Adaptive Tests of Significance Using Permutations of Residuals with R and SAS®

Adaptive Tests of Significance Using Permutations of Residuals with R and SAS®

Thomas W. O'Gorman
Northern Illinois University
Division of Statistics
DeKalb, IL

WILEY
A JOHN WILEY & SONS, INC., PUBLICATION

Published by John Wiley & Sons, Inc., Hoboken, New Jersey.
Published simultaneously in Canada.

Library of Congress Cataloging-in-Publication Data:

O'Gorman, Thomas W.
Adaptive tests of significance using permutations of residuals with R and SAS / Thomas W. O'Gorman.
p. cm.
Includes bibliographical references and index.
ISBN 978-0-470-92225-5 (hardback)
1. Regression analysis. 2. Computer adaptive testing. 3. R (Computer program language) 4. SAS (Computer file) I. Title.
QA278.2.O35 2012
519.5'36—dc23 2011038049

To my wife, Martha;
and my children, Kelly
and Tim.

CONTENTS

PREFACE

This book was written to introduce researchers to adaptive tests of significance and to describe the advantages of using these testing methods. Traditional tests of significance, such as the two-sample t test and the t test for slope, are robust in the sense that non-normality of the error distribution often does not dramatically change the level of significance. So, why should we use adaptive tests?

Adaptive tests are used to increase the power when the errors are not normally distributed. In real-world testing situations we rarely know the distribution of the errors, so it is important to know just how the traditional tests compare to adaptive tests with a variety of normal and non-normal error distributions. The power comparisons, which are displayed throughout the book, show that the adaptive test is often much more powerful than the traditional test with many non-normal error distribution.

Adaptive tests use the data to adjust the test procedures. For example, if a researcher wants to perform a two-sample test and the data suggests that the error distributions may be normally distributed, the traditional test procedure is not modified a great deal, so that the resulting test will approximate a two-sample t test. However, if the data contain a few outliers, then the test procedure will be modified to downweight the importance of those outliers.

At first glance these adaptive tests of significance are suspicious. It does not seem right to use the data to modify the test procedure and then use the data again to perform the test. If a test is not properly constructed, it may not maintain its level of significance; but if a test is properly constructed, it will maintain its level of significance. All of the adaptive tests in this book maintain their significance level because they use permutations methods. These permutation methods have become practical in the last few years because fast computers are now readily available. Most of the adaptive tests in this book can be performed in just a few minutes.

Because permutation methods are used to compute the p-value of adaptive test, some software is necessary. In this book we use R functions and SAS® macros to perform the adaptive weighting and permutation methods. It is not expected that the reader be familiar with both of these languages; either language is sufficient to perform the adaptive tests, confidence intervals, and estimates. Those readers who are not interested in the computational aspects of these tests may choose to skip those sections that describe software.

The basic adaptive testing method is described in the first four chapters. Applications of this method to tests for slope in a regression, tests for main effects and interaction effects in a two-way design, tests for randomized complete block designs, and tests for the analysis of paired data are explained in Chapters 5 through 7. The use of adaptive tests in multicenter clinical trials and in cross-over trials are described in Chapter 8. Chapters 9 and 10 concern multivariate tests and their application to the analysis of repeated measures data. Adaptive confidence intervals, which tend to be narrower than traditional confidence intervals when the errors have a non-normal distribution, are described in Chapter 12.

On a personal note, I became interested in adaptive testing around 1985 while working as a statistical consultant at the University of Iowa. A few years later, while working on variable selection methods in case-control studies with Robert (Skip) Woolson, I became convinced that variables selection methods could be improved if the performance of the testing methods could be improved. After publishing several rank-based adaptive tests, I decided that a more general approach was necessary. In the last ten years I have developed an adaptive test, which is described in this book, that can be used in many testing situations to provide increased power.

I am indebted to many individuals who have contributed to the literature on adaptive tests. The important paper by Hogg, Fisher, and Randles (1975) showed that an adaptive test could maintain its significance level and be relatively powerful. The papers by Büning (1996, 1999, 2002) extended the methods proposed by Hogg. Both Hogg and Büning proposed rank-based adaptive tests that did not use a permutation method.

I am also indebted to Freedman and Lane (1983), who proposed the permutation of residuals method that is used in this book. The papers by Anderson and Legendre (1999) and Anderson and Robinson (2001) increased my understanding of the importance of distinguishing between the various kinds of permutation methods. Most of the ideas concerning permutation tests, which are described in Chapter 4, come from those papers.

All of the simulation studies in this book were programmed in FORTRAN because they could not have been executed in a reasonable amount of time in either R or SAS. In general, programs written in FORTRAN execute much more quickly than those using R code or SAS macros. However, FORTRAN is not usually used for the analysis of data because it takes much longer to write FORTRAN code than R or SAS code. Hence, in most consulting situations, it is necessary to use either R or SAS to perform the adaptive test. The author would be happy to supply the FORTRAN code for any of the simulations performed in this book to any researcher who is familiar with the use of adjustable dimension arrays in FORTRAN. Please contact the author at *ogorman@math.niu.edu* and describe your interests or concerns to obtain the FORTRAN source code for the simulation study.

I want to acknowledge the support I have been given by my family and colleagues. I would also like to thank Professor Alan Polansky for his encouragement to develop these methods and for his assistance with the R language.

T. W. O'GORMAN

Dekalb, Illinois
November 2011

CHAPTER 1

INTRODUCTION

1.1 WHY USE ADAPTIVE TESTS?

Many adaptive tests have been developed in an effort to improve the performance of tests of significance. We will consider a test of significance to be "adaptive" if the test procedure is modified after the data have been collected and examined. For example, if we are using a certain kind of two-sample adaptive test we would collect the data and calculate selection statistics to determine which two-sample test procedure should be used. If the data appear to be normally distributed, then a Wilcoxon rank-sum test would be used; but if the data appear to contain outliers, then a median test would be used.

Adaptive tests of significance have several advantages over traditional tests. They are usually more powerful than traditional tests when used with linear models having long-tailed or skewed distributions of errors. In addition, they are carefully constructed so that they maintain their level of significance. That is, a properly constructed adaptive test that is designed to maintain a significance level of α will have a probability of rejection of the null hypothesis at or near α when the null hypothesis is true. Hence, adaptive tests are recommended because their statistical properties are often superior to those of traditional tests.

Adaptive Tests of Significance Using Permutations of Residuals with R and SAS®.
By Thomas W. O'Gorman

The adaptive tests described in this book have the following properties:

- The actual level of significance is maintained at or near the nominal significance level of α.
- If the error distribution is long-tailed or skewed, the adaptive test is usually more powerful than the traditional test, sometimes much more powerful.
- If the error distribution is normal, there is little power loss compared to the traditional tests.
- Adaptive test are practical. R code and SAS® macros are available to quickly perform these tests.

The adaptive tests in this book automatically reduce the influence of outliers. They are sometimes said to be robust; but to be clear about robustness, we should describe the two kinds of robustness. A test is said to be robust for size if its actual significance level is quite close to the nominal significance level, even when the usual assumptions are not met. For example, a test that is derived by assuming normality of the error distribution would be robust for size if it maintains its level of significance with non-normal errors. A test is said to be robust for power if it has high power relative to other tests when the usual distributional assumptions are not met. Many traditional tests are robust for size with non-normal errors but are not robust for power. Our objective is to develop adaptive tests that are robust for size and robust for power.

In this chapter we will give a brief history of adaptive tests and will present some of the procedures that are used to develop adaptive tests. We will also show the power advantages of several two-sample adaptive tests. Subsequent chapters will describe the advantages of the adaptive approach for a test of any subset of coefficients in a linear model. In Chapters 3 through 8 we show how adaptive tests can be used in almost all testing situations and that these tests have better properties than the traditional tests. In Chapters 9 and 10 we develop a multivariate adaptive test and show how it can be used to analyze repeated measures data. In Chapter 11 we describe several rank-based approaches to testing and in Chapter 12 we show how adaptive confidence intervals and estimates can be computed. In most chapters, R code and SAS macros will be used to perform the tests.

1.2 A BRIEF HISTORY OF ADAPTIVE TESTS

1.2.1 Early Tests and Estimators

The first two-sample adaptive test that was practical and relatively powerful was proposed by Hogg, Fisher, and Randles (1975). Prior to 1975, the adaptive tests were interesting but not too practical. For example, the test proposed by Hájek (1962) was designed to improve the power by finding scores that would produce a locally most powerful rank test. The test required an estimate of the density function (f) and the first derivative of the function (f'). The problem with this approach is that f and f'

are difficult to estimate unless the samples are very large. Hence, these adaptive tests are not practical and do not appear to be used. See Hogg (1974) for a discussion of the problems associated with this approach.

In order to avoid estimating densities and their derivatives, Hogg (1967) proposed an adaptive procedure that used the sample kurtosis to select one of four estimators of the mean of a symmetric distribution. In that research, four symmetric distributions were considered having various amounts of kurtosis. The idea was to use the selection statistic to select an estimator that would have low variance for samples from that distribution. One difficulty with this approach is that the sample kurtosis is highly variable, so it may sometimes fail to select the correct estimator for that symmetric distribution. In spite of this problem, the robust adaptive estimator had excellent performance with $n = 25$ observations that were generated from the four distributions that were used in that study. In arguing for greater use of these robust methods, Hogg (1967) stated "In this age of excellent computing devices, the statistician should take a broader view and not select a narrow model prior to observing the sample items." Over the following years this estimator has been modified and the more recent version of this adaptive estimator, as described by Hogg and Lenth (1984), has excellent properties.

1.2.2 Rank Tests

After the paper by Hogg was published in 1967, the idea of using a selection statistic to modify a statistical method was developed further in a paper by Randles and Hogg (1973), which included an adaptive one-sample test and an adaptive two-sample test. In these tests the rank scores were selected based on selection statistics. Instead of using the sample kurtosis as a selection statistic, they used a measure of tailweight. Although these tests were adaptive in nature, they were not as powerful as traditional tests.

However, just two years later Hogg, Fisher, and Randles (1975) published an improved two-sample adaptive test. This test, which will be described in the next section, was the first practical and effective two-sample adaptive test. Although the test attracted considerable attention from statisticians, it appears to be rarely used by researchers. One problem with their adaptive test is that, because it is a rank-based test, it is not easy to generalize this approach to tests of significance of regression coefficients in more complex models.

In a series of articles published more than 20 years after the paper by Hogg, Fisher, and Randles (1975), several researchers used the same selection statistics to construct tests for a variety of situations. Büning (1996) proposed an adaptive test of equality of medians using data from a one-way layout. This test was based on an extension of Hogg's method of using selection statistics to select a set of rank scores. Two years later, Büning and Kossler (1998) proposed an adaptive test for umbrella alternatives and, in the following year, Büning (1999) proposed a test for ordered alternatives. Further extensions of the adaptive approach were made by Büning and Thadewald (2000), who proposed a location-scale test and by Büning (2002), who proposed a test that could be used to test the null hypothesis that the distributions are equal against the general alternative that the distributions are not equal.

The tests proposed by Hogg and by Büning used selection statistics to determine the set of rank scores for the two-sample test. One small problem with this approach is that, if the selection statistics fall near the edge of a region corresponding to a set of rank scores, any small change in the data could change the selection statistics slightly, and this could result in the selection of an entirely different set of rank scores. This is undesirable because a small change in a single data value could result in a large change in the p-value. To remedy this situation, Ruberg (1986) proposed a continuously adaptive two-sample test and O'Gorman (1997) proposed a continuously adaptive test for the one-way layout. Using a different approach, Hall and Padmanabhan (1997) proposed several adaptive tests for the two-sample scale problem. They used a bootstrap testing approach with adaptively trimmed sample variances.

We should note that in the last 40 years there has also been work in the area of adaptive estimation. Yuh and Hogg (1988) proposed two adaptive regression estimators that rely on selection statistics to choose one of several robust regression estimators. Further work in the area of adaptive estimation was published by Hill and Padmanabhan (1991), who described the performance of some adaptive estimators when they were used with real data.

Although some material on adaptive confidence intervals and estimates is included in the last chapter of this book, we are primarily concerned with adaptive tests of significance. Further, we will focus on methods that can be used in a variety of testing situations. One such adaptive test utilizes a weighted least squares approach that we will now describe.

1.2.3 The Weighted Least Squares Approach

Before the year 2000, all of the adaptive tests were rank-based so they were limited to one-sample tests, two-sample tests, and tests for the one-way layout. A different approach was taken in 2001 when O'Gorman (2001) proposed a test that used an adaptive weighting method to assign weights to the observations so that the weighted observations could be used to test a subset of coefficients in a linear model. An improved version of this approach was proposed by O'Gorman (2002), and the book by O'Gorman (2004) described various applications of this method. With this approach the p-value was computed by using a permutation method. A few years later an adaptive multivariate test was proposed by O'Gorman (2006a), and this multivariate adaptive test was used in the analysis of repeated measures data by O'Gorman (2008a, 2010).

The adaptive tests proposed by O'Gorman prior to 2006 used a permutation method that required permutations of independent variables. An adaptive test that used permutation of residuals was proposed by O'Gorman (2006b), and this method was shown to be as effective as the test based on the permutation of independent variables. The advantages and disadvantages of various permutation methods will be addressed in Chapter 4. An improved adaptive weighting method that could be used with univariate and multivariate data will be described in Chapters 2 and 3, and it will be used for most of the tests of significance.

1.2.4 Recent Rank-Based Tests

While most of the adaptive testing literature prior to 2000 focused on two-sample tests, some recent research has been published on one-sample adaptive tests. Lemmer (1993) suggested two adaptive tests for the median. Freidlin, Miao, and Gastwirth (2003) proposed an interesting and effective adaptive test for paired data. These authors use the p-value from a test of normality, rather than a measure of skewness or tailweight, as the selection statistic. They showed that their test is reasonably effective for moderate sample sizes. Baklizi (2005) proposed a continuously adaptive test for the median when the symmetry of the distribution is in doubt. It maintains its size and is relatively powerful. Most recently, Miao and Gastwirth (2009) proposed a test that uses the same score functions that were used by Freidlin, Miao, and Gastwirth (2003), but the test uses a measure of tail-heaviness as the selection statistic. This test will be described and evaluated in Chapter 11.

A different approach to robustifying and improving two-sample tests was taken by Neuhäuser, Büning, and Hothorn (2004). To construct their test, they used four sets of rank scores to produce four standardized linear rank statistics. Next, they computed the maximum of those four statistics as the overall test statistic, which is then used with a permutation method to compute the p-value. This test maintains its level of significance and has higher power than many of the traditional parametric and nonparametric tests. In addition, it has the advantage of not using any selection statistic. While it is not always classified as an adaptive test, it does achieve the same objective as the adaptive test. In Chapter 11 we will give a detailed description of this promising test.

1.3 THE ADAPTIVE TEST OF HOGG, FISHER, AND RANDLES

We will now consider the two-sample adaptive test that was proposed by Hogg, Fisher, and Randles (1975), which will be called the HFR test. Much of the early work in adaptive testing was based on the selection of an appropriate score function, which was a function of a selection statistic. Although Hogg (1967) used the sample kurtosis as a selection statistic, robust measures of asymmetry and tailweight were used in the HFR test. Their robust measure of asymmetry is calculated by combining the observations over both samples, sorting the observations, and then computing

$$Q_3 = \frac{\overline{U}_{.05} - \overline{M}_{.5}}{\overline{M}_{.5} - \overline{L}_{.05}},$$

where $\overline{U}_{.05}$ is the average of the upper 5%, $\overline{M}_{.5}$ is the average of the middle 50%, and $\overline{L}_{.05}$ is the average of the lower 5% of the observations. It should be noted that while Q_3 may be more robust than some other measures, $\overline{U}_{.05}$ and $\overline{L}_{.05}$ may be sensitive to outliers. The robust measure of tail length is given by

$$Q_4 = \frac{\overline{U}_{.05} - \overline{L}_{.05}}{\overline{U}_{.5} - \overline{L}_{.5}},$$

where $\overline{U}_{.5}$ and $\overline{L}_{.5}$ are the averages of the upper and lower 50% of the combined samples, respectively. These two selection statistics are used to determine the most appropriate set of rank scores, as shown in the selection regions in Figure 1.1.

The general procedures for the HFR test are:

- Compute the selection statistics Q_3 and Q_4 using the combined observations.
- Use Q_3 and Q_4 to select the most effective rank scores based on the selection regions that are shown in Figure 1.1.
- Compute the test statistic using the selected scores and then find the p-value for the test based on those scores.

Hogg, Fisher, and Randles (1975) demonstrated that this test would maintain its significance level. They also showed, with a simulation study, that this adaptive test was often more powerful than traditional parametric and non-parametric tests.

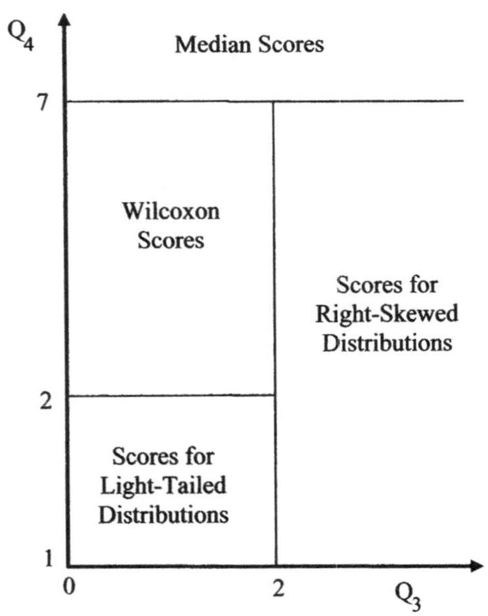

Figure 1.1 Selection regions for the HFR test.

To illustrate the selection method, suppose we obtained a data set for a two-sample test, with each sample having 20 observations, and we calculated the measure of asymmetry to be $Q_3 = 1.1$, which indicated that the distribution was nearly symmetric. Next, we calculated the measure of tailweight as $Q_4 = 7.8$, which indicated that the distribution was long-tailed. With this set of selection statistics, we would have used the selection regions in Figure 1.1 to select the median scores. If the data were slightly skewed and we obtained $Q_3 = 1.5$ and $Q_4 = 3.5$ as selection statistics, the Wilcoxon scores would have been selected.

1.3.1 Level of Significance of the HFR Test

Hogg, Fisher, and Randles (1975) carefully constructed the adaptive HFR test so that it maintains its level of significance. To demonstrate this important property, let E_j be the event that the vector of selection statistics falls in region j and let R be the event that the null hypothesis is rejected. Then, for the jth region $P(R|E_j) \leq \alpha$ because these rank tests are distribution-free. Consequently, the HFR test will maintain its size because

$$P(R) = \sum P(R|E_j)P(E_j) \leq \sum \alpha P(E_j) = \alpha,$$

where the summations are over the four regions. Hence, even though we use the data to determine the rank scores, we find that the test has a significance level less than or equal to α. In addition, Hogg, Fisher, and Randles (1975) showed, using a simulation study with 15 observations per group, that the empirical significance level closely approximates α.

1.3.2 Comparison of Power of the HFR Test to the t Test

The real advantage of adaptive tests is that they often have greater power than the traditional test for many non-normal error distributions. To demonstrate this, we will compare the power of the HFR test to that of the pooled two-sample t test for several error distributions. The error distributions are listed in Table 1.1. We will compare the tests by means of a simulation study that uses 100,000 data sets for each distribution. This simulation program was written in FORTRAN to estimate the power of these tests. By writing the simulation program in FORTRAN we are able, in a reasonable amount of time, to analyze 100,000 data sets to obtain accurate estimates of the power.

For each data set we use 15 observations in each sample. For the first sample the observations are generated from one of the error distributions. For the second sample we add a constant δ to a random variable generated from the same distribution. After each data set is generated we perform the pooled two-sample t test and the HFR test. We count the number of times that the null hypothesis is rejected and calculate the proportion of rejections, which is the empirical power of the test.

These distributions will be described in much greater detail in Chapter 2. We note that the first four distributions are symmetric and that the other five are skewed. Some of these distributions have short tails while most of the others are long-tailed, and two are bimodal.

Table 1.1 Error distributions used in the simulation studies

Distribution	Description
Uniform	Uniform[0,1]
Normal	Standard normal $N(0,1)$
t_4	t distribution with d.f. = 4
Bimodal symmetric	Mixture of normals
RST $\alpha_3 = 1, \alpha_4 = 4.2$	Moderate skewness, low kurtosis
RST $\alpha_3 = 1, \alpha_4 = 8.4$	Moderate skewness, high kurtosis
RST $\alpha_3 = 2, \alpha_4 = 11.4$	High skewness, low kurtosis
RST $\alpha_3 = 2, \alpha_4 = 15.6$	High skewness, high kurtosis
Bimodal skewed	Mixture of normals

In order to make the power comparisons meaningful, we have set the shift (δ) of the random variables in the second population relative to those in the first population so that the powers of the t test would be near 80% for the nine error distributions. To meet this objective, we used $\delta = 1.0$ in a simulation study, with 15 observation in each sample for each of the 100,000 data sets that were generated. We obtained powers, which are displayed in Figure 1.2, for the t test and the HFR test with the nine error distributions. We note that the HFR test has greater power than the t test for most of the distributions but it has lower power for three of the distributions.

It is encouraging that the HFR test maintains its significance level and has greater power than the t test for several distributions, but it is discouraging to see a moderate power loss with the uniform, the normal, and the bimodal error distributions.

1.4 LIMITATIONS OF RANK-BASED TESTS

One problem with the Wilcoxon test, the adaptive HFR test, and other rank-based tests is that the testing methods make sense only when the data can be ranked. Consequently, rank-based tests are usually used to test equality of medians with two populations or to test equality of medians in a one-way layout. For example, Büning (1996) proposed an adaptive test for the one-way layout that used the HFR method of selecting rank scores. However, rank tests are difficult or impossible to perform with more complex models. For example, if we want to compare two groups and need to include one covariate, it may be difficult to find a suitable rank test.

To overcome this problem, O'Gorman (2001) proposed an adaptive test that did not use ranks. Instead, it uses an adaptive weighting scheme to improve the power of the tests. More recently, several variants of that method have been proposed to increase the power of the test and to allow it to be used with a wider variety of models. In this book we describe a new adaptive test that is powerful and flexible. The next section gives a brief description of this approach for the two-sample test. Most of

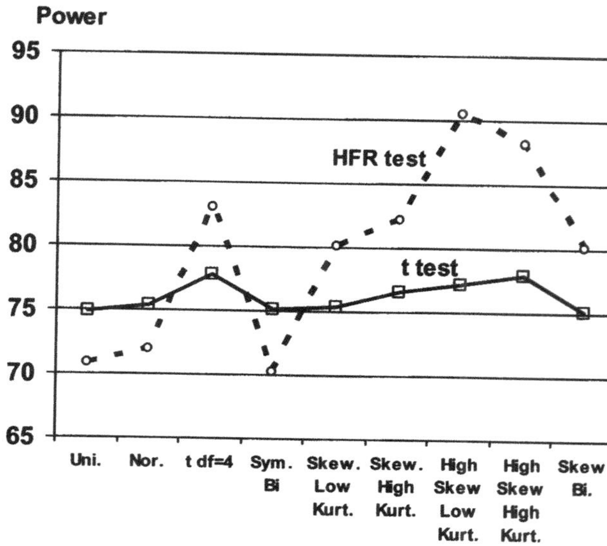

Figure 1.2 Power of the two-sample t test and the HFR test when 15 observations are generated for each sample. The t test is indicated by hollow squares and the HFR test is indicated by hollow circles.

the rest of this book is devoted to generalizing this testing procedure so that this new adaptive test can be used for most of the common testing situations.

1.5 THE ADAPTIVE WEIGHTED LEAST SQUARES APPROACH

The adaptive test that will be developed in this book is performed in two steps. In the first step the observations are weighed in a manner that produces residuals, in the weighted model, that are roughly normally distributed. In the second step a permutation method is used to compute a p-value. The details of the adaptive test will be described in subsequent chapters.

It may seem strange to use the weighted least squares method to normalize errors in regression models. In many books on regression analysis, the weighted least squares method is used to ensure that the errors have the same variability. See Rawlings, Pantula, and Dickey (1998, Chapter 12) for a description of the weighted least squares method in regression. In this book we are primarily interested in adaptive methods for non-normal data, so we are interested in weighting the observations to normalize the errors.

An example may illustrate the basic approach used in the adaptive weighted least squares test. In an experiment that was described by Koziol *et al.* (1981), mice were injected with colon carcinoma cells in order to determine the effectiveness of several immunotherapy regimens. Five days later the mice were randomly assigned to Group

1 or to Group 2. Mice in Group 1 received injections of tissue culture medium around the tumor while mice in Group 2 received injections of normal spleen cells, immune RNA, and tumor antigen. The tumor volumes (mm^3) at day 13 are given in Table 1.2 for the 10 mice in these two groups. In Koziol *et al.* (1981), Group 1 and Group 2 are labeled as Group A and Group C, respectively.

Table 1.2 Tumor volumes, in mm^3, for 2 groups of mice measured on day 13 of the experiment and the adaptive weights. Group 1 is the group that received the tissue culture medium and Group 2 is the group that received the spleen cells, immune RNA, and tumor antigen.

	Group 1			Group 2	
Obs	Volume	Weight	Obs	Volume	Weight
1	217.6	1.118	11	186.2	1.107
2	176.6	1.044	12	196.6	1.143
3	196.1	1.142	13	191.3	1.129
4	205.9	1.147	14	129.6	0.625
5	196.0	1.142	15	420.0	0.400
6	225.1	1.081	16	32.0	0.511
7	274.7	0.800	17	55.0	0.466
8	202.5	1.149	18	84.7	0.470
9	205.8	1.147	19	258.8	0.867
10	225.0	1.082	20	176.4	1.043

To show the relative effectiveness of these regimens, the volumes are displayed in Figure 1.3. This dot plot shows that there are several outliers present in Group 2 and that the tumor volumes in Group 1 tend to be larger than those in Group 2. However, if we perform a two-sample test, the pooled t test gives a p-value of $p = 0.292$ and the unequal variance t test gives a p-value of $p = 0.304$. These large p-values are due, in part, to the large variability that is present in Group 2.

In the adaptive WLS test we give reduced weights to the observations that are extreme. The adaptive weights are given in Table 1.2 for the 20 mice. Note that mouse 15 was given a weight of 0.400 because the tumor volume was so large, and mice 16, 17, and 18 were given low weights because the tumor volumes were quite small. In this way the adaptive test reduces the influence of the outliers. We then used a permutation method to find a p-value of $p = 0.061$, which is much smaller than the p-values found with the pooled and unequal variance t tests.

However, it is important not to place too much importance on the difference in p-values for this one data set because we do not know if the distributions are different. The weighting procedures and permutation methods will be described in great detail in Chapters 2 and 3. In the remainder of this chapter we will briefly describe the significance level and the power of this adaptive test.

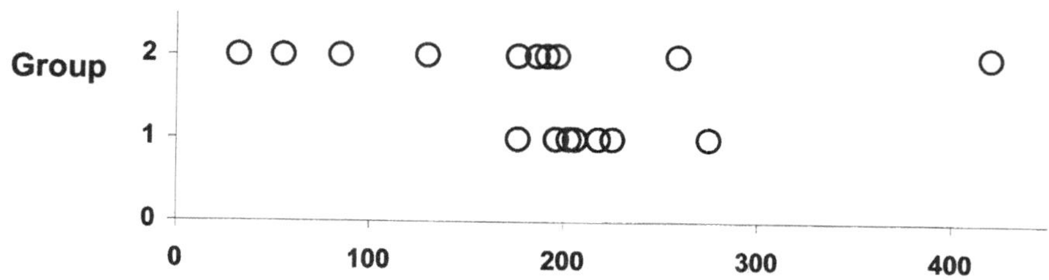

Figure 1.3 Tumor volumes, in mm^3, measured on day 13 of the experiment that was described by Koziol *et al.* (1981). Group 1 is the group that received the tissue culture medium, and Group 2 is the group that received the spleen cells, immune RNA, and tumor antigen.

1.5.1 Level of Significance

We can estimate the relative performance of this adaptive test in the same manner as we evaluated the performance of the HFR test. It will be shown in Chapter 3 that the two-sample adaptive test will maintain its level of significance. It maintains its level of significance because it is based on a permutation test. To further illustrate this important point, we show, in Figure 1.4, the empirical significance level of the t test, the HFR test, and the adaptive test. Note that the adaptive test and the HFR test maintain their level of significance near the nominal level of $\alpha = 0.05$ with 15 observations in each of the two samples. The empirical significance level of the t test does not exceed 5%, but it does not reject often enough for non-normal errors. Since 100,000 data sets were used in the simulation for each error distribution, we achieved a standard error of 0.07% in the estimation of the level of significance, so the estimates in Figure 1.4 are quite accurate.

1.5.2 Comparison of Power of the Adaptive WLS Test to the t Test and the HFR Test

In a simulation study of the relative power of two-sample tests, we set the difference (δ) between the populations to $\delta = 1.0$. With 15 observations in each sample we obtained the empirical powers shown in Figure 1.5 when we used 100,000 data sets in the simulations for each of the nine error distributions.

The simulation results show that the adaptive WLS test has greater power than the t test for all non-normal distributions. Also, it is clear that, for skewed distributions, the HFR and the adaptive WLS tests have similar power. However, there are two advantages the adaptive WLS test has over the HFR test. A minor advantage is that the adaptive test is slightly more powerful than the HFR test for some distributions. The major advantage of the adaptive WLS test is that the same general approach that is used for the two-sample adaptive WLS test can be used for testing a subset of

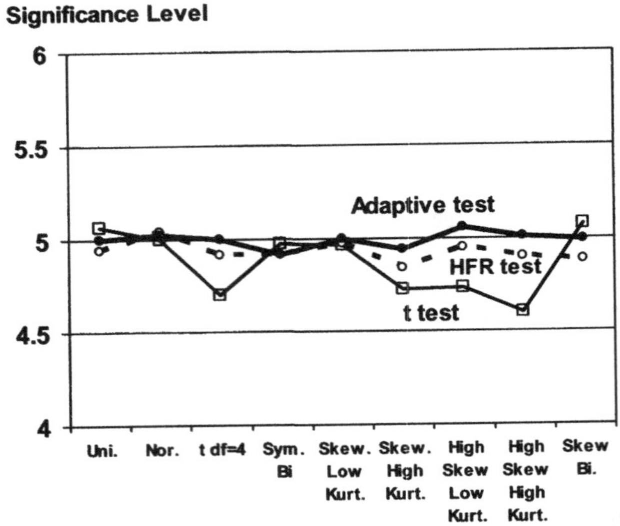

Figure 1.4 Empirical significance level in percent of the adaptive WLS test, the t test, and the HFR test with 15 observations in each sample. For each of the nine error distributions we generated 100,000 data sets. The t test is indicated by hollow squares, the HFR test is indicated by hollow circles, and the adaptive WLS test is indicated by filled circles.

coefficients in a linear model. Hence, the adaptive WLS test can be used for virtually all tests of significance.

1.6 DEVELOPMENT OF THE ADAPTIVE WLS TEST

We saw that the adaptive WLS two-sample test has a power advantage over the t test for many distributions. Can we create adaptive tests for regression models and factorial designs? Will these adaptive tests have a power advantage over traditional tests? Also, what is the justification for the adaptive weighting and how are these tests performed?

In the next seven chapters we will describe the advantages of the adaptive WLS test. Before we describe the adaptive test in detail, we will need to explain, in the next chapter, smoothing methods and their use in normalizing transformations. In Chapter 3 we return to the two-sample test and describe the adaptive test in great detail. This two-sample test will form the basis for most of the adaptive tests used in this book. To extend this adaptive WLS approach to tests for a subset of coefficients in a linear model, we will need to perform permutation tests with linear models. Chapter 4 gives an introduction to permutation methods that can be used with regression models. Chapter 5 gives the general adaptive testing method for any subset of coefficients in a linear model and Chapter 6 describes many of applications

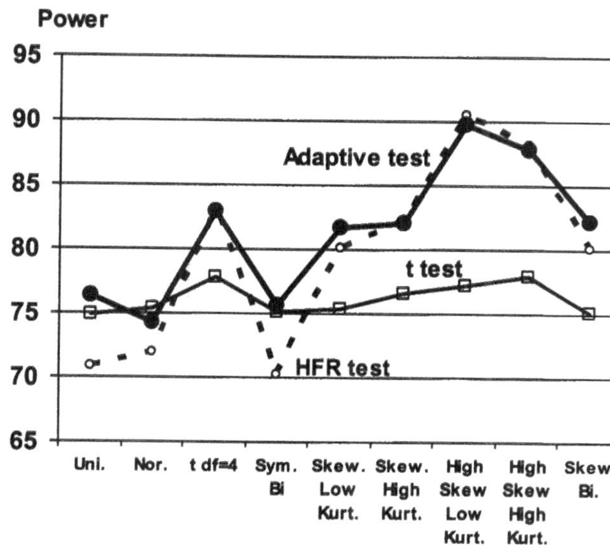

Figure 1.5 Empirical power in percent for the adaptive WLS test, the t test, and the HFR test with 15 observations in each sample of 100,000 data sets for each of the nine error distributions. The difference between the mean of these two population was set to $\delta = 1.0$. The t test is indicated by hollow squares, the HFR test is indicated by hollow circles, and the adaptive WLS test is indicated by filled circles.

of adaptive testing. In Chapters 7 and 8 we describe how adaptive tests can be used with paired data, data from multicenter trials, and data from crossover trials. Chapters 9 and 10 concern multivariate tests and their use in repeated measures studies. In Chapter 11 we describe rank-based approaches to testing. Adaptive confidence intervals and estimates are developed in Chapter 12.

Most of the adaptive testing methods in this book use a weighted least squares approach combined with a permutation method. The weighting procedure is intended to be a normalizing transformation that will reduce the influence of outliers. In the next chapter we discuss these normalizing procedures.

CHAPTER 2

SMOOTHING METHODS AND NORMALIZING TRANSFORMATIONS

In this chapter we will describe many of the fundamental techniques that will be used to perform adaptive tests of significance. We will focus our attention on a weighting procedure that is used to transform the data so that the transformed data is approximately normally distributed. We will see in subsequent chapters how this transformed data is used in the adaptive tests.

2.1 TRADITIONAL ESTIMATORS OF THE MEDIAN AND THE INTERQUARTILE RANGE

The adaptive weighting procedure relies heavily on percentile estimators. In this section we describe the traditional estimator of a percentile and in the next section we describe an improved estimator that is based on the smoothed cumulative distribution function (c.d.f.).

Let $x_1, x_2, \ldots, x_n$ be a random sample from a continuous distribution having a c.d.f. of $F(x)$ and let $x_{(1)}, x_{(2)}, \ldots, x_{(n)}$ be the set of ordered observations, in ascending order, that we will call the order statistics. If n is odd, the usual estimator of the median is $x_{((n+1)/2)}$; and if n is even, the usual estimator is $\frac{1}{2}[x_{(n/2)} + x_{(n/2+1)}]$.

Adaptive Tests of Significance Using Permutations of Residuals with R and SAS®.
By Thomas W. O'Gorman

Now suppose we want to estimate the pth percentile of the continuous distribution. Let x_p be the pth percentile of the distribution and let $\hat{x}_{t,p}$ be the traditional estimator of x_p. To clearly define $\hat{x}_{t,p}$ we need some notation. Let $\lfloor x \rfloor$ represent the greatest integer less than or equal to x and let $k = \lfloor p(n+1) \rfloor$ be the greatest integer less than or equal to $p(n+1)$. Now suppose we let $f = p(n+1) - k$. Although many percentile estimators have been proposed, one traditional estimator of x_p is $\hat{x}_{t,p} = x_{(k)} + f(x_{(k+1)} - x_{(k)})$, and this is the traditional estimator that we will use for $1 \leq p(n+1) \leq n$. With this notation, f represents the fraction between $x_{(k)}$ and $x_{(k+1)}$. If $p(n+1) < 1$, we set $\hat{x}_{t,p} = x_{(1)}$. If $p(n+1) > n$, we set $\hat{x}_{t,p} = x_{(n)}$.

In the adaptive weighting method we will use the interquartile range (IQR) as an estimate of the variability in the data. Using the traditional estimates of the 25th and 75th percentiles, the estimate of the IQR is $\widehat{IQR} = \hat{x}_{t,.75} - \hat{x}_{t,.25}$.

One problem with these traditional estimators of the percentiles (or quantiles) is that they are sensitive to one or two order statistics near $p(n+1)$. To address this problem, several authors, including Kaigh and Lachenbruch (1982) and Harrell and Davis (1982), have proposed quantile estimators that are linear combinations of order statistics. These estimators generally have less variability than the traditional estimators but are more complicated. Another approach is to estimate the c.d.f. of the observations and then use it to determine the quantile. One advantage of using the estimator that is based on the smoothed c.d.f. is that the variability of this estimator is less than the variability of the corresponding traditional estimator. Another advantage is that, because the smoothed c.d.f. will also be used in the normalization procedure, little extra effort will be required to compute these estimates. This is the approach we will take in the next section.

2.2 PERCENTILE ESTIMATORS THAT USE THE SMOOTH CUMULATIVE DISTRIBUTION FUNCTION

2.2.1 Smoothing the Cumulative Distribution Function

In this section we will estimate the cumulative distribution function ($F(x)$) and then use our smoothed estimate of $F(x)$ to estimate the percentiles. In our adaptive test we will estimate percentiles of residuals, so we will assume that we have obtained residuals from some linear regression. Now suppose we have $n = 7$ residuals with order statistics $\{-4, -3, -2, -1, 1, 2, 7\}$. The empirical c.d.f. is a function of x that roughly estimates $F(x)$ by computing the proportion in the sample that are less than or equal to x. The empirical distribution function for these residuals, which increases by $1/n$ at each value in the sample, is plotted in Figure 2.1.

Let $\hat{F}_{empirical}(x)$ be the empirical c.d.f. obtained from the order statistics of a set of values. To estimate the median using the empirical c.d.f., we find $\hat{x}_{.5}$ such that $\hat{F}_{empirical}(\hat{x}_{.5}) = .5$, which, for this data set, produces $\hat{x}_{.5} = -1$. We can also use $\hat{F}_{empirical}(x)$ in Figure 2.1 to find other percentiles. We find the 75th percentile to be $\hat{x}_{.75} = 2$ and the 25th percentile to be $\hat{x}_{.25} = -3$. Hence, the estimate of the interquartile range is $\widehat{IQR} = \hat{x}_{.75} - \hat{x}_{.25} = 2 - (-3) = 5$. One problem with

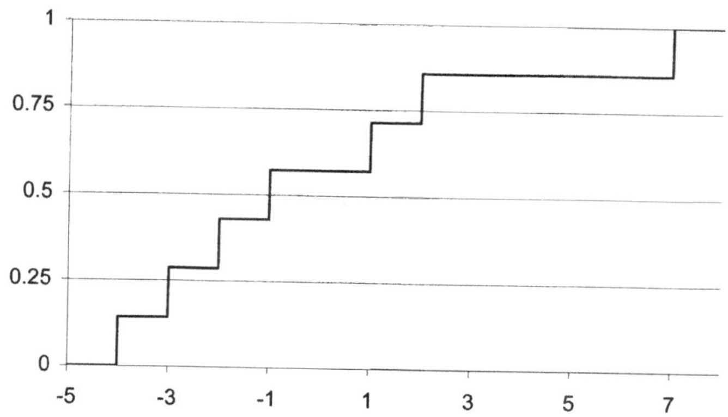

Figure 2.1 Empirical cumulative distribution function for a small data set.

this approach is that the estimate of the median is not unique if n is even because $\hat{F}_{empirical}$ is not a strictly increasing function. However, if we made some rules to define the sample percentiles uniquely, we would find that these percentiles are similar to the traditional percentiles. Hence, there is no reason to use the empirical c.d.f. to estimate the percentiles.

However, if we smooth the empirical c.d.f. and use the smoothed c.d.f. to find the percentiles, we can reduce the variability of the estimator. Before we smooth the c.d.f., note that the empirical c.d.f. increases by $1/n$ at each order statistic so that the empirical c.d.f. can be written as

$$\hat{F}_{empirical}(x) = \frac{1}{n}\sum_{i=1}^{n} I_{[x_i \le x]},$$

where

$$I_{[x_i \le x]} = \begin{cases} 1 & \text{if } x_i \le x, \\ 0 & \text{if } x_i > x \end{cases}$$

is an indicator function. To construct a smoothed c.d.f., we use the same general approach but substitute a smooth nondecreasing function in place of the indicator function. Many smoothing functions have been proposed but we will use the c.d.f. of the normal distribution in place of the indicator function. Let $\Phi(x)$ be the c.d.f. of the standard normal distribution. We compute the smoothed c.d.f. as

$$\hat{F}_h(x) = \frac{1}{n}\sum_{i=1}^{n} \Phi\left(\frac{x - x_i}{h}\right), \tag{2.1}$$

where h is a smoothing constant (or bandwidth) that must be specified. If h is set to a very small value relative to the smallest distance between any two observations, then $\Phi((x - x_i)/h)$ will approximate the indicator function so that $\hat{F}_h(x)$ will approximate $\hat{F}_{empirical}(x)$ for $-\infty < x < \infty$. If h is set to a moderate value, say $h = 1$, with our $n = 7$ observations, we obtain the smoothed c.d.f. that is shown in Figure 2.2. It is important to select the bandwidth (h) so that the empirical c.d.f. is smooth, but not so smooth that it no longer fits the empirical cumulative distribution function. Note that, because we have used the c.d.f. of the normal distribution in the smoothing, the smoothed c.d.f. will be increasing. Thus, the percentiles will be uniquely determined. We will discuss the selection of the smoothing constant later in this chapter.

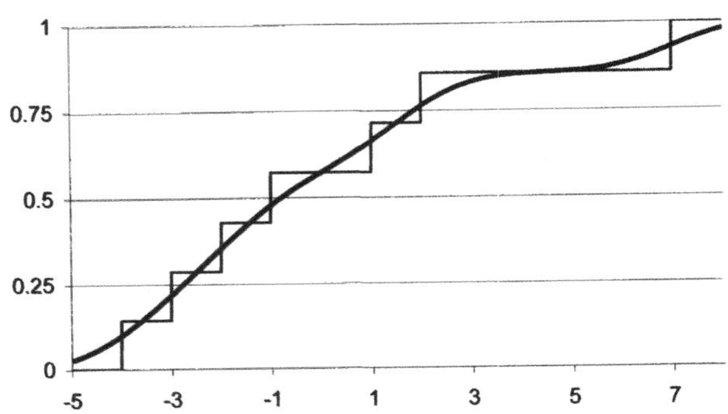

Figure 2.2 The empirical c.d.f. (thin line) and the smoothed c.d.f. (bold line), using a bandwidth of $h = 1$, for a small data set of $n = 7$ residuals.

The smoothed c.d.f. is important because it will be used in a procedure to normalize the data. For this normalization method we will need to find $\hat{F}_h(x_i)$ for $x_i = 1, \ldots, n$. In addition, we will use the smoothed cumulative distribution function to compute estimates of the 25th, 50th, and 75th percentiles.

2.2.2 Using the Smoothed c.d.f. to Compute Percentiles

To compute the median from the smoothed c.d.f. we use an iterative search method to find $\hat{x}_{.5}$ such that $\hat{F}_h(\hat{x}_{.5}) = .5$. Because $\hat{F}_h(x)$ is an increasing function of x, we can determine a point, which we will denote by x_{low}, such that $\hat{F}_h(x_{low}) < .5$, and we can find an upper bound x_{high} such that $\hat{F}_h(x_{high}) > .5$. The median must be in the interval (x_{low}, x_{high}) and must be unique because $\hat{F}_h(x)$ is increasing.

A variety of methods could be used to find the median. We could use a complicated root finding method, but it is quite easy and fairly efficient to find the root using the bisection method. That is, once we have obtained x_{low} and x_{high}, we compute the midpoint $x_{middle} = (x_{low} + x_{high})/2$ and $\hat{F}_h(x_{middle})$. If $\hat{F}_h(x_{middle}) > .5$, we use x_{middle} as the new upper bound but if $\hat{F}_h(x_{middle}) < .5$ we use x_{middle} as the new lower bound. With these new bounds we compute a new midpoint and we repeated this procedure. We continue this process of finding new upper and lower bounds until the difference between the upper and lower bounds is less than a small value, which is the tolerance. This bisection method is easy to program and the method always converges in a reasonable number of iterations. Interpolation methods may be used instead of bisection, but these are not always faster than bisection because the relative speed depends on the shape of the c.d.f. and the initial bounds. We used this bisection method to find the median for the data whose c.d.f. is displayed in Figure 2.2. The iterative procedure quickly produced an estimate of $\hat{x}_{.5} = -0.78$ as the median.

Now suppose we use this bisection method to obtain the 25th percentile as $\hat{x}_{.25} = \hat{F}_h^{-1}(.25)$ and the 75th percentile as $\hat{x}_{.75} = \hat{F}_h^{-1}(.75)$. Then we can use $\hat{x}_{.75} - \hat{x}_{.25}$ as a robust estimator of the IQR. Because this estimator uses the smoothed c.d.f. instead of the empirical c.d.f., it will be less sensitive to individual order statistics than the traditional estimator of the IQR. Because it is difficult to compute these estimates by manual calculation, we will describe how these can be computed using R.

2.2.3 R Code for Smoothing the c.d.f.

The R code given below is used to compute the smoothed c.d.f. at a specified point. Those readers who are not familiar with R may skip this section. We begin by defining a function, `cdfhat`, that computes the smoothed c.d.f. at a point (`xpoint`) using the observations that are contained in a vector (`xvector`). Later in this chapter we will learn how to determine a reasonable value for the bandwidth, but for this function it must be specified as `h`. The normal c.d.f. is produced by the function `pnorm` that is built-in as part of the R language. Because the R language uses vectors, rather than individual elements of vectors, the vector `phi` contains the values that immediately follow the summation sign in Equation (2.1). Next, the `mean` function computes the estimate of the c.d.f. as the scalar `cdf`, which is returned to the calling program by the function `cdfhat`. Thus, for any vector of values (`xvector`) and any smoothing parameter (`h`) this R code will obtain the value of the c.d.f. at a specified point (`xpoint`).

```
> cdfhat <- function(xvector,h,xpoint){
   phi <- pnorm((xpoint-xvector)/h)
   cdf <- mean(phi)
   return(cdf)
   }
```

We use the small data set of $n = 7$ residuals to illustrate how this R function can be used to evaluate the c.d.f. at a point. Suppose we have already typed the `cdfhat` function and now want to evaluate the smoothed c.d.f. at the point $x = -0.78$ with a smoothing parameter of $h = 1$. We would type the following R code:

```
> xvector <- c(-4,-3,-2,-1,1,2,7)
> xpoint <- c(-0.78)
> h <- 1
```

```
> fhat <- cdfhat(xvector,h,xpoint)
> print(fhat)
[1] 0.5003197
```

The program produces a value of $\hat{F}_h(-0.78) = .5003197$, which implies that the median, as evaluated by the smoothed c.d.f., is near -0.78. By using many values of `xpoint` in the function we could obtain the smooth c.d.f. that was shown in Figure 2.2.

2.2.4 R Code for Finding Percentiles

Instead of using the traditional estimator of the median we can obtain a more accurate estimate by means of a root finding algorithm that uses bisection. In the R code that is given below the function `rootcdf` is used to find an estimate of a percentile using the data in vector `x` along with the bandwidth `h`. The proportion less than the percentile is specified as `p`, so to obtain the median we would set the parameter `p` to 0.5. Because we are using bisection we must give a lower bound and an upper bound for the interval that contains the percentile. We specify the lower bound as `xlow` and the upper bound as `xhigh`. We continue the iterative process of the bisection method until the difference between `xlow` and `xhigh` is less than `tolerance`, which is usually specified as a small positive number. The following code uses the bisection method to find the percentiles:

```
> rootcdf <- function(x,h,p,xlow,xhigh,tolerance) {
    flow  <-  cdfhat(x,h,xlow)
    fhigh  <-  cdfhat(x,h,xhigh)
    for (i in seq(1:60) ) {
      xmiddle <- (xlow+xhigh)/2
      fmiddle <- cdfhat(x,h,xmiddle)
      if( (fmiddle == p) ) break
      if( (fmiddle < p) ) {xlow  <- xmiddle;  flow <- fmiddle}
      if( (fmiddle > p) ) {xhigh <- xmiddle; fhigh <- fmiddle}
      if( ( (xhigh-xlow) <= tolerance ) ) break
      if( (i==60) ){print (" Over 60 iterations");q()}
    }
    return(xmiddle)
    }
```

Note that `flow` and `fhigh` are values of the c.d.f. at `xlow` and `xhigh`, respectively. The value of the c.d.f. at `xmiddle` is computed to be `fmiddle`, which is then used in the bisection logic. The `for` loop is used to perform the bisection. If more than 60 bisections are required the program will print a warning and then will stop.

Now that we have defined the function `rootcdf`, we can use it to find the median. Because we want the median, we specify the lower proportion as `p=0.5`. Our bounds for the bisection will be specified as `xlow=-4` and `xhigh=7`. The iterations will continue until the difference between the upper and lower estimate is less than `tolerance=0.0000001`. The following R code will quickly find the median.

```
> xvector <- c(-4,-3,-2,-1,1,2,7)
> h <- 1
> p <- 0.5
> xlow <- -4
> xhigh <- 7
> tolerance <- 0.0000001
> median <- rootcdf(xvector,h,p,xlow,xhigh,tolerance)
> print(median)
[1] -0.7831708
```

The root finding function found the median to be approximately -0.7831708, based on the smoothed c.d.f. This function can also be used to find the 25th percentile by setting `p=0.25` and it can be used to find the 75th percentile by setting `p=0.75`, so it is easy to calculate the interquartile range. This root finding procedure will be used to find an estimate of the standard deviation that will be used in the adaptive weighting procedure.

2.3 ESTIMATING THE BANDWIDTH

In the adaptive testing procedure we will need to smooth the empirical c.d.f. of the residuals in order to obtain the adaptive weights. To perform the smoothing, we will use the method that we described in the last section. However, the c.d.f. is a function of the bandwidth (h) that has not yet been specified. In this section we describe how to calculate the bandwidth.

Any reasonable value for the bandwidth will be a function of the variability that we observe in the residuals, but the variability is sometimes difficult to estimate. The problem is that real-world data sets often contain outliers, and the sample standard deviation can be heavily influenced by the presence of a single outlier. Hence, we will use a function of the interquartile range as a robust estimate of variability. In this section we will describe how we can use the traditional estimates of percentiles to obtain our robust estimate of the standard deviation and, once an estimate of the bandwidth is obtained, how to use the smoothed c.d.f. to obtain an improved estimate of the standard deviation.

2.3.1 An Estimator of Variability Based on Traditional Percentiles

One robust estimator of the standard deviation is

$$\hat{\sigma}_t = \frac{\hat{x}_{t,.75} - \hat{x}_{t,.25}}{1.349} = \frac{\widehat{IQR}}{1.349}, \tag{2.2}$$

where $\widehat{IQR}$ is an estimator of the interquartile range. This estimator is a reasonable estimator of the population standard deviation (σ) if n is large and the observations are normally distributed. To see why this is a reasonable estimator, we note that in sampling from a standard normal distribution we should obtain a value of $(\hat{x}_{t,.75} - \hat{x}_{t,.25})$ near 1.349, because $\Phi^{-1}(.75) - \Phi^{-1}(.25) = 0.6745 - (-0.6745) = 1.349$. Hence, $\hat{\sigma}_t$ should approximate one when sampling from a standard normal distribution. This estimator of variability was suggested by Fenstad, Kjaernes, and Walløe (1980) when traditional percentiles are used. This estimator will be used in a formula to determine the bandwidth.

In this book we will use the function $h = 1.587\hat{\sigma}_t n^{-1/3}$ to compute the bandwidth. Note that h is directly proportional to the standard deviation and inversely proportional to the cube root of the sample size. This functional form was suggested by Polansky (1998), who used it as the bandwidth to estimate the distribution function. He showed that this bandwidth would minimize the mean integrated squared error with normally distributed variables.

In our work we are interested in the power of the adaptive test, which depends on the adaptive weights. Hence, we are interested in the accuracy of the adaptive weights. We will use $h = 1.587\hat{\sigma}_t n^{-1/3}$ as the bandwidth in this book, but later in this chapter we will investigate the functional form of the bandwidth to determine if the exponent of n and the constant 1.587 will produce accurate weights. Although there is a vast literature on smoothing methods, much of the literature on smoothing concerns estimation of the density function, not the cumulative distribution function. It should be pointed out that if we wanted to smooth the density function, rather than the distribution function, we would consider using $(-1/5)$ as the exponent of n. See Silverman (1986, Chapter 3) for an introduction to density estimation.

2.3.2 R Code for Finding the Bandwidth

The traditional estimates of the percentiles can be found using the R function `quantile`. The `type=6` option in the `quantile` function will give the traditional estimates that were described in Section 2.1. If we have a vector x and we want to find $\hat{x}_{t,0.25}$, $\hat{x}_{t,0.75}$, and $\hat{\sigma}_t$ we can use the following code:

```
q25 <- quantile(x,0.25,type=6)
q75 <- quantile(x,0.75,type=6)
sigma <- (q75-q25)/1.349
```

Now that we have a measure of variability, we can determine a reasonable value for the bandwidth. If the vector x is an $n \times 1$ vector and we wanted to use the function $h = 1.587\hat{\sigma}_t n^{-1/3}$, we could use the following code to determine the bandwidth:

```
n <- length(x)
h <- 1.587*sigma*n^(-0.33333333333)
```

The `length` function returns the number of elements in the vector x. This bandwidth will now be used to smooth the c.d.f., which will be used in the adaptive weighting.

2.3.3 An Estimator of Variability Based on Percentiles from the Smoothed Distribution Function

After the bandwidth is determined, a measure of variability can be defined by using percentiles that are found from the smoothed distribution function. We will assume that h has been determined, that the smoothed c.d.f. was obtained, and that a root finding method was used to compute the percentiles $\hat{x}_{.25}$ and $\hat{x}_{.75}$. We can now compute an improved estimate of the variability as

$$\hat{\sigma} = \frac{\hat{x}_{.75} - \hat{x}_{.25}}{1.349}.$$

This improved estimate of variability will be used in the adaptive weighting procedure.

2.4 NORMALIZING TRANSFORMATIONS

Tests of significance concerning parameters in a linear model usually require that the errors be normally distributed and independent and have constant variance. In our development of the adaptive test we will assume that the errors are independent and that they have constant variance, but we will not assume that they are normally distributed. Our approach will be to transform the model in such a way as to obtain a new model whose residuals are approximately normally distributed. Before we describe our approach, we will describe some of the traditional methods that have been suggested to normalize the errors.

2.4.1 Traditional Normalizing Methods

Many transformations have been proposed for normalizing the errors in models. (See Bartlett (1947) for examples of the square root and logarithmic transformations.) In regression models these transformations could be applied to the dependent variable, the independent variables, or all variables. One of the most popular transformations is due to Box and Cox (1964), who proposed a transformation that could be applied to the dependent variable. The Box–Cox method attempts to estimate a parameter (λ) that defines the transformation. For the ith observation let the observed dependent variable be Y_i and let the new transformed variable be $Y_i^{(\lambda)}$. Let $\dot{Y} = (Y_1 Y_2 \cdots Y_n)^{1/n}$ be the geometric mean of the dependent variable. The Box–Cox transformation is

$$Y_i^{(\lambda)} = \begin{cases} \dfrac{Y_i^{\lambda} - 1}{\lambda \dot{Y}^{(\lambda-1)}} & \text{for } \lambda \neq 0\,, \\ \dot{Y} \ln(Y_i) & \text{for } \lambda = 0 \end{cases}$$

for $i = 1, \ldots, n$. To use this transformation we must have $Y_i > 0$ for $i = 1, \ldots, n$. In their paper they used a likelihood method and a Bayesian method to estimate λ. The objectives of this transformation were to obtain the simplest possible regression model, to achieve a constant error variance, and to normalize the errors. Because this transformation is used for all of the observations we will call it a "global" transformation.

There are several problems with this approach. One problem is that the Box–Cox transformation is defined for $Y > 0$, but we often have negative values in models. For example, a researcher studying changes in tumor volumes from one time period to the next may have positive values for most observations but negative values for some observations. We are seeking a general method of normalizing residuals, so a method that is limited to positive data is not too useful for our purposes.

Another problem with Box–Cox transformations is that, if the researcher believes that the regression model, before transformation, is the correct model, then the transformed model will not be the correctly specified model. Because researchers are often interested in predicting the mean values for the untransformed data, the regression coefficients obtained from the transformed data may not be too useful. Also, Hernandez and Johnson (1980) pointed out that the transformed random variable cannot be normal unless the original random variable is lognormal.

A further difficulty in applying these global transformations is that non-normal distributions can depart from the normal distribution in many ways. Some examples of non-normal distributions include bimodal symmetric distributions, unimodal skewed distributions, and other unusual and bizarre distributions. Real-world distributions are often non-normal and may not approximate any of the textbook distributions. For example, consider the dot plot of the data in Figure 2.3. These data appear to be normally distributed except for two outliers, so it is difficult to see how a global transformation could normalize these data. To normalize these data, we need a "local" transformation that can transform only the two outliers.

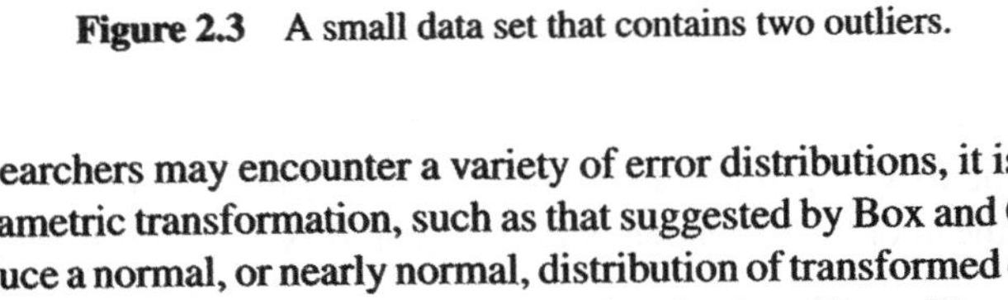

Figure 2.3 A small data set that contains two outliers.

Since researchers may encounter a variety of error distributions, it is unrealistic to expect a parametric transformation, such as that suggested by Box and Cox (1964), to always produce a normal, or nearly normal, distribution of transformed errors. Hence, we will not consider the Box–Cox transformation further. Nor will we consider any other global transformation, such as the square root transformation. Instead, we will use a weighting procedure that is flexible enough to normalize any continuous distribution.

Our approach to normalization of the error terms will be to transform the dependent and independent variables in such a way that the original structure of the model is maintained. We will use a transform-both-sides approach in all of our adaptive tests that will allow us to use the original model specification. The advantages of the transform-both-sides approach have been described by Carroll and Ruppert (1988). We will use an adaptive weighting procedure to normalize the errors. This will be accomplished by a local transformation that can be used with any data set.

2.4.2 Normalizing Data by Weighting

We describe a weighting method that has the flexibility to adequately normalize continuous distributions. The basic idea is to use the probability integral transform approach to normalize the data, with the c.d.f. approximated by the smoothed empirical c.d.f. To describe this procedure more completely, let us assume that the c.d.f. is known and is strictly increasing. Let X be a random variable that we wish to normalize and let $F(x)$ be the c.d.f. of X. Consider the random variable $U = F(X)$ that has c.d.f. $F_U(u)$. Since $F(x)$ is strictly increasing it follows that

$$F_U(u) = P(U \leq u) = P[F(X) \leq u] = P[X \leq F^{-1}(u)] = F[F^{-1}(u)] = u,$$

which implies that U is uniformly distributed on the interval [0,1]. The transformation $U = F(X)$ is the probability integral transform, as described by DeGroot (1986, p. 154). Let $\Phi(\cdot)$ be the c.d.f. of a standard normal random variable. Now consider the random variable $Z = \Phi^{-1}(U) = \Phi^{-1}[F(X)]$ with c.d.f. $F_Z(z)$. Then

$$F_Z(z) = P(Z \leq z) = P\{\Phi^{-1}[F(X)] \leq z\} = P[F(X) \leq \Phi(z)].$$

But, since $U = F(X)$ and $F_U(u) = u$,

$$F_Z(z) = P[U \leq \Phi(z)] = F_U[\Phi(z)] = \Phi(z).$$

Hence, $Z = \Phi^{-1}[F(X)]$ has a standard normal distribution.

We will write $X \sim N(\mu, \sigma^2)$ to indicate that X is distributed as a normal random variable with a mean of μ and a standard deviation of σ. Now suppose we want a normalized variable to have a standard deviation of σ. Then $\sigma Z = \sigma\Phi^{-1}[F(X)] \sim N(0, \sigma^2)$. So, if we know σ and $F(x)$, we could compute the transformed value for the ith observation as $x_i^* = \sigma\Phi^{-1}[F(x_i)]$. In our weighting procedure we will find it useful to write the transformed value as a product of a weight and a centered untransformed value. For the ith observation, if $x_i \neq x_{.5}$, we define the weight as

$$w_i = \frac{\sigma\Phi^{-1}[F(x_i)]}{x_i - x_{.5}}. \tag{2.3}$$

If $x_i = x_{.5}$, we set $w_i = 1$. We can now write the transformed value x_i^* as

$$x_i^* = \sigma\Phi^{-1}[F(x_i)] = \frac{\sigma\Phi^{-1}[F(x_i)]}{x_i - x_{.5}}(x_i - x_{.5}) = w_i(x_i - x_{.5}), \tag{2.4}$$

for $x_i \neq x_{.5}$. The transformed value x_i^* will be a normally distributed random variable with mean zero and standard deviation σ.

To obtain a better understanding of the normalization method, let us assume that an observation x is from a population that has a median of zero and a standard deviation of one. To normalize these random variables, we would use Equation (2.4) with $x_{.5} = 0$ and $\sigma = 1$ so that $x^* = \sigma\Phi^{-1}[F(x)]$. Suppose we have a random variable X from a right skewed distribution. In Figure 2.4 the c.d.f. for the right skewed distribution ($F(x)$) is displayed by a dotted line and the c.d.f. of the standard normal distribution is displayed as a thick line. Now let us consider a point, $x = 2.10$, such that $F(x) = 0.9$. To normalize this value, we compute $x^* = \Phi^{-1}[F(x)] = \Phi^{-1}[.9] = 1.28$, which implies that $\Phi(x^*) = 0.9$. That is, because x is at the 90th percentile of the right skewed distribution, we seek the standard normal variate that is at the 90th percentile, as shown in Figure 2.4. In this example we find $w_i = \Phi^{-1}[F(x)]/x_i = 1.28/2.10 = 0.61$. Hence, we are downweighting the observation to normalize the distribution of the transformed data. In Formula (2.4) we multiply $\Phi^{-1}[F(x_i)]$ by σ to restore the standard deviation of the original data.

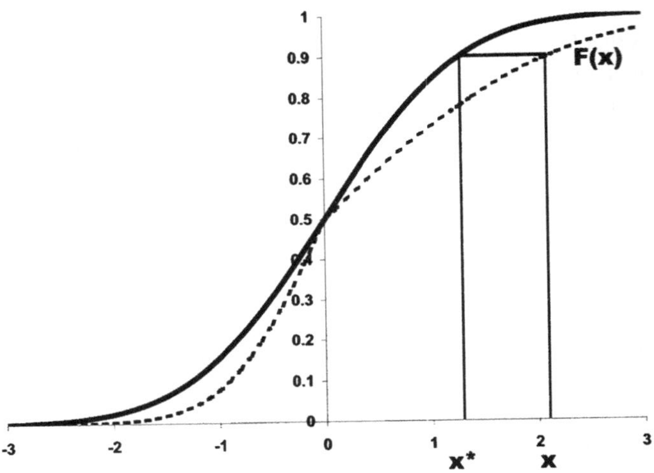

Figure 2.4 An illustration of the normalization process. The c.d.f. for the right skewed distribution is displayed by a dotted line and the c.d.f. of the standard normal distribution is displayed as a thick line. For this example the raw data value is $x = 2.10$ and the transformed data value is $x^* = 1.28$.

In practice, we do not know σ, $F(x)$, and $x_{.5}$, so we cannot compute the weights w_i for $i = 1, \ldots, n$. Our approach will be to smooth the empirical c.d.f. using a normal kernel smoothing function with bandwidth h to obtain $\hat{F}_h(x)$, which will be an estimate of $F(x)$. We can also use $\hat{F}_h(x)$ to find an estimate of the median ($\hat{x}_{.5}$), which is obtained by an iterative root-finding algorithm. To ensure that the variability in the transformed data is appropriate, we use the robust estimator $\hat{\sigma} = (\hat{x}_{.75} - \hat{x}_{.25})/1.349$ that was described in Section 2.3. In the next two sections we will give details concerning this weighting method and the specification of the bandwidth.

2.5 THE WEIGHTING ALGORITHM

In this section we describe the adaptive weighting procedure that we will use in all of the adaptive tests of significance in this book. We begin by computing the traditional estimates of the 25th and 75th percentiles and the initial estimate of the standard deviation using Equation (2.2). Next, we compute the bandwidth as $h = 1.587\hat{\sigma}_t n^{-1/3}$. With this bandwidth we can now smooth the c.d.f. and use an iterative procedure to find $\hat{x}_{.25}$, $\hat{x}_{.5}$, and $\hat{x}_{.75}$. Next, we compute the final estimate of variability as $\hat{\sigma} = (\hat{x}_{.75} - \hat{x}_{.25})/1.349$. This estimate will be used to standardize the normally distributed data so that it has roughly the same variability as the original data.

After we have determined the bandwidth, we use the weighting method that was described in Section 2.4 to normalize the data. Let $z_i = \Phi^{-1}[\hat{F}_h(x_i)]$ be the normal score for the ith observation and let $s_i = (x_i - \hat{x}_{.5})/\hat{\sigma}$ be the standardized value. Note that these are computed using the smoothed c.d.f. and the estimates of the median and σ. The normalizing weight for the ith observation, if $x_i \neq \hat{x}_{.5}$, is given by

$$\hat{w}_i = \frac{\hat{\sigma}\Phi^{-1}[\hat{F}_h(x_i)]}{x_i - \hat{x}_{.5}} = \frac{\Phi^{-1}[\hat{F}_h(x_i)]}{(x_i - \hat{x}_{.5})/\hat{\sigma}} = \frac{z_i}{s_i}. \tag{2.5}$$

If $x_i = \hat{x}_{.5}$, we set $\hat{w}_i = 1$. To obtain the transformed values, which will be centered at the median, we will compute $x_i^* = \hat{w}_i(x_i - \hat{x}_{.5})$ for $i = 1, \ldots, n$. Because the weighting procedure will often be used in this book, we provide a summary of the steps involved in the weighting.

The steps in the adaptive weighting algorithm are:

1. Use traditional percentile estimators to compute $\hat{\sigma}_t = (\hat{x}_{t,.75} - \hat{x}_{t,.25})/1.349$.
2. Compute the bandwidth $h = 1.587\hat{\sigma}_t n^{-1/3}$.
3. Smooth the data to obtain $\hat{F}_h(x_i)$ for $i = 1, \ldots, n$.
4. Use $\hat{F}_h(x)$ in an iterative procedure to compute $\hat{x}_{.25}$, $\hat{x}_{.5}$, and $\hat{x}_{.75}$. Then compute $\hat{\sigma} = (\hat{x}_{.75} - \hat{x}_{.25})/1.349$.
5. Compute the normal scores $z_i = \Phi^{-1}[\hat{F}_h(x_i)]$ and the standardized values $s_i = (x_i - \hat{x}_{.5})/\hat{\sigma}$ for $i = 1, \ldots, n$.
6. Compute the estimates of the normalizing weights

$$\hat{w}_i = \begin{cases} z_i/s_i & \text{if } s_i \neq 0, \\ 1 & \text{if } s_i = 0 \end{cases}$$

for $i = 1, \ldots, n$.

In most of the adaptive tests of significance, we will use this algorithm on the residuals to normalize the residuals and to obtain the adaptive weights. These weights will be used along with a weighted least squares estimation method to obtain a linear model with normalized residuals. An example will illustrate the computations involved.

2.5.1 An Example of the Weighing Procedure

In order to describe the computations that are required for the adaptive weighting and to provide some insight into the effectiveness of the normalization method, we will consider a data set containing the weights (in pounds) of $n = 11$ undergraduate males at a college. The ordered weights are 121, 125, 144, 148, 149, 156, 162, 167, 182, 195, and 318 pounds. These values are displayed in the second column of Table 2.1.

It is easy to verify that $\hat{x}_{t,.25} = x_{(3)} = 144$, $\hat{x}_{t,.75} = x_{(9)} = 182$, and $\hat{\sigma}_t = 28.17$. Hence, we use $h = 1.587(28.17)11^{-1/3} = 20.10$ as the bandwidth. At this point a root finding algorithm is used to find the median $\hat{x}_{.5} = 158.20$ and the final estimate of the variability as $\hat{\sigma} = 34.64$. The smoothing formula is used to find $\hat{F}_h(x)$ for each of the 11 values, which are shown in the third column of Table 2.1. The fourth column contains $z_i = \Phi^{-1}[\hat{F}_h(x_i)]$. Of course, a computer would usually be used for most of these calculations, but it is possible to find z_i from tables of the normal distribution. The standardized values $s_i = (x_i - \hat{x}_{.5})/\hat{\sigma}$ are in the fifth column. The adaptive weights, which are in the sixth column, can then be calculated as z_i/s_i. Note that most of the weights are between 0.873 and 1.106, except for the relatively low adaptive weight given to the last observation, which was a bit of an outlier. Next, we compute $x_i^* = \hat{w}_i(x_i - \hat{x}_{.5})$ for $i = 1, \ldots, n$, and in the last column we give $x_i^* + \hat{x}_{.5}$, which are the normalized observations scaled to have a standard deviation of $\hat{\sigma} = 34.64$ and a median of $\hat{x}_{.5} = 158.20$. Note that the normalized values closely approximate the original observations except for the outlier.

2.5.2 R Code for Weighting the Observations

In this section we give the R functions that will be used in the weighting process. We will assume that the bandwidth h has already been computed. We now need to smooth the empirical c.d.f. and determine, using a root-finding algorithm, the final estimates for the percentiles. The root-finding method `rootcdf`, which was described earlier in this chapter, has 6 arguments. The first argument is simply the vector that is to be smoothed. The bandwidth is the second argument. The third argument is the percentile, the fourth argument is a lower bound on the percentile, and the fifth argument is the upper bound on the percentile. The last argument is the tolerance, which must be greater than the width between the lower and upper limits after many iterations of the bisection method have been performed.

For the fourth argument we have specified a lower limit that equals the minimum minus the range and for the fifth argument we have specified an upper limit that equals the maximum plus the range. This may appear to be much too broad but it

Table 2.1 Weights (in pounds) of $n = 11$ undergraduates along with the intermediate calculations for the adaptive weights

Observation	Pounds	$\hat{F}_h(x)$	z_i	s_i	w_i	$x_i^* + \hat{x}_{.50}$
1	121	0.118	-1.187	-1.074	1.106	117.06
2	125	0.146	-1.053	-0.958	1.098	121.733
3	144	0.331	-0.436	-0.41	1.064	143.098
4	148	0.378	-0.311	-0.294	1.055	147.44
5	149	0.39	-0.279	-0.266	1.053	148.517
6	156	0.474	-0.066	-0.063	1.035	155.924
7	162	0.544	0.112	0.11	1.017	162.063
8	167	0.6	0.254	0.254	0.999	166.995
9	182	0.74	0.645	0.687	0.938	180.532
10	195	0.823	0.928	1.062	0.873	190.344
11	318	0.955	1.691	4.613	0.367	216.766

does not take the bisection algorithm many iterations to narrow that interval. The user could specify a more efficient method but it is not easy to specify a narrow initial interval that will bracket the root for all data sets. Because we do need great accuracy for the determination of the 50th percentile, we have specified the tolerance to be one-millionth of the range. In the last line of the R code we compute our final estimate of variability.

```
range <- max(x)-min(x)
lower <- min(x)  - range
upper <- max(x)  + range
tolerance <- 0.000001*range
cdf25 <- rootcdf(x,h,0.25,lower,upper,tolerance)
cdf50 <- rootcdf(x,h,0.50,lower,upper,tolerance)
cdf75 <- rootcdf(x,h,0.75,lower,upper,tolerance)
sigma <- (cdf75-cdf25)/1.349
```

If we used this R code on our data set of the weights of $n = 11$ undergraduate males, we would find, using $h = 20.10$, that $\hat{\sigma} = 34.64$.

Because we have an improved estimate of the 50th percentile and a good final estimate of variability, we can find the standardized values $s_i, i = 1, \ldots, n$, using the first line of the following code. Next, we define the weight vector `w` to have length n. In the loop over all of the observations we first compute $\hat{F}_h(x_i)$ using the smoothing function that was described earlier in this section and assign it to the variable `fhat`. Next we find $z_i = \Phi^{-1}[\hat{F}_h(x_i)]$ for $i = 1, \ldots, n$ by using the R function `qnorm` and assign it to the variable `z`. The weights are assigned in the `if` statement as we have described earlier in this section, but we check to see if s_i is close to zero. If it is then the weight is set to one, but usually all of the weights will be set to z_i/s_i.

The closeness to zero will be determined by `smallvalue`. This will typically be set to a small value, but to avoid the possibility of a bizarre weight being computed when s_i is close to zero, we recommend that `smallvalue` be much larger than `tolerance`.

```
s <- (x-cdf50)/sigma
w  <- matrix(0,n,1)
smallvalue <- 100*tolerance
for (i in 1:n) {
  fhat <- cdfhat(x,h,x[i])
  z <- qnorm(fhat)
  if( (abs(s[i]) >= smallvalue) )
     w[i] <- z/s[i]
    else
     w[i] <- 1
  }
```

This code, when used with the example data, gives the values for $\hat{F}_h(x)$, z_i, s_i, and w_i that are shown in Table 2.1. Although we have used this R code to normalize a set of observations, we will usually use this code to normalize a set of residuals that we obtain from a linear model. These procedures will be used in a two-sample test in the next chapter.

2.6 COMPUTING THE BANDWIDTH

Since we are going to use $h = 1.587\hat{\sigma}_t n^{-1/3}$ as the smoothing parameter in the adaptive weighting procedure, we need to determine if it is a reasonable formula for the bandwidth. Is the exponent of n the correct exponent? Will the constant 1.587 be effective in normalizing the data? In this section we describe the results from several simulation studies to verify that the formula for the bandwidth is a reasonable formula. In these studies data were generated from one of nine distributions and the bandwidth was determined by the formula $h = K\hat{\sigma}_t n^{-1/3}$ for a large number of values of K and n. For each data set a statistic that measures the accuracy of the adaptive weighting was computed, and these were averaged over many data sets to produce an estimate of the effectiveness of that bandwidth. These results were used to determine if the constant and the exponent of n are close to optimal values. In addition, we will see that the weighting method is not too sensitive to h. We begin by describing the distributions that we will use in this book.

2.6.1 Error Distributions

In simulation studies we will use nine error distributions in the linear models. In addition to the normal error distribution, we will use 8 other non-normal error distributions, which are listed in Table 2.2. All nine distributions have means equal to zero. The four symmetric error distributions include the uniform, the standard normal, the t (with 4 degrees of freedom), and a symmetric bimodal distribution. Let $N(\mu, \sigma^2)$ denote the normal distribution with a mean of μ and a variance of σ^2. The symmetric bimodal errors are generated from a $N(-\sqrt{9/13}, 4/13)$ distribution with probability $1/2$ and from a $N(\sqrt{9/13}, 4/13)$ distribution with probability $1/2$.

Four of the five skewed error distributions were members of the generalized lambda family of distributions (see Ramberg *et al.* (1979)), which has a four-parameter inverse c.d.f. of

$$F^{-1}(u) = \lambda_1 + \left[u^{\lambda_3} - (1-u)^{\lambda_4}\right] / \lambda_2.$$

The location parameter is λ_1 and the scale parameter is λ_2 . The four parameters are calculated, using the formulas in Ramberg *et al.* (1979), to obtain the desired mean, variance, skewness, and kurtosis. In this book, skewness (α_3) is defined to be $E[(x-\mu)^3/\sigma^3]$ and kurtosis (α_4) is defined to be $E[(x-\mu)^4/\sigma^4]$.

This family of distributions is popular because it can be used to generate random numbers from a distribution having a specified mean, variance, skewness, and kurtosis. For all four of the generalized lambda distributions that we used in these simulations we set $\mu = 0$ and $\sigma^2 = 1$. An example will illustrate how we generated random numbers from one of these distributions. If we want to generate random variables from a skewed distribution having $\alpha_3 = 1$ with a low kurtosis of $\alpha_4 = 4.2$, we use the tables in Ramberg *et al.* (1979) to obtain the parameters $\lambda_1 = -0.7869$, $\lambda_2 = 0.1142$, $\lambda_3 = 0.0212$, and $\lambda_4 = 0.1243$. To generate a random variable from this skewed, low kurtosis distribution, we first generate a uniform [0,1] random number (U) and then compute $\lambda_1 + [U^{\lambda_3} - (1-U)^{\lambda_4}]/\lambda_2$ as the random number. If we want to generate random numbers from the skewed, high kurtosis generalized lambda distribution we set $\alpha_3 = 1$ and $\alpha_4 = 8.4$, which is used to find λ_1, λ_2, λ_3, and λ_4, and we generate the random numbers in a similar fashion. For highly skewed distributions having $\alpha_3 = 2$, we set $\alpha_4 = 11.4$ for the low kurtosis distribution and $\alpha_4 = 15.6$ for the high kurtosis distribution.

In addition to the four generalized lambda distributions, we also used a skewed bimodal distribution that has errors generated from a $N(-1/2, 1/4)$ distribution with probability $3/4$ and from a $N(3/2, 1/4)$ distribution with probability $1/4$. For details and graphs of these distributions see O'Gorman (2006b).

Note that there are more skewed distributions than symmetric distributions and more long-tailed distributions than short-tailed distributions. Also, two of the nine are bimodal distributions. The uniform, normal and the t distribution with 4 degrees of freedom are familiar to statisticians so we do not need to describe them further. The density of the symmetric bimodal distribution is displayed in Figure 2.5 and the density of the bimodal skewed distribution is shown in Figure 2.6. The generalized lambda distributions, while commonly used in simulation studies, may not be familiar to most statisticians. The highly skewed, low kurtosis distribution is displayed in Figure 2.7 and the highly skewed, high kurtosis distribution is displayed in Figure 2.8. We expect that the wide variety of distributions include some that are reasonably close to what are found in the real world.

Table 2.2 Nine distributions that will be used in the simulation studies

Distribution	Description
Uniform	Uniform on the interval $[-\sqrt{3}, \sqrt{3}]$
Normal	Standard normal
t with $d.f. = 4$	t distribution with 4 degrees of freedom
Symmetric bimodal	$.5N(-\sqrt{9/13}, 4/13) + .5N(\sqrt{9/13}, 4/13)$
Skewed, low kurtosis	Generalized lambda with $\alpha_3 = 1$ and $\alpha_4 = 4.2$
Skewed, high kurtosis	Generalized lambda with $\alpha_3 = 1$ and $\alpha_4 = 8.4$
Highly skewed, low kurtosis	Generalized lambda with $\alpha_3 = 2$ and $\alpha_4 = 11.4$
Highly skewed, high kurtosis	Generalized lambda with $\alpha_3 = 2$ and $\alpha_4 = 15.6$
Skewed bimodal	$.75\ N(-1/2, 1/4) + .25N(3/2, 1/4)$

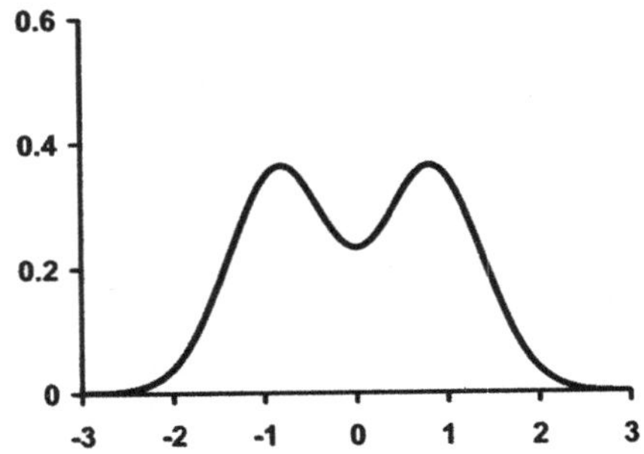

Figure 2.5 The symmetric bimodal distribution.

2.6.2 Measuring Errors in Adaptive Weighting

In this section our objective is to determine the bandwidth h that will most accurately normalize a variety of non-normal distributions. In simulation studies we know σ and $F(x)$, so we can easily compute the population weights for $w_i, i = 1, \ldots, n$, using Formula (2.3). We can also use the simulation data to compute the estimated weights $\hat{w}_i, i = 1, \ldots, n$, using Formula (2.5). Our objective in these simulation studies is to find a bandwidth that produces estimated weights $\hat{w}_i$ that are close to the population weights w_i for $i = 1, \ldots, n$. The idea is that if the estimated weights approximate the population weights, then the transformed random variables will be approximately normal. We will use $h = K\hat{\sigma}n^{-1/3}$ as the bandwidth, where K is a constant to be specified. Later we will address the appropriateness of the exponent of n. The approach to determining the correct value of K uses the following steps for each data set:

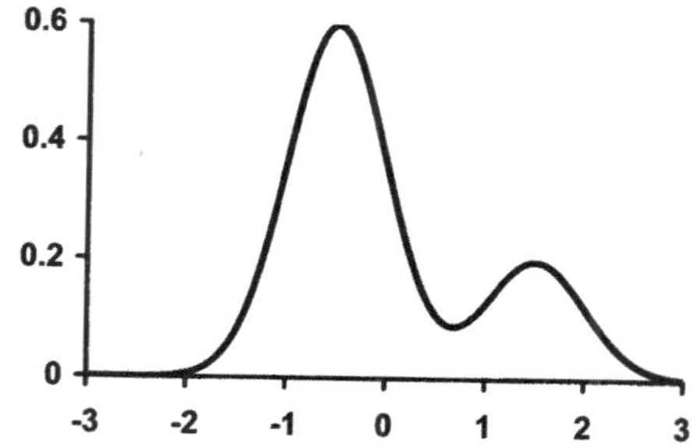

Figure 2.6 The skewed bimodal distribution.

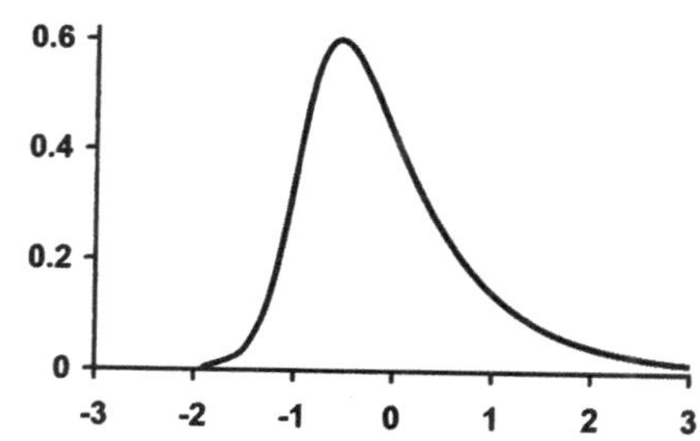

Figure 2.7 The highly skewed, low kurtosis generalized lambda distribution.

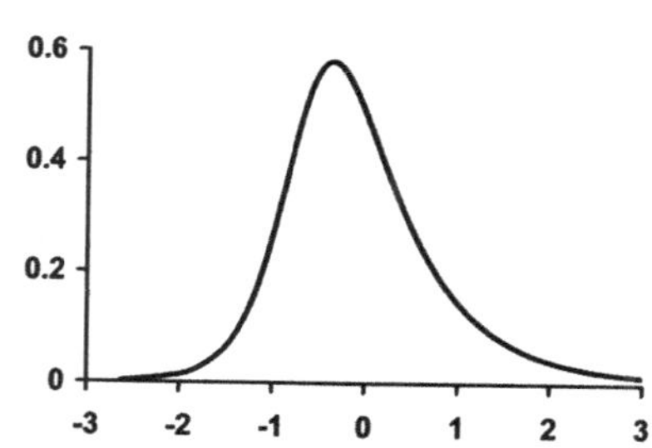

Figure 2.8 The highly skewed, high kurtosis generalized lambda distribution.

1. For a specified value of K, compute the bandwidth and, using the methods described in the previous section, obtain the normalizing weights as

$$\hat{w}_i = \frac{\hat{\sigma}\Phi^{-1}[\hat{F}_h(x_i)]}{x_i - \hat{x}_{.5}}$$

for $i = 1, \ldots, n$.

2. In simulation studies we know $F(x)$ so we can compute $x_{.25}$, $x_{.5}$, $x_{.75}$ and the population weights as

$$w_i = \frac{\sigma\Phi^{-1}[F(x_i)]}{x_i - x_{.5}},$$

where σ is computed as $\sigma = (x_{.75} - x_{.25})/1.349$.

3. Compute the mean square error of weights (MSEW) as

$$\text{MSEW} = \frac{1}{n}\sum_{i=1}^{n}(\hat{w}_i - w_i)^2.$$

The MSEW is a measure of the inaccuracy of the weighting procedure used to normalize the data.

In a simulation study designed to determine the most accurate bandwidth, we used 21 equally spaced values of K ranging from 1 to 3 in the formula $h = K\hat{\sigma}n^{-1/3}$. The nine error distributions were used in the study with sample sizes of $n = 25, 50, 100, 200$, and 400. For each value of K and n we generated 100,000 data sets and for each data set the MSEW was computed. The average of these MSEW over all of the data sets was computed as a measure of the inaccuracy in the weighting for each value of K and n. We will usually tabulate the square root of the average of the MSEW, which will be abbreviated as RMSEW.

We also attempted to determine the proper value of K by considering a slightly different measure of inaccurate weighting. Recall that in weighted least squares regression we will obtain the same test statistic if we use $\hat{w}_i = 2$, for $i = 1, \ldots, n$ or $\hat{w}_i = 1$, for $i = 1, \ldots, n$. Because in Chapters 3 to 8 we will be using the weights in a weighted least squares analysis with univariate dependent variables, we will not be too interested in the values of $\hat{w}_i$, for $i = 1, \ldots, n$, but we will be interested in their relative values. Consequently, for most of the adaptive tests a more relevant statistic is the mean squared error of the mean adjusted weights (MSEMAW), which we define as

$$\text{MSEMAW} = \frac{1}{n}\sum_{i=1}^{n}\left(\hat{w}_i/\bar{\hat{w}} - w_i/\bar{w}\right)^2,$$

where $\bar{\hat{w}}$ is the mean of the estimated weights and $\bar{w}$ is the mean of the population weights. Note that, if the distribution is normal so that $w_i = 1$ for $i = 1, \ldots, n$, then we would obtain MSEMAW $= 0$ if $\hat{w}_i = 2$ for $i = 1, \ldots, n$ or if $\hat{w}_i = 1$ for $i = 1, \ldots, n$. We will use the square root of the average of the MSEMAW, which will be called the RMSEMAW, in the figures to determine a reasonable value of K. For the multivariate adaptive tests and the tests for repeated measures that we will consider in Chapters 9 and 10 we are concerned with the absolute accuracy of the weights, so we will use both RMSEW and RMSEMAW to evaluate the effectiveness of the adaptive weighting procedure.

2.6.3 Simulation Studies

In Figure 2.9 we plot the RMSEW as a function of K in the formula $h = K\hat{\sigma}n^{-1/3}$, with $n = 50$ observations, for nine distributions. The results show that:

- The RMSEW decreases in the interval $[1.0, 1.2]$ for observations that were generated from all distributions. Thus, it is not necessary to consider values of K in that interval as minimizing the RMSEW.

- With data from 6 distributions the RMSEW increases for $K > 2.0$. With data from the normal distribution, the t distribution with $df = 4$, and for one skewed distribution, the RMSEW decreases for $K > 2.0$.

Hence, no single value of K will minimize the RMSEW. Although there is no single value of K that minimizes the RMSEW over these nine distributions, there may be a value of K that produces a low value of the RMSEW for most of the distributions.

In sampling from a standard normal distribution, we noted in Figure 2.9 that RMSEW decreases as h increases. If h is so large that there is clearly oversmoothing, then $\hat{F}_h(x)$ will not approximate $\hat{F}_{empirical}(x)$ for $i = 1, \ldots, n$. In this situation z_i will tend to be much smaller than x_i in absolute value. However, $\hat{\sigma} = (\hat{x}_{.75} - \hat{x}_{.25})/1.349$ will greatly exceed one because of the oversmoothing, so s_i will tend to be smaller than x_i in absolute value. Thus, it is possible with large values of h to obtain weights near one. See Exercise 2.11.

Because a researcher, in a real-world situation, would not know the distribution and because the distribution would rarely be any of these nine distributions, our objective should be to minimize the RMSEW over the types of distributions that might be used in a test of significance. Since these nine distributions were selected to represent a wide variety of distributions, it seems reasonable to average the RMSEW over the nine distributions to obtain an overall measure of inaccuracy of estimation of the adaptive weights.

Figure 2.10 displays the average, over all nine distributions, of the RMSEW for values of K in the interval $[1.0, 3.0]$ for $n = 25, 50, 100, 200$, and 400. This figure illustrates the following points:

- Values of K outside the interval $[1.4, 2.0]$ will not minimize the average RMSEW.

- For each sample size the minimum is in the narrow range of $[1.5, 1.7]$. These minimums are indicated by filled circles on the lines in Figure 2.10.

- A value of K near 1.6 will give an average RMSEW near the minimum for all sample sizes.

We see that the average RMSEW is not too sensitive to K and that the relationship between K and the average RMSEW is consistent over the values of n that we have used. We can draw several important conclusions from this simulation:

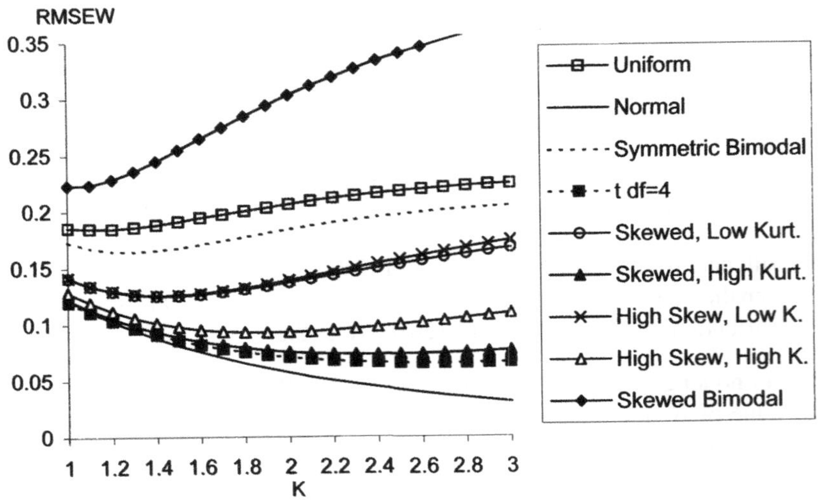

Figure 2.9 The root mean squared error of weights (RMSEW) for data having $n = 50$ observations generated from the nine distributions. The horizontal axis is the value of K in the formula for the bandwidth $h = K\hat{\sigma}n^{-1/3}$.

- Because the minimums are near $K = 1.6$ for samples sizes from $n = 25$ to $n = 400$, we can be assured that the exponent of n is not too far from an optimum value. [See Exercise 2.10.] As we noted earlier in the chapter, several authors have recommended that $(-1/3)$ be used as the exponent to smooth an empirical distribution function. Hence, we will continue to use $n^{-1/3}$ in the bandwidth formula.

- The bandwidth formula $1.587\hat{\sigma}n^{-1/3}$, which was proposed by Polansky (1998) to estimate the c.d.f., produces values of h that are close to those that would minimize the average RMSEW for each sample size.

Another approach to finding the right bandwidth is to find the value of h that will minimize the root mean square of the mean adjusted weights (RMSEMAW). Most of the adaptive tests of significance in this book use the adaptive weights in a weighted least squares regression. For those tests the relative weights are more important than the absolute weights so the RMSEMAW should be minimized. In Figure 2.11 we display the RMSEMAW for the same sample sizes and values of K that we used in Figure 2.10. The minimum value of the average RMSEMAW is shown by the filled circles on the lines for each sample size. These results are similar to those found for the average RMSEW. We conclude that the bandwidth formula $h = 1.587\hat{\sigma}n^{-1/3}$ will produce a value of h that will be close to minimizing the relative mean squared error of the weights. Of course, in the adaptive tests that will be described in subsequent chapters, the researcher could use other values of h, but

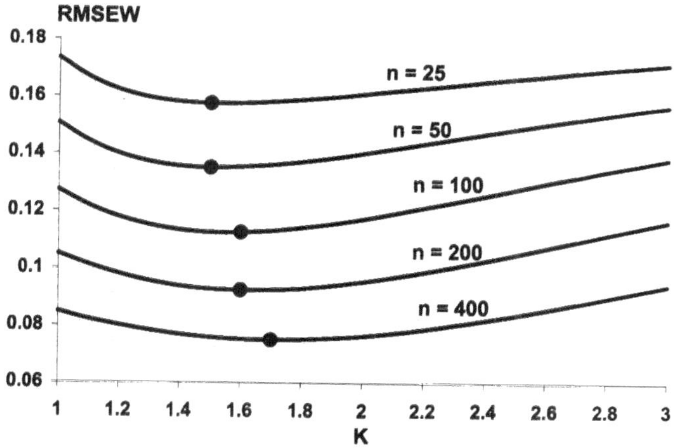

Figure 2.10 The average of the root mean squared error of weights (RMSEW) averaged over the nine distributions for several sample sizes. The horizontal axis is the value of K in the formula for the bandwidth $h = K\hat{\sigma}n^{-1/3}$. The minimum value of RMSEW for each sample size is indicated by a filled circle.

the results shown in this chapter should provide some guidance concerning the choice of the bandwidth.

2.7 EXAMPLES OF TRANSFORMED DATA

We explained in Section 2.4 that if we knew σ and $F(x)$ we could normalize the observations by computing

$$w_i = \frac{\sigma\Phi^{-1}[F(x_i)]}{x_i - x_{.5}}$$

for $i = 1, \ldots, n$ and then find the transformed values $x_i^* = w_i(x_i - x_{.5})$ for $i = 1, \ldots, n$. With real data we use our method with $h = 1.587\hat{\sigma}_t n^{-1/3}$ to compute the normalizing weights

$$\hat{w}_i = \frac{\hat{\sigma}\Phi^{-1}[\hat{F}_h(x_i)]}{x_i - \hat{x}_{.5}}$$

for $i = 1, \ldots, n$ and then find the transformed values $\hat{x}_i^* = \hat{w}_i(x_i - \hat{x}_{.5})$ for $i = 1, \ldots, n$.

One way to evaluate the adequacy of the normalizing transformation is to see if it appears to normalize a non-normal distribution. For example, consider the histogram shown in Figure 2.12 of $n = 400$ observation that were generated from a highly skewed, low kurtosis distribution that was described earlier in this chapter. As expected, the distribution of the generated data appears to be quite skewed with a

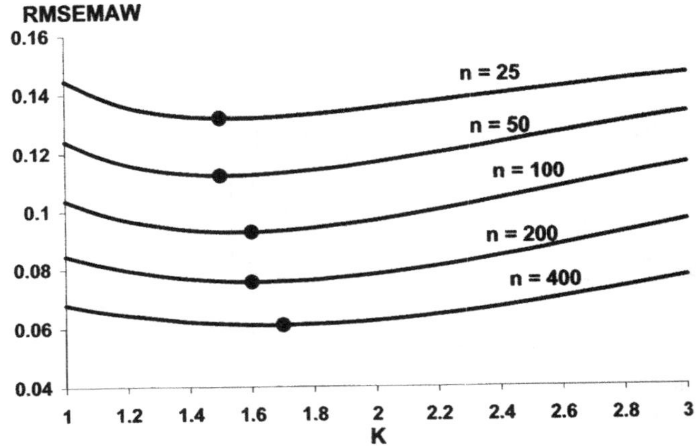

Figure 2.11 The average of the root mean squared error of adjusted weights (RMSEMAW) averaged over all nine distributions for several sample sizes. The horizontal axis is the value of K in the formula for the bandwidth $h = K\hat{\sigma}n^{-1/3}$. The minimum value of RMSEMAW is indicated by a filled circle for each sample size.

long right tail. However, after we compute the weights and the transformed data ($\hat{x}_i^*$ for $i = 1, \ldots, n$), we obtain a distribution that is shown in Figure 2.13, which is not skewed. Also, its overall shape approximates that of a normal distribution. Hence, the transformation was quite successful with this skewed data set.

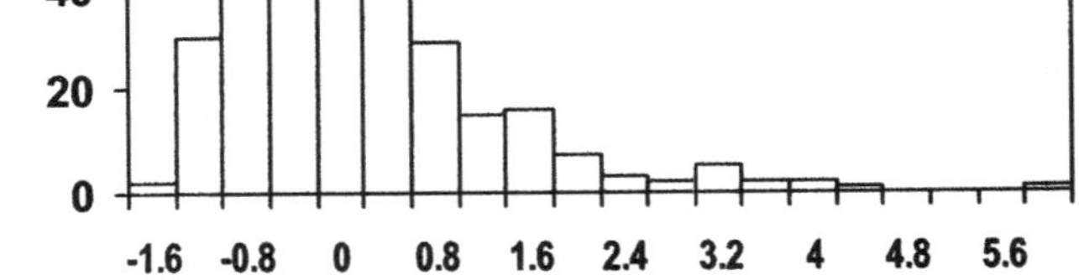

Figure 2.12 Histogram of the raw data on one data set of size $n = 400$ generated from a highly skewed, low kurtosis distribution.

A greater challenge was posed by the $n = 400$ observations displayed by the histogram in Fig 2.14. These data were generated from the skewed bimodal distribution

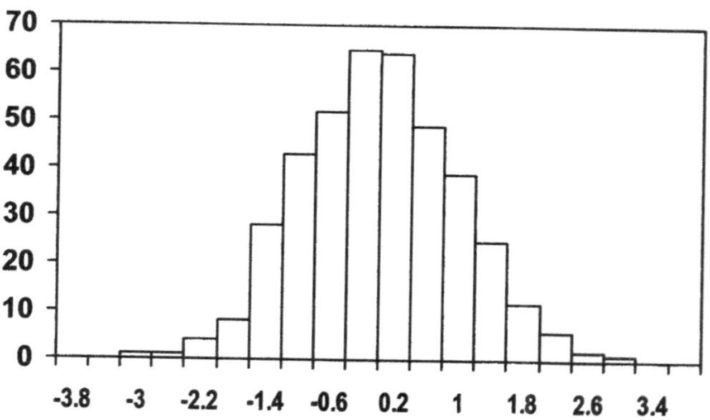

Figure 2.13 Histogram of the normalized data on one data set of size $n = 400$ that was obtained by weighting the highly skewed data that is displayed in Figure 2.12.

and, as expected, these data show one mode near -0.75 and another mode near 1.50. A histogram of the transformed data is shown in Figure 2.15. This histogram shows little evidence of bimodality and appears to be approximately normal.

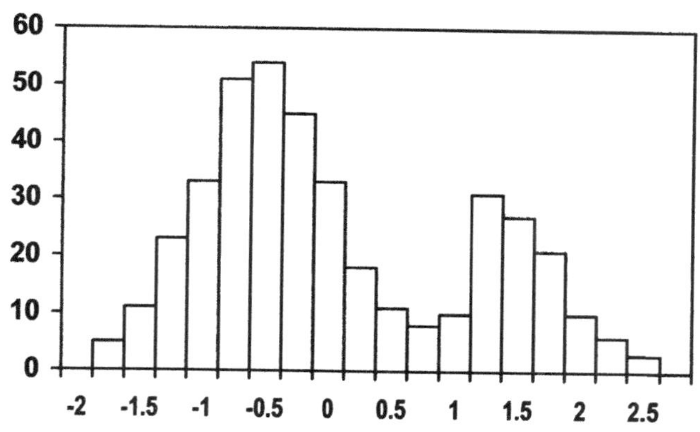

Figure 2.14 Histogram of the raw data on one data set of size $n = 400$ generated from a bimodal skewed distribution.

Data sets generated from the other seven distributions also demonstrated that for $n = 400$ the weighting procedure did an excellent job of normalizing the non-normal distributions. In addition, the procedure did not greatly distort the data that was generated from a normal distribution.

The smoothing method and the normalization method will be used to perform many of the tests in this book. In the next chapter we will illustrate the use of these

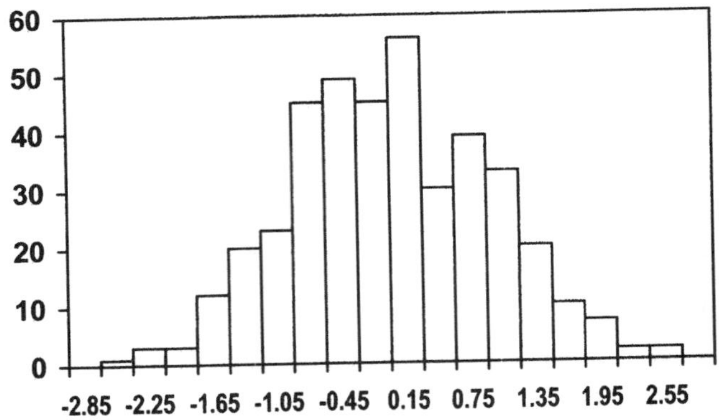

Figure 2.15 Histogram of the normalized data on one data set of size $n = 400$ that was obtained by weighting the bimodal data that is displayed in Figure 2.14.

methods with two-sample data. The objective is to construct an adaptive two-sample test that is more powerful, for a variety of error distributions, than the traditional two-sample *t* test.

EXERCISES

2.1 Suppose we have obtained weights (in pounds) for $n = 11$ soccer players. The order statistics are $\{125, 132, 141, 147, 148, 156, 161, 162, 166, 175, 189\}$.

- **a)** Compute the 25th, 50th, and 75th percentiles using the traditional method of computing percentiles.
- **b)** Compute the interquartile range, using the traditional estimates, and $\hat{\sigma}_t$.
- **c)** Suppose $x_{(11)} = 310$ instead of $x_{(11)} = 189$. Will this change the percentiles that were computed in part a?

2.2 Suppose a researcher obtained the following weight changes (post - pre) for $n = 19$ individuals who were enrolled in a program of diet and exercise.

```
-37, -15, -13, -12, -8, -6, -5, -4, -3, -1, 0,
  1,   3,   4,   5,  7, 10, 14,  28
```

Using the traditional method of estimating percentiles, compute

- **a)** $\hat{x}_{t,.25}$
- **b)** $\hat{x}_{t,.75}$
- **c)** $\hat{\sigma}_t$

2.3 Suppose the $n = 4$ residuals from a regression model are $x_{(1)} = -3$, $x_{(2)} = -2$, $x_{(3)} = -1$, and $x_{(4)} = +6$

- **a)** Let $h = 0.000001$. Without doing complex calculations, what are the approximate values of $\hat{F}_h(-4)$, $\hat{F}_h(-2.5)$, $\hat{F}_h(5.9)$, and $\hat{F}_h(6.1)$? Make a rough sketch of this smoothed c.d.f. on the interval [-4, 7].

b) Let $h = 10000$. Without doing complex calculations, what are the approximate values of $\hat{F}_h(-4)$, $\hat{F}_h(-2.5)$, $\hat{F}_h(5.9)$, and $\hat{F}_h(6.1)$? Make a rough sketch of this smoothed c.d.f. on the interval [-4, 7].

2.4 (R code) For this exercise use the data from the previous exercise and the R function `cdfhat` that was described in this chapter.

a) Suppose we use the bandwidth $h = 1$. Find the value of the smoothed c.d.f. for $x = -4, -3, -2, -1, 0, 1, 2, 3, 4, 5, 6$. Draw a rough sketch of the c.d.f.

b) Suppose we use the bandwidth $h = 10$. Find the value of the smoothed c.d.f. for $x = -4, -3, -2, -1, 0, 1, 2, 3, 4, 5, 6$. Draw a rough sketch of the c.d.f.

c) Which of the two smoothed c.d.f.s best fits the data? Explain.

2.5 (R code) Use the same data as we used in Exercise 2.2.

a) Using $h = 1.587\hat{\sigma}_t n^{-1/3}$ compute the bandwidth.

b) Compute $\hat{x}_{.25}$, $\hat{x}_{.50}$, $\hat{x}_{.75}$, and $\hat{\sigma}$.

2.6 Suppose we have $n = 7$ observations with order statistics $\{0, 3, 4, 5, 6, 7, 30\}$. Because there is one outlier, we would like to normalize the observations. Using hand calculations, find the following:

a) $\hat{\sigma}_t$.

b) The value of h produced by the recommended formula for the bandwidth.

2.7 Now suppose we have the same data as in the previous problem and we computed the smoothed c.d.f. using a bandwidth of $h = 2.46$. Further, suppose we found the median to be $\hat{x}_{.50} = 5.0295$. In addition, we obtained the estimates of the c.d.f., using the smoothed c.d.f. values, that are shown in the third column of following table.

Obs	X	$\hat{F}_h(x)$
1	0	0.099
2	3	0.300
3	4	0.395
4	5	0.497
5	6	0.596
6	7	0.683
7	30	0.929

a) Compute z_i for $i = 1, \ldots, 7$.

b) Compute s_i for $i = 1, \ldots, 7$.

c) Compute $\hat{w}_i$ for $i = 1, \ldots, 7$.

d) Compute $\hat{w}_i(x_i - \hat{x}_{.5}) + \hat{x}_{.5}$ for $i = 1, \ldots, 7$.

e) Compare x_i to $\hat{w}_i(x_i - \hat{x}_{.5}) + \hat{x}_{.5}$ for $i = 1, \ldots, 7$.

2.8 Consider the data that we have used in the last two exercises. Perhaps another transformation method could be used to transform this data. Which of the following transformations are possible? Explain?

a) The square root transformation.

b) The logarithmic transformation.

c) The Box–Cox transformation.

2.9 Suppose a researcher used the approach described in this chapter to normalize data, but used $h = 3\hat{\sigma}n^{-1/3}$ for the bandwidth. Suppose the researcher always had data sets of size $n = 50$. Use the information in Figure 2.9 to answer these questions.

a) If the researcher always had data that was normally distributed, would there be advantage or disadvantage of using this value of the bandwidth?

b) If the researcher always had highly skewed data, would there be any advantage or disadvantage of using this value of the bandwidth?

2.10 Suppose a researcher used a bandwidth formula of $h = \hat{\sigma}n^{-1/5}$ instead of the formula recommended in this chapter. [Note that the exponent of n differs from what was recommended.] Further, suppose that $\sigma = 1$ and that the samples are large enough so that, for purposes of this exercise, we can assume that $\hat{\sigma} = 1$. Using the information in Figure 2.11, answer the following questions.

a) Roughly estimate the RMSEMAW for $n = 400$ observations using $h = \hat{\sigma}n^{-1/5}$. [Hint: Compute h and then, using the recommended formula, find K that would give that value of h.]

b) If the researcher always had $n = 400$ observations what would be the consequences in terms of the RMSEMAW?

c) In your opinion, would the formula $h = \hat{\sigma}n^{-1/5}$ cause a great deal of inaccuracy in estimating the relative weights if $25 \leq n \leq 50$? Explain.

2.11 Suppose we have $n = 4$ observations from a standard normal distribution having a mean of zero and a standard deviation of one. The order statistics are $x_{(1)} = -1.6$, $x_{(2)} = -0.7$, $x_{(3)} = 1.1$, and $x_{(4)} = 1.9$. Further, suppose we use $h = 1000$ as the smoothing parameter. Draw a rough figure that includes $F(x)$ and $\hat{F}_h(x)$ in the interval [-2.0, 2.0]. [Hint: The c.d.f. of the standard normal is widely tabulated. To find $\hat{F}_h(x)$ write out a few terms in Equation 2.1.] Does $\hat{F}_h(x)$ approximate $F_{empirical}(x)$? Explain.

CHAPTER 3

A TWO-SAMPLE ADAPTIVE TEST

One of the more common objectives in research is to determine if two populations differ, so it is not surprising that a variety of tests of significance have been developed to perform two-sample tests for location. The pooled t test and the unequal variance t test are the most popular of the parametric tests, but many others have been proposed, including several rank-based tests. One of these rank tests is the Wilcoxon rank-sum test, which is popular because it is reasonably powerful and robust. It was one of the rank-based tests that were described in Section 1.3. In Chapter 1 we also described the adaptive test of Hogg, Fisher, and Randles (1975), which we called the HFR test, and we compared the power of that test to the power of the t test.

In this chapter we will describe a modern adaptive two-sample test. Unlike the HFR test, this test can be generalized to perform adaptive tests of significance for any subset of coefficients in a linear model. Hence, the approach used in this chapter will also be used in subsequent chapters to perform many tests, including a test of slope in a linear regression, a test of interaction in a two-way layout, and a test of a main effect in a multifactorial design.

Adaptive Tests of Significance Using Permutations of Residuals with R and SAS®.
By Thomas W. O'Gorman

3.1 A TWO-SAMPLE MODEL

Now suppose we have n observations and we want to perform a test of location. Let y_i be the dependent (or response) variable and let x_i be the independent (or predictor) variable for the ith observation. For a test of location we can fit the complete model

$$y_i = \beta_0 + \beta_1 x_i + \varepsilon_i,$$

for $i = 1, \ldots, n$, where

$$x_i = \begin{cases} 1 & \text{if } i\text{th observation is from population 2,} \\ 0 & \text{if } i\text{th observation is from population 1.} \end{cases}$$

To perform a two-sample test, we test the null hypothesis $H_0 : \beta_1 = 0$ versus the alternative hypothesis $H_a : \beta_1 \neq 0$. Under the null hypothesis the model simplifies to

$$y_i = \beta_0 + \varepsilon_i,$$

for $i = 1, \ldots, n$, which we will call the reduced model. We would usually use ordinary least squares to obtain estimates of β_0 and β_1 for the complete model and β_0 for the reduced model, and from these we would calculate residuals under the complete model and under the reduced model.

The traditional test is based on the F test statistic, which is a function of the sum of the squared residuals from the complete model and the sum of the squared residuals from the reduced model. Let $\hat{\varepsilon}_{C,i}$ be the ordinary residual for the ith observation from the complete model and let $\hat{\varepsilon}_{R,i}$ be the ordinary residual for the ith observation from the reduced model. The sum of the squared residuals from the complete model is $SSE_C = \sum \hat{\varepsilon}^2_{C,i}$, where the summation is over $i = 1, \ldots, n$ and the sum of the squared residuals from the reduced model is $SSE_R = \sum \hat{\varepsilon}^2_{R,i}$. The F test statistic is given by

$$F = \frac{SSE_R - SSE_C}{SSE_C/(n-2)}.$$

If the errors are normally distributed, independent, and have constant variance, the F test statistic is distributed as an F variate with $\nu_1 = 1$ and $\nu_2 = n - 2$ degrees of freedom. Typically, the p-value (p) is obtained from that distribution and the null hypothesis is rejected if the p-value is less than or equal to the significance level (α).

This p-value is the same p-value that would be obtained from the pooled two-sample test with a two-sided alternative. This test maintains its level of significance if the errors are normally distributed, independent, and have constant variance. In addition, we shall see that if the errors are not normally distributed, the F test maintains its level of significance at, or slightly below, the nominal value. These characteristics explain why the two-sample t test has been popular among researchers.

A major problem with the F test is that it is not as powerful as many other tests when the errors are not normally distributed. As we saw in Figure 1.5, the F test (or t test) is often 5% to 15% less powerful than the Wilcoxon test and the adaptive test.

The adaptive test relies on an adaptive weighting procedure combined with a permutation method to estimate the p-value. Because this weighting and permutation method will be used throughout this book, we will carefully and completely describe this adaptive two-sample test in this chapter.

3.2 COMPUTING THE ADAPTIVE WEIGHTS

The adaptive weighting method attempts to normalize the residuals from the reduced model so that, after weighting, the distribution of the residuals for the weighted model will be approximately normal. The calculation of the weights uses the same general method that was used in Chapter 2 to normalize the identically and independently distributed random variables. Instead of normalizing random variables, we will normalize the residuals from the reduced model $\hat{\varepsilon}_{R,i}$ for $i = 1, \ldots, n$. For large samples the residuals may closely approximate the errors, so it seems reasonable to use the smoothed c.d.f. of the residuals as an estimate of the c.d.f. of the errors.

An example will help explain the weighting procedures. To obtain the weights we begin by computing, using the ordinary residuals from the reduced model, the traditional estimates of the 25th percentile ($\hat{\varepsilon}_{t,.25}$) and the 75th percentile ($\hat{\varepsilon}_{t,.75}$). Next, we compute $\hat{\sigma}_t = (\hat{\varepsilon}_{t,.75} - \hat{\varepsilon}_{t,.25})/1.349$ as an estimate of the standard deviation of the residuals. This estimate is then used to compute the bandwidth as $h = 1.587\hat{\sigma}_t n^{-1/3}$.

Once the bandwidth is determined, the smoothing methods that were described in Chapter 2 are used to find the smoothed c.d.f. of the set of residuals. The smoothed c.d.f. is also used along with a root-finding algorithm to find improved estimates of the 25th, 50th, and 75th percentiles as $\hat{\varepsilon}_{.25}$, $\hat{\varepsilon}_{.5}$, and $\hat{\varepsilon}_{.75}$, respectively, and to compute $\hat{\sigma} = (\hat{\varepsilon}_{.75} - \hat{\varepsilon}_{.25})/1.349$ as the final robust estimate of the standard deviation. Next, for the ith observation we compute the normal score $z_i = \Phi^{-1}[\hat{F}_h(\hat{\varepsilon}_i)]$ and the standard score $s_i = (\hat{\varepsilon}_i - \hat{\varepsilon}_{.5})/\hat{\sigma}$. For the ith observation we obtain the weight as

$$\hat{w}_i = \begin{cases} z_i/s_i & \text{if } s_i \neq 0, \\ 1 & \text{if } s_i = 0. \end{cases}$$

Note that if the residuals are nearly normal, then z_i should approximate s_i for $i = 1, \ldots, n$. Also, if an outlier is present, then s_i will be quite large but z_i will be moderate, which will result in a small weight for that observation. Because $s_i > 0$ whenever $\hat{\varepsilon}_i > \hat{\varepsilon}_{.5}$ we can see that z_i and s_i will always have the same sign so that the weights will always be positive. Because we will use this method for most of the tests in this book, it will be convenient to summarize the weighting procedures.

The steps in the adaptive weighting algorithm are:

1. Use the set of ordinary residuals from the reduced model to compute $\hat{\sigma}_t = (\hat{\varepsilon}_{t,.75} - \hat{\varepsilon}_{t,.25})/1.349$.
2. Compute the bandwidth $h = 1.587\hat{\sigma}_t n^{-1/3}$.
3. Smooth the c.d.f. of the residuals to obtain $\hat{F}_h(\hat{\varepsilon}_i)$ for $i = 1, \ldots, n$.
4. Use $\hat{F}_h(\hat{\varepsilon})$ in an iterative procedure to compute $\hat{\varepsilon}_{.25}$, $\hat{\varepsilon}_{.5}$, and $\hat{\varepsilon}_{.75}$. Then compute $\hat{\sigma} = (\hat{\varepsilon}_{.75} - \hat{\varepsilon}_{.25})/1.349$.
5. Compute $z_i = \Phi^{-1}[\hat{F}_h(\hat{\varepsilon}_i)]$ and $s_i = (\hat{\varepsilon}_i - \hat{\varepsilon}_{.5})/\hat{\sigma}$ for $i = 1, \ldots, n$.
6. For $i = 1, \ldots, n$ compute the estimates of the weights as

$$\hat{w}_i = \begin{cases} z_i/s_i & \text{if } s_i \neq 0, \\ 1 & \text{if } s_i = 0. \end{cases}$$

Although the weighting procedure is not too complicated, the smoothing method would usually be performed using statistical software. A SAS macro will be used later in this chapter to calculate these weights. We now describe the R code for computing these weights.

3.2.1 R Code for Computing the Weights

In the following R code we begin the weighting process by fitting the reduced model, which in this case contains only the intercept. The reduced model is specified as `y ~ 1`. The fitted model object (`red`) for the reduced model is obtained by calling the linear model function `lm()`. The residuals, which determine the weights, are obtained from the object `red` and are stored in the vector `resid`. Next, the traditional percentiles (or quantiles) are computed and these are used to obtain the bandwidth h. The smoothed c.d.f. of the residuals is used in the `rootcdf` function that was described in Chapter 2 to find the final estimates of the quantiles.

```
red <- lm(y~1)
resid <- residuals(red)
#                          compute traditional quantiles
q25 <- quantile(resid,0.25,type=6)
q75 <- quantile(resid,0.75,type=6)
sigmat <- (q75-q25)/1.349
#                                compute bandwidth (h)
n <- length(resid)
h <- 1.587*sigmat*n^(-0.33333333333)
#                    compute final estimates of quantiles
range <- max(resid) - min(resid)
```

```
lower <- min(resid) - range
upper <- max(resid) + range
tol <- 0.00000001*sigmat
cdf25 <- rootcdf(resid,h,0.25,lower,upper,tol)
cdf50 <- rootcdf(resid,h,0.50,lower,upper,tol)
cdf75 <- rootcdf(resid,h,0.75,lower,upper,tol)
```

Our final estimate of variability ($\hat{\sigma}$), which is called `sigma` in the R code, is used to compute the vector of standardized values, which is placed in the vector `s`. The weighting loop is slightly different than that described in Chapter 2 because we also compute w_i^2 for $i = 1, \ldots, n$, which will be used later to compute the test statistic. Also, the tolerances for the root-finding function and the weighting function are proportional to the initial estimate of the variability.

```
sigma <- (cdf75-cdf25)/1.349
s <- (resid-cdf50)/sigma
tolerance <- 100*tol
w  <- matrix(0,n,1)
w2 <- matrix(0,n,1)
for (i in 1:n) {
  fhat <- cdfhat(resid,h,resid[i])
  z <- qnorm(fhat)
  if( (abs(s[i]) >= tolerance) )
     w[i] <- z/s[i]
    else
     w[i] <- 1
    w2[i] <- w[i]^2
  }
```

After execution of this R code, we now have the adaptive weights $\hat{w}_i$, for $i = 1, \ldots, n$, and $\hat{w}_i^2$, for $i = 1, \ldots, n$. These weights will be used in a weighted least squares regression to perform an adaptive two-sample test.

3.3 THE TEST STATISTICS FOR ADAPTIVE TESTS

Once the weights have been determined for all observations, we then use a weighted least squares (WLS) approach to fit the weighted model. For the ith observation the unweighted complete model is

$$y_i = \beta_0 + \beta_1 x_i + \varepsilon_i,$$

for $i = 1, \ldots, n$, where

$$x_i = \begin{cases} 1 & \text{if } i\text{th observation is from population 2,} \\ 0 & \text{if } i\text{th observation is from population 1.} \end{cases}$$

Let $\hat{w}_i$ be the weight for the ith observation. To obtain the WLS estimates, one approach is to find $\hat{\beta}_0$ and $\hat{\beta}_1$ by minimizing the sum of the squares of the weighted errors. That is, we minimize $\sum(\hat{w}_i\hat{\varepsilon}_i)^2 = \sum \hat{w}_i^2\hat{\varepsilon}_i^2$ to find $\hat{\beta}_0$ and $\hat{\beta}_1$. (Note that in some presentations of weighted least squares estimation the weight for the ith observation is given by $\hat{w}_i^2$, but we will continue to use $\hat{w}_i$ to indicate the weight.)

Another approach to fitting the model is to transform both sides of the original model by multiplying by the weights to obtain a weighted model. For the ith observation the weighted model is

$$\hat{w}_i y_i = \hat{w}_i \beta_0 + \hat{w}_i \beta_1 x_i + \hat{w}_i \varepsilon_i.$$

In order to simplify the presentation, we will define several new variables. For the ith observation let $y_i^* = \hat{w}_i y_i$ be the transformed dependent variable. Our new independent variable is $x_i^* = \hat{w}_i x_i$ and our new error variable is $\varepsilon_i^* = \hat{w}_i \varepsilon_i$. With this notation our model can be written as

$$y_i^* = \hat{w}_i \beta_0 + \beta_1 x_i^* + \varepsilon_i^*, \tag{3.1}$$

for $i = 1, \ldots, n$. After we use the adaptive weights, we hope that the empirical c.d.f. of the residuals from this model will approximate the c.d.f. of a normal distribution. Note that in Equation (3.1) there is no intercept term because the multiplier of β_0 does not equal one. To find $\hat{\beta}_0$ and $\hat{\beta}_1$, we can use an ordinary least squares (OLS) regression program with y_i^* as the dependent variable and $\hat{w}_i$ and x_i^* as the independent variables. To see that this approach will give the same values of $\hat{\beta}_0$ and $\hat{\beta}_1$ that we would obtain if we minimized $\sum \hat{w}_i^2\hat{\varepsilon}_i^2$, we note that with our first approach we find $\hat{\beta}_0$ and $\hat{\beta}_1$ to minimize

$$\begin{aligned}\sum \hat{w}_i^2\hat{\varepsilon}_i^2 &= \sum \{\hat{w}_i[y_i - (\hat{\beta}_0 + \hat{\beta}_1 x_i)]\}^2 \\ &= \sum [y_i^* - (w_i\hat{\beta}_0 + \hat{\beta}_1 x_i^*)]^2 \\ &= \sum (\hat{\varepsilon}_i^*)^2.\end{aligned}$$

Hence, $\hat{\beta}_0$ and $\hat{\beta}_1$ will minimize the sum of squared residuals in a regression of y_i^* on $\hat{w}_i$ and x_i^*.

We will use WLS on the complete model and the reduced model to perform the test. For the ith observation the residual for the weighted complete model will be denoted by $\hat{\varepsilon}_{C,i}^*$ and the residual for the weighted reduced model will be denoted by $\hat{\varepsilon}_{R,i}^*$. Regardless of the way we compute $\hat{\beta}_0$ and $\hat{\beta}_1$, for the weighted complete model, we find residuals as

$$\hat{\varepsilon}_{C,i}^* = \hat{w}_i[y_i - (\hat{\beta}_0 + \hat{\beta}_1 x_i)] = y_i^* - (\hat{w}_i\hat{\beta}_0 + \hat{\beta}_1 x_i^*),$$

for $i = 1, \ldots, n$. A measure of the lack of fit is given by $SSE_C^* = \sum(\hat{\varepsilon}_{C,i}^*)^2 = \sum(\hat{w}_i\hat{\varepsilon}_{C,i})^2$.

In a test of significance we will compare the fit under the null hypothesis to the fit under the alternative hypothesis. Under the null hypothesis we will fit the reduced model, which in this case is $y_i = \beta_{R,0} + \varepsilon_i$, for $i = 1, \ldots, n$. We will find $\hat{\beta}_{R,0}$ either by minimizing the sum of squares of the weighted errors or by using OLS on the transformed model $\hat{w}_i y_i = \hat{w}_i \beta_{R,0} + \hat{w}_i \varepsilon_i$. The residuals from this weighted reduced model are

$$\hat{\varepsilon}^*_{R,i} = \hat{w}_i(y_i - \hat{\beta}_{R,0}) = y^*_i - \hat{w}_i \hat{\beta}_{R,0},$$

for $i = 1, \ldots, n$, and the measure of lack of fit is $SSE^*_R = \sum(\hat{\varepsilon}^*_{R,i})^2 = \sum(\hat{w}_i \hat{\varepsilon}_{R,i})^2$.

To compare the two populations we will test the null hypothesis $H_0 : \beta_1 = 0$ versus the alternative hypothesis $H_a : \beta_1 \neq 0$ by comparing SSE^*_C to SSE^*_R. If SSE^*_R approximates SSE^*_C then there is no need to include β_1 in the model. However, if SSE^*_R greatly exceeds SSE^*_C, then the inclusion of β_1 is necessary. In this book we will perform a test of $H_0 : \beta_1 = 0$ versus $H_a : \beta_1 \neq 0$ using Wilks' lambda test statistic

$$\Lambda^{2/n,*} = \frac{SSE^*_C}{SSE^*_R}. \tag{3.2}$$

We will reject the null hypothesis for small values of $\Lambda^{2/n,*}$. This test will be equivalent to that based on the F statistic because

$$\begin{aligned} F^* &= \frac{SSE^*_R - SSE^*_C}{SSE^*_C/(n-2)} = \left(\frac{SSE^*_R}{SSE^*_C} - 1\right)\left(\frac{n-2}{1}\right) \\ &= \left(\frac{1}{\Lambda^{2/n,*}} - 1\right)\left(\frac{n-2}{1}\right). \end{aligned}$$

If we use the F^* test statistic we will reject the null hypothesis for large values of F^*. With traditional tests having no adaptive weighting, the F test statistic is often used because, if the errors are normally distributed, the F test statistic follows an F distribution with 1 and $n-2$ degrees of freedom under the null hypothesis. However, it is important to remember that this F^* test statistic may not follow the F distribution with 1 and $n-2$ degrees of freedom because the weights have been determined from the data. Hence, there is no need to use the F^* statistic in the adaptive test; we could use any statistic that is a strictly increasing (or strictly decreasing) function of F^*.

However, for adaptive tests, there are several reasons for using Wilks' lambda instead of F^* as the test statistic. One reason is that $\Lambda^{2/n,*}$ can be used with multivariate regression models that we will discuss in Chapter 9. Another reason is that the $\Lambda^{2/n,*}$ statistic is easier to conceptualize than the F^* statistic because $\Lambda^{2/n,*}$ is simply the ratio of the sum of squared residuals for two models. Note also that $\Lambda^{2/n,*} = 1 - R^{2,*}$, where $R^{2,*}$ is the coefficient of determination for the weighted model. When we perform an adaptive test with a model having only one dependent variable, we will be using permutation methods so that either the F^* statistic, the $R^{2,*}$ statistic, or the $\Lambda^{2/n,*}$ statistic could be used. The p-values will be identical.

We have described the adaptive test statistic for the two-sided test. If a one-sided test is appropriate, this approach can be modified in the manner that will be described in Section 3.11.

Now that we have computed the test statistic we need to compute a p-value. Because the distribution of the test statistic is not known, we will perform a test by finding the permutation distribution of the test statistic under the null hypothesis. This will be described in Section 3.4. We will now describe how to compute $\Lambda^{2/n,*}$ using R.

3.3.1 R Code to Compute the Test Statistic

The following code uses the squares of the weights, which have already been computed and are in the vector `w2`, to compute the test statistic `lambda`. In the first line of code the fitted model object `compw` is obtained for the weighted complete model using the `lm()` function. To obtain the WLS estimates, we specify the squares of the weights in the vector `w2`. We use the squares of the weights because the `lm()` function requires the squares of the weights, rather than the weights themselves. The residuals are placed in `residc`, and the squares of the residuals are computed and stored in `residc2`. In the fourth line the sum $\sum \hat{w}_i^2 \hat{\varepsilon}_{C,i}^2$ is computed by using the `t()` function to transpose the `w2` vector and by using `%*%` to multiple two matrices. The result is stored in `ssec`. We now perform the same calculations on the reduced model that is specified as `y~1`. The next three lines are used to compute $\sum \hat{w}_i^2 \hat{\varepsilon}_{R,i}^2$ and it is placed in `sser`. Finally, one test statistic is computed as `lambda`. Next, we obtain $n-2$ by using the `df.residual()` function on the fitted model object `compw` for the complete model and we compute `f` as the other test statistic.

```
compw <-lm(y~group, weights=w2)
residc <-residuals(compw)
residc2 <- residc^2
ssec <- t(w2)%*%residc2
redw <- lm(y~1,weights=w2)
residr <- residuals(redw)
residr2 <- residr^2
sser <- t(w2)%*%residr2
lambda <- ssec/sser
dfc <- df.residual(compw)
f <- (sser-ssec)/(ssec/dfc)
```

Now that we have obtained a reasonable method of weighting the observations and have described how the test statistic is computed, we need to perform the test. To obtain a p-value for this adaptive test, we will use a permutation method.

3.4 PERMUTATION METHODS FOR TWO-SAMPLE TESTS

3.4.1 Permutation of Observations

In this section we will find the permutation distribution of the test statistic $\Lambda^{2/n,*}$. The distribution is obtained by permuting (or shuffling) the observations between the two samples and then computing the test statistic after each permutation. The p-value equals the proportion of all permutations that produce test statistics that are less than or equal to the test statistic obtained from the data. An example should clarify this idea.

Suppose we have 6 subjects who have been randomly assigned to either the Drug A or Drug B treatment groups and that a symptom severity score is obtained from each subject. Suppose the scores are as follows:

	Group A			Group B		
Scores	3	1	7	5	8	6

With this assignment of subjects to treatments we can compute the adaptive weights and, using the methods of the last section, compute the test statistic to be $\Lambda^{2/n,*} = 0.6967$. [Recall that we reject the null hypothesis for small values of $\Lambda^{2/n,*}$.] To find the probability of observing a test statistic that is less than or equal to 0.6967, we perform a permutation test. To construct the permutation distribution of the test statistic, we begin by noting that, under the null hypothesis, the drugs are equally effective so the first subject in the Drug A group, whose score was 3, would have had that score if that subject were assigned to Drug B. Also, the first subject in the Drug B group, whose score was 5, would have had that score if that subject were assigned to the Drug A group. So, under the null hypothesis we are just as likely to observe this assignment:

	Group A			Group B		
Scores	5	1	7	3	8	6

To find the test statistics we compute the weights and then use the weighted model to obtain 0.9212 as the test statistic. [For this test we can simply permute the original weights because we compute the weights from the reduced model that does not include the group indicator.] Because any combination of three subjects from these six subjects could have been assigned to group A, there are $\binom{6}{3} = 20$ possible assignments of the subjects to the groups. Now, to construct the full permutation distribution we would compute $\Lambda^{2/n,*}$ for each of these $\binom{6}{3} = 20$ assignments of subjects to treatments.

For small samples it is possible to compute the full permutation distribution. If 20 animals are assigned equally to two treatment groups, there are only $\binom{20}{10} = 184756$ assignments of animals to treatments, so a modern computer should be able to quickly evaluate the full permutation distribution. For larger samples a full enumeration is impractical or impossible. For example, if 60 subjects are assigned to two groups so that 30 are in each group, there are $\binom{60}{30} = 1.18 \times 10^{17}$ assignments. If a computer could evaluate 20 million of these permutation per second, it would take over 187 years to determine the full permutation distribution! Clearly, another approach is needed.

Instead of attempting to find the full permutation distribution, we will take a large random sample from the permutation distribution. Fortunately, computers are so fast that a very large sample from the full permutation distribution can be obtained very quickly. Let R equal the number of sample permutations, not including the one assignment that we observed. If we select $R = 2000$ sample permutations, from the population of all possible permutations, we can obtain a reasonably accurate estimate of the full permutation distribution. If efficient software is available, samples of $R = 10{,}000$ may be quite practical.

If we have a full permutation distribution the p-value is the proportion of assignments that produce test statistics that are less than or equal to the observed test statistic. Let $\Lambda_u^{2/n,*}$ be the observed test statistic for the unpermuted data and let E equal the number of permutation test statistics that are less than or equal to $\Lambda_u^{2/n,*}$. Since the observed test statistic is from one of the possible permutations, there are $R+1$ test statistics and $E+1$ of these test statistics are less than or equal to $\Lambda_u^{2/n,*}$. Hence, if we use a sample permutation test the p-value is $p = (E+1)/(R+1)$. This estimator of the p-value, which was recommended by Davison and Hinkley (1997), will be compared to the estimator E/R in Section 3.7. If $p \leq \alpha$, we will reject the null hypothesis.

3.4.2 Permutation of Residuals

It is also possible to permute the residuals from the reduced model to perform a permutation test. In a two-sample test we would permute the residuals under the reduced model, which is a model that includes only the intercept. Under the reduced model, the estimate of the intercept is simply the mean of the observations over both samples, which is $\bar{y} = 5$ in the example. The unpermuted residuals are:

	Group A			Group B		
Residuals	-2	-4	2	0	3	1

Now suppose we permute the residuals in the same manner that we permuted the observations to obtain:

	Group A			Group B		
Residuals	0	-4	2	-2	3	1

We now find the y-values that correspond to those residuals by adding the permuted residuals to the predicted value, which was $\bar{y} = 5$. These new y-values are:

	Group A			Group B		
Residuals	5	1	7	3	8	6

These new y-values are identical to what we obtained by permuting the original y-values. For a two-sample test it can be seen if we compute the new y-values using the permutation of residuals approach, they will be identical to the permuted y-values. Because the permutation of residuals method is a more complex way of doing what is very easy to do, the reader may wonder why anyone would use the permutation of residuals approach.

We will use the permutation of residuals approach in this book because it can be generalized to multiple regression models. A detailed discussion of the many permutation methods that could be considered for multiple regression models will be presented in the next chapter. For the two-sample test it really does not matter which method is used because they are equivalent, but the SAS macros and the R code will use the permute-residuals method for all tests in this book.

3.4.3 R Code for Permutations

The permutation method is performed with the R code that is given below. We begin by setting the counter `e` equal to zero. It is the number of permutation test statistics that are less than or equal to the unpermuted test statistic `lambdau`. Next, we use the R function `lm(y~1)` to find the predicted values `yhat` and the residuals `yresidual` from the reduced model. To perform the permutations test we loop over the `r` permutations. In the first line in the loop, the `sample()` function will perform a permutation of the vector `yesidual` using the default random number generator. This `sample()` function is built into R and, when used with a vector as the only argument, it produces a permutation of the elements in the vector. This vector of permuted residuals is then added to the predicted values (`yhat`) to obtain a new vector of y-values, which we call `ynew`.

To obtain the test statistic we need to specify the complete and reduced models. The complete model is specified as

```
complete <- c("y~group")
```

This character string indicates that the intercept and the group indicator variable should be used in the complete model. It will be converted to a formula when it is used in the `lm()` function that is called by the `teststats` function. The reduced model is specified as

```
reduced  <- c("y~1")
```

This character string, which specifies that only the intercept should be used in the reduced model, will be converted to a formula when it is used by a `lm()` function that is called by the `teststats` function. On the second line of the loop the vector `statsp` is obtained from the `teststats` function that finds the test statistic, based on the new values of the dependent variable and on the specified complete and reduced models. This `teststats` function, which will be described in Section 3.6, performs the weighting and the computation of the test statistic. The `teststats` function produces the vector `statsp`, which contains $\lambda^{2/n,*}$ and F^*. On the next line the permutation test statistic `lambdap` is obtained from the `statsp` vector. In the last line of the loop we determine if `e` should be increased by one. After the loop is complete we compute the p-value as $(E+1)/(R+1)$.

```
e <- 0
redu <- lm(y~1)
yhat <- predict(redu)
yresidual <- residuals(redu)
for (k in 1:r) {
  ynew <- yhat+sample(yresidual)
```

```
  statsp <- teststats(ynew,complete,reduced)
  lambdap <- statsp[1]
  if( (lambdap <= lambdau) ) e  <-  e + 1
  }
p <- (e+1)/(r+1)
```

Note that F^*, which is stored at `statsp[2]`, is not used in the computation of the p-value, but some researchers may be more comfortable with an F statistic, so F^* will be printed on some of the computer outputs. We noted in Section 3.3 that if we had used F^* instead of $\lambda^{2/n,*}$ in the logic we would obtain the same p-value.

We have described the R code that is necessary to compute the weights and perform the permutations necessary to estimate the p-value. In Section 3.6 we will put these pieces together so the reader can understand how to perform an adaptive two-sample test using R. However, before we proceed further we should describe a data set that could be used in an adaptive two-sample test.

3.5 AN EXAMPLE OF A TWO-SAMPLE TEST

The data tabulated in Table 3.1 represent the total number of seizures reported over an eight-week period by 59 epileptics who were enrolled in a study to determine the effectiveness of a drug to reduce the number of seizures. The data, which were published by Thall and Vail (1990), originate from an experiment described by Leppik et al. (1985). In that experiment, subjects who were suffering from seizures were randomized to receive either the drug progabide or a placebo. Although the study was designed so that each patient would be crossed over to the other drug, the seizure counts in Table 3.1 are the precrossover responses. Note that the distribution of seizure counts is highly skewed. The adaptive weights were computed, as described in Section 3.2, and these weights were between 0.87 and 1.54 for seizure counts less than 35. Note that the subject who had a seizure count of 302 was given a low weight of 0.185, so that it would have little influence on the estimate of the treatment effect.

To determine if the weighting method was effective in normalizing the residuals, we display in Figure 3.1 a histogram of the residuals from the reduced model, which are highly skewed, and in Figure 3.2 the residuals from the adaptively weighted model. It appears that the weighting method was effective in reducing the skewness of this set of residuals.

A two-tailed t test produced a p-value of $p = 0.8320$, while the adaptive p-value, using $R = 2000$ sample permutations, was $p = 0.1669$. The difference in the p-values is largely due to the fact that the highest seizure count of 302 was in the progabide group, which tended to have lower seizure counts. We now show how to perform this test using R.

Table 3.1 Total seizure counts and their weights

Progabide			Placebo		
Obs	Total	Weight	Obs	Total	Weight
1	42	0.729	32	14	1.323
2	28	0.977	33	14	1.323
3	7	1.483	34	11	1.396
4	13	1.348	35	13	1.348
5	19	1.195	36	55	0.605
6	11	1.396	37	22	1.118
7	74	0.547	38	12	1.372
8	20	1.169	39	95	0.454
9	10	1.419	40	22	1.118
10	24	1.069	41	33	0.876
11	29	0.956	42	66	0.566
12	4	1.541	43	30	0.935
13	6	1.503	44	16	1.273
14	12	1.372	45	42	0.729
15	65	0.568	46	59	0.586
16	26	1.022	47	16	1.273
17	39	0.773	48	6	1.503
18	7	1.483	49	123	0.365
19	32	0.895	50	15	1.298
20	3	1.559	51	16	1.273
21	302	0.185	52	14	1.323
22	13	1.345	53	14	1.323
23	26	1.022	54	13	1.348
24	10	1.419	55	30	0.935
25	70	0.558	56	143	0.346
26	13	1.348	57	6	1.503
27	15	1.298	58	10	1.419
28	51	0.632	59	53	0.617
29	6	1.503			
30	0	1.609			
31	10	1.419			

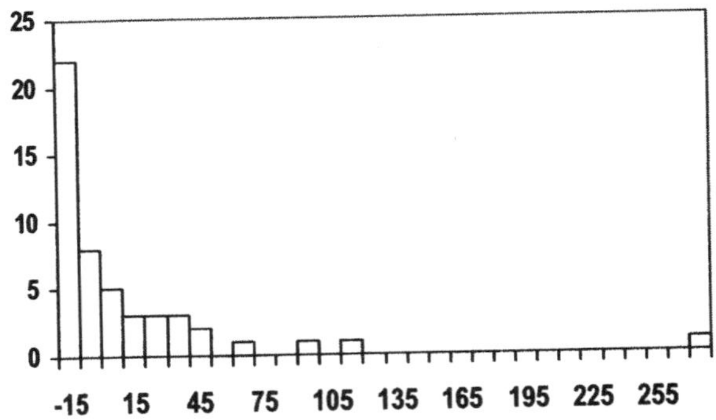

Figure 3.1 Histogram of the residuals from the reduced model for the seizure count data.

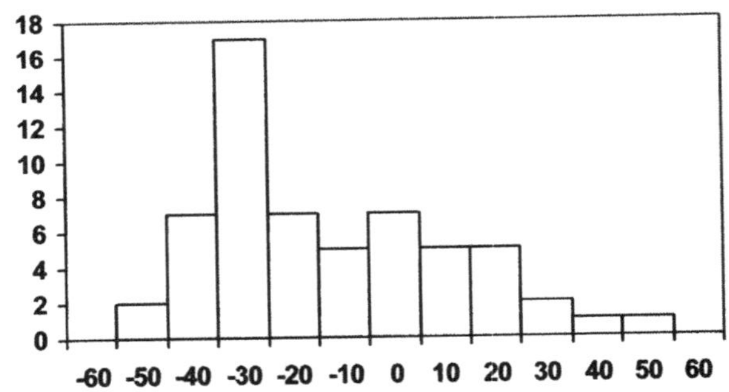

Figure 3.2 Histogram of the residuals from the adaptively weighted reduced model for the seizure count data.

3.6 R CODE FOR THE TWO-SAMPLE TEST

3.6.1 R Code for Computing the Test Statistics

The `teststats` function computes the $\lambda^{2/n,*}$ and F^* test statistics for the adaptive test. This function, which is given below, has three arguments. The first argument in the function is the dependent variable. The second and third arguments are character strings that specify the complete and reduced models using the syntax that is used in the `lm()` function. We will show how to specify these arguments later in this section.

The second line of the `teststats` function fits the reduced model. The `as.formula` function converts the character string in `reduced` into a formula. In the third line the vector of residuals is placed in `resid`. Next, the traditional quantiles are computed, and these are used to compute the first estimate of the standard deviation. In the next few lines the bandwidth h is computed and it is used in the rootcdf function to compute $\hat{\varepsilon}_{.25}$, $\hat{\varepsilon}_{.5}$, $\hat{\varepsilon}_{.75}$, and $\hat{\sigma}$. Next, the weights are computed in the same manner that we described in Section 3.2. After the weights are computed they are used in the lm() function to compute the test statistic lambda and f. The vector teststats contains both of these test statistics and that vector is returned by the teststats function.

```
teststats <- function(y,complete,reduced) {
red <- lm(as.formula(reduced))
resid <- residuals(red)
#                compute traditional quantiles and bandwidth
q25 <- quantile(resid,0.25,type=6)
q75 <- quantile(resid,0.75,type=6)
sigmat <- (q75-q25)/1.349
n <- length(resid)
h <- 1.587*sigmat*n^(-0.33333333333)
#                     compute improved estimate of sigma
range <- max(resid) - min(resid)
lower <- min(resid) - range
upper <- max(resid) + range
tol <- 0.00000001*sigmat
cdf25 <- rootcdf(resid,h,0.25,lower,upper,tol)
cdf50 <- rootcdf(resid,h,0.50,lower,upper,tol)
cdf75 <- rootcdf(resid,h,0.75,lower,upper,tol)
sigma <- (cdf75-cdf25)/1.349
#                          compute adaptive weights
s <- (resid-cdf50)/sigma
tolerance <- 100*tol
w  <- matrix(0,n,1)
w2 <- matrix(0,n,1)
for (i in 1:n) {
  fhat <- cdfhat(resid,h,resid[i])
  z <- qnorm(fhat)
  if( (abs(s[i]) >= tolerance) )
     w[i] <- z/s[i]
    else
     w[i] <- 1
    w2[i] <- w[i]^2
  }
#                            compute the test statistics
compw <-lm(as.formula(complete), weights=w2)
residc <-residuals(compw)
residc2 <- residc^2
ssec <- t(w2)%*%residc2
```

```
redw <- lm(as.formula(reduced),weights=w2)
residr <- residuals(redw)
residr2 <- residr^2
sser <- t(w2)%*%residr2
lambda <- ssec/sser
dfc <- df.residual(compw)
dfr <- df.residual(redw)
f <- ( (sser-ssec)/(dfr-dfc) )/ (ssec/dfc)
teststats <- c(lambda,f)
return(teststats)
  }
```

The `teststats` function is a very general function that will be used to compute the test statistics for most of the adaptive tests in this book. In this chapter the complete model will include an intercept term and a group indicator while the reduced model will include only the intercept.

3.6.2 R Code to Compute the Traditional F Test Statistic and p-Value

For some data sets we will want to compare the p-value from the adaptive test to the p-value from the traditional F (or t) test. The function `trad.ftest`, which is listed below, returns the p-value for the traditional nonadaptive F test. The user must specify the data set, the complete model, and the reduced model as character variables. The `attach` statement makes the variables in the `testdata` data frame to be available for use in the `lm()` function. The `trad.ftest` function uses two calls to the `lm()` function to obtain `comp.model` and `red.model` as the unweighted complete and reduced fitted model objects. The `deviance` function is used to compute the residual sum of squares. Thus, `ssec` and `sser` contain the sum of squares of the residuals for the complete and reduced models, respectively. The `df.residual` function computes the residual degrees of freedom so that `dfc` and `dfr` contain the residual degrees of freedom for the complete and reduced models, respectively. The traditional F test statistic is computed and the p-value, which is obtained from the c.d.f. of the F distribution, is returned. The `detach` statement removes the variables in the `testdata` data frame from the search path.

```
trad.ftest <- function(dataset,complete,reduced) {
  testdata <- read.table(dataset)
  attach(testdata)
  comp.model <- lm(as.formula(complete) )
  red.model  <- lm(as.formula(reduced) )
  ssec <- deviance(comp.model)
  sser <- deviance(red.model)
  dfc <- df.residual(comp.model)
  dfr <- df.residual(red.model)
```

```
numdf <- dfr-dfc
f <- ( (sser-ssec)/numdf)/(ssec/dfc)
p <- 1-pf(f,numdf,dfc)
detach(testdata)
return(p)
}
```

3.6.3 An R Function that Computes the p-Value for the Adaptive Test

We now have all the R functions that we need to perform the test. The `adapt` function, which is given below, calls these functions to perform the test. It uses the permutation of residuals approach to obtain a p-value for the adaptive test. The `adapt` function will be used to perform most of the adaptive tests in this book. It begins by using the `trad.ftest` function to obtain the p-value for the nonadaptive test. Next, the `teststats` function is called to obtain `stats`, which is the vector of test statistics. The first element in that vector is $\Lambda_u^{2/n,*}$, which is called `lambdau`. The vector of predicted values from the reduced model is `yhat` and the vector of residuals is `yresidual`. The `for` loop performs `r` permutations of the residuals, as we have described in Section 3.4, and the p-value is computed. The `set.seed` function is used to initialize the random number generator. The remainder of this function prints the data set name, the models, and the adaptive p-value.

```
adapt <- function(dataset,complete,reduced,r=2000,seed=1492)
                    {
tradpvalue <- trad.ftest(dataset,complete,reduced)
testdata <- read.table(dataset)
print(testdata)
attach(testdata)
statsu<- teststats(y,complete,reduced)
lambdau <- statsu[1]
fu <- statsu[2]
e <- 0
redu <- lm(as.formula(reduced))
yhat <- predict(redu)
yresidual <- residuals(redu)
set.seed(seed)
for (k in 1:r) {
  ynew <- yhat+sample(yresidual)
  statsp <- teststats(ynew,complete,reduced)
  lambdap <- statsp[1]
  if( (lambdap <= lambdau) ) e  <-  e + 1
  }
p <- (e+1)/(r+1)
detach(testdata)
cat(" Output from Adaptive Test:","\n","\n")
```

```
cat("    Unpermuted lambda = ",round(lambdau,4),"\n","\n")
cat("    Unpermuted F      = ",round(fu,4),"\n","\n")
cat("    p-value           = ",round(p,4),"\n","\n","\n")
cat(" Traditional F test p-value = ",round(tradpvalue,4),"\n")
return(p)
}
```

3.6.4 R Code to Perform the Adaptive Test

To execute the `adapt` function, we need only define the appropriate arguments. The `adapt` function and all of the functions that it calls must be placed before the function is called. In our example, we have used the data reported by Thall and Vail (1990), which is in Table 3.1, and we have analyzed the total seizure count, which we have put in a dataset `thall.total.dat`.

In the following code we execute the `adapt` function after specifying the five arguments that are used in the function call. The first argument is the dataset name that is specified by the character string

```
dataset <- c("thall.total.dat")
```

The test of significance is determined by the specification of the complete and reduced models using the syntax that the `lm()` function requires. For this two-sample test we specify the complete model as a character string

```
complete <- c("y~group")
```

This character variable, which is the second argument in the function call, must be a valid R formula for the `lm()` function. The third argument is the character variable `reduced` that specifies the model that contains only the intercept, which can be specified as

```
reduced  <- c("y~1")
```

The `as.formula` function is used in the `lm()` function to allow these character strings to describe the models. By defining the adaptall function in this manner, the statistician is able to perform a wide variety of statistical tests using the standard model syntax for the `lm()` function. For the fourth and fifth arguments we have specified that 2000 permutations be used and that the initial random number seed be set to 71492.

R code can be executed in a variety of ways. If we have the `adapt` function on a file named `adapt.r` we can use the `source` statement to include this code before we call the function. The `adapt.r` file, which is listed in Appendix A, can be downloaded from *www.math.niu.edu/~ogorman*. The only statements that are needed to execute this program are:

```
source("adapt.r")
dataset <- c("thall.total.dat")
```

```
complete <- c("y~group")
reduced  <- c("y~1")
r <- 2000
seed <- c(71492)
pvalue <- adapt(dataset,complete,reduced,r,seed)
```

This program computed the adaptive p-value for the two-sample test as $p = 0.1669$. The test is quite practical; it can execute in less than a minute on an inexpensive computer purchased in 2008. If greater accuracy is desired, the number of permutations could be increased.

3.7 LEVEL OF SIGNIFICANCE OF THE ADAPTIVE TEST

The permutation test statistics will be equally likely under the null hypothesis when used with data that are obtained from subjects who were randomly assigned to one of two treatment groups. Once the p-value is obtained, permutation tests are performed in the same manner that traditional test are performed; we reject the null hypothesis if the p-value is less than or equal to α. Recall that we will be using $\Lambda^{2/n,*}$ as a test statistic, so we will reject the null hypothesis for small values of $\Lambda^{2/n,*}$. Now suppose we have constructed the full permutation distribution and have determined the critical value, which is the largest test statistic that has a p-value less than or equal to α. Although the permutation distribution is discrete, the p-value for the critical value will be very close to α for any reasonable sample size. For our example we have $\binom{59}{31} = 5.53 \times 10^{16}$ permutations so that, even with some ties, there will be a large number of distinct test statistics. Thus, if we were to tabulate the full permutation distribution and use it to perform the test, we could find a critical value that would produce a Type I error close to α. However, we will usually take a sample of R permutations to estimate the p-value. Hence, we would like to know the probability of a Type I error when we take a random sample with replacement from a set of all the permutations and we reject the null hypothesis whenever $p = (E+1)/(R+1) \leq \alpha$.

Suppose there are N test statistics and all the test statistics are distinct. Let Λ be the set of all test statistics. Suppose we order the test statistics so that they satisfy $\Lambda_{(1)} < \Lambda_{(2)} < \ldots < \Lambda_{(N)}$. Thus, $\Lambda_{(l)}$ is greater than $l-1$ test statistics. As before, we will take a random sample of R permutations, with replacement, from the set of all permutations and we will let E be the number of test statistics that are less than or equal to the observed test statistic.

Because we are taking a random sample of R permutations, we cannot be certain that a specified test statistic will lead to rejection of the null hypothesis. The total probability of rejection of the null hypothesis is

$$P(Rejection) = \sum_{l=1}^{N} P(\Lambda_{(l)})P(Rejection|\Lambda_{(l)}). \tag{3.3}$$

Under the null hypothesis all test statistics are equally likely, so $P(\Lambda_{(l)}) = 1/N$ for $l = 1, \ldots, N$.

Now suppose we consider the probability of rejection for a specific $\Lambda_{(l)}$. For that value of $\Lambda_{(l)}$ we may or may not reject the null hypothesis, depending on the R test statistics that were selected. If we take a sample of R permutations, then for any $\Lambda_{(l)}$

$$\begin{aligned} P(Rejection|\Lambda_{(l)}) &= P(p \leq \alpha|\Lambda_{(l)}) \\ &= P\left(\frac{E+1}{R+1} \leq \alpha|\Lambda_{(l)}\right) \\ &= P(E \leq \alpha(R+1) - 1|\Lambda_{(l)}). \end{aligned}$$

Let $p_l = l/N$ be the proportion of test statistics that are less than or equal to the test statistic $\Lambda_{(l)}$. Because the permutations are selected independently with replacement, the probability of obtaining a permutation test statistic less than or equal to $\Lambda_{(l)}$ is equal to p_l for each of the permutations that are selected. Hence, E is a binomial with R trials and probability p_l. Thus,

$$P(Rejection|\Lambda_{(l)}) = \sum_{i=0}^{\lfloor \alpha(R+1)-1 \rfloor} \binom{R}{i} p_l^i (1-p_l)^{R-i}$$

For a two-sample test having $N = \binom{n_1+n_2}{n_1}$ distinct test statistics we find that $P(\Lambda_{(l)}) = 1/\binom{n_1+n_2}{n_1}$. Consequently, using Equation (3.3), we can compute the probability of rejection as

$$P(Rejection) = \frac{1}{\binom{n_1+n_2}{n_1}} \sum_{l=1}^{N} \sum_{i=0}^{\lfloor \alpha(R+1)-1 \rfloor} \binom{R}{i} p_l^i (1-p_l)^{R-i} \quad (3.4)$$

Computing this probability directly can be problematic because of the underflows and overflows that will be encountered, but the cumulative binomial can be computed by using the relationship with the incomplete Beta distribution or by using a recursive relationship between terms in a binomial sum. Using Equation (3.4), we were able to compute, for several sample sizes, the probability of rejection that are shown in the fourth column of Table 3.2. It can be seen that, except for very small sample sizes, the actual probabilities of rejection are very close to the nominal significance level. Consequently, when $R = 2000$ randomly selected permutations are used, the two-sample adaptive test will have a level of significance close to α if $n_1 \geq 6$ and $n_2 \geq 6$.

Instead of using $p = (E+1)/(R+1)$, some researchers use $p = E/R$ as the p-value in the test procedure. The empirical significance level of the test when $p = E/R$ is used is given in the last column of Table 3.2. These results show that the test procedure that uses $p = E/R$ will not maintain its level of significance, but they also demonstrate that, with $R = 2000$ permutations, there is little difference between the significance levels of these two test procedures.

I believe that most people would consider $p = E/R$ to be the most natural formula to use in the test procedure. However, it should be kept in mind that our objective is to perform a test, not to estimate a proportion. Also, in this test procedure we

Table 3.2 Exact probability of rejection under the null hypothesis using a sample permutation test with $\alpha = 0.05$ and $R = 2000$ permutations

n_1	n_2	$\binom{n_1+n_2}{n_1}$	$P(Rejection)$ Using $p = (E+1)/(R+1)$	$P(Rejection)$ Using $p = E/R$
4	4	70	0.04298	0.04337
6	6	924	0.04943	0.04993
9	5	2002	0.04973	0.05023
8	8	12870	0.04988	0.05044
12	12	2704156	0.04997	0.05047

are not estimating the population proportion that exceed a specified value. In our test procedure we have already obtained one sample and have obtained R additional samples from the population, and these samples are equally likely. Consequently, it seems reasonable to use $p = (E+1)/(R+1)$ in the test because it will maintain its level of significance. For a more complete discussion of this topic see Davison and Hinkley (1997, p. 140).

Because our results apply to two-sample permutation tests, including the adaptive test, we have shown that when $p = (E+1)/(R+1)$ is used, the adaptive test will maintain its level of significance with continuous error distributions. If the errors are normally distributed, have equal variance, and are independent, the t test will maintain its size because the distribution of the test statistic is known to be a t distribution with $n-1$ degrees of freedom. However, if the errors are not normally distributed, the empirical significance level of the traditional t test may depart from the nominal value. Hence, we performed an extensive simulation study using 100000 data sets to accurately estimate the significance level of the t test and the adaptive test. The standard error is approximately 0.07% when 100000 data sets are used. These empirical significance levels were shown in Figure 1.4 for the two-sample test with $n_1 = n_2 = 15$ observations. As expected, the empirical significance level of the adaptive test was in the range $[.05 \pm 2(.0007)]$ for all distributions. Note that the significance level of the adaptive test was very close to the nominal value but the empirical significance level of the t-test was somewhat below the nominal value for several non-normal error distributions.

3.8 POWER OF THE ADAPTIVE TEST

Because the adaptive test and the t test maintain their significance level, it is possible to fairly compare their power. To obtain accurate estimates of the power, a simulation program was written in FORTRAN. By writing the simulation in FORTRAN, we are able to analyze many data sets in a reasonable amount of time, so that we can obtain accurate estimates of the power. All of the simulation studies in this book that are used to estimate the level of significance and the power are written in FORTRAN by the author, except for the mixed model test in Chapter 10 that uses a procedure in SAS. Any researcher who is interested in a specific simulation study and who is proficient in FORTRAN may contact the author at *ogorman@math.niu.edu* to obtain the source code. It should be noted that these programs make extensive use of adjustable array dimensions in FORTRAN-77. Researchers who are more familiar with R or SAS should be able to verify the simulation results by generating many data sets, in R or in SAS, and then using the appropriate R function or SAS macro to obtain the p-value for the adaptive test.

The power of these tests with $n_1 = n_2 = 25$ observations, when we set the difference between the distributions equal to $\delta = 0.792$, is shown in Figure 3.3 based on a simulation study that used 10000 data sets for each error distribution. The value of δ was selected to produce probabilities of rejection near 80% for the t test. We note that with normal errors the adaptive test is nearly as powerful as the t test. With the other three symmetric distributions the adaptive test is clearly more powerful than the t test, including the long-tailed $t_{df=4}$ distribution. With the five skewed distributions the adaptive test was much more powerful than the t test. These results are important because many of the real-world distributions are skewed or are long-tailed symmetric. With $n_1 = n_2 = 25$, if we use the adaptive test there is little to lose in power if the error distribution is normal, while there is much to gain in power if the error distribution is skewed or long-tailed.

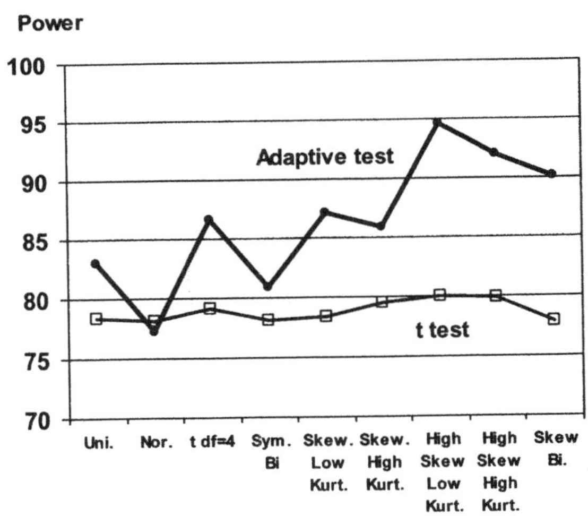

Figure 3.3 The empirical power, in percent, for the adaptive test and the t test for several error distributions with $n_1 = n_2 = 25$ observations. The difference between the two populations was set to $\delta = 0.792$ to give powers near 80% for the t test.

The relative power for these two tests that we have shown in Figure 3.3 also roughly describes the relationship between the power of the tests for many other sample sizes. In Figure 1.5 we gave the empirical power of the adaptive test and the t test for $n_1 = n_2 = 15$ observations with $\delta = 1.0$ as the shift parameter, and we observed the same power advantage of the adaptive test with non-normal errors. The results that we observe for the balanced two-sample test can also be observed with unbalanced sample sizes. If we use unbalanced data sets with $n_1 = 40$ and $n_2 = 10$ observations and use a shift of $\delta = 0.991$, we find the powers that are displayed in Figure 3.4. We again see that there is only a small loss in power with normal errors, but there is a sizable gain in power with all of the non-normal distributions.

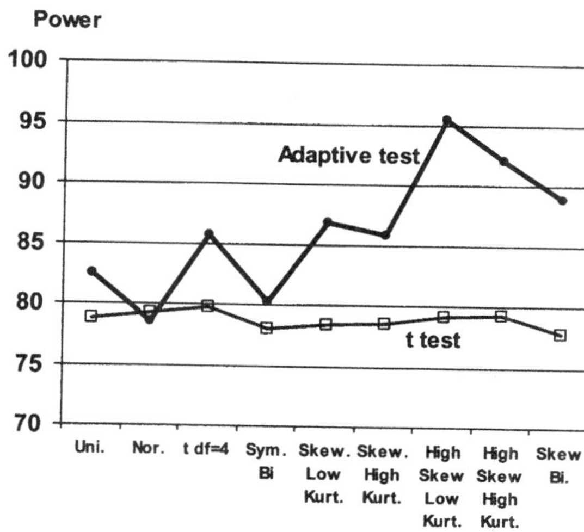

Figure 3.4 The empirical power, in percent, for the adaptive test and the t test for several error distributions with $n_1 = 40$ and $n_2 = 10$ observations. The difference between the two populations was set to $\delta = 0.991$ to give powers near 80% for the t test.

3.9 SAMPLE SIZE ESTIMATION

The power advantage of the adaptive test is important for researchers who are analyzing data that have already been collected. However, if the study is in the process of being designed, the researcher may be more interested in the sample size reductions that are possible when adaptive tests are employed in the analysis of the data. Since, for a fixed sample size, we have found the adaptive test to be more powerful than the t test for non-normal errors, we expect to find significant reductions in sample sizes with non-normal errors. To determine the sample size reductions that are possible when adaptive tests are used we will perform a simulation study. We will assume that the mean difference between the groups (δ) is fixed. We will find the sample size needed to obtain 80% power when the t test is used and the sample size needed to obtain the same power when the adaptive test is used.

To find the sample size needed to achieve 80% power for a specified value of $\delta = 0.75$, we begin by using an initial guess of the sample size and then we perform a simulation study to estimate the actual power. Based on that result we adjust the sample size upward or downward and then perform another simulation study to estimate the power. We continue this process until we find the sample size that gives power nearest to 80%. The resulting sample sizes for the t test and the adaptive test are shown in Figure 3.5. As expected, we observe little difference in sample size requirements with normal errors, but with non-normal errors the t test clearly requires a larger sample size.

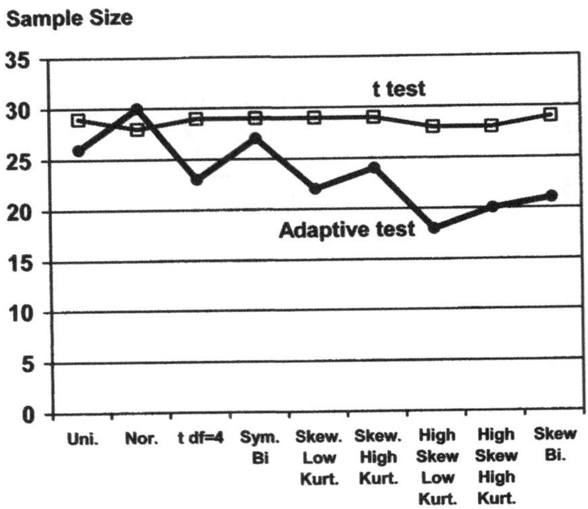

Figure 3.5 The sample size in each group that are required for a two-sample test to achieve 80% power, assuming a difference between the populations of $\delta = 0.75$ and assuming a standard deviation of $\sigma = 1$ for the errors.

In an actual experiment we would rarely know the distribution of the errors. Also, it is unlikely that any of these nine distributions will exactly match the distribution of the errors. However, based on a pilot study or experience, it may be possible to make an educated guess about the overall shape of the error distribution.

For example, if a variable measures the triglyceride level, the researcher may believe that the distribution will be skewed to the right, because the triglyceride level cannot be negative and because other studies have shown that most of the subjects will have moderate or low triglyceride levels, but some may have very large values. In that situation a linear model might be expected to have a skewed distribution of errors. Hence, it is reasonable to assume that the sample size requirements will be less than those obtained under the assumption of normality.

To obtain the sample size required for an adaptive test to achieve a specified power, we would usually proceed in two steps:

1. Calculate the sample size requirements for a given δ and σ assuming the errors are normal.

2. Adjust the sample size that was found in step 1 according to the expected shape of the distribution of the errors.

Let the power of the test be $1 - \beta$. To calculate the sample size requirements when the errors are normal, we use the standard formula for the sample size in each group:

$$n_1 = n_2 = 2[\Phi^{-1}(1 - \alpha/2) + \Phi^{-1}(1 - \beta)]^2 \left(\frac{\sigma^2}{\delta^2}\right). \tag{3.5}$$

To adjust these sample sizes, we consider the ratios of the sample size needed with the adaptive test to the sample size determined by the standard formula, based on a simulation study that used $\delta = 0.75$ to achieve approximately 80% power. These ratios are shown in Figure 3.6.

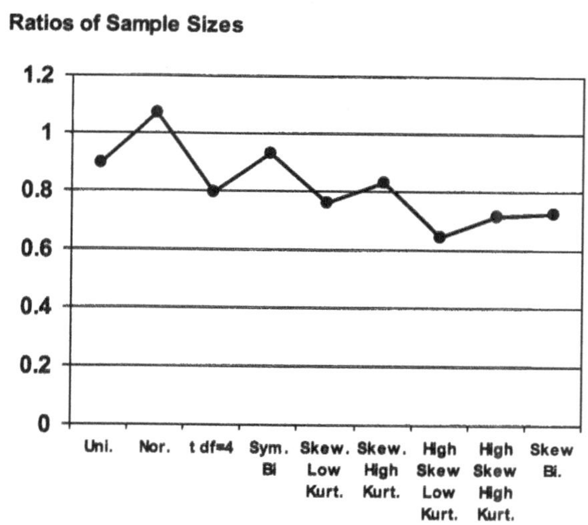

Figure 3.6 The ratio of the sample size needed for the adaptive test to the sample size needed for the t test with balanced sample sizes. We used $\delta = 0.75$ for these simulations to give powers near 80% for the t test and adaptive test. The t test required samples sizes near 30 in each group.

Based on the ratios in Figure 3.6, we have produced the sample size multipliers that are given in Table 3.3. Note that we have used only four categories of shapes of distributions because it is unlikely that the researcher would know a great deal about the exact shape of the distribution. If the errors are approximately normally distributed, then no adjustment is necessary. If the distribution is likely to be symmetric and long-tailed, but not normal, then a multiplier of 0.85 would seem to be appropriate so that there would be a 15% reduction in the number of experimental units needed. If the distribution is expected to be skewed, then a greater reduction may be possible. The multiplier is near 0.80 for moderately skewed distributions, which are quite common. Although these multipliers were based on one simulation study having $n_1 + n_2$ near 60, they can be applied to other studies whose power estimates are similar to those shown in Figure 3.3.

Table 3.3 Multipliers of the sample size that is obtained from the standard formula.[a]

Distribution	Multiplier
Normal	1.05
Long-tailed symmetric	0.85
Moderately skewed	0.80
Highly skewed	0.70

[a]These multipliers are appropriate if $n_1 + n_2$ approximates 60.

3.10 A SAS MACRO FOR THE ADAPTIVE TEST

Adaptive testing can easily be performed in SAS by using a SAS macro that can be downloaded from the author's web site (*www.math.niu.edu/~ogorman*). This macro can be used for a wide variety of tests, ranging from very simple univariate tests to complex tests used in multivariate regression models having several dependent and many independent variables. For tests that are based on simple models, only a few parameters need to be specified.

Suppose we want a two-sample adaptive test of the null hypothesis of group effect with the seizure data that we have analyzed earlier in the chapter. To perform the adaptive test, we need to do the following:

- Create a SAS data set that includes the dependent variable and the group indicator variable.
- Include the SAS macro "adaptall" into the code, either by placing it at the beginning of the program or by using an "include" statement, so that SAS can find the macro.
- Specify the appropriate parameters in the call to the macro.

The SAS macro uses the GLM procedure to obtain the residuals, so we must specify the dependent and independent variable in a manner that conforms to the syntax for the MODEL and CLASSES statements in PROC GLM. In the macro call we must specify the dependent variable as the keyword parameter (`y`) and the independent variable as the keyword parameter (`xa`). In a two-sample test we would want SAS to use the independent variable as a class variable, so we specify the class variable as the keyword parameter (`classa`). In the code given below, the data set name is `seizure`, the dependent variable is `count`, and the independent variable is group, which is also specified as a class variable. When we invoke the macro, the keyword parameter y will be given the value count, the keyword parameter xa will be given the value group, and the keyword parameter classa will be given the value group. The random number generator that was proposed by Wichmann and Hill (1982) is used to select the permutations. That generator uses three seeds that are specified as seed1, seed2, and seed3. These seeds must be positive integers less than 30269. So, to perform a two sample adaptive test, if the adaptall macro was located in the directory "c:\research" on a personal computer, we would use this SAS code:

```
include "c:\research\adaptall.sas";
data seizure;
  input count group;
  datalines;
  42 progabide
  .
  .
  53 placebo
;
%adaptall(dataset=seizure, y=count, xa=group, classa=group,
          seed1=17833, seed2=18632, seed3= 18972);
```

The adaptall.sas file, which is listed in Appendix B, can be downloaded from *www.math.niu.edu/~ogorman*. The SAS macro produced the output below.

```
Dependent variables  :  count

Independent variables in reduced model  : Intercept
      Class variables in reduced model  :

Independent variables in complete model : Intercept   group
      Class variables in complete model :   group

Adaptive test statistics for unpermuted data:

      Wilks Lambda =   0.9849
      F statistic  =   0.8711

      p-value for adaptive test =   0.1714
```

Note that the p-value for the adaptive test is $p = 0.1714$, which varies slightly from the p-value obtained by the R code, which was $p = 0.1669$. The difference in p-values is caused by differences in the random number generators that are used in the two programs. We will have much more to say about this SAS macro in subsequent chapters.

3.11 MODIFICATIONS FOR ONE-TAILED TESTS

We have used the F^* and $\Lambda^{2/n,*}$ test statistics to perform two-tailed tests. Often, two-tailed tests are more appropriate than one-tailed tests; but if a one-tailed test is needed, it is not difficult to modify the two-tailed procedure to produce a one-tailed p-value.

Suppose we want to perform the one-sided test of $H_0 : \beta_1 = 0$ versus $H_a : \beta_1 > 0$. Let t_ν denote the test statistic obtained from the pooled version of the two-sample t test with $\nu = n_1 + n_2 - 2$ degrees of freedom. It can be shown that $t_\nu^2 = F$ using OLS and that $(t_\nu^*)^2 = F^*$ when the observations are adaptively weighted. Hence, we can easily compute the t test statistics as $t_\nu^* = (F^*)^{1/2}$ if $\beta_1 \geq 0$ or as $t_\nu^* = -(F^*)^{1/2}$ if $\beta_1 < 0$. For this one-tailed test of $H_0 : \beta_1 = 0$ versus $H_a : \beta_1 > 0$, we would perform the R permutations of the residuals and would let E equal the number of permutations with t test statistics that exceed the observed t test statistic. The one-tailed p-value would be computed as $p = (E + 1)/(R + 1)$. So, while we will use two-sided tests in this book, the same weighting approach could be used to perform one-sided tests, and the R functions and SAS macros could be modified to perform these tests.

3.12 JUSTIFICATION OF THE WEIGHTING METHOD

We described, in Chapter 2, a weighting method of normalizing data. In this book we will use that weighting method to normalize the residuals. For two-sample tests the reduced linear models, which are used to compute the residuals, contain only the intercept. In this section we show that the weighting formula can be used to normalize the errors in these linear models and that the weights do not depend on our estimate of β_0.

In the adaptive tests we use the residuals from the reduced model to compute the weights as

$$\hat{w}_i = \frac{z_i}{s_i} = \frac{\hat{\sigma}\Phi^{-1}[\hat{F}_h(\hat{\varepsilon}_i)]}{\hat{\varepsilon}_i - \hat{\varepsilon}_{.5}} \tag{3.6}$$

for $i = 1, \ldots, n$. For the two-sample test the reduced model for the ith observation is $y_i = \beta_0 + \varepsilon_i$ and the weighted reduced model is

$$\hat{w}_i y_i = \hat{w}_i \beta_0 + \hat{w}_i \varepsilon_i.$$

After applying the adaptive weights we expect that $\varepsilon_i^* = \hat{w}_i \varepsilon_i$ will be approximately normally distributed.

To see why these adaptive weights are appropriate for the two-sample test consider the reduced model

$$y_i = \beta_0 + \varepsilon_i,$$

for $i = 1, \ldots, n$. Now, suppose the c.d.f. of the errors, $F(\varepsilon)$, is known and that β_0 is known so that the errors, $\varepsilon_i, i = 1, \ldots, n$ are known. Because we know $F(\varepsilon)$ we could compute $\varepsilon_{.25}$, $\varepsilon_{.5}$, and $\varepsilon_{.75}$. If we let $\beta_{0,.5} = \beta_0 + \varepsilon_{.5}$ we can write the reduced model as

$$y_i = (\beta_0 + \varepsilon_{.5}) + (\varepsilon_i - \varepsilon_{.5}) = \beta_{0,.5} + (\varepsilon_i - \varepsilon_{.5}).$$

With $\sigma = (\varepsilon_{.75} - \varepsilon_{.25})/1.349$ we could compute the population weights as

$$w_i = \frac{\sigma \Phi^{-1}[F(\varepsilon_i)]}{\varepsilon_i - \varepsilon_{.5}} \tag{3.7}$$

for $i = 1, \ldots, n$. Hence, the weighted reduced model for the ith observation, using the population weights, becomes

$$w_i y_i = w_i \beta_{0,.5} + w_i(\varepsilon_i - \varepsilon_{.5}).$$

If $(\varepsilon_i - \varepsilon_{.5})$ has a continuous distribution, then, using the same approach that we described in Chapter 2, the random variable $w_i(\varepsilon_i - \varepsilon_{.5}) = \sigma\Phi^{-1}[F(\varepsilon_i)]$ will be normally distributed. Hence, if we knew β_0 the weights in (3.7) will normalize the errors.

Note that, if we add a constant (c) to each residual so that the shifted residuals are $\hat{\varepsilon}_i + c$, for $i = 1, \ldots, n$, the sample weights will not change. The addition of the constant will not change $\hat{\sigma}$ or h so that $\hat{F}_h(\hat{\varepsilon}_i)$, when smoothed over the set of residuals, will equal $\hat{F}_h(\hat{\varepsilon}_i + c)$, when smoothed over the set of shifted residuals. Hence, the numerator in (3.6) is location invariant. The denominator is also location invariant because $\hat{\varepsilon}_i - \hat{\varepsilon}_{.5} = (\hat{\varepsilon}_i + c) - (\hat{\varepsilon}_{.5} + c)$ for $i = 1, \ldots, n$. Consequently, the sample weights given in (3.6) are location invariant. This is important because it shows that the sample weights do not depend on our estimate of β_0. Thus, the only source of inaccuracy in the weighting is due to the inaccuracy in the estimation of the c.d.f. of the errors. Using a smoothed c.d.f. with a reasonable bandwidth can reduce this inaccuracy but cannot eliminate it. With large n we expect to have a good estimate of $F(\cdot)$, so the normalization should be effective. With small sample sizes the weighting may produce residuals that are only roughly normally distributed due to the inaccuracies involved in the estimation of the c.d.f. of the errors. Although we would like to normalize the errors as accurately as possible, in this book we are primarily concerned with the level of significance and the power of the tests that are obtained after weighting.

3.13 COMMENTS ON THE ADAPTIVE TWO-SAMPLE TEST

We have seen that the adaptive two-sample test has excellent statistical properties. It maintains its significance level, is more powerful than the t test for non-normal errors, and has only a small power loss with normal errors. In addition, using either R or SAS, it is possible to perform the test in a few minutes. Consequently, there is little to lose and much to gain by routinely using this adaptive two-sample test.

But what make these results so exciting is that the WLS approach can also be used for much more complex models. These basic procedures can be extended to perform a test of any subset of coefficients in most linear models. We will see in subsequent chapters how this approach can be used to test main effects and interaction effects with multifactorial designs. In addition, we shall see that the same adaptive testing method can be used with multiple regression models and with models for paired data.

Before we consider more complex models, we need to learn more about the various types of permutations that can be used with those models. We saw in this chapter that the permute-residuals method was equivalent to the permutation of the group indicator for a two-sample test. However, with multiple regression models there are several methods that can be considered and they are not equivalent. These permutation methods will be discussed in the next chapter.

EXERCISES

3.1 How does the set of ordinary residuals differ from a set of errors? Are there any restrictions on the residuals?

3.2 Explain the difference between WLS and OLS. What is minimized when OLS is used? What is minimized when WLS is used?

3.3 Suppose we obtain $\hat{w}_i = 2$ for $i = 1, \ldots, n$ and we use WLS to obtain estimates of β_0 and β_1. Will these estimates equal the OLS estimates? Why or why not?

3.4 Suppose a researcher always performed experiments using two groups of animals and always used the two-sample adaptive test to analyze the data.

- **a)** Under permutations of residuals will SSE_R^* change? Explain.
- **b)** Could SSE_C^* be used as the test statistic instead of $\Lambda^{2/n,*}$? Why or why not?

3.5 A researcher preferred to use $R^{2,*}$ as the test statistic instead of $\Lambda^{2/n,*}$.

- **a)** How should E be defined if $R^{2,*}$ is used?
- **b)** Will this test statistic produce the same results that would be obtained if $\Lambda^{2/n,*}$ were used as the test statistic? Explain.

3.6 Suppose 16 rats are randomly assigned to one of two treatment groups so that 8 rats are assigned to each group. How many different assignments are possible?

3.7 Suppose 68 rats are randomly assigned to one of two treatment groups so that 34 rats are assigned to each group.

- **a)** How many different assignments are possible?
- **b)** If a single processor on a computer could compute the test statistic for 20 million assignments per second and there are four processors on a computer that could be programmed to use all four processors simultaneously, how much time would it take to evaluate the full permutation distribution?

3.8 A nutritionist performed an experiment to determine how much weight would be lost by people who followed one of two diets. To compare these diets nine subjects were randomly assigned to one of two diet groups. The weight loss in pounds were recorded as:

	Group A					Group B			
Weight Loss	12	2	-3	7	4	6	11	-2	8

a) Compute the residuals from the reduced model.

b) Using the permutation of residuals method with one permutation of the residuals, compute a new vector of observations by adding the predicted value (from the reduced model) to the permuted residuals.

c) Did you obtain a permutation of the original data? Will the permutation of residuals method always produce a permutation of the y-values when two-sample tests are used? Explain.

3.9 Suppose, based on the full permutation distribution, that the p-value is 0.05 for a certain test statistic. Now, suppose we obtain that test statistic in a sample permutation test that uses R samples. Calculate the standard error of the estimate of the p-value for:

a) a sample of $R = 2000$.

b) a sample of $R = 10000$.

3.10 Suppose we want to perform a two-sample adaptive test on data that we have already collected. We compute the adaptive weights from the residuals of the reduced model on the unpermuted data. Assume that we perform one permutation of the residuals to obtain a new vector of dependent variables and then we calculate the adaptive weights based on the new residuals. Can the weights for the permuted data be found easily without doing any calculations? Why or why not?

3.11 Blumberg, West, and Ellis (1954) gave data on a psychological test that was given to two groups of cancer patients. The scores for the patients in the two groups were:

Group 1	-25	-24	-22	-22	-21	-18	-18	-18	-18
	-17	-16	-14	-14	-13	-13	-13	-13	-9
	-8	-7	-5	1	3	7	7		
Group 2	-21	-18	-16	-16	-16	-14	-13	-13	-12
	-11	-11	-11	-9	-9	-9	-9	-7	-6
	-3	-2	3	10					

Using R code find the p-value for the adaptive test and the p-value for the F test. If there are differences between these p-values, explain what may have caused the difference.

3.12 Using the SAS macro on the data given in Exercise 3.11 find the p-value for the adaptive test and the p-value for the F test. If there is a difference between these p-values, explain what may have caused the difference.

3.13 Consider the data given in Exercise 3.11. Suppose the subject who had the largest value in Group 2 was incorrectly classified into that group. If we move that subject into Group 1, what will be the effect on the adaptive weights? Explain.

3.14 (SAS) A group of researchers obtained the weights w_i, for $i = 1, \ldots, n$ and wanted to obtain estimates of β_0 and β_1 using WLS. For the ith observation they created a new variable `WY` that equaled $w_i y_i$, and a variable `WX` that equaled $w_i x_i$. Then they used the following SAS code to obtain the estimates:

```
proc glm;
model wy=wx;
```

What is wrong with this approach? How could they correct the problem?

3.15 Suppose a researcher wanted to design a study to compare the weight loss of subjects over an 8-week period on one diet to the weight loss on another diet. Further, suppose the design objective was to have 80% power to detect a specified δ using a level of significance of $\alpha = 0.05$. The traditional sample size formula indicated that the design should call for $n_1 = n_2 = 32$ subjects. Previous studies of weight loss indicate that the responses tend to be skewed. If an adaptive test is used to analyze the data, what sample size would be required?

CHAPTER 4

PERMUTATION TESTS WITH LINEAR MODELS

4.1 INTRODUCTION

In Chapter 3 we discussed permutation methods for two-sample tests and we permuted the residuals to perform that test. We also showed that permutations of residuals was equivalent to permutations of observations in that two-sample test. Both of the methods are valid, so there is little controversy about the permutation method.

In contrast, when different permutation methods are used to test a coefficient in a linear regression model the results are not equivalent and it is not clear which method is superior. Some of the methods that have been proposed are:

- Permute the residuals from the reduced model and add these to the predicted values from the reduced model to form a new vector of predicted values.
- Permute the independent variable for the coefficient that is to be tested.
- Permute the dependent variable.
- Permute the residuals from the complete model and add these to the predicted values from the complete model to form a new vector of predicted values.

Adaptive Tests of Significance Using Permutations of Residuals with R and SAS®.
By Thomas W. O'Gorman

We will describe the first three of these methods in some detail in this chapter. For details concerning the last method see Anderson and Legendre (1999).

Usually we are not able to tabulate the full permutation distribution, so we take a sample of permutations from a population of all possible permutations. We could use sampling with replacement or sampling without replacement. Sampling without replacement could slightly reduce the variability in the p-value, but it is more difficult to program. Also, sampling without replacement takes much more computer time to execute because each new permutation that is generated must be checked against all of the others that have been generated. We use sampling with replacement, but there is little difference between sampling with replacement and sampling without replacement when we take samples of size $R = 2000$. For example, if we have $n = 20$ observations that we use in a test for slope in a simple linear regression, there are $20! = 2.43 \times 10^{18}$ permutations; so if we sample with replacement, the probability of obtaining any duplicates is approximately 8.2×10^{-13}. For very small values of n it may be slightly more efficient to sample without replacement; but since adaptive tests are not generally recommended if n is much smaller than 20, there appears to be no practical difference between sampling with replacement and sampling without replacement.

In the next chapter we will show how these permutation methods can be used to perform adaptive tests. In this chapter we will limit our discussion to nonadaptive tests. We will describe the permutation methods and will show how permutation methods can be used for more complex experiments. Before we describe these permutation methods it will be helpful to introduce some notation.

4.2 NOTATION

Suppose we have an $n \times 1$ vector $\mathbf{Y}$ and we want to permute (or shuffle) the rows of $\mathbf{Y}$. To permute the rows, we can premultiply $\mathbf{Y}$ by a permutation (or shuffling) matrix $\mathbf{S}$. The $\mathbf{S}$ matrix is an $n \times n$ matrix having only ones and zeros, with each row and column having exactly one of its elements equaling one. Let s_{ij} be the element of $\mathbf{S}$ in the ith row and jth column. If $s_{ij} = 1$, the jth row of $\mathbf{Y}$ will be permuted to the ith row of $\mathbf{SY}$. For example, let $\mathbf{Y} = \begin{bmatrix} 3 & 1 & 7 & 5 \end{bmatrix}'$ and let

$$\mathbf{S} = \begin{bmatrix} 0 & 0 & 0 & 1 \\ 0 & 1 & 0 & 0 \\ 1 & 0 & 0 & 0 \\ 0 & 0 & 1 & 0 \end{bmatrix}$$

be a permutation matrix. Then

$$\mathbf{SY} = \begin{bmatrix} 0 & 0 & 0 & 1 \\ 0 & 1 & 0 & 0 \\ 1 & 0 & 0 & 0 \\ 0 & 0 & 1 & 0 \end{bmatrix} \begin{bmatrix} 3 \\ 1 \\ 7 \\ 5 \end{bmatrix} = \begin{bmatrix} 5 \\ 1 \\ 3 \\ 7 \end{bmatrix}$$

is a permutation of the rows of $\mathbf{Y}$. It can be shown that the inverse of $\mathbf{S}$ is the transpose of $\mathbf{S}$. Hence, to restore the original order we can multiply $\mathbf{SY}$ by the transpose of $\mathbf{S}$. With our example we find

$$\mathbf{S}'(\mathbf{SY}) = \begin{bmatrix} 0 & 0 & 1 & 0 \\ 0 & 1 & 0 & 0 \\ 0 & 0 & 0 & 1 \\ 1 & 0 & 0 & 0 \end{bmatrix} \begin{bmatrix} 5 \\ 1 \\ 3 \\ 7 \end{bmatrix} = \begin{bmatrix} 3 \\ 1 \\ 7 \\ 5 \end{bmatrix} = \mathbf{Y}.$$

We will sometimes call $\mathbf{S}' = \mathbf{S}^{-1}$ the inverse shuffle because it restores the original order.

4.3 PERMUTATIONS WITH BLOCKING

In some situations we may want to permute only within a region (or block). For example, if we have performed an experiment by randomly assigning subjects to treatments at two locations, we may want to permute the subjects within the locations but not between the locations. Suppose we have n_1 observations from location 1 in the first n_1 rows of $\mathbf{Y}$ and n_2 observations from location 2 in the last n_2 rows of $\mathbf{Y}$. To perform within-block permutations the permutation matrix would have this form

$$\mathbf{S} = \begin{bmatrix} \mathbf{S}_1 & \mathbf{0} \\ \mathbf{0} & \mathbf{S}_2 \end{bmatrix},$$

where $\mathbf{S}_1$ and $\mathbf{S}_2$ are the permutation matrices for the first and second locations. We can then perform a within-block permutation as $\mathbf{SY}$. Permutation tests that use this approach will be called blocked permutation tests.

4.4 LINEAR MODELS IN MATRIX FORM

In subsequent chapters we will use adaptive tests for a subset of coefficients in regression models. Suppose we have a regression model with q independent variables. For the ith observation the regression model can be written as

$$y_i = \beta_1 x_{i,1} + \beta_2 x_{i,2} + \cdots + \beta_q x_{i,q} + \varepsilon_i,$$

where $\beta_1, \beta_2, \ldots, \beta_q$ are the regression coefficients and ε_i is the error term. It will be more convenient to use the matrix form of this model. Let the $n \times 1$ vector of dependent variables be

$$\mathbf{Y} = \begin{bmatrix} y_1 \\ y_2 \\ \cdot \\ \cdot \\ \cdot \\ y_n \end{bmatrix}.$$

The $n \times q$ matrix of independent variables is

$$\mathbf{X} = \begin{bmatrix} x_{1,1} & x_{1,2} & \cdots & x_{1,q} \\ x_{2,1} & x_{2,2} & \cdots & x_{2,q} \\ \cdot & \cdot & \cdot & \cdot \\ x_{n,1} & x_{n,2} & \cdots & x_{n,q} \end{bmatrix},$$

and the $q \times 1$ vector of regression coefficients is $\boldsymbol{\beta}$. The error terms are in the $n \times 1$ vector $\boldsymbol{\varepsilon}$. Typically, we will include an intercept term in the model so we will usually set $x_{i,1} = 1$ for $i = 1, \ldots, n$.

With this notation we can write the model in matrix form as $\mathbf{Y} = \mathbf{X}\boldsymbol{\beta} + \boldsymbol{\varepsilon}$. Usually, we will be testing a subset of coefficients in a linear model, so we will partition the model into two parts. The partitioned model can be written as

$$\mathbf{Y} = \mathbf{X}_R\boldsymbol{\beta}_R + \mathbf{X}_A\boldsymbol{\beta}_A + \boldsymbol{\varepsilon} = [\mathbf{X}_R|\mathbf{X}_A] \begin{bmatrix} \boldsymbol{\beta}_R \\ \boldsymbol{\beta}_A \end{bmatrix} + \boldsymbol{\varepsilon},$$

where $\mathbf{X}_R$ is the $n \times r$ matrix containing the first r columns of $\mathbf{X}$, $\mathbf{X}_A$ is the $n \times (q-r)$ matrix containing the last $(q-r)$ columns of $\mathbf{X}$, $\boldsymbol{\beta}_R$ is the $r \times 1$ vector consisting of the first r elements of $\boldsymbol{\beta}$, and $\boldsymbol{\beta}_A$ is the $(q-r) \times 1$ matrix containing the last $q-r$ elements of $\boldsymbol{\beta}$.

We will test $H_0 : \boldsymbol{\beta}_A = \mathbf{0}$ versus $H_a : \boldsymbol{\beta}_A \neq \mathbf{0}$, where $\mathbf{0}$ is a $(q-r) \times 1$ matrix of zeros. Under the null hypothesis, $\mathbf{X}_A\boldsymbol{\beta}_A = \mathbf{0}$, so our model becomes

$$\mathbf{Y} = \mathbf{X}_R\boldsymbol{\beta}_R + \boldsymbol{\varepsilon}$$

We will call this model the "reduced" model because the complete model is reduced to it under the null hypothesis. Our basic strategy is to fit the complete model

$$\mathbf{Y} = \mathbf{X}_R\boldsymbol{\beta}_R + \mathbf{X}_A\boldsymbol{\beta}_A + \boldsymbol{\varepsilon}$$

to the data and the reduced model to the same data to see if the addition of the terms in $\mathbf{X}_A\boldsymbol{\beta}_A$ improves the fit.

4.5 PERMUTATION METHODS

Several permutation methods have been proposed for regression models. For certain simple models some of these permutation methods are equivalent, but for more complex models these methods are not equivalent. Also, it is not obvious which of these methods is superior. So, in this section we will briefly describe some of the permutation methods that can be used to test a subset of coefficients in multiple regression models.

4.5.1 The Permute-Errors Method

Before we discuss some practical permutation methods we will describe an impractical, but important, permutation method. Suppose we know $\boldsymbol{\beta}_R$ and we are interested in testing $H_0 : \boldsymbol{\beta}_A = 0$. Under the null hypothesis, the errors could be calculated as $\boldsymbol{\varepsilon} = \mathbf{Y} - \mathbf{X}_R\boldsymbol{\beta}_R$. Note that, because the errors are independent any permutation of these errors is as likely as any other permutation. For the kth permutation let $\mathbf{S}_k$ be a permutation matrix so that $\mathbf{S}_k\boldsymbol{\varepsilon}$ are the permuted errors. We could, for the kth permutation, compute a new vector of dependent variables as

$$\mathbf{Y}_k = \mathbf{X}_R\boldsymbol{\beta}_R + \mathbf{S}_k\boldsymbol{\varepsilon}.$$

Using $\mathbf{Y}_k$ we could fit the complete and reduced model to obtain a test statistic for testing $H_0 : \boldsymbol{\beta}_A = \mathbf{0}$ for the kth permutation. So, if we know β_R we could construct the permutation distribution of the test statistic, and because the errors are independent, the test statistics are equally likely. Although this method is impractical to use because $\boldsymbol{\beta}_A$ is usually unknown, we will use this method to justify other, more practical, approaches. We will refer to this method as the permute-errors method.

4.5.2 The Permute-Residuals Method

One of the most promising approaches was proposed by Freedman and Lane (1983). With this method we permute the residuals from the reduced model and add these permuted residuals to the predicted values, which are also based on the reduced model, to obtain a new vector of dependent variables. This new $\mathbf{Y}$ vector is used to compute the test statistic for that permutation. After a large number of these permutations are performed, the p-value is calculated. This is called the permute-residuals method.

In matrix notation, let the permutation matrix for the kth permutation be $\mathbf{S}_k$. With the permute-residual approach we first compute residuals from the reduced model, $\mathbf{Y} = \boldsymbol{\beta}_R\mathbf{X}_R + \boldsymbol{\varepsilon}$. These residuals will be denoted by $\hat{\boldsymbol{\varepsilon}}_R$ and the predicted values will be denoted by $\hat{Y}_R$. Next, for the kth permutation, we permute this vector to obtain $\mathbf{S}_k\hat{\boldsymbol{\varepsilon}}_R$, and compute the new vector of dependent variables as

$$\mathbf{Y}_k = \hat{\mathbf{Y}}_R + \mathbf{S}_k\hat{\boldsymbol{\varepsilon}}_R.$$

We then fit the reduced model $\mathbf{Y}_k = \mathbf{X}_R\boldsymbol{\beta}_R + \boldsymbol{\varepsilon}$ and the complete model $\mathbf{Y}_k = \mathbf{X}_R\boldsymbol{\beta}_R + \mathbf{X}_A\boldsymbol{\beta}_A + \boldsymbol{\varepsilon}$ to obtain the sum of squared errors for the reduced and complete models. These are used to compute the test statistic for the kth permutation.

With this permute-residuals method, we perform R permutations of the residuals and compute the test statistic for each permutation. If we reject the null hypothesis for small values of the test statistic, let E equal the number of permutation test statistics that are less than or equal to the test statistic from the unpermuted data. The p-value for this test is given by $p = (E+1)/(R+1)$. This is a non-adaptive permute-residuals permutation test. We will discuss the adaptive version of this test in the next chapter.

Now suppose the reduced model contains only an intercept term so that $\mathbf{X}_R = \mathbf{1}$, where $\mathbf{1}$ is an $n \times 1$ vector of ones. Then $\hat{\beta}_R = \bar{Y}$ and the new vector of dependent variables for the kth permutation is

$$\mathbf{Y}_k = \mathbf{1}\hat{\boldsymbol{\beta}}_R + \mathbf{S}_k\hat{\boldsymbol{\varepsilon}}_R.$$

But, since $\hat{\boldsymbol{\varepsilon}}_R = \mathbf{Y} - \mathbf{1}\hat{\boldsymbol{\beta}}_R$ and $\mathbf{S}_k\mathbf{1} = \mathbf{1}$,

$$\mathbf{Y}_k = \mathbf{1}\hat{\boldsymbol{\beta}}_R + \mathbf{S}_k(\mathbf{Y} - \mathbf{1}\hat{\beta}_R) = \mathbf{S}_k\mathbf{Y}.$$

Hence, the permute-residuals method simply permutes the rows of $\mathbf{Y}$ if $\mathbf{X}_R = 1$. Also, since $\mathbf{Y} = \mathbf{1}\boldsymbol{\beta}_R + \boldsymbol{\varepsilon}$,

$$\mathbf{Y}_k = \mathbf{S}_k\mathbf{Y} = \mathbf{S}_k(\mathbf{1}\boldsymbol{\beta}_R + \boldsymbol{\varepsilon}) = \mathbf{1}\boldsymbol{\beta}_R + \mathbf{S}_k\boldsymbol{\varepsilon}.$$

Consequently, if $\mathbf{X}_R = \mathbf{1}$, the permute-residuals method is equivalent to the permute-error method. That is, if we had known the errors, which are independent, and permuted them, we would obtain the same test statistic that we would obtain by permuting the residuals. Because the test statistic obtained from the permute-errors method are equally likely, it follows that the test statistics obtained from the permute-residuals method are also equally likely.

However, if $\mathbf{X}_R \neq \mathbf{1}$ the permute-residuals method gives the new vector of dependent variables as

$$\begin{aligned}
\mathbf{Y}_k &= \mathbf{X}_R\hat{\boldsymbol{\beta}}_R + \mathbf{S}_k\hat{\boldsymbol{\varepsilon}}_R \\
&= \mathbf{X}_R\hat{\boldsymbol{\beta}}_R + \mathbf{S}_k(\mathbf{Y} - \mathbf{X}_R\hat{\boldsymbol{\beta}}_R) \\
&= \mathbf{X}_R\hat{\boldsymbol{\beta}}_R + \mathbf{S}_k\mathbf{Y} - \mathbf{S}_k\mathbf{X}_R\hat{\boldsymbol{\beta}}_R.
\end{aligned}$$

But no simplification is possible because, in general, $\mathbf{S}_k\mathbf{X}_R \neq \mathbf{X}_R$. Hence, if $\mathbf{X}_R \neq \mathbf{1}$ the permute-residuals method is not equivalent to the permute-errors method. So, while it makes good sense to permute the residuals because they may, in large samples, approximate the errors, we cannot claim that the test statistics are equally likely.

This permute-residuals method is the method that we will use in subsequent chapters. However, there are two other permutation methods that some researchers have used to perform permutation tests, and these will now be described.

4.5.3 The Permutation of Independent Variables Method

Another method that can be used to test the null hypothesis $H_0 : \boldsymbol{\beta}_A = \mathbf{0}$ versus the alternative hypothesis $H_A : \boldsymbol{\beta}_A \neq \mathbf{0}$ uses row permutations of the $\mathbf{X}_A$ matrix. Many years ago this procedure was used by Draper and Stoneman (1966) to analyze a data set. More recently, Oja (1987) and Collins (1987) described the permutation distribution of test statistics obtained by this method. Kennedy and Cade (1996) used a small simulation study and O'Gorman (2005) used a larger simulation study to investigate the properties of this test.

With this method, for the kth permutation we compute the sum of squared errors from the model

$$\mathbf{Y} = \mathbf{X}_R\boldsymbol{\beta}_R + (\mathbf{S}_k\mathbf{X}_A)\boldsymbol{\beta}_A + \varepsilon.$$

Note that, with this permutation method, the reduced model $\mathbf{Y} = \mathbf{X}_R\boldsymbol{\beta}_R + \varepsilon$ does not change. With each permutation we calculate a permutation test statistic and compute the p-value in the usual manner.

This method is appropriate if the rows of $\mathbf{X}_A$ have been assigned at random. In the two-sample test that we described in the previous chapter, the treatments were assigned at random which implies that, under the null hypothesis, the permutation test statistics obtained by permuting the indicator variables are equally likely. Hence, this permutation test can be justified because the rows of $\mathbf{X}_A$ have been assigned at random. However, in many regression models the rows of $\mathbf{X}_A$ have not been assigned at random; so, for these models, this permutation method is difficult to justify. Also, in most regression models the variables in $\mathbf{X}_R$ are related to those in $\mathbf{X}_A$. For these regression models a permutation of the rows of $\mathbf{X}_A$ will change the relationship between the variables in $\mathbf{X}_A$ and those in $\mathbf{X}_R$. However, even with these problems the method usually produces tests that have good properties. In a large simulation study, O'Gorman (2005) showed that, even when independent variables are highly correlated, the test maintained its level of significance.

Because this approach is not intellectually appealing, it is not widely used. Although this method has been used by O'Gorman (2002, 2004) for adaptive tests, it will not be used in this book. The permutation of residuals method is as effective and is applicable to a wider variety of regression models.

4.5.4 The Permutation of Dependent Variables Method

Another approach, which was proposed by Manly (1997), is to permute the dependent variables. For the kth permutation the test statistic is based on the fit of the complete model $\mathbf{S}_k\mathbf{Y} = \mathbf{X}_R\boldsymbol{\beta}_R + \mathbf{X}_A\boldsymbol{\beta}_A + \varepsilon$ and the reduced model $\mathbf{S}_k\mathbf{Y} = \mathbf{X}_R\boldsymbol{\beta}_R + \varepsilon$. This method was evaluated by Anderson and Legendre (1999), who found that, while it generally had properties similar to that of the permute-residuals method, the Type I error was somewhat erratic when an extreme outlier was included as a covariate. They concluded that the permute-residuals method of Freedman and Lane has somewhat better performance than the permutation of dependent variables method. We will not use the permutation of dependent variables method in this book.

4.6 PERMUTATION TEST STATISTICS

Permutation tests use the permutation distribution of a test statistic to determine the p-value. For a nonadaptive two-sample test we could use a t test statistic, an F test statistic, or even the difference between the two samples to construct the permutation distribution. We saw, in Chapter 3, that $\Lambda^{2/n} = SSE_C/SSE_R$ was a strictly decreasing function of F, so it could be used as the test statistic. A permutation test that used F as the test statistic will give the same p-value as a permutation test that used $\Lambda^{2/n}$ as the test statistic. However, not all test statistics will give the same p-value, and some of these do not have good statistical properties. For example, we might consider using the regression coefficient as the test statistic, but Kennedy and Cade (1996) have shown that using it may produce a test that does not maintain its level of significance. Hence, in this book we will use $\Lambda^{2/n} = SSE_C/SSE_R$ as the test statistic. In regression models that have only one dependent variable our results will also describe the properties of the tests that use the F test statistic.

4.7 AN IMPORTANT RULE OF TEST CONSTRUCTION

Even if we use $\Lambda^{2/n}$ (or F) as the test statistic, the permutation test may not be valid if the test statistic is not computed correctly. For the permutation test to be valid, the test statistics must be computed for all permutations in the same manner that they were computed for the unpermuted (or raw) data. The idea is that, under the null hypothesis, the observed data is just one of the many permutations of the data, so we should use identical procedures for calculating the test statistics. We would violate this rule if we used, for the kth permutation, the permute-residuals method to find Y_k and then calculated SSE_C from Y_k and calculated SSE_R from Y. Since we calculated SSE_C and SSE_R using Y with the unpermuted data, we must, for the kth permutation, use Y_k to compute SSE_C and SSE_R. The general rule is that we must calculate each permutation test statistic using precisely the same procedures that were used to calculate the test statistic for the unpermuted data.

4.8 A PERMUTATION ALGORITHM

Fisher and Yates (1938) proposed a permutation algorithm and Durstenfeld (1964) proposed a more efficient permutation algorithm that is known as the Durstenfeld shuffle. The basic idea can be explained with a small example. Suppose we want to generate a permutation of the integers:

1	2	3	4

To start the Durstenfeld shuffle, we generate a random number between 1 and 4, then we swap the number in that position with the number in the last position. Suppose we generate a "2". After swapping the number in the second position with the number in the last position, we obtain

1	4	3	2

Next, we generate a random number between 1 and 3 and swap the number in that place with the 3rd element in the list. (Note that, after the first step, the last number in the list cannot be changed.) Suppose we generate a "1". After swapping the number in the first position with the number in the 3rd position, we obtain

3	4	1	2

Next, we generate a random number between 1 and 2. Suppose we generate a "2". The number in the 2nd place swaps with itself so we obtain the final permutation of

3	4	1	2

The FORTRAN computer programs, which are used to perform all the simulation studies in this book, use the Durstenfeld shuffle, as does the SAS macros. In those languages the random numbers are generated by using the pseudo-random number generator that was proposed by Wichmann and Hill (1982). In the R language we use the `sample()` function to perform the permutation. This function does not use the Wichmann–Hill pseudo-random number generator, so it is not possible to obtain exact agreement between permutation tests in SAS and in R.

4.9 A PERFORMANCE COMPARISON OF THE PERMUTATION METHODS

The most surprising aspect of permutation tests is that many of the methods appear to have good properties. If extreme situations are avoided, they usually maintain their level of significance and they often have power comparable to that of the traditional t and F tests.

However, there are some important differences between these methods. Anderson and Legendre (1999) compared several permutation methods for Type I error probabilities and for power. They found that the method of permuting the dependent variables sometimes resulted in inflating the probability of a Type I error. In addition, they found that the permutation methods generally had slightly higher power than the t test with non-normal errors. Anderson and Robinson (2001) described the asymptotic properties of several permutation tests. They showed that all of the permutation tests that they considered had the same asymptotic power. In the studies by Anderson and Legendre (1999) and by Anderson and Robinson (2001), they did not include the permute-independent variables method in their analysis.

In an effort to provide a more comprehensive comparison of permutation methods, O'Gorman (2005) compared the performance of the permute-residual method, the permute independent variable method, and the F test. The extensive simulation studies showed that all of the permutation tests maintained their level of significance near $\alpha = 0.05$, except for simulations that used the exponential cubed distribution, which is an extremely skewed distribution. For the exponential cubed distribution the permutation tests tended to have significance levels closer to the nominal than the F test. The simulation studies also showed that the power of the permutation test approximates that of the F test.

To illustrate the fact that the power of the permute-residuals test approximates the power of the t test, consider the power of the two-sample tests that are shown in Figure 4.1. These results were from the same simulation study that was used in Chapter 3, having 25 observations in each sample, to show the relative performance of the adaptive test compared to that of the t test, except in this simulation we did not use adaptive weights. That is, this is the same algorithm that we used for adaptive testing except the weights were set to $w_i = 1.0$ for $i = 1, \ldots, n$. It can be seen that there is very little difference in power between the tests, but the permutation test is slightly more powerful than the t test with some highly skewed errors. These results are in agreement with those of Anderson and Legendre (1999).

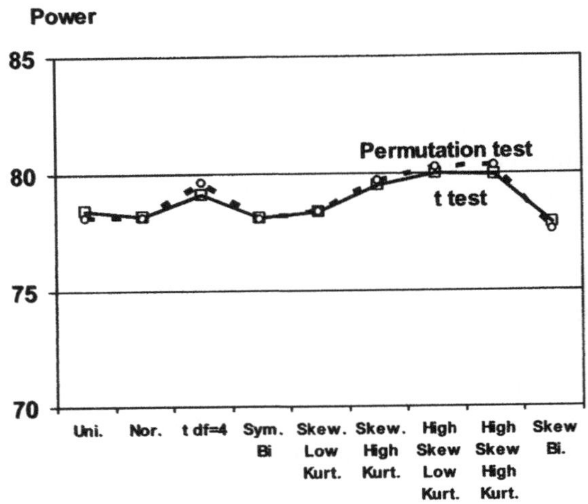

Figure 4.1 The empirical power, in percent, for the non-adaptive permute-residual test and the two-sample t test for several error distributions with 25 observations in each sample. The difference between the populations was set to $\delta = 0.792$ to achieve approximately 80% power.

4.10 DISCUSSION

If the permutation tests have power near that of the traditional F test, why would anyone bother using them? The simulation studies by Anderson and Legendre (1999) and by O'Gorman (2005) clearly show little difference between the permutation tests and the traditional tests. The traditional tests are well known and the permutation tests are not. In addition, the permutation tests are computationally intensive. So, the direct answer to this question is that there appears to be no good reason to prefer a permutation test over the traditional F test.

The reason that we have described permutation tests is to show the reader how to produce p-values for a test statistic when the sampling distribution is not known. When we use the normalizing weights to compute the test statistic, the permutation method is used to compute the p-value. This adaptive test maintains its level of significance and often has greater power than traditional tests. In addition, the adaptive permutation tests are quite practical; the p-values can be computed in a few minutes. So, there are good reasons to use permutation tests with adaptive weighting methods to increase the power of tests. In the next chapter we describe an adaptive test for any subset of coefficients in any linear model that contains fixed effects.

EXERCISES

4.1 Consider the vector of observation

$$\mathbf{Y} = \begin{bmatrix} 2 \\ 6 \\ 10 \\ 22 \\ 7 \end{bmatrix}.$$

Suppose the kth permutation is

$$\mathbf{Y_k} = \begin{bmatrix} 22 \\ 10 \\ 6 \\ 2 \\ 7 \end{bmatrix}.$$

a) Find the shuffling matrix S_k.

b) What permutation matrix will restore the original order? That is, find $S_{restore}$ such that $S_{restore}Y_k = Y$.

4.2 Suppose four patients are enrolled at each of three treatment centers and that these patients are assigned within each center to one of two treatment groups with 2 patients assigned to each group. Let Y_1, Y_2, Y_3, Y_4 be the observations from center 1, let Y_5, Y_6, Y_7, Y_8 be the observations from center 2, and let $Y_9, Y_{10}, Y_{11}, Y_{12}$ be the observations from center 3. Give a permutation matrix that will perform a within-center permutation of these observations.

4.3 Suppose a researcher had only enough resources to use $n = 4$ animals in an experiment. The analysis was based on a nonadaptive test of slope in a simple linear regression model, using the permute-residuals method. Because there are only 4 animals the researcher will use the full permutation distribution to compute the p-value.

a) How many distinct values of $\Lambda^{2/n} = SSE_C/SSE_R$ are possible? You may assume that there are no tied test statistics.

b) Suppose the researcher performed many experiments that are similar to the one that we described, with each experiment having $n = 4$ animals, and used $\alpha = 0.05$ in the test procedure. If the null hypothesis is true for each experiment, how often will the researcher reject the null hypothesis?

c) Suppose the researcher performed many experiments that are similar to the one that we described, with each experiment having $n = 4$ animals, and used $\alpha = 0.01$ in the test procedure. If the null hypothesis is true for each experiment, how often will the researcher reject the null hypothesis?

4.4 Suppose we use the permutation of independent variables method and, for the kth permutation, we fit the model

$$\mathbf{Y} = \mathbf{X}_R\beta_R + (\mathbf{S}_k\mathbf{X}_A)\boldsymbol{\beta}_A + \varepsilon. \tag{4.1}$$

Suppose we premultiply both sides of the model (4.1) by $\mathbf{S}_k^{-1}$. Does this demonstrate that the permutation of the dependent variables method is equivalent to the permutation of independent variables method? Why or why not?

4.5 Suppose we use the permutation of dependent variables method to test $H_0 : \boldsymbol{\beta}_A = \mathbf{0}$ versus $H_a : \boldsymbol{\beta}_A \neq \mathbf{0}$ in a linear model

$$\mathbf{Y} = \mathbf{1}\boldsymbol{\beta}_R + \mathbf{X}_A\boldsymbol{\beta}_A + \boldsymbol{\varepsilon}.$$

Note that the reduced model contains only the intercept.

a) For this test, will the permutation of dependent variables method be equivalent to the permutation of residuals method? Why or why not?

b) For this test, will the permutation of dependent variables method be equivalent to the permutation of independent variables method? Why or why not?

4.6 Suppose we use the permutation of dependent variables method to test $H_0 : \boldsymbol{\beta}_A = \mathbf{0}$ versus $H_a : \boldsymbol{\beta}_A \neq \mathbf{0}$ in a linear model

$$\mathbf{Y} = \mathbf{X}_R\boldsymbol{\beta}_R + \mathbf{X}_A\boldsymbol{\beta}_A + \boldsymbol{\varepsilon}.$$

Note that the reduced model may contain several independent variables.

a) For this test, will the permutation of dependent variables method be equivalent to the permutation of residuals method? Why or why not?

b) For this test, will the permutation of dependent variables method be equivalent to the permutation of independent variables method? Why or why not?

4.7 Suppose we have a data set that has $n = 52$ observations and we want to test $H_0 : \boldsymbol{\beta}_A = \mathbf{0}$ versus $H_a : \boldsymbol{\beta}_A \neq \mathbf{0}$ in a linear model by computing the full permutation distribution using the permutation of residuals method. The linear model is

$$\mathbf{Y} = \mathbf{X}_R\boldsymbol{\beta}_R + \mathbf{X}_A\boldsymbol{\beta}_A + \boldsymbol{\varepsilon}.$$

Suppose every man, woman, and child on Earth has a personal computer with 100 processors in each computer. We will assume that approximately 7 billion people live on Earth. If each processor can evaluate 20 million permutations per second and all of the processors can be programmed to evaluate the permutation distribution, how much time would be required to evaluate the full permutation distribution?

CHAPTER 5

AN ADAPTIVE TEST FOR A SUBSET OF COEFFICIENTS IN A LINEAR MODEL

5.1 THE GENERAL ADAPTIVE TESTING METHOD

In Chapter 3 we demonstrated that the two-sample test was more powerful than the traditional test when the errors were not normally distributed. In this chapter we consider a test for a subset of coefficients in a regression model. Will the performance that we saw with two-sample tests carry over to the tests that are used with more complex models?

An adaptive test for a subset of coefficients in a linear regression model will be used for all tests in this chapter and the next three chapters. The adaptive test, which is a generalization of the two-sample test that was described in Chapter 3, requires two steps:

1. Compute the weights that will normalize the residuals.

2. Compute the p-value by using a permutation method.

In this chapter we will carefully estimate the level of significance and the relative power of this adaptive test for testing a slope in a simple linear regression and for testing coefficients in a multiple linear regression. We will also show how these can

Adaptive Tests of Significance Using Permutations of Residuals with R and SAS®.
By Thomas W. O'Gorman

be performed in R and in SAS. We begin by describing the weighting procedure in detail.

5.1.1 Weighting Step

In this section we will use matrix notation to describe the linear models. Let $\mathbf{Y}$ be the $n \times 1$ vector for the dependent variable, let $\mathbf{X}$ be the $n \times q$ matrix of independent variables, and let $\boldsymbol{\beta}$ be the $q \times 1$ vector of regression coefficients. With this notation the linear model can be written as

$$\mathbf{Y} = \mathbf{X}\boldsymbol{\beta} + \boldsymbol{\varepsilon} = \mathbf{X}_R\boldsymbol{\beta}_R + \mathbf{X}_A\boldsymbol{\beta}_A + \boldsymbol{\varepsilon}, \tag{5.1}$$

where $\mathbf{X}_R$ is the $n \times r$ matrix containing the first r columns of $\mathbf{X}$, $\mathbf{X}_A$ is the $n \times (q-r)$ matrix containing the last $(q-r)$ columns of $\mathbf{X}$, $\boldsymbol{\beta}_R$ is the $r \times 1$ vector consisting of the first r elements of $\boldsymbol{\beta}$, $\boldsymbol{\beta}_A$ is the $(q-r) \times 1$ matrix containing the last $q-r$ elements of $\boldsymbol{\beta}$, and $\boldsymbol{\varepsilon}$ is the $n \times 1$ vector of errors. We wish to test $H_0 : \boldsymbol{\beta}_A = \mathbf{0}$ versus $H_a : \boldsymbol{\beta}_A \neq \mathbf{0}$. We begin by fitting the reduced model

$$\mathbf{Y} = \mathbf{X}_R\boldsymbol{\beta}_R + \boldsymbol{\varepsilon} \tag{5.2}$$

to obtain the vector of ordinary residuals ($\hat{\boldsymbol{\varepsilon}}_R$). We then use the general procedures that we described in Chapter 3 to obtain the adaptive weights.

In the two-sample problem we started the weighting procedure by using the percentiles to determine an estimate of the standard deviation, and then we used it to find a reasonable value for the bandwidth. Next, the smoothed c.d.f. was used to find a more precise estimate of variability and a precise estimate of the median. Then we computed the normalizing weights. Because we will use this weighting method extensively in this chapter, we will give the steps here for easy reference.

The steps in the adaptive weighting algorithm are:

1. Use the set of ordinary residuals from the reduced model to compute $\hat{\sigma}_t = (\hat{\varepsilon}_{t,.75} - \hat{\varepsilon}_{t,.25})/1.349$.
2. Compute the bandwidth $h = 1.587\hat{\sigma}_t n^{-1/3}$.
3. Smooth the c.d.f. of the residuals to obtain $\hat{F}_h(\hat{\varepsilon}_i)$ for $i = 1, \ldots, n$.
4. Use $\hat{F}_h(\hat{\varepsilon})$ in an iterative procedure to compute $\hat{\varepsilon}_{.25}$, $\hat{\varepsilon}_{.5}$, and $\hat{\varepsilon}_{.75}$. Then compute $\hat{\sigma} = (\hat{\varepsilon}_{.75} - \hat{\varepsilon}_{.25})/1.349$.
5. Compute $z_i = \Phi^{-1}[\hat{F}_h(\hat{\varepsilon}_i)]$ and $s_i = (\hat{\varepsilon}_i - \hat{\varepsilon}_{.5})/\hat{\sigma}$ for $i = 1, \ldots, n$.
6. For $i = 1, \ldots, n$ compute the estimates of the weights as

$$\hat{w}_i = \begin{cases} z_i/s_i & \text{if } s_i \neq 0, \\ 1 & \text{if } s_i = 0. \end{cases}$$

After we compute the weights, we define $\mathbf{W}$ to be the diagonal matrix of weights having w_i in the ith row and ith column of $\mathbf{W}$. One way of finding estimates of $\boldsymbol{\beta}_R$ and $\boldsymbol{\beta}_A$ is to minimize the sum of the squares of the weighted residuals. The other way of finding the estimates is to use the adaptive weights to transform the model. With this approach we premultiply both sides of the linear model by $\mathbf{W}$. The weighted complete model becomes

$$\mathbf{WY} = \mathbf{WX}_R\boldsymbol{\beta}_R + \mathbf{WX}_A\boldsymbol{\beta}_A + \mathbf{W}\boldsymbol{\varepsilon}.$$

Let the transformed variables be denoted by $\mathbf{Y}^* = \mathbf{WY}$, $\mathbf{X}_R^* = \mathbf{WX}_R$, and $\mathbf{X}_A^* = \mathbf{WX}_A$. Using this notation, the transformed complete model can be written as

$$\mathbf{Y}^* = \mathbf{X}_R^*\boldsymbol{\beta}_R + \mathbf{X}_A^*\boldsymbol{\beta}_A + \mathbf{W}\boldsymbol{\varepsilon}.$$

With this transformed model we can use ordinary least squares methods to find estimates of $\boldsymbol{\beta}_R$ and $\boldsymbol{\beta}_A$, which are denoted by $\hat{\boldsymbol{\beta}}_R$ and $\hat{\boldsymbol{\beta}}_A$. Note that the vector of residuals is

$$\hat{\boldsymbol{\varepsilon}}_C^* = \mathbf{Y}^* - (\mathbf{X}_R^*\hat{\boldsymbol{\beta}}_R + \mathbf{X}_A^*\hat{\boldsymbol{\beta}}_A) = \mathbf{W}[\mathbf{Y} - (\mathbf{X}_R\hat{\boldsymbol{\beta}}_R + \mathbf{X}_A\hat{\boldsymbol{\beta}}_A)].$$

We measure the lack of fit by the sum of the squared residuals from the weighted model, which is $SSE_C^* = \sum_{i=1}^{n}(\hat{\varepsilon}_{C,i}^*)^2 = (\hat{\boldsymbol{\varepsilon}}_C^*)'\hat{\boldsymbol{\varepsilon}}_C^*$

We use the same approach and the same weights with the reduced model, which becomes, after weighting,

$$\mathbf{WY} = \mathbf{WX}_R\boldsymbol{\beta}_R + \mathbf{W}\boldsymbol{\varepsilon}.$$

Using the notation defined above, the transformed reduced model can be written as

$$\mathbf{Y}^* = \mathbf{X}_R^*\boldsymbol{\beta}_R + \mathbf{W}\boldsymbol{\varepsilon}.$$

We use ordinary least squares to compute an estimate of $\boldsymbol{\beta}_R$ that will be denoted by $\hat{\boldsymbol{\beta}}_{R,0}$. The residuals from the reduced model are $\hat{\boldsymbol{\varepsilon}}_R^* = \mathbf{Y}^* - \mathbf{X}_R^*\hat{\boldsymbol{\beta}}_{R,0} = \mathbf{W}(\mathbf{Y} - \mathbf{X}_R\hat{\boldsymbol{\beta}}_{R,0})$. The lack of fit is measured by $SSE_R^* = (\hat{\boldsymbol{\varepsilon}}_R^*)'\hat{\boldsymbol{\varepsilon}}_R^*$. Next, for the unpermuted data, we compute the test statistic

$$\Lambda_u^{2/n,*} = \frac{SSE_C^*}{SSE_R^*}. \tag{5.3}$$

If the additional terms given in $\mathbf{X}_A^*\boldsymbol{\beta}_A$ do not greatly improve the fit, then SSE_C^* will not be much smaller than SSE_R^*, so $\Lambda_u^{2/n,*}$ will approximate one. However, if the terms in $\mathbf{X}_A^*\boldsymbol{\beta}_A$ are important, then SSE_C^* will be much smaller than SSE_R^*, so $\Lambda_u^{2/n,*}$ will be much smaller than one.

In our notation we use q to denote the number of independent variables in the complete model and r to denote the number of independent variables in the reduced model, so the number of parameters in $\boldsymbol{\beta}_A$ is $(q - r)$. Instead of using $\Lambda_u^{2/n,*}$ as the test statistic we could compute the traditional F test statistic as

$$F_u^* = \frac{(SSE_R^* - SSE_C^*)/(q-r)}{SSE_C^*/(n-q)} = \left(\frac{SSE_R^*}{SSE_C^*} - 1\right)\left(\frac{n-q}{q-r}\right)$$
$$= \left(\frac{1}{\Lambda_u^{2/n,*}} - 1\right)\left(\frac{n-q}{q-r}\right).$$

Because F_u^* is a strictly decreasing function of $\Lambda_u^{2/n,*}$ we will obtain the same p-value regardless of which test statistic is used in the permutation test.

5.1.2 Permutation Step

We will assume that we have already computed $\Lambda_u^{2/n,*}$, using Equation (5.3), for the unpermuted data. To obtain a p-value we perform a permutation test using the permutation of residuals method that was described in the previous chapter. The steps are as follows:

1. Compute, from the unweighted reduced model $\mathbf{Y} = \mathbf{X}_R\beta_R + \varepsilon$, the ordinary residuals $\hat{\varepsilon}_R$ and the vector of predicted values $\hat{\mathbf{Y}}$.

2. Generate, using a random number generator, a permutation of the elements of $\hat{\varepsilon}_R$. For the kth permutation the permuted residuals are $\mathbf{S}_k\hat{\varepsilon}_R$, where $\mathbf{S}_k$ is a permutation matrix.

3. Compute the new dependent variable as $\mathbf{Y}_k = \hat{\mathbf{Y}} + \mathbf{S}_k\hat{\varepsilon}_R$ for the kth permutation.

4. Using the reduced model $\mathbf{Y}_k = \mathbf{X}_R\boldsymbol{\beta}_R + \varepsilon$, with the new vector as the dependent variable, calculate the adaptive weights and use these as the diagonal elements in the weighting matrix $\mathbf{W}_k$. Use the weighted complete model $\mathbf{W}_k\mathbf{Y} = \mathbf{W}_k\mathbf{X}_R\boldsymbol{\beta}_R + \mathbf{W}_k\mathbf{X}_A\boldsymbol{\beta}_A + \mathbf{W}_k\boldsymbol{\varepsilon}$ and the weighted reduced model $\mathbf{W}_k\mathbf{Y} = \mathbf{W}_k\mathbf{X}_R\boldsymbol{\beta}_R + \mathbf{W}_k\boldsymbol{\varepsilon}$ to compute the test statistic in the same manner that was used with the unpermuted data. For the kth permutation the test statistic will be denoted by $\Lambda_k^{2/n,*}$.

5. Perform R permutations of the residuals. Let E equal the number of permutations that have $\Lambda_k^{2/n,*} \leq \Lambda_u^{2/n,*}$. The adaptive p-value is $p = (E+1)/(R+1)$.

6. If $p \leq \alpha$, reject the null hypothesis.

Typically, we use $R = 2000$ permutations, but if the objective is to estimate the p-value more precisely then a larger value of R would be necessary. Although the weights must be computed for each permutation, the p-value can be obtained quickly with a modern computer.

5.2 SIMPLE LINEAR REGRESSION

5.2.1 The Significance of the Adaptive Test

The simple linear regression model is a special case of the model specified in Equation (5.1). The simple linear regression model can be expressed as

$$\mathbf{Y} = \mathbf{1}\beta_R + \mathbf{X}_A\beta_A + \boldsymbol{\varepsilon},$$

where $\mathbf{X}_A$ is an $n \times 1$ matrix. Under $H_0 : \beta_A = \mathbf{0}$, if we permute the errors, we would obtain a permutation of $\mathbf{Y}$, since $\mathbf{1}\beta_R + \mathbf{S}_k\boldsymbol{\varepsilon} = \mathbf{S}_k(\mathbf{1}\beta_R + \boldsymbol{\varepsilon}) = \mathbf{S}_k\mathbf{Y}$. If we permute the residuals, we would also obtain a permutation of $\mathbf{Y}$ because $\mathbf{1}\hat{\beta}_R + \mathbf{S}_k\hat{\boldsymbol{\varepsilon}}_R = \mathbf{S}_k(\mathbf{1}\hat{\beta}_R + \hat{\boldsymbol{\varepsilon}}_R) = \mathbf{S}_k\mathbf{Y}$. Since any permutation of the errors is as likely as any other permutation of the errors, the permute-residuals method will, under the null hypothesis, produce equally likely test statistics.

We must be careful, in constructing our permutation tests, that we use the same computational procedures with each permutation and that this procedure is the same procedure that is used with the unpermuted data. If a permutation test has this property, then we will say that the test uses identical procedures under permutations.

In this chapter we are concerned with the properties of adaptive tests. Our adaptive test statistics are more complex functions of the data than the test statistics that are not adaptively weighted. However, if a permutation test uses identical procedures under permutations and if the probabilities of obtaining the permutations of $\mathbf{Y}$ are equal, then the adaptive test statistics will be equally likely.

With the simple linear regression model, under the null hypothesis, the permute-residuals method will produce equally likely adaptive test statistics. This means that if we performed a permutation test using all $n!$ permutations of the residuals, the p-value would be the probability of observing a test statistic as extreme, or more extreme, than that observed. Hence, the adaptive test for slope in a simple linear regression will maintain its level of significance, and it is reasonable to compare the power of the adaptive test to the power of the traditional t (or F) test.

5.2.2 The Power of the Adaptive Test

We need to compare the power of three tests for slope. The first is the traditional t test for slope. The second is the permutation t test that uses the unweighted t test statistic in a permutation test. The third is the adaptive test that uses the adaptively weighted test statistic.

The relative powers of these tests for $n = 25$ observations are displayed in Figure 5.1 for a simulation study that used 10,000 data sets for each of the nine error distributions. These power estimates were obtained by a FORTRAN program that was written by the author. See Section 3.8 for more information about the FORTRAN code. The slope was set at $\beta_1 = 0.58$ for all error distributions to obtain powers near 80% for the t test. The adaptive test is more powerful than the t test for the long-tailed symmetric distribution and all of the skewed distributions. There is little power loss with normal errors. This is the same pattern of results that we observed in Chapter 3 with the two-sample tests. We note that the t test and the unweighted permutation t test give nearly indentical results.

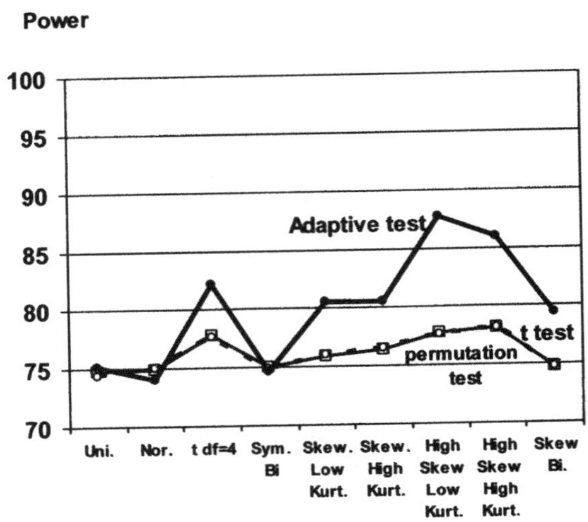

Figure 5.1 Empirical power in percent for the t test for slope, the permutation test, and the adaptive test for $n = 25$ observations. The slope was set to $\beta_1 = 0.58$ for all error distributions. The t test is indicated by hollow squares, the permutation t test by hollow circles, and the adaptive test by filled circles.

To determine if the relative powers of these tests change when larger samples are used, we ran another simulation using $n = 100$ observations, which produced the power estimate that are shown in Figure 5.2. We used a slope of $\beta_1 = 0.29$ to achieve powers near 80% for the t test. It is clear from these simulation results that there is very little power loss when the adaptive test is used and the errors are normally distributed. With $n = 100$ observations the power loss is less than 1% with normally distributed errors while the power gain is generally between 5% and 15% with non-normal errors.

5.2.3 Justification of the Weighting Method

We demonstrated, in Section 3.12, that the adaptive weights are approximations of the population weights that could be computed if we knew the c.d.f of the errors. Those results for the two-sample test can be directly applied to the simple linear regression because the reduced model consists of an intercept term for the two-sample test and for the test for slope in a simple linear regression. Hence, it is appropriate to use the methods that were described in Chapter 2 to normalize $\varepsilon_i - \varepsilon_{.5}$, for $i = 1, \ldots, n$, where $\varepsilon_{.5}$ is the population median of the errors. Of course, with small samples the sample estimates of the c.d.f. of the errors may not be very accurate, which may result in the residuals being only roughly normal. However, our primary objective

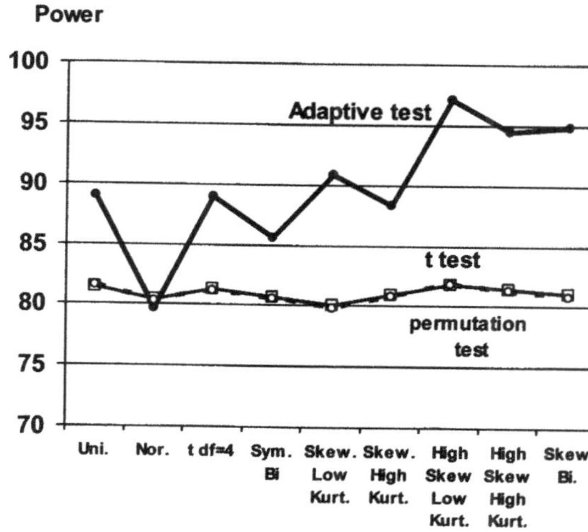

Figure 5.2 Empirical power in percent for the t test for slope, the permutation test, and the adaptive test for $n = 100$ observations. We used a slope of $\beta_1 = 0.29$ to achieve powers near 80% for the t test. The t test is indicated by hollow squares, the permutation t test by hollow circles, and the adaptive test by filled circles.

was to increase the power of the test of slope and this appears to have been achieved even with samples of $n = 25$ observations.

5.3 AN EXAMPLE OF A SIMPLE LINEAR REGRESSION

As part of a larger study of the nutritional status of the elderly in the Dominican Republic, Simon (2001) collected height, weight, and blood pressure data on $n = 181$ men and women. The researcher obtained the body mass index, which is the weight (in kilograms) divided by the height (in meters) squared, diastolic blood pressure, and many other variables. The relationship between the body mass index (BMI) and diastolic blood pressure is shown in Figure 5.3 for the $n = 108$ females in the study. These data can be obtained at *www.math.niu.edu/~ogorman.*

The scatterplot does not show a clear relationship between these variables. The ordinary least squares fit indicated a slope of $\hat{\beta}_1 = 0.267$. The traditional t test that gives equal weights to the observations produced a p-value of $p = 0.2469$. Hence, the relationship between BMI and diastolic blood pressure in the population has not been demonstrated. However, there is one outlier, which is an individual who has a BMI of 30.5 and a diastolic blood pressure of 22. We would like to greatly reduce the influence of this observation.

If we perform an adaptive test, the outlier is given a weight of 0.460, while the weights for the other observations are in the interval [0.835, 1.062]. The adaptive test produced a p-value of 0.0825, based on 2000 permutations, which is much smaller

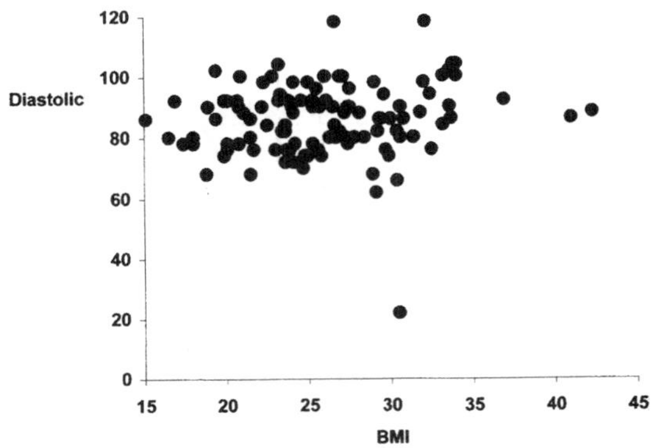

Figure 5.3 Scatterplot of body mass index (BMI) and diastolic blood pressure for $n = 108$ females.

than the p-value obtained with the traditional test. With this adaptive test there is some evidence that there is a relationship between BMI and diastolic blood pressure, but we would not reject $H_0 : \beta_1 = 0$ if we used $\alpha = 0.05$. However, we do not know the value of β_1. Thus, it is not clear that the adaptive test did a better job with this data set. We must rely on the simulation studies that we described in the previous section to determine the effectiveness of the adaptive weighting procedure.

5.3.1 Using R Code to Perform the Adaptive Test for Slope

In order to analyze these data with R, we must first create a data set that conforms to the standard R data frame format, with the first column of data containing the observation number. The variable names are given in the first row. The first four rows of the data set, which contain the first three observations, are given as follows:

	sex	age	bmi	y	systolic
74	Female	60	20.7	92	118
75	Female	90	28.1	88	138
76	Female	60	42.3	88	130
.	.	.	.	.	.

Note that we have used `y` as the dependent variable, which in this data set is the diastolic blood pressure. To perform the adaptive test using R, we need to include the function `adapt ()` and all of the functions that it uses at the beginning of the program, and then we must specify the data set name, the complete model, the reduced model, and the number of permutations that are to be used in the test. The `adapt` function, which is listed in Appendix A, can be downloaded from *www.math.niu.edu/~ogorman.*

Suppose the data set is called `simon.females.data`, the body mass index variable is indicated by `bmi`, and we want 2000 permutations. The R code that will execute the `adapt` function is:

```
dataset <- c("simon.females.data")
complete <- c("y~bmi")
reduced  <- c("y~1")
r <- 2000
seed <-c(71492)
pvalue <- adapt(dataset,complete,reduced,r,seed)
```

Note that `y~1` specifies the linear model for the reduced model, which includes only the intercept. The output of the R program is given below, which includes the two-tailed *p*-value for the traditional test in addition to the *p*-value for the adaptive test.

```
Output from Adaptive Test:
   Unpermuted lambda =  0.9732
   Unpermuted F      =  2.9237
   p-value           =  0.0825

Traditional F test p-value =  0.2469
```

We used $R = 2000$ permutations and found that the program executed in less than one minute on an inexpensive personal computer purchased in 2008. Because the estimated *p*-value is not too far from $\alpha = 0.05$, the researcher may want a more precise estimate of the *p*-value. When we used $R = 10{,}000$ permutations, it executed in less than 4 minutes on the same computer.

5.3.2 Using a SAS Macro to Perform the Adaptive Test

The adaptive test for slope can easily be performed by invoking the `adaptall` macro. This macro, which is listed in Appendix B, can be downloaded from *www.math.niu.edu/~ogorman*. (The macro could be placed earlier in the SAS program or it could be referenced in an `include` statement.) To execute this macro, the SAS data set must have already been created. Suppose that we have already created a data set called `simon` that contains a dependent variable named `diastol`, and an independent variable named `bmi`. After we invoke this macro, the keyword parameter for the dependent variable `y` will be replaced by `diastol`, and the keyword parameter for the independent variables `xa` will be replaced by `bmi`. We specified three random number seeds, which can be any integers between 1 and 30269, and the number of permutations, which is specified by the macro parameter `nperm`. This SAS code will invoke the macro:

```
%adaptall(dataset=simon, y=diastol, xa=bmi,
          seed1=10670, seed2=20588, seed3=1932, nperm=2000);
```

The SAS output, which is not shown here, gives all of the information contained in the macro call as well as the p-value for the adaptive test, which was 0.0835. (Later in this chapter a SAS output will be given.) Because the SAS macro uses a different random number generator than that used in R, the p-value will usually be slightly different from the one obtained by the R program. The SAS macro took 4 minutes to execute on an inexpensive personal computer purchased in 2008. It is expected to execute in less time on a more modern computer.

5.4 MULTIPLE LINEAR REGRESSION

In Section 5.1 we described how the multiple regression model could be written as

$$\mathbf{Y} = \mathbf{X}_R\boldsymbol{\beta}_R + \mathbf{X}_A\boldsymbol{\beta}_A + \boldsymbol{\varepsilon}. \tag{5.4}$$

In this section we will test $H_0 : \boldsymbol{\beta}_A = \mathbf{0}$ versus $H_a : \boldsymbol{\beta}_A \neq \mathbf{0}$. To perform an adaptive test of a subset of coefficients in a multiple linear regression, we use the residuals from the reduced model $\mathbf{Y} = \mathbf{X}_R\boldsymbol{\beta}_R + \boldsymbol{\varepsilon}$ to compute the weights. Also, the residuals from the reduced model are the residuals that we permute. In Chapter 4 we found that we could not conclude, as we did with simple linear regression, that permutation of residuals is equivalent to permutation of errors, because X_R is not a column of ones.

If we knew the errors, we would permute the errors from the reduced model because the errors are independent. However, we usually do not know the errors, but we do know the residuals. It seems reasonable to permute the residuals because, in large samples, our estimates of the elements in $\boldsymbol{\beta}_R$ may be reasonably accurate so the residuals may approximate the errors. We noted in Chapter 4 that Anderson and Robinson (2001) investigated the asymptotic distribution of partial correlations and concluded that the permutation of residuals method was as effective as, or more effective than, the other permutation methods. But these asymptotic results may not accurately describe the performance with small samples. In addition, these asymptotic results may not be applicable to this adaptive test.

Hence, when we use the permutation of residuals procedure with our adaptive weighting method, we will need to carefully document the performance of the test. Our first concern is with the significance level of the test. Will the adaptive test maintain its significance level?

Before we attempt to answer that question, we should note that there is no guarantee that the traditional F test will maintain its level of significance for non-normal errors. Of course, with normal errors the traditional F statistic follows an F distribution, so we would expect the test to maintain its level of significance for normal errors. But the performance of the F test with non-normal errors may not be satisfactory. Consequently, we will use extensive simulation studies to determine the significance level and power of the adaptive test and the F test. Before we describe these studies, we should briefly comment on the use of the adaptive weighting methods in multiple regression.

5.4.1 Comments on the Weighting Method

For the two-sample test the reduced model contained only an intercept term. For those models we did not need an accurate estimate of the intercept because the weighing procedure is location invariant. We showed in Section 3.12 that the inaccuracy in the weighting was due to the inaccuracy in the estimation of the c.d.f. of the errors.

In these multiple regression models we will assume that there is at least one independent variable, not including the intercept. If we perform an adaptive test for a subset of coefficients in a linear model, we will show that the inaccuracy in the weighting is due to the inaccuracy in the estimation of the c.d.f. of the errors and the inaccuracy in the estimation of the elements in $\boldsymbol{\beta}_R$.

To see this, consider the reduced model $\mathbf{Y} = \mathbf{X}_R\boldsymbol{\beta}_R + \boldsymbol{\varepsilon}$. If $F(\cdot)$ is known, and if $\boldsymbol{\beta}_R$ is known, then the errors are known so the population weights can be computed as

$$w_i = \frac{\sigma\Phi^{-1}[F(\varepsilon_i)]}{\varepsilon_i - \varepsilon_{.5}}.$$

For the two-sample test we showed that we did not need to know β_R because the sample weights

$$\hat{w}_i = \frac{z_i}{s_i} = \frac{\hat{\sigma}\Phi^{-1}[\hat{F}_h(\hat{\varepsilon}_i)]}{\hat{\varepsilon}_i - \hat{\varepsilon}_{.5}}$$

are location invariant. However, with multiple regression models the relationship between the residuals and the errors can be expressed as

$$\begin{aligned}\hat{\boldsymbol{\varepsilon}}_R &= \mathbf{Y} - \mathbf{X}_R\hat{\boldsymbol{\beta}}_R \\ &= \mathbf{Y} - \mathbf{X}_R\boldsymbol{\beta}_R + \mathbf{X}_R\boldsymbol{\beta}_R - \mathbf{X}_R\hat{\boldsymbol{\beta}}_R \\ &= \boldsymbol{\varepsilon} + \mathbf{X}_R(\boldsymbol{\beta}_R - \hat{\boldsymbol{\beta}}_R).\end{aligned}$$

Hence, if $\hat{\boldsymbol{\beta}}_R \neq \boldsymbol{\beta}_R$ we cannot, in general, claim that $\hat{\boldsymbol{\varepsilon}}_R$ will equal $\boldsymbol{\varepsilon} + \mathbf{1}c$, where c is a constant. So, the inaccuracies in the weighting will be due to the inaccuracies in the estimation of $F(\cdot)$ and inaccuracies in the estimation of $\boldsymbol{\beta}_R$. Consequently, we expect that the normalization method will not be too effective if the estimate of β_R is not too accurate.

5.4.2 Significance Level of the Adaptive Test

A simulation study was designed to accurately estimate the significance level of the test of $H_0 : \boldsymbol{\beta}_A = 0$ in the model $\mathbf{Y} = \mathbf{X}_R\boldsymbol{\beta}_R + \mathbf{X}_A\boldsymbol{\beta}_A + \boldsymbol{\varepsilon}$. In our simulations we have used two variables in $\mathbf{X}_R$, in addition to the intercept, and two variables in $\mathbf{X}_A$. For the ith observation the model is

$$y_i = \beta_0 + \beta_1 x_{i1} + \beta_2 x_{i2} + \beta_3 x_{i3} + \beta_4 x_{i4} + \varepsilon_i.$$

We will test $H_0 : \beta_3 = \beta_4 = 0$ versus $H_a : \beta_3 \neq 0$ or $\beta_4 \neq 0$. To make the simulations more interesting and informative, we generated the pairs (x_{i1}, x_{i3}) from a bivariate normal distribution having a correlation of $\rho = 0.8$ and generated the pairs (x_{i2}, x_{i4}) in the same manner. That is, we generated correlated independent variables to introduce a fair amount of multicollinearity into the regression model.

In the simulation that we used to estimate the significance level, we set $\beta_3 = \beta_4 = 0$ and observed how often we rejected the null hypothesis. We used 100,000 data sets with $n = 25$ observations in each data set to obtain the empirical probabilities of rejecting the null hypothesis, which are shown in Figure 5.4 for the nine error distributions. The standard error of these estimates is approximately 0.07%. Note that the empirical significance level is very close to the nominal level of 5% for the F test and the permutation test for the nine error distributions. With the adaptive test the small departures from the nominal value that we observe indicate that the test may not reject often enough, but the proportion of rejections, even in this sample of $n = 25$ observations with several covariates, was always greater than 4.48% and less than 5.09%. Generally, we are more concerned if a test rejects too often rather than not often enough, so the adaptive test appears to be quite acceptable with sample sizes of $n = 25$.

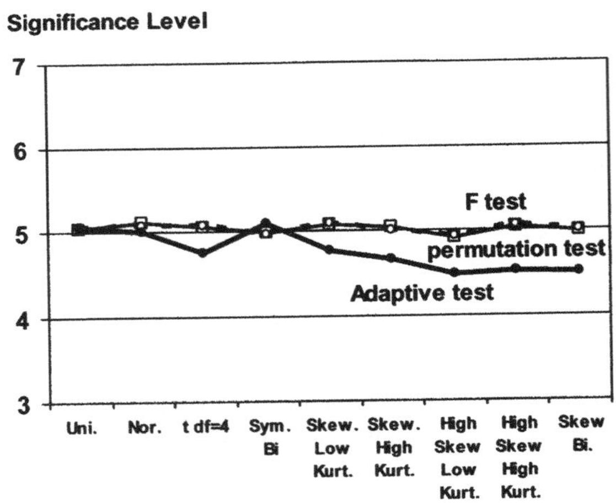

Figure 5.4 Empirical level of significance in percent for the F test and the adaptive test of $H_0 : \beta_3 = \beta_4 = 0$ in a multiple regression model having $n = 25$ observations with correlated predictor variables. The F test is indicated by hollow squares, the permutation test by hollow circles, and the adaptive test by filled circles.

Similar estimates of the empirical significance level are shown in Figure 5.5 for $n = 100$ observations. Again, we used 100,000 data sets for these simulations to obtain an accurate estimate of the level of significance. We observe that the significance level of the adaptive test is in the interval [4.71%, 5.01%]. Hence, we conclude that with $n = 100$ observations the significance level of these tests are at or near $\alpha = 0.05$ for the nine error distributions.

Power results for other simulations that have uncorrelated independent variables, which are not shown here, show the same pattern. Because the significance level of the adaptive test is quite close to the nominal value for $n = 100$, and because researchers would seldom use an adaptive test with fewer than $n = 25$ observations in a four-variable regression model, we conclude that the adaptive test will maintain its significance level near, or slightly below, the nominal level.

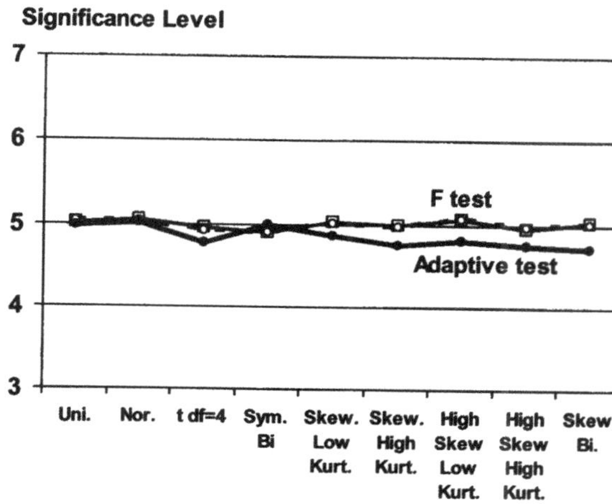

Figure 5.5 Empirical level of significance in percent for the *F* test and the adaptive test of $H_0 : \beta_3 = \beta_4 = 0$ in a multiple regression model having $n = 100$ observations with correlated predictor variables. The *F* test is indicated by hollow squares, the permutation test by hollow circles, and the adaptive test by filled circles.

5.4.3 Power of the Adaptive Test

We wanted to compare the power of these tests, so we used a simulation study with 10,000 data sets with $\beta_3 = \beta_4 = 0.8$ to achieve power near 80%. With $n = 25$ observations we obtain the results displayed in Figure 5.6, which show a moderate power advantage of 2% to 7% for the adaptive test with skewed errors and a power disadvantage of 0.91% with normal errors. So, except for highly skewed errors, there is little difference in power between the tests with $n = 25$ observations.

However, for larger sample sizes we do see differences in power between these tests. The results for $n = 50$ observations that are displayed in Figure 5.7 show a clear advantage of using the adaptive test for non-normal errors and that there is a small power loss of approximately 1% with normal errors. We used $\beta_3 = \beta_4 = 0.556$ in these simulations. So it appears, even with a fairly complex model having $n = 50$ observations, that there is much to gain and little to lose if the adaptive test is used.

The power results for simulations that used $n = 100$ observations are displayed in Figure 5.8. We used $\beta_3 = \beta_4 = 0.4$ in these simulations. These results show a

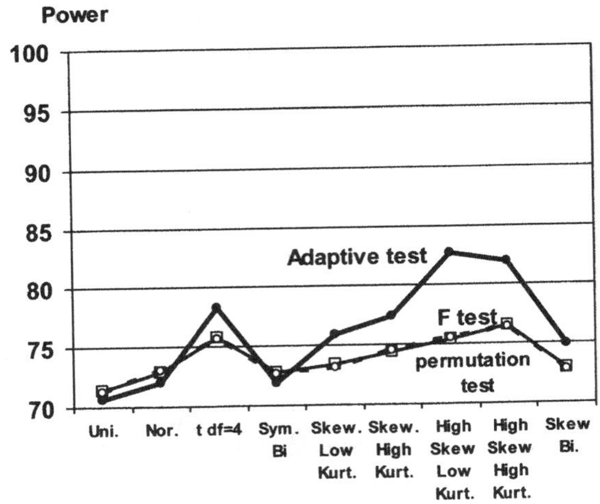

Figure 5.6 Empirical power in percent for the F test, the permutation test, and the adaptive test of $H_0 : \beta_3 = \beta_4 = 0$ in a multiple regression model with $n = 25$ observations and correlated predictor variables. The F test is indicated by hollow squares, the permutation test by hollow circles, and the adaptive test by filled circles.

large power advantage for the adaptive test with all non-normal errors and a loss of less than 1% with normal errors.

These results clearly demonstrate that the adaptive test has a greater power advantage if the number of observations is large. The modest advantage in power for the adaptive test with small n is due to the difficulty of accurately normalizing the error distribution when there are few residuals. Thus, if we have four independent variables, the adaptive test appears to be preferable if n is large, but for $n \leq 25$ either test could be used.

5.5 AN EXAMPLE OF A TEST IN MULTIPLE REGRESSION

The cigarette consumption data set was analyzed by Chatterjee, Hadi, and Price (2000) to estimate the relationship between several social and economic predictor variables and the demand for cigarettes in the United States during the year 1970. This data set is given in Tables 5.1 and 5.2.

Suppose we believe that price and income are related to the demand for cigarettes, but it is not clear if gender is related to demand. To test the null hypothesis we specify the complete model for the ith observation as

$$y_i = \beta_0 + \beta_1 x_{i1} + \beta_2 x_{i2} + \beta_3 x_{i3} + \varepsilon_i,$$

where y_i is the per capita cigarette sales (in packs), x_{i1} is the average price per pack (in cents), x_{i2} is the per capita personal income (in dollars), and x_{i3} is the percent of

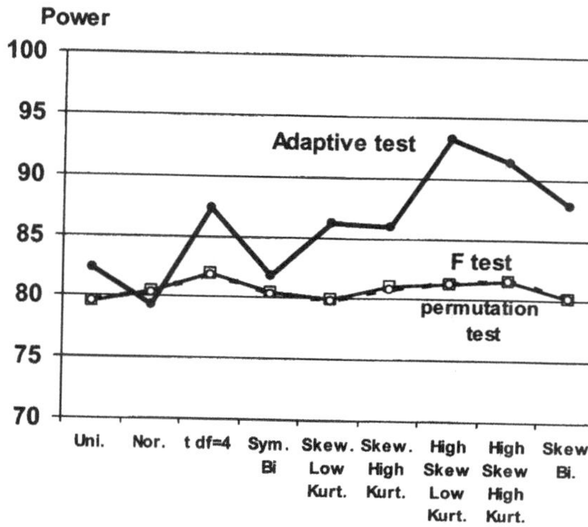

Figure 5.7 Empirical power in percent for the F test, the permutation test, and the adaptive test of $H_0 : \beta_3 = \beta_4 = 0$ in a multiple regression model with $n = 50$ observations and correlated predictor variables. The F test is indicated by hollow squares, the permutation test by hollow circles, and the adaptive test by filled circles.

females living in the state. We want to test $H_0 : \beta_3 = 0$ to see if gender is related to cigarette sales after adjusting for price and income. Thus, the reduced model is

$$y_i = \beta_0 + \beta_1 x_{i1} + \beta_2 x_{i2} + \varepsilon_i.$$

The traditional F test produced a p-value of $p = 0.143$, but the adaptive test gave a p-value of $p = 0.036$ based on 2000 permutations of the residuals from the reduced model. So, when the adaptive test is used, it appears that gender is related to the demand for cigarettes after adjusting for price and income. The difference between these p-values is due to the downweighting of the observations having the largest residuals.

The weights that we computed from the residuals of the reduced model are given in the last column of Tables 5.1 and 5.2. New Hampshire had the highest demand for cigarettes, and this was not expected based on the price or income, so the residual was large and this produced a weight of 0.256 with the unpermuted data. Nevada also had high cigarette sales so the adaptive weighing procedure gave it a weight of 0.392, while the weights for the other observations were in the range of 0.604 to 1.141 with the unpermuted data.

5.5.1 Example Using R Code

As we explained earlier in this chapter, to use the `adapt` function a data frame must first be created with the appropriate variable names. The dependent variable must

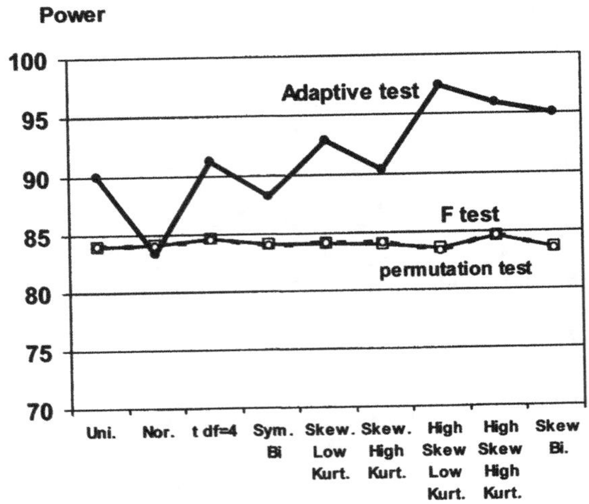

Figure 5.8 Empirical power in percent for the F test, the permutation test, and the adaptive test of $H_0 : \beta_3 = \beta_4 = 0$ in a multiple regression model with $n = 100$ observations and correlated predictor variables. The F test is indicated by hollow squares, the permutation test by hollow circles, and the adaptive test by filled circles.

be given the variable name of `y`. The independent variables will be given the names of `price`, `income`, and `gender`. The statistician must include the R code for the `adapt` function before these statements, either by placing them before this code or by using a `source` statement. The `adapt` function, which is listed in Appendix A, can be downloaded from *www.math.niu.edu/~ogorman*. To perform the adaptive test for gender, the complete model includes price, income, and gender while the reduced model includes only the price and income. The following code will execute the `adapt` function to perform the adaptive test:

```
dataset <- c("cigs.data")
complete <- c("y~price + income + gender")
reduced  <- c("y~price + income")
r <- 2000
seed <-c(91492)
pvalue <- adapt(dataset,complete,reduced,r,seed)
```

The standard output will include the p-value for the traditional F test, which was 0.143, and the adaptive p-value, which was 0.036. The R program can easily be modified to print the residuals from the reduced model and the weights from the unpermuted reduced model.

Table 5.1 Cigarette data

State	Sales	Price	Income	Age	Gender	Weight
AL	89.8	42.7	2948	27	51.7	1.069
AK	121.3	41.8	4644	22.9	45.7	1.141
AZ	115.2	38.5	3665	26.3	50.8	1.112
AR	100.3	38.8	2878	29.1	51.5	1.064
CA	123	39.7	4493	28.1	50.8	1.141
CO	124.8	31.1	3855	26.2	50.7	1.071
CT	120	45.5	4917	29.1	51.5	1.127
DE	155	41.3	4524	26.8	51.3	0.633
DC	200.4	32.6	5079	28.4	53.5	0.604
FL	123.6	43.8	3738	32.3	51.8	0.666
GA	109.9	35.8	3354	25.9	51.4	1.141
HI	82.1	36.7	4623	25	48	0.599
ID	102.4	33.6	3290	26.4	50.1	1.052
IL	124.8	41.4	4507	28.6	51.5	1.114
IN	134.6	32.2	3772	27.2	51.3	1.128
IO	108.5	38.5	3751	28.8	51.4	1.137
KA	114	38.9	3853	28.7	51	1.138
KY	155.8	30.1	3112	27.5	50.9	0.641
LA	115.9	39.3	3090	24.8	51.4	0.752
ME	128.5	38.8	3302	28	51.3	0.670
MD	123.5	34.2	4309	27.1	51.1	1.055
MA	124.3	41	4340	29	52.2	1.091
MI	128.6	39.2	4180	26.3	51	1.049
MN	104.3	40.1	3859	26.8	51	1.128
MS	93.4	37.5	2626	25.1	51.6	1.128
MO	121.3	36.8	3781	29.4	51.8	1.126
MT	111.2	34.7	3500	27.1	50	1.117

5.5.2 Example Using a SAS Macro

It is also easy to use SAS to perform this adaptive test. We first must include the adaptive macro and create the SAS data set before we invoke the SAS macro. The `adaptall` function, which is listed in Appendix B, can be downloaded from *www.math.niu.edu/~ogorman*. If the `adaptall` macro is in the same subdirectory as the source program, the following code will perform the adaptive test:

```
%include 'adaptall.sas';
data cigs;
```

Table 5.2 Cigarette data (Continued)

State	Sales	Price	Income	Age	Gender	Weight
NB	108.1	34.7	3789	28.6	51.2	1.037
NV	189.5	44	4563	27.8	49.3	0.392
NH	265.7	34.1	3737	28	51.1	0.256
NJ	120.7	41.7	4701	30.1	51.6	1.140
NM	90	41.7	3077	23.9	50.7	1.132
NY	119	41.7	4712	30.3	52.2	1.132
NC	172.4	29.4	3252	26.5	51	0.594
ND	93.8	38.9	3086	26.4	49.5	1.140
OH	121.6	38.1	4020	27.7	51.5	1.132
OK	108.4	39.8	3387	29.4	51.3	1.066
OR	157	29	3719	29	51	0.843
PA	107.3	44.7	3971	30.7	52	1.051
RI	123.9	40.2	3959	29.2	50.9	0.972
SC	103.6	34.3	2990	24.8	50.9	1.134
SD	92.7	38.5	3123	27.4	50.3	1.127
TN	99.8	41.6	3119	28.1	51.6	1.007
TX	106.4	42	3606	26.4	51	1.070
UT	65.5	36.6	3227	23.1	50.6	0.638
VT	122.6	39.5	3468	26.8	51.1	0.774
VT	124.3	30.2	3712	26.8	50.6	1.070
VA	96.7	40.3	4053	27.5	50.3	1.034
WA	114.5	41.6	3061	30	51.6	0.671
WV	106.4	40.2	3812	27.2	50.9	1.141
WI	132.2	34.4	3815	27.2	50	1.097

```
  input  State $ Age Income Gender Price Sales ;
  datalines;
AL 27.0  2948.0  51.7 42.7 89.8
AK 22.9  4644.0  45.7 41.8 121.3
.   .    .     .   .   .   .   .
;
%adaptall(dataset=cigs, y=sales, xr=price income, xa = gender,
        seed1=27293, seed2=18923, seed3=29349, nperm=2000);
```

We have listed only the first two observations of the $n = 51$ observations in the data set. Note that the variables in the X_R matrix are listed in the `xr` keyword parameter. The intercept is always included in X_R, so it should not be specified in the list for the keyword parameter `xr`. The variables in the X_A matrix are listed in the `xa` keyword parameter. After the macro is invoked, it will produce the following output:

```
Number of permutations = 2000
Starting seeds = 27293  18923  29349
SAS data set: cigs   Number of observations = 51

Dependent variables  :  sales

Independent variables in reduced model :Intercept price income
      Class variables in reduced model :

Independent variables in complete model:Intercept price income
                                        gender
      Class variables in complete model:

Adaptive test statistics for unpermuted data:
      Wilks Lambda =    0.9389
      F statistic  =    3.0562

p-value for adaptive test =    0.0390
```

This output gives the information that would be needed to document exactly what hypothesis was tested. In this multiple regression program there were no class variables, but we will be using many of these in the next chapter. This SAS code will execute in approximately 5 minutes on an inexpensive PC purchased in 2008. Newer computers should be able to execute this program much more quickly. Although it appears that the adaptive test was useful in determining whether gender should be included in the model, we cannot determine that it was useful because we do not know the relationship between the variables.

5.6 CONCLUSIONS

In this chapter we have investigated the size and the power of the adaptive test for slope in a simple linear regression model and an adaptive test for two coefficients in a multiple regression model. We have learned that:

- The adaptive test for slope in a simple linear regression will maintain its significance level.
- The adaptive test for slope in a simple linear regression is more powerful than the traditional test for long-tailed symmetric and skewed distributions.
- The adaptive test for a subset of coefficients in a multiple linear regression will maintain its significance level, at or slightly below the nominal value of $\alpha = 0.05$, even when the independent variables are highly correlated.
- If there are at least 50 observations, the adaptive test has a power advantage, with non-normal errors, when used in a multiple regression model.

- The adaptive test is quite practical; it can be performed quickly using either R code or SAS code.

The multiple regression results suggest that, for moderate and large data sets, the adaptive approach can improve the performance of a test for any subset of coefficients in a linear model. In the next three chapters we will use this adaptive test in a variety of testing situations.

EXERCISES

5.1 Suppose we have a simple linear regression model and we perform an adaptive test for slope. Could we use SSE_C^* instead of $\Lambda^{2/n,*}$ as the test statistic? Explain fully.

5.2 Suppose we have a multiple linear regression model with a reduced model that contains an independent variable in addition to the intercept. If we wanted to perform an adaptive test, could we use SSE_C^* instead of $\Lambda^{2/n,*}$ as the test statistic? Explain fully.

5.3 Suppose we have a multiple linear regression model with a reduced model that contains an independent variable in addition to the intercept. We used the permutation of residuals method to obtain $\mathbf{Y}_k$ for the *k*th permutation, but used the weights for the unpermuted data in the analysis. That is, for the *i*th observation we continue to use, for every permutation, the weight (w_i) that was computed from the unpermuted residuals. Does this procedure make sense? Why or why not?

5.4 If a researcher used F^*, instead of $\Lambda^{2/n,*}$, as the test statistic, what changes would be needed in the permutation steps to compute the *p*-value?

5.5 With the data collected by Simon (2001), we used the body mass index (bmi) to predict diastolic blood pressure. Using either R code or SAS code with the blood pressure data set, perform an adaptive test of significance for slope in a simple linear regression that uses the body mass index to predict the systolic blood pressure. The blood pressure data set can be found at *www.math.niu.edu/~ogorman*.

5.6 Suppose we perform an adaptive test for slope in a simple linear regression model using a data set that has $n = 43$ observations. If there are no tied values for the dependent variable and the independent variable, how many distinct test statistics are possible? Will this test maintain its level of significance? Why or why not?

5.7 Use the power estimates for a test for slope in a simple linear regression model that are shown Figures 5.1 and 5.2 to give approximate answers to the following questions.

a) What is the average increase in power of the adaptive test over the *t* test with $n = 25$ observations when the average is taken over all nine distributions?

b) What is the average increase in power of the adaptive test over the *t* test with $n = 100$ observations when the average is taken over all nine distributions?

5.8 For the multiple regression model that was used in the simulation studies to determine the empirical significance level of the test, we indicated that we are more concerned if the test rejects too often rather than not often enough. Do you agree? Explain?

5.9 With the cigarette data from Chatterjee, Hadi, and Price (2000), test the null hypothesis that both age and gender are unrelated to cigarette sales, after controlling for price and income. Use either R code or SAS code to perform an adaptive test of significance. The cigarette data set can be found at *www.math.niu.edu/~ogorman.*

5.10 Suppose we had a multiple regression model and attempted to perform an adaptive test of significance, but calculated $\hat{\sigma}$ incorrectly so that it equaled three times the correct value. This miscalculation was made for the unpermuted data set and for all of the permutation data sets. Further, suppose all of the other statistics, including $\hat{\sigma}_t$, that were calculated in the first four steps of the weighing process were correctly calculated. How would the weights be changed with this miscalculation? What effect would this have on the estimation of the elements of $\boldsymbol{\beta}_R$ and $\boldsymbol{\beta}_A$? What effect would this have on the test statistic?

5.11 Suppose two students were analyzing the same data set that has $n = 67$ observations using a simple linear regression with $R = 2000$ permutations of the residuals, but one of the students used with replacement sampling while the other used without replacement sampling of the vector of residuals. Suppose they both used the same random number generator and algorithm to select the permutation. If the student who used without replacement sampling checked to see if the next permutation had already been selected, what is the chance that the second permutation would match the first permutation selected? What is the chance that the third permutation would match either the first or second permutations? Without doing any further calculations, do you expect that their answers would be identical? Explain.

5.12 Consumers Union (2010) published data on 21 small cars. This data included engine size (in liters), weights (in thousands of pounds), and EPA city mileage. These data, which can be found at *www.math.niu.edu/~ogorman*, are listed in Table 5.3. Using a SAS macro or R code, perform an adaptive test of the null hypothesis that weight is unrelated to mileage, after controlling for engine size.

5.13 Coleman *et al.* (1966), as part of a larger study, obtained data on sixth graders at 20 schools in the United States. The outcome variable was the average verbal test scores. The potential predictor variables include the average staff salaries per pupil, the percent of white collar fathers, the socioeconomic status composite deviation (SED), the teachers' verbal test scores, and the average of the mothers' education level. The data, which can be obtained at *www.math.niu.edu/~ogorman*, is listed in Table 5.4.

Table 5.3 Small car mileage data

Car	Make	Model	Engine size	Weight (1000lbs.)	MPG city
1	Chevrolet	Cruze	1.8	3.275	24
2	Chevrolet	HHR	2.4	3.295	22
3	Dodge	Caliber	2.0	3.185	23
4	Ford	Focus	2.0	2.715	25
5	Honda	Civic	1.8	2.810	25
6	Hyundai	Elantra	1.8	2.755	29
7	Kia	Forte	2.0	2.875	25
8	Kia	Soul	2.0	2.950	24
9	Mazda	3	2.0	2.915	24
10	Mitsubishi	Lancer	2.0	3.015	25
11	Nissan	Cube	1.8	2.855	27
12	Nissan	Sentra	2.0	2.965	27
13	Nissan	Versa	1.8	2.805	28
14	Scion	tC	2.5	3.060	23
15	Scion	xB	2.4	3.120	22
16	Scion	xD	1.8	2.645	26
17	Subaru	Impreza	2.5	3.155	20
18	Suzuki	SX4	2.0	2.990	23
19	Toyota	Corolla	1.8	2.850	26
20	Toyota	Matrix	2.4	3.000	21
21	Volkwagen	Golf	2.5	3.045	24

Use SAS or R to perform adaptive tests of significance.

a) Perform an adaptive test for slope in a simple linear regression using SED as the sole predictor variable.

b) Perform an adaptive test of the null hypothesis that salary, percent white collar fathers, teachers' verbal scores, and mothers' education level are unrelated to the verbal scores, after controlling for SED.

c) Perform an adaptive test of the null hypothesis that salary is unrelated to the verbal scores, after controlling for all of the other predictor variables.

Table 5.4 Coleman data used to model the verbal test scores of sixth-grade students at 20 schools in the United States

School	Salary per Pupil	Percent White Collar	SED	Teachers' Verbal Scores	Mothers' Education Level	Verbal Scores
1	3.83	28.87	7.20	26.60	6.19	37.01
2	2.89	20.10	-11.71	24.40	5.17	26.51
3	2.86	69.05	12.32	25.70	7.04	36.51
4	2.92	65.40	14.28	25.70	7.10	40.70
5	3.06	29.59	6.31	25.40	6.15	37.10
6	2.07	44.82	6.16	21.60	6.41	33.90
7	2.52	77.37	12.70	24.90	6.86	41.80
8	2.45	24.67	-0.17	25.01	5.78	33.40
9	3.13	65.01	9.85	26.60	6.51	41.01
10	2.44	9.99	-0.05	28.01	5.57	37.20
11	2.09	12.20	-12.86	23.51	5.62	23.30
12	2.52	22.55	0.92	23.60	5.34	35.20
13	2.22	14.30	4.77	24.51	5.80	34.90
14	2.67	31.79	-0.96	25.80	6.19	33.10
15	2.71	11.60	-16.04	25.20	5.62	22.70
16	3.14	68.47	10.62	25.01	6.94	39.70
17	3.54	42.64	2.66	25.01	6.33	31.80
18	2.52	16.70	-10.99	24.80	6.01	31.70
19	2.68	86.27	15.03	25.51	7.51	43.10
20	2.37	76.73	12.77	24.51	6.96	41.01

CHAPTER 6

MORE APPLICATIONS OF ADAPTIVE TESTS

The general adaptive testing procedure that we described in the last chapter for the multiple regression problem can be used to perform a test of significance for any subset of coefficients in a linear model containing fixed effects. In this chapter we show how an adaptive approach could be used in most of the common testing situations to increase the power of the tests. We will analyze data from a one-way layout, a randomized complete block design, and a two-way layout. We will also show how the residuals can be permuted within blocks in a design that uses blocking. For each of these applications we will give the R and SAS code to perform the test.

6.1 THE COMPLETELY RANDOMIZED DESIGN

6.1.1 Model Specification

In this section we will describe the adaptive test for group effect in a completely randomized design having one factor with g levels. This is the standard one-way layout with n_j observations at level j. The model can be specified as

$$y_i = \beta_0 + \beta_1 x_{i,1} + \cdots + \beta_{g-1} x_{i,g-1} + \varepsilon_i, \tag{6.1}$$

By Thomas W. O'Gorman

for $i = 1, \ldots, n$, where

$$x_{i,j} = \begin{cases} 1 & \text{if } i\text{th observation is from population } j, \\ 0 & \text{otherwise.} \end{cases}$$

Other parameterizations of the model could be used. With this parameterization the mean of the gth level is the intercept. To determine if any of the level means are different from any of the other treatment means, we will test $H_0 : \beta_1 = \beta_2 = \ldots = \beta_{g-1} = 0$ against the alternative $H_a : \beta_j \neq 0$ for some $j = 1, \ldots, g-1$. Although this model was intended for the analysis of data from a completely randomized design, it often is used to analyze observational or experimental data that conform to a one-way layout.

To perform the tests, we will assume that the errors are independent and identically distributed. The two-sample adaptive test that was described in Chapter 3 is a special case of this test with $g = 2$. To perform the test, the complete model is given in Equation (6.1) and the reduced model is $y_i = \beta_0 + \varepsilon_i$ for $i = 1, \ldots, n$. No special weighting method is needed; the adaptive test is performed using the weighting algorithm and the permutation method that were described in Chapter 5.

6.1.2 Level of Significance and Power of the Adaptive Test

Because the reduced model contains only an intercept term, the adaptive test will maintain its level of significance. The reasoning is the same as that given in Chapter 3 for the two-sample test. Under the null hypothesis the test statistics that result from the permutations of the residuals are equally likely, so the resulting test will maintain its level of significance.

In order to compare the power of the adaptive test to other tests, we used four simulation studies, with one having $g = 2$ groups and the others having $g = 4$ groups. The power estimates were obtained by a FORTRAN program that was written by the author. FORTRAN was used in the simulations because it executed more quickly than R or SAS. See Section 3.8 for more information about the FORTRAN code. In the simulation with $g = 2$ groups we estimated the power of the pooled t test, the Wilcoxon rank-sum test, and the adaptive test. In this simulation we set the difference between the population means to $\delta = 0.80$ to achieve power near 80% for the t test. The simulation results that are shown in Figure 6.1 indicate that the adaptive test is more powerful than the t test with all non-normal error distributions, but has a power loss of 1.14% with the normal error distribution. The adaptive test is more powerful than the Wilcoxon test for 7 of the 9 distributions, but has a 1.69% power loss with the t_4 error distribution and a 0.85% power loss with the skewed high kurtoisis error distribution.

In general terms, the Wilcoxon test is approximately as powerful as the adaptive test for long-tailed and skewed distributions, but is not as powerful as the adaptive test for the uniform and the bimodal error distributions. Despite its popularity, the t test has relatively low power when compared to the Wilcoxon and adaptive tests for the long-tailed symmetric distribution and the skewed distributions.

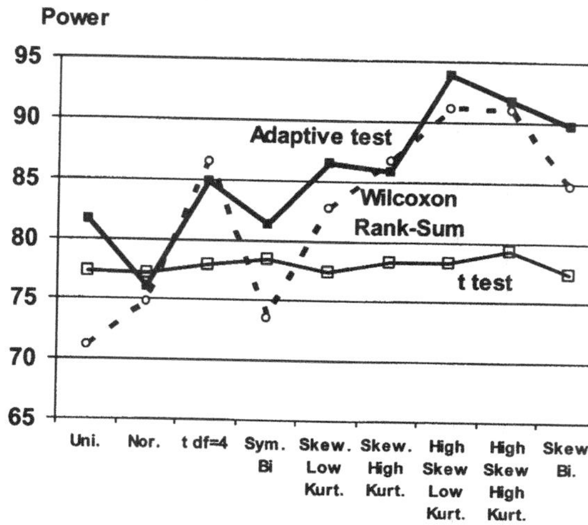

Figure 6.1 Empirical power in percent for the two-sample t test, the Wilcoxon rank-sum test, and the adaptive test for equality of group means for a balanced design having sample sizes of $n_1 = n_2 = 24$. The t test is indicated by hollow squares, the Wilcoxon rank-sum test by hollow circles, and the adaptive test by filled squares.

The other simulations compared the F test, the Kruskal–Wallis test, and the adaptive test in completely randomized designs having $g = 4$ groups. The first simulation study had balanced group sizes of $n_1 = n_2 = n_3 = n_4 = 8$. The second study had balanced group sizes of $n_1 = n_2 = n_3 = n_4 = 12$. The third study had unbalanced group sizes of $n_1 = n_2 = n_3 = 8$ and $n_4 = 24$. These small sample sizes were used because we wanted to determine the effectiveness of the adaptive test with the small studies that are sometimes used in animal experiments.

The empirical power in percent for the F test, the Kruskal–Wallis test, and the adaptive test are shown in Figure 6.2 for the balanced design having $n_1 = n_2 = n_3 = n_4 = 8$. To achieve powers near 80%, we used $\beta_1 = \beta_2 = \beta_3 = 1.4$ in these simulations. When we compare the adaptive test to the F test, we see that the adaptive test has a slightly lower power with the normal and the symmetric bimodal distributions and we see that the adaptive test has somewhat greater power with many of the other distributions. Only with highly skewed errors do we see dramatic differences between the two methods.

In contrast, we see large differences in power between the Kruskal–Wallis test and the F test with many of the error distributions in Figure 6.2. The Kruskal–Wallis test is less powerful than the adaptive test for all distributions. For these sample sizes the Kruskal–Wallis test cannot be recommended.

If larger sample sizes are used, the relative power of these tests change. In Figure 6.3 we see the empirical power for these three tests with balanced data sets having group sizes of $n_1 = n_2 = n_3 = n_4 = 12$. To achieve powers near 80%, we used

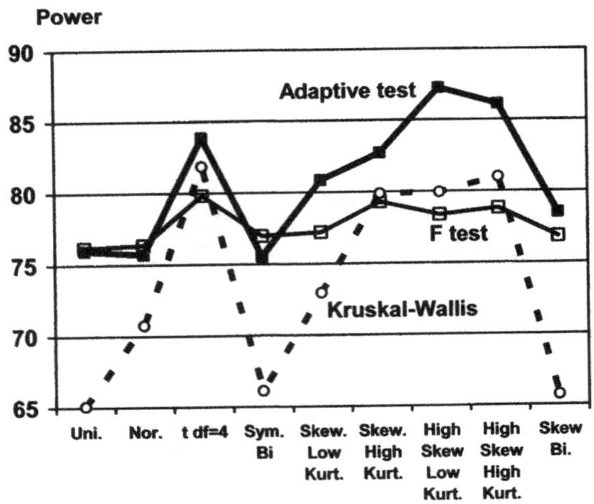

Figure 6.2 Empirical power in percent for the F test, the Kruskal–Wallis test, and the adaptive test for equality of group means for a balanced design having sample sizes of $n_1 = n_2 = n_3 = n_4 = 8$. The F test is indicated by hollow squares, the Kruskal–Wallis test by hollow circles, and the adaptive test by filled squares.

$\beta_1 = \beta_2 = \beta_3 = 1.1$ in these simulations. For studies of this size, the adaptive test is often considerably more powerful than the F test. The Kruskal–Wallis test is much more powerful than the F test with the t_4 distribution and the highly skewed error distributions. The adaptive test was more powerful than the other tests for 7 distributions and had power within 2% of the most powerful test for the other 2 distributions. The same overall pattern is seen in Figure 6.4 for an unbalanced design having sample sizes of $n_1 = n_2 = n_3 = 8$ and $n_4 = 24$. To achieve powers near 80% we used $\beta_1 = \beta_2 = \beta_3 = 1.0$ in these simulations. By comparing the results in Figure 6.3 to those in Figure 6.4 it appears that the differences between the power of these test is primarily due to the increased sample size and is relatively insensitive to the lack of balance in the design. Unless the researcher knew that the errors were normally distributed the adaptive test would be recommended for these sample sizes.

6.1.3 An Example of a Completely Randomized Design

The biomass of ants in the stomach contents of the eastern horned lizard, *Phrynosoma douglassi brevirostre*, was recorded by Powell and Russell (1984). These data were published by Linton, Edgington, and Davies (1989) and were analyzed by Manly (1997). The data were obtained for $g = 4$ months for adult males and yearling females. The biomass (in milligrams) are displayed in Table 6.1.

We will test the null hypothesis that the biomass is unrelated to the month. Note that this is observational data and that the distribution of biomass is highly skewed.

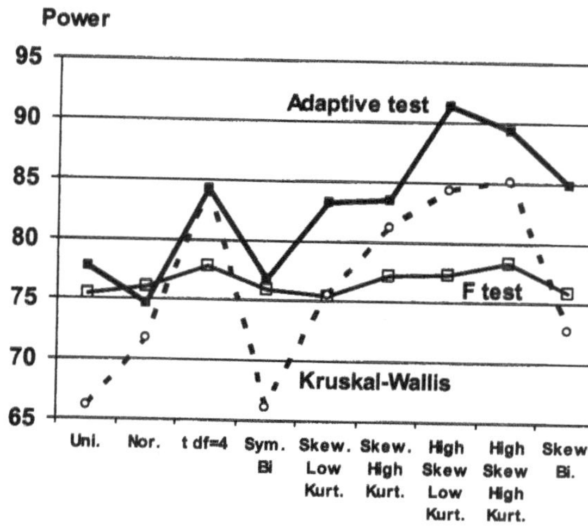

Figure 6.3 Empirical power in percent for the F test, the Kruskal–Wallis test, and the adaptive test for equality of group means for a balanced design having sample sizes of $n_1 = n_2 = n_3 = n_4 = 12$. The F test is indicated by hollow squares, the Kruskal–Wallis test by hollow circles, and the adaptive test by filled squares.

The traditional F test produced a p-value of $p_F = 0.211$, whereas the adaptive test, using an R program, produced a p-value of $p_{adaptive} = 0.038$. We would reject the null hypothesis if we used the adaptive test with $\alpha = 0.05$. This is reasonable because it does appear that the August mean is greater than the September and July means. The weights shown in Table 6.1 indicate how much downweighting was performed for those observations that are on the right of the distribution. If there really is a difference between months then the downweighting of the extreme observations was helpful in detecting the difference.

6.1.3.1 Using R to Perform the Adaptive Test

We can easily perform this adaptive test using R. First we make a data set that has the month as a character variable and the biomass as a numeric variable. The first two lines and last line of the data set are:

```
        month     y
1       June     13
2       June    105
.         .       .
24  September    44
```

Now suppose that this data set is called "biomass.dat". It is important that character values be used for the `month` variable so that it will be defined as a factor and used as a factor in the calls to the `lm` function. We will assume that the `adapt` function is on a file `adapt.r`. The `adapt` function, which is listed in Appendix A, can

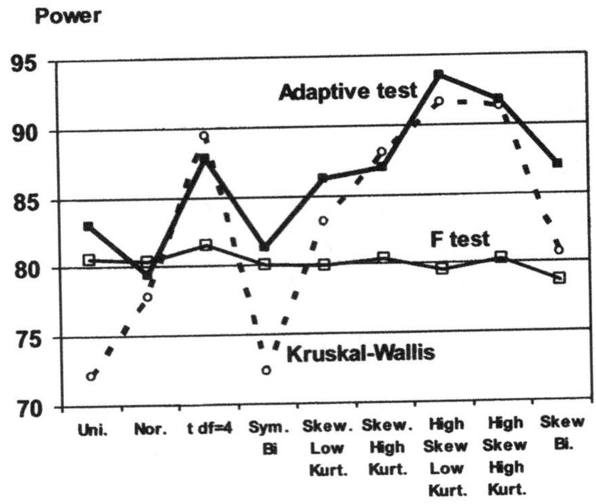

Figure 6.4 Empirical power in percent for the *F* test, the Kruskal–Wallis test, and the adaptive test for equality of group means with an unbalanced design having sample sizes $n_1 = n_2 = n_3 = 8$ and $n_4 = 24$. The *F* test is indicated by hollow squares, the Kruskal–Wallis test by hollow circles, and the adaptive test by filled squares.

be downloaded from *www.math.niu.edu/∼ogorman*. Instead of copying that code to the beginning of the program, we will include it by specifying that file in a `source` statement. We can also specify the number of permutations and the random number seed in the call to the `adapt` function, as shown in the following R code:

```
source("adapt.r")
dataset <- c("biomass.dat")
complete <- c("y~month")
reduced  <- c("y~1")
pvalue <- adapt(dataset,complete,reduced,r=2000,seed=91492)
```

It is important to remember that `complete` and `reduced` are character strings that are used as formulas in the `lm` function that is called from the `adapt` function. This R code produced the following output:

```
Output from Adaptive Test:
  Unpermuted lambda =  0.8536
  Unpermuted F      =  1.1438
  p-value           =  0.038

Traditional F test p-value =  0.211
```

If a great deal of precision is needed in the *p*-value the number of permutations should be increased. With 10,000 permutations the test should take only a few minutes to execute on most computers.

Table 6.1 The ant biomass data and the adaptive weights

Observation	Month	Dry Biomass	Weight
1	June	13	1.306
2	June	105	1.104
3	June	242	0.780
4	July	2	1.326
5	July	8	1.315
6	July	20	1.293
7	July	59	1.213
8	July	245	0.774
9	August	40	1.253
10	August	50	1.232
11	August	52	1.228
12	August	82	1.160
13	August	88	1.145
14	August	233	0.797
15	August	488	0.492
16	August	515	0.487
17	August	600	0.484
18	August	1889	0.195
19	September	0	1.329
20	September	5	1.320
21	September	6	1.319
22	September	18	1.297
23	September	21	1.291
24	September	44	1.245

6.1.3.2 Using SAS to Perform the Adaptive Test This data set can easily be analyzed using the SAS macro `adaptall.sas`. The `adaptall` macro, which is listed in Appendix B, can be downloaded from *www.math.niu.edu/~ogorman*. Before we invoke the macro we must include the SAS macro and create a temporary SAS data set. If the macro is in the same directory as the SAS program, we would use the code:

```
%include 'adaptall.sas';
data lizards;
  input observation month $ biomass;
  datalines;
 1      June    13
 2      June   105
```

```
.    .    .
24  September  44
;
%adaptall(dataset=lizards, y=biomass, xa=month, classa=month,
          seed1=4321, seed2=15497, seed3=24733);
```

The SAS output produced a p-value of $p = 0.0455$ with 2000 permutations, which is the default. This p-value differs slightly from that produced by the R code because the R code uses the Mersenne-Twister random number generator while the SAS code uses the Wichmann–Hill random number generator. In SAS we must specify which variables to use as class variables and for these variables SAS will generate indicator variables. The independent variables in X_A are specified in the `xa` keyword parameter, and those that are class variables are specified in the `classa` keyword parameter.

6.1.4 Multiple Comparison Procedures

When analyzing data from a one-way layout having g levels of one factor, it is often necessary to make several comparisons between the levels. If we perform all pairwise comparisons of means, then we would have $g(g-1)/2$ comparisons. So, if $g = 4$ and we wanted to perform many two-sample tests to compare all levels to all other levels, it would require $4(3)/2 = 6$ two-sample tests. If each of these tests were performed using a significance level of α, the chance of rejecting one or more of these hypotheses would be much greater than α. To ensure that the overall (or experimentwise) error rate does not exceed a specified significance level of α, a variety of multiple comparison procedures have been proposed. For an explanation of the reasoning behind these methods see Miller (1980). For a description of many of the more popular methods see Ott and Longnecker (2001, Chapter 9). Using adaptive tests that are different from those considered here, Hill, Padmanabhan, and Puri (1988) proposed several multiple comparison procedures.

The Bonferroni multiple comparison procedure is popular because it can be used in many situations. Using this method, if we want to make c comparisons and we want to control the overall (or experimentwise) error rate we would use an adjusted $\alpha' = \alpha/c$ for each of the comparisons. Thus, if we wanted the experimentwise error rate not to exceed $\alpha = 0.05$ and we performed $c = 6$ comparisons, then we would use $\alpha' = \alpha/c = 0.05/6 = 0.0083$ for each of the individual comparisons.

To perform an adaptive multiple comparison procedure with data in a one-way layout, we would perform an adaptive test of $H_0 : \mu_i = \mu_j$ to compare the ith group to the jth group. We would obtain the p-value for the adaptive test and reject the null hypothesis if the p-value was less than or equal to α'.

To illustrate how one of these comparisons could be performed, we will use the lizard data to compare the mean of August to the mean of September. One way to do this is to write the complete model as

$$y_i = \beta_1 x_{i,1} + \beta_2 x_{i,2} + \beta_3 x_{i,3} + \beta_4 x_{i,4} + \varepsilon_i,$$

for $i = 1, \ldots, n$, where

$$x_{i,j} = \begin{cases} 1 & \text{if the } i\text{th observation is from the } j\text{th month,} \\ 0 & \text{otherwise.} \end{cases}$$

The reduced model, under the null hypothesis of no difference between the August and September means, is

$$y_i = \beta_1 x_{i,1} + \beta_2 x_{i,2} + \beta_{34} x_{i,34} + \varepsilon_i,$$

for $i = 1, \ldots, n$, where

$$x_{i,34} = \begin{cases} 1 & \text{if the } i\text{th observation is from August or September,} \\ 0 & \text{otherwise.} \end{cases}$$

By comparing the fit after adaptively weighting the complete and reduced models we can compute the test statistic in the usual way. This is the method that will be used in the SAS macro.

Another approach that is sometimes used is to perform a two-sample test using only the data from August and September. However, by using this complete and reduced model, we include all of the data in the analysis, which is helpful to estimate the c.d.f. of the residuals. We will now use a SAS macro that will perform all of these pairwise comparisons.

6.1.4.1 An Example of a Multiple Comparison Procedure Using a SAS Macro

Because multiple comparison procedures are popular, a SAS macro was written to facilitate the use of the Bonferroni multiple comparison procedure. The SAS macro `adaptmc` sets up the appropriate indicator variables to perform these adaptive tests, as we have described in this section, and then it calls the `adaptall` macro. The `adaptmc` macro, which is listed in Appendix C, can be downloaded from *www.math.niu.edu/~ogorman*. To perform all the comparisons between months with the lizard data set, we would use this code:

```
%include 'adaptall.sas';
%include 'adaptmc.sas';
data lizards;
  input observation month$ biomass;
  datalines;
 1      June     13
 2      June    105
.     .      .
24   September  44
;
%adaptmc(dataset=lizards, y=biomass,
         xa=month, classa=month,
         seed1=27823, seed2=25332, seed3=19234, nperm= 2000);
```

The `adaptmc` macro is designed to perform all the comparisons in a one-way layout. It was not designed to be used with covariates or with multifactorial designs. Hence, the only variable in the model must be a class variable. In other respects the macro call is similar to the macro call for the `adaptall` macro. The macro sets up the indicator variables, as described above, to perform the test of $H_0 : \mu_i = \mu_j$ versus $H_a : \mu_i \neq \mu_j$ for $i = 1, \ldots, g-1$ and $j = i+1, \ldots, g$. This multiple comparison macro produced the following output:

```
Compare August  to July  for variable month
     with 2000 permutations, p-value =    0.0605

Compare August  to June  for variable month
     with 2000 permutations, p-value =    0.2479

Compare July  to June  for variable month
     with 2000 permutations, p-value =    0.6097

Compare August  to September  for variable month
     with 2000 permutations, p-value =    0.0205

Compare July  to September  for variable month
     with 2000 permutations, p-value =    0.5807

Compare June  to September  for variable month
     with 2000 permutations, p-value =    0.3223
```

Note that the p-value for the test of the null hypothesis that the August mean equals the September mean is $p_{adaptive} = 0.0205$. If only two comparisons were of interest and the experimentwise error rate was $\alpha = 0.05$ then that null hypothesis would be rejected because $p_{adaptive} \leq \alpha/2$. However, if three tests were to be performed and if the Bonferroni multiple comparison procedure is used, this p-value would not exceed $\alpha/3$, so the null hypothesis would not be rejected.

6.2 TESTS FOR RANDOMIZED COMPLETE BLOCK DESIGNS

In many experimental settings there is so much variability in the experimental units that it is difficult to detect a difference between the levels of a factor. One way to increase the efficiency of a test of significance is to assign all of the levels of a factor to each experimental unit. For example, it may be difficult to compare two varieties of corn because the fertility of the soil may vary by location. To increase the chance of detecting a difference between the two varieties, we might use several blocks (locations) and then within each block randomly assign part of a block to one variety and the other part to the other variety. In the analysis of these data there is a blocking factor and a treatment factor, but usually we are only interested in the treatment factor. These designs are called randomized complete block (RCB) designs.

Suppose we have B blocks and wish to compare g treatments. To test the null hypothesis of no treatment effect we specify the complete model as

$$y_{ij} = \mu + \beta_i + \tau_j + \varepsilon_{ij}, \tag{6.2}$$

where μ is the overall effect, $\beta_1, \ldots, \beta_B$ are the block effects, and $\tau_1, \ldots, \tau_g$ are the treatment effects. This is an overspecified model. Usually, to estimate the effects we add the constraints $\sum_{i=1}^{B} \beta_i = 0$ and $\sum_{j=1}^{g} \tau_j = 0$. Sometimes several replications for each block-treatment combination are used, but for this example we will assume that there is only one measurement for each treatment in each block. Note that we have not included an interaction between block and treatment. If an interaction is suspected then we would need more than one measurement for each treatment in each block.

To test $H_0 : \tau_1 = \ldots = \tau_g = 0$ against $H_a : \tau_j \neq 0$ for at least one $j = 1, \ldots, g$, we fit the complete model, as specified by Equation (6.2), and the reduced model

$$y_{ij} = \mu + \beta_i + \varepsilon_{ij}. \tag{6.3}$$

Several methods are commonly used to test the null hypothesis of no treatment effect in data from randomized complete block designs. The traditional test is an F test that is based on the fit of the complete model, which includes the block and treatment effects, and the fit of the reduced model, which includes only the block effects. Another approach was proposed by Friedman (1937), who used a test statistic that is a function of the within-block ranks. Yet another rank test was proposed by Sen (1968). This aligned rank test was based on a test statistic that is a function of the ranks, after ranking over all blocks and treatments, of the "aligned" observations that are obtained by subtracting the block mean (or median) from all observations in that block.

Although the model formulation in Equation (6.2) appears to be similar to what would be used in a two-way analysis of variance, there is a major difference between the randomized block analysis and the two-way analysis of variance. In the randomized block designs we believe that, under the null hypothesis, we are justified in permuting observations within blocks. In our adaptive test we will permute the observations within blocks by permuting the residuals within each block and then adding these to the estimated block effect, which is the mean of the observations in the block. Hence, in the evaluation of the properties of this adaptive test we will use within-block permutation of residuals.

To describe the permutation method, it is helpful to write the model in matrix form. Let the dependent variable be arranged as

$$\mathbf{Y}_k = \begin{bmatrix} y_{11} \\ y_{12} \\ . \\ y_{1g} \\ . \\ . \\ y_{Bg} \end{bmatrix},$$

so that the first block of g rows contains the data from the first block, the second block of g rows contains the data from the second block, etc., and the vector of residuals from the reduced model ($\hat{\varepsilon}^*_R$) are arranged in the same manner. Let $\hat{\mathbf{Y}}$ be the vector of estimated values, which consists of the within-block means, and let $\mathbf{S}_{k,j}$ be a $(g \times g)$ permutation matrix for the kth permutation of the residuals in the jth block. For the kth permutation the new vector of dependent variables is $\mathbf{Y}_k = \hat{\mathbf{Y}} + \mathbf{S}_k\hat{\varepsilon}^*_R$, where

$$\mathbf{S}_k = \begin{bmatrix} \mathbf{S}_{k,1} & 0 & 0 & \dots & 0 \\ 0 & \mathbf{S}_{k,2} & 0 & \dots & 0 \\ . & . & . & . & . \\ 0 & 0 & 0 & \dots & \mathbf{S}_{k,B} \end{bmatrix}.$$

By setting up the permutation matrix in this manner, we ensure that the permutations will be performed within each block. Although the permutation method differs from our previous method, the adaptive weighting procedure is the same as that used in Section 5.1 for the general adaptive testing method.

6.2.1 The Significance Level and Power of the Adaptive RCB Test

Two simulation studies were used to estimate the empirical level of significance and power of the F test, a permutation test, and the adaptive test. The first simulation study used 8 blocks with 4 treatments per block. We generated 10,000 data sets from one of the error distributions for each of the studies. With 10,000 data sets the standard error of the empirical significance level is 0.0022, so that a test that maintains its size should fall in the range $0.05 \pm 3(0.0022) = [4.34\%, 5.66\%]$. As we expected, the permutation and adaptive tests maintained their level of significance in that range for all distributions. This was expected because, under the null hypothesis, the within-block permutations of the residuals produce within-block permutations of the dependent variables that are equally likely. However, the empirical size of the F test was 4.23% for the skewed, high kurtosis distribution. A second simulation study was performed using 8 blocks with 8 treatments per block, and it demonstrated that all three tests maintained their significance level within the range.

In the simulations for power the treatment effects were set so that the empirical power would be near 80%. For the simulations that used 4 treatments, the effect for the jth treatment was set to $\tau_j = 0.566j$, for $j = 1, \ldots, 4$ to achieve that power for the F test. The empirical power estimates are given in Figure 6.5 for the RCB model having 8 blocks and 4 treatments. These power estimates show only a modest power advantage of the adaptive test over the traditional test.

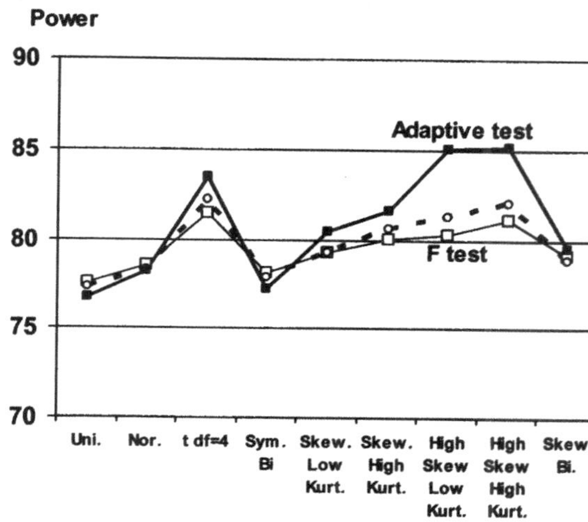

Figure 6.5 Empirical power in percent for the F test, the permutation test, and the adaptive test for equality of means in a randomized complete block design having 8 blocks and 4 treatments. The F test is indicated by hollow squares, the permutation test by hollow circles, and the adaptive test by filled squares.

For the second simulation we used 8 treatments in a RCB model having 8 blocks. We used treatment effects of $\tau_j = 0.212j$, for $j = 1, \ldots, 8$, to achieve power near 80%. The empirical power estimates, which are displayed in Figure 6.6, show that with 8 treatments the adaptive test is more powerful than the other tests for non-normal distributions. The results for 4 treatments and for 8 treatments suggest that the relative power of these tests is closely related to the number of treatments per block. This makes sense because if there are 4 treatments per block the block means will be sensitive to the presence of an outlier, so it becomes difficult to determine what observations are outliers. Hence, with 4 treatments per block it is difficult to accurately downweight outliers.

6.2.2 An Example of the Analysis of RCB Design Data

Damaser *et al.* (1963) obtained data on the skin potential (in millivolts) for 8 subjects in an experiment, each of whom were asked to feel four emotions while under hypnosis. The data are recorded in Table 6.2. We can use the RCB model to analyze

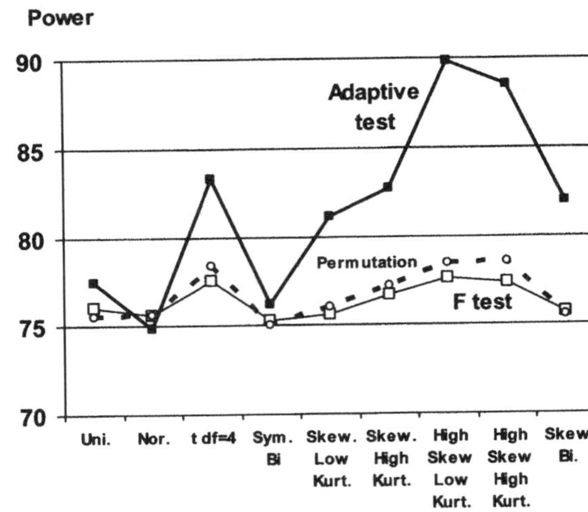

Figure 6.6 Empirical power in percent for the F test, the permutation test, and the adaptive test for equality of means in a randomized complete block design having 8 blocks and 8 treatments. The F test is indicated by hollow squares, the permutation test by hollow circles, and the adaptive test by filled squares.

this data, using the subjects as blocks. The traditional F test produced a p-value of 0.0344, while the adaptive p-value, using the SAS macro, was somewhat lower at 0.0270. Both of these p-values indicate that there is a some difference between these emotions. In this section we describe how to perform this test using R and SAS.

Table 6.2 Skin potential (in millivolts) for eight subjects under hypnosis for four requested emotions

Subject	Fear	Happiness	Depression	Calmness
1	23.13	22.74	22.52	22.64
2	57.60	53.18	53.71	53.11
3	10.49	9.72	10.80	8.29
4	23.61	19.60	21.09	21.62
5	11.89	13.80	13.67	13.30
6	54.57	47.06	39.16	36.96
7	21.01	13.55	13.66	14.78
8	20.30	23.56	16.28	14.82

6.2.2.1 Using R to Analyze RCB Data To perform a test of the null hypothesis of no treatment effect with a RCB design, we need to permute the residuals within each block. To restrict the permutations to those within each block we subset the vector of residuals and then use the `sample` function to perform the permutations. The R function `adaptblock`, which is shown below, performs the same adaptive analysis as the R function `adapt` that was described in Chapter 3, except that in `adaptblock` the residuals are randomized within the blocks. The `adaptblock` function, which is listed in Appendix D, can be downloaded from *www.math.niu.edu/~ogorman*. Note that the user must include a numeric variable `block` in the data frame and the number of distinct blocks must be specified in the function call as `nblocks`.

```
adaptblock <- function(dataset,complete,reduced,
                                r,seed,nblocks) {
tradpvalue <- trad.ftest(dataset,complete,reduced)
testdata <- read.table(dataset)
print(testdata)
attach(testdata)
statsu<- teststats(y,complete,reduced)
lambdau <- statsu[1]
e <- 0
countperm <- 0
redu <- lm(as.formula(reduced))
yhat <- predict(redu)
yresidual <- residuals(redu)
set.seed(seed)
n <- length(yresidual)
for (k in 1:r) {
  for (b in 1:nblocks[1]) {
    permres <- yresidual[block==b]
    lengthperm <- length(permres)
    if ( (lengthperm>1) )permres <- sample(permres)
    countperm[1] <- 0
    for (i in 1:n) {
      if( (block[i]==b) )  {
        countperm[1] <- countperm[1] + 1
        yresidual[i] <- permres[countperm[1]]
                            }
                  }
                            }
  ynew <- yhat + yresidual
  statsp <- teststats(ynew,complete,reduced)
  lambdap <- statsp[1]
  if( (lambdap <= lambdau) ) e  <-  e + 1
  }
p <- (e+1)/(r+1)
detach(testdata)
return(p)
}
```

Note that `permres` is a vector that contains the within-block residuals. The `sample` function is used to permute these residuals. In the `for` loop that is indexed by `i`, we place the permuted residuals for block `b` into the correct place in vector `yresidual`. After the residuals in all blocks have been permuted, they are added to `yhat` to obtain `ynew`, which equals $\mathbf{Y}_k$ for the kth permutation. We use `ynew` in the `teststats` function and compute the p-value in the usual manner.

The data set `damaser.dat` included the skin potential as the dependent variable having the variable name `y`. The emotions were recorded as the character variable `emotion`, and the `block` variable was entered as a numeric vector. The first few lines and the last line of the data set are

```
   emotion    y    block
1     f     23.13    1
2     h     22.74    1
3     d     22.52    1
4     c     22.64    1
.     .       .      .
.     .       .      .
32    c     14.82    8
```

To perform the adaptive test we use $\lambda^{2/n,*}$ as the test statistic and we count the number of times that the permutation test statistic is less than or equal to the unpermuted test statistic. We then use that count to estimate the p-value.

To invoke this function we use the following R code:

```
dataset <- c("damaser.dat")
complete <- c("y~as.factor(block) + emotion")
reduced  <- c("y~as.factor(block)")
r <- 2000
seed <-c(71492)
nblocks <- c(8)
pvalue <- adaptblock(dataset,complete,reduced,r,seed,nblocks)
```

Note that the blocking factor in this example is numeric, but we want the `lm` function to treat it as a factor. This is accomplished by specifying it with `as.factor(block)` in the complete and reduced models. This R program produced an adaptive p-value of $p = 0.028$ in less than one minute on a personal computer purchased in 2008.

6.2.2.2 Using SAS to Analyze RCB data The following SAS macro can be used to perform the adaptive test for a RCB design. Note that the blocking variable is specified as `block`. The X_R matrix includes the indicator variables for the blocking factor and the X_A matrix includes the indicator variables for the treatment factor. This is the same macro that we have used in previous chapters for the adaptive test. The only difference is that the blocking variable is specified so that the residuals are permuted within that blocking variable. If we place the `adaptall` macro in the same directory as the program, the SAS code will be

```
%include 'adaptall.sas';
data main;
  input obs treat y block;
  datalines;
1  1  23.13 1
2  2  22.74 1
3  3  22.52 1
4  4  22.64 1
.  .   .    .
.  .   .    .
32  4  14.82 8
;
%adaptall(dataset=main, y=y, xr=block, classr=block,
          xa=treat,  classa=treat,  blkvars=block,
          seed1=14920, seed2=22187, seed3=7878,
          nperm=2000);
```

The SAS code produced the following output. Note that the blocking variable is included in the output in order to document the fact that the residuals were permuted within the levels of `block`.

```
Blocking variable:  block
Dependent variables  :  y

Independent variables in reduced model   : Intercept block
      Class variables in reduced model   : block

Independent variables in complete model : Intercept block treat
      Class variables in complete model : block  treat

Adaptive test statistics for unpermuted data:

      Wilks Lambda =   0.7945
      F statistic  =   1.8107

p-value for adaptive test =   0.0270
```

6.3 ADAPTIVE TESTS FOR TWO-WAY DESIGNS

In order to reduce the number of experimental units needed to evaluate the effect of two factors, researchers often use a two-way design. In this section we present an approach to adaptive tests for data obtained from these designs. Suppose factor A has a levels, factor B has b levels, and that n_{ij} is the number of observations at the *i*th level of factor A and the *j*th level of factor B. Let μ be the overall effect, let α_i be the fixed effect of the *i*th level of factor A, and let β_j be the fixed effect of the *j*th level of factor B. The interaction of the *i*th level of factor A and the *j*th level of factor B is given by γ_{ij}. Let ε_{ijk} be the error for the *k*th observation at the *i*th level of factor A and the *j*th level of factor B. With this notation the model that includes an interaction term is

$$y_{ijk} = \mu + \alpha_i + \beta_j + \gamma_{ij} + \varepsilon_{ijk} \tag{6.4}$$

for $i = 1, \ldots, a$, $j = 1, \ldots, b$, and $k = 1, \ldots, n_{ij}$.

Typically, in the analysis of data from two-way layouts, we test the interaction effect first. If we conclude that the interaction effect is negligible, we then proceed to test one or both of the main effects. If the interaction effect is significant, we proceed to compare levels of factor A at specified levels of factor B. We will begin our analysis of two-factor designs with the adaptive test for interaction.

6.3.1 Tests for Interaction

To test for an interaction effect we compare the fit of the complete model, which contains an intercept term, to the fit of the reduced model

$$y_{ijk} = \mu + \alpha_i + \beta_j + \varepsilon_{ijk}, \tag{6.5}$$

which contains only the main effects. The traditional *F* test is based on the sum of the squared errors for the reduced model specified in Equation (6.5) and the sum of the squared errors for the complete model, which includes an interaction term, that is specified in Equation (6.4).

To perform the adaptive test we find the residuals from the reduced model, which includes only the main effects, and we use these residuals to compute the adaptive weights and the test statistics for the unpermuted data. Next, the residuals from the reduced model are permuted and then added to the values predicted by the reduced model to produce a new set of observations, which are used to compute the adaptive weights and the permutation test statistics. Unlike many of the models we have used, the weights will need to be recomputed with each permutation. Also, unlike many of the models that we have considered, the adaptive test may not maintain its level of significance because it cannot be shown that the vector of new observations are equally likely under permutations of residuals, so the resulting permutation test statistics may not be equally likely under the null hypothesis. However, we can still argue that, for large samples, the residuals should approximate the errors so the resulting permutation distributions should approximate the permutation distribution that would have been obtained if we had been able to permute the errors, rather than the residuals. So, although we expect the test to have acceptable properties, we need to carefully document the size and power of the test for interaction with finite sample sizes.

6.3.1.1 Level of Significance and Power The significance level of the F test, the permutation test, and the adaptive test were evaluated using the nominal significance level of $\alpha = 0.05$. These simulation studies used 100,000 data sets for each error distribution. In Figure 6.7 the empirical significance level is displayed for a test of interaction in a two-way layout having 2 levels for factor A and 4 levels for factor B, in a balanced design that had 10 observations assigned to each combination of factor A and factor B.

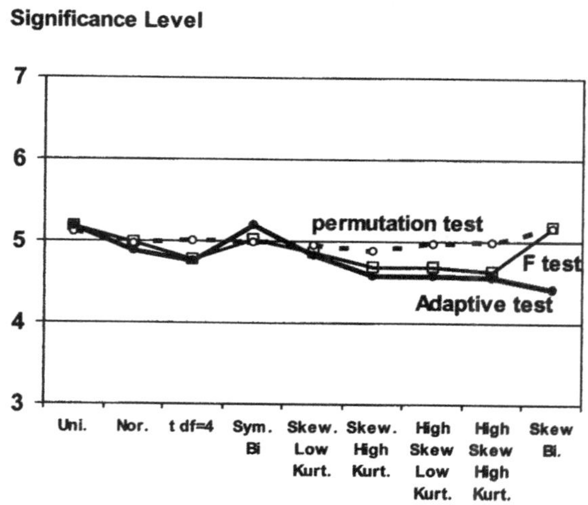

Figure 6.7 Empirical level of significance in percent for the F test, the permutation test, and the adaptive test for the interaction effect in a 2×4 two-way layout. A balanced design having 10 observations for each combination of factor A and factor B was used. The F test is indicated by hollow squares, the permutation test by hollow circles, and the adaptive test by filled circles.

As we have seen in Chapter 5 with the multiple regression model, the adaptive test and the F test do not reject often enough with skewed error distributions. However, in this simulation study we note that all of the empirical significance levels are between 4.4% and 5.2%. Also, because none of the empirical significance levels of these tests greatly exceed their nominal level of significance of $\alpha = 0.05$, we can say that they all maintain their level of significance and we can proceed to evaluate their power.

The power of these tests of interaction in a 2×4 layout is given in Figure 6.8. The interaction effect, which was a constant of 1.60 that was added to the mean at the 2nd, 3rd, and 4th levels of factor B at the second level of factor A, was adjusted to produce powers near 80%. The simulation results show that for this test of interaction in a balanced design there was only a 1% loss in power with normal errors, but there was a gain in power between 4% and 18% with non-normal errors. This large increase in power was due to the effectiveness of the normalizing transformation when $n = 80$ observations are used in a linear model that has only 5 parameters in the reduced model. In addition, it should be noted that researchers have few other tests to choose from because nonparametric test are usually not used for tests of interaction. Consequently, unless the researcher was quite confident that the errors were normally distributed, it would seem prudent to use the adaptive test to analyze data that are similar in structure to those used in this simulation.

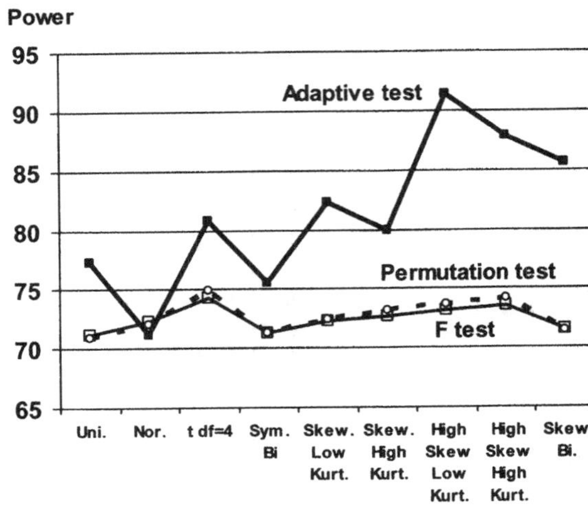

Figure 6.8 Empirical power in percent for the F test, the permutation test, and the adaptive test for the interaction effect in a 2×4 two-way layout. A balanced design having 10 observations for each combination of factor A and factor B was used. The F test is indicated by hollow squares, the permutation test by hollow circles, and the adaptive test by filled squares.

6.3.2 An Example

To show how to perform this adaptive test for interaction, consider the insect count data listed in Table 6.3 that was analyzed in Simpson, Roe, and Lewontin (1960, p. 284). The data are the number of insects found in a square-foot bottom sampler in two creeks during four months. One factor is the month and the other factor is the creek. Thus, we have a two-way layout with four levels of month and two levels of creek. This is a balanced design with 6 observations in each of the $2 \times 4 = 8$ cells. We begin our analysis with a test of interaction between month and creek.

The traditional F test produced a p-value of $p_F = 0.0467$, while the adaptive test produced a p-value of $p_{adaptive} = 0.0230$, using a SAS macro. Hence, it appears that the interaction effect is important and it would not be appropriate to perform a test of either main effect in the presence of this interaction. While both p-values are less than $\alpha = 0.05$, the adaptive test more clearly indicates that there is an interaction. The adaptive test for interaction is easy to perform using either R or SAS.

Table 6.3 Number of insects in a square foot bottom sampler

Obs.	Month	Creek	Insects	Obs.	Month	Creek	Insects
1	Dec.	Shope	7	25	June	Shope	124
2	Dec.	Shope	9	26	June	Shope	51
3	Dec.	Shope	19	27	June	Shope	63
4	Dec.	Shope	1	28	June	Shope	81
5	Dec.	Shope	18	29	June	Shope	83
6	Dec.	Shope	15	30	June	Shope	106
7	Dec.	Ball	25	31	June	Ball	20
8	Dec.	Ball	9	32	June	Ball	44
9	Dec.	Ball	16	33	June	Ball	26
10	Dec.	Ball	28	34	June	Ball	127
11	Dec.	Ball	10	35	June	Ball	38
12	Dec.	Ball	14	36	June	Ball	52
13	March	Shope	29	37	Sept.	Shope	72
14	March	Shope	37	38	Sept.	Shope	87
15	March	Shope	114	39	Sept.	Shope	100
16	March	Shope	49	40	Sept.	Shope	68
17	March	Shope	24	41	Sept.	Shope	67
18	March	Shope	64	42	Sept.	Shope	9
19	March	Ball	35	43	Sept.	Ball	40
20	March	Ball	45	44	Sept.	Ball	45
21	March	Ball	22	45	Sept.	Ball	263
22	March	Ball	29	46	Sept.	Ball	100
23	March	Ball	18	47	Sept.	Ball	129
24	March	Ball	27	48	Sept.	Ball	115

6.3.2.1 Using R Code to Perform the Test of Interaction Suppose that, on the file `creek.dat`, the variables `month` and `creek` are recorded as character variables. Because they are character variables, when this file is used to create a data frame these variables become factor, rather than numeric, variables. Consequently, when `month` and `creek` are used in the linear model function the software will automatically create the appropriate indicator variables. Note that the interaction of the two factors is specified by `month:creek` in the R code.

```
source("adapt.r")
dataset <- c("creek.dat")
complete <- c("y~month+creek+month:creek")
reduced  <- c("y~month+creek")
pvalue <- adapt(dataset,complete,reduced,r=2000,seed=7142)
```

The R output is not shown but the p-value for the adaptive test is $p_{adaptive} = 0.025$. Although this is a more complex model than those we have previously used and it requires the adaptive weights to be computed for each permutation, it executed in less than one minute on a personal computer purchased in 2008.

6.3.2.2 Using a SAS Macro to Perform the Test of Interaction The traditional F test can be obtained with the following SAS code. We will assume that the data is on the SAS data set `dcreek`. In `PROC GLM` we will identify the dependent variable as `insects` and the two factors as `month` and `creek`. Because these independent variables are character variables we will need SAS to create indicator variables for them, so these two variables are listed in the `class` statement. The interaction effect is specified by `month*creek` in PROC GLM. The SAS code to produce the nonadaptive F test is

```
proc glm data=dcreek;
  class month creek;
  model  insects = month creek month*creek;
```

If we keep in mind that the adaptive macro uses PROC GLM to fit the models, it is easy to perform the adaptive test of interaction using the SAS macro. This is done by specifying `month` and `creek` as the `xr` variables so that these effects are included in the reduced model. Both of these variables are class variables so they must also be specified in the `classr` argument. The interaction effect is what we want to test, so we will enter it as the `xa` macro variable, but there is no need to specify the variables in the interaction as class variables; SAS already recognizes them as class variable because these were previously specified as class variables. The following code invokes the SAS macro:

```
%adaptall(dataset=dcreek, y=insects,
         xr=month creek, classr=month creek, xa=month*creek,
         seed1=18973, seed2=8732, seed3=23433, nperm=2000);
```

The output of this SAS macro is given below. In addition to providing the adaptive p-value, the macro documents the random number seeds, the variables in the complete and reduced model, and the class variables.

```
Number of permutations = 2000
Starting seeds = 18973  8732  23433
SAS data set: dcreek   Number of observations = 48

Dependent variables   :  insects

Independent variables in reduced model  : Intercept month creek
      Class variables in reduced model  :  month creek

Independent variables in complete model : Intercept month creek
                                          month*creek
```

```
      Class variables in complete model : month creek

Adaptive test statistics for unpermuted data:
      Wilks Lambda =   0.8521
      F statistic  =   2.3151

p-value for adaptive test =    0.0230
```

With 2000 permutations this macro executes in a few minutes on a personal computer purchased in 2008.

6.3.3 Tests for Main Effects

If the interaction effect can be ignored, based on the interaction test and other information, we can proceed to analyze the main effects in a two-way layout. To investigate the statistical properties of the test for main effect we generated data for a 2×4 layout using the main effects but without an interaction effect. We generated 10 observations in each of the cells in the 2×4 layout.

To determine the significance level of the test, we generated 100,000 data sets for each error distribution. We used the traditional F test, the permutation test, and the adaptive test to perform a test of the null hypothesis that factor B had no effect. The empirical significance levels, which are shown in Figure 6.9, indicate that all three tests maintained their level of significance near the nominal value of $\alpha = 0.05$. The empirical significance level was below 5.2% for all error distributions. Hence, it is possible to fairly compare the power of the tests.

Next, a simulation study was performed to determine the relative power of these tests. We used 10,000 data sets with each error distribution to obtain the empirical powers displayed in Figure 6.10. In these simulations the main effect, which was adjusted to give powers near 80%, was set to 0.8 for the second, third, and fourth levels of factor B while it was set to zero for the first level. The power results show that the adaptive test is much more powerful than the other tests for all non-normal errors.

6.3.3.1 Using R Code to Perform a Test of a Main Effect On the file `creek.dat` the variables are the same as the variables that we used in the test for interaction. The following R code can be used to perform an adaptive test for the effect of creek:

```
source("adapt.r")
dataset <- c("creek.dat")
complete <- c("y~month+creek")
reduced  <- c("y~month")
pvalue <- adapt(dataset,complete,reduced,r=2000,seed=7142)
```

This code produced an adaptive p-value of 0.3023, which was much smaller than the traditional F test p-value of 0.9423.

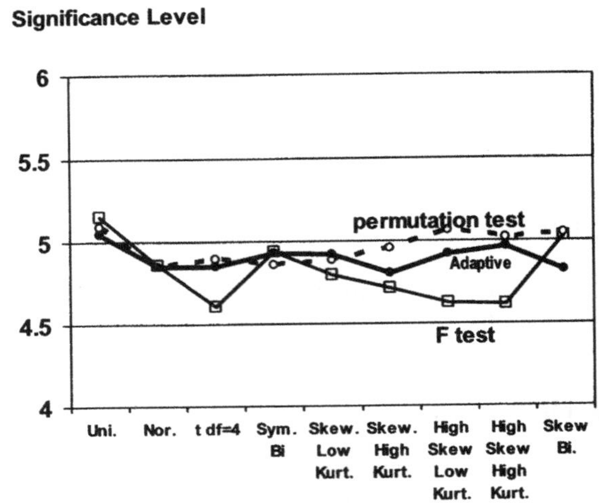

Figure 6.9 Empirical level of significance in percent for the F test, the permutation test, and the adaptive test for the main effect of factor B in a 2×4 two-way layout. A balanced design having 10 observations for each combination of factor A and factor B was used. The F test is indicated by hollow squares, the permutation test by hollow circles, and the adaptive test by filled circles.

6.3.3.2 *Using a SAS Macro to Perform a Test of a Main Effect*

The SAS macro can be used to perform a test of the main effect of creek. The data set name and the variable names are the same as we have used to perform the test for interaction. To perform the test for the main effect of creek, we specify `month` as the only effect in the reduced model. In the complete model we include `month` and `creek`, so the SAS variable `xa` contains only the `creek` variable. Both `month` and `creek` are specified as class variables in the following SAS code:

```
%adaptall(dataset=dcreek, y=insects,
        xr=month, classr=month, xa=creek, classa=creek,
        seed1=18973, seed2=28732, seed3=2343, nperm=2000);
```

The output of this macro is similar in appearance to that shown for the test of interaction. The SAS macro produced a p-value of 0.2874 for the adaptive test.

6.4 DEALING WITH UNEQUAL VARIANCES

The adaptive test was designed to give increased power with non-normal distributions. However, it is not clear if it will be reasonably powerful when the variances are not equal. In this section we investigate the properties of a few tests that might be used to deal with unequal variances. We will restrict our attention to the two-sample test

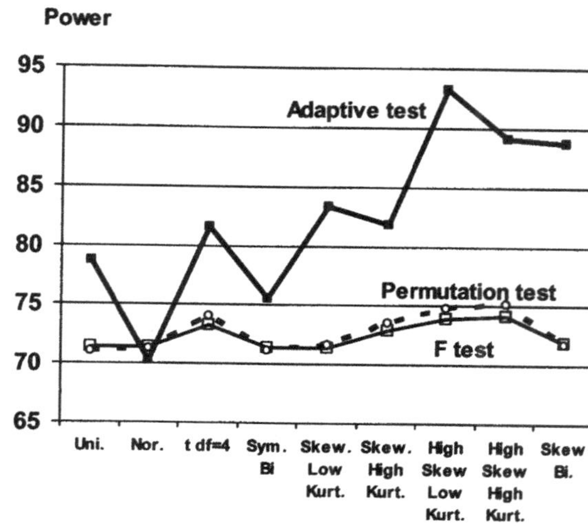

Figure 6.10 Empirical power in percent for the F test, the permutation test, and the adaptive test for the main effect of factor B in a 2×4 two-way layout. A balanced design having 10 observations for each combination of factor A and factor B was used. The F test is indicated by hollow squares, the permutation test by hollow circles, and the adaptive test by filled squares.

because many testing situations involve only two samples and because many methods have been proposed to deal with unequal variances.

In order to clarify the discussion, it will be helpful to describe the relationship between the distribution functions of the random variables in the two populations. It is possible that the distribution function of a random variable from one population is shifted relative to the distribution function of the random variable from the other population, but is otherwise identical. This will occur if the treatment effect is constant. On the other extreme, the distribution function of one of the random variables may cross the distribution function of the other. Another possibility is that the distribution function of one of the random variables is less than or equal to the distribution function of the other random variable for all values of the random variables. If this is true, we will say the random variables are stochastically ordered.

Unequal variances can arise in a variety of situations with data from two populations. Some of the more popular situations are:

1. With experimental data the treatment effect may not be constant but is a positive random variable. This will cause the treatment group to have greater variability than the control group. Because the treatment is always positive, the random variables from these two populations will be stochastically ordered.

2. With observational data we may feel that the populations have different variances under the alternative hypothesis; but if we observe a difference between

the populations, the random variables will be stochastically ordered. Under the null hypothesis we assume that the distribution functions are equal.

3. With either observational or experimental data we may want to perform a two-sample test but we are not willing to assume that the variances are equal under the null hypothesis. We may believe that the distributions have the same shape, but under the alternative hypothesis the means or variances may not be equal. In this situation we are not assuming that the random variables are stochastically ordered.

4. In some situations we may not be willing to make any assumptions about the means, variances, or distributional shapes of the random variables. In these situations we are looking for any difference in the distributions of the random variables.

Note that in the first two situations the variances are assumed to be equal under the null hypothesis, but are not equal under the alternative hypothesis. In these situations the random variables are stochastically ordered. In the last two situations the random variables may not be stochastically ordered. In this section we describe several tests that could be used in these situations. We begin by seeing how to deal with stochastically ordered random variables.

6.4.1 Tests with Stochastically Ordered Random Variables

6.4.1.1 Tests with Experimental Data We now concentrate on the statistical properties of a two-sample test when the treatment effect is variable but always positive. Let $y_{1,i}$ be the ith observation in the first (control) group and let $y_{2,i}$ be the ith observation in the second (treatment) group. Let $y_{2,i}^*$ be the unobserved value of the ith observation in the treatment group that would have been recorded if the treatment had not been used. With a constant treatment effect of Δ the observed values are $y_{2,i} = y_{2,i}^* + \Delta$, for $i = 1, \ldots, n_2$.

To model variable treatment effects we will add a random variable, instead of a constant, to $y_{2,i}^*$. In previous research we found that using treatment effects from a $U[\frac{1}{2}\Delta, \frac{3}{2}\Delta]$ distribution produced results that were very similar to those obtained by using a constant treatment effect, so we have not used those treatment effects here. To introduce greater variability we generated treatment effects from a $U[0, 2\Delta]$ distribution in the simulation study.

In our simulation study we compare the size and power of the pooled *t* test, the Welch *t* test, the Wilcoxon rank-sum test, and the adaptive test. The Welch *t* test is often recommended as a test procedure when the variances may be unequal. For this test the degrees of freedom are calculated using the Welch–Satterthwaite equation. We used $n_1 = n_2 = 18$ for the balanced design, $n_1 = 24$ and $n_2 = 12$ for the unbalanced design having the smaller number of observations in the sample that has the larger variance, and $n_1 = 12$ and $n_2 = 24$ for the design having the larger number of observations in the sample that has the larger variance.

The simulations for size showed that the three tests maintained their significance level near the nominal value. The simulations for power using $n_1 = n_2 = 18$ observations, which are shown in Figure 6.11, demonstrate that the adaptive test has greater power than the pooled *t* test and Welch's *t* test for most non-normal errors. In this simulation the shift was generated from a U[0,2] distribution. We also note that the Wilcoxon test has high power for many skewed and long-tailed distributions, but has low power for the uniform and symmetric bimodal distributions. Because the shift is a random variable having variance equal to 4/12, and since $y^*_{2,i}$ has a variance of one, the variance of $y_{2,i}$ will equal $1 + 4/12$, so the ratio of the standard deviations is 1.16. It is not surprising that the power of Welch's *t* test closely approximates the power of the pooled *t* test because with equal sample sizes the test statistics will be identical, although the degrees of freedom will be slightly different.

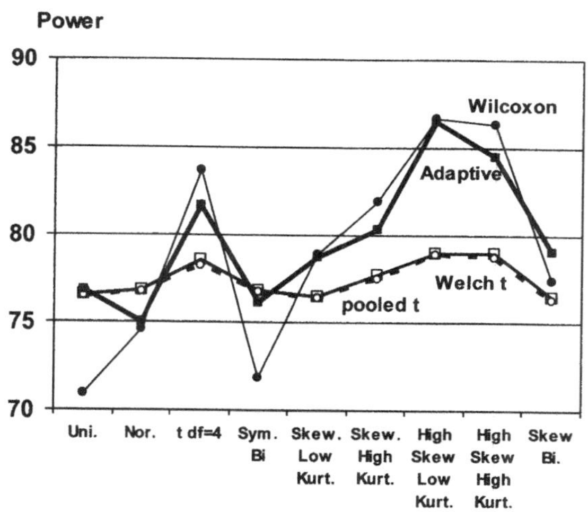

Figure 6.11 Empirical power in percent for the pooled *t* test, the Welch *t* test, and the adaptive test for equality of means for a balanced design having sample sizes of $n_1 = n_2 = 18$. The adaptive test is indicated by solid squares, the Wilcoxon test by solid circles, the pooled *t* test by hollow squares, and the Welch *t* test by hollow circles.

The results for an unbalanced design having $n_1 = 24$ and $n_2 = 12$ observations, which are given in Figure 6.12, show a different pattern than the results for the balanced design. The most striking result of this simulation is that the Welch *t* test has relatively low power for all distributions. This low power is due to the fact that the smaller group has the larger variance. The adaptive test has relatively high power for all distributions. Again, the Wilcoxon test has excellent power for most distributions that are skewed and for the long-tailed distribution.

We also used another unbalanced design having $n_1 = 12$ and $n_2 = 24$ so that the larger group would have the larger variance. These power estimates are given in Figure 6.13. In these simulations the power of the Welch *t* test exceeded that of the

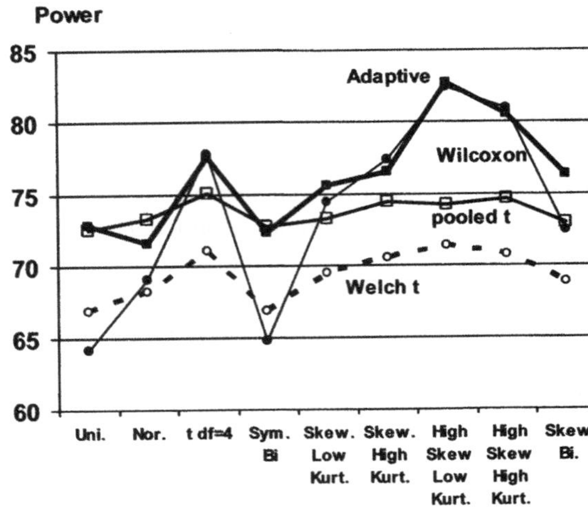

Figure 6.12 Empirical power in percent for the pooled t test, the Welch t test, and the adaptive test for equality of means for an unbalanced design having sample sizes of $n_1 = 24$ and $n_2 = 12$. In this study the smaller sample had the larger variance. The adaptive test is indicated by solid squares, the Wilcoxon test by solid circles, the pooled t test by hollow squares, and the Welch t test by hollow circles.

pooled t test for all distributions. The adaptive and Wilcoxon tests had greater power than both t tests for skewed distributions but had lower power than the Welch t test with normally distributed errors. This seems to imply that if you knew that the errors were normally distributed and you knew that the larger sample would be taken from the population with the larger variance then the Welch t test would be preferred.

These power results clearly show that the Wilcoxon and adaptive tests have reasonable power for most skewed and long-tailed symmetric distributions. They also show erratic performance for the two versions of the t test. It should be pointed out that many other tests have been proposed, some of which may have better performance than the adaptive test. Our sole objective in this section has been to show that the adaptive test has reasonable power when used to analyze experimental data when the treatment effect is quite variable.

6.4.1.2 Tests with Observational Data With observational data the researcher may not be willing to assume that the variances are equal, but may be confident that any difference between the samples would arise from random variables that are stochastically ordered. For example, suppose an epidemiologist wanted to compare the body mass index (BMI) of males in one region to the BMI of males in another region. The epidemiologist may suspect that both distributions are skewed but may not have a clear idea of the relationship between these populations. However, she may believe that if there is a difference between the populations the BMI in one region

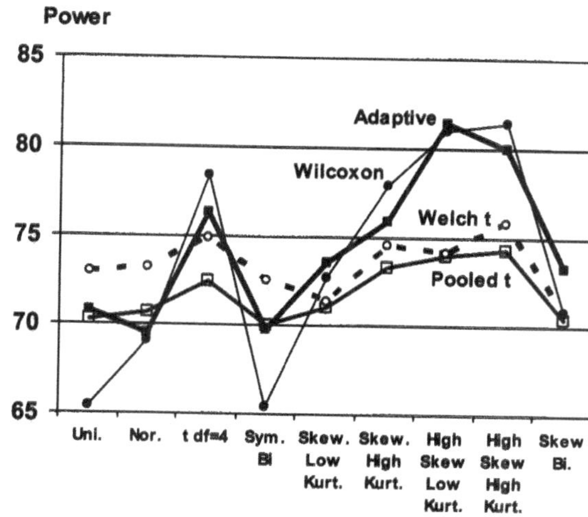

Figure 6.13 Empirical power in percent for the pooled t test, the Welch t test, and the adaptive test for equality of means for an unbalanced design having sample sizes of $n_1 = 12$ and $n_2 = 24$. In this study the larger sample had the larger variance. The adaptive test is indicated by solid squares, the Wilcoxon test by solid circles, the pooled t test by hollow squares, and the Welch t test by hollow circles.

will tend to be larger than the BMI in another region. Hence, it may be reasonable to assume that the random variables are stochastically ordered.

For the results given in Figures 6.11, 6.12, and 6.13, the ratio of the standard deviations is 1.16. Hence, if the random variables are stochastically ordered and the ratio of the standard deviations does not exceed 1.2, it seems reasonable to rely on the results given above for the variable shift model. However, if there are large differences in variability, the researcher might want to avoid a test for location. In this situation it may be more valuable to describe the differences in variability, in addition to the differences in location.

6.4.2 Tests when the Random Variables Are Not Stochastically Ordered

6.4.2.1 Tests for Location with Unequal Variances A variety of procedures have been proposed for testing the null hypothesis that the distributions are equal against the alternative hypothesis that either the means are different or the variances are different, or both are different. One approach is to use a version of the Wilcoxon test that has been modified so that it will maintain its level of significance when there are large differences between the variances in the populations. Fligner and Policello (1981) proposed a modification of the Wilcoxon test and showed that it maintains its significance level for a variety of error distributions where there are differences in variability. Another two-sample test was proposed by Brunner and Munzel (2000) for use with unequal variances.

6.4.2.2 *Tests for Location and Scale* One nonparametric test of location and scale was proposed by Lepage (1971). It uses a test statistic that combines information from a Wilcoxon test, which is used to detect changes in location, and an Ansari–Bradley test, which is used to detect differences in variability. Lepage showed that the asymptotic distribution of the test statistic is chi-squared with 2 degrees of freedom. This test will be described in Chapter 11 along with other rank tests. Another test for location and scale, which does not require any special software, was proposed by O'Brien (1988).

An adaptive test of location and scale was proposed by Büning and Thadewald (2000). This test uses a measure of tailweight as the selection statistic. It also will be described in Chapter 11 with the rank tests.

6.4.2.3 *Tests for General Alternatives* In some circumstances it may not be a good idea to make an assumption about the relationship between the two distribution functions. For example, in an experiment designed to evaluate the duration of sleep of subjects that were given an experimental drug, the researcher may believe that many subjects may get much more sleep on the medication while many other subjects may get much less sleep on the medication. In this situation the researcher may be unwilling to assume that the random variables are stochastically ordered. The objective may be to identify any difference in the distributions of these random variables, so the researcher may want a test for a general alternative. The Kolmogorov–Smirnov test or the Cramer–von Mises test are sometimes used to detect any differences between the distributions. (See, for example, Randles and Wolfe (1979, Chapter 11).) An adaptive test for the general alternative has been published by Büning (2002). In this adaptive test either the Kolmogorov–Smirnov test or one of the version of the Cramer–von Mises test is performed depending on selection statistics that are measures of skewness and tailweight.

6.5 EXTENSIONS TO MORE COMPLEX DESIGNS

Although the adaptive test can be used to test any subset of coefficients in a linear model that contains only fixed effects, it is not possible to give examples, with R and SAS code, for every experimental model. However, in this section we will discuss a few of the more common designs and describe how the adaptive test can be performed. Because the adaptive test can be used to test any subset of coefficients in a linear model, it is generally only necessary to identify which variables should be included in the reduced and complete models.

6.5.1 Analysis of Covariance

One common research design is the one-way layout with one covariate. For example, a researcher may want to compare the achievement of second grade students at three schools, but the researcher may want to use the age of the student as a covariate. In this model we use age in the reduced model and use age and school in the complete model. The complete model is

$$y_i = \beta_0 + \beta_1 x_{i1} + \beta_2 x_{i2} + \beta_3 x_{i3} + \varepsilon_i,$$

for $i = 1, \ldots, n$, where x_{i1} is the age of the *i*th student, and x_{i2} and x_{i3} are indicator variables for the second and third schools, respectively.

In R we would perform this test by first establishing a data file that could be read as a data frame called `achievement.dat` that has the achievement score recorded as the `y` variable and the age recorded as the `age` variable. Because we want R to set up indicator variables for the schools we will use a character variable for `school` so that it will be a factor vector object. The R code for the test of equality of school achievement, after controlling for age, is

```
source("adapt.r")
dataset <- c("achievement.dat")
complete <- c("y~age + school")
reduced  <- c("y~age")
pvalue <- adapt(dataset,complete,reduced,r=2000,seed=7293)
```

In SAS we can easily invoke the SAS macro `adaptall` after we create a SAS data set. Suppose the SAS data set `achievement` has been created. If it contains the dependent variable (`scores`), the covariate (`age`), and the school name (`school`), the following statement will invoke the macro:

```
%adaptall(dataset=achievement, y=scores,
         xr=age, xa=school, classa=school,
         seed1=22172, seed2=14572, seed3=7872);
```

Note that we have not included `nperm` in the macro call, so the macro will use the default of 2000 permutations. If there are several covariates they can be included in the `xr` list. The adaptall macro can be used for any adaptive test by including the appropriate variables in the `xr` and `xa` lists.

6.5.2 Multifactorial Designs

The basic approach that we used in the analysis of two-way designs can be extended to multifactorial designs. To illustrate this approach, suppose that 72 adults were enrolled in a weight loss program that was designed to evaluate the effectiveness of diet, exercise, and medication on weight loss over a two-month period. Further, suppose that the experimenter used a complete factorial design with 3 levels of diet (usual, low fat, very low fat), 3 levels of exercise (none, 30 minutes per day, 60 minutes per day), and 2 levels of medication (placebo, appetite suppressant). The outcome measure is the weight change over the two-month period. This is a 3-factor design with $3 \times 3 \times 2 = 18$ combinations of the three factors. Suppose that four subjects were assigned to each of the treatment combinations so that $18 \times 4 = 72$ weight change values were used.

The analysis of the data would proceed by testing the higher-order interactions using the adaptive test. Suppose we want to use 5000 permutations of the residuals and that the data is on a dataset called `weightloss`. In R there are many ways of specifying main effects and interaction effects. One way to use R to test the null hypothesis of no three-factor interaction is

```
source("adapt.r")
dataset <- c("weightloss.dat")
complete <- c("y~diet + exercise + medication
    + diet:exercise + diet:medication + exercise:medication
    + diet:exercise:medication")
reduced  <- c("y~diet + exercise + medication
    + diet:exercise + diet:medication + exercise:medication")
pvalue <- adapt(dataset,complete,reduced,r=5000,seed=5157)
```

If the interaction is not significant, then we might proceed to test a first-order interaction. To test the interaction of diet and medication we would use the following code:

```
source("adapt.r")
dataset <- c("weightloss.dat")
complete <- c("y~diet + exercise + medication
    + diet:exercise + diet:medication + exercise:medication")
reduced  <- c("y~diet + exercise + medication
    + diet:exercise + exercise:medication")
pvalue <- adapt(dataset,complete,reduced,r=5000,seed=5157)
```

Another approach to the analysis of data from multifactorial designs is to test for all interactions of a certain order. To make the models easier to specify, we can use the model syntax options described in Chambers and Hastie (1992, p. 29). We can simplify the expressions by using the formula operator "`*`" to include all interactions. For example, we might want to test the null hypothesis of no interaction effects by using the following code:

```
source("adapt.r")
dataset <- c("weightloss.dat")
complete <- c("y~diet*exercise*medication")
reduced  <- c("y~diet + exercise + medication")
pvalue <- adapt(dataset,complete,reduced,r=5000,seed=5157)
```

It is also easy to analyze multifactorial data using the SAS macro. Suppose we have already created a SAS data set `weightloss` that has a dependent variable `loss` along with factor variables of `diet`, `exercise`, and `medication` on 72 subjects. The SAS macro can be invoked to perform a test of no three factor interaction with this code:

```
%adaptall(dataset=weightloss, y=loss,
      xr=diet exercise medication
```

```
diet*exercise diet*medication exercise*medication,
classr=diet exercise medication,
xa=diet*exercise*medication,
seed1=15732, seed2=22955, seed3=5732, nperm=5000);
```

Note that the model syntax for SAS differs from that of R in several respects. In SAS the class variables must be explicitly stated, whereas in R the `lm` function recognizes factor objects and sets up the appropriate indicator variables. Also, "*" is used to indicate an interaction in SAS but is used to indicate all interactions in R.

6.5.3 Other Designs

Many complex designs can be handled by the SAS macro and the R function for models that have fixed effects. The user only needs to know how to specify the appropriate complete and reduced models for the null hypothesis of interest. The properties of most adaptive tests should follow the properties of the tests that we have described if the number of regression coefficients is not too large. Although we have seen high relative power for the adaptive test in a wide variety of models, we have made no attempt to document the properties of the adaptive test in linear models that have more than 12 regression coefficients in the full model.

One important test that we have not described yet is the adaptive test for paired data. It will be shown in the next chapter that the adaptive test that we have described for blocked data could be used with paired data, but we will develop some computational simplifications that will make it a practical test. Also, we will compare this adaptive test to some adaptive tests for paired data that have been proposed.

We have postponed the discussion of the tests used in multicenter clinical trials and the tests that are used in crossover trials. These topics will be addressed in Chapter 8.

In addition, the reader may have noticed that we have not yet described an adaptive test that could be used to analyze repeated measures data. This important topic will be deferred to Chapter 10, after we have developed an adaptive test for multivariate regression models. In that chapter we will compare our approach to the mixed model approach.

EXERCISES

6.1 Suppose a researcher used 32 rats in an experiment to compare four drugs (A, B, C, and D) with 8 animals assigned to each drug group. In addition, suppose drug Group B contains an extreme outlier, which is given a weight of 0.12, and the other observations had weights in the interval $[0.9, 1.1]$.

a) Suppose we want to test $H_0 : \mu_A = \mu_B = \mu_C = \mu_D$ against the alternative that at least two of the means are different. When the residuals are permuted, if the residual for the outlier in Group B is permuted to an observation in Group C, will that observations be given that weight? Why or why not?

b) If the researcher wanted to base the comparison between Group A and Group C only on the observations in Group A and Group C, what statistical analysis could be performed to accomplish that objective?

c) If, to avoid the outlier in Group B, the researcher used a rank transformation over all of the observations in all 4 groups, would the comparison between Group A and Group C be influence by the observations in Group B?

6.2 Use the power estimates given in the text for the one-way layout, with 4 groups, to compare the Kruskal–Wallis test to the F test.

a) If $n_1 = n_2 = n_3 = n_4 = 8$, for which error distributions does the power of the Kruskal–Wallis test exceed the power of the F test by more than 5 percent?

b) If $n_1 = n_2 = n_3 = n_4 = 12$, for which error distributions does the power of the Kruskal–Wallis test exceed the power of the F test by more than 5 percent?

6.3 Use the power estimates given in the text for the one-way layout, with 4 groups, to compare the adaptive test to the F test.

a) If $n_1 = n_2 = n_3 = n_4 = 12$, for which error distributions does the power of the adaptive test exceed the power of the F test by more than 5 percent?

b) If $n_1 = n_2 = n_3 = n_4 = 12$, for which error distributions does the power of the F test exceed the power of the adaptive test by more than 5 percent?

6.4 Some researchers use logarithmic transformations to reduce the skewness in data. Suppose a researcher used a logarithmic transformation of the dry biomass in the stomach contents of the eastern horned lizards that are given in Table 6.1. For the following questions you may assume that the researcher is performing a traditional F test.

a) What problems would be encountered in taking the logarithmic transformation?

b) Suppose one researcher added 0.5 to each count in the table before performing the logarithmic transformation, while another researcher added 10,000 to each count before performing the transformation. Do you think they would obtain the same test statistic?

6.5 A registered dietitian was interested in determining the effectiveness of diet and exercise on weight loss with overweight adults. He randomly assigned 24 adults to a calorie restricted diet and 24 adults to a diet that did not have any calorie restriction. In each of these diet groups, 12 subjects were asked to perform 30 minutes of additional walking per day and 12 were asked to perform no additional walking. The weight loss (in pounds per day) over the 8-week period were recorded as follows:

Diet	Walking	Weight loss
No	No	0.04, 0.11, 0.08, 0.01, 0.02, 0.04
		-0.03, 0.05, 0.02, 0.00, 0.02, 0.03
No	Yes	0.11, 0.16, 0.09, 0.14, 0.17, 0.13
		0.13, 0.12, 0.11, 0.08, 0.52, 0.15
Yes	No	0.08, 0.12, 0.18, 0.14, 0.07, 0.09
		0.08, 0.11, 0.14, 0.06, 0.08, 0.14
Yes	Yes	0.22, 0.18, 0.28, 0.19, 0.17, 0.23
		0.25, 0.26, 0.22, 0.27, 0.21, 0.18

a) Using R or SAS, perform a test of interaction using the traditional test and the adaptive test. Note any difference between the results.

b) Were there any outliers? If so, how were these weighted in the adaptive test?

c) Is it reasonable to perform a test of the main effect of exercise?

d) If it is reasonable to do so, perform a test of the effect of exercise. Use the traditional and adaptive tests and note any differences.

6.6 Golfers use "Drivers" to hit golf balls long distances. As part of a larger study to compare several brands of drivers, a consumer magazine attempted to determine if some of these drivers produced greater distances. Ten experienced golfers volunteered to participate in a study and each golfer used 4 brands of drivers. The yards driven by each driver were recorded as follows:

	Driver			
Golfer	A	B	C	D
1	223	218	207	231
2	227	208	219	227
3	205	185	193	202
4	226	162	218	215
5	283	235	223	244
6	232	216	112	235
7	205	195	202	217
8	205	192	214	211
9	221	205	215	206
10	209	201	193	216

a) Using R or SAS, are there any differences between these drivers? Use the traditional test and the adaptive test.

b) Are there any outliers? If so, how are these weighted?

c) Is is appropriate to permute the residuals for each golfer? Why or why not?

d) Describe one possible permutation for golfer 3.

e) Give the shuffling matrix for that permutation for golfer 3.

f) Describe, using matrix notation, the shuffling matrix. Give the overall dimensions of the shuffling matrix and the dimensions of the submatrices.

6.7 Suppose two researchers obtain data from a randomized complete block design with 10 blocks with each block having 8 treatments. We will assume that both researchers will use a permutation test to analyze the data. One researcher treats the blocking factor as a fixed effect, so the analysis is equivalent to that of a two-way layout. The other researcher treats the block as a blocking factor and analyzes the data with the methods that are appropriate for randomized complete block designs.

a) Are the unpermuted test statistics identical?

b) Will the residuals from the reduced model be the same for these two researchers?

c) Will the permutation distributions be identical? Why or why not?

6.8 Golf balls have small indentations, call "dimples", that are used to improve the performance of the ball. Many American golf balls have 336 dimples and many British golf balls have 330 dimples. In an effort to determine if there is a difference between the balls that have 336 dimples and those that have 330 dimples, a researcher enlisted 8 golfers to drive both kinds of golf balls. The yards driven by these golfers for these golf balls are:

	Number of Dimples	
Golfer	British (330)	American (336)
1	215	227
2	241	232
3	234	218
4	196	208
5	210	243
6	237	233
7	217	242
8	243	228

a) For golfer 3, what is the residual for the British golf ball? No computer program should be necessary for this calculation.

b) For golfer 3, what is the residual for the American golf ball?

c) For the *i*th golfer, what is the relationship between the residual for the British golf ball and the residual for the American golf ball?

d) Will the distribution of the residuals be skewed or symmetric? Explain?

6.9 Suppose a researcher used three factors (A, B, and C) in a complete balanced factorial design. She was confident that the only interaction between these factors would occur between factors B and C, so she wanted to test the null hypothesis that there is no interaction between factors B and C against the alternative that there is an interaction.

a) State the complete model.

b) State the reduced model.

c) Give either the R code or the SAS code to compute $p_{adaptive}$ with 5000 permutations of the residuals.

6.10 Several varieties of barley were evaluated to determine which had the greatest yield. Because the fertility of the soil varies dramatically from place to place, the agronomist planted each of 7 varieties of barley within each of the plots of land. The number of bushels per acre are

Plot	A	B	C	D	E	F	G
1	47	53	51	47	52	45	42
2	67	68	71	64	63	69	70
3	52	59	61	63	64	58	59
4	51	61	47	55	31	49	48
5	63	68	57	59	62	61	64

a) Using R or SAS, test the null hypothesis that the varieties have equal yields. Use the traditional test and the adaptive test.

b) Identify any outliers and comment on the weighting.

c) In this experiment is it appropriate to permute the residuals within each plot? Why or why not?

6.11 An educator wanted to compare student achievement at the 6th grade level in three schools. The researcher wanted to include a covariate (age) into the analysis to more accurately compare the schools.

a) Give the R or SAS code that would perform the adaptive test for school effect, after controlling for the age of the student.

b) Suppose the researcher wanted to use all of the data to compare school 2 to school 3. Give the complete and reduced model for the appropriate test. Describe how this test would be performed with R or SAS.

CHAPTER 7

THE ADAPTIVE ANALYSIS OF PAIRED DATA

7.1 INTRODUCTION

One of the most common tests of significance is the test for paired data. This test can be used to analyze data collected on individuals before and after a treatment, or data from experiments where the subjects are paired. For example, Mitchell, Burroughs, and Beadles (1936) collected data on ten pairs of rats, with one rat from each pair assigned to a raw peanut diet and the other member of the pair to a roasted peanut diet. Usually, in the analysis of paired data, the experimenter will compute the differences between the pairs and the analysis will be based solely on these differences. These differences are assumed to have a symmetric distribution. Under the null hypothesis the distribution will be symmetric about zero.

To demonstrate that the distribution of differences will be symmetric about zero, suppose that we are trying to evaluate the effectiveness of a medication to lower diastolic blood pressure and we observe a blood pressure of 85 (mm Hg) before the drug is taken and 80 after the drug is taken. Under the null hypothesis, we are just as likely to obtain 80 as the first reading and 85 as the second reading. If we define the difference as the posttreatment blood pressure minus the pretreatment blood pressure we are just as likely to record a +5 as a -5. Because the same argument

Adaptive Tests of Significance Using Permutations of Residuals with R and SAS®.
By Thomas W. O'Gorman

can be made for all measurements made under the null hypothesis, it can be seen that the distribution of the differences is symmetric about zero. For a more detailed explanation see Lehmann (1975, p. 157).

Let d_i be the difference between the measurements for the ith pair and let B equal the number of pairs. The usual t test statistic is

$$t = \frac{\bar{d}}{(s/\sqrt{n})}$$

where $\bar{d}$ is the average of the differences and s^2 is the usual unbiased estimator of the variance of the differences. If the differences are normally distributed then, under the null hypothesis, the test statistic t will be distributed as a t distribution with $B - 1$ degrees of freedom. This test is popular because it is the most powerful test if the differences are normally distributed.

However, if the differences are not normally distributed, the t test may not be more powerful than the other tests. As we shall see, the Wilcoxon signed-rank (SR) test is more powerful than the t test for many distributions. It is based on the sum of the ranks of the absolute values of the differences, over all positive differences. For small samples it is possible to use a table of the distribution of the sum to determine the exact p-value. If there are a large number of pairs, a normal approximation is usually used to approximate the p-value.

The Wilcoxon SR test is only one of many rank tests that have been proposed. For rank-based tests we let R_i^+ be the rank of $|d_i|$ among $|d_1|, \ldots, |d_B|$ and we let $\Psi_i = 1$ if $d_i \geq 0$ and $\Psi_i = 0$ if $d_i < 0$. With this notation the SR test is based on $\sum \Psi_i R_i^+$, where the sum is over the B differences.

In the SR test we sum the ranks, but it is easy to create other rank-based tests that use the sum of a function of the ranks. If we define a general function of R_i^+ as $a(R_i^+)$, the sum over the positive differences is $\sum \Psi_i a(R_i^+)$. We perform the test by first calculating the test statistic

$$Z = \frac{\sum_{i=1}^{B} \Psi_i a(R_i^+) - \frac{1}{2} \sum_{i=1}^{B} a(i)}{\sqrt{\frac{1}{4} \sum_{i=1}^{B} a^2(i)}}. \tag{7.1}$$

Because Randles and Wolfe (1979, Chapter 10) show that Z has a limiting standard normal distribution, we usually use a normal distribution to approximate the p-value. If we set $a(i) = i$, for $i = 1, \ldots, B$, we obtain the SR test. If we set $a(i) = 1$, for $i = 1, \ldots, B$, we obtain the sign test.

In addition to the t test and the Wilcoxon SR test, many other tests have been proposed. Randles and Hogg (1973) proposed an adaptive test for paired data that uses a selection statistic to select a set of rank scores. More recently, Friedlin, Miao, and Gastwirth (2003) proposed an adaptive test that uses the p-value produced by the Shapiro–Wilks test of normality to select one of several sets of rank scores. O'Gorman (2008b) compared the performance of these adaptive tests to the t test and the SR test and found that the test proposed by Friedlin *et al.* (2003) had relatively high power for small and moderate sample sizes, but not for $B = 200$ pairs. When there are a large number of pairs the test may often select a set of scores for a non-normal distribution even when there is little departure from normality. This selection method results in reduced power when B is large.

To address this problem, Miao and Gastwirth (2009) proposed an adaptive rank test. This test uses several of the same score functions as the test by Friedlin *et al.* (2003), but it uses a measure of tail-heaviness to select an appropriate set of rank scores. Miao and Gastwirth (2009) showed that their test has higher power than the test of Friedlin *et al.* (2003) for $B = 200$ pairs. The test of Miao and Gastwirth (2009) will be described in the next section and will be compared to other tests later in this chapter.

An adaptively weighted least squares approach was used by O'Gorman (2004, Chapter 5) to develop a test for paired data. That test used a weighting procedure that was specifically designed for the analysis of paired data. That is, the adaptive weighting procedure that was used for linear regression models was not used in the development of the test for paired data. We will take a different approach. In this chapter we will describe an adaptive test that uses the adaptive weighting and permutation methods that we developed in Chapter 2 through Chapter 5.

In addition to the adaptive tests for paired data, several other tests have been proposed for testing the null hypothesis that the median equals zero when the distribution is not symmetric. Lemmer (1993) suggested two adaptive tests for the median. Baklizi (2005) proposed a continuously adaptive test for the median when the symmetry of the distribution is in doubt. It maintains it size and is relatively powerful. Another adaptive location test was proposed by Bandyopadhyay and Dutta (2007), which used a measure of symmetry to select either the sign test or the SR test. We will not consider these tests further because we will restrict our attention to the analysis of paired data, which should be symmetrically distributed.

7.2 THE ADAPTIVE TEST OF MIAO AND GASTWIRTH

7.2.1 A Measure of Tail-Heaviness

Miao and Gastwirth (2009) published an adaptive rank-based test that used a measure of tail-heaviness to determine the set of rank scores to be used to perform the test. Under the null hypothesis the differences are symmetric about zero, so one measure of variability is $\tilde{s} = [(\sum_{i=1}^{B} d_i^2)/n]^{1/2}$. Note that $\tilde{s}$ will be sensitive to outliers. Another measure of variability, which is not sensitive to the presence of a few outliers, is

$$\widetilde{M} = \frac{median(|d_1|,\ldots,|d_B|)}{\Phi^{-1}(.75)} = \frac{median(|d_1|,\ldots,|d_B|)}{0.6745}.$$

Miao and Gastwirth (2009) proposed that the tail heaviness measure $TH = \tilde{s}/\widetilde{M}$ be used as a measure of tail-heaviness. If the differences are normally distributed about zero with a standard deviation of σ, then $\tilde{s}$ and $\widetilde{M}$ should approximate σ, so TH should be close to one. If the distribution has heavy tails, then $\tilde{s}$ will be large compared to $\widetilde{M}$, which will produce a ratio that will greatly exceed one. This test, which will be called the MG test, uses TH to select a set of rank scores.

7.2.2 Rank Score Functions

In constructing the MG test, the next step is to use TH to select a set of scores that give high power for distributions with that amount of tail-heaviness. In the MG test, if the data appear to come from a light-tailed distribution the MG test selects the normal scores

$$a_1(R_i^+) = \Phi^{-1}\left[\frac{1}{2} + \frac{1}{2}\left(\frac{R_i^+}{B+1}\right)\right], \tag{7.2}$$

for $i = 1, \ldots, B$. If the distribution has moderate tail-heaviness, such as a normal or logistic distribution, then the Wilcoxon scores

$$a_2(R_i^+) = \frac{R_i^+}{B+1}, \tag{7.3}$$

for $i = 1, \ldots, B$, will be selected. If the distribution has heavy tails then the t_2 scores

$$a_3(R_i^+) = \frac{3\sqrt{2}}{2}\left(\frac{R_i^+}{B+1}\right)\sqrt{1-\left(\frac{R_i^+}{B+1}\right)^2}, \tag{7.4}$$

for $i = 1, \ldots, B$, will be selected. If the distribution has very heavy tails, then the Cauchy scores

$$a_4(R_i^+) = \frac{2\tan\left[\frac{1}{2}\pi\left(\frac{R_i^+}{B+1}\right)\right]}{1+\tan^2\left[\frac{1}{2}\pi\left(\frac{R_i^+}{B+1}\right)\right]}, \tag{7.5}$$

for $i = 1, \ldots, B$, will be selected.

In Figure 7.1 we have graphed these four functions. Note that these are functions of $u = R_i^+/(B+1)$, so that u is in the interval $[0,1]$. The plots in Figure 7.1 show that the Wilcoxon signed-rank (SR) scores and the normal scores are strictly increasing functions of the ranks, whereas the scores for the long-tailed distributions are decreasing with R_i^+ for large values of the signed ranks.

7.2.3 Selecting the Score Function

The MG test uses the tail heaviness measure measure TH to select the score function. The selection strategy is:

- If $TM < 1.02$, use the normal scores $a_1(R_i^+)$ in Equation (7.2).
- If $1.02 \leq TM < 1.2$, use the Wilcoxon scores $a_2(R_i^+)$ in Equation (7.3).
- If $1.2 \leq TM < 2.7$, use the t_2 scores $a_3(R_i^+)$ in Equation (7.4).
- If $TM \geq 2.7$, use the Cauchy scores $a_4(R_i^+)$ in Equation (7.5).

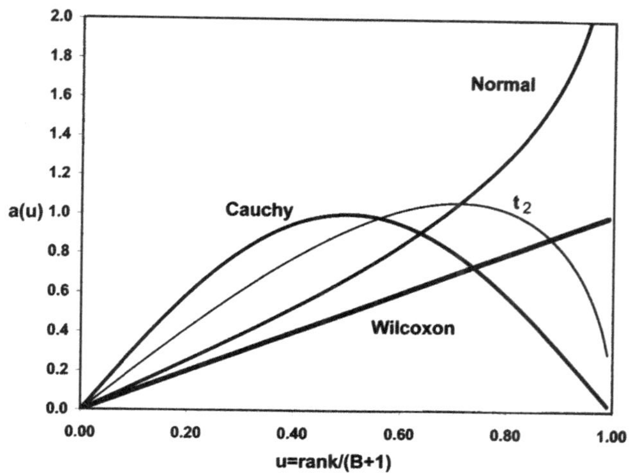

Figure 7.1 Rank scores used in the Miao–Gastwirth adaptive test.

After the scores are selected, Z is computed using Equation (7.1) and the p-value (p_{MG}) is obtained from the normal distribution. If $p_{MG} \leq \alpha$ we reject the null hypothesis. See Miao and Gastwirth (2009) for a justification of the cutpoints used in the test selection.

The MG test maintains its size because the signed ranks, which are used in the test statistic, are independent of the absolute order statistics, which are used in the selection statistic. A proof of independence is given in Randles and Wolfe (1979, Chapters 8 and 11). The power of this test will be compared to the power of other tests in Section 7.5.

7.3 AN ADAPTIVE WEIGHTED LEAST SQUARES TEST

7.3.1 The Unweighted Test Statistic

To perform an adaptive test for paired data, we usually will use the differences to perform the test. However, it will be useful to consider how a linear model could be used when both measurements are used for each pair. By writing the model for the differences as a traditional linear model we can use the weighting and permutation methods that we have already developed. Suppose we have B pairs and within each pair there is a first member and a second member. For example, in a pretest-posttest design the first member is the pretest score and the second member is the posttest score. Let Y_{ij} be the observed value for the jth member of the ith pair. Let μ_i be the effect of pair i and let m_j be the member effect.

The linear model can be written as

$$Y_{ij} = \mu_i + m_j + \varepsilon_{ij}, \tag{7.6}$$

for $i = 1, \ldots, B$ and $j = 1, 2$. In this model ε_{ij} is the error for the jth member of the ith pair. We will assume that these errors are independent and identically distributed.

It will be convenient to parameterize the model using $m_1 = -\delta/2$ and $m_2 = \delta/2$. With this notation the first members of the pairs are

$$Y_{i1} = \mu_i - \frac{1}{2}\delta + \varepsilon_{i1}, \tag{7.7}$$

for $i = 1, \ldots, B$, and the second members of the pairs are

$$Y_{i2} = \mu_i + \frac{1}{2}\delta + \varepsilon_{i2}, \tag{7.8}$$

for $i = 1, \ldots, B$. With this notation the differences between the members can be written as

$$d_i = Y_{i2} - Y_{i1} = (\mu_i + \tfrac{1}{2}\delta + \varepsilon_{i2}) - (\mu_i - \tfrac{1}{2}\delta + \varepsilon_{i1})$$

$$= \delta + (\varepsilon_{i2} - \varepsilon_{i1})$$

for $i = 1, \ldots, B$. In order to perform the traditional nonadaptive test, we need to compute SSE_R and SSE_C. In the linear model shown in Equations (7.7) and (7.8) we are interested in testing $H_0 : \delta = 0$ versus $H_a : \delta \neq 0$. Under the null hypothesis, without any adaptive weighting, we can write the SSE_R as

$$SSE_R = \sum_{i=1}^{B} \sum_{j=1}^{2} (Y_{ij} - \hat{Y}_{ij})^2.$$

To write this as a function of the differences, we note that, under the null hypothesis,

$$\hat{Y}_{ij} = \bar{Y}_i = (Y_{i1} + Y_{i2})/2,$$

for $i = 1, \ldots, B$ and $j = 1, 2$. We also note that $Y_{i1} - \bar{Y}_i = Y_{i1} - (Y_{i1} + Y_{i2})/2 = -d_i/2$ and $Y_{i2} - \bar{Y}_i = d_i/2$, for $i = 1, \ldots, B$. Hence, we can write SSE_R as

$$SSE_R = \sum_{i=1}^{B} \sum_{j=1}^{2} (Y_{ij} - \bar{Y}_i)^2 = \sum_{i=1}^{B} \left[\left(-\frac{d_i}{2}\right)^2 + \left(\frac{d_i}{2}\right)^2 \right] = \sum_{i=1}^{B} \frac{d_i^2}{2},$$

which can be calculated directly from the differences.

To compute SSE under the alternative hypothesis, we need to find estimates for Y_{ij}. Because the design is balanced, the estimate of μ_i is $\bar{Y}_i = (Y_{i1} + Y_{i2})/2$, for $i = 1, \ldots, B$, and the estimate of δ is $\bar{d} = (\sum_{i=1}^{B} d_i)/B$. Thus, for the first member of the ith pair, we have $\hat{Y}_{i1} = \bar{Y}_i - \bar{d}/2$, and for the second member of the ith pair, we have $\hat{Y}_{i2} = \bar{Y}_i + \bar{d}/2$.

Thus, under the alternative hypothesis,

$$\begin{aligned} SSE_C &= \sum_{i=1}^{B}\sum_{j=1}^{2}(Y_{ij} - \hat{Y}_{ij})^2 \\ &= \sum_{i=1}^{B}\left\{[Y_{i1} - (\bar{Y}_i - \bar{d}/2)]^2 + [Y_{i2} - (\bar{Y}_i + \bar{d}/2)]^2\right\} \\ &= \sum_{i=1}^{B}\left[\left(\frac{-d_i}{2} + \frac{\bar{d}}{2}\right)^2 + \left(\frac{d_i}{2} - \frac{\bar{d}}{2}\right)^2\right] \\ &= \sum_{i=1}^{B}[(d_i - \bar{d})^2]/2. \end{aligned}$$

Consequently, it is easy to compute SSE_R, SSE_C, and all of the usual test statistics using only the differences. This is an important result because, for studies having a large number of pairs, we do not need to deal with large matrices that contain indicator variables for the pairs.

7.3.2 Adaptive Weighting and Permutations

The reason we use a linear model to perform the test is that we can use the adaptive weighting method that we have already developed. The residual, under the null hypothesis, for the first member of the ith pair is

$$\hat{\varepsilon}_{i1} = Y_{i1} - \hat{Y}_{i1} = Y_{i1} - \hat{\mu}_i = Y_{i1} - \bar{Y}_i = -\frac{d_i}{2}.$$

The residual for the second member of the ith pair is

$$\hat{\varepsilon}_{i2} = Y_{i2} - \hat{Y}_{i2} = Y_{i2} - \hat{\mu}_i = Y_{i2} - \bar{Y}_i = \frac{d_i}{2}.$$

Hence, if we know the B differences, we can compute the residuals as

$$\left\{-\frac{d_1}{2}, \frac{d_1}{2}, -\frac{d_2}{2}, \frac{d_2}{2}, \ldots\right\}.$$

These residuals are then used to compute the weights using the adaptive weighting method that was described in Chapter 3. It can be shown that the weights for the two members of a pair are equal. To see this, note that the distribution of the residuals will be symmetric about zero. Thus, if the weight for one member of a pair is Z_i/s_i then the weight for the other member of that pair will be $-Z_i/(-s_i)$.

Now, if we permute the residuals within each pair and add them to the predicted values that are obtained under the null hypothesis, we will obtain the new vector of dependent variables. Let $\{\hat{\varepsilon}_{11,k}, \hat{\varepsilon}_{12,k}, \hat{\varepsilon}_{21,k}, \hat{\varepsilon}_{22,k}, \ldots\}$ be the vector obtained by shuffling the residuals within each pair. Now suppose, for the kth permutation, the residuals are permuted for the ith pair so that the residual for the second member is permuted to the first member. For that permutation, the first member of the pair becomes

$$Y_{i1,k} = \hat{\mu}_i + \hat{\varepsilon}_{i2} = \bar{Y}_i + \hat{\varepsilon}_{i2} = Y_{i2},$$

and the second member of the pair becomes

$$Y_{i2,k} = \bar{Y}_i + \hat{\varepsilon}_{i1} = Y_{i1}.$$

Therefore, the members of each pair will be permuted in the same way that the residuals were permuted. If the kth permutation results in a permutation of the residuals for the ith pair, the difference becomes

$$d_{i,k} = Y_{i2,k} - Y_{i1,k} = Y_{i1} - Y_{i2} = -d_i.$$

Hence, a permutation of the residuals for the ith pair results in a sign change for that difference. This implies that a permutation of the residuals will produce a sign change of the difference.

7.3.3 The Test Statistic for the Adaptive WLS Test

Now, suppose we compute the adaptive weights for all $2B$ observations. Since $Y_{i1} - \bar{Y}_i = -d_i/2$ and $Y_{i2} - \bar{Y}_i = d_i/2$, the SSE for the weighted model, under the null hypothesis, is

$$SSE_R^* = \sum_{i=1}^{B}\sum_{j=1}^{2}\left[w_i(Y_{ij} - \bar{Y}_i)\right]^2 = \sum_{i=1}^{B} w_i^2 \frac{d_i^2}{2}.$$

Under the alternative hypothesis we compute the SSE for the weighted model as

$$SSE_C^* = \sum_{i=1}^{B}\sum_{j=1}^{2}\left[w_i(Y_{ij} - \hat{Y}_{ij})\right]^2.$$

However, to compute $\hat{Y}_{ij}$ we need estimates of μ_i and δ using the adaptively weighted model. We now will proceed to find those estimates.

It will be helpful if we rewrite the models into their linear model form using standard matrix notation. Suppose we write, using matrix notation, the full unweighted model, based on the model in Equations (7.7) and (7.8), as

$$\mathbf{Y} = \mathbf{X}\boldsymbol{\beta} + \boldsymbol{\varepsilon},$$

where the vector of dependent variables is

$$\mathbf{Y} = \begin{bmatrix} Y_{11} \\ Y_{12} \\ Y_{21} \\ Y_{22} \\ \vdots \end{bmatrix},$$

and the predictor variables are

$$\mathbf{X} = \begin{bmatrix} 1 & 0 & \dots & -1/2 \\ 1 & 0 & \dots & +1/2 \\ 0 & 1 & \dots & -1/2 \\ 0 & 1 & \dots & +1/2 \\ \vdots & \vdots & \dots & \vdots \end{bmatrix},$$

and the regression coefficients are

$$\boldsymbol{\beta} = \begin{bmatrix} \mu_1 \\ \mu_1 \\ \mu_2 \\ \mu_2 \\ \vdots \\ \delta \end{bmatrix}.$$

After the residuals are computed, they are used to calculate the weights, which are placed in the diagonal matrix $\mathbf{W}$. The weighted model is

$$\mathbf{WY} = \mathbf{WX}\boldsymbol{\beta} + \mathbf{W}\boldsymbol{\varepsilon}.$$

Using $\mathbf{Y}^* = \mathbf{WY}$ and $\mathbf{X}^* = \mathbf{WX}$, the transformed model can be written as

$$\mathbf{Y}^* = \mathbf{X}^*\boldsymbol{\beta} + \mathbf{W}\boldsymbol{\varepsilon},$$

where

$$\mathbf{Y}^* = \begin{bmatrix} w_1 & 0 & 0 & 0 & \dots \\ 0 & w_1 & 0 & 0 & \dots \\ 0 & 0 & w_2 & 0 & \dots \\ 0 & 0 & 0 & w_2 & \dots \\ \cdot & \cdot & \cdot & \cdot & \cdot \\ \cdot & \cdot & \cdot & \cdot & \cdot \\ \cdot & \cdot & \cdot & \cdot & \cdot \end{bmatrix} \begin{bmatrix} Y_{11} \\ Y_{12} \\ Y_{21} \\ Y_{22} \\ \vdots \\ \vdots \end{bmatrix} = \begin{bmatrix} w_1 Y_{11} \\ w_1 Y_{12} \\ w_2 Y_{21} \\ w_2 Y_{22} \\ \vdots \\ \vdots \end{bmatrix}$$

and

$$\mathbf{X}^* = \begin{bmatrix} w_1 & 0 & 0 & 0 & \cdots \\ 0 & w_1 & 0 & 0 & \cdots \\ 0 & 0 & w_2 & 0 & \cdots \\ 0 & 0 & 0 & w_2 & \cdots \\ . & . & . & . & . \\ . & . & . & . & . \\ . & . & . & . & . \end{bmatrix} \begin{bmatrix} 1 & 0 & \cdots & -1/2 \\ 1 & 0 & \cdots & +1/2 \\ 0 & 1 & \cdots & -1/2 \\ 0 & 1 & \cdots & +1/2 \\ 0 & 0 & \cdots & . \\ 0 & 0 & \cdots & . \\ . & . & \cdots & . \end{bmatrix}$$

$$= \begin{bmatrix} w_1 & 0 & \cdots & -w_1/2 \\ w_1 & 0 & \cdots & +w_1/2 \\ 0 & w_2 & \cdots & -w_2/2 \\ 0 & w_2 & \cdots & +w_2/2 \\ . & . & \cdots & . \\ . & . & \cdots & . \\ . & . & \cdots & . \end{bmatrix}.$$

Using the transformed model, the least squares solution is

$$\hat{\beta} = (\mathbf{X}^{*'}\mathbf{X}^*)^{-1}\mathbf{X}^{*'}\mathbf{Y}^*$$

$$= \begin{bmatrix} 2w_1^2 & 0 & \cdots \\ 0 & 2w_2^2 & \cdots \\ \vdots & \vdots & \vdots \\ 0 & \cdots & \sum_{i=1}^B w_i^2/2 \end{bmatrix}^{-1} \begin{bmatrix} w_1^2(Y_{11}+Y_{12}) \\ w_2^2(Y_{21}+Y_{22}) \\ \vdots \\ \sum_{i=1}^B w_i^2(Y_{i2}-Y_{i1})/2 \end{bmatrix}.$$

It is easy to compute the inverse of $(\mathbf{X}^{*'}\mathbf{X}^*)$ because it is a diagonal matrix. For example, the first element of $\hat{\beta}$ is

$$(2w_1^2)^{-1}[w_1^2(Y_{11}+Y_{12})] = (Y_{11}+Y_{12})/2 = \bar{Y}_1.$$

Consequently,

$$\hat{\beta} = \begin{bmatrix} \bar{Y}_1 \\ \bar{Y}_2 \\ \vdots \\ \sum_{i=1}^B w_i^2 d_i / \sum_{i=1}^B w_i^2 \end{bmatrix}.$$

This shows that the estimate of μ_i is simply $\bar{Y}_i$ and the estimate of δ is $\bar{d}_w = (\sum_{i=1}^{B} w_i^2 d_i)/(\sum_{i=1}^{B} w_i^2)$, which is the weighted average of the differences. Hence, using the weighted model, we find the predicted values

$$\hat{Y}_{i1} = \bar{Y}_i - \frac{1}{2}\bar{d}_w,$$

and

$$\hat{Y}_{i2} = \bar{Y}_i + \frac{1}{2}\bar{d}_w,$$

for $i = 1, \ldots, B$. Recalling that $Y_{i1} - \bar{Y}_i = -d_i/2$ and $Y_{i2} - \bar{Y}_i = d_i/2$, the SSE for the complete weighted model is

$$\begin{aligned} SSE_C^* &= \sum_{i=1}^{B}\sum_{j=1}^{2}[w_i(Y_{ij} - \hat{Y}_{ij})]^2 \\ &= \sum_{i=1}^{B}\left(\{w_i[Y_{i1} - (\bar{Y}_i - \bar{d}_w/2)]\}^2 + \{w_i[Y_{i2} - (\bar{Y}_i + \bar{d}_w/2)]\}^2\right) \\ &= \sum_{i=1}^{B}\left\{\left[w_i\left(-\frac{d_i}{2} + \frac{\bar{d}_w}{2}\right)\right]^2 + \left[w_i\left(\frac{d_i}{2} - \frac{\bar{d}_w}{2}\right)\right]^2\right\} \\ &= \sum_{i=1}^{B} w_i^2(d_i - \bar{d}_w)^2/2. \end{aligned}$$

Let $SSE_{C,k}^*$ denote the SSE of the complete weighted linear model using the kth permutation of residuals with the linear model (or under k random sign changes of the differences) and let $SSE_{C,u}^*$ denote the SSE of the complete weighted linear model of the unpermuted data. We could use $\Lambda_k^{2/n,*}$ or F_k^* as the test statistic for the kth permutation. However, since SSE_R^* does not change with random sign changes of the differences, we can define E to equal the number of sign changes having $SSE_{C,k}^* \leq SSE_{C,u}^*$. Then we compute the p-value as $p = (E+1)/(R+1)$. This test will be called the adaptive WLS test for paired data.

To calculate the p-value we could use the original observations in the linear model or the differences in the equations for $SSE_{C,u}^*$ and $SSE_{C,k}^*$. However, there are several good reasons for using the differences:

- The linear model approach may be effective for small samples, but it becomes a computational burden for large samples because $X'X$ will be large.

- Researchers are accustomed to using the differences in the t test and Wilcoxon SR test, so it may be easier to continue using differences.

So, while it is true that we could have used the linear model approach that we developed in previous chapters to perform the adaptive test for paired data, we have found from experience that it is best to perform this test using the differences. Also, the software that is described in the next section uses the differences.

7.4 AN EXAMPLE USING PAIRED DATA

7.4.1 Data from Twins

In an effort to study the effect of environment on intelligence, Farber (1981) collected data on 33 twins. For each of these twins, one of the twins was raised by someone other than the parents and one was raised by the parents. An IQ test was administered to every child in the study. The data is the IQ of the child that was raised by a parent minus the IQ of the child that was not raised by a parent. These differences were recorded as follows:

4	-6	13	-3	-14	-15	-6	-3	-18
-14	-15	1	1	-1	-12	12	-4	9
4	11	-5	-11	1	-5	-11	-4	-9
-19	-2	5	8	3	-1			

If we analyze the data using the traditional t test we obtain a two-tailed p-value of $p_t = 0.0441$.

If we use the adaptive MG test we find that the tail-heaviness measure is $TH = 9.244/8.895 = 1.039$. Using this TH measure as the selection statistic we determine that Wilcoxon scores should be used. The Wilcoxon SR test produced a p-value of $p_{SR} = 0.0463$, which is slightly larger than the p-value obtained from the t test. We now will show how to use R functions and SAS macros to perform the adaptive WLS test.

7.4.2 The Adaptive WLS Test Using R

To analyze the data from Farber (1981) using R, we first obtain the R function `adaptpaired` that is on the file `adaptpaired.r`. This function, which is listed in Appendix E, can be downloaded from *www.math.niu.edu/~ogorman*. This function computes the weights and performs the random sign changes for the permutation test. Next, we set up a data frame on the file `farber.dat` that contains the differences as the variable `differences` on the data frame. Suppose we want 100,000 random sign changes and want to set our own seed for the random number generator. The R code to perform the test is:

```
source("adaptpaired.r")
dataset <- c("farber.dat")
r <- 100000
seed <- 31416
pvalue <- adaptpaired(dataset,r,seed)
```

The variable to be analyzed should not be included in the function call but must be given the variable name of `differences` on the data frame. This R program produced $p_{adaptive} = 0.0372$ as the two-tailed p-value for the adaptive test.

7.4.3 The Adaptive WLS Test Using SAS

A SAS macro, which can be found on the file `adaptpaired.sas`, can be used to perform the adaptive WLS test. The `adaptpaired` macro, which is listed in Appendix F, can be downloaded from *www.math.niu.edu/~ogorman*. For this example, we create the data set `farber` that has the variable `diffvar` as the difference between the IQ scores for the twins. The difference variable is specified in the macro call by the `diff` keyword parameter. The number of random sign changes must be specified by the `nsignchs` keyword parameter. If we want to perform 10,000 random sign changes the following SAS program could be used:

```
%include 'adaptpaired.sas';
data farber;
input diffvar;
datalines;
4
.
-1
;
%adaptpaired(dataset=farber, diff=diffvar,
         seed1=2237, seed2= 8767, seed3=23787, nsignchs=10000);
```

This invocation of the SAS macro produced a p-value of $p_{adaptive} = 0.0334$. Note that the adaptive WLS p-values produced by the R function and the SAS macro are slightly smaller than the p-value from the t test. However, to evaluate the effectiveness of the adaptive WLS test and the adaptive MG test we need to perform a simulation study.

7.5 SIMULATION STUDY

An extensive simulation study was used to document the finite sample properties of the t test, the Wilcoxon signed-rank (SR) test, the adaptive MG test, and the adaptive WLS test for sample sizes of $B = 15$, 25, 50, 100, and 200. Ten symmetric distributions were used in the simulations to generate the differences. These distributions are briefly described in Table 7.1. The uniform and beta distributions are short-tailed, the normal and logistic distributions have moderate tail lengths, and the distributions from the t_4 distribution to the 30% Extreme Outlier distribution are long-tailed. The bimodal distribution, with modes at -1.5 and +1.5, seems to have the characteristics of a uniform distribution.

As expected, the SR, MG, and adaptive WLS tests maintained their significance level near the nominal value of $\alpha = 0.05$, but the t test, for small samples of $B \leq 25$ differences, had empirical levels of significance near 0.03 for some distributions. Because none of the tests had empirical levels of significance that exceeded 5.41%, and because almost all of the significance levels exceeded 4.0%, it is reasonable to compare the power of these tests.

Table 7.1 Ten distributions of differences used in the simulation studies

Distribution	Description
Uniform	Uniform on the interval $[-\sqrt{3}, \sqrt{3}]$
Beta	Beta distribution with parameters $p = q = 2$
Normal	Standard normal
Logistic	Logistic distribution
t with $d.f. = 4$	t distribution with 4 degrees of freedom
10% Outliers	.9 from $N(0,1)$ and .1 from $N(0,3^2)$
30% Outliers	.7 from $N(0,1)$ and .3 from $N(0,3^2)$
10% Extreme Outliers	.9 from $N(0,1)$ and .1 from $N(0,10^2)$
30% Extreme Outliers	.7 from $N(0,1)$ and .3 from $N(0,10^2)$
Bimodal	.5 from $N(-1.5,1)$ and .5 from $N(1.5,1)$

The power results for $B = 15$ pairs are shown in Figure 7.2. For these simulations the distributions were shifted to the right by δ, which was determined to give 60% to 80% power for the Wilcoxon SR test. The power estimates show little difference in power between the tests for the normal and logistic distributions, but do show that the t test is less powerful than the other tests with long-tailed distributions. For the short-tailed and bimodal distributions the t test and the adaptive WLS test appear to be more powerful than the SR and the adaptive MG tests. The MG test is more powerful than the other test with the 30% Extreme Outlier distribution. For the other distributions the adaptive WLS test has relatively high power.

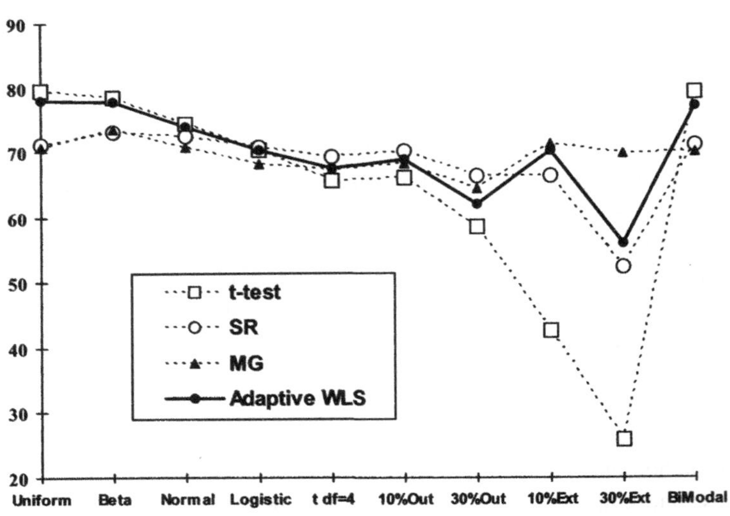

Figure 7.2 Empirical power in percent for four tests with $B = 15$ pairs.

The results for $B = 25$ pairs, which are given in Figure 7.3, demonstrate that the t test is not as powerful as the other tests with long-tailed distributions. The results also show that the SR test has relatively high power, except for short-tailed and extremely long-tailed distributions. The MG test has high power overall and has the highest power with the 30% Extreme Outlier distribution. The adaptive WLS test also has high overall power for short-tailed and moderate-tailed distributions, but not for the 30% Extreme Outlier distribution.

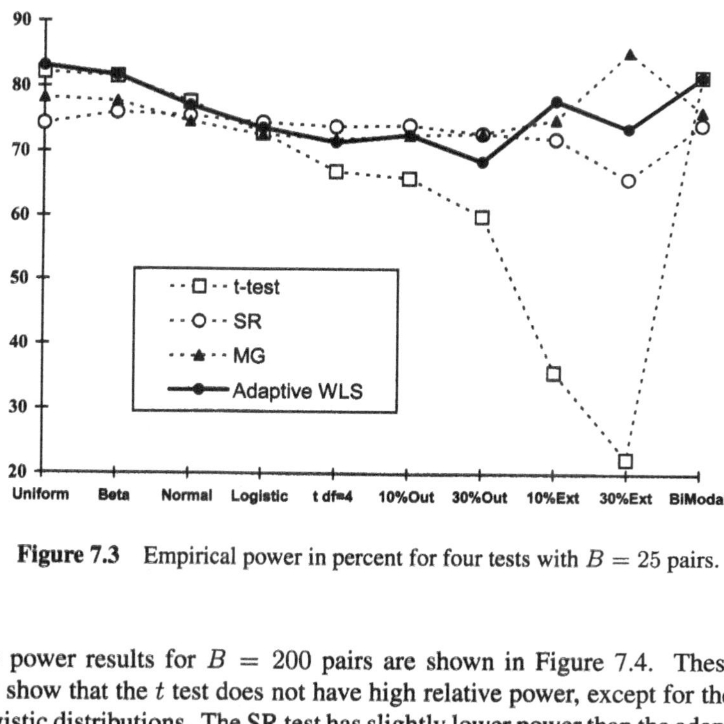

Figure 7.3 Empirical power in percent for four tests with $B = 25$ pairs.

The power results for $B = 200$ pairs are shown in Figure 7.4. These results clearly show that the t test does not have high relative power, except for the normal and logistic distributions. The SR test has slightly lower power than the adaptive tests for short-tailed and some long-tailed distributions. The adaptive WLS test and the adaptive MG test have high power for all distributions. Results for the simulations having $B = 50$ and $B = 100$ are not shown because they are similar to those shown for $B = 25$ and $B = 200$.

7.6 SAMPLE SIZE ESTIMATION

We have seen that, for non-normal error distributions, there is a power loss if the t test is used. We would like to approach the same subject from another point of view. What sample size reductions are possible if we use the Wilcoxon SR test or an adaptive test instead of the t test?

Let the power of the test be $1 - \beta$. Let σ be the standard deviation and δ be the mean of the population of differences. For a one-sample t test the standard formula

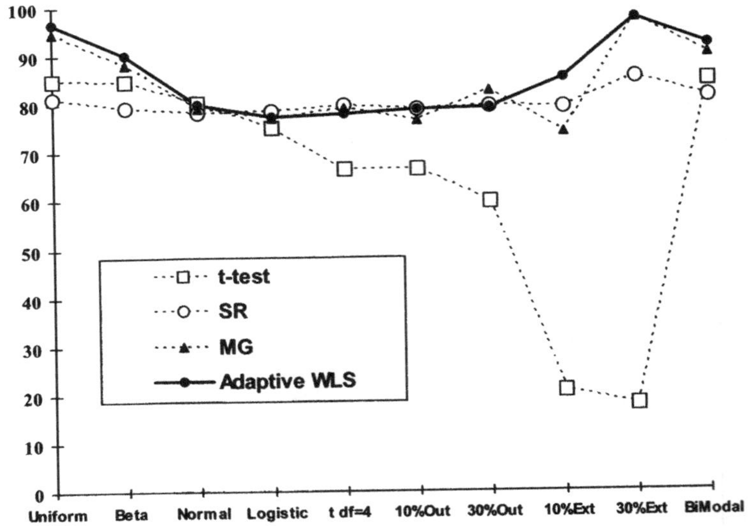

Figure 7.4 Empirical power in percent for four tests with $B = 200$ pairs.

for the sample size with normal errors is

$$B = [\Phi^{-1}(1 - \alpha/2) + \Phi^{-1}(1 - \beta)]^2 \left(\frac{\sigma^2}{\delta^2}\right). \tag{7.9}$$

However, this formula may not be accurate if the errors are not normally distributed or if the *t* test is not used. To investigate these sample size requirements a computer program was used to determine what sample size would be needed. We assumed that we would use a significance level of $\alpha = 0.05$ and generated differences from a population having a standard deviation of $\sigma = 1.0$ and a mean of $\delta = 0.4$. We then found the sample sizes that are necessary to achieve 80% power. In this simulation it was necessary to standardize each of the distributions so that the standard deviation would equal one. For example, the 10% Outlier distribution generates random variables that have a standard deviation of 1.3416, so we divided the generated values by 1.3416 so that the distribution would have a standard deviation of one.

The standard formula in Equation (7.9) was used to find the initial sample size of $B = 49$, assuming the errors are normal. For each test and for each distribution we ran a series of simulation studies that started with $B = 49$ pairs. We then adjusted the sample size up or down and then ran another simulation study with the adjusted sample size. We continued to adjust the sample size until the power was near 80% based on 10,000 data sets for that test with that distribution. The results of this simulation study is shown in Figure 7.5 for the 10 distributions described in Table 7.1. For example, with the 10% Outlier distribution the *t* test requires 50 pairs while the other tests require only 40 pairs.

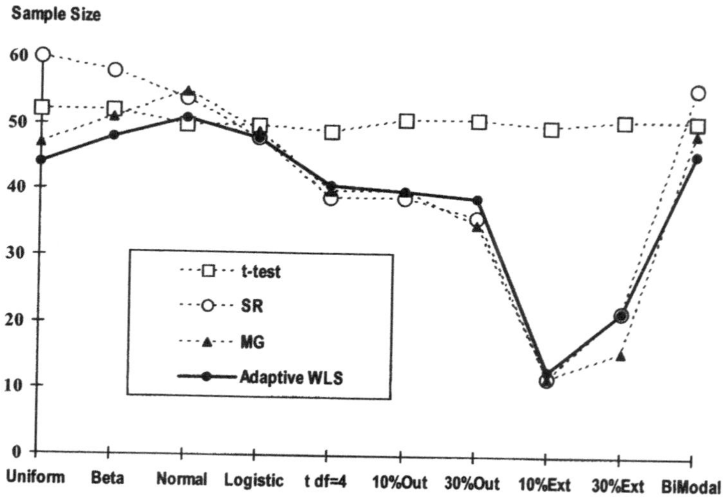

Figure 7.5 Sample sizes need to achieve 80% power with a shift of $\delta = 0.4$.

The results clearly show that the t test, as expected, requires a sample size near $B = 49$ to achieve 80% power for all distributions. However, the results also show that far fewer observations are necessary if another test is used. For long-tailed symmetric distributions the Wilcoxon signed-rank test, the MG test, and the adaptive test require fewer samples.

When designing a study how should the researcher proceed? One approach is to use the standard formula to find the sample size for normal errors. If previous data tend to be long-tailed, then it appears that the sample size could be reduced by 20% if the Wilcoxon test or one of the adaptive tests were used. This approach could save a considerable amount of money and time.

Another approach is to use the standard formulas, assuming the distribution of differences will approximate a normal, and specify that the sample size be used without modification. With that sample size, use the adaptive test to analyze the data. When the study is completed, if it appears that the data came from a normal distribution then there is little loss in power. However, if the data appeared to be from a long-tailed distribution, the power should be somewhat larger than 80%. This approach does not save any money or time but it does greatly increase the chance of rejecting the null hypothesis while maintaining the probability of a Type I error at $\alpha = 0.05$.

7.7 DISCUSSION OF TESTS FOR PAIRED DATA

The simulation results show that the t test would be a poor choice if the distribution was expected to be long-tailed. The simulation results also show that, for large samples, the SR test would be preferred to the t test because it is more powerful than the t test for all long-tailed distributions and is about as powerful as the t test for the normal distribution. The Wilcoxon SR test is surprisingly powerful for such a simple test. The adaptive Miao–Gastwirth test appears to be an improvement over the SR test for short-tailed and long-tailed distributions.

The adaptive WLS test has power roughly equal to that of the adaptive MG test for most distributions. The major disadvantage of the adaptive WLS test is that it requires a computer to estimate the permutation distribution of the test statistic. However, the adaptive WLS test has a number of practical advantages over the adaptive MG test.

One advantage of the WLS test is that small changes to the data produce only small changes to the p-values. With the adaptive Miao–Gastwirth test, if the tail-heaviness measure TH is near 1.2, then a small change in the data could cause a switch from Wilcoxon scores to t_2 scores, which could result in a large change in the p-value. This undesirable property of adaptive tests that are based on selection statistics was first observed by Hüsler(1987).

Another advantage of the adaptive WLS test is that it is a special case of a general approach to adaptive testing. If data were obtained from a pretest–posttest study, the adaptive WLS test could be used if the researcher wanted to use the pretest score as a covariate. Also, the adaptive WLS test is easy to understand because it is a special case of the general adaptive WLS test.

EXERCISES

7.1 Suppose we have data from a small study that has only $B = 3$ pairs and the differences are $d_1 = +7$, $d_2 = +9$, and $d_3 = -4$.

- **a)** List all the sign changes that are possible.
- **b)** If you have B blocks how many distinct sign changes are possible?

7.2 Suppose we have $B = 3$ pairs with differences of $d_1 = +7$, $d_2 = +9$, and $d_3 = -4$.

- **a)** Compute the unweighted SSE_C for this data.
- **b)** If the kth permutation of the residuals results in differences of $d_1 = -7$, $d_2 = -9$, and $d_3 = +4$, compute the unweighted $SSE_{C,k}$.
- **c)** Comment on the relationship between the $SSE_{C,k}$ (that you just computed for the kth permutation) and SSE_C.

7.3 For this exercise assume that during each second a computer can compute the test statistic for 100 million data sets that were created by sign changes.

- **a)** If we have $B = 8$ pairs of observations, how many sign changes are possible? How long would it take to find the complete permutation distribution of the test statistic?
- **b)** If we have $B = 25$ pairs of observations, how many sign changes are possible? How long would it take to find the complete permutation distribution of the test statistic?

c) If we have $B = 187$ pairs of observations, how many sign changes are possible? How long would it take to find the complete permutation distribution of the test statistic?

7.4 Five individuals were enrolled in a weight loss program that consisted of diet and exercise over an eight week period. The weight loss, in pounds, for these individuals were

2	-4	7	8	25

Using the Miao–Gastwirth test, compute the tail-heaviness measure TH. In the Miao–Gastwirth test what set of rank scores would be selected?

7.5 Suppose the compression strength of bricks must exceed 2800 pounds per square inch. Further, suppose we obtain a sample of $B = 9$ bricks having compression strengths of

3200	2850	2450	3100	2600
3950	2900	3450	4900	

If we want to test $H_0 : \mu = 2800$ versus $H_a : \mu \neq 2800$, is it appropriate to use random sign changes in a permutation test? Explain.

7.6 Mitchell, Burroughs, and Beadles (1936) obtained data from $B = 10$ pairs of rats, with one member of the pair assigned to a diet of raw peanut protein, and the other member of the pair assigned to a diet of roasted peanut protein. The differences in biological values for these rat pairs were

6	6	9	4	5
2	2	2	-18	3

a) Perform the t test to obtain p_t.

b) Using either R or SAS, perform the adaptive WLS test using 200 random sign changes to obtain $p_{adaptive}$.

c) Compare p_t to $p_{adaptive}$ and comment on any difference that you see between these results.

7.7 The adaptive weighting procedure requires the median to be estimated. For paired data can the median be estimated easily? Explain.

7.8 Suppose we have obtained reading scores for 18 third-grade students who participated in an enrichment program for 4 weeks. The reading scores were obtained on each student before and after the program, and the improvement in reading scores was calculated as the after score minus the before score. The improvement scores were:

8	12	-6	4	-2	4
5	7	-5	4	12	27
-3	9	11	-5	13	2

a) Compute the t test statistic and find the p-value p_t.

b) Perform the adaptive WLS test, using either the R code or the SAS macro, and find the p-value.

c) Do the p-values differ? If so, why?

7.9 To answer the following questions, refer to the material in the first section of Chapter 1 concerning the topic of robustness.

a) Is the t test robust for size?

b) Is the t test robust for power?

7.10 Based on the empirical power shown in Figure 7.3 for $B = 25$ pairs, for which distributions does the power of the SR test exceed the power of the t test? For which distributions does the power of the t test exceed the power of the SR test?

7.11 The data that were collected by Farber (1981) were used in this chapter to compute the p-values for the adaptive WLS test, the Wilcoxon test, and the t test. Using either R or SAS, find the weights that were assigned to each pair in the adaptive WLS test.

CHAPTER 8

MULTICENTER AND CROSS-OVER TRIALS

This chapter concerns the analysis of data from two important study designs that are often used in the development of new drugs. We will demonstrate how adaptive methods can be used in both multicenter trials and cross-over trials to increase the power of the tests.

In a multicenter clinical trial subjects are randomized to treatments within each center, and the subjects at one center may differ from the subjects at another center. Consequently, we will have a treatment effect and a center effect, but we are only interested in the treatment effect. Because subjects are randomized within center we will need to modify our permutation test methods so that the residuals are permuted within each center. We will see that this can easily be accomplished with existing software and that the resulting test is more powerful than the traditional test for non-normal error distributions.

Cross-over trials are also used to compare two treatments, but the approach is to give both treatments to each individual in the study. If we denote one treatment by A and the other treatment by B, a cross-over design gives some of the subjects treatment A first and treatment B second while it gives the other subjects treatment B first and treatment A second. These designs are appealing because this test of treatment equality may be much more powerful than one based on a design that

By Thomas W. O'Gorman

gives some subjects treatment A and other subjects treatment B. Because, in the cross-over design the between-subject variability may be reduced or eliminated, it has a greater chance of detecting a treatment effect. However, we shall see that the analysis of cross-over trials can be problematic because a carry-over effect may bias the estimates of treatment effect. We will show how adaptive test can be used to evaluate the treatment and carry-over effects.

8.1 TESTS IN MULTICENTER CLINICAL TRIALS

Researchers often need to compare the effectiveness of two drugs on a large sample of patients, but they cannot accrue enough patients at any one hospital or clinic to complete the study in a reasonable amount of time. So, to obtain many patients quickly, a number of centers may be enrolled in the study. In this section we will consider the properties of an adaptive test in the analysis of data from multicenter clinical trials.

Let C be the number of centers. For the ith center let n_{i1} be the number of patients given the first treatment and the let n_{i2} be the number of patients given the second treatment. Let c_i be the center effect for the ith center and let t_j be the treatment effect for the jth treatment. The complete model for a multicenter trial of two treatments is

$$y_{ijl} = \mu + c_i + t_j + \varepsilon_{ijl}, \tag{8.1}$$

for $i = 1, \ldots, C$, $j = 1, 2$, and $l = 1, \ldots, n_{ij}$. This model considers the center effect to be a fixed effect. In the simulation studies we will have C indicator variables for the C centers and one variable as an indicator of the second treatment. We will test the null hypothesis $H_0 : t_1 = t_2$ with a traditional F test, a permutation test, and the adaptive test. The reduced model for the multicenter two-sample test is

$$y_{ijl} = \mu + c_i + \varepsilon_{ijl}.$$

It will sometimes be convenient to write the model in matrix form as

$$\mathbf{Y} = \mathbf{X}_R \boldsymbol{\beta}_R + \mathbf{X}_A \boldsymbol{\beta}_A + \boldsymbol{\varepsilon}, \tag{8.2}$$

where $\mathbf{X}_R$ is a matrix of indicator variables for the center effect and $\mathbf{X}_A$ is a column vector that contains an indicator variable for the treatment effect. We will assume that the vector containing the outcomes is arranged as follows:

$$Y = \begin{bmatrix} Y_{111} \\ Y_{112} \\ \vdots \\ Y_{121} \\ Y_{122} \\ \vdots \\ Y_{211} \\ Y_{212} \\ \vdots \end{bmatrix},$$

so that the first $n_{11} + n_{12}$ rows contain the data for subjects from the first center, the next $n_{21} + n_{22}$ rows contain the data for subjects from the second center, etc. We will now investigate the properties of the test for treatment effect.

8.1.1 Level of Significance and Power

With multicenter clinical trial data we obtain the adaptive weights from the residuals of the reduced model $\mathbf{Y} = \mathbf{X}_R\boldsymbol{\beta}_R + \boldsymbol{\varepsilon}$. Under the reduced model the predicted values for the observations within a center are set equal to the mean for that center. Thus, the residuals are differences between the y-values and the within-center means.

The adaptive weights are obtained from the residuals by using the steps described in Section 5.1. The weights are used to compute SSE_R^* for the weighted reduced model. These weights are also used to compute SSE_C^* for the complete model that includes the treatment effect. The test statistic for the unpermuted data ($\Lambda_u^{2/n,*}$) is computed in the usual manner.

The permutation method permutes residuals within the centers and produces the within-center permutations of the dependent variable that would be obtained from a permutation of the errors. To see this we note that for the *k*th permutation the full permutation matrix can be written as

$$\mathbf{S}_k = \begin{bmatrix} \mathbf{S}_{1,k} & 0 & 0 & \dots & 0 \\ 0 & \mathbf{S}_{2,k} & 0 & \dots & 0 \\ . & . & . & . & . \\ 0 & 0 & 0 & \dots & \mathbf{S}_{C,k} \end{bmatrix},$$

where $\mathbf{S}_{i,k}$ is the permutation matrix for the *i*th center. The new vector of dependent variables for the *k*th permutation is

$$\mathbf{Y}_k = \mathbf{X}_\mathbf{R}\hat{\boldsymbol{\beta}}_R + \mathbf{S}_k\hat{\boldsymbol{\varepsilon}}_R,$$

where $\hat{\varepsilon}_R$ is the vector of residuals from the reduced model. Because the rows of $\mathbf{X}_R$ are identical for the observations that are in the same center, $\mathbf{S}_k\mathbf{X}_R = \mathbf{X}_R$. Since $\hat{\varepsilon}_R = \mathbf{Y} - \mathbf{X}_R\hat{\boldsymbol{\beta}}_R$ it follows that

$$\mathbf{Y}_k = \mathbf{X}_R\hat{\boldsymbol{\beta}}_R + \mathbf{S}_k(\mathbf{Y} - \mathbf{X}_R\hat{\boldsymbol{\beta}}_R) = \mathbf{S}_k\mathbf{Y}.$$

Thus, under the null hypothesis, the kth within-center permutation of residuals results in a within-center permutation of the dependent variables. We can use the same logic with the permutation of the errors, rather than the permutation of the residuals. Since $\mathbf{Y} = \mathbf{X}_R\boldsymbol{\beta}_R + \varepsilon$, a within-center permutation of the errors would produce a within-center permutation of the dependent variable because

$$\mathbf{Y}_k = \mathbf{X}_R\boldsymbol{\beta}_R + \mathbf{S}_k\varepsilon = \mathbf{X}_R\boldsymbol{\beta}_R + \mathbf{S}_k(\mathbf{Y} - \mathbf{X}_R\boldsymbol{\beta}_R) = \mathbf{S}_k\mathbf{Y}.$$

Because within-center permutations of errors are equally likely, and these will produce the same results as within-center permutations of the residuals, the resulting test statistics will be equally likely. Therefore, when the reduced model contains only a center effect and the permutations are performed within each center, the adaptive test will maintain its size.

Two simulation studies were designed to estimate the relative power of the traditional F test, the permutation test, and the adaptive test in a small and a large multicenter clinical trial. For the smaller study we used 5 centers, with 10 observations assigned to each of two treatments within each center. We chose a value for the treatment effect to obtain powers near 80%, and we generated 10,000 data sets with errors generated from the same nine distributions that we have used in Chapters 5 and 6. The power results for the smaller study are shown in Figure 8.1. These show a power advantage of the adaptive test over the other tests of 5% to 18% for the non-normal distributions.

For the larger study we used 10 centers with 100 observations in each of the two treatment groups in each center, so that there were a total of 2000 observations. Because these simulation studies required more calculations, we generated only 1000 data sets for the power estimates. The treatment effect was set to a value to obtain powers near 80% for the F test. The results, which are shown in Figure 8.2, demonstrate that the adaptive test is much more powerful than the other tests for all non-normal error distributions. The large differences in power between these tests is due to the large sample sizes that were used in these studies, which allows the adaptive test to accurately normalize the residuals. Based on the results from these simulation studies, we can say that there is little to lose and much to gain by using the adaptive test.

8.1.2 An Example of the Analysis of Data from a Multicenter Clinical Trial

Consider a hypothetical example of a clinical trial of a drug to reduce moderate hypertension. Suppose the study was conducted at 10 centers and at each center 20 subjects were enrolled in the trial. Within each center, 10 of the subjects received

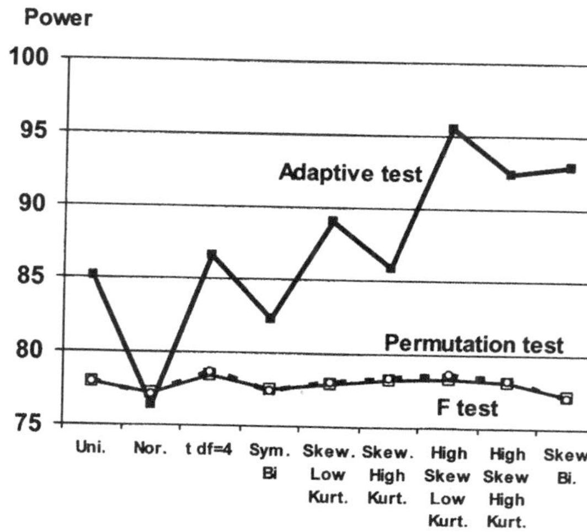

Figure 8.1 Empirical power in percent for the F test, the permutation test, and the adaptive test for the treatment effect in a multicenter clinical trial having 10 observations for each treatment in each of 5 centers. The F test is indicated by hollow squares, the permutation test by hollow circles, and the adaptive test by filled squares.

the drug and 10 received a placebo. The primary endpoint was the change in seated diastolic blood pressure over the 8 week duration of the trial. The change was defined as the diastolic blood pressure immediately after the treatment period minus the baseline diastolic blood pressure.

The change from baseline is recorded in Table 8.1 for the drug and placebo at the 10 treatment centers. The traditional F test gave a p-value of $p_F = 0.096$ and the adaptive test gave a p-value of $p_{adaptive} = 0.032$. The difference between these p-values is primarily due to the low weight of 0.45 given to the outlier in the tenth center. The other weights were in the range of $[0.69, 1.15]$.

8.1.2.1 Using R Code to Perform a Test of Treatment Effect We will use the `adaptblock` function to analyze this blocked data. This function was described in Section 6.2, where it was used with randomized complete block designs. The `adaptblock` function, which is listed in Appendix D, can be downloaded from *www.math.niu.edu/∼ogorman*. The data set `changebp.dat` contains the data from Table 8.1, and we will permute the residuals within each level of the variable `block`. In the adaptive analysis the center will be labeled as `block` and, for this example, the number of blocks equals 10. The R code is

```
source("adaptblock.r")
dataset <- c("changebp.dat")
complete <- c("y~as.factor(block)+treatment")
reduced  <- c("y~as.factor(block)")
```

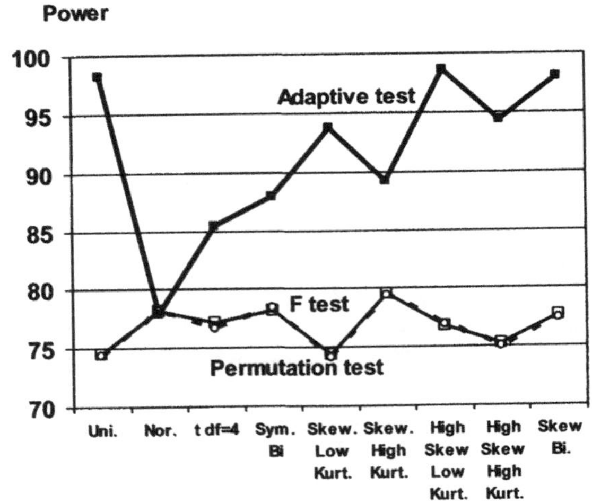

Figure 8.2 Empirical power in percent for the *F* test, the permutation test, and the adaptive test for the treatment effect in a multicenter clinical trial having 100 observations for each of two treatments in each of 10 centers. The *F* test is indicated by hollow squares, the permutation test by hollow circles, and the adaptive test by filled squares.

```
r <- 2000
seed <-c(51492)
nblocks <- c(10)
pvalue<-adaptblock(dataset,complete,reduced,r,seed,nblocks)
```

Note that the `as.factor` function must be used in the specification of the complete and reduced models so that R will set up indicator variables for the centers. That R code produced the following output:

```
Input:
    Data set: changebp.dat
    Complete model:  y~as.factor(block)+treatment
    Reduced model :  y~as.factor(block)
    Number of permutations =  2000
    Random seed =  51492

  Output:
    Unpermuted lambda =  0.9766
    Adaptive p-value  =  0.036
    Traditional F test p-value =  0.096
```

8.1.2.2 Using a SAS Macro to Perform a Test of Treatment Effect The SAS macro for performing this blocked permutation adaptive test is the same macro that has been used for all of the other tests in this book. The dependent variable is the

Table 8.1 The change from baseline for diastolic blood pressure over the eight weeks of treatments

Center	Treatment	Mean Change
1	placebo	-5, -8, -2, 3, -1, 3, 7, -2, 3, -5
	drug	-3, -6, 13, 2, -11, -3, -9, -7, -8, 5
2	placebo	-6, -7, -4, -3, 6, -8, -3, 5, -4, -5
	drug	-9, 8, -8, 3, 7, -4, -7, 3, -6, -4
3	placebo	-3, 2, -3, 5, -4, -9, -4, 3, -1, 3
	drug	-2, -5, 3, -1, -2, -26, -13, 3, -1, -3
4	placebo	-6, 7, -3, 4, -4, 5, 4, -5, 6, 8
	drug	7, 8, -2, -6, 3, -1, -8, 4, -12, -2
5	placebo	4, -8, -3, 3, -3, 2, -4, 6, -1, 3
	drug	-2, -7, -12, 2, -2, -5, 3, -1, -10, -3
6	placebo	-6, 7, -4, -2, 6, 8, -3, 3, -4, 5
	drug	-9, 8, -6, 23, -4, -2, -7, 3, -1, -15
7	placebo	-4, -7, -4, 5, -1, 3, -7, -5, 2, -4
	drug	-1, -8, -3, 3, 10, -3, 9, -3, -5, 3
8	placebo	5, -3, 4, -4, 5, 7, -4, -5, 6, 8
	drug	-9, 8, -8, 5, -7, -11, 7, 3, -2, -14
9	placebo	-6, -7, -1, -5, 2, -4, 5, 3, -1, -2
	drug	-3, -2, -4, 3, -1, -6, -12, 2, -11, -3
10	placebo	2, 7, -2, -3, -3, 2, -4, 7, 6, -38
	drug	-9, 8, 3, -2, -5, 3, -1, 1, -4, -6

change in the blood pressure, which is recorded as the `changbp` variable on the SAS dataset `multicenter`. When a blocking variable is used, the blocking variable must be specified as the `blkvars` keyword parameter. The `center` variable will be a class variable in the X_R matrix and it will be the blocking variable. The `treatment` variable is also a class variable and is the only variable in X_A. The following SAS code will perform the adaptive test on this data:

```
%adaptall(dataset=multicenter, y=changebp,
    xr=center, classr=center, xa=treatment, classa=treatment,
    blkvars=center,
    seed1=18521, seed2=28332, seed3=22343, nperm=2000);
```

The SAS output gives the blocking variable so that there is documentation that the residuals were permuted within centers. The SAS macro produced the following output:

```
Number of permutations = 2000
Starting seeds = 18521  28332  22343
```

```
SAS data set: main   Number of observations = 200
Blocking variable:  center

Dependent variables  :  changebp
Independent variables in reduced model   : Intercept center
      Class variables in reduced model   :  center
Independent variables in complete model  : Intercept center
                                           treatment
      Class variables in complete model  : center treatment

Adaptive test statistics for unpermuted data:
      Wilks Lambda =    0.9766
      F statistic  =    4.5297
p-value for adaptive test =    0.0320
```

With the adaptive test the treatment effect is significant, although it must be emphasized that this was a hypothetical data set. If the researcher fails to enter the blocking variable, the program will execute but it will permute the residuals between the centers, which usually is not appropriate.

8.2 ADAPTIVE ANALYSIS OF CROSS-OVER TRIALS

When researchers are interested in comparing the effectiveness of two treatments, a major problem is the large variability that is observed between the subjects. This between-subject variability makes it difficult to determine if there really is a difference between the treatments, especially if the sample sizes are small.

In order to make these studies more powerful, cross-over designs are sometimes used. In these designs those subjects who are given treatment A in the first period are then given treatment B in the second period, while those subjects who were given treatment B in the first period are then given treatment A in the second period. If the treatments are drugs which may have an effect that lasts longer than the treatment period, the researchers may have a "wash-out" period between the two treatments so as to reduce the potential bias that may be introduced into the analysis. If possible, a wash-out period of at least five times the half-life of the drug is sometimes recommended.

If a cross-over design is used the test for treatment effect may be much more powerful than a test that would be used with data obtained from a parallel design. This increase in efficiency is due to the removal of the between-subject variability. However, there are several disadvantages of this design. One problem is that cross-over studies are only suitable for the treatment of conditions that, after the wash-out period, have the same mean response. Consequently, cross-over studies are often used to evaluate the effectiveness of treatments for chronic conditions. However, they are not suitable to evaluate the effectiveness of curative treatments. For example, they usually would not be considered to evaluate the effectiveness of two antibiotics for the treatment of a specific infection because, after the treatment and wash-out period, the mean responses may not be equal.

The most serious problem in cross-over trials is the chance that a treatment by period interaction may not be detected. If the interaction is present, but is ignored in the analysis, the estimate of the treatment effect could be biased. We will address this issue by performing tests of interaction effects. The specific test depends on the study design. Cross-over designs sometimes include only the measurements that are taken immediately after the completion of the treatments. In these designs, for each subject there are two measurements and we are usually interested in the test for the effect of treatment. However, it is often desirable to collect data on each subject before the treatments are given. That is, a measurement is made at the beginning of the study before the first period treatment and another measurement is obtained after the wash-out period but before the second period treatment. These baseline measurements are used to more effectively determine if there is a group by treatment interaction.

While it is possible to compare three or more treatments in a cross-over design, we will confine our attention to two-period cross-over designs. In this chapter we will first describe how adaptive tests can be used with the basic two-period cross-over design, without baseline measurements. Next, we will perform the additional adaptive tests that can be used when baseline data are available.

8.2.1 Tests for Two-Period Cross-Over Trials without Baseline Measurements

Hills and Armitage (1979) showed how the treatment effect and the treatment by period interaction effect could be tested by using simple sums and differences. To show how these tests can be performed we will introduce some notation. Let μ be the overall mean and let s_k be the subject effect for the *k*th subject. The subject effect is a random variable with mean zero and variance σ_s^2. The subjects are randomly assigned to groups and γ_i is the group effect for the *i*th group. The effect of the *j*th period is given by π_j. We are primarily interested in the treatment effect, which is denoted by τ_1 for period 1 in group 1 and for period 2 in group 2. For period 2 in group 1 and period 1 in group 2 the treatment effect is denoted by τ_2. The treatment by period interaction for the *i*th group at period 2 is given by λ_1 for group 1 and λ_2 for group 2, but equals zero for the first period. Let μ_{ij} be the mean for the *i*th group at the *j*th period. The fixed effects are given in Table 8.2. With this notation we can write the model as

$$y_{ijk} = \mu_{ij} + s_k + \varepsilon_{ijk}, \tag{8.3}$$

where ε_{ijk} is the random error for the *k*th observation in the *i*th group at the *j* period. We will assume that the random errors have mean zero and variance σ^2.

With a two-period cross-over design we cannot estimate all of the effects in the model. Since we usually randomly assign subjects to groups, we can set $\gamma_1 = \gamma_2 = 0$. Because the random effects have zero means, the fixed effects in the table equal their means.

Table 8.2 Fixed effects for the cross-over trial without baseline measurements

	Group	
Period	1	2
1	$\mu + \gamma_1 + \pi_1 + \tau_1$	$\mu + \gamma_2 + \pi_1 + \tau_2$
2	$\mu + \gamma_1 + \pi_2 + \tau_2 + \lambda_1$	$\mu + \gamma_2 + \pi_2 + \tau_1 + \lambda_2$

Hills and Armitage (1979) proposed using the sum of the two measurements for each individual in a two-sample test to perform a test of interaction of treatment and period. We compute the sum as $t_{1k} = y_{11k} + y_{12k}$ for the kth subject in the first group and $t_{2k} = y_{21k} + y_{22k}$ for the kth subject in the second group. We note that for the kth subject in the first group the expected value is

$$E[t_{1k}] = 2\mu + \pi_1 + \pi_2 + \tau_1 + \tau_2 + \lambda_1,$$

and for the kth subject in the second group the expected value is

$$E[t_{2k}] = 2\mu + \pi_1 + \pi_2 + \tau_1 + \tau_2 + \lambda_2.$$

Note that if $\lambda_1 = \lambda_2$ then $E[t_{1k}] = E[t_{2k}]$. Hence, one approach is to use these sums to perform a test of $H_0 : \lambda_1 = \lambda_2$ with a two-sample test. If the errors are normally distributed, we could use a t test; or if they are not normally distributed, we could use an adaptive test or a nonparametric test. However, the test for interaction of treatment and period will have low power because the subject totals will include between-subject variability and within-subject variability. So, while a test for treatment by period interaction could be performed, it usually has insufficient power to detect the interaction. See Jones and Kenward (2003, Chapter 2) for details concerning this test.

An example may illustrate this point. Consider the hypothetical data on LDL cholesterol levels from a two-period crossover study. Suppose the data were used to determine the effect of lowering LDL cholesterol after taking one of two drugs, either drug A or drug B, for 90 days, with a wash-out period between the treatments of 30 days. In this study, 15 subjects were randomly assigned, with 8 assigned to the AB sequence and 7 assigned to the BA sequence, and their LDL concentrations (in mg/dL) were recorded. The hypothetical data, along with the sums of the two measurements, are displayed in Table 8.3.

The sums of the responses, which are given in the fifth column of Table 8.3, are plotted in Figure 8.3. With the overlap that we observe in these two groups (sequences) we cannot hope to reject $H_0 : \lambda_1 = \lambda_2$. Even if we had a large treatment by period interaction with $\lambda_1 = 10$ and $\lambda_2 = -10$, the effect would be to shift the sums from sequence 1 by 20 relative to those from sequence 2. But with the large variability we observe in the sums that we see in Figure 8.3 the two-sample test would not lead to rejection of the null hypothesis $H_0 : \lambda_1 = \lambda_2$. This low power is due to the large between-subject variability in the data set.

If interaction is suspected, it would be important to design the study to include baseline measurements because they could be used to perform a powerful test for

Table 8.3 LDL cholesterol (mg/dL) recorded after 90 days on either drug A or drug B.[a]

Subject	Sequence	After First Treatment	After Second Treatment	Sum of Responses	Difference Between Responses
1	AB	111	105	216	6
2	AB	182	171	353	11
3	AB	147	151	298	-4
4	AB	136	127	263	9
5	AB	172	169	341	3
6	AB	195	182	377	13
7	AB	210	208	418	2
8	AB	129	124	253	5
9	BA	163	169	332	-6
10	BA	138	141	279	-3
11	BA	191	185	376	6
12	BA	203	216	419	-13
13	BA	143	145	288	-2
14	BA	163	171	334	-8
15	BA	177	189	366	-12

[a]Each patient was randomly assigned to either the AB or BA sequence.

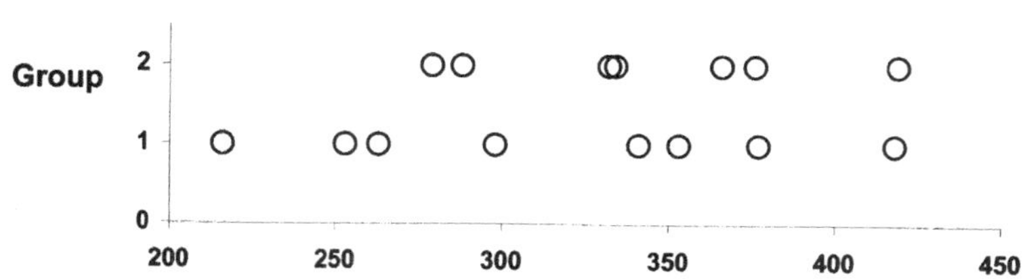

Figure 8.3 The sum of the LDL measurements for the subjects in group 1 (AB) and group 2 (BA).

interaction. This test for treatment by period interaction will be fully evaluated in Section 8.2.2. If we do not have baseline measurements, we usually assume that there is no interaction and proceed directly to test the treatment effect.

To test $H_0 : \tau_1 = \tau_2$ we compute the differences between the two measurements as $d_{1k} = y_{11k} - y_{12k}$ for the kth subject in Group 1 and $d_{2k} = y_{21k} - y_{22k}$ for the kth subject in Group 2. Using the fixed effects in Table 8.2, we find that the mean values are

$$E[d_{1k}] = \pi_1 - \pi_2 + \tau_1 - \tau_2$$

for the kth subject in the first group, and

$$E[d_{2k}] = \pi_1 - \pi_2 + \tau_2 - \tau_1$$

for the kth subject in the second group. Hence, we can test $H_0 : \tau_1 = \tau_2$ using a two-sample test on the differences. Note that the variance of these differences is not inflated by the between-subject variability. This is important because it is possible, with relatively small samples, to detect a treatment difference. In our example the differences between the responses that are given in the last column of Table 8.3, which are plotted in Figure 8.4, exhibit little variability because the between-subject variability has been removed by the subtraction. This allows the test for treatment effect with cross-over trials to be relatively powerful, even for trials that have few subjects. For the LDL data that are shown in Figure 8.4 we see that there is little overlap in the differences, so we would reject the null hypothesis of equality of treatments.

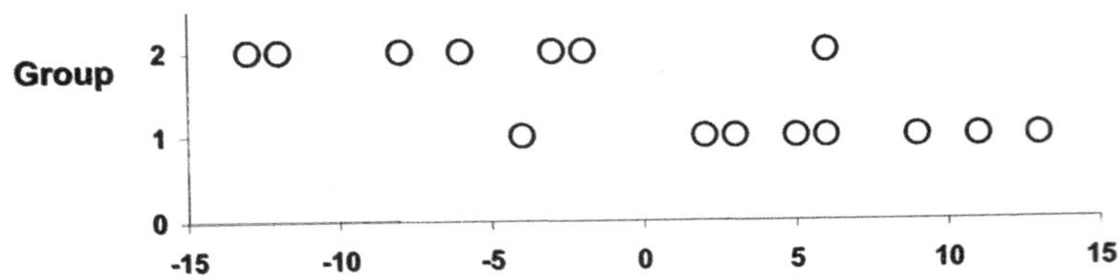

Figure 8.4 The difference of the LDL measurements for the subjects in group 1 (AB) and group 2 (BA).

To evaluate the power of the test for treatment effect, we used a simulation study with $n_1 = 16$ subjects in Group 1 and $n_2 = 16$ subjects in Group 2. Note that, because we use the differences between the measurements in the test, the distribution of the differences will not, in general, equal the distribution of the errors. In the beginning of the previous chapter we showed that the differences are symmetrically distributed, even if the random errors are highly skewed. Hence, the results from the two-sample simulations that were performed in Chapter 3 do not directly apply to cross-over studies.

Also, note that the difference for the kth subject is not related to the overall mean μ or the subject effect. We will assume in the simulation study that the group effect and the interaction effect is zero. Thus, it can be seen that the power of the test is a function of $(\tau_1 - \tau_2)$. In the simulation study we used the same symmetric distributions that were used for paired data. These were described in Table 7.1. The differences d_{1k} and d_{2k} were generated from one of these symmetric distributions, and the mean difference was selected to give powers between 60% and 80% for the Wilcoxon rank-sum test. These tests are two-sample tests for populations that have symmetrically distributed random variables. The empirical power is shown in Figure 8.5 for a cross-over study having 16 subjects in each group.

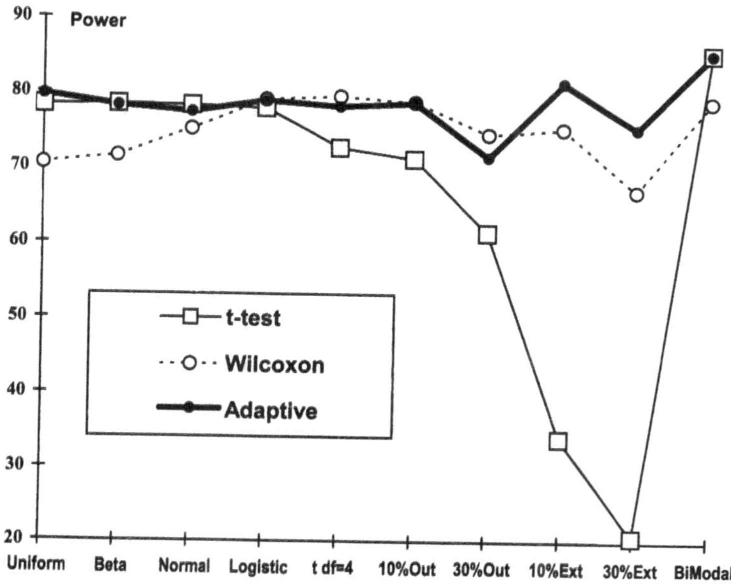

Figure 8.5 Empirical power in percent for the direct test of treatment effect with 16 subjects assigned to each group. The power of the t test is indicated by hollow squares, the Wilcoxon rank-sum test by hollow circles, and the adaptive test by filled circles.

These results in Figure 8.5 show that the Wilcoxon rank-sum test has lower power than the other tests with short-tailed distributions, which include the Uniform, Beta, and Bimodal distributions when there are 16 subjects assigned to each group. The t test has lower power than the other tests for all long-tailed distributions. Only the adaptive test has relatively high power for all distributions.

Most cross-over studies have a small number of subjects, but we wanted to compare the power of these tests when used with moderate sample sizes. A simulation study using 48 subjects in each group was performed. The results are not shown because they are nearly identical to those shown in Figure 8.5. Thus, it appears that the adaptive test has relatively high power for all distributions and sample sizes in the range of 16 to 48 subjects per group.

8.2.2 Tests for a Two-Period Cross-Over Design with Baseline Measurements

The major problem in the analysis of data obtained from the basic cross-over experiment (without baseline measurements) is that the test for treatment by period interaction often does not have high power. Recall that the test is based on the sum of the two measurements. This sum will be quite variable if the between-subject variability is large. Hence, the test often has low power. This is disturbing because, in the presence of a treatment by period interaction, the estimate of the treatment effect may be biased.

To avoid this problem, researchers often take measurements before and after each treatment. With this design each subject has four measurements: one before the first treatment, one after the first treatment, one after the "wash-out" period between the treatments, and one after the second treatment. These four measurements will be denoted by the vector $(x_{11k}, y_{11k}, x_{12k}, y_{12k})$ for the kth subject in the first group. In the second group the vector $(x_{21k}, y_{21k}, x_{22k}, y_{22k})$ will denote the measurements for the kth subject.

In many cases the subjects are randomly assigned to groups so it is reasonable to assume that the group effect is zero. The period effects are given by $\pi_1, \pi_2, \pi_3, \pi_4$ where $\pi_1 + \pi_2 + \pi_3 + \pi_4 = 0$. We will use τ to denote the treatment effect for treatment B and will use $-\tau$ to denote the treatment effect for treatment A. Ideally, the wash-out period is specified to be of sufficient duration so that, regardless of what treatment was administered in the first period, the within-group mean responses after the wash-out period will be equal. However, due to a first-order carry-over effect, the means may differ. This first-order carry-over effect will be specified as θ for Group 2 and as $-\theta$ for Group 1. It is also possible that the order of the treatments matters. That is, patients given treatment A and then treatment B may see greater improvement than those who were given treatment B then treatment A. This second-order carry-over effect is denoted by λ for Group 2 and by $-\lambda$ for Group 1. Using the notation of Kenward and Jones (1987), we can write the fixed effects for the 4 periods in Table 8.4 for both groups.

Table 8.4 Fixed effects for the two-period cross-over design with baseline measurements

	Group	
Period	1 (AB)	2 (BA)
Before first treatment	$\mu + \pi_1$	$\mu + \pi_1$
After first treatment	$\mu + \pi_2 - \tau$	$\mu + \pi_2 + \tau$
After wash-out	$\mu + \pi_3 - \theta$	$\mu + \pi_3 + \theta$
After second treatment	$\mu + \pi_4 + \tau - \lambda$	$\mu + \pi_4 - \tau + \lambda$

Kenward and Jones (1987) proposed a sequential procedure to perform the tests. They recommended performing the test for first-order effect and follow it with the test for second-order effect. If they fail to reject the null hypothesis for both the first-order and second-order effects, they will then perform the test for treatment effect. They begin by testing $H_o : \theta = 0$ versus $H_a : \theta \neq 0$. This is a two-sample test on the differences $x_{11k} - x_{12k}$ for the kth subject in the first group and $x_{21k} - x_{22k}$ for the kth subject in the second group. This is the same procedure that we used to test the treatment effect in the two-period cross-over trial without baseline measurements. These differences between baseline values will also be symmetrically distributed. Hence, the relative powers of this first-order test are identical to those shown in Figure 8.5 for 10 symmetric distributions of the differences.

If we fail to reject $H_0 : \theta = 0$ and have no reason to believe that a first-order carry-over effect exists, then we would proceed to test the second-order carry-over effect. Assuming that $\theta = 0$ we begin by testing $H_0 : \lambda = 0$ versus $H_a : \lambda \neq 0$ by computing $s_{1k} = [(x_{11k} - y_{11k}) + (x_{12k} - y_{12k})]/2$ for the *k*th subject in group 1 and $s_{2k} = [(x_{21k} - y_{21k}) + (x_{22k} - y_{22k})]/2$ for the *k*th subject in group 2. Let $b_{11k} = x_{11k} - y_{11k}$ and $b_{12k} = x_{12k} - y_{12k}$. Using this notation, we can see that $s_{1k} = (b_{11k} + b_{12k})/2$ is the mean of the differences between the baseline and the treatment measurements for the *k*th subject in Group 1. In Group 2 s_{2k} has the same interpretation for the *k*th subject. We then perform a two-sample test of $H_0 : \lambda = 0$ using the values of s_{1k} from Group 1 and the values of s_{2k} from Group 2.

In our simulation study, for the *k*th subject in Group 1, we generate b_{11k} and b_{12k} from a symmetric distribution, and then compute s_{1k}. Because s_{1k} is the average of b_{11k} and b_{12k} the distribution of s_{1k} will also be symmetric, but will generally have lower kurtosis than the distribution of b_{11k} and b_{12k}. We used 10,000 data sets having 16 subjects in each group with the difference between the baseline and the treatment measurements generated from one of the 10 symmetric distributions that were described in Table 7.1. We adjusted λ to obtain powers near 80% for the Wilcoxon test. The relative power of the tests for second-order effect are given in Figure 8.6.

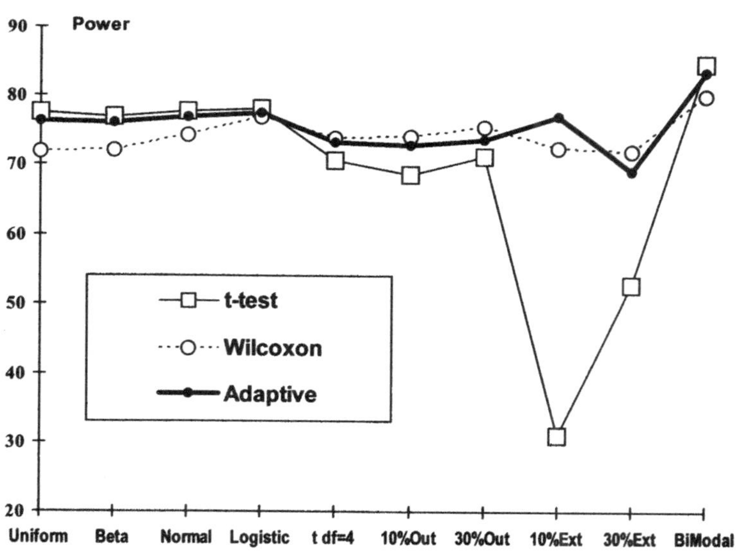

Figure 8.6 Empirical power in percent for the test of second-order effect with 16 subjects assigned to each group. The power of the *t* test is indicated by hollow squares, the Wilcoxon rank-sum test by hollow circles, and the adaptive test by filled circles.

The results show that the adaptive test has high power overall, but there is little difference between these tests for short-tailed symmetric distributions and for 10%

and 30% outlier distributions. The only meaningful differences between these tests occur with very long-tailed distributions, where the t test has much lower power than the other tests.

If we conclude, based on the tests for first-order and second-order carry-over effects, that $\theta = 0$ and $\lambda = 0$, we proceed to perform the test for the treatment effect. We test $H_0 : \tau = 0$ versus $H_a : \tau \neq 0$ by performing a two-sample test of the difference between treatments. This test is identical to the test that we described earlier for data sets that did not have baseline measurements. Hence, the power of these tests are shown in Figure 8.5. In these cross-over trials we use baseline measurements only to evaluate the first-order and second-order carry-over effects. We will now show how adaptive tests can be used to perform these tests in the example.

8.2.3 An Example

To illustrate how to perform an adaptive analysis, we will use a data set from a cross-over trial of two drugs that were used to treat bronchial asthma. This data set, which is given in Table 8.5, was published by Patel (1983) and analyzed by Kenward and Jones (1987). The measurements were of forced expiratory volume in one second (FEV1) taken at four times: once before the first treatment, once after the administration of a single oral dose of either drug A or drug B, once after the wash-out period, and once after the administration of the other drug. In this study, 17 subjects who had mild to moderate bronchial asthma were randomly assigned to either the AB sequence or the BA sequence. The first subject in the BA sequence had unusual responses, but the observation was retained in the analysis.

To analyze these data, we used the traditional t test and the adaptive test for the first-order, second-order, and treatment effects. The t test produced a p-value of $p_t = 0.5301$ for the first-order carry-over effect while the adaptive test, using a SAS macro, produced a p-value of $p_{AD} = 0.6737$. Based on these large p-values, it appears to be safe to assume that $\theta = 0$ and we can proceed with the analysis of the second-order effects. We performed the test of the second-order carry-over effect and obtained p-values of $p_t = 0.4696$ from the t test and $p_{AD} = 0.6397$ from the adaptive test. Consequently, we will assume that there is no carry-over effect and will proceed to test the treatment effect. The t test for the treatment effect produced a p-value of $p_t = 0.0472$, while the adaptive test, using a SAS macro, gave a p-value of $p_{AD} = 0.0260$. The difference between the p-values for the t test and the adaptive test is due to the downweighting of the first observation in the BA sequence. That observation was given a weight of $w_9 = 0.60$, while the other weights were in the range [0.82, 1.11]. So, with this data, the adaptive test more clearly detects a treatment difference.

8.2.3.1 Adaptive Tests for Cross-Over Trials Using SAS The SAS code in Table 8.6 was used to perform the tests for carry-over effects and the test for treatment effect. The variable `first` was computed in the data step as the difference between the baseline scores for each subject. It was used to perform the adaptive test of the

Table 8.5 Forced expiratory volume in one second (FEV1), measured in liters, from patients with bronchial asthma.[a]

Subject	Sequence	Before Treatment	After Period 1	After Washout	After Period 2
1	AB	1.09	1.28	1.24	1.33
2	AB	1.38	1.60	1.90	2.21
3	AB	2.27	2.46	2.19	2.43
4	AB	1.34	1.41	1.47	1.81
5	AB	1.31	1.40	0.85	0.85
6	AB	0.96	1.12	1.12	1.20
7	AB	0.66	0.90	0.78	0.90
8	AB	1.69	2.41	1.90	2.79
9	BA	1.74	3.06	1.54	1.38
10	BA	2.41	2.68	2.13	2.10
11	BA	3.05	2.60	2.18	2.32
12	BA	1.20	1.48	1.41	1.30
13	BA	1.70	2.08	2.21	2.34
14	BA	1.89	2.72	2.05	2.48
15	BA	0.89	1.94	0.72	1.11
16	BA	2.41	3.35	2.83	3.23
17	BA	0.96	1.16	1.01	1.25

Source: Patel, H. I. (1983) Data are used with permission of Taylor & Francis.

[a]Each patient was randomly assigned to either the AB or BA sequence.

first-order carry-over effect. In order to test the second-order carry-over effect, the variable `second` was computed as the sum, over the two treatments, of the differences between the baseline and treatment responses for each subject. This will be twice the mean of the differences between the baseline the treatment measurements, so the test using the sum will produce the same p-value as the test using the mean. We also computed the variable `treatment` as the difference between first and second treatments for each subject. These variables were used in the three calls to the `adaptall` macro to perform the three tests.

8.2.3.2 *Adaptive Tests for Cross-Over Trials Using R* The R code in Table 8.7 was used to analyze the FEV1 values on the asthma data set. Before we executed this code, we made a data frame `fev1.dat` that contained the information in Table 8.5 and used `x1, y1, x2, y2` as variable names for the four measurements for each subject. Recall that, for the R code, the dependent variable is always given `y` as the variable name. So, before we perform each test we compute the appropriate `y` variable that is then used in a two-sample test. The samples are denoted by the variable `sequence`.

Table 8.6 SAS code for three tests with the asthma data set

```
%include "adaptall.sas";
data fev1;
   input obs sequence $ x1 y1 x2 y2;
   first = x1 - x2;
   second = (x1 - y1) + (x2 - y2);
   treatment = y1 - y2;
   datalines;
1 AB 1.09 1.28 1.24 1.33
.
;
%adaptall(dataset=fev1, y=first, xa = sequence, classa = sequence,
   seed1 = 3475, seed2 = 7859, seed3 = 27873);
%adaptall(dataset=fev1, y=second, xa = sequence, classa = sequence,
   seed1 = 6985, seed2 = 3541, seed3 = 18765);
%adaptall(dataset=fev1, y=treatment, xa = sequence, classa=sequence,
   seed1 = 9723, seed2 = 8452, seed3 = 29543);
```

This example produced a p-value of $p_{AD} = 0.6797$ for the adaptive test of first-order effect, $p_{AD} = 0.6341$ for the adaptive test of second-order effect, and $p_{AD} = 0.0328$ for the adaptive test of treatment effect. These three tests can be executed quickly on a modern computer.

8.2.4 Recommendations for Cross-Over Trials

In the analysis of cross-over trial data, it is important to distinguish between data sets that have baseline measurements and those that lack these measurements. We have already noted that, if baseline measurements are not available, then the test for carry-over effect will have low power because there will be between-subject variability in the sum of the treatment measurements.

In addition, there is a potential problem because the testing procedure is sequential in nature. Without baseline measurements the first step is to test for carry-over. If we fail to reject the null hypothesis, we perform a test for treatment effect using the differences between treatment measurements. If we do reject the null hypothesis, we would perform a two-sample test using only the data from the first period. This is the approach recommended by Grizzle (1965) and followed by Hills and Armitage (1979). Although this appears to be a logical approach, Freeman (1989) showed that this sequential procedure, when used on data sets that did not contain baseline measurements, inflated the probability of a Type I error. For a detailed analysis of this topic see Jones and Kenward (2003, Section 2.7). They recommend that, to

Table 8.7 R code for three tests with the asthma data set

```
source("adapt.r")
dataset <- c("fev1.dat")
crossover <- read.table(dataset)
attach(crossover)
complete <- c("y ~ sequence")
reduced <- c("y ~ 1")
r <- 5000
seed <- 7882
y <- x1 - x2
pvalue <- adapt(dataset, complete, reduced, r, seed)
y <- x1 - y1 + x2 - y2
pvalue <- adapt(dataset, complete, reduced, r, seed)
y <- y1 - y2
pvalue <- adapt(dataset, complete, reduced, r, seed)
```

avoid these problems, the study should be designed to minimize the possibility of a carry-over effect. See Senn (2007, Chapter 17) for an excellent discussion of the carry-over problem.

When baseline measurements are available the sequential procedure given by Kenward and Jones (1987) is recommended. In their approach the test for the first-order carry-over effect avoids the between-subject variability by using the differences between baseline measurements. Hence, it should be reasonably powerful. However, caution should be used in interpreting the results because the overall approach is sequential in nature. See Jones and Kenward (2003, Chapter 2) for discussions concerning the use of baseline measurements in cross-over trials and for the use of covariates in the analysis of cross-over data.

It should also be emphasized that the tests that are used to evaluate treatment effect and carry-over effects are two-sample tests. It is possible to use any two-sample test to evaluate these effects. We have included, in Figures 8.5 and 8.6, the results for the Wilcoxon rank-sum test, which is easy to perform and is familiar to most researchers. It has the advantage of being more robust than the t test for long-tailed distributions. The normal scores test could also be used to perform the test. The two-sample adaptive test has the advantage of being relatively powerful for many symmetric distributions and can be generalized to more complex designs.

EXERCISES

8.1 In Table 8.1 there are $C = 10$ centers with 10 subjects assigned to each of two treatments. Suppose the reduced model contains the center effect, but we shuffle the residuals over all centers. If we add the shuffled residuals to the predicted values would the Y_k vector be equal to the same shuffle of the original observations? Explain.

8.2 In this chapter we have assumed that we have a multicenter trial with C centers and our objective is to compare the effectiveness of two treatments. This model was described by Equation (8.1). Now suppose we want to include a covariate in the model.

a) Write the complete model for a multicenter trial that includes one covariate.

b) If the permutations of residuals are performed within each center, will the resulting vector of dependent variables Y_k equal the same permutation of Y? Why or why not?

8.3 Suppose a small multicenter trial has only $C = 2$ centers and in each of these centers there are only 4 subjects who are randomly assigned to two treatments such that exactly 2 are given each treatment.

a) How many different permutations of the residuals could be obtained? Note that different permutations sometimes yield identical test statistics, so there will be fewer distinct test statistics. We are only interested in the number of permutations of the residuals.

b) If we have $C = 5$ centers and assign 10 subjects randomly such that exactly 5 are given each treatment, how many different permutations of the residuals could be obtained?

8.4 Suppose three centers were enlisted to participate in a clinical trial of the effectiveness of two statin drugs to lower total cholesterol in patients who were diagnosed as hypercholesteremic. The percent reduction in total cholesterol over an 8-week period of the trial are given below for the patients in the trial. Positive values indicate reductions in cholesterol while negative values indicate an increase in total cholesterol. Using R code, compare the drugs in this trial. Use center as a blocking factor.

	Drug	
Center	Statin A	Statin B
1	23, 6, 18, 12, 15	22, 27, 31, 13, -24
2	12, 29, 17, 25, 9	28, 43, 15, 21, 37
3	26, 21, 7, 13, 18, 22, 31	-27, 38, 16, 25, 11, 31, 35

8.5 Using the data from the previous exercise, use SAS to test the null hypothesis that the drugs are equally effective.

8.6 In Table 8.1 we displayed the data for the change from baseline for diastolic blood pressure over eight weeks of treatment in a multicenter trial. Suppose the researcher also obtained the age and gender of each subject and wanted to include age and gender as covariates in the analysis. Give the R code or the SAS code that would be required to test the null hypothesis of no difference between treatments after controlling for age and gender.

8.7 In a test for treatment effect in a multicenter clinical trial, why does the within-center permutation of residuals make more sense than permuting the residuals over all centers?

8.8 For testing the direct effect in a cross-over study we found, in Figure 8.5, the power of the t test, the Wilcoxon test, and the adaptive test. These results were shown for studies having 16 subject assigned to each group. Suppose you felt, prior to collecting the data, that the distribution of the treatment differences could follow any of the 10 distributions and that these distributions were equally likely. Roughly evaluate the average differences between the tests when averaged over the 10 distributions.

8.9 As stated in the text, in order to test the first-order carry-over effect in a cross-over trial that has baseline measurements, we compute the differences $x_{11k} - x_{12k}$ for the kth subject in the first group and $x_{21k} - x_{22k}$ for the kth subject in the second group.

- **a)** What is the expected value of $x_{11k} - x_{12k}$ when the expectation is taken over all subjects and all possible responses for those subjects? Recall that the subject effect is a random variable with a mean of zero.
- **b)** What is the expected value of $x_{21k} - x_{22k}$ over all subjects and responses?
- **c)** Explain why it makes sense to use these differences in a two-sample test of $H_0 : \theta = 0$.

8.10 Using the data in Table 8.3 for a cross-over study, perform the test for treatment by period interaction, using any one of the two-sample tests. Are you able to reject the null hypothesis? Why or why not?

8.11 In Table 8.5 we have data on the FEV1 measurements for 17 subjects. Now suppose we had, for the 17th subject, mistakenly recorded a value of 4.01 instead of 1.01 for the "after wash-out" measurement. This observation is clearly an outlier.

- **a)** How would the inclusion of this outlier affect the test for treatment effect?
- **b)** How would this affect the test for first-order carry-over effect?
- **c)** What approach should be used in the analysis of this data if it included such a outlier? Identify what tests would be performed.

8.12 Suppose a researcher routinely performed cross-over studies, and each time he performed a study he failed to obtain baseline measurements. He performed a test for treatment by period interaction by using a two-sample test on the sum of the two measurements for each individual in the study, in the manner that we have described in the text.

The researcher claims that he never observed an interaction so there is no reason to make the effort to obtain baseline measurements. Does this approach make sense? Explain.

CHAPTER 9

ADAPTIVE MULTIVARIATE TESTS

9.1 THE TRADITIONAL LIKELIHOOD RATIO TEST

In the first eight chapters we have restricted our attention to adaptive tests for a subset of coefficients in a linear model having one dependent variable. A test for a subset of coefficients in a linear model that has only one dependent variable will be called a univariate test. In this chapter we extend our methods to multivariate linear models that can include several dependent variables, and we develop an adaptive test for a subset of coefficients in these multivariate linear models. In this section we introduce the notation for multivariate linear models and describe one of the test statistics for traditional (non-adaptive) multivariate tests.

Consider the multivariate linear model in matrix form as $\mathbf{Y} = \mathbf{X}\boldsymbol{\beta} + \boldsymbol{\varepsilon}$, where $\mathbf{Y}$ is an $n \times p$ matrix of dependent variables, $\mathbf{X}$ is an $n \times q$ matrix of constants, $\boldsymbol{\beta}$ is a $q \times p$ matrix of regression coefficients, and $\boldsymbol{\varepsilon}$ is an $n \times p$ error matrix. The error vectors are assumed to be independent and identically distributed. However, for an individual observation the error for one variable may be correlated with the error for another variable. To test a subset of regression coefficients, we will use a reduced model that includes r independent variables with $1 \leq r < q$. Let $\mathbf{X}_R$ be the $n \times r$ matrix consisting of the first r columns of $\mathbf{X}$ and let $\boldsymbol{\beta}_R$ be the $r \times p$ matrix consisting

Adaptive Tests of Significance Using Permutations of Residuals with R and SAS®.
By Thomas W. O'Gorman

of the first r rows of $\boldsymbol{\beta}$. Let $\mathbf{X}_A$ be the $n \times (q-r)$ matrix of the last $q-r$ columns of $\mathbf{X}$, and let $\boldsymbol{\beta}_A$ be the $(q-r) \times p$ matrix of the last $q-r$ rows of $\boldsymbol{\beta}$. With this notation the full model can be written as

$$\mathbf{Y} = \mathbf{X}_R\boldsymbol{\beta}_R + \mathbf{X}_A\boldsymbol{\beta}_A + \boldsymbol{\varepsilon} = [\mathbf{X}_R|\mathbf{X}_A] \begin{bmatrix} \boldsymbol{\beta}_R \\ \boldsymbol{\beta}_A \end{bmatrix} + \boldsymbol{\varepsilon}.$$

We will test $H_0 : \boldsymbol{\beta}_A = \mathbf{0}$ versus $H_a : \boldsymbol{\beta}_A \neq \mathbf{0}$, where $\mathbf{0}$ is a $(q-r) \times p$ matrix of zeros.

With multivariate data having $p > 1$ we often use Wilks' lambda as the test statistic, which is

$$\Lambda^{2/n} = \frac{\left|\hat{\boldsymbol{\Sigma}}_C\right|}{\left|\hat{\boldsymbol{\Sigma}}_R\right|} = \frac{|\hat{\boldsymbol{\varepsilon}}'_C\hat{\boldsymbol{\varepsilon}}_C|}{|\hat{\boldsymbol{\varepsilon}}'_R\hat{\boldsymbol{\varepsilon}}_R|},$$

where $\hat{\boldsymbol{\varepsilon}}_C$ and $\hat{\boldsymbol{\varepsilon}}_R$ are the $n \times p$ matrices of residuals from the complete and reduced models, respectively, with $\hat{\boldsymbol{\Sigma}}_C = (1/n)\hat{\boldsymbol{\varepsilon}}'_C\hat{\boldsymbol{\varepsilon}}_C$ and $\hat{\boldsymbol{\Sigma}}_R = (1/n)\hat{\boldsymbol{\varepsilon}}'_R\hat{\boldsymbol{\varepsilon}}_R$. Note that $\Lambda^{2/n}$ is a generalization of the test statistic that we used earlier in this book with models having one dependent variable.

With multivariate tests we cannot use the F test statistic, but will continue to use $\Lambda^{2/n}$ as the test statistic for both the multivariate and univariate tests. For a traditional likelihood ratio (LR) test, under the assumption that the errors are distributed as a multivariate normal distribution, it is often possible to calculate the LR p-value (p_{LR}) based on its relation to the F distribution. For example, if $q-r=1$ or $q-r=2$, we can calculate the p-value using the relationship of the Λ distribution to the F distribution that is described by Mardia, Kent, and Bibby (1979, p. 83). We reject the null hypothesis if $p_{LR} \leq \alpha$. As we shall see, the LR test has excellent statistical properties when the error distribution is a multivariate normal distribution. However, multivariate normal error distributions may be rare, so we will attempt to develop an adaptive test that has higher power than the traditional LR test with non-normal errors.

9.2 AN ADAPTIVE MULTIVARIATE TEST

9.2.1 The Projection Method

Earlier in this book, when we performed tests of significance with one dependent variable, it was relatively easy to identify and downweight outliers. However, multivariate tests are more complex. With two dependent variables a visual inspection of a scatter plot may be sufficient to enable a researcher to identify the outliers, but, if the number of dependent variables exceeds two, it is not easy to identify outliers.

The general approach we will take is to use a projection method to evaluate how far an observation is from the center of the cloud of points. The projection method is not new. Wilcox (2003) used it to approximate the Tukey depth of an observation in p-dimensional space. By projecting the points in p-dimensional space onto a line we reduce the problem to identification of an outlier in one dimension. After we have projected the points, the adaptive weighting scheme is similar to the method that we used with univariate tests.

With multivariate dependent variables having $p > 1$ the weighting procedure uses the matrix $\hat{\varepsilon}$, which is an $n \times p$ matrix consisting of the ordinary residuals from the reduced model $\mathbf{Y} = \mathbf{X}_R\boldsymbol{\beta}_R + \boldsymbol{\varepsilon}_R$. Because the p variables could be measured on p different scales, it is desirable to standardize the residuals. To standardize these residuals we use robust estimates of the standard deviations of the residuals that are in the p columns of $\hat{\varepsilon}$. Let $\hat{\varepsilon}_{.25,j}$ denote the traditional estimator of the 25th percentile of the jth residual and let $\hat{\varepsilon}_{.75,j}$ denote the traditional estimator of the 75th percentile of the jth residual. We calculate $\hat{\sigma}_{\varepsilon,j} = (\hat{\varepsilon}_{.75,j} - \hat{\varepsilon}_{.25,j})/1.349$ for $j =, \ldots, p$ as the robust estimates of the standard deviations of the p residuals. Next, the standardized residuals are computed as $v_{i,j} = \hat{\varepsilon}_{i,j}/\hat{\sigma}_{\varepsilon,j}$ for $i = 1, \ldots, n$ and $j = 1, \ldots, p$, so that the pth column of the $n \times p$ matrix $\mathbf{v}$ consists of the standardized residuals for the pth variable. We now will center these standardized residuals.

Let the mediancenter be the $1 \times p$ row vector $\tilde{\mathbf{v}}$, where the mediancenter is defined to be the point that minimizes the sum of the Euclidean distances from the point to the n observations in p-dimensional space. Haldane (1948) showed that the mediancenter is unique if $p > 1$. The values in v are centered to obtain the matrix of centered standardized residuals $\mathbf{c} = \mathbf{v} - \mathbf{1}\tilde{\mathbf{v}}$, where $\mathbf{1}$ is an $n \times 1$ vector containing ones. After centering, the vector $\mathbf{c}_i$, which is the ith row of $\mathbf{c}$, contains the centered standardized residuals for the ith observation.

To obtain the weight for the ith observation, we begin by projecting the centered standardized residuals onto the line (L_i) that connects $\mathbf{0}$ to $\mathbf{c}_i$, as shown in Figure 9.1. The orthogonal projection of $\mathbf{c}_j$ onto L_i is $p_{ij} = [(\mathbf{c}_i\mathbf{c}_j')/(\mathbf{c}_i\mathbf{c}_i')]\mathbf{c}_i$, for $j = 1, \ldots, n$. (See Mardia, Kent, and Bibby (1979, p. 482.) Next, we compute the signed Euclidean distance (D_{ij}) from p_{ij} to zero, as $D_{ij} =$ sign $(\mathbf{c}_i\mathbf{c}_j')||p_{ij}||$ for $j = 1, \ldots, n$, where sign$(\mathbf{c}_i\mathbf{c}_j')$ is the sign of $\mathbf{c}_i\mathbf{c}_j'$ and $||p_{ij}||$ is the Euclidean distance from p_{ij} to zero. We project all of the points onto line L_i so that we can use a univariate weighting method to weight the ith observation. The distance D_{ij} can be interpreted as the distance of the jth observation from the mediancenter in the direction of the ith observation. If the jth observation is far from zero it is possible that D_{ij} could be close to zero because the projection of $\mathbf{c}_j$ onto L_i could be close to zero. Hence, D_{ij} is a measure that takes distance and direction into account.

9.2.2 Adaptive Weighting

To compute the weight for the ith observation we use the set of signed distances of projected points from the origin for the ith observation. Let $D_i = \{D_{ij} : j = 1, \ldots, n\}$ denote that set of points. The weights are computed in the same manner that we used with the residuals in the univariate regression. That is, we use the procedure in Section 5.1, with the set D_i in place of the residuals from the univariate model and perform all of the smoothing and centering in the same way. These signed distances are shown in Figure 9.2 for the points in Figure 9.1. For the ith observation

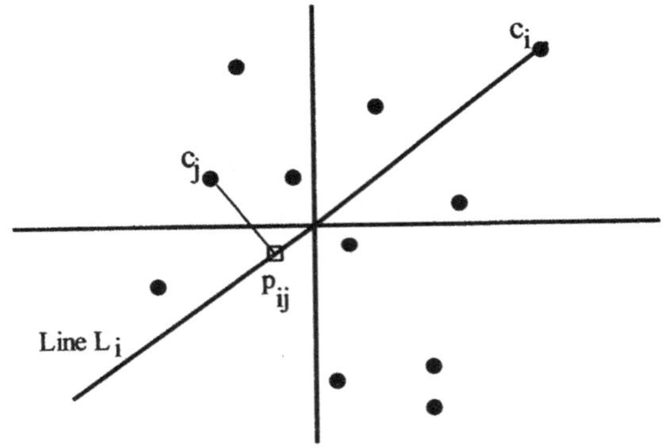

Figure 9.1 The projection of the points onto the line L_i.

we compute the sample weight $\hat{w}_i$, but do not compute the weights for the $n-1$ other points that were projected onto line L_i.

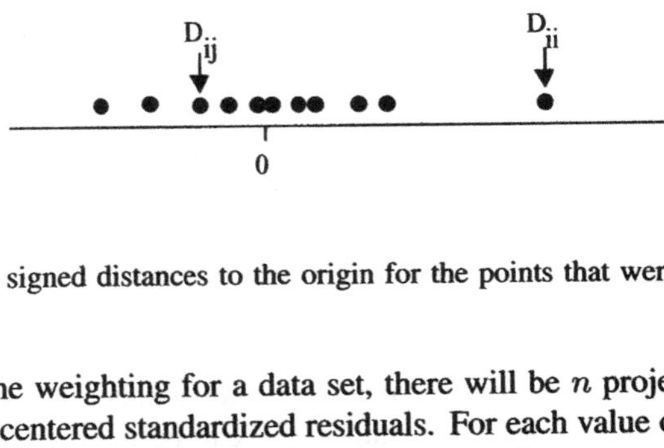

Figure 9.2 The signed distances to the origin for the points that were projected onto the line L_i.

To perform the weighting for a data set, there will be n projections for each of the n vectors of centered standardized residuals. For each value of $i = 1, \ldots, n$ we must project the points onto the line L_i to determine the set D_i and the weight w_i. Note that, if we used this multivariate method for univariate regression models having $p = 1$, the projected values would equal the standardized and centered residuals so the weights would be identical to those obtained from the univariate method. Hence, the multivariate weighting procedure is a generalization of the univariate weighting procedure.

To test $H_0 : \boldsymbol{\beta}_A = \mathbf{0}$ versus $H_a : \boldsymbol{\beta}_A \neq \mathbf{0}$, we will use the WLS method that was used in the previous chapters. As before, let $\mathbf{W}$ be an $n \times n$ matrix having weights w_i for $i = 1, \ldots, n$ on the diagonal and zeros off the diagonal. One easy way to find the test statistic is to premultiply the matrices in the model matrix by $\mathbf{W}$ to obtain

$$\mathbf{WY} = \mathbf{WX}_R\boldsymbol{\beta}_R + \mathbf{WX}_A\boldsymbol{\beta}_A + \mathbf{W}\boldsymbol{\varepsilon}.$$

We can express this full model using $\mathbf{Y}^* = \mathbf{WY}, \mathbf{X}_R^* = \mathbf{WX}_R, \mathbf{X}_A^* = \mathbf{WX}_A$, and $\boldsymbol{\varepsilon}^* = \mathbf{W}\boldsymbol{\varepsilon}$, as

$$\mathbf{Y}^* = \mathbf{X}_R^*\boldsymbol{\beta}_R + \boldsymbol{X}_A^*\boldsymbol{\beta}_A + \boldsymbol{\varepsilon}^*.$$

This is identical to the weighted transformed model that we used in Chapter 5, except that $\mathbf{Y}^*$, $\boldsymbol{\beta}_R$, $\boldsymbol{\beta}_A$, and $\boldsymbol{\varepsilon}^*$ now have p columns. Next, we use ordinary least squares methods to obtain the residuals $(\hat{\boldsymbol{\varepsilon}}_C^*)$ and the covariance matrix of the residuals $\hat{\boldsymbol{\Sigma}}_C^* = (1/n)(\hat{\boldsymbol{\varepsilon}}_C^*)'\hat{\boldsymbol{\varepsilon}}_C^*$ for this adaptively weighted complete model.

The reduced model is weighted in the same manner to obtain $\mathbf{Y}^* = \mathbf{X}_R^*\boldsymbol{\beta}_R + \boldsymbol{\varepsilon}_R^*$. The ordinary least squares method are used to obtain the residuals and the $\hat{\boldsymbol{\Sigma}}_R^*$ matrix. Based on these estimates of the covariance matrix of the errors we compute Wilks' lambda

$$\Lambda_u^{2/n,*} = \frac{|\hat{\boldsymbol{\Sigma}}_C^*|}{|\hat{\boldsymbol{\Sigma}}_R^*|}$$

as the adaptive test statistic for the unpermuted data. Because the distribution of this adaptive test statistic is not known, we will use a permutation method to compute the p-value.

9.2.3 Permutation Method

The permutation test uses the multivariate version of the Freedman and Lane (1983) method of permuting the residuals from the reduced model $(\hat{\boldsymbol{\varepsilon}}_R)$. Let $\mathbf{S}_k$ be the permutation matrix for the kth permutation. The permuted residuals for the kth permutation are $S_k\hat{\boldsymbol{\varepsilon}}_R$, which is a permutation of the rows of the $\hat{\boldsymbol{\varepsilon}}_R$ matrix. This is used to calculate the "new" Y matrix for the kth permutation as $\mathbf{Y}_k = \hat{\mathbf{Y}}_R + \mathbf{S}_k\hat{\boldsymbol{\varepsilon}}_R$, where $\hat{\mathbf{Y}}_R = \mathbf{X}_R\hat{\boldsymbol{\beta}}_R$ is the $n \times p$ matrix of predicted values from the reduced model. Next, we calculate the weights for the kth permutation, which are based on the reduced model $\mathbf{Y}_k = \mathbf{X}_R\boldsymbol{\beta}_R + \boldsymbol{\varepsilon}_{R,k}$, in the same manner that was used for the unpermuted data. After the weights are computed, we compute $\hat{\boldsymbol{\Sigma}}_{C,k}^*$ from the weighted complete model and we compute $\hat{\boldsymbol{\Sigma}}_{R,k}^*$ from the weighted reduced model. Next, we compute the permutation test statistic $\Lambda_k^{2/n,*}$ for the kth permutation. As we have done with other permutation tests, we let E equal the total number of permutations with $\Lambda_k^{2/n,*} \leq \Lambda_u^{2/n,*}$ and we compute the p-value as $p_{AD} = (E+1)/(R+1)$. We reject the null hypothesis if $p_{AD} \leq \alpha$.

Many of the permutation results that we have derived in previous chapters can be applied to multivariate regressions. For models having one or more covariates in the reduced model the weights will need to be computed with each permutation of the residuals, but if $\mathbf{X}_R = \mathbf{1}$ a permutation of the residuals will result in a permutation of the rows of $\mathbf{Y}$, so the weights for the kth permutation of the data will equal the kth permutation of the weights obtained from the unpermuted data.

9.2.4 Justification of the Projection Method

The projection method, while not difficult, may appear to be overly complex. However, the identification of outliers and the proper weighting of observations is much more difficult in several dimensions than it is in one dimension. Because we will be using this multivariate testing method to analyze repeated measures data, we need a procedure that can weight observations in three or more dimensions. Several approaches could be considered.

One approach is to identify and weight observations in each dimension and then combine the results to obtain an overall weight. However, this approach is problematic because a multivariate outlier may not appear to be an outlier in any of the dimensions. For example, in Figure 9.1, the point $\mathbf{c}_i$ is not an outlier in either the first or second dimensions, but it is separated from the rest of the data in the direction defined by the line L_i. Although variants of this approach could overcome some of the problems, it was not clear how this method would produce an appropriate adaptive weight for each observation, so this method was not considered further.

Another approach is to use a method related to convex hull trimming to downweight outliers. A convex hull method identifies all of the vertices of the smallest convex polyhedron that includes all of the residuals, but it is not clear what weight should be assigned to those points on the vertices of the convex hull. In addition, this method does not appear to be suited to our objective because we do not want to downweight observations that are not really outliers. For example, if we have residuals in three dimensions and they appear to be from a multivariate normal distribution we would not want to downweight any of the observations. Hence, we have not used the convex hull method.

We have used a projection method to weight the observations in a manner to normalize residuals along each line that was determined by connecting each point to the mediancenter. In this manner we hope to normalize the residuals in each of n directions defined by these lines. After weighting each point, we would like the empirical c.d.f. of the multivariate residuals from the weighted model to approximate the c.d.f. of a multivariate normal distribution.

9.3 AN EXAMPLE WITH TWO DEPENDENT VARIABLES

Data was collected by Simon (2001) on the nutritional status of an elderly population in the Dominican Republic. As part of that study, she obtained data on the age, body mass index (BMI), systolic blood pressure, and diastolic blood pressure for 108 females. Suppose we wanted to determine if the systolic and diastolic blood pressure measurements are related to BMI after controlling for age. We could perform a traditional multivariate test after fitting a full model that includes age and BMI as predictor variables and a reduced model that includes only the age variable. This approach gives a p-value of $p_{LR} = 0.2002$ for the non-adaptive likelihood ratio test. Hence, we would not reject the null hypothesis $H_0 : \boldsymbol{\beta}_A = \mathbf{0}$. However, we would like to see if an outlier influenced the p-value.

A scatterplot of the centered residuals for both dependent variables under the reduced model is shown in Figure 9.3. The scatterplot shows one outlier that may adversely affect the performance of a traditional test. If we use an adaptive test it may identify and downweight outliers so that the overall performance of the test might be improved.

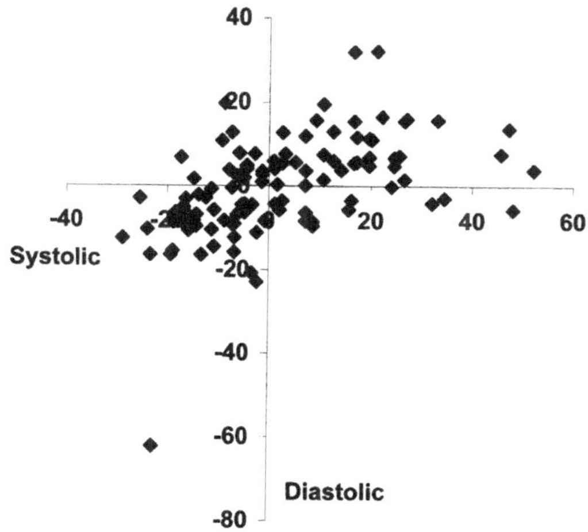

Figure 9.3 Scatterplot of the ordinary centered residuals for the systolic and diastolic blood pressure in a model that uses age as a predictor variable.

9.3.1 Using the SAS Macro

Suppose we have placed data for the diastolic blood pressure, the systolic blood pressure, the age, and the BMI on a SAS data set named `bpdata`. The SAS macro `adaptall` will be used to perform the multivariate adaptive test. This is the same macro that we used in previous chapters. The `adaptall` function, which is listed in Appendix B, can be downloaded from *www.math.niu.edu/~ogorman*. However, with multivariate data, the number of dependent variables will exceed one so the `adaptall` macro will perform the weighting by using the projection method. To invoke this macro we must specify the dataset, the dependent variables, and the number of dependent variables (`nvars`). The macro variables `xr` and `xa` are used to specify the variables in the X_A and X_R matrices, respectively. The other parameters are used to specify the random number seeds and the number of permutations. To perform an adaptive test of the null hypothesis that body mass index is unrelated to systolic and diastolic blood pressure, after controlling for age, we use the following statement to invoke the macro:

```
%adaptall(dataset=bpdata, y=systolic diastol, nvars=2,
        xr=age, xa=bmi,
        seed1=10670, seed2=20588, seed3=1932, nperm=2000);
```

The `adaptall` macro produced the following output, which should be sufficient to document the details of the test. Note that because `nvars` exceeds one the F test statistic is not printed.

```
Number of permutations = 2000
```

```
Starting seeds = 10670  20588  1932
SAS data set: simon   Number of observations = 108

Dependent variables   :  systolic diastol

Independent variables in reduced model   : Intercept  age
      Class variables in reduced model   :

Independent variables in complete model : Intercept age  bmi
      Class variables in complete model :

Adaptive test statistics for unpermuted data:

      Wilks Lambda =    0.9539

p-value for adaptive test =    0.0735
```

We observe that the p-value for the adaptive test ($p_{AD} = 0.0735$) is much smaller than the p-value for the LR test. This is due, in part, to the downweighting of the outlier, which was given a weight of 0.51 for the unpermuted data. Most the other weights were near 1.0 and all of the other weights were in the interval [0.74, 1.27]. This macro executed in under 10 minutes on an inexpensive personal computer purchased in 2008.

9.3.2 Using an R Function

An R function (`adaptmult`) computes the weights by standardizing the residuals, finding the mediancenter, centering the standardized residuals, and performing the projection method. This `adaptmult` function is more complex than the R functions that were used in the previous univariate tests. The `adaptmult` function, which is listed in Appendix G, can be downloaded from *www.math.niu.edu/~ogorman*. To test the null hypothesis that diastolic and systolic blood pressure are not related to the body mass index after controlling for age, we use the following code:

```
source("adaptmult.r")
dataset <- c("bp.dat")
testdata <- read.table(dataset)
attach(testdata)
yvars <- c("diastolic,systolic")
y <- cbind(diastolic,systolic)
complete <- c("y ~ age + bmi")
reduced <-  c("y ~ age")
r <- 1000
seed <- c(61413)
pvalue <- adaptmult(y,complete,reduced,r,seed)
```

The `source` statement loads the `adaptmult` function and the `dataset` variable contains the character string to identify the file containing the blood pressure data, which is `bp.dat`. In order to fully document the test, we must specify the two dependent variables in `yvars`. Because the `y` matrix is used as the matrix of dependent variables, we must use `cbind` to put them into the matrix `y`. We define the complete and reduced models by their character strings and specify the number of permutations and the random number seed. The last statement calls the `adaptmult` function that will perform the test. The following output is produced:

```
Input:
    Data set: multivariate.simon.r.dat
    Dependent variables:  diastolic,systolic
    Complete model:  y ~ age + bmi
    Reduced model :  y ~ age
    Number of permutations =  2000
    Random seed =  61413

  Output from Adaptive Test:
    Unpermuted lambda =  0.9539
    p-value           =  0.069
```

Unfortunately, the R code takes about 20 minutes to execute on an inexpensive personal computer purchased in 2008. As computers become faster, it is expected that this procedure will be quite practical.

9.4 PERFORMANCE OF THE ADAPTIVE TEST

9.4.1 Significance Level of the Tests

The tests that we have used in previous chapters have generally maintained their level of significance. For these tests, we used a linear model with one independent variable. However, in this chapter we will see that the LR test does not maintain its significance level in some situations with multivariate linear models. Hence, we need to carefully document the performance of these tests. As we did in Chapter 5, it will be convenient to classify reduced models as "simple" models if they have only the intercept in the reduced model. We will classify reduced models as "complex" models if they have at least one covariate.

If we have a simple model and we perform a test that is based on permutations of residuals, then the resulting test should maintain its level of significance. To see this, we can use the same argument that we used for univariate tests in Chapter 5. Recall that, if we knew the errors, we could perform a permutation test using those errors and the permutation test statistics would be equally likely because the errors are independent and identically distributed. If we have a simple model then, under the null hypothesis, the permutation of errors would produce a permutation of the rows in the Y matrix. Hence, we could perform a valid test by permuting the residuals under the reduced model, which also would produce a permutation of the rows in the Y matrix.

To check the level of significance of the adaptive test and the LR test, we performed a simulation study with $n = 50$ observations having $p = 2$ dependent variables. We used the complete model $\mathbf{Y} = \mathbf{1}\boldsymbol{\beta}_0 + \mathbf{x}_1\boldsymbol{\beta}_1 + \mathbf{x}_2\boldsymbol{\beta}_2 + \boldsymbol{\varepsilon}$ and the reduced model $\mathbf{Y} = \mathbf{1}\boldsymbol{\beta}_0 + \boldsymbol{\varepsilon}$ to test the hypothesis $H_0 : \boldsymbol{\beta}_1 = \boldsymbol{\beta}_2 = \mathbf{0}$ versus $H_a : \boldsymbol{\beta}_1 \neq \mathbf{0}$ or $\boldsymbol{\beta}_2 \neq \mathbf{0}$. In this model $\mathbf{x}_1$ and $\mathbf{x}_2$ are $n \times 1$ vectors, while $\boldsymbol{\beta}_1$ and $\boldsymbol{\beta}_2$ are 1×2 row vectors. We generated the two independent variables from a multivariate log-normal distribution with a correlation of 0.4 between the variables. We generated the errors from three different multivariate distributions each having a correlation of 0.4 between the errors. To find the empirical significance level of the test, we set $\boldsymbol{\beta}_1 = \boldsymbol{\beta}_2 = \mathbf{0}$. The empirical significance levels, based on a simulation study that used 10,000 data sets, are displayed in Figure 9.4.

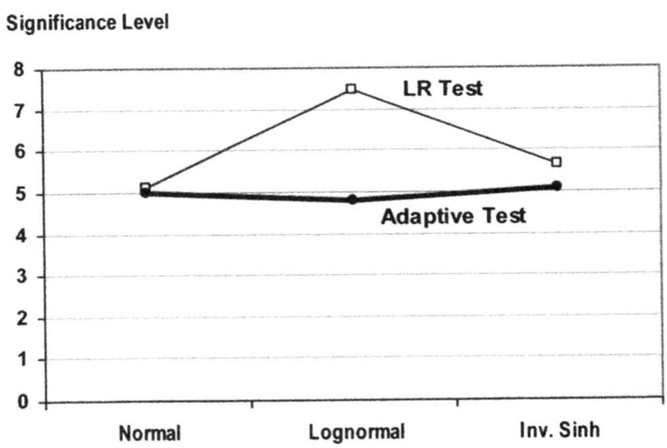

Figure 9.4 The empirical significance level, in percent, for the LR test and the multivariate adaptive test for $n = 50$ observations generated from a model $\mathbf{Y} = \mathbf{1}\boldsymbol{\beta}_0 + \mathbf{x}_1\boldsymbol{\beta}_1 + \mathbf{x}_2\boldsymbol{\beta}_2 + \boldsymbol{\varepsilon}$. We set $\boldsymbol{\beta}_1 = \boldsymbol{\beta}_2 = 0$ for these estimates.

We also used a simulation study with $n = 100$ observations having $p = 4$ dependent variables to check the empirical significance level of a test when the reduced model is not a simple model. We used the complete model $\mathbf{Y} = \mathbf{1}\boldsymbol{\beta}_0 + \mathbf{x}_1\boldsymbol{\beta}_1 + \mathbf{x}_2\boldsymbol{\beta}_2 + \mathbf{x}_3\boldsymbol{\beta}_3 + \mathbf{x}_4\boldsymbol{\beta}_4 + \boldsymbol{\varepsilon}$ and the reduced model $\mathbf{Y} = \mathbf{1}\boldsymbol{\beta}_0 + \mathbf{x}_1\boldsymbol{\beta}_1 + \mathbf{x}_2\boldsymbol{\beta}_2 + \boldsymbol{\varepsilon}$ to test the hypothesis $H_0 : \boldsymbol{\beta}_3 = \boldsymbol{\beta}_4 = \mathbf{0}$ versus $H_a : \boldsymbol{\beta}_3 \neq \mathbf{0}$ or $\boldsymbol{\beta}_4 \neq \mathbf{0}$. We generated the first and third covariates from a multivariate log-normal distribution with a correlation of 0.4 between the covariates and generated the second and the fourth covariates in a similar way. We generated the $1 \times p$ vector of errors from three different multivariate distributions, with each having a correlation of 0.4 between the errors. The empirical significance level, based on a simulation study that used 10000 data sets, are displayed in Figure 9.5.

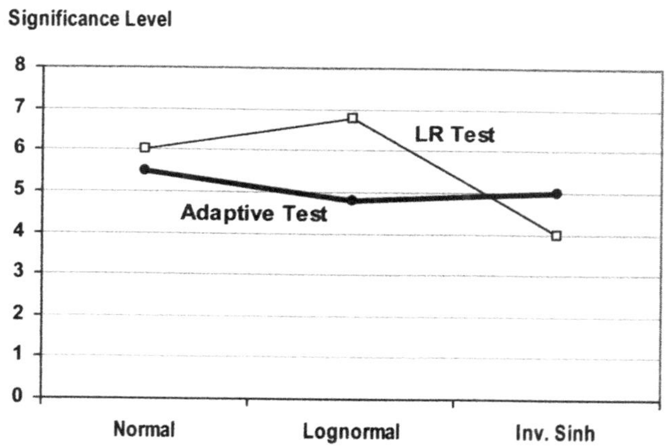

Figure 9.5 The empirical significance level, in percent, for the LR test and the multivariate adaptive test with $n = 100$ observations from a model $\mathbf{Y} = \mathbf{1}\boldsymbol{\beta}_0 + \mathbf{x}_1\boldsymbol{\beta}_1 + \mathbf{x}_2\boldsymbol{\beta}_2 + \mathbf{x}_3\boldsymbol{\beta}_3 + \mathbf{x}_4\boldsymbol{\beta}_4 + \boldsymbol{\varepsilon}$. We set $\boldsymbol{\beta}_3 = \boldsymbol{\beta}_4 = 0$ for these estimates.

Based on the results shown in Figure 9.4 for a simple model with $n = 50$ observations, we see that the LR test appears to reject too often with multivariate lognormal error distributions. As we expected, the adaptive test maintains its level of significance. Based on the simulation results shown in Figure 9.5, we conclude that the adaptive test maintains its level of significance close to the nominal value with this complex model. However, the LR test rejects too often with lognormal errors. This failure of the LR test to maintain its level of significance should be considered when evaluating the relative power of these tests.

9.4.2 Power of the Tests

The power estimates for the LR test and the adaptive test are given in Figure 9.6. The simulation study was identical to that described above for $n = 50$ observations generated from the model $\mathbf{Y} = \mathbf{1}\boldsymbol{\beta}_0 + \mathbf{x}_1\boldsymbol{\beta}_1 + \mathbf{x}_2\boldsymbol{\beta}_2 + \boldsymbol{\varepsilon}$, except we used $\boldsymbol{\beta}_1 = \boldsymbol{\beta}_2 = [1.2/\sqrt{n}, 1.2/\sqrt{n}]$ to obtain powers near 50% for the LR test. It can be seen that these tests are nearly equal in power with the normal error distribution and the inverse hyperbolic sine error distribution, but the adaptive test is much more powerful with the lognormal error distribution.

To evaluate the power of these tests with a complex model we used the same simulation study that we have described with $n = 100$ observations having $p = 4$ dependent variables from the model $\mathbf{Y} = \mathbf{1}\boldsymbol{\beta}_0 + \mathbf{x}_1\boldsymbol{\beta}_1 + \mathbf{x}_2\boldsymbol{\beta}_2 + \mathbf{x}_3\boldsymbol{\beta}_3 + \mathbf{x}_4\boldsymbol{\beta}_4 + \boldsymbol{\varepsilon}$, except we set $\boldsymbol{\beta}_3 = \boldsymbol{\beta}_4 = [1.5/\sqrt{n}, 1.5/\sqrt{n}]$ to achieve powers near 70% for the LR test. The independent variables and errors were correlated in the same manner that they were for the simulations used to estimate the significance level of the test. The power estimates, which are displayed in Figure 9.7, show that the adaptive test

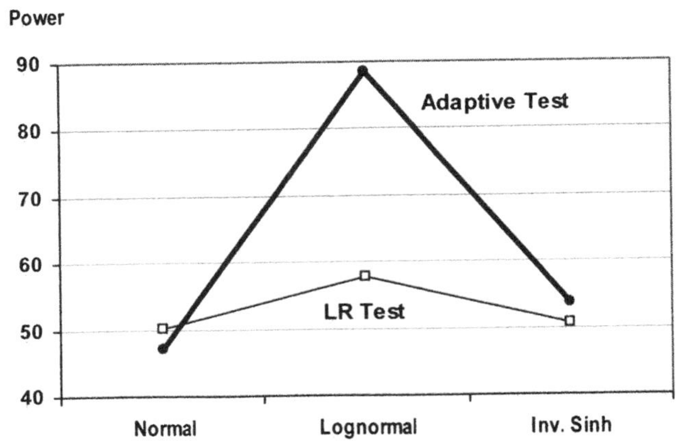

Figure 9.6 The empirical power, in percent, for the LR test and the multivariate adaptive test with $n = 50$ observations from the model $\mathbf{Y} = \mathbf{1}\boldsymbol{\beta}_0 + \mathbf{x}_1\boldsymbol{\beta}_1 + \mathbf{x}_2\boldsymbol{\beta}_2 + \boldsymbol{\varepsilon}$. We set $\boldsymbol{\beta}_1 = \boldsymbol{\beta}_2 = [1.2/\sqrt{n}, 1.2/\sqrt{n}]$ for these power estimates.

has a large power advantage over the LR test with the lognormal error distribution, a small power advantage with the inverse hyperbolic sine error distribution, and a small power disadvantage with the normal error distribution.

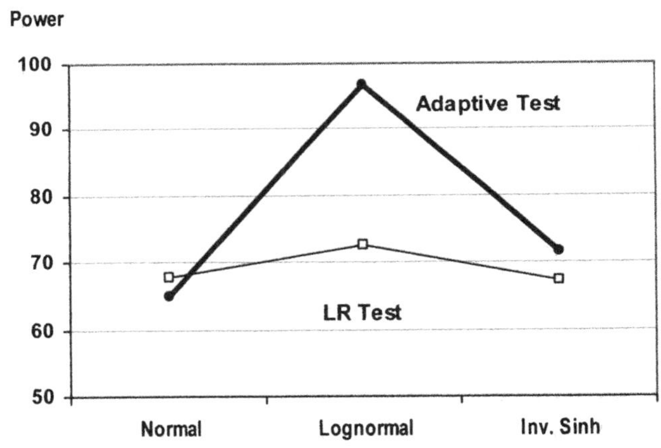

Figure 9.7 The empirical power, in percent, for the LR test and the multivariate adaptive test with $n = 100$ observations from the model $\mathbf{Y} = \mathbf{1}\boldsymbol{\beta}_0 + \mathbf{x}_1\boldsymbol{\beta}_1 + \mathbf{x}_2\boldsymbol{\beta}_2 + \mathbf{x}_3\boldsymbol{\beta}_3 + \mathbf{x}_4\boldsymbol{\beta}_4 + \boldsymbol{\varepsilon}$. We set $\boldsymbol{\beta}_3 = \boldsymbol{\beta}_4 = [1.5/\sqrt{n}, 1.5/\sqrt{n}]$ for these power estimates.

9.5 CONCLUSIONS FOR MULTIVARIATE TESTS

Several conclusions about the LR test can be drawn from the simulation studies. First, the LR test does not always maintain its level of significance. Second, with non-normal errors the LR test is often not as powerful as the adaptive test. So the LR test appears to have some undesirable properties when compared to the adaptive test.

In contrast, the projection method that is used for adaptive weighting appears to produce a test that has excellent properties. We argued that, for simple models having only an intercept term in the reduced model, the adaptive test will maintain its level of significance. For complex models it does appear, based on some simulation results, that the adaptive test maintains its level of significance. Further, the projection method produces a test that has little loss in power compared to the LR test with multivariate normal error distributions. With non-normal error distributions the adaptive test appears to have a power advantage over the LR test.

However, these simulation studies were performed on only a few of the many possible situations that could arise in practice. We used $n = 50$ observations with a simple model and $n = 100$ observations with a complex model. We reported in our figures the results for only one value of the correlation between the independent variables and for only one value of the correlation between the errors, but other simulations that were performed with several other correlation structures show the same pattern that we have shown in our figures.

We will, in the next chapter, learn more about the properties of this adaptive multivariate test. We will see how the adaptive multivariate test can be used to perform tests of significance with repeated measures data and we will compare it to the commonly used tests of significance with repeated measures data. So, we will get further insight into the properties of the adaptive multivariate test in the next chapter.

EXERCISES

9.1 A scatterplot of the standardized and centered residuals for two variables is shown in this figure.

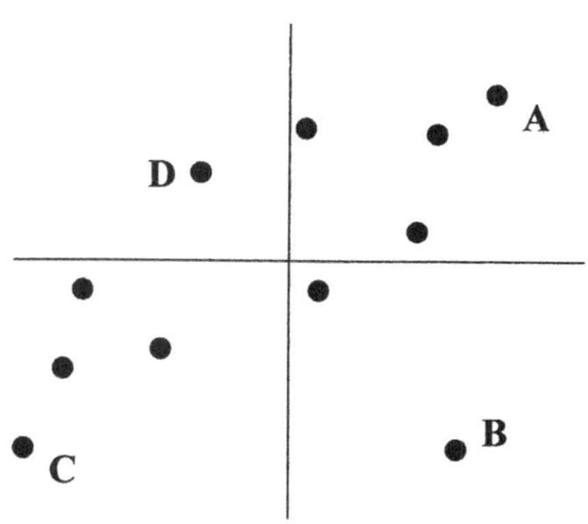

Which one of the points A, B, C, or D will be given the lowest weight? Explain why.

9.2 Suppose 32 volunteers, who had chronic lower back pain, were randomly assigned to one of four drug treatment groups. After six weeks of treatment the researchers measured their pain level on a visual analog scale and their range of motion. Suppose they wanted to test the null hypothesis that the drugs are equally effective and to do this they wanted to use the adaptive multivariate test. Can they permute the weights when they permute the residuals? Why or why not?

9.3 We permute the rows of the $n \times p$ matrix $\hat{\varepsilon}_R$ to obtain $\mathbf{S}_k\hat{\varepsilon}_R$ as the permuted residuals for the kth permutation. With this method, each row of $\hat{\varepsilon}_R$ is moved to a row of $\mathbf{S}_k\hat{\varepsilon}_R$. A different approach would be to use p different permutations for the p columns of $\hat{\varepsilon}_R$. Thus, with this method, for the first observation the residual for the first variable might be shuffled to row 4, while the residual for the second variable might be shuffled to row 6. Are there any advantages or disadvantages of using this method?

9.4
We defined the mediancenter as that point that minimizes the sum of the Euclidean distances to the n points.

a) In the following plot where is the mediancenter for these four points? No calculations should be necessary.

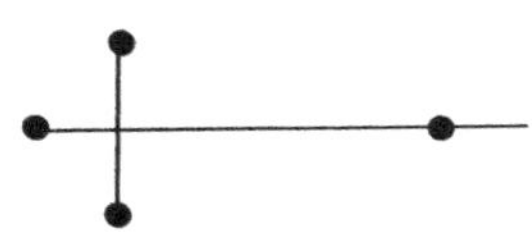

b) In the following plot where is the mediancenter? No calculations should be necessary.

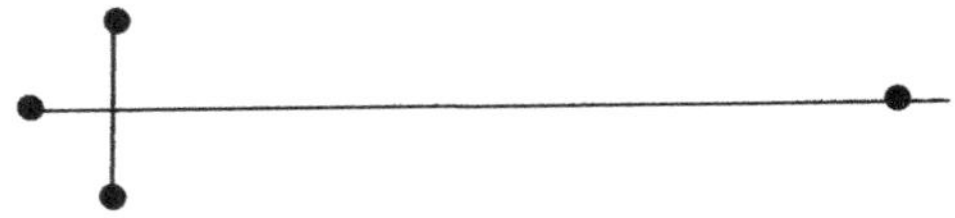

c) Is the mediancenter a robust measure of center? Explain.

9.5 Suppose that, in an effort to determine if the addition of barley to oat cereal would improve the health status of malnourished children, an experiment was conducted on 24 malnourished children. These children were randomly assigned to either a barley supplement group or a oat supplement (control) group. Each child was followed for 150 days and the gains in height and weight were used as the primary outcome measures. In Table 9.1 the treatment, the age of the child (in months), the gender, the height gain per day (centimeters per day), and the weight gain per day (kilograms per day) are given for the children. The two outcome measures are the height gain per day and the weight gain per day. This data set can be found at *www.math.niu.edu/~ogorman*.

Use a SAS macro to test the null hypothesis that the supplement was not effective, after controlling for age and gender.

Table 9.1 Data on 24 malnourished children

Supplement	Age	Gender	Height Gain (cm. per day)	Weight Gain (kg. per day)
Barley	28	male	0.035	0.018
Barley	32	female	0.028	0.015
Barley	37	female	0.031	0.022
Barley	29	female	0.033	0.028
Barley	34	male	0.029	0.025
Barley	36	male	0.034	0.029
Barley	41	female	0.026	0.019
Barley	31	male	0.031	0.025
Barley	35	male	0.033	0.021
Barley	38	female	0.036	0.028
Barley	37	female	0.037	0.027
Barley	35	female	0.032	0.029
Oatmeal	28	male	0.025	0.014
Oatmeal	30	male	0.027	0.025
Oatmeal	33	female	0.053	0.022
Oatmeal	27	female	0.023	0.018
Oatmeal	36	female	0.021	0.015
Oatmeal	37	male	0.031	0.019
Oatmeal	31	female	0.035	0.028
Oatmeal	38	male	0.021	0.015
Oatmeal	39	female	0.023	0.016
Oatmeal	33	female	0.026	0.018
Oatmeal	33	female	0.027	0.017
Oatmeal	32	female	0.022	0.019

9.6 Use the data in the previous problem along with the R function to test the null hypothesis that the supplement was not effective, after controlling for age and gender.

CHAPTER 10

ANALYSIS OF REPEATED MEASURES DATA

10.1 INTRODUCTION

Researchers often collect data on experimental units at several points in time because the additional cost of data collection is small compared to the benefits. For example, individuals who were randomly assigned to a weight loss treatment may be weighed every 2 weeks for 10 weeks. These repeated measurements could be used to evaluate the effectiveness of the treatment over the course of the experiment. By taking measurements every two weeks, instead of one measurement at the end of the 10 week study period, the researchers may obtain much more useful information with little additional cost.

In studies involving animals, researcher often obtain data at regular intervals after an intervention. Because animal experiments are costly and because researchers would like to use few animals in the experiment, they try to obtain as much information as possible from the animals that are used by taking several measurements.

Hence, statisticians often need to analyze data sets that have repeated measurements. This is not always an easy task because the analysis of repeated measurements can become complicated if the difference between the treatment groups varies over time. If the difference does change over time, we conclude that there is an interaction

Adaptive Tests of Significance Using Permutations of Residuals with R and SAS®.
By Thomas W. O'Gorman

of treatment and time. If the interaction is significant then the analysis proceeds by using the treatment differences at each time, or at a subset of the times. For example, if two groups of animals are measured at one hour intervals and the interaction is significant we might compare the groups at each time.

However, if the interaction is not significant then it may be reasonable to evaluate an overall effect of the treatment. The analysis might proceed by computing the average of the responses over the measurement times for each subject. For example, if animals are measured at one hour, two hours, and three hours after the treatment, we could compute the average response over those three times and use the averages for each animal in a two-sample test of significance to determine if the treatment is effective.

Thus, the analysis of repeated measurements usually requires two steps. The first step is a test for interaction between treatment and time. If that test is not significant, it may, in some situations, be reasonable to conclude that the responses are parallel and then perform the test for an overall treatment effect. If the test for interaction is significant, it makes little sense to evaluate an overall treatment effect, so we would compare the treatments at each time.

In the next few sections we will describe the likelihood ratio test, an adaptive test, and the mixed model test. In Section 10.6 we will consider several two-sample tests of parallelism when there are no missing values. We will compare the significance level and the power of the adaptive test to two non-adaptive tests. In Section 10.7 we will describe how a test for group (or treatment) effect could be performed in the absence of an interaction effect.

One common problem with repeated measures data is that some of the data may be missing at some of the measurement times. To address this problem, in Section 10.9 we will describe how these tests could be used to analyze repeated measurements having certain kinds of missing data.

Before we begin dealing with repeated measurements, we should point out that repeated measurements are a special kind of multivariate data. One important difference is that repeated measures data are presented in a time order whereas multivariate data can usually be put into any order. For example, if one variable is systolic blood pressure and the other is diastolic blood pressure, the ordering is not important.

Another important difference is that repeated measurements are usually on the same scale. Hence, with repeated measures data we can compute and interpret the differences between adjacent values as the change in that measurement over a time period. In contrast, with multivariate data it would usually make little sense to calculate the differences between variables. For example, if one variable is household income and another is the number of televisions in the home, the difference between these numbers would be difficult to interpret.

Because of the many challenges posed by repeated measures data, it is not surprising that a number of approaches have been developed to analyze these data. For a detailed description of many of these tests see Davis (2002). We will limit ourselves to three tests, and will begin by describing the LR test, the adaptive test, and the mixed model test.

10.2 THE MULTIVARIATE LR TEST

The traditional likelihood ratio (LR) test can be used to analyze repeated measures data. Suppose we have a nonmissing observation on each subject for each of t time points. Let y_{ij} denote the observed value of the ith subject at the jth time point. With the LR approach we view the t measurements as a vector and analyze the data using a multivariate regression model. That is, for the ith subject we record a vector of observations that becomes the ith row in the $\mathbf{Y}$ matrix. The multivariate tests use the model

$$\mathbf{Y} = \mathbf{X}\boldsymbol{\beta} + \boldsymbol{\varepsilon} = \mathbf{X}_R\boldsymbol{\beta}_R + \mathbf{X}_A\boldsymbol{\beta}_A + \boldsymbol{\varepsilon}. \quad (10.1)$$

This model is identical to that used in the previous chapter except that we have t observations instead of p dependent variables. In this model we define $\mathbf{X}_R$ to be the $n \times r$ matrix consisting of the first r columns of $\mathbf{X}$ and we define $\boldsymbol{\beta}_R$ as the $r \times t$ matrix consisting of the first r rows of $\boldsymbol{\beta}$. For the additional terms in the complete model we define $\mathbf{X}_A$ as the $n \times (q - r)$ matrix consisting of the last $q - r$ columns of $\mathbf{X}$, and we define $\boldsymbol{\beta}_A$ as the $(q - r) \times t$ matrix of the last $q - r$ rows of $\boldsymbol{\beta}$. The errors are denoted by $\boldsymbol{\varepsilon}$, which is an $n \times t$ matrix.

With this general model we perform a test of $H_0 : \boldsymbol{\beta}_A = \mathbf{0}$ versus $H_a : \boldsymbol{\beta}_A \neq \mathbf{0}$, where $\mathbf{0}$ is a $(q - r) \times t$ matrix of zeros. This is the same test that we performed in the last chapter. However, in this chapter we will confine our attention to the two-sample test of parallelism and the two-sample test of group effect. This test is performed by computing Wilks' lambda and the p-value associated with it by the method that was described in Chapter 9.

10.3 THE ADAPTIVE TEST

The adaptive test for repeated measures data uses the multivariate adaptive test that was described in the last chapter. In this chapter we will be performing two-sample tests, so the null hypothesis will be that there is no difference between the populations and we will compute the multivariate residuals under this null hypothesis. Next, observations that appear to be outliers in the t dimensional space will be downweighted and the permutation of residuals method will be used to produce a p-value.

With repeated measures data the adaptive method will downweight those observations that have an unusual growth pattern. For example, if we measure the heights (in inches) of males at 12, 13, and 14 years of age, we would expect these values to increase with age. Suppose we have a sample of 20 males and 19 of those individuals have increasing values for height. However, if the vector of heights for one individual is $[60, 66, 63]$ we would suspect that at least one of the measurements was in error. This individual would be downweighted because the residuals would be set apart from the the rest of the data in three dimensions.

So, in the application of the adaptive multivariate test to repeated measurements we will be concerned with differences between groups in the patterns of growth.

Individuals whose growth pattern is unusual will be downweighted so that it is easier to detect differences between groups.

10.4 THE MIXED MODEL TEST

Some researchers model the vector of responses as a function of the treatment effect, the subject effect, and the error. The treatment effect is assumed to be a fixed effect while the subject effect is assumed to be a random effect. Models that use fixed effects and random effects are called mixed models. This approach is popular because is allows very flexible modeling of fixed and random effects. Also, it is a practical approach because software is available to analyze repeated measures data. In much of the literature on mixed models the responses for the ith subject are written as a column vector. However, to be consistent with our notation for multivariate tests, we will use the $1 \times t$ row vector as the vector of responses. The fixed effects use a $1 \times b$ vector of coefficients denoted by $\boldsymbol{\beta}$. For the ith subject we will use a $b \times t$ model matrix of constants, which is denoted by $\mathbf{X}_i$, to indicate the treatment given to that subject. With this notation, for the ith subject the contributions of the fixed effects is $\boldsymbol{\beta}\mathbf{X}_i$, which is a $1 \times t$ vector.

For example, in a two-sample test we might let $\boldsymbol{\beta} = \begin{bmatrix} \mu & \Delta \end{bmatrix}$, where μ is the overall mean and Δ is the treatment effect for subjects in the second sample. If the treatment effect is constant over $t = 3$ times we would use $\mathbf{X}_i = \left[\begin{smallmatrix} 1 & 1 & 1 \\ 0 & 0 & 0 \end{smallmatrix}\right]$ if the ith subject is in sample 1 and $\mathbf{X}_i = \left[\begin{smallmatrix} 1 & 1 & 1 \\ 1 & 1 & 1 \end{smallmatrix}\right]$ if the ith subject is in sample 2. Thus, $\boldsymbol{\beta}\mathbf{X}_i = \begin{bmatrix} \mu & \mu & \mu \end{bmatrix}$ if the ith subject is in sample 1 and $\boldsymbol{\beta}\mathbf{X}_i = \begin{bmatrix} \mu + \Delta & \mu + \Delta & \mu + \Delta \end{bmatrix}$ if the ith subject is in sample 2.

The random effect for the ith subject is γ_i. This is a univariate random variable and we will assume that $E(\gamma_i) = 0$ and $V(\gamma_i) = \sigma_\gamma^2$. The $1 \times t$ model matrix $\mathbf{Z}_i$ will model the time effects so that $\gamma_i \mathbf{Z}_i$ is the $1 \times t$ vector of random effects. In our work with the mixed model we will add the same random effect to each time. Hence, we will use $\mathbf{Z}_i = \mathbf{1}$, where $\mathbf{1}$ is a $1 \times t$ vector of ones.

In addition to the fixed and random effects there is random error. Let the vector $\boldsymbol{\varepsilon}_i$ denote the $1 \times t$ random error vector. For the within-subject errors we will assume that $E(\boldsymbol{\varepsilon}_i') = \mathbf{0}$. The within-subject covariance matrix is $V(\boldsymbol{\varepsilon}_i') = E(\boldsymbol{\varepsilon}_i'\boldsymbol{\varepsilon}_i) = \mathbf{W}_i$. Although there may be high correlations between the within-subject errors, we will assume that the error vectors are independent, which implies that $E(\boldsymbol{\varepsilon}_i'\boldsymbol{\varepsilon}_j) = \mathbf{0}$ for $i \neq j$. With this notation the mixed model for the ith subject is

$$\mathbf{y}_i = \boldsymbol{\beta}\mathbf{X}_i + \gamma_i \mathbf{1} + \boldsymbol{\varepsilon}_i. \tag{10.2}$$

In addition, we will assume that the random effects and the within-subject errors are independent. Hence, the covariance matrix for y_i' is

$$\mathbf{1}'V(\gamma_i)\mathbf{1} + V(\boldsymbol{\varepsilon}_i') = \sigma_\gamma^2 \mathbf{1}'\mathbf{1} + \mathbf{W}_i. \tag{10.3}$$

Typically, the user specifies the correlation structure of W_i and the software uses either maximum likelihood or restricted maximum likelihood (REML) methods to estimate the parameters. See Davis (2002), or Fitzmaurice, Laird, and Ware (2004) for an explanation of these estimation methods.

One big problem with the mixed model method is that several authors, including Park, Park, and Davis (2001), and Overall *et al.* (1999), have shown that some tests of significance fail to maintain their level of significance. Kowalchuk *et al.* (2004) found that if the adjustment proposed by Kenward and Roger (1997) is used with the mixed model, the resulting test maintains its level of significance. Hence, we will use the Kenward–Roger method in our simulation study so that we can compare the power of the mixed model test to the LR test and the adaptive test.

Another difficulty with the mixed model approach is that the covariance structure in $\mathbf{W}_i$ must be specified. The MIXED procedure in SAS (2003) allows the user to specify one of several correlation structures. For example, if the user felt that an autoregressive (AR) correlation structure was the correct correlation structure, then that could be specified and the parameters would be estimated based on that structure. The problem is that the user would rarely know the correlation structure.

This problem has received a great deal of attention in the literature. We will assume in our simulations studies that the user is somehow able to correctly specify the correlation structure. This may be an unrealistic assumption, but if any bias is introduced by this procedure, it should be in the direction of improving the performance of the mixed model test.

10.5 TWO-SAMPLE TESTS

The LR test, the adaptive test, and the mixed model test can be used in a variety of situations with repeated measures data. In this chapter we will limit our discussion to two-sample tests. For the two-sample test we will let μ_{hj} be the mean of population h at time j. The difference between the population means will be denoted by $\Delta_j = \mu_{2j} - \mu_{1j}$ for $j = 1, \ldots, t$. Let $\Delta = [\Delta_1, \ldots, \Delta_t]$ be the vector of differences and let $\mu = [\mu_{11}, \ldots, \mu_{1t}]$ be the vector of means for the first population. To indicate that the subject is from the first or second sample, we define the indicator function such that $I_2(1) = 0$ and $I_2(2) = 1$. Let $\mathbf{y}_{hi}$ denote the $1 \times t$ vector of responses for the ith individual and let n_h be the sample size in sample h. With this notation we can write the model for the ith individual in sample h as

$$y_{hi} = \mu + I_2(h)\Delta + \gamma_i \mathbf{1} + \boldsymbol{\varepsilon}_i, \tag{10.4}$$

for $h = 1, 2$ and $i = 1, \ldots, n_h$.

For the simulations that are used in this chapter, we will use $t = 4$ time periods and will, for the means in the first group, set $\mu_{1j} = 0$ for $j = 1, \ldots, 4$. The means in the second group, which equal $\mu + \Delta$, could be set to a variety of values in the simulations for power, but we will use a simulation design that is similar to that used by Park, Park, and Davis (2001). We will set $\Delta_j = \delta_0 + j\delta_1$ for $j = 1, \ldots, 4$, which will have the effect of setting the means in the second group to $\mu_{2j} = \delta_0 + j\delta_1$ for $j = 1, \ldots, 4$. To investigate the power of a test of parallelism we will set δ_1 to a nonzero value. If we set $\delta_0 = \delta_1 = 0$, the groups means will be equal. This is used to test the level of significance for the group effect. If we set $\delta_1 = 0$ and set $\delta_0 \neq 0$, we can evaluate the power of the test for group effect.

In the model described by Equation (10.4) we generate the random effects using univariate distributions. We use the multivariate normal, log-normal, $\sinh^{-1}$, and outlier distributions to generate the multivariate errors. Because we want to investigate the power of these tests with a variety of correlation structures, we use the approach given in Johnson (1987) to generate these multivariate errors.

Although the data are generated using Equation (10.4), which includes fixed effects and random effects, we will use a multivariate linear model to compute the traditional LR test statistics. Let $n = n_1 + n_2$. The test for the group effect can be performed by writing the complete model as

$$\mathbf{Y} = \mathbf{1}_{n\times 1}[\mu_{11} \cdots \mu_{1t}] + \begin{bmatrix} \mathbf{0}_{n_1\times 1} \\ \mathbf{1}_{n_2\times 1} \end{bmatrix} [\Delta_1 \cdots \Delta_t] + \mathbf{E}, \tag{10.5}$$

where $\mathbf{E}$ is an $n \times t$ matrix of errors. Note that these errors include the random effects and the random errors. With this parameterization we test for the group effect by testing $H_0 : \Delta_1 = \ldots = \Delta_t = 0$ against $H_a : \Delta_j \neq 0$ for some j. For this test the reduced model is $\mathbf{Y} = \mathbf{1}_{n\times 1}[\mu_{11} \cdots \mu_{1t}] + \mathbf{E}_R$. To perform the test we calculate $\hat{\Sigma}_R$ from the reduced model, $\hat{\Sigma}_C$ from the complete model, and $\Lambda^{2/n}$ as the test statistic. This is the form of the LR test that we will use in this chapter for the test of parallelism and for the test of group effect.

10.6 TWO-SAMPLE TESTS FOR PARALLELISM

The analysis of two-sample repeated measures data usually is performed in two steps. In the first step, we check to see if the responses are consistent with the hypothesis that the means are parallel. If the parallelism hypothesis is reasonable, we perform a two-sample multivariate test of the null hypothesis that the mean vectors are equal. Consequently, it makes sense to first consider the performance of the LR, adaptive, and the mixed model test for parallelism.

10.6.1 Traditional LR Test for Parallelism

It can be shown that one way to perform the test of parallelism is to compute the adjacent differences for each subject and then use those differences in a two-sample test. For details concerning the use of adjacent differences, see Davis (2002, Chapter 4). Using this approach, for the ith subject the dependent variables are $[y_{i2} - y_{i1}, \cdots, y_{it} - y_{i(t-1)}]$. These differences are then used to compute the test statistic $\Lambda^{2/n}$. Because $\Lambda^{2/n}$ is a two-sample test statistic a p-value can easily be obtained from the F distribution, as described by Mardia, Kent, and Bibby (1979, p. 83).

In this model, which uses differences between adjacent values in the $\mathbf{Y}$ matrix, the null hypothesis implies that the means of the differences between adjacent values are the same in both treatment groups. If we fail to reject the null hypothesis, we might consider it reasonable to believe that there are no differences between the groups in the mean change between adjacent time points. So, under the null hypothesis we can say that profiles of the mean responses will be parallel.

10.6.2 Adaptive Test for Parallelism

The adaptive test for parallelism is performed by using the adaptive multivariate test that was described in the last chapter. We compute the $t-1$ adjacent differences $[y_{i2} - y_{i1}, \cdots, y_{it} - y_{i(t-1)}]$ and use them in a two-sample adaptive multivariate test. If the vector of adjacent difference is an outlier, that observation will be downweighted. A vector of adjacent differences will be an outlier if it shows an unusual growth pattern in the data. For the adaptive test the permutation of residuals approach is used, with the vector of adjacent differences, to estimate the p-value. The SAS and R code that can be used to perform this test will be given in Section 10.8.

10.6.3 Mixed Model Test for Parallelism

The test of parallelism can also be performed by using the MIXED procedure in SAS. The model is specified so that the group, time, and the interaction of group and time are fixed effects, and the subjects effects are considered random effects. The subject random effect is nested within the group. The REPEATED statement is used to specify the correlation structure and the REML method is used to estimate the parameters. If the p-value associated with the interaction of group and time is less than $\alpha = 0.05$ we reject the null hypothesis.

10.6.4 An Example

For example, consider the data of Potthoff and Roy (1964) on 16 boys and 11 girls that are shown in Table 10.1. For these children the distance from the center of the pituitary to the pteryomaxillary fissure was measured at 8, 10, 12, and 14 years of age. This data can be found at *www.math.niu.edu/~ogorman*. The objective was to determine if gender differences exists. These data are used directly in the mixed model test while the LR and adaptive tests are based on the three adjacent differences for each subject.

The LR test of parallelism was performed as a two-sample test on the three adjacent differences obtained from the 27 subjects. We obtained a Wilks' lambda test statistic of 0.74, which gave a p-value of $p_{LR} = 0.696$. This result suggests that the parallelism hypothesis is reasonable.

The adaptive test, using the three adjacent differences, gave a p-value of $p_{AD} = .1034$, which is much lower than the p-value for the LR test. The difference between the p-values was due to the downweighting of several observations. The adaptive weights are based on the projection to the lines connecting the standardized residuals to the mediancenter for the 27 adjacent differences in 3-dimensions. To help identify the outliers we show a scatterplot of the second and third differences in Figure 10.1.

Table 10.1 Distances, in millimeters, from the center of the pituitary to the pteryomaxillary fissure for 11 girls and 16 boys at four ages

		Age			
Individuals	Gender	8	10	12	14
1	F	21.0	20.0	21.5	23.0
2	F	21.0	21.5	24.0	25.5
3	F	20.5	24.0	24.5	26.0
4	F	23.5	24.5	25.0	26.5
5	F	21.5	23.0	22.5	23.5
6	F	20.0	21.0	21.0	22.5
7	F	21.5	22.5	23.0	25.0
8	F	23.0	23.0	23.5	24.0
9	F	20.0	21.0	22.0	21.5
10	F	16.5	19.0	19.0	19.5
11	F	24.5	25.0	28.0	28.0
12	M	26.0	25.0	29.0	31.0
13	M	21.5	22.5	23.0	26.5
14	M	23.0	22.5	24.0	27.5
15	M	25.5	27.5	26.5	27.0
16	M	20.0	23.5	22.5	26.0
17	M	24.5	25.5	27.0	28.5
18	M	22.0	22.0	24.5	26.5
19	M	24.0	21.5	24.5	25.5
20	M	23.0	20.5	31.0	26.0
21	M	27.5	28.0	31.0	31.5
22	M	23.0	23.0	23.5	25.0
23	M	21.5	23.5	24.0	28.0
24	M	17.0	24.5	26.0	29.5
25	M	22.5	25.5	25.5	26.0
26	M	23.0	24.5	26.0	30.0
27	M	22.0	21.5	23.5	25.0

Source: Potthoff, R. F. and Roy, S. N. (1964). Data are used with permission of Oxford University Press.

The second difference is the difference between the age 12 measurements and the age 10 measurements. The third difference is the difference between the age 14 measurement and the age 12 measurement. The scatterplot shows one clear outlier, which was given a weight of 0.0324. This outlier was observation 20 in Table 10.1, which had decreases in distances between 8 and 10 years of age and between 12 and

14 years of age. Another observation that was given a low weight was observation 24, which had a weight of 0.506. This low weight was primarily due to the large difference between 8 and 10 years of age. The other weights were in the interval [0.624, 1.170].

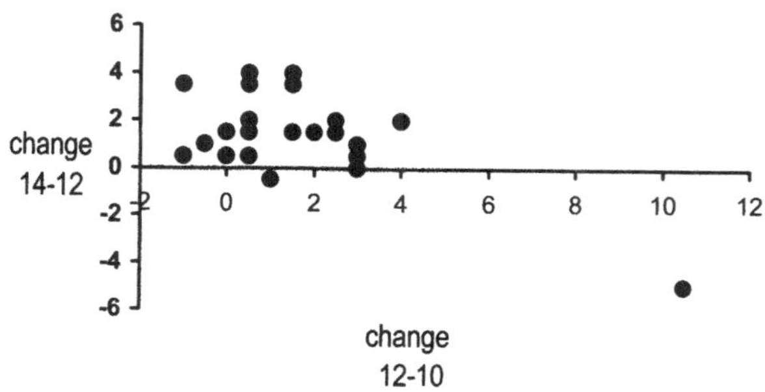

Figure 10.1 Scatterplot of the adjacent differences. The difference between the age 12 measurement and the age 10 measurement is plotted on the horizontal axis; the difference between the age 14 measurement and the age 12 measurement is plotted on the vertical axis.

The MIXED procedure in SAS, when used on the repeated measures data shown in Table 10.1 with the default REML estimation method, an autoregressive correlation structure, and the Kenward–Roger degrees of freedom, produced a p-value of $p_{MM} = 0.0746$ for the gender-age interaction.

With these results, it is not possible to reject the parallelism hypothesis, so we would proceed with a test of group effect. However, before we evaluate the test for group effect, we need to evaluate the significance level and the power of the tests of parallelism.

10.6.5 Comparison of the Tests for Parallelism

10.6.5.1 Simulation Design For these simulations the data are generated using Equation (10.4), with μ and Δ set to the values that are appropriate for the evaluation of the size or power of the test. We used $t = 4$ time points in these simulations with sample size configurations of $(n_1, n_2) = (25, 25)$, (20,80), and (100,100). The univariate random effects (γ_i) were generated from the normal, lognormal, and the inverse hyperbolic sine distributions. The lognormal, and $\sinh^{-1}$ variates were generated in a manner described by Johnson (1987) and were translated and rescaled so that they had a mean of zero and a standard deviation of one.

To generate the vectors of random errors we used three correlation structures. To describe these structures let ρ_{jk} be the correlation between the jth and the kth random error. For the uncorrelated structure we set $\rho_{jk} = 0$ for $j \neq k$. For the autoregressive (AR) structure with $\rho = .3$ we set $\rho_{jk} = .3^{|j-k|}$ for $j \neq k$. For the autoregressive (AR) structure with $\rho = .7$ we set $\rho_{jk} = .7^{|j-k|}$ for $j \neq k$.

We should note that these correlation structures describe the relationship between the random errors, so they represent the within-subject correlations. It can be shown that if the random errors are uncorrelated and the errors have equal variances then the within-group correlations for different variates will be $\sigma_\gamma^2/(\sigma_\gamma^2 + \sigma_\varepsilon^2)$, where σ_γ^2 is the variance of the random errors and σ_ε^2 is the common variance of the errors. Thus, if the errors are uncorrelated, the resulting within-group correlation matrix will have a compound symmetry structure. See Davis (2002, Section 6.3) for a discussion of covariance structures.

For these simulations we generated random error vectors from the multivariate normal, lognormal, and $\sinh^{-1}$ distributions with the uncorrelated, AR($\rho = .3$), and AR($\rho = .7$) correlation structures. The multivariate lognormal and $\sinh^{-1}$ errors are obtained by transforming multivariate normal random variables. The generation of these errors for the lognormal and $\sinh^{-1}$ distributions requires the computation of the appropriate correlation matrices for the multivariate normal distribution so that, after transformation, the random variables will have the desired correlation structure. The formulas in Johnson (1987, Chapter 5) were used in these calculations. We also generated random error vectors from an Outlier distribution, which were generated in the same manner as the multivariate normal errors, except each observation had a probability of 0.025 of being assigned a value of +5, and the same probability of being assigned a value of -5.

After the data are generated, we compute the adjacent differences to be used in the LR and adaptive tests. Note that these adjacent differences are not functions of the random effects.

The data were generated by a FORTRAN computer program. For each data set the LR and adaptive p-values were also obtained from that FORTRAN computer program. Later, after each data set was generated, it was stored and a separate SAS program analyzed each data set using PROC MIXED to obtain the p-value for the mixed model test. These p-values were then used to determine the proportion of rejections. Thus, for the results in the figures, the LR, adaptive, and mixed model tests used the same data sets, but the result for the LR and adaptive tests were produced by the FORTRAN program while the results for the mixed model test were produced by PROC MIXED in SAS.

10.6.5.2 Significance Level of Tests for Parallelism Because the reduced model is a simple model that contains only a vector of constants, the permutation test statistics will be equally likely under the null hypothesis. This implies that the adaptive test will maintain its level of significance. However, the LR and mixed model tests may not maintain the significance level, so will will perform a simulation study.

To estimate the level of significance of these tests of parallelism we set $\mu_{1j} = 0$ and $\mu_{2j} = \delta_0$ for $j = 1, \ldots, 4$ in Equation (10.4). For $n_1 = n_2 = 25$ with the AR($\rho = .3$) correlation structure we obtain the empirical significance levels that are shown in Figure 10.2. Because we used 10000 data sets in the simulations the standard error is approximately 0.22%. Figure 10.2 shows that the LR and mixed model tests do not maintain their level of significance with the AR($\rho = .3$) correlation structure with $n_1 = n_2 = 25$ observations. However, with $n_1 = n_2 = 100$ observations all of the tests have a level of significance near the nominal level of $\alpha = 0.05$ and none exceed the nominal level. For $n_1 = n_2 = 100$ observations, over the 3 error distributions and the 3 correlation structures the empirical significance levels were in the interval [4.1%, 5.2%] for the LR test, in the interval [4.7%, 5.2%] for the adaptive test, and in the interval [3.5%, 5.0%] for the mixed model test. There is no indication that the significance level of these tests greatly exceeds the nominal value. Hence, we can proceed to compare the power of these tests of parallelism.

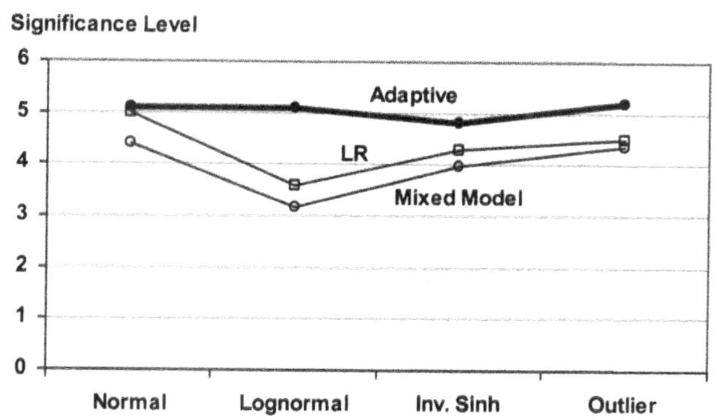

Figure 10.2 Empirical level of significance, in percent, for the test of parallelism with $n_1 = n_2 = 25$ observations measured at $t = 4$ times, using an AR($\rho = 0.3$) correlation structure. The multivariate error distributions are shown on the horizontal axis.

10.6.5.3 Power of Tests of Parallelism To make a meaningful comparison of the three tests of parallelism, we set the simulation parameters to values that would give approximately 80% power. Thus, for the second sample the means are $\mu_{2j} = \delta_0 + j\delta_1$, with δ_1 set to values that give approximately 80% power for the LR test. The empirical power estimates are shown in Figure 10.3 for the three tests with 25 observations in each group and with uncorrelated errors. For normally distributed errors the power of the mixed model test is slightly greater than the power of the LR test, and the power of the adaptive test is slightly less than the power of the LR test. However, for the non-normal errors the adaptive test is much more powerful than the other tests.

We obtain somewhat different results with correlated errors. These power results are shown in Figure 10.4 for $n_1 = n_2 = 25$ observations with an AR($\rho = .7$) correlation structure. We see that the mixed model test has lower power than the other tests and that the adaptive test is more powerful than the other tests for non-normal errors. The low power of the mixed models test cannot be explained by an improper specification of the model because the AR correlation structure was specified in the MIXED procedure.

A similar pattern was observed with larger sample sizes. The power results in Figure 10.5 are from simulation studies that used 100 observations in each group with an AR($\rho = .3$) correlation structure. In this simulation the LR test and the

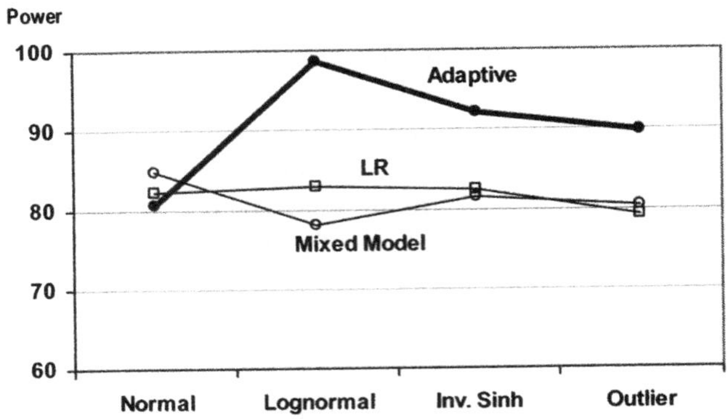

Figure 10.3 Empirical power, in percent, for the test of parallelism with $n_1 = n_2 = 25$ observations measured at $t = 4$ times, using an uncorrelated correlation structure. The multivariate error distributions are shown on the horizontal axis.

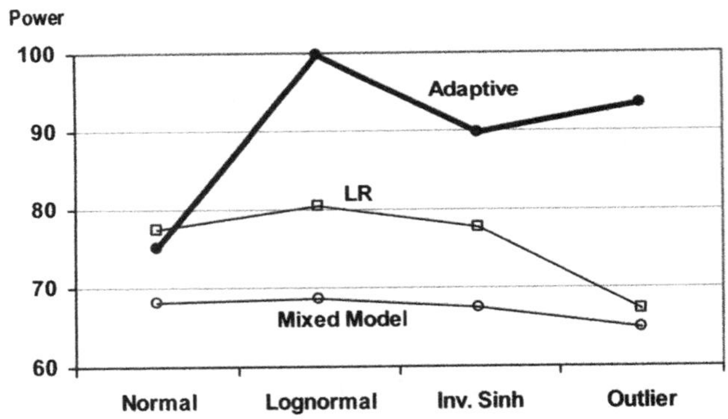

Figure 10.4 Empirical power, in percent, for the test of parallelism with $n_1 = n_2 = 25$ observations measured at $t = 4$ times, using an AR($\rho = 0.7$) correlation structure. The multivariate error distributions are shown on the horizontal axis.

mixed model test had similar power for all error distributions, while the adaptive test was much more powerful for non-normal errors. These results were similar to the simulation results with uncorrelated errors, which are not shown. The simulations results with the AR($\rho = .7$) correlation structure showed a slightly larger difference between the adaptive test and the other tests than that shown in Figure 10.5. The power estimates for the unbalanced design with $n_1 = 20$ and $n_2 = 80$ with an AR($\rho = .3$) are shown in Figure 10.6. It can be seen that these results are similar to those found for $n_1 = n_2 = 100$ shown in Figure 10.5.

We conclude that, for tests of parallelism, the adaptive test is more powerful than the other tests with non-normal errors. In addition, if the errors are multivariate normal, the three tests are nearly equal in power. Thus, there appears to be little to lose and much to gain by using the adaptive test. We will give the SAS and R

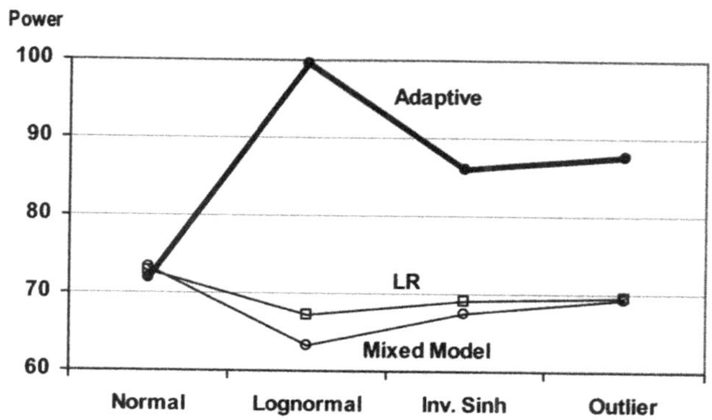

Figure 10.5 Empirical power, in percent, for the test of parallelism with $n_1 = n_2 = 100$ observations measured at $t = 4$ times, using an AR($\rho = 0.3$) correlation structure. The multivariate error distributions are shown on the horizontal axis.

code for these tests in Section 10.8. In the next section will turn our attention to the testing procedures that could be used if we do not reject the null hypothesis of parallel responses.

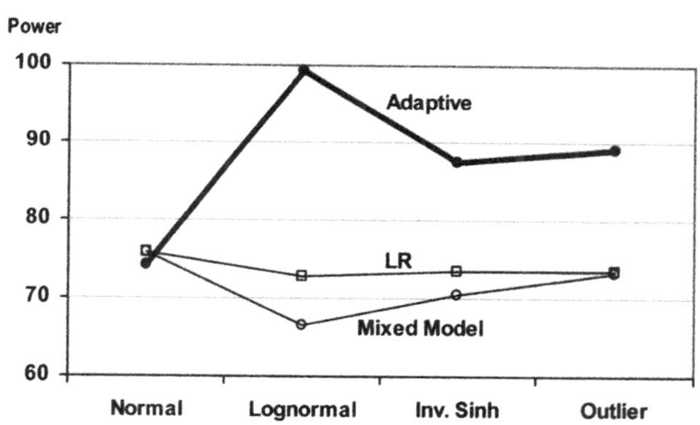

Figure 10.6 Empirical power, in percent, for the test of parallelism with $n_1 = 20$ and $n_2 = 80$ observations measured at $t = 4$ times, using an AR($\rho = 0.3$) correlation structure. The multivariate error distributions are shown on the horizontal axis.

10.7 TWO-SAMPLE TESTS FOR GROUP EFFECT

In this section we assume that the researcher has already concluded that there is no interaction between the group and time effect. If a test of parallelism is performed and we obtain a large p-value we may conclude that there is no interaction. In these situations it makes sense to test for the group effect.

There are two approaches to these tests for group effect. One approach is to use the vector of responses at the t time points, using a multivariate or mixed model test. The other approach is to compute an average of the t responses and then use these averages in a univariate two-sample test. Because there are several multivariate tests and several univariate tests, there are many test procedures that could reasonably be used. The five tests that we will compare are:

1. The multivariate two-sample test that was described in Chapter 9. We use the t observations obtained on each subject to perform the test, rather than using the $t-1$ adjacent differences that we used in the test of parallelism.

2. The multivariate adaptive test, which uses the projection, weighting, and permutation procedures that were described in Chapter 9 on the t observations obtained from the subjects. The adaptive weighting procedure should downweight the vectors that are set apart from the rest of the data in t-dimensional space.

3. The mixed model test for group effect that uses the procedures described in Section 10.4. We used the MIXED procedure in SAS and included only the group and time effects in the model. The p-value associated with the group effect was used to perform the test.

4. For each subject compute the average response over the t time periods, then use these averages in a two-sample pooled t test.

5. For each subject compute the average response over the t time periods, then use these averages in an adaptive two-sample test.

Note that the univariate tests are much easier to perform. In the absence of an interaction between time and group it is not clear that the multivariate tests will be more powerful than the univariate tests. Also, researchers may be more comfortable with univariate tests.

10.7.1 Simulation Results for Group Effects

10.7.1.1 Simulation Design The simulation studies for group effects were very similar to the simulations that were used for the tests of parallelism. We used the same random effect and error distributions, the same correlation structures, and the same sample size configurations. However, in these tests we assumed that the effects were parallel, so that in the model specified by Equation (10.4) we used $\mu_{2j} = \delta_0$ in the formula for the treatment differences $\Delta_j = \mu_{2j}$ for $j = 1, \ldots, 4$.

Unlike the tests of parallelism, the tests for group effect depend on the random effect distribution. Hence, we used 3 random effect distributions, 4 error distributions, and 3 correlation structures so there are $3 \times 4 \times 3 = 36$ simulations for each of the three sample size configuration. Because of the large number of simulation studies, we used only 1000 data sets for each of the $3 \times 36 = 108$ simulations.

10.7.1.2 Significance Level of the Test for Group Effect For the simulations that used $n_1 = n_2 = 25$ observations, all five tests had empirical significance levels at or near the nominal level of $\alpha = 0.05$. Because we used 1000 data sets in these simulations the standard error is 0.7%, so we expect most of the simulations to have empirical significance levels in the interval [3.6%, 6.4%]. Only two of the 36 simulations studies had empirical significance levels above 6.4%, and these two had significance levels of 6.6%. Hence, for this sample size, all of the tests maintained their level of significance at or near 5.0%.

As expected, the simulation studies that used other sample size configurations also produced empirical significance levels near the nominal significance level. The largest empirical significance level was 7.0%, which was observed with the mixed model test with $n_1 = 20$ and $n_2 = 80$ observations with normal random effects and AR($\rho = 0.7$) correlated errors generated from the $\sinh^{-1}$ distribution. Since all five tests appear to have significance levels that approximate the nominal significance level, we can fairly compare the power of these five tests.

10.7.1.3 Power of the Test for Group Effect The relative power of these five tests are shown in Figure 10.7 for simulations having $n_1 = n_2 = 25$ observations with errors generated using an AR($\rho = 0.3$) correlation structure with normally distributed random effects. It can be seen that the MIXED model test, the t test, and the univariate adaptive test have approximately the same power and that their power is greater than the power of the LR and the multivariate adaptive tests.

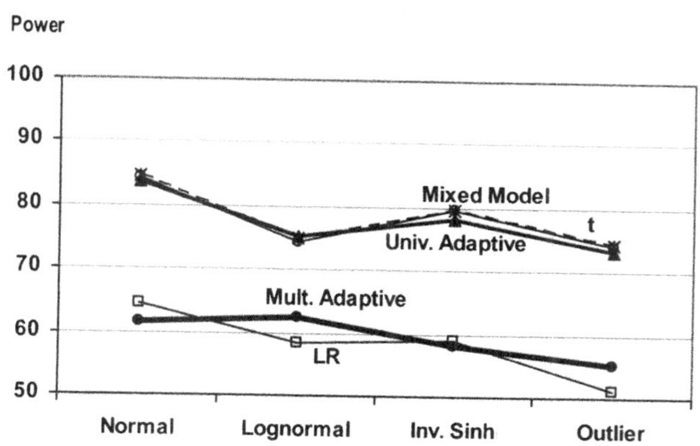

Figure 10.7 Empirical power, in percent, for the group effect with $n_1 = 25$, $n_2 = 25$ observations measured at $t = 4$ times, using an AR($\rho = 0.3$) correlation structure. The random effects were generated from a normal distribution. The multivariate error distributions are shown on the horizontal axis.

However, when we use lognormal random effects instead of normal random effects a different pattern can be seen. These results are shown in Figure 10.8. With lognormal random effects the univariate adaptive test is the most powerful test and the multivariate LR test is clearly the least powerful test for all error distributions. The MIXED model test and the t test appear to have similar power characteristics, while the multivariate adaptive test always has power less than the univariate adaptive test.

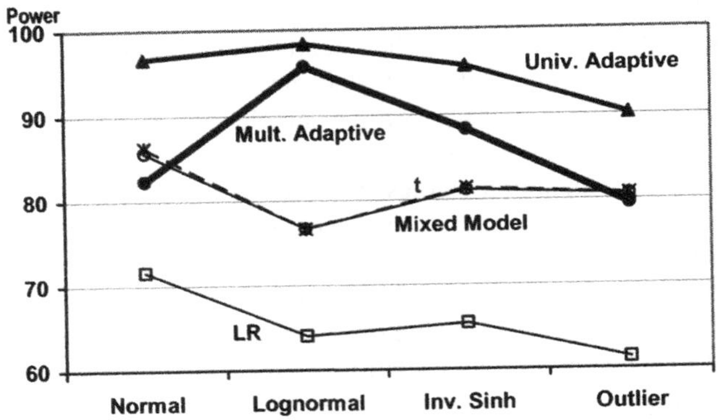

Figure 10.8 Empirical power, in percent, for the group effect with $n_1 = 25$, $n_2 = 25$ observations measured at $t = 4$ times, using an AR($\rho = 0.3$) correlation structure. The random effects were generated from a lognormal distribution. The multivariate error distributions are shown on the horizontal axis.

The power results from simulations that used the $\sinh^{-1}$ distribution are shown in Figure 10.9. These show the same overall pattern that we observed with lognormal random effects, but the differences in powers between the tests are reduced. We note in all of these simulation results that the multivariate LR test has lower power than the univariate t test and the multivariate adaptive test has lower power than the univariate adaptive test.

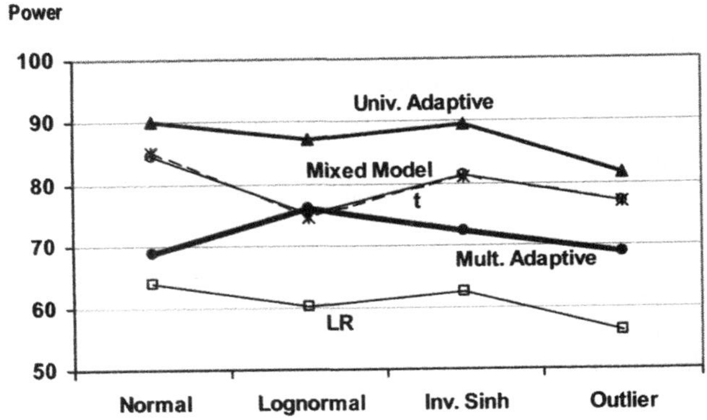

Figure 10.9 Empirical power, in percent, for the group effect with $n_1 = 25$, $n_2 = 25$ observations measured at $t = 4$ times, using an AR($\rho = 0.3$) correlation structure. The random effects were generated from an inverse hyperbolic sine distribution. The multivariate error distributions are shown on the horizontal axis.

These general results for $n_1 = n_2 = 25$ also apply to larger samples. Power estimates are shown in Figure 10.10 for simulations that use $n_1 = n_2 = 100$ observations with normally distributed random effects. Again, we see that the univariate tests outperform the multivariate tests and, with normal random effects, the univariate adaptive test, the MIXED model test, and the t test have approximately the same power for all error distributions.

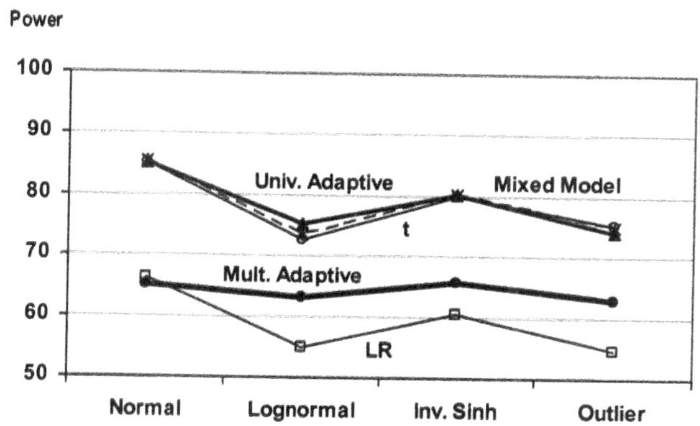

Figure 10.10 Empirical power, in percent, for the group effect with $n_1 = 100$, $n_2 = 100$ observations measured at $t = 4$ times, using an AR($\rho = 0.3$) correlation structure. The random effects were generated from a normal distribution. The multivariate error distributions are shown on the horizontal axis.

For $n_1 = n_2 = 100$ observations with random effects that are generated from the lognormal distribution, the power results are shown in Figure 10.11. These results clearly show that the adaptive tests have higher power than the other tests and that the univariate adaptive test almost always rejects the null hypothesis when the t test and MIXED model test reject the null hypothesis for only 80% of the datasets.

The results in Figures 10.7 through Figure 10.11 demonstrate that the relative power of these tests depends primarily on the random effect and that the univariate adaptive test has the greatest power, or nearly the greatest power, for all random effect distributions and error distributions. In addition, this univariate adaptive test can easily be performed.

10.8 AN EXAMPLE OF REPEATED MEASURES DATA

We now consider the data of Koziol *et al.* (1981) that concerns tumor sizes in mice who were injected with mouse colon carcinoma tumor cells. Thirty mice were used in the experiment but we will only analyze the data from the 20 mice who were injected with tissue culture median (Group A) or with normal spleen cells, immune RNA, and tumor antigen (Group C). The tumor volumes (in mm^3) were recorded for several days, but because some mice died at day 18 we used only the data from day 11, 13, 15, and 17. These data are displayed in Table 10.2.

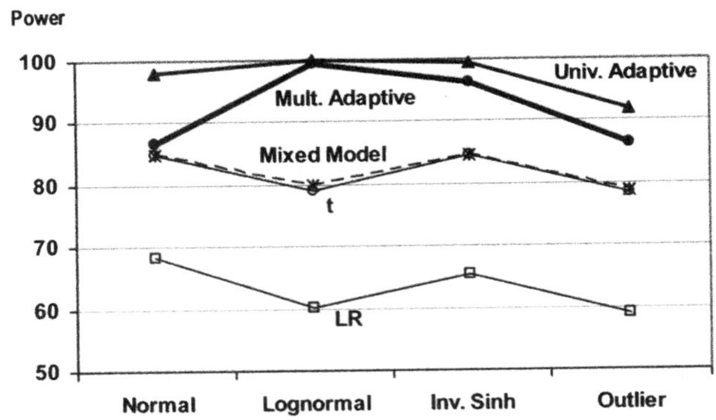

Figure 10.11 Empirical power, in percent, for the group effect with $n_1 = 100$, $n_2 = 100$ observations measured at $t = 4$ times, using an AR($\rho = 0.3$) correlation structure. The random effects were generated from a lognormal distribution. The multivariate error distributions are shown on the horizontal axis.

The test of parallelism was performed with the LR test, the multivariate adaptive test, and the mixed model test. The LR test produced $p_{LR} = 0.998$, the multivariate adaptive test produced $p_{MAD} = 0.964$, and the mixed model test produced $p_{MM} = 0.996$. We concluded that the parallelism hypothesis is reasonable and we proceeded to perform the test for group effect.

The multivariate two-sample LR test for group effect produced $p_{LR} = 0.210$, the multivariate adaptive test produced $p_{MAD} = 0.108$, and the mixed model test gave $p_{MM} = 0.341$. Next, we computed the average, over the four days, of the tumor volumes for the twenty mice. These averages were used with a univariate t test that gave $p_t = 0.313$ and with an adaptive test that produced $p_{AD} = 0.059$. The large difference in these p-values was due to the relatively small weights of 0.47, 0.50, and 0.49 that were given to observation 5, 6, and 7 in group C, while the weights for the other mice were in the range $[0.98, 1.02]$. Because p_{AD} is close to $\alpha = 0.05$, these results suggest, but do not demonstrate, that the means are not equal.

In this study we do not know if there really is a difference between the populations, so it is difficult to determine if the adaptive test was more effective than the non-adaptive test.

10.8.1 Using the SAS Macro

We can perform the test of parallelism with one call to the `adaptall` macro. Suppose we compute the adjacent differences as `diff1`, `diff2`, and `diff3`. If these differences are on the `koziol` dataset we can use the following code to perform the test:

```
%adaptall(dataset=koziol, y=diff1 diff2 diff3, nvars=3,
     xa=group, classa=group,
     seed1=23471, seed2=7839, seed3=11941, nperm=2000);
```

Table 10.2 Tumor volumes (in mm^3) recorded on four days for mice who have been injected with mouse colon carcinoma tumor cells

		Days			
Group	Mouse	11	13	15	17
A	1	157.1	217.6	379.0	556.6
	2	152.2	176.6	317.9	356.4
	3	122.4	196.1	388.9	469.3
	4	95.0	205.9	307.3	405.1
	5	168.8	196.0	340.4	507.3
	6	85.0	225.1	289.0	317.9
	7	129.8	274.7	340.3	507.2
	8	157.0	202.5	307.2	320.1
	9	129.7	205.8	419.1	421.2
	10	156.9	225.0	372.6	379.2
C	1	108.0	186.2	213.8	379.1
	2	129.6	196.6	397.1	500.0
	3	65.0	191.3	274.6	405.0
	4	52.9	129.6	303.5	415.0
	5	147.0	420.0	653.4	806.4
	6	115.2	32.0	3.2	1.4
	7	55.0	55.0	118.8	118.3
	8	156.8	84.7	291.5	400.0
	9	44.6	258.8	405.0	372.6
	10	118.3	176.4	340.2	361.0

With 2000 permutations the SAS macro produced a p-value of $p_{MAD} = 0.9640$ for the multivariate adaptive test of parallelism.

To perform the multivariate adaptive test of group effect we use the tumor volumes on the four days, which are `d11`, `d13`, `d15`, and `d17`. The SAS code is:

```
%adaptall(dataset=koziol, y=d11 d13 d15 d17, nvars=4,
      xa=group, classa=group,
      seed1=2671, seed2=17839, seed3=21941, nperm=2000);
```

This multivariate adaptive test gave a p-value of $p_{MAD} = 0.1084$.

To perform the univariate adaptive test we first compute the `average`, which is the average tumor volume over the 4 days, and include it on the `koziol` data set. Then we perform a univariate adaptive test with the following SAS code:

```
%adaptall(dataset=koziol, y=average, nvars=1,
```

```
        xa=group, classa=group,
        seed1=671, seed2=22839, seed3=7941, nperm=2000);
```

The univariate test produced a p-value of $p_{AD} = 0.0585$. While this is not statistically significant at the $\alpha = 0.05$ level, it does suggest that there may be a difference between the groups.

10.8.2 Using R Code

Suppose we have the data that was given in Table 10.2 on a file `koziol.dat`, and suppose the data is entered in a way that can be read by the `read.table()` function. To perform the test of parallelism we compute the adjacent differences `diff11`, `diff13`, and `diff15` and use the `cbind` function to build the `y` matrix. The following code performed the multivariate adaptive test of parallelism:

```
source("adaptmult.r")
dataset <- c("koziol.dat")
testdata <- read.table(dataset)
attach(testdata)
diff11 = d13-d11
diff13 = d15-d13
diff15 = d17-d15
yvars <- c("diff11, diff13, diff15")
y <- cbind(diff11,diff13,diff15)
complete <- c("y ~ group")
reduced <- c("y ~ 1")
r <- 2000
seed <- c(61413)
pvalue <- adaptmult(y,complete,reduced,r,seed)
```

This produced a p-value of $p_{AD} = 0.9685$, so there is no evidence of interaction of group and time.

To perform the multivariate test of group effect we use the same multinomial adaptive test, but define `y` to include the original data rather than the adjacent differences. The `R` code is

```
source("adaptmult.r")
dataset <- c("koziol.dat")
testdata <- read.table(dataset)
attach(testdata)
y <- cbind(d11, d13, d15, d17)
complete <- c("y ~ group")
reduced <- c("y ~ 1")
r <- 2000
seed <- c(7413)
pvalue <- adaptmult(y,complete,reduced,r,seed)
```

We obtained a p-value of $p_{MAD} = 0.105$, which suggests that there may be a difference between the groups. However, we know, from our simulation results, that univariate tests are generally more powerful than the univariate tests so we will complete the analysis with a univariate test.

To perform the univariate test we computed the averages over the 4 days and stored the averages in the y vector on the file koziol.univ.dat. The R code to perform the univariate test is

```
source("adapt.r")
dataset <- c("koziol.univ.dat")
complete <- c("y ~ group")
reduced <-  c("y ~ 1")
r <- 2000
seed <- c(61413)
pvalue <- adapt(dataset,complete,reduced,r,seed)
```

This univariate two-sample test produced a p-value of $p_{AD} = 0.054$. These results from the R code closely approximate the results from the SAS macro. This is not surprising because the only difference between the tests in R and in SAS is the random number generator.

10.9 DEALING WITH MISSING DATA

When several measurements are required from each subject, researchers often find that some of these observations are missing. For example, if subjects in a longitudinal study must have blood tests every two weeks for eight weeks, many subjects may miss one or more of the blood tests. When data are missing it is not always clear what analysis is appropriate.

Before a researcher analyzes a data set that contains missing data, the reasons for the missing data should be understood. One possibility is that data may be missing because an extraneous event happened to occur. For example, a subject may have a missing value at an observation time because the subject needed to travel for a family emergency. On the other hand, data may be missing because the outcome, which is unobserved, is extremely negative. For example, suppose some mice that were being treated for a certain cancer developed a large tumor that caused the mice to die. For those mice the missing values are informative, so the missing values should not be ignored.

A variety of terms have been used to distinguish the kinds of missingness. We will use the classifications described by Little and Rubin (1987). Depending on the missing data mechanism, the missing data can be classified as follows:

- Missing data are considered to be missing completely at random (MCAR) if the probability of a response is not related to the outcome, which is not observed, nor is the probability related to any of the other outcomes or the covariates.

- Missing data are considered to be missing at random (MAR) if the probability of a response is not related to the outcome, which is not observed, but is related to one or more covariates or to the outcomes that are observed.

- Missing data are considered to be missing not at random (MNAR) if the probability of response depends on the unobserved outcome. Data that are MNAR are sometimes called informative or nonignorable.

In general, if the data are missing completely at random the analysis can proceed using the nonmissing data without too much concern about creating bias. However, if the data are MCAR and all subjects that have any missing values are removed from the analysis, there can be a meaningful loss in power depending on the number of subject that would be removed. On the other extreme, if the missing data are nonignorable then the researcher needs to use the model for the missing data mechanism in the analysis.

In studies involving repeated measurements the missing data can occur in two patterns. One pattern of missing data occurs when some of the repeated measurements are missing but subsequent observations are present. These missing values are called intermittently missing. Another pattern of missing data occurs when a subject stops being measured so that all measurements are missing after the first missing value. This pattern is said to be a monotone missing pattern and the subjects are sometimes called dropouts.

Many approaches can be used to deal with missing data. One approach is to use a mixed model to analyze the data. This approach is popular because it allows the researcher to use all of the nonmissing data and it does not require any imputation of missing data.

Another approach is to impute reasonable values for all of the missing data and continue with the analysis as if the dataset contained complete data. We will now describe a simple and effective imputation strategy that was designed for repeated measures studies that use a permutation test.

For the ith subject at the jth time the adjacent difference is $d_{ij} = y_{i(j+1)} - y_{ij}$. Let $\tilde{d}_j$ be the median of the adjacent differences at the jth time over both groups. The imputation method uses these median differences to impute a reasonable quantity for the missing value. For $i = 1, \ldots, n$ and $j = 1, \ldots, t$ we impute values as follows:

1. If y_{ij} is missing but $y_{i(j-1)}$ and $y_{i(j+1)}$ are observed, impute the value

$$y_{ij} = \frac{1}{2}\{(y_{i(j-1)} + \tilde{d}_{j-1}) + (y_{i(j+1)} - \tilde{d}_j)\}$$

2. If y_{ij} is missing and $y_{i(j+1)}$ is missing, but $y_{i(j-1)}$ is observed, or, if $j = t$ and y_{it} is missing but $y_{i(t-1)}$ is observed, then impute $y_{ij} = y_{i(j-1)} + \tilde{d}_{j-1}$.

3. If y_{ij} is missing and $y_{i(j-1)}$ is missing but $y_{i(j+1)}$ is observed, or, if $j = 1$ and y_{i1} is missing but y_{i2} is observed, then impute $y_{ij} = y_{i(j+1)} - \tilde{d}_j$.

If these three steps are used and there are still some missing values in the data matrix, this procedure should be used again. This procedure is repeated until all of the missing values are replaced with imputed values.

The advantage of this imputation method is that the LR test and the adaptive test can be used on the data set after imputation. Because the imputation method does not depend on the sample indicator, any permutation test, including the adaptive test, should maintain its level of significance. In addition, O'Gorman (2010) showed that if the data were MAR or MCAR and this imputation method was used then an adaptive test had a power advantage over the LR and mixed model tests. However, the adaptive test used in O'Gorman (2008a, 2010) is not identical to the multivariate adaptive test that was proposed in Chapter 9 and used in this chapter. In addition, the simulations studies by O'Gorman (2010) assumed that no more than 20% of the data were missing. For details concerning those simulations and the imputation method see O'Gorman (2010).

Whatever approach is used, the researcher should be aware that the results may be very sensitive to the assumptions about the missing data mechanism. If the missing data are MCAR, the approach described above may be sufficient. However, if the data are MNAR, then the missing data model must be specified and used in the analysis. For a discussion of missing data in longitudinal studies see the review paper by Ibrahim and Molenberghs (2009).

The biggest challenge to researchers who have missing data is that the missing data mechanism is usually not known. Further, assumptions about the mechanism cannot be verified because the data are not observed. It also should be kept in mind that the tests may be sensitive to the assumption about the missing data mechanism. Hence, analysis of repeated measures data with missing values continues to be a challenge.

Researchers who have missing data and who want to use an adaptive test might consider using the imputation method described in O'Gorman (2010). Next, the adaptive weighting and permutation methods that were described in this chapter could be used to perform the adaptive test for parallelism and the adaptive test for group effect.

10.10 CONCLUSIONS AND RECOMMENDATIONS

In the simulation studies we attempted to use a variety of distributions, correlation structures, and sample sizes. However, a severe limitation of these simulation studies is that only three non-normal multivariate distributions were used to generate the non-normal errors. It is easy to generate non-normal data with a specified correlation structure for the lognormal and the $\sinh^{-1}$ distributions by using procedures described in Johnson (1987, Chapter 5), but the number of non-normal distributions whose correlation structure can be specified appears to be limited. Further research using a greater variety of distributions may be able to shed more light on the performance of these tests.

We discovered with these simulation studies that the adaptive test for parallelism is often much more powerful than the LR and mixed model tests with three non-normal distributions. We also found that if the errors are normal the adaptive test is nearly as powerful as the other tests. Because, with small samples, departures from multivariate normality are difficult to detect, it seems reasonable to use a test for parallelism that performs well for normal and non-normal distributions.

We also found clear evidence that the power of the test for group effect can be increased by computing the subject averages over time and then using those averages in a two-sample test. We found that an adaptive univariate two-sample test had greater power than the pooled t test in tests for the group effect, but we did not include many of the other two-sample univariate tests, such as the Wilcoxon rank-sum test, in the comparisons. See Section 6.1 for a comparison of the Wilcoxon rank-sum test, the univariate adaptive test, and the pooled t test. In addition, a univariate test for group effect could be performed using one of the rank-based tests that we will describe in Chapter 11.

EXERCISES

10.1 Suppose, in an effort to compare two reading methods, we have randomly assigned six students to one of two reading groups. Consider the following scores on reading tests taken every week for three weeks:

	Week		
Group	1	2	3
1	82	85	86
1	91	92	95
1	76	81	80
2	81	82	85
2	68	73	72
2	72	75	76

a) Compute the adjacent differences for the six observations.

b) Will the parallelism hypothesis be rejected? Explain. Few calculations should be required.

10.2 A weight loss experiment was conducted by a nutritionist to determine if overweight adult males would lose more weight with a diet treatment (Treatment A) or with a diet and exercise program (Treatment B). Suppose that 100 subjects were randomly assigned to either Treatment A or Treatment B and the weight of each subject was recorded before and after the eight-week treatment period. Thus, there were $t = 2$ measurements for each subject.

a) How would an adaptive test of parallelism be performed?

b) Could we have used an adaptive test that we introduced earlier in this book to test the null hypothesis of no difference in mean weight loss between the treatments? Explain.

10.3 Write the SAS program for performing the adaptive test for parallelism on the Potthoff and Roy (1964) data given in Table 10.1.

10.4 Write the R code for performing the adaptive test for parallelism on the Potthoff and Roy (1964) data given in Table 10.1.

10.5 Suppose a researcher analyzes a data set and finds, for the LR test of parallelism, a p-value of $p_{LR} = 0.127$. For the same data the p-value for the multivariate adaptive test of parallelism is $p_{MAD} = 0.021$. The researcher then computes the average of the recorded values over the times and then performs a test of group effect. The researcher obtains $p_t = 0.045$ for the t test and $p_{AD} = 0.028$ for the univariate adaptive test. The researcher concludes that the group effect is significant. Comment on the procedure that was used. Are these tests appropriate? Why or why not?

10.6 Suppose six patients were randomly assigned to one of two surgical treatment groups and their strength measurements were recorded as follows for three weeks of post-surgical follow-up:

	Week		
Group	1	2	3
1	18	27	32
1	32	43	47
1	13	24	30
2	46	57	61
2	35	40	49
2	51	60	66

a) Compute the adjacent differences.
b) Make a scatterplot of the adjacent differences.
c) Which observation will be given the lowest weight in a test of parallelism? No calculations should be necessary.

10.7 Write the SAS program for performing the adaptive test for group effect on the Potthoff and Roy (1964) data given in Table 10.1.

10.8 Write the R code for performing the adaptive test for group effect on the Potthoff and Roy (1964) data given in Table 10.1.

10.9 Suppose a dietitian randomly assigned subjects to one of two weight loss diets, which are labeled as A or B, and then followed them for 8 weeks. He obtained the baseline weight and the weights after 4 and 8 weeks on a diet for 8 subjects. However, the dietitian was not able to obtain two observations. The weights are:

		Week		
Subject	Group	0	4	8
1	A	152	150	150
2	A	173		171
3	A	187	182	179
4	A	178	174	170
5	B	164	162	158
6	B	157	151	
7	B	195	191	188
8	B	171	168	167

a) Calculate the adjacent differences and $\tilde{d}_j$ for $j = 1, 2$.

b) Find the imputed value for y_{22}. Is this reasonable? Why or why not?

c) Find the imputed value for y_{63}. Is this reasonable? Why or why not?

10.10 Zerbe (1979) reported the plasma inorganic phosphate measurements of 13 control and 20 obese patients at several times after a standard oral glucose challenge. The measurements at 30 minutes, 60 minutes, 90 minutes, and 120 minutes after the challenge are given in Table 10.3. This data can be found at *www.math.niu.edu/~ogorman*.

a) Perform an adaptive test of parallelism using either SAS or R.

b) Perform an adaptive test of group effect using either SAS or R.

c) State your conclusions.

10.11 We noted in the text that the imputation method does not depend on what group the individual is in.

a) Under the null hypothesis are the permutations of the y-vectors equally likely?

b) Why will this test, when used with imputed data, maintain its level of significance?

10.12 The data listed in Table 10.2 gives the tumor volumes at day 11, 13, 15, and 17 for 20 mice. If mouse 5 in Group C had missing values at day 15 and day 17, and none of the other observations were missing, do you believe that the missing data mechanism would be MCAR? Explain.

Table 10.3 Plasma inorganic phosphate (mg/dl) after glucose challenge

		Minutes after Glucose Challenge			
Group	Patient	30	60	90	120
Control	1	3.3	3.0	2.6	2.2
	2	2.6	2.6	1.9	2.9
	3	4.1	3.1	2.3	2.9
	4	3.0	2.2	2.8	2.9
	5	3.8	2.1	3.0	3.6
	6	2.2	2.0	2.6	3.8
	7	3.0	2.4	2.5	3.1
	8	3.9	2.8	2.1	3.6
	9	4.0	3.4	3.4	3.3
	10	3.1	2.9	2.2	1.5
	11	2.6	2.6	2.3	2.9
	12	3.7	3.1	3.2	3.7
	13	3.1	3.2	3.3	3.2
Obese	14	3.3	3.0	2.6	2.2
	15	4.9	4.1	3.7	3.7
	16	4.4	3.9	3.9	3.7
	17	3.9	3.1	3.1	3.1
	18	3.1	3.3	2.6	2.6
	19	5.0	2.9	2.8	2.2
	20	3.1	3.3	2.8	2.9
	21	4.7	3.9	3.1	2.8
	22	2.5	2.3	2.2	2.1
	23	5.0	4.1	3.7	3.7
	24	4.3	4.0	4.0	3.3
	25	4.2	4.2	3.4	3.5
	26	4.3	4.0	4.0	3.3
	27	4.1	4.6	4.1	3.4
	28	4.6	4.6	4.4	4.1
	29	3.5	3.8	3.6	3.3
	30	6.1	4.2	4.1	4.3
	31	3.4	3.1	2.8	2.1
	32	4.0	3.7	3.3	2.4
	33	4.4	3.8	3.8	3.8

CHAPTER 11

RANK-BASED TESTS OF SIGNIFICANCE

11.1 THE QUEST FOR POWER

In previous chapters we have said little about the rank-based adaptive tests, preferring instead to rely on tests that used weighted least squares and permutation methods. The adaptive WLS tests were emphasized because they can be used to test any subset of coefficients in a linear model. However, the earliest two-sample adaptive tests, which were based on ranks, had greater power than the t test and the Wilcoxon test for many error distributions. Therefore, we will attempt to describe these methods in some detail in this chapter.

Rank-based tests use test statistics that are functions of the data only through the ranks. For example, the Wilcoxon rank-sum test uses the sum of the ranks in the second sample as the test statistic, where the observations are ranked over all observations in both samples. Because the test statistic is a function of the ranks, for small samples it is easy to tabulate the c.d.f. of the permutation distribution and this permutation distribution can be used for any sample taken from a continuous distribution. Hence, the Wilcoxon test can be used with samples from a wide variety of distributions and, because it is based on ranks, it is naturally robust with respect to the presence of an outlier.

Adaptive Tests of Significance Using Permutations of Residuals with R and SAS®.
By Thomas W. O'Gorman

The Wilcoxon test statistic, which uses the sum of the ranks in the second sample, is only one of many rank-based two-sample tests. Instead of computing the sum of the ranks in the second sample, we could compute the sum of the scores obtained from a function of the ranks. For example, we could count the number of observations in the second sample whose rank exceeds the median rank. This is the two-sample median test, which has been shown to have greater power than the Wilcoxon test to detect a shift with distributions that are symmetric and heavy-tailed. In addition to the rank-sum and median tests, a variety of other rank tests have been devised that have greater power to detect a shift when the sample is taken from populations that have other distributions. See Randles and Wolfe (1979, Chapter 9) for a comparison of the asymptotic relative efficiencies of many of these rank tests.

So, while all of the rank-based two-sample tests maintain their significance level for all distributions, their relative power to detect a shift depends on the distribution. This fact led Randles and Hogg (1973) to propose several adaptive tests that use selection statistics to select a set of rank scores in order to produce a powerful test. In their early work in adaptive testing the emphasis was on the choice of an appropriate selection statistic and how often the selection statistic selected the correct set of rank scores. The paper by Hogg, Fisher, and Randles (1975) showed that it is possible to find selection statistics that give the two-sample adaptive test relatively high power compared to the traditional tests. This test, which will be called the HFR test, was briefly described in Chapter 1. They showed that their adaptive test would maintain its level of significance and that it was relatively powerful with many non-normal error distributions.

In the next section we will describe several rank tests and their properties. These tests are used in the HFR test that will be described in Section 11.3. Later in this chapter we will describe several more recent adaptive tests. The objective of these tests is to increase the power of the two-sample test.

11.2 TWO-SAMPLE RANK TESTS

The most common two-sample tests are the pooled t test and the unequal variance t test. The pooled t test is popular, has good statistical properties, and is easy to compute. Although it is the most powerful test if the errors are normally distributed and the variances are equal, it may not be as powerful as the Wilcoxon rank-sum test if the errors are not normally distributed. The unequal variance t test is another popular test that may maintain its level of significance when the errors do not have equal variance and they are normally distributed, but it too may be less powerful than rank-based tests with non-normal errors. See O'Gorman (1995) for details concerning the performance of these tests with unequal variance. The major problem with t tests is that they are not robust for power. That is, they are not as powerful as other tests when the errors are non-normal because the test statistics are functions of means and standard deviations that are sensitive to the presence of outliers.

Several rank-based test have been proposed in order to reduce the sensitivity of the test to outliers. The most common rank-based two-sample test is the Wilcoxon rank-sum test. This test, which is based on the sum of the ranks in one of the samples, is often more powerful than the t test for many non-normal errors. However, we are not restricted to using the sum of the ranks as the test statistic. We could use the sum of a function of the ranks as the test statistic.

Rank-based tests can be described most generally using rank scores, which give a score for each observation as a function of the rank of the observation. Suppose we let $X_1, \ldots, X_m$ be a simple random sample of size m from a continuous population with a cumulative distribution function (c.d.f.) of $F(x)$ and let $Y_1, \ldots, Y_n$ be a simple random sample of size n from a population with a c.d.f. of $F(x - \Delta)$. We combine the observations from these samples to compute the ranks $Q_1, \ldots, Q_m$ for the observations in the first group and the ranks $R_1, \ldots, R_n$ for the observations in the second group. With this notation the Wilcoxon rank-sum is $S = R_1 + \cdots + R_n$. The p-value for the test can be based on the permutation distribution of S if n is small or on a large sample normal approximation if n is large.

The general two-sample rank statistic can be written as

$$S = \sum_{j=1}^{n} a(R_j), \tag{11.1}$$

where the summation is over the scores for the observations in the second sample and $a(1), \ldots, a(n)$ are the rank scores that we assign to those observations. Usually the scores are nondecreasing so that in a test of $H_0 : \Delta = 0$ versus $H_a : \Delta > 0$ we reject the null hypothesis for large values of S. Let $N = m + n$. For the Wilcoxon test we use the rank scores

$$a_W(i) = i, \qquad i = 1, \ldots, N. \tag{11.2}$$

For the Wilcoxon test the jth observation in the second sample will have a rank of R_j and a score of $a_W(R_j) = R_j$. Thus, the scores are simply their ranks. For example, if Y_2 has rank 5 in the combined sample then $R_2 = 5$, and for the Wilcoxon test we use $a_W(R_2) = a_W(5) = 5$ for the score.

For the median test we use the scores

$$a_M(i) = \begin{cases} 1 & \text{if } i \geq (N+1)/2, \\ 0 & \text{otherwise} \end{cases} \tag{11.3}$$

so that S is simply the number of observations in the second sample that equal or exceed the median. This test could be expected to distinguish between two populations that have peaks in the center of their densities, as shown in Figure 11.1. Since S is the count of the observations in the second sample above the median, and because observations that are just above the median get the same weight as those that are far above the median, this test tends to be sensitive to shifts in location for symmetric distributions that have central peaks and longtails.

Another set of rank scores that are effective with long-tailed distributions is

$$a_{LT}(i) = \begin{cases} -(N+1)/4 & \text{if } i < (N+1)/4, \\ i - (N+1)/2 & \text{if } (N+1)/4 \leq i \leq 3(N+1)/4, \\ (N+1)/4 & \text{if } i > 3(N+1)/4. \end{cases} \tag{11.4}$$

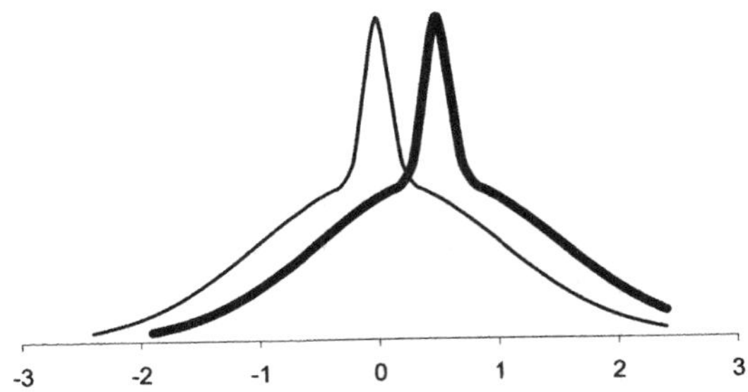

Figure 11.1 Two density functions that have central peaks. The distribution for the second population has a density function (bold line) that is shifted to the right of density function for the first population (thin line).

These scores were proposed by Randles and Wolfe (1979, p. 293) for detecting a shift in a population whose distribution is long-tailed. The score function for $a_{LT}(i)$ is shown in Figure 11.2. These scores downweight the extreme values that can arise when sampling from long-tailed distributions.

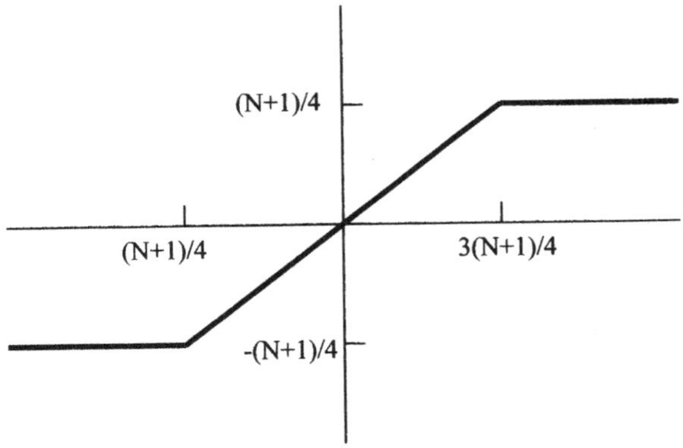

Figure 11.2 Rank scores for long-tailed densities.

If the errors have a short-tailed distribution, the density functions for the two populations might look like those in Figure 11.3. To detect a shift in these populations we would need to emphasize the observations near the tails, which will have the smallest and largest ranks. Thus, we might use the scores

$$a_{ST}(i) = \begin{cases} i - (N+1)/4 & \text{if } i < (N+1)/4, \\ 0 & \text{if } (N+1)/4 \le i \le 3(N+1)/4, \\ i - 3(N+1)/4 & \text{if } i > 3(N+1)/4. \end{cases} \tag{11.5}$$

This score function is displayed in Figure 11.4. These scores, which were suggested by Gastwirth (1965), are effective in detecting a shift in location for short-tailed distributions.

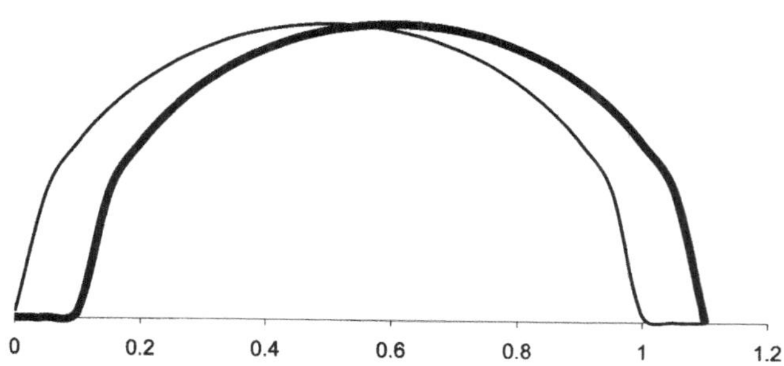

Figure 11.3 Density functions having short tails. The distribution for the second population has a density function (bold line) that is shifted to the right of density function for the first population (thin line).

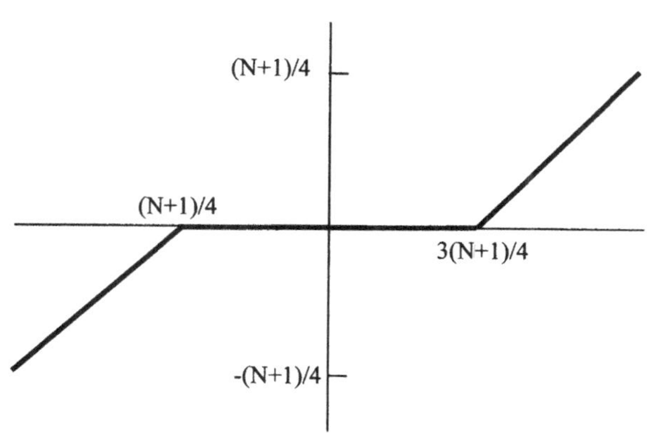

Figure 11.4 Rank scores for short-tailed densities.

Many real-world distributions are skewed to the right, so their density functions may appear to be similar to those shown in Figure 11.5. We would want to emphasize the smallest observations and give little weight to the largest observations. For these distributions the scores

$$a_{RS}(i) = \begin{cases} i - (N+1)/2 & \text{if } i \leq (N+1)/2, \\ 0 & \text{if } i > (N+1)/2 \end{cases} \tag{11.6}$$

were used by Hogg, Fisher, and Randles (1975) to detect a shift in right-tailed distributions. This score function is displayed in Figure 11.6. The test statistic that uses these scores should be sensitive to the smallest observations.

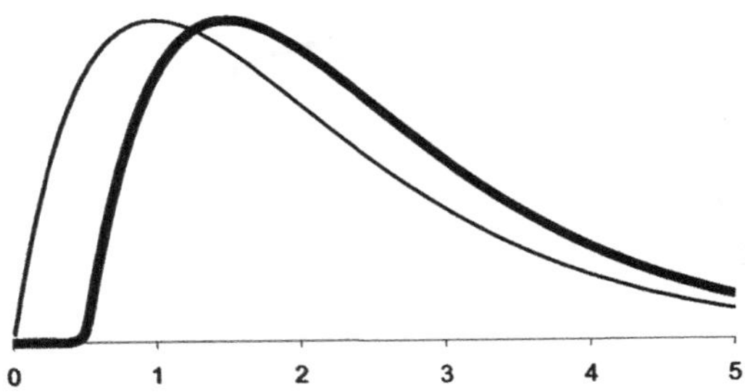

Figure 11.5 Density functions that are skewed to the right. The distribution for the second population has a density function (bold line) that is shifted to the right of the density function for the first population (thin line).

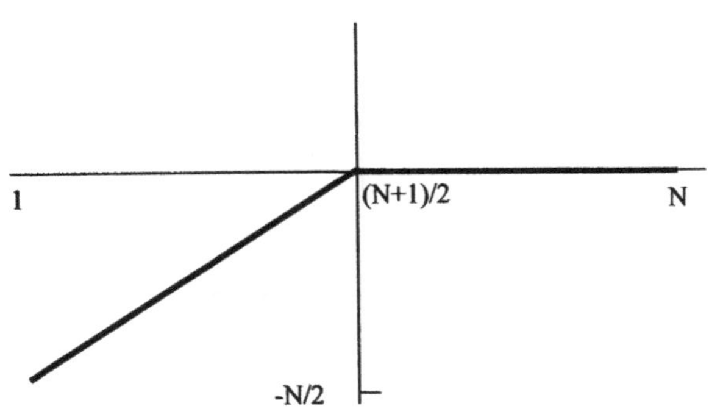

Figure 11.6 Rank scores for right-skewed densities.

With some data sets we may want the rank scores to be normally distributed. To make these scores approximate a normal distribution we could use the normal scores

$$a_{NS}(i) = \Phi^{-1}[i/(N+1)], \qquad i = 1, \ldots, N, \tag{11.7}$$

where $\Phi(\bullet)$ is the c.d.f. of the standard normal distribution. The test that uses this score function was introduced by Van der Waerden (1952, 1953). Note that, if n is large and the original observations were distributed as a standard normal random variable then these normal scores should approximate the observations. If the original observations were highly skewed, this would have the effect of producing normally distributed scores.

Once we have selected a set of scores we could perform a permutation test to determine the appropriate p-value. However, for large samples it may not be feasible to determine the full permutation distribution so we usually use a normal approximation for $m \geq 10$ and $n \geq 10$. For large samples we calculate the test statistic

$$Z = \frac{S - n\bar{a}}{\sigma} \tag{11.8}$$

where

$$\bar{a} = \sum_{i=1}^{N} a(i)/N \tag{11.9}$$

and

$$\sigma = \left\{ \frac{mn}{N(N-1)} \sum_{i=1}^{N} [a(i) - \bar{a}]^{2} \right\}^{\frac{1}{2}}. \tag{11.10}$$

The mean and variance formulas are derived in Randles and Wolfe (1979, Chapter 8).

To perform a two-sample rank test for large samples, we would first choose a set of scores and then rank the observations and compute S, σ, and Z. Next, we would obtain the p-value from the c.d.f. of the standard normal distribution.

These rank scores can be illustrated by a data set containing the measurements of total ventilation (liters of air per minute per square meter of body area) for $m = 8$ subjects in a control group and $n = 8$ subjects in an experimental group that were obtained by Agosti and Camerota (1965). The ventilation rates from both samples are tabulated in Table 11.1 along with the rank scores, the test statistics, and the two-tailed p-values from each test. Using $\alpha = 0.05$, we would reject the null hypothesis if we had used the Wilcoxon test, but would not reject the null hypothesis if we had used the median test.

A researcher would like to have the greatest chance of rejecting the null hypothesis while maintaining the stated level of significance. But which set of scores should be used? If we knew the distribution of the errors, we might use that knowledge to select an appropriate set of rank scores. Because we rarely know the distribution of the errors, we would like to have some procedure to select the rank scores. The adaptive test that we will describe in the next section will give a fully automatic selection of scores for this test.

Table 11.1 Ventilation rates for 8 control subject and 8 experimental subjects.[a]

Obs	Sample	Rates	Rank	Wilcoxon	Median	NS	ST	RS
1	1	4.50	1.00	1.00	0.00	-1.56	-3.25	-7.50
2	1	4.78	2.00	2.00	0.00	-1.19	-2.25	-6.50
3	1	4.79	3.00	3.00	0.00	-0.93	-1.25	-5.50
4	1	4.86	4.00	4.00	0.00	-0.72	-0.25	-4.50
5	1	5.41	6.00	6.00	0.00	-0.38	0.00	-2.50
6	1	5.70	8.00	8.00	0.00	-0.07	0.00	-0.50
7	1	6.08	11.00	11.00	1.00	0.38	0.00	0.00
8	1	6.21	12.00	12.00	1.00	0.54	0.00	0.00
9	2	5.32	5.00	5.00	0.00	-0.54	0.00	-3.50
10	2	5.60	7.00	7.00	0.00	-0.22	0.00	-1.50
11	2	5.74	9.00	9.00	1.00	0.07	0.00	0.00
12	2	6.06	10.00	10.00	1.00	0.22	0.00	0.00
13	2	6.32	13.00	13.00	1.00	0.72	0.25	0.00
14	2	6.34	14.00	14.00	1.00	0.93	1.25	0.00
15	2	6.79	15.00	15.00	1.00	1.19	2.25	0.00
16	2	7.18	16.00	16.00	1.00	1.56	3.25	0.00
			$S=$	89.00	6.00	3.93	7.00	-5.00
			$Z=$	2.21	1.94	2.25	2.31	2.07
			$p=$	0.027	0.053	0.024	0.021	0.039

[a] The last three columns contain the normal scores (NS), the scores for short-tailed distributions (ST), and the scores for right skewed distributions (RS).

11.3 THE HFR TEST

The HFR test was proposed by Hogg, Fisher, and Randles (1975) to increase the chance of detecting a difference between the populations. This adaptive test uses two selection statistics to determine which set of rank scores to use. The first of these selection statistics is obtained by combining both samples, sorting the observations, and then computing

$$Q_3 = \frac{\bar{U}_{.05} - \bar{M}_{.5}}{\bar{M}_{.5} - \bar{L}_{.05}},$$

where $\bar{U}_{.05}$ is the average of the upper 5%, $\bar{M}_{.5}$ is the average of the middle 50%, and $\bar{L}_{.05}$ is the average of the lower 5% of the observations. The second selection statistic, which is a robust measure of tail length, is given by

$$Q_4 = \frac{\bar{U}_{.05} - \bar{L}_{.05}}{\bar{U}_{.5} - \bar{L}_{.5}},$$

where $\bar{U}_{.5}$ and $\bar{L}_{.5}$ are the averages of the upper and lower 50% of the combined samples, respectively. These two selection statistics are then used to select a set of rank scores that are sensitive to a shift in location. Figure 11.7 shows the selection regions in the (Q_3, Q_4) plane for the adaptive test. We showed, in Chapter 1, that the HFR test has reasonable power for many of the error distributions.

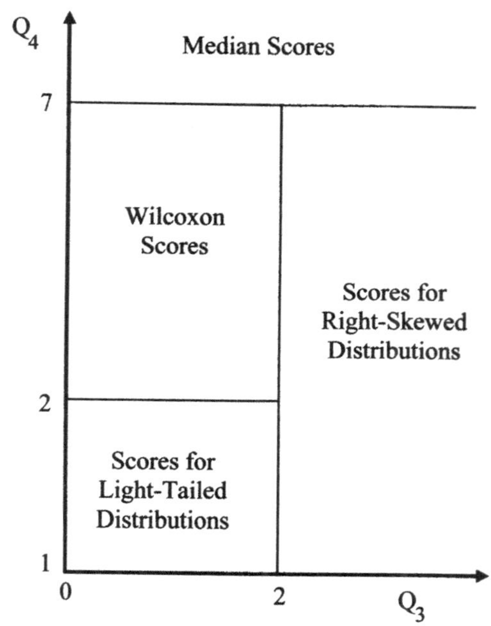

Figure 11.7 Selection regions for the HFR test.

Even with all of their careful work, the adaptive test seems, at first glance, to be questionable. It does not appear to be appropriate to use the data to determine exactly what test statistic should be used. Why does the HFR test maintain its level of significance? Will other rank tests also maintain their level of significance? We will explore these topics in the next section.

11.4 SIGNIFICANCE LEVEL OF ADAPTIVE TESTS

We will limit our discussion of the significance level of the adaptive test to the two-sample situation with m observations $X_1, X_2, \ldots, X_m$ from the first sample and n observations $Y_1, Y_2, \ldots, Y_n$ from the second sample. Let $Z_{(1)} \leq \ldots \leq Z_{(m+n)}$ be the order statistics from the combined sample. Let $R^* = \{Q_1, \ldots, Q_m, R_1, \ldots, R_n\}$ be the vector of ranks, which is obtained by ranking over both samples. Note that the elements in R^* are not ordered. Under the null hypothesis, it can be shown that the vector of ranks is independent of the set of order statistic. For a short proof of this property see Randles and Wolfe (1979, p. 273).

The basic approach taken in many adaptive tests is to use selection statistics to select a set of rank scores that are then used in the calculation of the test statistic. Since the order statistics are independent of the vector of ranks, the selection statistics are independent of the rank test statistics. Hence, the resulting test should maintain its level of significance.

To see why this is true, consider the adaptive HFR test that uses Q_3 and Q_4 as selection statistics. The vector $Q = (Q_3, Q_4)$ is used to determine the set of rank scores that will be used, as shown in Figure 11.7. Note that a distribution-free rank test could be performed using any one of these 4 sets of rank scores. Now suppose S_j is the sum, over the second sample, of the rank scores assigned to the *j*th region, and suppose that c_j is the critical value for the *j*th region, such that $P(S_j > c_j) \leq \alpha$ for all distributions. Then, for all distributions, the overall significance level of the adaptive test is

$$P(Rejection) = P\left[\bigcup_{j=1}^{4}(Q \in Region_j \text{ and } S_j > c_j)\right].$$

Since Q is a function of the order statistics and S_j is a function of the ranks, Q and S_j are independent under the null hypothesis. Hence, for every distribution, the significance level of the test is

$$\begin{aligned} P(Rejection) &= \sum_{j=1}^{4}[P(Q \in Region_j) \cdot P(S_j > c_j)] \\ &\leq \sum_{j=1}^{4} P(Q \in Region_j)\alpha = \alpha. \end{aligned}$$

Thus, the HFR test maintains its level of significance. In general, if the selection statistic is a function of the order statistics and the test statistic is a function of the ranks, then the test will maintain its level of significance for all distributions.

However, not all rank-based adaptive tests maintain their level of significance. For example, Hüsler (1987) proposed a two-sample test whose selection statistics were weighted averages of the within-sample values of Q_3 and Q_4. These selection statistics were not independent of the ranks. Hence, this test did not maintain its level of significance for small samples. In this chapter all of the rank-based adaptive tests use selection statistics that are obtained from the combined rather than the within-group order statistics, so they all maintain their level of significance.

11.5 BÜNING'S ADAPTIVE TEST FOR LOCATION

Much of the early literature on adaptive testing was limited to one-sample and two-sample tests. In an effort to make these methods more widely applicable, Büning

(1996) extended adaptive methods to the analysis of data from a one-way layout, using the same general approach that was used in the HFR test. Büning defined Q_3 and Q_4 in the same manner that they were in the HFR test, but he used a different rank score selection scheme. The selection regions for the test are shown in Figure 11.8 for this test and these are similar to those shown in Figure 11.7 for the HFR test. However, if $Q_4 > 3$ the scores for long-tailed distributions are used instead of the median scores that were used in the HFR test.

After the rank scores are selected they are used in a one-way rank test. Suppose we have K samples from K populations. The rank of the jth observation in the ith sample is R_{ij} and the rank score will be denoted by $a(R_{ij})$. For these one-way tests we will let $\bar{a}_i$ be the average rank score in the ith sample, and let $\bar{a}$ be the average of the rank scores over all samples. Let n_i be the number of observations in the ith group and let N be the total sample size. The rank-based test statistic for any set of rank scores is

$$S_i = \frac{\sum_{i=1}^{K} n_i \left(\bar{a}_i - \bar{a}\right)^2}{\left[\sum_{i=1}^{K} \sum_{j=1}^{n_i} \left\{a\left(R_{ij}\right) - \bar{a}\right\}^2\right] \Big/ (N-1)}.$$

This statistic has an approximate chi-square distribution with $K - 1$ degrees of freedom provided the average of the scores are used for ties. See Conover (1973) for details concerning this approximation. If the Wilcoxon scores are selected then the resulting test is the Kruskal–Wallis test. Büning's adaptive test maintains its level of significance for the same reason that the HFR test maintains its level of significance. In a small simulation study having $K = 3$ samples with $n_1 = n_2 = n_3 = 10$, Büning (1996) demonstrated that this test had slightly less power than that of the F test for normal errors, but it had higher power than the F test for five non-normal errors.

11.6 AN ADAPTIVE TEST FOR LOCATION AND SCALE

With experimental data it is common to find that the treatment group has shifted to the right or left of the control group and that the observations in the experimental group are more variable than those in the control group. To detect the differences in these populations, Lepage (1971) proposed a test that is sensitive to changes in location and scale. Let Z_l be the test statistic obtained from a set of rank scores that are used to perform a test for location. This test statistic would usually be computed using Equation (11.8), with the means and standard deviations computed from Equations (11.9) and (11.10). Let Z_s be a test statistic computed from a different set of rank scores that would be sensitive to changes in scale. Lepage (1971) suggested creating a new test statistic that is a function of the location and scale test statistics. The proposed test statistic is

$$L = Z_l^2 + Z_s^2 \qquad (11.11)$$

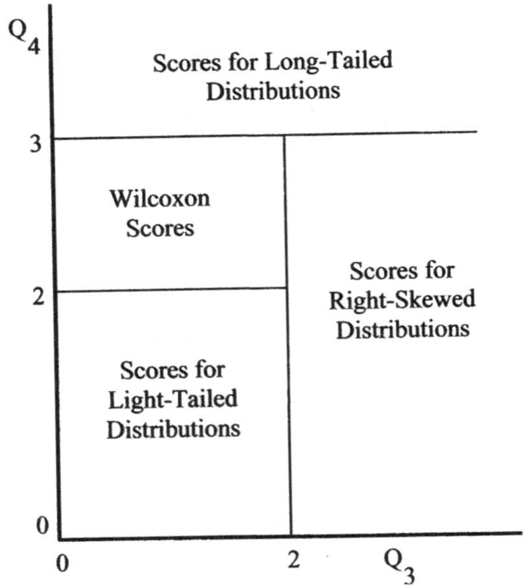

Figure 11.8 Selection regions for Büning's test.

The rank scores for the test of location and the rank scores for the tests of scale are usually selected so that they are uncorrelated under the null hypothesis. If they are uncorrelated under the null hypothesis then the distribution of L should be approximately chi-squared with two degrees of freedom.

Lepage (1971) proposed using the Wilcoxon scores given in Equation (11.2) for the location test and the Ansari–Bradley scores for the scale test. The scores for the Ansari–Bradley test are

$$a_{AB}(i) = \frac{N+1}{2} - \left| i - \frac{N+1}{2} \right|, i = 1, \ldots, N. \tag{11.12}$$

Both Z_l and Z_s can be computed in the usual manner using Equations (11.8), (11.9), and (11.10). If there are no ties, the formulas for the mean and variance of the Ansari–Bradley test statistic can be simplified. These formulas can be found in Gibbons and Chakraborti (2003, Chapter 9). Lepage computes Z_l, Z_s, and the L statistic, which is then used to perform the test for location and scale. This test appears to have excellent power to detect a change in location or a change in scale, or both, with symmetric distributions having tails of medium length.

For short-tailed or long-tailed symmetric distributions other scores could be used in place of the Wilcoxon and Ansari–Bradley scores. Büning and Thadewald (2000) suggested three sets of scores for these tests. The first test uses the scores that were described in Equation (11.5) for the location test. These scores, which are denoted by $a_{ST}(i)$, are effective with short-tailed distributions. For the scale test with short-tailed distributions, they proposed using $a_{ST,s}(i) = |a_{ST}(i)|$ for $i = 1, \ldots, N$. Using these scores, the values of Z_l and Z_s are computed and the Lepage-type test statistic (L_{ST}) is calculated.

For median-tailed distributions, Büning and Thadewald (2000) proposed using normal scores $a_{NS}(i)$ that were defined in (11.7) for the location test and the rank scores $a_{NS,s}(i) = [a_{NS}(i)]^2$ for $i = 1, \ldots, N$ for the scale test. Next, the Lepage-type test statistic is computed for medium tailed distributions, which will be called L_{MT}.

For long-tailed distributions Büning and Thadewald (2000) proposed using the rank scores $a_{LT}(i)$ that were defined in (11.4) for the location test. The scores for the test for scale are given by $a_{LT,s}(i) = [i - (N+1)/2]^2$ for $i = 1, \ldots, N$. The corresponding Lepage-type test statistic for long-tailed distributions will be denoted by L_{LT}.

Büning and Thadewald (2000) proposed a selection statistic to determine which of these three sets of test scores to use. The selection statistic, which is a location and scale invariant measure of tail length, is

$$\hat{T} = \frac{\hat{x}_{t,0.975} - \hat{x}_{t,0.025}}{\hat{x}_{t,0.875} - \hat{x}_{t,0.125}}, \tag{11.13}$$

where $\hat{x}_{t,p}$ is the traditional estimator of the pth percentile from the combined data. It can be shown that a sample from a normal distribution will have $\hat{T}$ near 1.704. Larger values of $\hat{T}$ suggest that the distribution has longer tails than that of the normal. In their adaptive test, they use L_{ST} if $\hat{T} \leq 1.5$ because that test should be appropriate for short-tailed distributions. For medium tailed distribution having $1.5 < \hat{T} \leq 1.9$ they use L_{MT} as the test statistic. They use L_{LT} as the test statistic if $\hat{T} > 1.9$ because those scores should be effective with long-tailed symmetric distributions. The adaptive test of Büning and Thadewald (2000) was shown, in simulation studies, to be effective in detecting shifts and changes in scale with a variety of symmetric distributions.

11.7 OTHER ADAPTIVE RANK TESTS

One desirable feature of rank tests, such as the Wilcoxon test, is that the p-value will change only by a small amount if a single observation is added to a large data set. For example, if $m = n = 50$ we would expect the p-value to change little if one observation is added to one of the groups. Nor would we expect the p-value to dramatically change if a small correction is made to one of the values in a data set. However, in the HFR test a small change in the data set could change Q_3 or Q_4 enough so that an entirely different set of rank scores would be used in the test. Thus, the p-value can be quite sensitive to small changes in the data.

To address this problem Ruberg (1986) proposed a continuously adaptive rank test. This test was shown by O'Gorman (1996) to have low power in some circumstances. Another continuously adaptive rank test was proposed by O'Gorman (1997), and it maintains its size and has high relative power. It avoids the problems associated with selection regions, but it introduces a fair amount of complexity into the testing procedure. The adaptive weighted least squares approach that we used in Chapter 3 for the two-sample test is continuously adaptive, so it also avoids the problems associated with selection regions.

11.8 MAXIMUM TEST

So far, we have described adaptive tests that use the approach taken by Hogg, Fisher, and Randles (1975). These tests use selection statistics to determine the set of rank scores that should be used. However, these methods require selection statistics to be computed and selection regions to be determined. Also, the p-value may be sensitive to small changes in the data. One way to avoid the difficulties associated with selection methods is to use a maximum test. The general approach is to compute the Z-values, using Equation (11.8), for several sets of rank scores and then to use the maximum of these Z-values as the test statistic. The distribution of the test statistic, and the p-value, is estimated by a permutation method.

A maximum test was proposed by Neuhäuser, Büning, and Hothorn (2004) that appears to have relatively high power for a variety of distributions. Their test is performed using the following procedure:

1. Compute the Wilcoxon test statistic (Z_W) using the scores from Equation (11.2).

2. Compute the test statistic for long-tailed distributions (Z_{LT}) using the scores from Equation (11.4).

3. Compute the test statistic for short-tailed distributions (Z_{ST}) using the scores from Equation (11.5).

4. Compute the test statistic for right skewed distributions (Z_{RS}) using the scores from Equation (11.6).

5. Let $Z_u = \max(Z_W, Z_{LT}, Z_{ST}, Z_{RS})$ be the test statistic for the maximum test.

6. Use a permutation test to find the p-value. This p-value can be obtained by permuting the residuals and adding them to the overall mean or by permuting the sample labels.

   (a) For the kth permutation compute the statistics $Z_{W,k}$, $Z_{LT,k}$, $Z_{ST,k}$, and $Z_{RS,k}$ from the permuted data.

   (b) Compute $Z_k = \max(Z_{W,k}, Z_{LT,k}, Z_{ST,k}, Z_{RS,k})$ as the maximum statistic for the kth permutation.

   (c) Perform R permutations and compute the maximum test statistic for each of these permutations. Let E equal the number of permutations that have

$Z_k \geq Z_u$. The p-value for the maximum test is $p_{max} = (E+1)/(R+1)$, where R is the number of permutations.

Neuhäuser, Büning, and Hothorn (2004) showed, using a simulation study, that the maximum test had power that was comparable to the HFR test for a variety of distributions. One advantage of the maximum test is that it does not need a vector of selection statistics. Another advantage of this test is that the p-value is not too sensitive to small changes in the data.

11.9 DISCUSSION

In contrast to the adaptive tests that are used in the rest of this book, the rank-based adaptive tests do not require a complicated weighting procedure. In addition, they have several other advantages:

- Because they are rank tests, it is easy to convince researchers that they will maintain their level of significance.
- They have high relative power for a variety of distributions.
- A permutation procedure is not required to perform the HFR test.

However, there are serious limitations with these rank-based adaptive methods that have discouraged their use. The disadvantages of the adaptive rank-based tests are:

- They are not easy to generalize to more complex models.
- The HFR test is not as powerful as the adaptive WLS test with the normal and uniform error distributions.

Depending on the complexity of the model the researcher may choose either the rank-based or the weighted least squares approach to adaptive testing. The rank methods seem to be suitable for those researchers who usually perform two-sample tests and rarely have the need to use tests of significance with more complex models. For researchers who are faced with a variety of testing situations the weighted least squares approach seems to more desirable because it can be used for most of the commonly encountered tests of significance.

EXERCISES

11.1 Blumberg, West, and Ellis (1954) collected data concerning psychological factors related to the progression of cancer. In their study they identified two groups. One group had a rapidly progressing form of the disease while the other group had a slowly progressing form. In the data shown below, negative values indicated a tendency to present the appearance of serenity in the presence of stress. The scores are:

Rapidly Progressing	-25, -24, -22, -22, -21, -18, -18, -18, -18
	-17, -16, -14, -14, -13, -13, -13, -13, -9
	-8, -7, -5, 1, 3, 7, 7
Slowly Progressing	-21, -18, -16, -16, -16, -14, -13, -13, -12
	-11, -11, -11, -9, -9, -9, -9, -7, -6
	-3, -2, 3, 10

Using SAS or R compute the p-value for the pooled t test, the Wilcoxon rank-sum test, and the median test. (With SAS, PROC NPAR1WAY will give these p-value.) Use the computer output to answer the following questions:

a) Give the two-tailed p-value for the pooled t test.

b) Give the two-tailed p-value for the Wilcoxon rank-sum test.

c) Explain the difference in p-values between the t test and the Wilcoxon test.

d) Give the two-tailed p-value for the median test.

e) Explain the difference in p-values between the median test and the other tests.

11.2 In Table 11.1 the ventilation rates for observation 13 was 6.32 and for observation 14 it was 6.34. These are closely spaced values, considering the range was from 4.50 to 7.18 for these ventilation rates. Yet, the rank scores for the normal scores were 0.72 and 0.93 for observation 13 and 14, respectively. Verify each of these normal scores and explain why these are more widely separated than the raw data.

11.3 In the previous problem it was noted, in Table 11.1, that the ventilation rates for observations 13 and 14 were closely spaced. If we had used the adaptive weighted least squares test that was described in Chapter 3 would the weights have been closely spaced? Why or why not?

11.4 Suppose a researcher routinely computed the p-value from a pooled t test (p_t) and the p-value from the Wilcoxon rank-sum test (p_W). Further, suppose the researcher used $p = \min(p_t, p_W)$ in a test of significance and that the null hypothesis was rejected if $p \leq \alpha$. For this exercise you may assume that the errors were normally distributed, independent, and had constant variance. Do you believe that this test maintained its level of significance? Why or why not?

11.5 Suppose a researcher used a minimum p-value test that is based on the same component tests that were used by Neuhäuser, Büning, and Hothorn (2004) in their maximum test. Let p_W, p_{LT}, p_{ST}, and p_{RS} be the p-values obtained by the normal approximation for the tests based on Wilcoxon, long-tailed, short-tailed, and right-skewed scores, respectively. Let $p = \min(p_W, p_{LT}, p_{ST}, p_{RS})$ be the p-value that is used in the test of significance. Further, suppose that a permutation method was used to ensure that the test would maintain its level of significance. Would the test have properties that are similar to those of the maximum test? Would the test have properties that are identical to those of the maximum test? Explain.

11.6 In the HFR test there are scores for right-skewed distributions, but there are no scores for left-skewed distributions. Suggest a method of expanding the HFR test so that the test would be powerful for detecting a location shift with left-skewed distributions. Propose a set of scores and suggest a modification of the selection regions.

11.7 Suppose a researcher, in an attempt to program the HFR test, made an error in the selection logic. Suppose he used the same selection regions that are shown in Figure 11.7, but the program used scores for right-skewed distributions for $Q_4 > 7$ and used the median scores for $Q_3 \geq 2$ and $Q_4 < 7$. Would the resulting test maintain its level of significance? Would you expect this test to be effective in detecting a shift in location with right-skewed populations?

CHAPTER 12

ADAPTIVE CONFIDENCE INTERVALS AND ESTIMATES

12.1 THE RELATIONSHIP BETWEEN TESTS AND CONFIDENCE INTERVALS

Often researchers publish confidence intervals, instead of p-values from tests of significance, because they provide more useful information to the reader. Tests are used to determine if a value c, when used in a test of $H_0 : \beta = c$, is reasonably consistent with the data, whereas confidence intervals give a range of values of c that are consistent with the data. Although tests and confidence intervals are often used in different situations, there is a close relationship between the two approaches.

One way to construct confidence intervals is to determine the interval that contains values of c that, when used in a test of significance, would lead to acceptance of the null hypothesis. To illustrate this idea, suppose we want to construct a $100(1-\alpha)\%$ confidence interval for a regression parameter β in a regression model. Let $[\hat{\beta}_L, \hat{\beta}_U]$ denote the confidence interval for β. If $c \in [\hat{\beta}_L, \hat{\beta}_U]$, then the test of $H_0 : \beta = c$ versus $H_a : \beta \neq c$ would produce a two-sided p-value that is greater than α. If $c \notin [\beta_L, \beta_U]$, then the p-value would be less than α.

In the first ten chapters of this book we have described how to calculate p-values for many kinds of adaptive tests, so we should be able to construct confidence intervals

By Thomas W. O'Gorman

based on those tests. It would be tedious, but we could compute those intervals by performing tests of significance for many values of c to determine which of those c values would produce p-values greater than α. Although it may be possible to use such a procedure to find confidence intervals, we will use an iterative search procedure that was proposed by Garthwaite (1996). This procedure is easy to program and is practical. Later in this chapter we will demonstrate how the confidence limits can be computed using this iterative search with R code or a SAS macro.

12.2 THE ITERATIVE PROCEDURE OF GARTHWAITE

We will describe the method proposed by Garthwaite (1996) to estimate the lower and upper limits of a confidence interval. This method uses a stochastic approximation method that was proposed by Robbins and Monro (1951) to estimate the quantiles of a distribution. Garthwaite and Buckland (1992) demonstrated how to use the Robbins–Monro method to calculate a confidence interval and a few years later Garthwaite (1996) showed how a permutation test could be used to compute the limits. This is the general approach that we will use in this chapter.

To construct the confidence interval we will work on each limit separately. To find the lower limit we begin by using the lower limit of the traditional interval as our first estimate. We then perform one permutation of the residuals and adjust our estimate of the lower limit either up or down, depending on the t test statistic that we obtain from an adaptively weighted model. In the next step this updated lower limit is used as the new estimate and it is then updated by a single permutation of the residuals. This stochastic process continues until the estimate converges. Typically it takes between 5000 and 10,000 steps to obtain an accurate estimate of the lower limit. A similar approach is used for the upper limit.

To describe the method in more detail, we first need to generalize our adaptive testing method. In previous chapters we have used an adaptive approach to test $H_0 : \boldsymbol{\beta}_A = 0$ versus $H_a : \boldsymbol{\beta}_A \neq 0$ in the model

$$\mathbf{Y} = \mathbf{X}_R\boldsymbol{\beta}_R + \mathbf{X}_A\boldsymbol{\beta}_A + \boldsymbol{\varepsilon},$$

where $\mathbf{Y}$ is an $n \times 1$ matrix, $\mathbf{X}_R$ is an $n \times r$ matrix, $\boldsymbol{\beta}_R$ is a $r \times 1$ vector, $\mathbf{X}_A$ is an $n \times 1$ vector, $\boldsymbol{\beta}_A$ is a 1×1 matrix, and $\boldsymbol{\varepsilon}$ is an $n \times 1$ vector. However, to compute a confidence interval we must test $H_0 : \boldsymbol{\beta}_A = c$, where c is a constant. We can perform the test of $H_0 : \boldsymbol{\beta}_A = c$ versus $H_a : \boldsymbol{\beta}_A \neq c$ with a simple adjustment of the $\mathbf{Y}$ vector. To perform this test, we subtract $c\mathbf{X}_A$ from both sides of the model to obtain

$$\mathbf{Y} - c\mathbf{X}_A = \mathbf{X}_R\boldsymbol{\beta}_R + \mathbf{X}_A(\boldsymbol{\beta}_A - c) + \boldsymbol{\varepsilon}.$$

To test $H_0 : \boldsymbol{\beta}_A - c = 0$ versus $H_a : \boldsymbol{\beta}_A - c \neq 0$, we first compute $\mathbf{Y}_{ADJ} = \mathbf{Y} - c\mathbf{X}_A$ and then use the methods we have previously described to perform the adaptive test using the model

$$\mathbf{Y}_{ADJ} = \mathbf{X}_R\boldsymbol{\beta}_R + \mathbf{X}_A(\boldsymbol{\beta}_{\mathbf{A}} - c) + \boldsymbol{\varepsilon}.$$

Note that $H_0 : \boldsymbol{\beta}_A - c = 0$ is equivalent to $H_0 : \boldsymbol{\beta}_A = c$. To simplify the test procedure, we let $\boldsymbol{\beta}_{ADJ} = \boldsymbol{\beta}_A - c$ and write the model as

$$\mathbf{Y}_{ADJ} = \mathbf{X}_R\boldsymbol{\beta}_R + \mathbf{X}_A\boldsymbol{\beta}_{ADJ} + \boldsymbol{\varepsilon}.$$

In the iterative procedure we will perform an adaptive test of $H_0 : \boldsymbol{\beta}_{ADJ} = 0$ versus $H_a : \boldsymbol{\beta}_{ADJ} \neq 0$. To compute a confidence interval for a parameter in a linear model we need to perform many tests, each with a different value of c.

We now describe the Garthwaite (1996) procedure for finding the upper limit of the confidence interval for $\boldsymbol{\beta}_A$, which will be denoted by $\hat{\boldsymbol{\beta}}_{A,U}$. The upper limit $\hat{\boldsymbol{\beta}}_{A,U}$ is unknown, but we will use a stochastic approach that will produce an accurate estimate of $\hat{\boldsymbol{\beta}}_{A,U}$. In order to make the notation consistent with the literature, we will use U_l to denote our estimate of the upper limit $(\hat{\boldsymbol{\beta}}_{A,U})$ at the lth step. We begin by finding an initial estimate of the upper limit that we will denote by U_1. This estimate is updated, one step at a time, to obtain estimates of the upper limit. Now, suppose at the lth step we have obtained U_l. We compute the t statistic from the adaptively weighted model for the test of $H_0 : \boldsymbol{\beta}_A = U_l$. This will require that we first compute $\mathbf{Y}_{ADJ} = \mathbf{Y} - U_l\mathbf{X}_A$. Next, we compute the weights and the residuals from the reduced model $\mathbf{Y}_{ADJ} = \mathbf{X}_R\boldsymbol{\beta}_R + \boldsymbol{\varepsilon}$. After the weights are applied to the complete and reduced models we compute the t statistic (t_{unperm}) for the unpermuted data. Next, one permutation of the ordinary residuals is performed and we compute a new $\mathbf{Y}$ vector as the sum of the predicted values and the permuted residuals. That is, we compute $\mathbf{Y}_{NEW} = \hat{\mathbf{Y}}_{ADJ} + \mathbf{S}\hat{\boldsymbol{\varepsilon}}_R$, where $\hat{\mathbf{Y}}_{ADJ}$ is the vector of predicted values, under the reduced model, $\mathbf{S}$ is a permutation matrix, and $\hat{\boldsymbol{\varepsilon}}_R$ is the vector of ordinary residuals obtained from the unweighted reduced model, which is $Y_{ADJ} = \mathbf{X}_R\boldsymbol{\beta}_R + \boldsymbol{\varepsilon}$. This $\mathbf{Y}_{NEW}$ vector is used to compute adaptive weights and the t statistic (t_{perm}) for the adaptively weighted permuted model. That is, t_{perm} is computed in the same way that t_{unperm} was computed, except Y_{NEW} is used instead of Y_{ADJ}. The surprising aspect of Garthwaite's method is that the updating method uses only one permutation at each step.

Now suppose we have, at the lth step, an estimate U_l of the upper limit and we have computed t_{unperm} and t_{perm}. Let k denote a positive steplength constant that will be described later. Our estimate of the upper limit at step $l + 1$ is

$$U_{l+1} = \begin{cases} U_l - k(\alpha/2)/l & \text{if } t_{perm} > t_{unperm}, \\ U_l + k(1-\alpha/2)/l & \text{if } t_{perm} \leq t_{unperm}. \end{cases} \tag{12.1}$$

We should note that the estimate of the upper limit can be adjusted up or down and that the amount of the adjustment will decrease as the step number increases. Also note that the downward adjustment is much smaller than the upward adjustment. While this search always converges, it does so slowly because the denominator of the step change is proportional to $1/l$.

To compute the limits for a 95% confidence interval, Garthwaite (1996) proposed using a steplength constant of $k = 17.46(U_l - \hat{\boldsymbol{\beta}}_{AD})$, where $\hat{\boldsymbol{\beta}}_{AD}$ is an adaptive estimate of the slope. However, their steplength constant varies with U_l and can become too small if U_l approximates $\hat{\beta}_{AD}$. We have taken a slightly different approach and have used the constant $k = 17.46(\hat{\beta}_{U,OLS} - \hat{\beta}_{OLS})$, where $\hat{\beta}_{OLS}$ is the OLS estimate of β_A and $\hat{\beta}_{U,OLS}$ is the OLS estimate of the upper confidence interval for β_A, assuming normally distributed errors. The value of k is not critical; any positive constant will ensure convergence. However, if the value of k is small the method will be slow to converge.

A similar procedure is used to obtain the lower limit. We start with an estimate of the lower limit L_l at the lth step and we update it using the formula

$$L_{l+1} = \begin{cases} L_l + k(\alpha/2)/l & \text{if } t_{perm} < t_{unperm}, \\ L_l - k(1-\alpha/2)/l & \text{if } t_{perm} \geq t_{unperm}. \end{cases} \tag{12.2}$$

The lower limit is obtained independently of the upper limit that was updated using Equation (12.1). Traditional confidence intervals are symmetric about the traditional OLS point estimate. These adaptive confidence intervals may not be symmetric about the adaptive estimate.

Robbins and Monro (1951) give a description of their stochastic approximation method and show that it will converge to the correct value. However, it may not be clear why it converges to one of the limits. The following example may illustrate the reasoning behind Equations (12.1) and (12.2).

Suppose we have 10 observations in each of two samples and we want to compute a 95% confidence interval for the population difference. We have plotted the y-values for this example in Figure 12.1. We will assume that the model is

$$y_i = \beta_0 + \beta_A x_i + \varepsilon_i, \tag{12.3}$$

where

$$x_i = \begin{cases} 0 & \text{if the } i\text{th observation is from population 1,} \\ 1 & \text{if the } i\text{th observation is from population 2} \end{cases}$$

for $i = 1, \ldots, n$. With this parameterization β_A represents the mean of population 2 minus the mean of population 1. Note that the values from population 2 tend to be greater than those from population 1.

Now suppose the upper 95% confidence interval for β_A is $\hat{\beta}_{A,U} = 20$. For purposes of illustration, let us assume that we happen to be given $U_l = 20$ for the lth step. In the Robbins–Munro procedure we test $H_0 : \beta_A = 20$ by computing $y_{i,ADJ} = y_i - 20x_i$ for $i = 1, \ldots, n$. A plot of these data are shown in Figure 12.2. Note that the y-values from sample 1 are unchanged while those from sample 2 are shifted 20 units to the left.

We can see from Figure 12.2 that t_{unperm} will be negative. Recall that we permute residuals under the null hypothesis and that for a reduced model containing only an intercept the permutation of residuals would result in a permutation of the values in the Y_{ADJ} vector. For a single permutation, with the adjusted data that are shown in Figure 12.2, it is likely that t_{perm} will be close to zero. Hence, for most permutations t_{perm} will exceed t_{unperm}. When we perform a one-sided test of $H_0 : \beta_A = 20$ versus $H_a : \beta_A < 20$ we will use the data shown in Figure 12.2 to compute t_{unperm}.

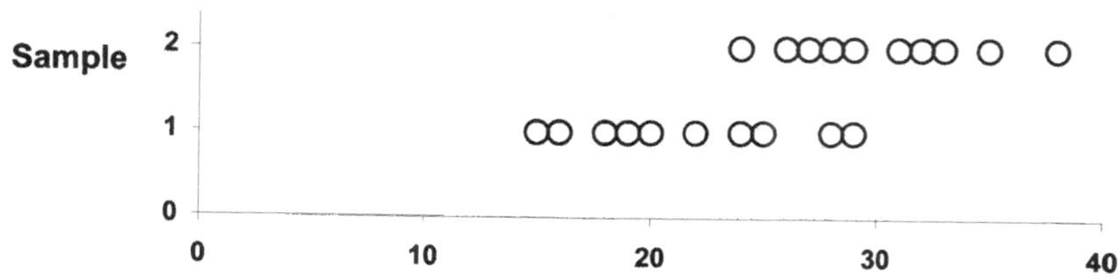

Figure 12.1 Hypothetical data from two populations. The y-values are plotted on the horizontal axis.

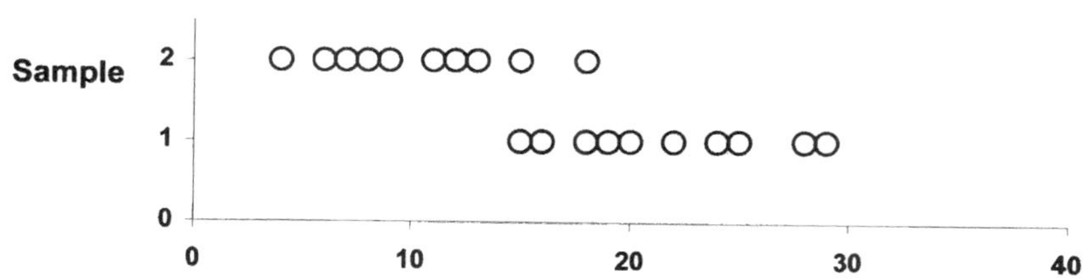

Figure 12.2 Adjusted y-values for the test of $H_0 : \beta_A = 20$ are plotted on the horizontal axis. For the second sample these are shifted 20 to the left of those in Figure 12.1.

Since U_l equals the upper limit of the $100(1-\alpha)\%$ confidence interval we know that $P(t_{perm} > t_{unperm}) = 1 - \alpha/2$ and that $P(t_{perm} \le t_{unperm}) = \alpha/2$. From Equation (12.1), if $U_l = \hat{\beta}_{A,U}$, the expected value of U_{l+1} is

$$\begin{aligned}E[U_{l+1}] &= P(t_{perm} > t_{unperm})[U_l - k(\alpha/2)/l] \\ &\quad + P(t_{perm} \le t_{unperm})[U_l + k(1-\alpha/2)/l] \\ &= (1-\alpha/2)[U_l - k(\alpha/2)/l] + (\alpha/2)[U_l + k(1-\alpha/2)/l] \\ &= (1-\alpha/2+\alpha/2)U_l - (1-\alpha/2)k(\alpha/2)/l + (\alpha/2)k(1-\alpha/2)/l \\ &= U_l.\end{aligned}$$

Hence, if l is large and U_l is near the upper limit then the estimate of the upper limit will change little if we take 1000 steps. If we guess some other value of the upper limit, say $U_l = 23$, we would obtain the adjusted values that are shown in Figure 12.3. Note that because this value of U_l is greater than the upper limit of the confidence interval we obtain smaller values for the adjusted y-values in the second sample. With this estimate of the upper limit, it can be seen from Figure 12.3 that there is now little overlap between these samples, which implies that $P(t_{perm} > t_{unperm}) > 1 - \alpha/2$. Suppose that there is a value $\alpha^* < \alpha$ such that $P(t_{perm} > t_{unperm}) = 1 - \alpha^*/2$. We would then find, if $U_l > \hat{\beta}_{A,U}$, the expected value for U_{l+1} to equal

$$\begin{aligned}E[U_{l+1}] &= (1-\alpha^*/2)[U_l - k(\alpha/2)/l] + (\alpha^*/2)[U_l + k(1-\alpha/2)/l]\\&= U_l - (1-\alpha^*/2)k(\alpha/2)/l + (\alpha^*/2)k(1-\alpha/2)/l\\&= U_l - k(\alpha/2)/l + (\alpha^*/2)k/l\\&= U_l + (\alpha^*/2 - \alpha/2)k/l.\end{aligned}$$

Hence, if $U_l > \hat{\beta}_{A,U}$ then $\alpha^*/2 < \alpha/2$ and $E(U_{l+1}) < U_l$. This implies that, if $U_l > \hat{\beta}_{A,U}$ the updated upper limit will probably decrease. Conversely, if $U_l < \hat{\beta}_{A,U}$ then $\alpha^*/2 > \alpha/2$ and it follows that $E(U_{l+1}) > U_l$. Consequently, if $U_l < \hat{\beta}_{A,U}$ the updated upper limit will probably be increased. Hopefully, this example will help explain why the search will converge. For a more detailed description of this iterative method see Robbins and Monro (1951), who state the conditions that will ensure convergence to the correct limit.

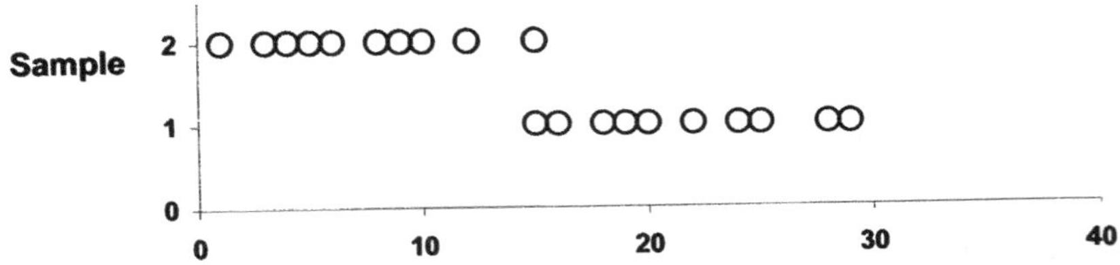

Figure 12.3 Adjusted y-values are plotted on the horizontal axis. For the second sample these are shifted 23 to the left of those in Figure 12.1.

We need not start the Robbins–Monro search at $l = 1$. In fact, starting at $l = 1$ may produce large changes to the initial estimate, which are undesirable because the search may be slow to converge. For 95% confidence intervals, Garthwaite (1996) recommends using $l = 23$ as the initial step, and we have used this initial value of l in the simulation studies.

How do we know when to stop the iterative process? Usually 5000 steps will produce reasonably accurate estimates of the limits, but in the simulation studies we usually took more steps to ensure that the estimates are stationary and have little variability. In the simulation studies we evaluate blocks of 1000 steps, starting at $l = 3000$, and compute the range of these estimates over those 1000 steps, where the range is the maximum of the estimates over the 1000 steps minus the minimum of the estimates over the 1000 steps. This range is one of the criteria we use to determine if the iterative process should be stopped.

Another criterion to consider is how often $t_{perm} \leq t_{unperm}$. Let $p_{empirical}$ denote the proportion of steps in a block of 1000 steps having $t_{perm} \leq t_{unperm}$. If the estimate is near the true upper limit, then $p_{empirical}$ should approximate $\alpha/2$. The standard error of $p_{empirical}$ should approximate $[\alpha/2(1-\alpha/2)/1000]^{1/2}$. We stop the iterative process if the range is less than 0.5% of the width of the traditional confidence interval and $p_{empirical}$ is in the interval $\alpha/2 \pm [\alpha/2(1-\alpha/2)/1000]^{1/2}$.

That is, we stop the iterations if the estimates are sufficiently stable and $p_{empirical}$ is reasonably close to $\alpha/2$. With this strict stopping rule we usually use between 8000 and 20,000 steps for each limit. For the 36,000 data sets that we analyzed in the simulation studies, the method proposed by Garthwaite (1996) always satisfied the strict stopping rules with fewer than 100,000 steps.

After it has been determined that the estimates have converged we can use either the last estimate of the limit or an average of previous estimates as our estimate. Because we stopped the iterative method after determining that a block of 1000 steps met our criteria, we have averaged the estimates over the last 1000 steps to provide our final estimate.

12.3 A 95% CONFIDENCE INTERVAL FOR THE DIFFERENCE BETWEEN POPULATION MEANS

In this section we describe the properties of adaptive confidence intervals for the mean difference between two population means. Because the calculation of the adaptive confidence intervals are computationally intensive, they take more effort to produce than the traditional confidence intervals. We would like to know if there is a real advantage of using adaptive confidence intervals instead of the traditional confidence intervals.

To investigate the properties of the two-sample confidence intervals we generate data using the model

$$y_i = \beta_0 + \beta_A x_i + \varepsilon_i, \tag{12.4}$$

where

$$x_i = \begin{cases} 0 & \text{if the } i\text{th observation is from population 1,} \\ 1 & \text{if the } i\text{th observation is from population 2} \end{cases}$$

for $i = 1, \ldots, n$. This is the same model that we used in Equation (12.3) in our explanation of the iterative method. With this model the mean of the first population is β_0, the mean of the second population is $\beta_0 + \beta_A$, and the difference is β_A. Our objective is to find a confidence interval for β_A.

For example, in an experiment designed to estimate the effects of diet on cholesterol, Anderson *et al.* (1984) randomly assigned 10 men to an oat-bran diet and 10 men to a bean diet. After 21 days on the diet the triglyceride levels were measured. These levels are shown in Table 12.1.

Table 12.1 Final triglyceride levels for men who were randomly assigned to either a bean diet or an oat-bran diet

Bean Diet	175, 173, 110, 116, 136, 209, 351, 176, 341, 476
Oat-Bran Diet	159, 155, 144, 101, 196, 150, 195, 270, 456, 523

It is important to note that there is a great deal of overlap between these two samples, as can be seen in the plot of this data in Figure 12.4, so that a test of significance would not lead to rejection of the null hypothesis. Also, the distributions of the values in both samples appear to be skewed to the right.

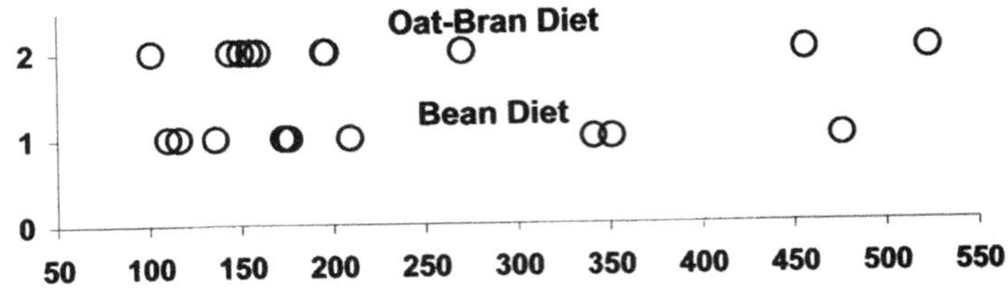

Figure 12.4 Triglyceride levels for men who were put on an oat-bran or bean diet. Diet 1 is the bean diet and diet 2 is the oat-bran diet.

Narrow confidence intervals are better than wide confidence intervals, so we will compare the average width of the adaptive confidence intervals to the average width of the traditional confidence confidence intervals for several error distributions and sample size configurations. But narrow confidence intervals are worthless if they do not usually include the population parameter. Hence, we will also carefully evaluate the coverage probabilities of the adaptive confidence intervals.

12.3.1 Comparison of Coverage Probabilities

We used, for this analysis of the coverage probabilities, the same nine error distributions that we have used for adaptive tests. To estimate the coverage probability, we designed a simulation study that used 1000 data sets with each data set having 15 observations in both samples. The coverage probabilities, in percent, are shown in Figure 12.5 for 95% confidence intervals for the difference between population means. Note that the traditional confidence interval and the adaptive confidence interval maintain their nominal coverage probability.

This pattern is also observed in studies having unbalanced sample sizes of $n_1 = 40$ and $n_2 = 10$ that produce empirical coverage probabilities for the adaptive confidence interval in the interval [93.8%, 95.5%]. Because we used 1000 data sets in these simulations, the standard error of a 95% confidence interval is 0.7%, so we see no evidence that the adaptive confidence interval departs from the nominal coverage probability. For a larger balanced design having $n_1 = n_2 = 50$ we observe coverage probabilities for the adaptive confidence interval in the interval [94.4%, 95.8%]. Consequently, we conclude that the confidence interval for the difference between means maintains its coverage probability. It now makes sense to compare the widths of the traditional confidence intervals to those of the adaptive confidence intervals.

12.3.2 Comparison of Average Width

In the same simulations that were used to estimate the coverage probabilities, we also computed the width of each interval and the average width of the intervals over

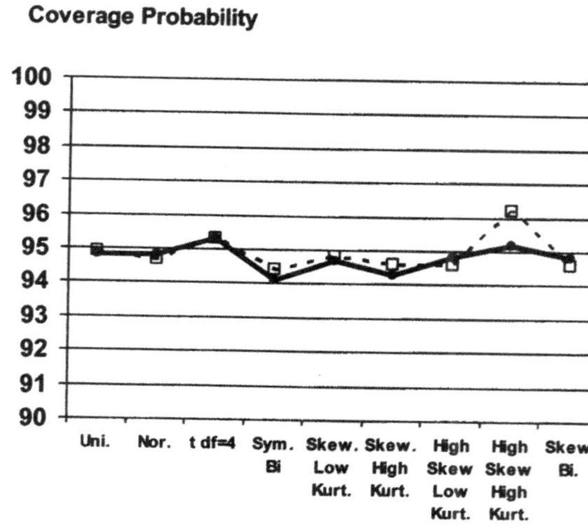

Figure 12.5 Empirical coverage probabilities, in percent, for the traditional confidence interval and the adaptive confidence interval for the difference between the means of two populations. The coverage probabilities for the traditional interval are shown by the hollow squares joined by the dashed line and the coverage probabilities for the adaptive interval are shown by the solid circles joined by the solid line.

the 1000 data sets that were generated. For the simulations that used 15 observations in each sample, we display these average widths in Figure 12.6. We observe that for non-normal errors the adaptive confidence intervals are shorter, on average, than the traditional confidence intervals. That is, we are able to obtain, for non-normal errors, a more precise estimate of the parameter while maintaining the coverage probability. These results are not too surprising because we found, in Chapter 3, that the adaptive tests have greater power than the t test for non-normal errors and because the confidence intervals were based on the same adaptive test procedure.

These results show, with the non-normal error distributions that we used in these simulation studies, that the adaptive confidence intervals tend to be narrower than the traditional confidence intervals. However, averages can be misleading. Could the average widths that are plotted in Figure 12.6 distort the true relationship between the widths of these confidence intervals?

To answer this question we have displayed a scatterplot in Figure 12.7 of the widths of the adaptive and traditional confidence intervals from data sets that had $n_1 = n_2 = 15$ with errors generated from a generalized lambda distribution with a skewness of $\alpha_3 = 2$ and a kurtosis of $\alpha_4 = 11.4$. This distribution is what we call the high-skewness, low-kurtosis distribution. Each point on the scatterplot gives the traditional confidence interval width on the horizontal axis and the adaptive confidence interval width on the vertical axis for one of the 1000 data sets that were generated. If a data set has a point below the line, then it has a narrower adaptive confidence interval than the traditional confidence interval. The scatterplot in Figure

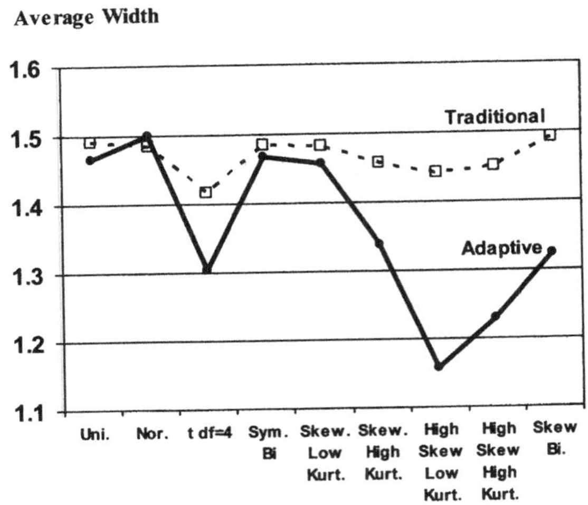

Figure 12.6 The average width of the traditional and the adaptive confidence intervals from samples of size $n_1 = n_2 = 15$ for the difference between the means of two populations. These simulations used 1000 data sets with $n_1 = n_2 = 15$. The average widths for the traditional confidence intervals are indicated by the hollow squares joined by the dashed line and the average widths for the adaptive confidence intervals are indicated by the filled circles joined by the solid line.

12.7 shows that the widths of the intervals are often nearly equal, but that when there are differences in width the adaptive interval is narrower. Thus, the average width appears to measure the overall performance of these confidence intervals.

We now compare the average widths of the traditional and adaptive confidence intervals for the unbalanced design having $n_1 = 40$ and $n_2 = 10$. These average widths are shown in Figure 12.8. We see that the adaptive confidence intervals are narrower, on average, than the traditional confidence intervals, for all non-normal errors, but the adaptive confidence interval is slightly wider than the traditional confidence interval for normal errors.

The same overall pattern is observed in Figure 12.9 for larger sample sizes of $n_1 = n_2 = 50$. Again, we see that the average width of the adaptive confidence interval is less than the average width of the traditional confidence interval. Some researchers may think that the central limit theorem should guarantee, with sample sizes of $n_1 = n_2 = 50$, that the traditional intervals will be almost as narrow as any other interval. I believe that the sampling distribution of the difference between sample means will be approximately normal, but the standard error, and hence the traditional confidence interval, will be influenced by outliers while those outliers will have less influence on the adaptive confidence interval. In this respect, the adaptive weighting acts as a variance reduction technique.

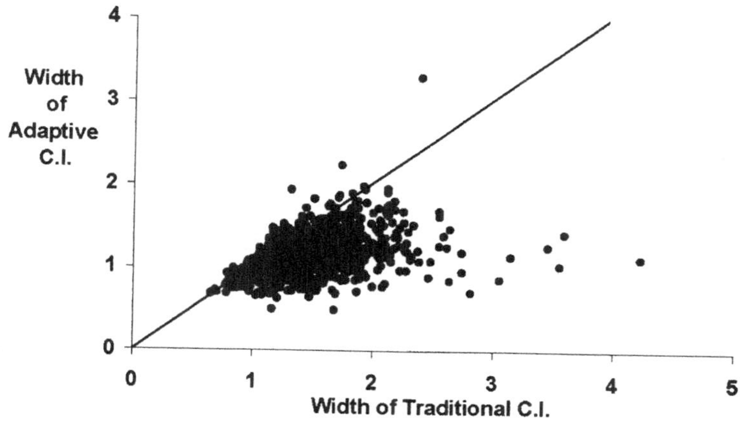

Figure 12.7 The scatterplot gives the width of the traditional confidence interval on the horizontal axis and the width of the adaptive width on the vertical axis for each of the 1000 data sets generated from a model having skewed errors with $n_1 = n_2 = 15$. If the width of the adaptive confidence interval equals the width of a traditional confidence interval, then the point will be on the line.

12.4 A 95% CONFIDENCE INTERVAL FOR SLOPE

In this section we describe the properties of a confidence interval for β_A in a model

$$y_i = \beta_0 + \beta_1 x_{i,1} + \beta_A x_{i,2} + \varepsilon_i$$

for $i = 1, \ldots, n$. In these simulations the independent variables $(x_{i,1}, x_{i,2})$ were generated from a bivariate normal distribution with a correlation of 0.8 between the variables. This high correlation was used to evaluate the effect of multicollinearity on the confidence interval for β_A.

As expected, the empirical coverage probabilities for the adaptive confidence intervals were between 93.7% and 95.7%, so the coverage probabilities appear to be maintained near 95%. The average widths for confidence intervals with data sets having $n = 50$ are shown in Figure 12.10. These results are similar to those shown for the two-sample tests, and are consistent with the relative power of the adaptive test to the traditional test. Because the adaptive confidence interval uses the same weights and permutation method as the adaptive test and because the results for the adaptive confidence intervals are consistent with the adaptive test results, it seems reasonable to conclude that, for non-normal error distributions, the higher power of the adaptive test will produce narrower confidence intervals when used with many of the other models that have been evaluated earlier in this book.

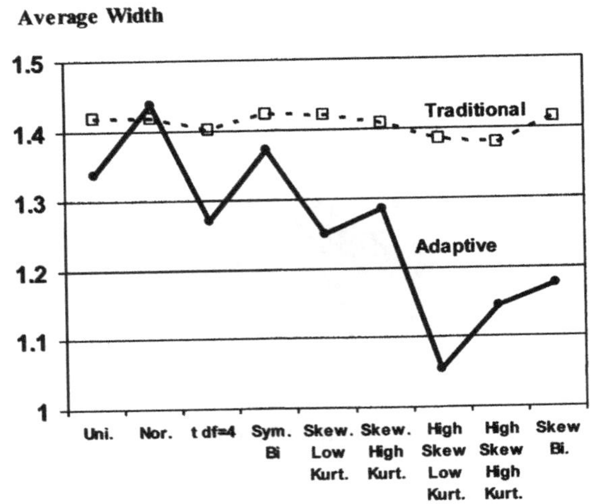

Figure 12.8 The average width of the traditional and the adaptive confidence intervals from samples of size $n_1 = 40$ and $n_2 = 10$ for the difference between the means of two populations. The average widths for the traditional confidence intervals are indicated by the hollow squares joined by the dashed line and the average widths for the adaptive intervals are indicated by the filled circles joined by the solid line.

12.5 A GENERAL FORMULA FOR CONFIDENCE LIMITS

We now describe a general formula for computing confidence limits. Let $\hat{\beta}_{A,L}$ and $\hat{\beta}_{A,U}$ be the lower and upper limits, respectively, which are unknown. We know that if Garthwaite's procedure is used and the upper limit is found, then at that upper limit $P(t_{perm} > t_{unperm}) = 1 - \alpha/2$. Using the same approach, we can see that when the lower limit is found that $P(t_{perm} < t_{unperm}) = 1 - \alpha/2$. For the lower limit let $p_{left} = P(t_{perm} \geq t_{unperm}) = \alpha/2$. To estimate $\hat{\beta}_{A,L}$ using the iterative procedure we can write the updating Formula (12.2) as

$$L_{l+1} = \begin{cases} L_l + k(p_{left})/l & \text{if } t_{perm} < t_{unperm}, \\ L_l - k(1 - p_{left})/l & \text{if } t_{perm} \geq t_{unperm}. \end{cases} \tag{12.5}$$

Now, suppose we want the upper limit. For the upper limit let $p_{left} = P(t_{perm} > t_{unperm}) = 1 - \alpha/2$, which implies that $\alpha/2 = 1 - p_{left}$. Thus, the updating Formula (12.1) for the upper limit becomes

$$U_{l+1} = \begin{cases} U_l - k(1 - p_{left})/l & \text{if } t_{perm} > t_{unperm}, \\ U_l + k(p_{left})/l & \text{if } t_{perm} \leq t_{unperm}. \end{cases} \tag{12.6}$$

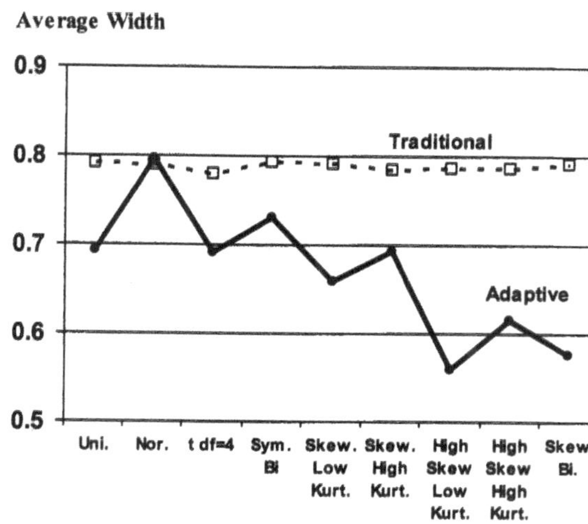

Figure 12.9 The average width of the traditional and the adaptive confidence intervals from samples of size $n_1 = n_2 = 50$ for the difference between the means of two populations. The average widths for the traditional intervals are indicated by the hollow squares joined by the dashed line and the average widths for the adaptive intervals are indicated by the filled circles joined by the solid line.

Usually we would set p_{left} to a small value, say 0.025, for the lower limit, and set p_{left} to a large value, say 0.975, for the upper limit. Note that equations (12.5) and (12.6) are nearly identical, differing only when $t_{perm} = t_{unperm}$.

In our simulations we have rarely found $t_{perm} = t_{unperm}$ because the sample sizes are large enough so that it is extremely unlikely that we would obtain, at random, the same permutation as the original, and because ties would rarely occur with these error distributions. However, ties are often encountered with real-world data sets, so we need to deal with them in a manner that is consistent with the method proposed by Garthwaite (1996). To combine these updating formulas into one general formula, let p_{left} denote the desired value of $P(t_{perm} > t_{unperm})$. The general updating formula that can produce estimates of the lower and the upper limits is

$$G_{l+1} = \begin{cases} G_l + \dfrac{k(p_{left})}{l} & \text{if } t_{perm} < t_{unperm} \text{ or} \\ & \quad \text{if } (t_{perm} = t_{unperm} \text{ and } p_{left} > .5), \\ G_l & \text{if } t_{perm} = t_{unperm} \text{ and } p_{left} = .5, \\ G_l - \dfrac{k(1 - p_{left})}{l} & \text{if } t_{perm} > t_{unperm} \text{ or} \\ & \quad \text{if } (t_{perm} = t_{unperm} \text{ and } p_{left} < .5). \end{cases} \tag{12.7}$$

For the lower limit we will have $p_{left} < .5$ so it will agree with updating Equation (12.2). For the upper limit we will have $p_{left} > .5$ so it will agree with updating

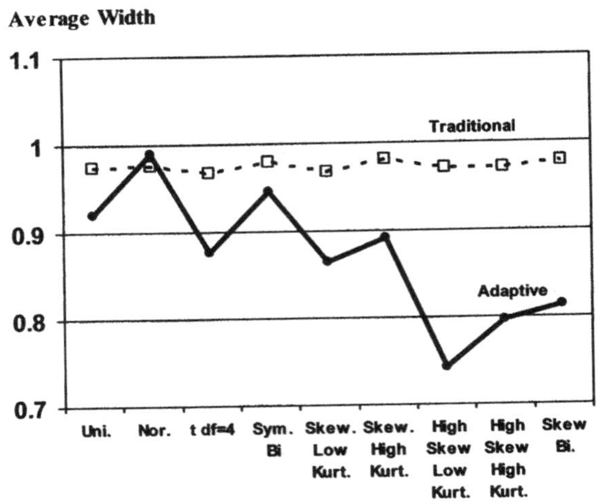

Figure 12.10 The average width for the traditional confidence interval and the adaptive confidence interval for β_A in a multiple regression model with data sets having $n = 50$ observations. The average widths for the traditional confidence interval are shown by the hollow squares joined by the dashed line and the average widths for the adaptive interval are shown by the filled circles joined by the solid line.

Equation (12.1). We will use this formula with $p_{left} = 0.025$ in the R code that is described in the next section to compute the lower limit of the adaptive confidence interval. We will also use it later in this chapter in the adaptive estimation procedure.

12.6 COMPUTING A CONFIDENCE INTERVAL USING R

An R function entitled `adaptest.r` is used to estimate the lower and upper limits of a confidence interval. This R function uses the general updating Equation (12.7) to perform the iterative method described by Garthwaite (1996). The `adaptest` function, which is listed in Appendix H, can be downloaded from *www.math.niu.edu/~ogorman*. To call this macro we must specify the complete model and the reduced model, in the same manner that they were specified in earlier chapters and we must specify the variable name (`varname`) corresponding to the parameter that we want to estimate. In this example we are interested in a 95% confidence interval for the parameter associated with the group indicator. Unlike the simulation studies, which had a specified stopping rule, this function will perform the number of steps specified by the `maxstep` parameter. The user will need to determine if the estimates are sufficiently accurate. An initial estimate of the limit must be specified by the parameter `initial` and p_{left} is specified as the `pleft` parameter in the function call. Typically, if a 95% confidence interval is desired, the researcher would call the `adaptest` function twice, once with an initial estimate of the lower limit with $p_{left} = 0.025$, and a second time with an initial estimate of the upper limit with $p_{left} = 0.975$.

We will use the triglyceride data set that we described in Section 12.3 to illustrate how easy it is to use R to compute these limits. If the data from Table 12.1 is on the file `tri.dat`, with the diet specified by the `group` variable, we can obtain the lower and upper limits using this R code:

```
source("adaptest.r")
dataset  <- c("tri.dat")
complete <- c("y~group")
reduced  <- c("y~1")
varname  <- c("group")
pleft    <- 0.025
initial  <- -120
maxsteps <- 5000
seed     <- c(71492)
lowerlimit <- adaptest(dataset, complete, reduced, varname,
                       pleft, initial, maxsteps, seed)
pleft    <- 0.975
initial  <- +130
upperlimit <- adaptest(dataset, complete, reduced, varname,
                       pleft, initial, maxsteps, seed)
```

In Table 12.2 we give the partial output for the lower limit of a 95% confidence interval. Note that, at each step that is a multiple of 1000, the average of the estimates, over the last 1000 steps, is printed. Because the averages are still somewhat erratic, I believe most researchers would continue this simulation, perhaps using 20,000 steps. The output for the upper limit, which is not shown, is similar to that for the lower limit.

Table 12.2 Partial output from R program that estimates the lower limit of the 95% confidence interval for the triglyceride data.

Step = 1 estimate = -117.6413
Step = 2 estimate = -115.3808
Step = 100 estimate = -99.11369
Step = 200 estimate = -79.27942
Step = 1000 estimate = -68.65023 average = -82.65953
Step = 2000 estimate = -78.74534 average = -76.05234
Step = 3000 estimate = -79.03844 average = -79.6314
Step = 4000 estimate = -80.70212 average = -79.22125
Step = 5000 estimate = -82.31379 average = -82.28741

This R function can be used to provide a variety of confidence intervals. If a 99% confidence interval is desired the value of p_{left} would be set to $p_{left} = .005$ for the lower limit and $p_{left} = .995$ for the upper limit. Also, if a researcher had used an initial estimate that appears to be inaccurate, the researcher could rerun the program using a new estimate that was obtained from the output of the preliminary analysis.

12.7 COMPUTING A 95% CONFIDENCE INTERVAL USING SAS

A SAS macro is also available for computing the 95% confidence interval. This program is less flexible than the R code because it always produces a 95% confidence interval and it uses traditional 95% confidence interval limits as the initial values in the Garthwaite (1996) procedure. This SAS macro, which is listed in Appendix I, can be downloaded from *www.math.niu.edu/~ogorman*.

The SAS code given here generates the 95% confidence interval for the mean difference between the two treatments in the triglyceride data set that we described in Section 12.3. We have used the indicator variable `group` in the call to the macro `adaptci`. Note that the parameters in the macro call are similar to those used in the tests of significance, but we must specify the maximum number of steps as the `maxsteps` variable. In the program we are assuming that the macro is on the file `adaptci.sas`.

```
%include 'adaptci.sas';
data dtri;
  input tri group @@;
  datalines;
159 1  155 1  144 1  101 1  196 1
150 1  195 1  270 1  456 1  523 1
175 0  173 0  110 0  116 0  136 0
209 0  351 0  176 0  341 0  476 0
;
%adaptci(dataset=dtri, y=tri, xa=group,
   seed1=3577, seed2=4755, seed3=18177, maxsteps= 5000);
```

This SAS macro produces output that is similar to that produced by the R code. The output is given in Table 12.3 for some of the iterations. Note that this SAS macro prints, for each step that is a multiple of 1000, the current estimate and the average of the last 1000 steps. No stopping rules are included in the program, but the researcher can use the output to determine if convergence has been achieved.

12.8 ADAPTIVE ESTIMATION

There is an extensive literature on robust estimation but there is little published on adaptive estimation. With robust estimation the idea is to reduce the influence of outliers by using some predetermined downweighting function. For example, the 10% trimmed mean might be used to estimate the mean of a population having a symmetric distribution. This trimming is not adaptive. The trimming is performed

Table 12.3 Partial output from a SAS program that gives the lower limit of the 95% confidence interval for the triglyceride data

after step 1 adaptive lower limit = -113.328
after step 2 adaptive lower limit = -111.068
after step 100 adaptive lower limit = -106.702
after step 200 adaptive lower limit = -74.3242
after step 1000 adaptive lower limit = -97.2957
after step 2000 adaptive lower limit = -89.8875 average = -93.16
after step 3000 adaptive lower limit = -84.2330 average = -85.81
after step 4000 adaptive lower limit = -84.3775 average = -84.81
after step 5000 adaptive lower limit = -84.6833 average = -83.74

by giving weights of zero to observations in the top and bottom 10% of the sample. The data are trimmed even if the data appear to be normally distributed.

In contrast, an adaptive estimator would perform the trimming only if it appeared to be necessary. For example, the adaptive estimator proposed by Hogg and Lenth (1984) uses a selection statistic that is a measure of tailweight to determine which downweighting function to use. This estimator uses the mean if the tails of the distribution are not long, it uses a 25% trimmed mean if the tails are moderately long, and it uses the median if the tails are very long.

In the following description of various robust methods we will assume that we are interested in estimating the slope in a simple linear regression. Many of the earliest robust estimators of the slope were M estimators. For these estimators a function is specified that reduces the influence of the observations that have large residuals. A variety of functions have been proposed and many of these functions contain a tuning parameter that must be specified. See the book by Hampel *et al.* (1986, Chapter 2) for details. One problem with these methods is that there does not appear to be agreement on what function and tuning parameters are most effective.

A very different approach to robust estimation of the slope was described by Rousseeuw (1984). In this method the slope is determined to be the value that minimizes the median of the squares of the residuals. This estimator, which is sometimes called the LMS estimator, is appealing because it will ignore a few unusual observations. However, Hettmansperger and Sheather (1992) and Stromberg (1993) pointed out that this estimator can be unstable in certain circumstances.

A trimming approach can also be used in regression problems by using the least trimmed squares (LTS) estimator. A 25% trimming with an LTS estimator gives a slope that minimizes the sum of the squared residuals for a subset of points of size $\lfloor .75n \rfloor$ that have the smallest squared residuals. See Rousseeuw and Leroy (1987) for details. With large data sets this method may be computationally intensive because there are many subsets that contain 75% of the observations. In addition to the LMS and LTS estimators there are many other robust estimators that have been proposed.

Because there is such a large literature in this area and because this book concerns adaptive tests of significance, we will use one estimator that can be determined directly from the adaptive test. The idea of using a test procedure to find an estimate is not new. Hodges and Lehmann (1963) proposed using a Wilcoxon two-sample test to determine an estimate of the difference between population means. Their approach is to determine a shift of one sample relative to the other sample that will give the "best possible agreement" of the two samples. Suppose that there are m observations in sample one and n observations in sample two and we compute each of the mn possible differences obtained by selecting one observation from sample one and one observation from sample two. The Hodges–Lehmann estimate of the difference between population means is the median of the set of all mn differences. See Lehmann (1975, Chapter 2) for an example of this procedure. We will extend this idea to the estimation of one parameter in a linear model. Thus, we want to find that value of the parameter that will produce the largest p-value. To find this estimate we use a modification of the search method that Garthwaite (1996) used to calculate confidence limits.

Consider the general updating Formula (12.7). Our estimator is the value that is produced when we set $p_{left} = .5$ and use the adaptive test to compute t_{perm} and t_{unperm}. That is, we find the value that, when used to adjust the y-values, would maximize the two-tailed p-value. Using $p_{left} = .5$, the updating formula becomes

$$G_{l+1} = \begin{cases} G_l + k/(2l) & \text{if } t_{perm} < t_{unperm}, \\ G_l & \text{if } t_{perm} = t_{unperm}, \\ G_l - k/(2l) & \text{if } t_{perm} > t_{unperm}. \end{cases} \tag{12.8}$$

Now suppose the sample size is large enough so that the probability that t_{perm} equals t_{unperm} is very small. Then, with $p_{left} = .5$ the estimates obtained from Equation (12.8) will converge to a value such that half of the permutations will have $t_{perm} < t_{unperm}$ and half will have $t_{perm} > t_{unperm}$. This appears to be a reasonable way of robustly estimating a parameter in a linear model.

Some intuition concerning this method can be obtained if we refer to the hypothetical data shown in Figure 12.1. Suppose we want to estimate the difference between the population means. We begin our estimation procedure by using an initial estimate of the difference, say $\hat{\beta}_A = 9$, and compute the adjusted y-values as $y_{i,ADJ} = y_i - 9x_i$ for $i = 1, \ldots, n$. The plot of these adjusted y-values that is given in Figure 12.11 indicates that the medians of the adjusted y-values are nearly equal, so t_{unperm} would be approximately zero. A permutation of the residuals would be almost as likely to produce $t_{perm} < t_{unperm}$ as it would be to produce $t_{perm} > t_{unperm}$. Thus $\hat{\beta}_A = 9$ will be close to the adaptive estimate. The updating Formula (12.8) would be used with several thousand steps to obtain a more precise estimate of the difference between the population means.

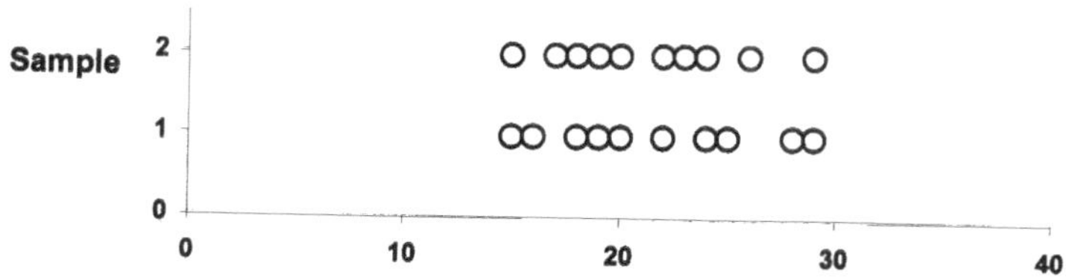

Figure 12.11 Adjusted y-values for the test of $H_0 : \beta_A = 9$ are plotted on the horizontal axis. For the second sample these are shifted 9 to the left of those in Figure 12.1.

12.9 ADAPTIVE ESTIMATION OF THE DIFFERENCE BETWEEN TWO POPULATION MEANS

In this section we will compare the traditional estimator of the difference between population means to the adaptive estimator. The model, as we specified earlier in the chapter, is

$$y_i = \beta_0 + \beta_A x_i + \varepsilon_i, \tag{12.9}$$

where

$$x_i = \begin{cases} 0 & \text{if the } i\text{th observation is from population 1,} \\ 1 & \text{if the } i\text{th observation is from population 2} \end{cases}$$

for $i = 1, \ldots, n$. This is the model that was used for the confidence interval for the difference between two population means. In this section our objective is to find a better estimator of the difference (β_A) than the traditional estimator. We will begin our evaluation of this estimator by checking the bias.

We used a simulation study with 1000 data sets to evaluate the bias. Each data set had 15 observations in both samples. The bias for both estimators is shown in Figure 12.12. It can be seen that the bias is close to zero for these nine error distributions. In these simulations we used $\beta_A = 3$ and the standard deviation of the errors was equal to one. Hence, the standard deviation of the difference between group means, without weighting, is $(1/15+1/15)^{1/2} = 0.365$, so the standard error of the average difference is approximately $.365/(1000)^{1/2} = 0.012$. With this standard error the bias estimates shown in Figure 12.12 are consistent with a bias of zero. A similar pattern was seen for simulation studies having $n_1 = 40$ and $n_2 = 10$ and for studies having $n_1 = n_2 = 50$. The bias of the adaptive estimator was in the interval [-0.020, 0.012] for studies having $n_1 = 40$ and $n_2 = 10$ and was in the interval [-0.008, 0.011] for studies having $n_1 = n_2 = 50$. Hence, we conclude that the estimator has little or no bias.

One measure of accuracy is the RMSE, which is the square root of the mean squared error (MSE) of an estimator. To compute the MSE we averaged $(\hat{\beta}_A - \beta_A)^2$ over the 1000 data sets that were used in the simulation. The RMSE for the simulations that used $n_1 = n_2 = 15$ are shown in Figure 12.13. We note that the RMSE for the adaptive estimator is slightly greater than the RMSE of the traditional estimator

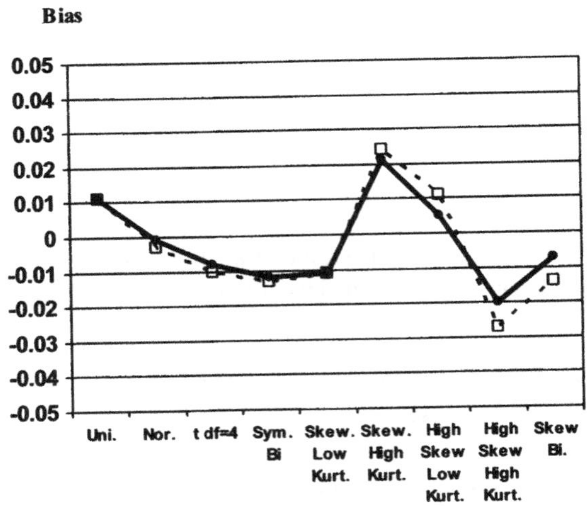

Figure 12.12 The bias of the adaptive estimator and the traditional estimator with samples of size $n_1 = n_2 = 15$. The bias for the traditional estimator is shown by hollow squares joined by the dashed line and the bias for the adaptive estimator is shown by the filled circles joined by the solid line.

for normally distributed errors. However, for the non-normal errors the adaptive estimator has a much smaller RMSE than the traditional estimator.

For a larger unbalanced sample size of $n_1 = 40$ and $n_2 = 10$ the simulation results are shown in Figure 12.14. Again, we see that the RMSE of the adaptive estimator is much less than the RMSE of the traditional estimator for non-normal error distributions. The results for simulations having $n_1 = n_2 = 50$, which are not shown, are similar to those for the smaller samples. These results are in agreement with those we found for the width of the confidence intervals.

There is a close relationship between the adaptive confidence intervals and the adaptive estimates. The adaptive estimator could be approximated as the midpoint of an adaptive confidence interval having a low coverage probability. Hence, the narrow confidence intervals produce more precise estimates.

12.10 ADAPTIVE ESTIMATION OF A SLOPE IN A MULTIPLE REGRESSION MODEL

Suppose we have a multiple regression model

$$y_i = \beta_0 + \beta_1 x_{i,1} + \beta_A x_{i,A} + \varepsilon_i, \tag{12.10}$$

for $i = 1, \ldots, n$, and we want to estimate β_A. We would like to compare the traditional OLS estimator to the adaptive estimator. To find the estimate we use the

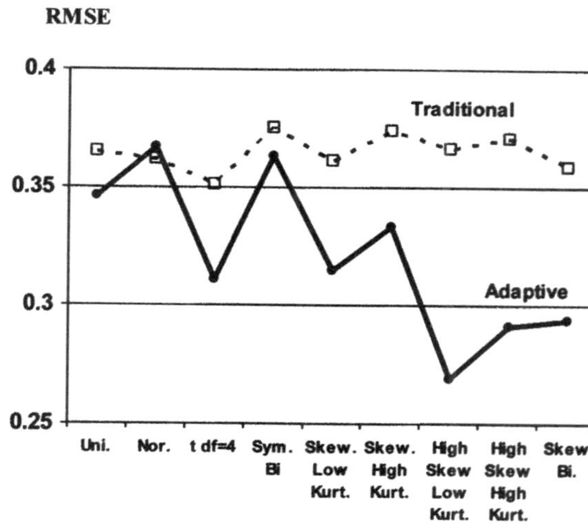

Figure 12.13 The RMSE for the adaptive estimator and the traditional estimator for the difference between the means of two populations with samples of size $n_1 = n_2 = 15$. The RMSE for the traditional estimator is shown by the hollow squares joined by the dashed line and the RMSE for the adaptive estimator is shown by filled circles joined by the solid line.

updating Formula (12.8). Let $x_{i,A}$ be the value of the predictor variable corresponding to β_A for the *ith* observation. At the *l*th step, if we have an estimate of G_l, we compute the adjusted *y*-values as $y_{i,ADJ} = y_i - G_l x_{i,A}$, for $i = 1, \ldots, n$. We then use Garthwaite's procedure, which involves computing t_{unperm} for the test of $H_0 : \beta_A = G_l$ and, after one permutation of the residuals, computing t_{perm}. Next, the updating Equation (12.8) is used to update the estimate of the slope.

In a simulation study we used $n = 50$ observations in a multiple regression model specified in Equation (12.10) along with errors generated from one of the usual error distributions to obtain y_i for $i = 1, \ldots, 50$. The two independent variables were generated from a bivariate normal distribution with a correlation of $\rho = 0.8$. We set $\beta_A = 3.0$ and generated 1000 data sets to evaluate the bias and RMSE of these estimators.

As expected, the bias estimates, which are shown in Figure 12.15, are close to zero. Note also that the small departures from zero of the bias of the adaptive estimator are matched by those of the traditional estimator that we know is unbiased. Hence, we conclude that the adaptive estimator has little or no bias when used in a multiple regression model with two correlated independent variables.

The RMSE of these estimators is shown in Figure 12.16 for the same simulations having $n = 50$ observations. With normally distributed errors we found a RMSE estimate of 0.254 for the traditional estimator, while the adaptive estimator had a RMSE of 0.256. Thus, for this sample size there appears to be little loss in accuracy if the adaptive estimator is used with normally distributed errors. With the non-

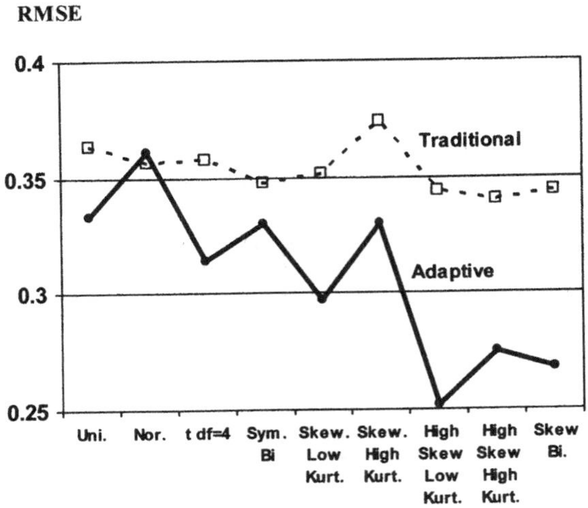

Figure 12.14 The RMSE for the adaptive estimator and the traditional estimator for the difference between the means of two populations with samples of size $n_1 = 40$, and $n_2 = 10$. The RMSE for the traditional estimator is shown by the hollow squares joined by the dashed line and the RMSE for the adaptive estimator is shown by filled circles joined by the solid line.

normal error distributions the adaptive estimator has a considerably smaller RMSE than the RMSE of the traditional estimator.

12.11 COMPUTING AN ADAPTIVE ESTIMATE USING R

The Scottish Hills Runners Association collected data on 35 races that were run in Scotland in 1984. These data, which were published by Atkinson (1986), contained a serious error that was not discovered for 2 years after publication. The data set, without the correction, is given in Table 12.4. The error occurred in the recording of the time for the Knock Hill race. The original data was given as 4719 seconds, but due to a tabulation error it should have been one hour less, so it should have been recorded as 1119 seconds.

Using this race data, our objective is to estimate the coefficient for the distance variable, which should indicate how many seconds would be required to run one mile in these races. Our multiple regression model is

$$y_i = \beta_0 + \beta_1 x_{i,1} + \beta_A x_{i,2} + \varepsilon_i,$$

where y_i is the time to complete the race (in seconds), $x_{i,1}$ is the climb (in feet), and $x_{i,2}$ is the distance (in miles) for the ith race.

The ordinary least squares estimate of β_A is 373.1, but we believe that this value may underestimate the true time per mile because the Knock Hill race had

Table 12.4 The Scottish Hills race data

Race	Name	Time (sec)	Climb (ft)	Distance (mi)
1	Greenmantle	965	650	2.5
2	Carnethy	2901	2500	6
3	Craig Dunain	2019	900	6
4	Ben Rha	2736	800	7.5
5	Ben Lomond	3736	3070	8
6	Goatfell	4393	2866	8
7	Bens of Jura	12277	7500	16
8	Cairnpapple	2182	800	6
9	Scolty	1785	800	5
10	Traprain Law	2385	650	6
11	Lairig Ghru	11560	2100	28
12	Dollar	2583	2000	5
13	Lomonds of Fife	3900	2200	9.5
14	Cairn Table	2648	500	6
15	Eildon Two	1616	1500	4.5
16	Cairngorm	4335	3000	10
17	Edinburgh	5905	2200	14
18	Knock Hill	4719	350	3
19	Black Hill	1045	1000	4.5
20	Creag Beag	1954	600	5.5
21	Kildoon	957	300	3
22	Meall Ant-Suiche	1674	1500	3.5
23	Half Ben Nevis	2859	2200	6
24	Cow Hill	1076	900	2
25	North Berwick Law	1121	600	3
26	Creag Dubh	1573	2000	4
27	Burnswark	2066	800	6
28	Largo	1714	950	5
29	Criffel	3030	1750	6.5
30	Achmony	1257	500	5
31	Ben Nevis	5135	4400	10
32	Knockfarrel	1943	600	6
33	Two Breweries Fell	10215	5200	18
34	Cockleroi	1686	850	4.5
35	Moffat Chase	9590	5000	20

Source: Atkinson, A. C. (1986) Data are used with permission of the Institute of Mathematical Statistics.

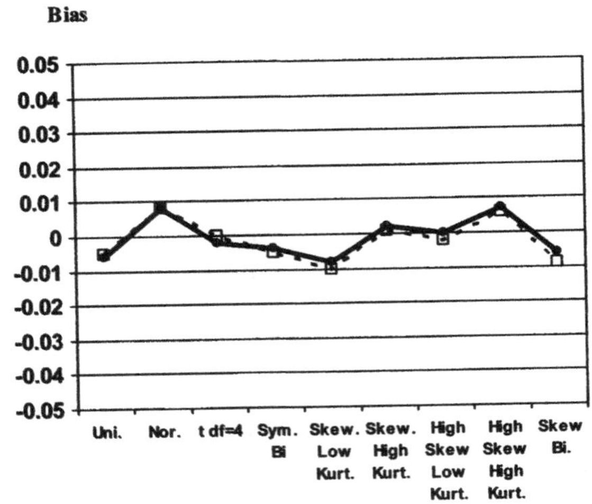

Figure 12.15 The bias of the adaptive estimator and the traditional estimator with a sample of size $n = 50$ in a multiple regression model. The bias for the traditional estimator is shown by the hollow squares joined by the dashed line and the bias for the adaptive estimator is shown by filled circles joined by the solid line.

erroneously recorded a long time for a short race. Because we anticipated that the adaptive estimate will downweight the erroneous data, we used the following R code to find the adaptive estimate:

```
source("adaptest.r")
dataset   <- c("scottish.dat")
complete <- c("y~climb+miles")
reduced   <- c("y~climb")
varname   <- c("miles")
pleft     <- 0.5
initial   <- 0.0
maxsteps <- 10000
seed      <- c(777)
beta      <- adaptest(dataset, complete, reduced, varname,
                          pleft, initial, maxsteps, seed )
```

The R code for estimation is on the file named `adaptest.r` and the data are on the file `scottish.dat`. The complete and reduced models are specified in the same manner that they were specified in earlier chapters. The predictor variable that corresponds to the parameter that is to be estimated is given in the `varname` variable. The value p_{left} is specified as `pleft`, which is always set to 0.5 for estimation. The initial estimate is specified by `initial` and the maximum number of steps is specified by `maxsteps`. Note that we have used an initial estimate of

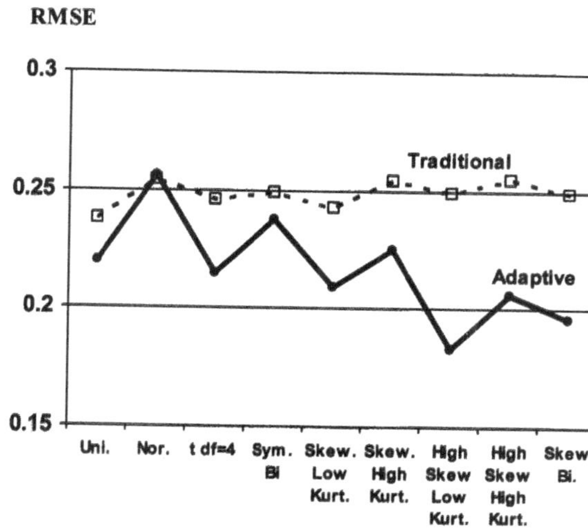

Figure 12.16 The RMSE for the adaptive estimator and the traditional estimator for the slope in a multiple regression model with a sample of size $n = 50$. The RMSE for the traditional estimator is shown by the hollow squares joined by the dashed line and the RMSE for the adaptive estimator is shown by filled circles joined by the solid line.

zero, which is far from a reasonable value, but the estimation routine quickly found a more reasonable value, as can be seen in the output that is shown in Table 12.5. A random number seed should be specified as `seed`.

Table 12.5 Output from an R program that gives the estimate for the number of seconds required to race one mile in the Scottish Hills races

```
Step = 1 estimate = 27.88625
Step = 2 estimate = 54.61058

Step = 10 estimate = 235.8253
Step = 20 estimate = 407.8750

Step = 100 estimate = 398.2935
Step = 200 estimate = 389.1547

Step = 1000 estimate = 396.2677 average = 391.2715
Step = 2000 estimate = 393.8002 average = 393.8414

Step = 10000 estimate = 393.61 average = 394.8544
```

The adaptive estimate, after 10,000 steps, was 394.9, which is somewhat larger than the traditional ordinary least squares estimate of 373.1. While almost all of the observations in the data set at the last step had weights in the interval [0.9, 1.1], the weight for the Knock Hill race was 0.262, which was the lowest weight in the data set.

Two years after the data were originally published, Atkinson (1988) corrected the time for the Knock Hill race. After that time was corrected to 1119 seconds the OLS estimate becomes 380.7 and the adaptive estimate becomes 387.9. Hence, the adaptive estimate correctly downweighted an erroneous observation. However, it is not possible to state which estimator is more accurate because we do not know the true value of the parameter.

12.12 COMPUTING AN ADAPTIVE ESTIMATE USING SAS

It is also possible to compute an adaptive estimate of the number of seconds required to run one mile using the `adaptest` macro, which is listed in Appendix J and which can be downloaded from *www.math.niu.edu/~ogorman*. We must call the macro `adaptest` to compute the estimate using `maxsteps` as the total number of steps. The initial estimate does not need to be specified because the macro always uses the ordinary least squares estimate as the initial estimate. We will assume the SAS macro `adaptest` is on the file `adaptest.sas`. The SAS code to obtain an estimate of the time per mile with the Scottish Hills race data is

```
%include "adaptest.sas";
data race;
   input Hill $   time   miles   climb;
   datalines;
Greenmantle    965    2.5     650
Carnethy      2901    6.0    2500
   .            .      .       .
MoffatChase   9590   20.0    5000
;
%adaptest(dataset=race, y=time, xr=climb, xa=miles,
    seed1=14554, seed2=10733, seed3=13349, maxsteps = 10000);
```

The output from this SAS macro is given in Table 12.6. Note that the estimate obtained from the SAS macro approximates the estimate obtained from the R code.

12.13 DISCUSSION

We have seen that the adaptive confidence intervals are often narrower than the traditional confidence intervals. The only problem with adaptive confidence intervals is that they tend to take longer to compute than adaptive tests, but this is a function of the software that this used and the speed of the computer. The R programs in this chapter take only a few minutes to run on an inexpensive personal computer purchased in 2008, so these are quite practical. The estimation procedure, because it uses the same approach that is used to find the limits of the confidence interval, can be executed in about half of the time that is necessary with the confidence interval.

Table 12.6 Output from a SAS program that gives the estimate for the number of seconds required to race one mile in the Scottish Hills races

after step 1 adaptive lower limit = 327.5441
after step 2 adaptive lower limit = 354.2367
after step 100 adaptive lower limit = 385.7765
after step 200 adaptive lower limit = 381.6242
after step 1000 adaptive lower limit = 394.7562
after step 2000 lower limit = 392.3199 average = 394.9524
after step 10000 lower limit = 393.7422 average = 393.8530

In the software we recompute the weights in each step. However, if the estimate changes little there may be no need to recompute the weights in each step. We might consider, after we have obtained a reasonable estimate of a limit, to use the same weights for 10 steps. Consequently, it may be possible to reduce the amount of computation needed in these confidence interval procedures. However, in an effort to keep the code as straightforward as possible we have not investigated these methods.

The estimation procedure is test-based. That is, the estimate is the value that maximizes the p-value. This idea was originated in the estimation procedures of Hodges and Lehmann (1963), where they found the estimate of the difference between two-samples such that the Wilcoxon rank-sum test would produce the largest two-sided p-value. We have used the same basic idea to estimate one parameter in a linear model.

Another approach to adaptive estimation was proposed by O'Gorman (2004, Chapter 7). That approach did not use a test based method, but used a different weighting procedure applied directly to the model. In simulation studies it generally had a slightly larger RMSE than the procedure described in this chapter, but it does have the advantage of producing adaptive estimates for all of the regression parameters. We have, in the chapter, concentrated on one regression parameter so that we could use the test-based approach.

We have seen in this chapter that the desirable properties of the adaptive tests carry over to the confidence intervals that are derived from them. The confidence intervals maintain their coverage probabilities because the tests maintained their level of significance. The adaptive confidence intervals, for non-normal error distributions, were often narrower than the traditional intervals because the corresponding adaptive test procedures were more powerful than the traditional test procedures. Hence, as we have observed with testing methods, there is little to lose and much to gain by using an adaptive procedure.

EXERCISES

12.1 Consider the triglyceride data that are displayed in Figure 12.4. Suppose we want an adaptive estimate of the upper limit of a 95% confidence interval. The traditional 95% confidence interval for the difference between means with the triglyceride data set is [-115.7,132.9]. Between step $l = 25$ and step $l = 26$ what is the maximum change in the estimate of the upper limit? Between step $l =$ 10,000 and step $l =$ 10,001 what is the maximum change in the estimate?

12.2 Suppose that at step $l =$ 10,000 we obtain an estimate of the lower limit of a 95% confidence interval. After we perform 1000 more steps we find that the lower limit estimate is the same at step $l =$ 11,000 as it was at step $l =$ 10,000. Roughly estimate how many of these 1000 steps produced $t_{unperm} < t_{perm}$. You may assume that $t_{unperm} \neq t_{perm}$ for all steps.

12.3 A skeptic believes that the statistics on the average widths of confidence intervals that are shown in some figures in this chapter do not represent the true widths because one or two large values of the widths could inflate the average width of the traditional confidence intervals. What information that was shown in this chapter could be used to address this concern?

12.4 Suppose that after $l =$ 10,000 steps we obtain an estimate for β_A and this estimate does not change over the next 1000 steps. For how many of these steps do we find $t_{unperm} < t_{perm}$? You may assume that $t_{unperm} \neq t_{perm}$ at all steps.

12.5 Refer to the Scottish Hills race data shown in Table 12.4. This data can be found at *www.math.niu.edu/~ogorman*. Using either an R program or a SAS program compute the 95% confidence interval for the distance parameter. *Warning*: If you use a SAS program do not label the distances as `distance` because the macro uses a variable called "distance" in its calculations.

12.6 Using the data on triglycerides found in Table 12.1, write an R program or a SAS program to calculate the adaptive estimate of the difference between the mean triglyceride level on the bean diet and the mean triglyceride level on the oat-bran diet.

12.7 Use the cigarette data that is listed in Table 5.1 and Table 5.2 for this exercise. Suppose we are using as the model for the ith observation

$$y_i = \beta_0 + \beta_1 x_{i1} + \beta_2 x_{i2} + \beta_3 x_{i3} + \varepsilon_i,$$

where y_i is the per capita cigarette sales (in packs), x_{i1} is the average price per pack (in cents), x_{i2} is the per capita personal income (in dollars), and x_{i3} is the percent of females living in the state. Use R code to compute the 95% confidence interval for β_1. Interpret.

12.8 Refer to the cigarette data and the model that is described in the previous exercise. Use the SAS macro to compute the 95% confidence interval for β_1. Interpret.

12.9 Data on gas mileage in small cars was given in Exercise 5.12 of Chapter 5. Use that data to give an adaptive estimate of the coefficient of weight in a model that includes engine size and weight to predict EPA city mileage.

12.10 Data concerning the average verbal scores for sixth grade students at 20 schools in the United States was described in Exercise 5.13 of Chapter 5 and was listed in Table 5.4. Use that data to answer the following questions:

a) Find an adaptive 95% confidence interval for the coefficient that measures the average salary per pupil, after including all of the other predictor variables into the model.

b) Is the confidence interval consistent with the results for the adaptive test of significance that you performed in Exercise 5.13 of Chapter 5?

c) Find an adaptive point estimate for the coefficient that relates the average salary per pupil to the average verbal score in a model that includes all five predictor variables.

APPENDIX A

R CODE FOR UNIVARIATE ADAPTIVE TESTS

```
adapt <- function(dataset,complete,reduced,r=2000,seed=1492){

tradpvalue <- trad.ftest(dataset,complete,reduced)
testdata <- read.table(dataset)
print(testdata)
attach(testdata)
statsu<- teststats(y,complete,reduced)
lambdau <- statsu[1]
fu <- statsu[2]
e <- 0
redu <- lm(as.formula(reduced))
yhat <- predict(redu)
yresidual <- residuals(redu)
set.seed(seed)
for (k in 1:r) {
  ynew <- yhat+sample(yresidual)
```


Adaptive Tests of Significance Using Permutations of Residuals with R and SAS®.
By Thomas W. O'Gorman

```
  statsp <- teststats(ynew,complete,reduced)
  lambdap <- statsp[1]
  if( (lambdap <= lambdau) ) e  <-  e + 1
  }
p <- (e+1)/(r+1)
detach(testdata)

cat("\n","*************************************","\n","\n")
cat("  Input:","\n","\n")
cat("    Data set:",dataset,"\n","\n")
cat("    Complete model: ",complete,"\n","\n")
cat("    Reduced model : ",reduced,"\n","\n")
cat("    Number of permutations = ",r,"\n","\n")
cat("    Random seed = ",seed,"\n","\n","\n","\n")
cat("  Output from Adaptive Test:","\n","\n")
cat("    Unpermuted lambda = ",round(lambdau,4),"\n","\n")
cat("    Unpermuted F      = ",round(fu,4),"\n","\n")
cat("    p-value           = ",round(p,4),"\n","\n","\n")
cat("  Trad. F test p-value = ",round(tradpvalue,4),"\n")

cat("\n","*************************************","\n","\n")
return(p)
}

teststats <- function(y,complete,reduced) {
red <- lm(as.formula(reduced))
resid <- residuals(red)
#                             compute traditional quantiles
q25 <- quantile(resid,0.25,type=6)
q75 <- quantile(resid,0.75,type=6)
sigmat <- (q75-q25)/1.349
#                             compute bandwidth (h)
n <- length(resid)
h <- 1.587*sigmat*n^(-0.3333333333)
range <- max(resid) - min(resid)
lower <- min(resid) - range
upper <- max(resid) + range
tol <- 0.00000001*sigmat
cdf25 <- rootcdf(resid,h,0.25,lower,upper,tol)
cdf50 <- rootcdf(resid,h,0.50,lower,upper,tol)
cdf75 <- rootcdf(resid,h,0.75,lower,upper,tol)
sigma <- (cdf75-cdf25)/1.349
#                             compute adaptive weights
s <- (resid-cdf50)/sigma
tolerance <- 100*tol
w  <- matrix(0,n,1)
w2 <- matrix(0,n,1)
for (i in 1:n) {
  fhat <- cdfhat(resid,h,resid[i])
```

```
    z <- qnorm(fhat)
    if( (abs(s[i]) >= tolerance) )
       w[i] <- z/s[i]
      else
       w[i] <- 1
      w2[i] <- w[i]^2
    }
  compw <-lm(as.formula(complete), weights=w2)
  residc <-residuals(compw)
  residc2 <- residc^2
  ssec <- t(w2)%*%residc2
  redw <- lm(as.formula(reduced),weights=w2)
  residr <- residuals(redw)
  residr2 <- residr^2
  sser <- t(w2)%*%residr2
  lambda <- ssec/sser
  dfc <- df.residual(compw)
  dfr <- df.residual(redw)
  f <- ( (sser-ssec)/(dfr-dfc) )/ (ssec/dfc)
  teststats <- c(lambda,f)
  return(teststats)
    }
  rootcdf <- function(x,h,p,xlow,xhigh,tolerance) {
    flow  <-  cdfhat(x,h,xlow)
    fhigh  <-  cdfhat(x,h,xhigh)
    if ( (flow >p) | (fhigh < p) ) {
      cat("Stop in rootcdf flow,fhigh=",flow,fhigh);q() }
    for (i in seq(1:60) ) {
      xmiddle <- (xlow+xhigh)/2
      fmiddle <- cdfhat(x,h,xmiddle)
      if( (fmiddle == p) ) break
      if( (fmiddle < p) ) {xlow   <- xmiddle; flow <- fmiddle}
      if( (fmiddle > p) ) {xhigh  <- xmiddle; fhigh <- fmiddle}
      if( ( (xhigh-xlow) <= tolerance ) ) break
      if((i==60)){print ("stop in rootcdf, over 60 iterations")
                   ;q()}
      }
    return(xmiddle)
    }
 cdfhat <- function(xvector,h,xpoint){
   phi <- pnorm((xpoint-xvector)/h)
   cdf <- mean(phi)
   return(cdf)
   }
 trad.ftest <- function(dataset,complete,reduced) {
   testdata <- read.table(dataset)
   attach(testdata)
   comp.model <- lm(as.formula(complete) )
   red.model  <- lm(as.formula(reduced) )
```

```
ssec <- deviance(comp.model)
sser <- deviance(red.model)
dfc <- df.residual(comp.model)
dfr <- df.residual(red.model)
numdf <- dfr-dfc
f <- ( (sser-ssec)/numdf)/(ssec/dfc)
p <- 1-pf(f,numdf,dfc)
detach(testdata)
return(p)
}
```

APPENDIX B

SAS MACRO FOR ADAPTIVE TESTS

```
/*      Notes:

1)  The variables that are used in the macro call are:

    Variable    Explanation                        Required (default)
    ------------------------------------------------------------------
    dataset     Name of SAS dataset                     required
    y           Name of dependent variable              required
    nvars       Number of dependent variables             (1)
    xr          Variables in Xr matrix                  optional
    classr      class variables in Xr matrix            optional
    xa          Variables in Xa matrix             required for tests
    classa      class variables in Xa matrix            optional
    blkvars     Blocking variables                      optional
    ici         Indicator for conf. interval         (0=no), 1=yes
    iestimat    Indicator for adaptive estimates     (0=no), 1=yes
```

Adaptive Tests of Significance Using Permutations of Residuals with R and SAS®.
By Thomas W. O'Gorman

```
iequalwt  Indicator for equal weights      (0=no), 1=yes
seed1     Random generator seed 1          (1492)
seed2     Random generator seed 2           (314)
seed3     Random generator seed 3          (2718)
nperm     Number of permutations           (2000)
short     Indicator for short printout     (0=no), 1=yes
_______________________________________________________

2)  No missing values are allowed in the dependent variables,
    independent variables, or blocking variables.

3)  If variables are formatted, then the formats will need to
    be placed before the macro.

4)  The number of observations must not exceed 10000.

5)  The number of dependent variables must not exceed 20.
   */
options nocenter linesize=80 nonotes ;

%macro adaptall(dataset=, y=, nvars=1,
     xr=, classr=, xa=, classa=,
     blkvars=, ici=0, iestimat=0, iequalwt=0,
     seed1=1492, seed2=314, seed3=2718, nperm=2000, short=0);
data _null_;
   /*  The code in this data step is used to obtain a list of
       the variables in the keep list by deleting
       the interaction (*) characters. */
   cxr=symget('xr');
   listxr = translate(cxr,'    ','*|()');
   call symput('mlistxr',listxr);
   cxa=symget('xa');
   listxa = translate(cxa,'    ','*|()');
   call symput('mlistxa',listxa);
data awlsdata(keep= &y &mlistxr &mlistxa);
   retain n 0;
   set &dataset;
     n=n+1;
     call symput('nobs',left(n));
     output awlsdata;
proc glm data=awlsdata   %if &short ne 0 %then noprint; ;
  classes  &classr &classa;
  model &y=&xr &xa ;
  %if &nvars ne 1 %then manova h=&xa / printe printh;;
  title2 ' traditional analysis -- no adaptation';
/*     Now begin adaptive analysis    */
proc glm data=awlsdata noprint;
  classes  &classr ;
  model &y=&xr;
```

```
  output out=dresid residual = res1-res&nvars
             predicted= yhat1-yhat&nvars;
  title2 'Adaptive weighted least squares analysis';
%if &short eq 0 %then %do;
  proc print data=dresid;
  %end;

%adaptwts(datares=dresid, resvars=res1-res&nvars,
          nobs= &nobs, nvars = &nvars, iequalwt=&iequalwt);
data dall;
  merge awlsdata dweights dresid;
%if &short eq 0 %then %do;
  proc print data=dall ;
  title2' Data with weights  SASWT = weights used by SAS';
    %end;
%if &iestimat = 1 %then %do;
proc glm data=dall ;
  classes  &classr ;
  model &y=&xr /solution;
  weight saswt;
%go to endestim;
    %end;
%adaptlam(dataall=dall, yvars=&y, nvars=&nvars,
         xrvars=&xr, clrvars=&classr,
         xavars=&xa, clavars=&classa, wt2=saswt);

%if &nvars =1 %then data dlamstar(keep=lamstar fstar count);;
%if &nvars gt 1 %then data dlamstar(keep=lamstar count); ;
  set dlambda;
  lamstar = lambda;
%if &nvars =  1 %then fstar = f; ;
  count=0;
data dseed;
   next1=mod(171*&seed1,30269);
   next2=mod(172*&seed2,30307);
   next3=mod(170*&seed3,30323);
/*    Now begin permutations        */
  %do i=1 %to &nperm;
     data dperm(keep= place &blkvars res1-res&nvars)
     dseed(keep=next1 next2 next3);
       array k{&nobs} _temporary_;
       retain tmpseed1 tmpseed2 tmpseed3;
       set dseed(in=inseed) dall(in=indall) end=last;
       if inseed then do;
         tmpseed1 = next1;
         tmpseed2 = next2;
         tmpseed3 = next3;
         do i = 1 to &nobs;
           k{i} = i;
```

```
          end;
        do l = &nobs to 2 by -1;
          /* This is the Wichmann-Hill random generator  */
          tmpseed1 = mod(171*tmpseed1,30269);
          tmpseed2 = mod(172*tmpseed2,30307);
          tmpseed3 = mod(170*tmpseed3,30323);
          random =tmpseed1/30269.0+tmpseed2/30307.0+
                  tmpseed3/30323;
          random = mod(random,1.0);
          /* This is the Fisher-Yates permutation algorithm,
             as improved by Durstenfeld (1964) Comm. of the
             ACM 1964, algorithm 235 */
          itrade = int(l*random) + 1;
          ktemp      = k{itrade};
          k{itrade} = k{l};
          k{l}       = ktemp;
          end;
        end;
      if indall then do;
        l=_n_-1;
        place = k{l};
        output dperm;
        end;
      if last then do;
        next1 = tmpseed1;
        next2 = tmpseed2;
        next3 = tmpseed3;
        output dseed;
        end;
    proc sort data=dperm; by &blkvars place;
    data dynew;
      array yhatval{&nvars} yhat1-yhat&nvars;
      array ynewval{&nvars} ynew1-ynew&nvars;
      array tempres{&nvars} res1-res&nvars;
      merge dall dperm;
      do  j = 1 to &nvars;
        ynewval{j} = yhatval{j} + tempres{j};
        end;
 proc glm data= dynew noprint;
    classes &classr &classa;
    model ynew1-ynew&nvars = &xr;
    output out=dresperm residual=resperm1-resperm&nvars;
%adaptwts(datares=dresperm, resvars=resperm1-resperm&nvars,
          nobs= &nobs, nvars = &nvars, iequalwt=&iequalwt);
data dallperm;
  merge dynew dweights;
%adaptlam(dataall=dallperm, yvars=ynew1-ynew&nvars,
          nvars=&nvars, xrvars=&xr, clrvars=&classr,
          xavars=&xa, clavars=&classa, wt2=saswt);
```

```
%if &nvars =1 %then data dlamstar(keep=lamstar fstar count);;
%if &nvars gt 1 %then data dlamstar(keep=lamstar count); ;
    merge dlamstar dlambda;
    if lambda le lamstar then count = count + 1;
 %end;

     data _null_;
       set dlamstar;
       put lamstar =;
%if &nvars =  1 %then put fstar =; ;
       p_value=(count+1)/(&nperm+1);
  if &short eq 0 then do;
    file print ;
    put '1';
    put ;
    put '   ********************************************** ';
    put ;
    %if &iequalwt eq 0 %then
        put "   Adaptive Weighted Least Squares Test";;
    %if &iequalwt eq 1 %then
        put "     Permutation Test Using Equal Weights";;
    put ;
    put " Number of permutations = &nperm ";
    put " Starting seeds = &seed1  &seed2  &seed3" ;
    put " SAS data set: &dataset   Number of obs. = &nobs";
    put ;
    put " Dependent variables   :  &y ";
    %if &blkvars ne %str() %then
        put " Blocking variable: &blkvars ";;
    put ;
    put "  Ind. vars. in reduced model   : Intercept  &xr";
    put " Class vars. in reduced model   :  &classr";
    put;
    put "  Ind. vars. in complete model : Intercept &xr &xa";
    put " Class vars. in complete model : &classr  &classa";
    put;
    put " Adaptive test statistics for unpermuted data:";
    put;
    put '       Wilks Lambda = ' lamstar 8.4;
    put;
    %if &nvars eq 1 %then put "   F statistic = " fstar 8.4;;
    put;
    put ' p-value for adaptive test = ' p_value 8.4 ;
    put;
    put '   ********************************************** ';
    put;
    end;
 else do;
```

```
    file print noprint;
    put " with &nperm permutations, p-value = " p_value 8.4 ;
    put;
    end;
run;
%endestim:;
%mend;

%macro adaptlam(dataall=, yvars=, nvars=, xrvars=, clrvars=,
                xavars=, clavars=, wt2=);
/* Find determinant of the error matrix for the full model */
proc glm data=&dataall outstat=dsigstar noprint;
  classes  &clrvars &clavars;
  model &yvars=&xrvars &xavars/ss1;
  weight &wt2;
%if &nvars = 1  %then  data dsigstar(keep =ss df); ;
%if &nvars gt 1 %then data dsigstar(keep = &yvars); ;
  set dsigstar;
  if _type_ = "ERROR";
%determin(datasig=dsigstar,yvars=&yvars,n=&nvars);
%if &nvars =  1 %then data dssstar(keep=ssstar dfc); ;
%if &nvars gt 1 %then data dssstar(keep=ssstar); ;
  set ddeter;
    ssstar=deter;
%if &nvars =  1 %then dfc=df; ;
/* Find determinant of the error matrix for reduced model*/
proc glm data=&dataall outstat=dsigstar noprint;
  classes  &clrvars ;
  model &yvars=&xrvars /ss1;
  weight &wt2;
%if &nvars = 1  %then  data dsigstar(keep =ss df); ;
%if &nvars gt 1 %then data dsigstar(keep = &yvars); ;
  set dsigstar;
  if _type_ = "ERROR";
%determin(datasig=dsigstar,yvars=&yvars,n=&nvars);
%if &nvars =  1 %then data dssrstar(keep=ssrstar dfr); ;
%if &nvars gt 1 %then data dssrstar(keep=ssrstar); ;
  set ddeter;
    ssrstar=deter;
%if &nvars =  1 %then  dfr=df; ;
/* Now find Wilks lambda */
%if &nvars =  1 %then data dlambda(keep = lambda f); ;
%if &nvars gt 1 %then data dlambda(keep = lambda); ;
  merge dssstar dssrstar;
  lambda = ssstar/ssrstar;
%if &nvars =  1 %then
    f = ( (ssrstar-ssstar)/(dfr-dfc) )/(ssstar/dfc); ;
%mend;
```

```
%macro determin(datasig,yvars=&yvars,n=);
%if &nvars =  1 %then data ddeter(keep=deter df); ;
%if &nvars gt 1 %then data ddeter(keep=deter); ;
    array a{&n,&n} _temporary_;
    array temp{&n} &yvars;
    set &datasig end=last;
    if &n = 1 then do;
      deter=ss;
      output;
      go to skiplu;
      end;
    do j = 1 to &n;
      a{_n_,j} = temp{j};
      end;
    if last then do;
      do j = 1 to &n;
        do i = 1 to j-1;
          sum = a(i,j);
          do k = 1 to i-1;
            sum = sum - a(i,k)*a(k,j);
            end;
          a(i,j) = sum;

          end;
        do i = j to &n;
          sum = a(i,j);
          do k = 1 to j-1;
            sum = sum - a(i,k)*a(k,j);
            end;
            a(i,j) = sum;
          end;
        if j ne &n then do;
          dum=1.0/a(j,j);
          do i = j + 1 to &n;
            a(i,j) = a(i,j)*dum;
            end;
          end;
        end;
      deter=1.0;
      do i = 1 to &n;
        deter=deter*a(i,i);
        end;
      output;
      end;
      skiplu:;
%mend;

%macro adaptwts(datares=, resvars=, nobs=, nvars=,
                iequalwt=0);
```

```
data dweights(keep= weight saswt );
  array d{&nobs,&nvars} _temporary_;
  array smcdf{*} f1-f10000;
  array tempres{&nvars} &resvars;
  array median{20} med1-med20;
  array tempmed{20} t1-t20;
  array p{&nobs,&nvars} _temporary_;
  array distance{&nobs} dst1-dst&nobs;
  array points{*} p1 p2 p3;
  array smpoint{*} smp1 smp2 smp3;
  set &datares nobs=rn end=last;
  retain h f1-f10000 ;
  do j = 1 to &nvars;
    d{_n_,j} = tempres{j};
    end;
  if last then do;
   if &iequalwt = 1 then do;
    do i = 1 to &nobs;
      weight = 1.0;
      saswt  = 1.0;
      output;
      end;
    go to skipadap;
    end;
    if &nvars = 1 then do;
      do j = 1 to &nobs;
        distance{j} = d{j,1};
         end;
      go to skipcntr;
      end;
      np1 = %eval(&nobs)+1;
      nlast = np1-1;
      lower = floor(.25*np1);
      lfrac = .25*np1 - lower;
      lowerp1 = lower + 1;
      upper = floor(.75*np1);
      ufrac = .75*np1 - upper;
      upperp1 = upper + 1;
  /*    standardize the residuals    */
      do j = 1 to &nvars;
        do i = 1 to &nobs;
          distance{i} = d{i,j};
          end;
        plow = ordinal(lower, of distance{*});
        plowp1 = ordinal(lowerp1, of distance{*});
        p25 = (1-lfrac)*plow + lfrac*plowp1;
        pupper = ordinal(upper, of distance{*});
        pupperp1 = ordinal(upperp1, of distance{*});
        p75 = (1-ufrac)*pupper + ufrac*pupperp1;
```

```
      sdvar = (p75 - p25)/1.349;
      do i = 1 to &nobs;
        d{i,j} = d{i,j}/sdvar;
        end;
      end;
/*   begin finding mediancentre of array d    */
  stepsize=0.1;
  oldsum=9.9e99;
  do jj = 1 to &nvars;
    median{jj} = 0.0;
    end;
  another:;
  do j=1 to &nvars;
    do jj = 1 to &nvars;
      tempmed{jj} = median{jj};
      end;
    do k = -1 to +1 by 2;
      tempmed{j} = median{j} + k*stepsize;
      sum=0.0;
      do i = 1 to &nobs;
        dist = 0.0;
        do jj = 1 to &nvars;
          dist = dist + (d{i,jj} - tempmed{jj})**2;
          end;
        dist = sqrt(dist);
        sum = sum + dist;
        end;
      if(sum < oldsum) then do;
        do jj = 1 to &nvars;
          median{jj} = tempmed{jj};
          end;
        oldsum=sum;
        go to another;
        end;
      end;
    end;
    if stepsize > .00000001 then do;
      stepsize = stepsize/3.0;
      go to another;
      end;
     else do;
      end;
       Now center the residuals
  do i = 1 to &nobs;
    do j = 1 to &nvars;
      d{i,j} = d{i,j} - median{j};
      end;
    end;
      Now begin the weighting loop
```

```
skipcntr:;
  do i = 1 to &nobs;
    if &nvars = 1 and i gt 1 then go to wt;
    if &nvars = 1 then go to standard;
      Begin projections for the ith observation
    cicip = 0.0;
    do jj = 1 to &nvars;
      cicip = cicip + d{i,jj}*d{i,jj};
      end;
    do j = 1 to &nobs;
      cicjp = 0.0;
      do jj = 1 to &nvars;
        cicjp = cicjp + d{i,jj}*d{j,jj};
        end;
      sqdist = 0.0;
      do jj = 1 to &nvars;
        pij=(cicjp/cicip)*d{i,jj};
        sqdist = sqdist + pij*pij;
        end;
      distance{j} = sign(cicjp)*sqrt(sqdist);
      end;
       Now standardize the distances
standard:;
     np1 = %eval(&nobs)+1;
     nlast = np1-1;
     lower = floor(.25*np1);
     fraction = .25*np1 - lower;
     lowerp1 = lower + 1;
     plow = ordinal(lower, of distance{*});
     plowp1 = ordinal(lowerp1, of distance{*});
     p25 = (1-fraction)*plow + fraction*plowp1;
     upper = floor(.75*np1);
     fraction = .75*np1 - upper;
     upperp1 = upper + 1;
     pupper = ordinal(upper, of distance{*});
     pupperp1 = ordinal(upperp1, of distance{*});
     p75 = (1-fraction)*pupper + fraction*pupperp1;
     sdinit = (p75 - p25)/1.349;
     h = 1.587*sdinit*rn**(-0.3333333333333333);
/*    Now find estimates for the 25th, 50th, and 75th
      percentiles with smoothed c.d.f.  */
     n = %eval(&nobs);
     range = ordinal(n, of distance{*})
            -ordinal(1, of distance{*});
     tol=0.000000001*range;

     do l = 1 to 3;
       if l = 1 then percent = 0.25;
       if l = 2 then percent = 0.50;
```

```
        if l = 3 then percent = 0.75;
        p1 = ordinal(1, of distance{*}) - range;
        p2 = 0.0;
        p3 = ordinal(n, of distance{*}) + range;
         do j = 1 to 3;
          sum = 0.0;
          do k = 1 to rn;
            z = (points{j} - distance{k})/h;
            sum = sum + probnorm(z);
            end;
          smpoint{j} = sum/rn;
          end;
  nextiter:;
        if smp2 lt percent then do;
          p1=p2;
          smp1=smp2;
          end;
        if smp2 gt percent then do;
          p3=p2;
          smp3=smp2;
          end;
        if abs(smp2 - percent) lt tol then go to over;
        p2=(p1+p3)/2;
        sum=0.0;
        do k = 1 to rn;
          z = (p2 - distance{k})/h;
          sum = sum + probnorm(z);
          end;
        smp2 = sum/rn;
        if abs(p3 -p1) lt tol then go to over;
                                  else go to nextiter;
  over:;
        if l = 1 then perc25 = p2;
        if l = 2 then perc50 = p2;
        if l = 3 then perc75 = p2;
        end;
        sigmahat = (perc75 - perc25)/1.349;
/*
       Now we have the distances standardized.
       Will compute weight for the ith observation.
 */
  wt:;
        sum = 0.0;
        do k = 1 to rn;
          z = (distance{i} - distance{k})/h;
          sum = sum + probnorm(z);
          end;
        fhat = sum/rn;
        z=(distance{i} - perc50)/sigmahat;
```

```
        if abs(z) gt 0.0001 then weight = probit(fhat)/z;
                              else weight = 1.0;
        saswt=weight*weight;
        output;
     /*
   End of weighting loop for the ith observation.
   */
 end;
 end;
 skipadap:;
%mend;
```

APPENDIX C

SAS MACRO FOR MULTIPLE COMPARISONS PROCEDURES

Note: The SAS macro adaptall.sas must also be included. It is in Appendix B and on the file adaptall.sas.

```
/*
Notes:

 1)  This macro must be included in a SAS program that also
     has the "adaptall" macro.

 2)  No missing values are allowed in the dependent or
     independent variables.

 3)  If variables are formatted, then the formats will need
     to be placed before the macro.

 4)  The macro is to be used with a one-way layout.  Do not
     specify anything for xr or classr; they are not used as
```


By Thomas W. O'Gorman

```
     inputs in this macro.

 5)  The indicator variable for the groups should be
     specified as the xa and classa variables.

 6)  The number of independent variables should not be
     specified; it will be calculated by this macro before
     the call to the awls macro.

 7)  The number of observations must not exceed 10,000.
   */
%macro adaptmc(dataset=, y=, xa=, classa=,
                    seed1=, seed2=, seed3=, nperm=2000);
 proc sort data=&dataset out=dsorted;
   by &classa;
 data dk;
   retain k 0 ;
   keep k ;
   set dsorted end=last;
     by &classa;
   if first.&classa then do;
     k=k+1;
     call symput ('km',k);
     end;
   if last then output;
   run;
  %do imc=2 %to &km;
  %do jmc=1 %to &imc-1;
   data expand;
     length lower upper $ 8;
     retain lower upper;
     retain c 0;
     file print noprint;
     keep  &y &xa newclass ia;
     set  dsorted end=last;
       by &classa;
     if first.&classa then do;
         c = c + 1;
         if c = &jmc then lower=&classa;
         end;
     if c = &imc then do;
                             newclass=lower;
                             ia=1;
                             upper=&classa;
                             end;
                       else do;
                             newclass=&classa;
                             ia=0;
                             end;
```

```
    if last then
 put "  Compare " lower " to " upper " for variable &classa";
  %adaptall(dataset=expand, y=&y, xr=newclass,
     classr=newclass, xa=ia, seed1=&seed1, seed2=&seed2,
     seed3=&seed3, nperm=&nperm, short=1);
  %end;
  %end;
%mend;
;
```

APPENDIX D

R CODE FOR ADAPTIVE TESTS WITH BLOCKING FACTORS

Note: The functions teststats, rootcdf, cdfhat, and trad.ftest must also be included. These are in Appendix A and on the file adapt.r .

```
adaptblock <- function(dataset,complete,reduced,r,seed,
                       nblocks) {
tradpvalue <- trad.ftest(dataset,complete,reduced)
testdata <- read.table(dataset)
print(testdata)
attach(testdata)
statsu<- teststats(y,complete,reduced)
lambdau <- statsu[1]
e <- 0
countperm <- 0
redu <- lm(as.formula(reduced))
yhat <- predict(redu)
yresidual <- residuals(redu)
set.seed(seed)
```


By Thomas W. O'Gorman

```
n <- length(yresidual)
for (k in 1:r) {
  for (b in 1:nblocks[1]) {
    permres <- yresidual[block==b]
    lengthperm <- length(permres)
    if ( (lengthperm>1) )permres <- sample(permres)
    countperm[1] <- 0
    for (i in 1:n) {
      if( (block[i]==b) )  {
        countperm[1] <- countperm[1] + 1
        yresidual[i] <- permres[countperm[1]]
                                   }
                   }
                                  }
  ynew <- yhat + yresidual
  statsp <- teststats(ynew,complete,reduced)
  lambdap <- statsp[1]
  if( (lambdap <= lambdau) ) e  <-  e + 1
  }
p <- (e+1)/(r+1)
detach(testdata)

cat("\n","****************************************","\n","\n")
cat("  Input:","\n","\n")
cat("    Data set:",dataset,"\n","\n")
cat("    Complete model: ",complete,"\n","\n")
cat("    Reduced model : ",reduced,"\n","\n")
cat("    Number of permutations = ",r,"\n","\n")
cat("    Random seed = ",seed,"\n","\n","\n","\n")
cat("  Output:","\n","\n")
cat("    Unpermuted lambda = ",round(lambdau,4),"\n","\n")
cat("    Adaptive p-value  = ",round(p,4),"\n","\n")
cat("Traditional F test p-value = ",round(tradpvalue,4),"\n")

cat("\n","****************************************","\n","\n")
return(p)
}
```

APPENDIX E

R CODE FOR ADAPTIVE TEST WITH PAIRED DATA

Note: The functions rootcdf and cdfhat must also be included. These are in Appendix A and on the file adapt.r .

```
adaptpaired <- function(dataset,r=2000,seed=1492) {
testdata <- read.table(dataset)
print(testdata)
attach(testdata)
pairs <- length(differences)
dneg <- -1.0*differences
n <- 2*pairs
resid <- matrix(0,n,1)
resid <- (rbind(differences,dneg))/2
#                                compute traditional quantiles
q25 <- quantile(resid,0.25,type=6)
q75 <- quantile(resid,0.75,type=6)
sigmat <- (q75-q25)/1.349
#                                compute bandwidth (h)
```


By Thomas W. O'Gorman

```
h <- 1.587*sigmat*n^(-0.33333333333)
range <- max(resid) - min(resid)
lower <- min(resid) - range
upper <- max(resid) + range
tol <- 0.000001*sigmat
cdf25 <- rootcdf(resid,h,0.25,lower,upper,tol)
cdf50 <- rootcdf(resid,h,0.50,lower,upper,tol)
cdf75 <- rootcdf(resid,h,0.75,lower,upper,tol)
sigma <- (cdf75-cdf25)/1.349
#                               compute adaptive weights
s <- (resid-cdf50)/sigma
tolerance <- 100*tol
w  <- matrix(0,n,1)
weights2 <- matrix(0,pairs,1)
for (i in 1:n) {
  fhat <- cdfhat(resid,h,resid[i])
  z <- qnorm(fhat)
  if( (abs(s[i]) >= tolerance) )
     w[i] <- z/s[i]
    else
     w[i] <- 1
  }
for (i in 1:pairs){
   j = 2*i-1
   weights2[i] <- w[j]^2
  }
print(weights2)
dwbar <- (t(weights2)%*%differences)/sum(weights2)
sseu <- (t(weights2)%*%((differences-dwbar)^2))/2
e <- 0
set.seed(seed)
diffsignch <- matrix(0,pairs,1)
for (k in 1:r) {
  unifnum <- runif(pairs)
  for (j in 1:pairs){
    if( (unifnum[j] <= 0.5) )
      diffsignch[j] <- differences[j]
     else
      diffsignch[j] <- -differences[j]
     }
  dwbarsignch <- (t(weights2)%*%diffsignch)/sum(weights2)
  dmdbar <- as.vector(diffsignch)-dwbarsignch
  ssesignch <- (t(weights2)%*%((dmdbar)^2))/2
  if( (ssesignch <= sseu) ) e  <-  e + 1
  }
p <- (e+1)/(r+1)
detach(testdata)
cat("\n","*************************************","\n","\n")
cat("  Input:","\n","\n")
```

```
cat("     Data set:",dataset,"\n","\n")
cat("     Number of sign changes = ",r,"\n","\n")
cat("     Random seed = ",seed,"\n","\n","\n","\n")
cat("  Output from Adaptive Test:","\n","\n")
cat("     Unpermuted sse     = ",round(sseu,4),"\n","\n")
cat("     p-value            = ",round(p,4),"\n","\n","\n")

cat("\n","****************************************","\n","\n")
return(p)
}
```

APPENDIX F

SAS MACRO FOR ADAPTIVE TEST WITH PAIRED DATA

Note: The macro adaptwts in Appendix B on file adaptall.sas must also be included.

```
      /*
Notes:

 1)   The variables that are used in the macro call are
      described as follows:

      Variable       Explanation                  Required (default)
      ________________________________________________________________
      dataset       Name of SAS dataset                      required
      diff          Name of variable                         required
      seed1         Random generator seed 1                    (1492)
      seed2         Random generator seed 2                     (314)
      seed3         Random generator seed 3                    (2718)
      nsignchs      Number of sign changes                     (2000)
      ________________________________________________________________
```

```
2)  No missing values are allowed in the differences.

3)  If variables are formatted, then the formats will need
    to be placed before the macro.

4)  The number of observations must not exceed 10,000.

5)  The number of dependent variables must not exceed 20.
  */

options nocenter linesize=80 nonotes;

%macro adaptpaired(dataset=, diff=, nvars=1,
         seed1=1492, seed2=314, seed3=2718, nsignchs=2000);
data dresid1 dresid2;
  set &dataset;
  res1 = &diff/2; output dresid1;
  res1 =-&diff/2; output dresid2;
data dresid;
  retain n 0;
  set dresid1 dresid2;
  n = n + 1;
  call symput('nobs',n);
  output dresid;
run;
%adaptwts(datares=dresid, resvars=res1-res&nvars,
          nobs= &nobs, nvars = &nvars);
data ddiffwts;
  merge &dataset(in=ind) dweights;
  if ind eq 1;
data ddbaru(keep=sumwt2 dbaru);
  retain sumwt2 0;
  retain sumwtd 0;
  set ddiffwts end=last;
  sumwt2 = sumwt2 + weight*weight;
  sumwtd = sumwtd + weight*weight*&diff;
  if last then do;
    dbaru=sumwtd/sumwt2;
    output;
    end;
data dsse(keep=sseu count);
  retain dbarur sumwt2r sseu 0;
  set ddbaru(in=indbar) ddiffwts(in=indiff) end=last;
  if indbar then dbarur = dbaru;
  if indiff then
     sseu=sseu+weight*weight*((&diff-dbarur)**2)/2.0;
  if last then do;
```

```
    count=0;
    output;
    end;
data dseed;
   next1=mod(171*&seed1,30269);
   next2=mod(172*&seed2,30307);
   next3=mod(170*&seed3,30323);
  %do i=1 %to &nsignchs;
    data dseed(keep=next1 next2 next3)
         ddiffsc(keep= weight diffsc);
       retain tmpseed1 tmpseed2 tmpseed3 0;
       set dseed(in=inseed) ddiffwts end=last;
       if inseed then do;
         tmpseed1 = next1;
         tmpseed2 = next2;
         tmpseed3 = next3;
         end;
        else do;
           tmpseed1 = mod(171*tmpseed1,30269);
           tmpseed2 = mod(172*tmpseed2,30307);
           tmpseed3 = mod(170*tmpseed3,30323);
           random=tmpseed1/30269.0+tmpseed2/30307.0
                  +tmpseed3/30323;
           random = mod(random,1.0);
           if random le 0.5 then diffsc =  &diff;
                                 else diffsc = -&diff;
           output ddiffsc;
           end;
      if last then do;
         next1 = tmpseed1;
         next2 = tmpseed2;
         next3 = tmpseed3;
         output dseed;
         end;
   data ddbarsc(keep=dbarsc);
     retain sumwtd sumwt2r 0;
     set ddbaru(in=inbaru) ddiffsc(in=indiff) end=last;
     if inbaru then do;
       sumwt2r=sumwt2;
       end;
     if indiff then do;
       sumwtd = sumwtd + weight*weight*diffsc;
       end;
     if last then do;
       dbarsc = sumwtd/sumwt2r;
       output;
       end;
 data dssesc(keep=ssesc);
   retain dbarscr ssesc 0;
```

```
    set ddbarsc(in=inbarsc) ddiffsc(in=indiffsc) end=last;
    if inbarsc then dbarscr = dbarsc;
    if indiffsc then
       ssesc = ssesc+weight*weight*((diffsc-dbarscr)**2)/2.0;
    if last then output;
  data dsse (keep = sseu count);
    merge dsse dssesc;
    if ssesc le sseu then count = count + 1;
%end;;
data _null_;
  set dsse;
  npairs = &nobs/2;
  p_value=(count+1)/(&nsignchs + 1);
     file print ;
    put '1';
    put ;
    put '   ********************************************** ';
    put ;
    put "   Adaptive Weighted Least Squares Test";;
    put ;
    put " Number of sign changes = &nsignchs ";
    put " Starting seeds = &seed1  &seed2  &seed3" ;
    put " Data set: &dataset  Number of differences=" npairs;
    put ;
    put " Difference variable  :  &diff ";
    put;
    put ' p-value for adaptive test = ' p_value 8.4 ;
    put;
    put '   ********************************************** ';
    put;
 run;
 %endestim:;
 %mend;
```

APPENDIX G

R CODE FOR MULTIVARIATE ADAPTIVE TESTS

Note: The functions cdfhat must also be included. This is in Appendix A and on the file adapt.r .

```
adaptmult <- function(y,complete,reduced,r=2000,seed=1492) {
lambdau<- teststat(y,complete,reduced)
e <- 0
redu <- lm(as.formula(reduced))
yhat <- predict(redu)
yresidual <- residuals(redu)
set.seed(seed)
n <- length(yhat[,1])
nvars <- length(yhat[1,])
y <- matrix(0,n,nvars)
observations <- c(1:n)
for (k in 1:r) {
  place <- sample(observations)
  for (i in 1:n) {
```

```
    row <- place[i]
    y[i,] <- yhat[i,] + yresidual[row,]
                  }
  lambdap <- teststat(y,complete,reduced)
  if( (lambdap <= lambdau) ) e  <-  e + 1
  }
p <- (e+1)/(r+1)
detach(testdata)

cat("\n","*************************************","\n","\n")
cat("  Input:","\n","\n")
cat("    Data set:",dataset,"\n","\n")
cat("    Dependent variables: ", yvars,"\n","\n")
cat("    Complete model: ",complete,"\n","\n")
cat("    Reduced model : ",reduced,"\n","\n")
cat("    Number of permutations = ",r,"\n","\n")
cat("    Random seed = ",seed,"\n","\n","\n","\n")
cat("  Output from Adaptive Test:","\n","\n")
cat("    Unpermuted lambda = ",round(lambdau,4),"\n","\n")
cat("    p-value           = ",round(p,4),"\n","\n","\n")
cat("\n","*************************************","\n","\n")
return(p)
}
teststat <- function(y,complete,reduced) {
red <- lm(as.formula(reduced))
resid <- residuals(red)
n <- length(resid[,1])
nvars <- length(resid[1,])
#                               standardize residuals
temp <- matrix(0,n,1)
for (j in 1:nvars) {
  temp <- resid[,j]
  q25 <- quantile(temp,0.25,type=6)
  q75 <- quantile(temp,0.75,type=6)
  sigmavar <- (q75-q25)/1.349
  resid[,j] <- resid[,j]/sigmavar
  }
#                               begin finding mediancentre
initial <- matrix(0,nvars,1)
for (j in 1:nvars) initial[j]<-quantile(resid[,j],0.5,type=6)
tempmed <- initial
mediancentre <- optim(initial,sumdistances,tempmed,resid,n)
median <- mediancentre$par
ones <- matrix(1,n,1)
c <- matrix(0,n,nvars)
c <- resid - ones%*%t(median)
#                               now begin projections
w  <- matrix(0,n,1)
w2 <- matrix(0,n,1)
```

```
 distance <- matrix(0,n,1)
 for (i in 1:n) {
   ci <- c[i,]
   cicip <- crossprod(ci)
   for (j  in 1:n) {
     cj <- c[j,]
     cicjp <- crossprod(ci,cj)
     cijdcii <- cicjp/cicip
     pij <- cijdcii*ci
     sqdist <- crossprod(pij)
     distance[j] <- sign(cicjp)*sqrt(sqdist)
                         }
#                                    now standardize the distances
q25 <- quantile(distance,0.25,type=6)
q50 <- quantile(distance,0.50,type=6)
q75 <- quantile(distance,0.75,type=6)
sigmat <- (q75-q25)/1.349
#                                    compute bandwidth (h)
h <- 1.587*sigmat*n^(-0.33333333333)
range <- max(distance) - min(distance)
lower <- min(distance) - range/10
upper <- max(distance) + range/10
tol   <- 0.00001*sigmat
cdf25 <- rootcdfm(distance,h,0.25,lower,q25,q50,tol)
cdf50 <- rootcdfm(distance,h,0.50,lower,q50,upper,tol)
cdf75 <- rootcdfm(distance,h,0.75,q50,q75,upper,tol)
sigma <- (cdf75-cdf25)/1.349
#                                    compute adaptive weights
s <- (distance[i]-cdf50)/sigma
tolerance <- 100*tol
fhat <- cdfhat(distance,h,distance[i])
z <- qnorm(fhat)
if( (abs(s) >= tolerance) )
   w[i] <- z/s
  else
   w[i] <- 1
   w2[i] <- w[i]^2
#                    completed weighting for the ith observation
compw <-lm(as.formula(complete), weights=w2)
residc <-residuals(compw)
wresidc <- diag(as.vector(w))%*%residc
ssec <- t(wresidc)%*%wresidc
redw <- lm(as.formula(reduced),weights=w2)
residr <- residuals(redw)
wresidr <- diag(as.vector(w))%*%residr
sser <- t(wresidr)%*%wresidr
lambda <- det(ssec)/det(sser)
#cat("lambda")
#print(lambda)
```

```
teststat <- c(lambda)
return(teststat)
  }
rootcdfm <- function(x,h,p,xlow,xmiddle,xhigh,tolerance) {
  flow     <-  cdfhat(x,h,xlow)
  fmiddle  <-  cdfhat(x,h,xmiddle)
  fhigh    <-  cdfhat(x,h,xhigh)
  if ( (flow >p) | (fhigh < p) ) {
    cat("Stop in rootcdfm flow,fhigh=",flow,fhigh);q() }
  for (i in seq(1:60) ) {
    slope <- (fhigh - flow)/(xhigh - xlow)
    xnew <- xmiddle - (fmiddle-p)/slope
    diff <- fmiddle - p
    adiff <- abs(diff)
    if( (adiff < tolerance) ) break
    if( (fmiddle < p) ) {
      midhigh  <- (xmiddle + xhigh)/2
      xlow <- xmiddle
      flow <- fmiddle
      xmiddle <- min(xnew,midhigh)
                            }
    if( (fmiddle > p) ) {
      midlow  <- (xlow + xmiddle)/2
      xhigh <- xmiddle
      fhigh <- fmiddle
      xmiddle <- max(xnew,midlow)
                            }
    fmiddle <- cdfhat(x,h,xmiddle)
    if( (i==60) ){
     print (" Stop in rootcdfm with over 60 iterations");q()}
    }
   return(xmiddle)
   }
 sumdistances <- function(center,points,n){
   sumdist <- 0
   for (i in 1:n) {
     diff <- points[i,] - center
     dist2 <- crossprod(diff)
     dist <- sqrt(dist2)
     sumdist <- sumdist + dist
                   }
     return(sumdist)
```

APPENDIX H

R CODE FOR CONFIDENCE INTERVALS AND ESTIMATES

Note: The functions rootcdf and cdfhat must also be included. These are in Appendix A and on the file adapt.r .

```
adaptest <- function(dataset, complete, reduced, varname,
                     plower, initial, maxsteps, seed)  {
testdata <- read.table(dataset)
cat("   Data Set ", "\n","\n")
print(testdata)
attach(testdata)
yvector <- y
xa <- as.vector(as.matrix(subset(testdata,select=varname)))
set.seed(seed)
olsmodel <- lm(as.formula(complete))
olsse <- summary(olsmodel)$coefficients[varname,2]
dfc <- df.residual(olsmodel)
tcrit <- abs(qt(0.975,dfc))
k <- 17.46*tcrit*olsse
```

```
estimate <- initial
estimatesum <- 0
cat("\n","\n","\n","******************************","\n","\n")
cat("  Input:","\n","\n")
cat("    Data set:",dataset,"\n","\n")
cat("    Complete model: ",complete,"\n","\n")
cat("    Reduced model : ",reduced,"\n","\n")
cat("    Variable estimated : ",varname, "\n","\n")
cat("    Maximum number of steps = ",maxsteps,"\n","\n")
cat("    Random seed = ",seed,"\n","\n")
cat("    Initial estimate = ", initial, "\n", "\n")
cat("    Proportion to left = ",plower,"\n","\n")
cat("    Steplength constant = ", k,"\n","\n")
cat("\n","\n","    Estimates:","\n","\n")
for (l in 1:maxsteps) {
  yadj  <- yvector - estimate*xa
  tunperm <- teststat(yadj,complete, reduced, varname)
  yadj <- permres(yadj,complete,reduced)
  tperm <- teststat(yadj,complete,reduced, varname)
  if( (tperm < tunperm) | ( (tperm==tunperm) & (plower > .5)))
    estimate <- estimate + k*plower/(l+22)
  if( (tperm > tunperm) | ( (tperm==tunperm) & (plower < .5)))
    estimate <- estimate - k*(1-plower)/(l+22)
  estimatesum <- estimatesum + estimate
  if( l <= 10 | (l <=100  & l==  10*floor(l/10)   )
              | (l < 1000 & l== 100*floor(l/100) ) )
     cat("      Step = ",l," estimate = ", estimate,"\n")
  if( l==1000*floor(l/1000) )                 {
    estimateaver=estimatesum/1000
    estimatesum=0
     cat("      Step = ",l," estimate = ",estimate,
         "  average = ",estimateaver,"\n")          }
                          }

cat("\n","****************************************","\n","\n")
detach(testdata)
                                                  }

permres <- function(y,complete,reduced) {
  redu     <- lm(as.formula(reduced))
  yhat     <- predict(redu)
  yresidual <- residuals(redu)
  ynew   <- yhat + sample(yresidual)
  return(ynew)
                                             }
teststat <- function(y,complete,reduced,varname) {
red <- lm(as.formula(reduced))
resid <- residuals(red)
#                              compute traditional quantiles
q25 <- quantile(resid,0.25,type=6)
```

```
q75 <- quantile(resid,0.75,type=6)
sigmat <- (q75-q25)/1.349
#                                       compute bandwidth (h)
n <- length(resid)
h <- 1.587*sigmat*n^(-0.33333333333)
range <- max(resid) - min(resid)
lower <- min(resid) - range
upper <- max(resid) + range
tol <- 0.00000001*sigmat
cdf25 <- rootcdf(resid,h,0.25,lower,upper,tol)
cdf50 <- rootcdf(resid,h,0.50,lower,upper,tol)
cdf75 <- rootcdf(resid,h,0.75,lower,upper,tol)
sigma <- (cdf75-cdf25)/1.349
#                                       compute adaptive weights
s <- (resid-cdf50)/sigma
tolerance <- 100*tol
w  <- matrix(0,n,1)
w2 <- matrix(0,n,1)
for (i in 1:n) {
  fhat <- cdfhat(resid,h,resid[i])
  z <- qnorm(fhat)
  if( (abs(s[i]) >= tolerance) )
     w[i] <- z/s[i]
    else
     w[i] <- 1
    w2[i] <- w[i]^2
  }
compw <-lm(as.formula(complete), weights=w2)
tvalue <- summary(compw)$coefficients[varname,3]
return(tvalue)
  }
```

APPENDIX I

SAS MACRO FOR CONFIDENCE INTERVALS

```
    /*
      Note: If variables are formatted then the formats will
            need to be placed before the macro.
    */
options nocenter linesize=80 nonotes ;
%macro adaptci(dataset=, y=, xr=, xa=, seed1=, seed2=,seed3=,
                 maxsteps=8000);
data dseed;
   next1=&seed1;
   next2=&seed2;
   next3=&seed3;
data dupdate;
  c=0;
data dgarth(keep=&y &xr &xa);
  retain n 0;
  set &dataset;
```

Adaptive Tests of Significance Using Permutations of Residuals with R and SAS®.
By Thomas W. O'Gorman

```
    n = n + 1;
    call symput('nobs',trim(left(put(n,8.) ) ) );
    output;
proc reg data=dgarth;
  model &y=&xr &xa/clb;
title1 ' Regression Analysis with 95% Confidence Intervals ';
 /*  Now begin finding the adaptive estimate for parameter */
proc glm data=dgarth noprint;
  model &y=&xr;
  output out=dresid predicted=yhat1 residual=res;
  title2 'Adaptive weighted least squares analysis';
%adaptwts1(datares=dresid, resvars=res, nobs=&nobs, nvars=1);
data dall;
  merge dgarth dresid dweights;

    proc print data=dall ;
     var &y &xr &xa res weight saswt;
 title2' Data, after adjustment   SASWT=weights used by SAS';
proc reg data=dall outest=dunperm outseb noprint;
  model &y = &xr &xa;
  weight saswt;
data dtunperm(keep=tunperm beta);
  set dunperm;
  retain beta;
  if _type_ eq 'PARMS' then beta=&xa;
  if _type_ eq 'SEB'   then do;
    tunperm=beta/&xa;
    output;
    end;
file print noprint;
title1 'Adaptive Weighted Least Squares Confidence Intervals';
title2 'Search method based on method proposed by ';
title3 'Garthwaite, P. H. , Biometrics, vol. 52,  1996.';
 put ;
proc reg data=dgarth outest=dtheta outseb noprint; ;
  model &y=&xr &xa;
data dupdate;
  set dtunperm(in=inunperm) dtheta;
  retain thetahat theta;
  if inunperm  then thetahat=beta;
  if _type_ eq 'PARMS' then theta=&xa;
  if _type_ eq 'SEB'   then do;
  file print noprint notitle;
  put ' ';
  put '  **************************************************';
  put '          now begin adaptive lower limit';
  put '  **************************************************';
  put ' ';
                        nm2=&nobs-2;
```

```
                              tcrit=tinv(0.975,nm2);
                              ui=theta+tcrit*&xa;
                              li=theta-tcrit*&xa;
                              olswidth=ui-li;
                              c=li;
                              iter=1;
                              output;
                              end;
 %do istep=1 %to &maxsteps;
 %nextstep( y=&y, xr=&xr, xa=&xa, nobs=&nobs);
 data dupdatet;
   merge dupdate dtunperm dtperm;
   if tperm lt tunperm then
          li=li+17.46*olswidth/2*0.025/(iter+22);
     else li=li-17.46*olswidth/2*0.975/(iter+22);
   file print noprint notitle;
   if iter le 10 or (iter le 1000 and
      iter eq 100*(int(iter/100)))
      then put ' after iteration ' iter 6.0
              ' adaptive lower limit = ' li 8.4;
   if (iter -1) eq 1000*int((iter-1)/1000) then average = li;
                                   else average = average + li;
   if iter gt 1000 and iter eq 1000*(int(iter/1000)) then do;
     average = average/1000;
     put ' after iteration ' iter 6.0
         ' adaptive lower limit = ' li 8.4
         ' average = ' average 8.2;
     end;
   iter=iter+1;
   c=li;
data dupdate;
  set dupdatet;
%end;
data dupdate;
  set dupdate;
  file print noprint notitle;
  put ' ';
  put '  ************************************************';
  put '          now begin adaptive upper limit';
  put '  ************************************************';
  put ' ';
  c=ui;
  iter=1;
%do istep=1 %to &maxsteps;
%nextstep( y=&y, xr=&xr, xa=&xa, nobs=&nobs);
data dupdatet;
  merge dupdate dtunperm dtperm;
  if tperm ge tunperm then
         ui=ui-17.46*olswidth/2*0.025/(iter+22);
```

```
    else ui=ui+17.46*olswidth/2*0.975/(iter+22);
  file print noprint notitle;

  if iter le 10 or (iter le 1000 and
     iter eq 100*(int(iter/100))) then
    put ' after iteration ' iter 6.0
        '  adaptive upper limit = ' ui 8.4;
  if (iter -1) eq 1000*int((iter-1)/1000) then average = ui;
                                   else average = average + ui;
  if iter gt 1000 and iter eq 1000*(int(iter/1000)) then do;
   average = average/1000;
   put ' after iteration ' iter 6.0
       '  adaptive upper limit = ' ui 8.4
       '  average = ' average 8.2;
   end;
  iter=iter+1;
  c=ui;
data dupdate;
  set dupdatet;
%end;
%mend;
%macro nextstep( y=, xr=, xa=, nobs=);
data awlsdata(keep= &y yadj &xr &xa) ;
   set dupdate(in=inup)  dgarth(in=indata);
   retain const;
   if inup then const=c;
   if indata then do;
     yadj=&y-const*&xa;
     output awlsdata;
     end;
proc glm data=awlsdata noprint;
  model yadj=&xr;
  output out=dresid predicted=yhat1 residual=res;
  title2 'adaptive weighted least squares analysis';
%adaptwts1(datares=dresid, resvars=res, nobs=&nobs, nvars=1);
data dall;
  merge awlsdata dresid dweights;
proc reg data=dall outest=dunperm outseb noprint;
  model yadj = &xr &xa;
  weight saswt;
data dtunperm(keep=tunperm beta);
  set dunperm;
  retain beta;
  if _type_ eq 'PARMS' then beta=&xa;
  if _type_ eq 'SEB'   then do;
    tunperm=beta/&xa;
    output;
    end;
/*    Now perform one permutation         */
```

```
    data dperm(keep= place  res)
         dseed(keep=next1 next2 next3);
      array k{&nobs} _temporary_;
      retain tmpseed1 tmpseed2 tmpseed3;
      set dseed(in=inseed) dall(in=indall) end=last;
      if inseed then do;
        tmpseed1 = next1;
        tmpseed2 = next2;
        tmpseed3 = next3;
  /* This is the Wichmann-Hill random number generator. */
          tmpseed1 = mod(171*tmpseed1,30269);
          tmpseed2 = mod(172*tmpseed2,30307);
          tmpseed3 = mod(170*tmpseed3,30323);
        do i = 1 to &nobs;
          k{i} = i;
          end;
        do l = &nobs to 2 by -1;
          tmpseed1 = mod(171*tmpseed1,30269);
          tmpseed2 = mod(172*tmpseed2,30307);
          tmpseed3 = mod(170*tmpseed3,30323);
          random=tmpseed1/30269.0+tmpseed2/30307.0
                 +tmpseed3/30323;
          random = mod(random,1.0);
  /* This is the Fisher-Yates permutation algorithm, as
     improved by Durstenfeld (1964) Comm. of the ACM 1964
     algorithm 235 */
          itrade = int(l*random) + 1;
          ktemp     = k{itrade};
          k{itrade} = k{l};
          k{l}      = ktemp;
          end;
        end;
      if indall then do;
        l=_n_-1;
        place = k{l};
        output dperm;
        end;
      if last then do;
        next1 = tmpseed1;
        next2 = tmpseed2;
        next3 = tmpseed3;
        output dseed;
        end;
   proc sort data=dperm out=dtemp; by place;
   data dynew;
     merge dall dtemp;
       ynew1 = yhat1 + res;
proc glm data=dynew noprint;
  model ynew1=&xr;
```

```
  output out=dresid residual=res;
%adaptwts1(datares=dresid, resvars=res, nobs=&nobs, nvars=1);
     data dallperm;
       merge dynew dweights;
     proc reg data=dallperm outest=dperm outseb noprint;
       model ynew1 = &xr &xa ;
       weight saswt;
     data dtperm(keep=tperm);
       set dperm;
       retain beta;
       if _type_ eq 'PARMS' then beta=&xa;
       if _type_ eq 'SEB'   then do;
         tperm=beta/&xa;
         output;
         end;
%mend;
;
%macro adaptwts1(datares=, resvars=, nobs=, nvars=1,
                 iequalwt=0);
data dweights(keep= weight saswt );
  array d{&nobs} _temporary_;
  array smcdf{*} f1-f&nobs;
  array distance{&nobs} dst1-dst&nobs;
  array points{*} p1 p2 p3;
  array smpoint{*} smp1 smp2 smp3;
  set &datares end=last;
  retain h f1-f&nobs;
  n = &nobs;
  d{_n_} = res;

  if last then do;
   if &iequalwt = 1 then do;
    do i = 1 to n;
      weight = 1.0;
      saswt  = 1.0;
      output;
      end;
    go to skipadap;
    end;
      do j = 1 to n;
        distance{j} = d{j};
         end;
 /*
         Now standardize the distances
 */

      np1 = n + 1;
      nlast = np1-1;
      lower = floor(.25*np1);
```

```
    fraction = .25*np1 - lower;
    lowerp1 = lower + 1;
    plow = ordinal(lower, of distance{*});
    plowp1 = ordinal(lowerp1, of distance{*});
    p25 = (1-fraction)*plow + fraction*plowp1;
    upper = floor(.75*np1);
    fraction = .75*np1 - upper;
    upperp1 = upper + 1;
    pupper = ordinal(upper, of distance{*});
    pupperp1 = ordinal(upperp1, of distance{*});
    p75 = (1-fraction)*pupper + fraction*pupperp1;
    sdinit = (p75 - p25)/1.349;
    h = 1.587*sdinit*n**(-0.3333333333333333);
/*
    Now find estimates for the 25th, 50th, and
    75th percentiles with smoothed c.d.f.
 */
    range=ordinal(n, of distance{*})
         -ordinal(1, of distance{*});
    tol=0.000000001*range;
    do l = 1 to 3;
      if l = 1 then percent = 0.25;
      if l = 2 then percent = 0.50;
      if l = 3 then percent = 0.75;
      p1 = ordinal(1, of distance{*}) - range;
      p2 = 0.0;
      p3 = ordinal(n, of distance{*}) + range;
       do j = 1 to 3;
        sum = 0.0;
        do k = 1 to n;
          z = (points{j} - distance{k})/h;
          sum = sum + probnorm(z);
          end;
        smpoint{j} = sum/n;
        end;
  nextiter:;
      if smp2 lt percent then do;
        p1=p2;
        smp1=smp2;
        end;
      if smp2 gt percent then do;
        p3=p2;
        smp3=smp2;
        end;
      if abs(smp2 - percent) lt tol then go to over;
      p2=(p1+p3)/2;
      sum=0.0;
      do k = 1 to n;
        z = (p2 - distance{k})/h;
```

```
          sum = sum + probnorm(z);
          end;
        smp2 = sum/n;
        if abs(p3 -p1) lt tol then go to over;
                               else go to nextiter;
 over:;
        if l = 1 then perc25 = p2;
        if l = 2 then perc50 = p2;
        if l = 3 then perc75 = p2;
        end;
        sigmahat = (perc75 - perc25)/1.349;
   do i = 1 to n;
        sum = 0.0;
        do k = 1 to n;
          z = (distance{i} - distance{k})/h;
          sum = sum + probnorm(z);
          end;
        fhat = sum/n;
        z=(distance{i} - perc50)/sigmahat;
        if abs(z) gt 0.0001 then weight = probit(fhat)/z;
                            else weight = 1.0;
        saswt=weight*weight;
        output;
        end;
   end;
 skipadap:;
%mend;
```

APPENDIX J

SAS MACRO FOR ESTIMATES

Note: The macros nextstep and adaptwts1 must be included. These are in Appendix I and on the file adaptci.sas.

```
  /*
  Note: If variables are formatted then the formats will
        need to be placed before the macro.
   */
options nocenter linesize=80 nonotes ;

%macro adaptest(dataset=, y=, xr=, xa=, seed1=, seed2=,
                seed3=, maxsteps=8000);
data dseed;
   next1=&seed1;
   next2=&seed2;
   next3=&seed3;
data dupdate;
  c=0;
```

Adaptive Tests of Significance Using Permutations of Residuals with R and SAS®.
By Thomas W. O'Gorman

```
data dgarth(keep=&y &xr &xa);
  retain n 0;
  set &dataset;
    n = n + 1;
    call symput('nobs',trim(left(put(n,8.) ) ) );
    output;
proc reg data=dgarth;
  model &y=&xr &xa/clb;
title1 ' Regression Analysis with no adaptation';
proc glm data=dgarth noprint;
  model &y=&xr;
  output out=dresid predicted=yhat1 residual=res;
  title2 'Adaptive weighted least squares analysis';
%adaptwts1(datares=dresid, resvars=res, nobs=&nobs, nvars=1);
data dall;
  merge dgarth dresid dweights;
    proc print data=dall ;
     var &y &xr &xa res weight saswt;
  title2' Data, after adjustment  SASWT=weights used by SAS';
proc reg data=dall outest=dunperm outseb noprint;
  model &y = &xr &xa;
  weight saswt;
data dtunperm(keep=tunperm beta);
  set dunperm;
  retain beta;
  if _type_ eq 'PARMS' then beta=&xa;
  if _type_ eq 'SEB'   then do;
    tunperm=beta/&xa;
    output;
    end;
file print noprint;
title1 ' Adaptive Weighted Least Squares Point Estimate';
title2 ' Search method based on method proposed by ';
title3 ' Garthwaite, P. H. , Biometrics, vol. 52, 1996.';
 put ;
proc reg data=dgarth outest=dtheta outseb noprint; ;
  model &y=&xr &xa;
data dupdate;
  set dtunperm(in=inunperm) dtheta;
  retain thetahat theta;
  if inunperm  then thetahat=beta;
  if _type_ eq 'PARMS' then theta=&xa;
  if _type_ eq 'SEB'   then do;
  file print noprint notitle;
  put ' ';
  put '  ***********************************************';
  put '          now begin adaptive point estimate';
  put '  ***********************************************';
  put ' ';
```

```
                                    nm2=&nobs-2;
                                    tcrit=tinv(0.975,nm2);
                                    ui=theta+tcrit*&xa;
                                    li=theta-tcrit*&xa;
                                    olswidth=ui-li;
                                    c=theta;
                                    iter=1;
                                    output;
                                    end;
%do istep=1 %to &maxsteps;
%nextstep( y=&y, xr=&xr, xa=&xa, nobs=&nobs);
data dupdatet;
  merge dupdate dtunperm dtperm;
  if tperm lt tunperm then
         li=li+17.46*olswidth/2*(0.5)/(iter+22);
    else li=li-17.46*olswidth/2*(0.5)/(iter+22);
  file print noprint notitle;
  if iter le 10 or (iter le 1000 and
     iter eq 100*(int(iter/100))) then
     put ' after step ' iter 6.0
     '  adaptive lower limit = ' li 8.4;
  if (iter -1) eq 1000*int((iter-1)/1000) then average = li;
                                     else average = average + li;
  if iter gt 1000 and iter eq 1000*(int(iter/1000)) then do;
    average = average/1000;
    put ' after step ' iter 6.0 ' lower limit = ' li 10.4
        '  average = ' average 10.4;
    end;
  iter=iter+1;
  c=li;
data dupdate;
  set dupdatet;
%end;
%mend;
```

REFERENCES

Agosti, E. and Camerota, G. (1965). Some effects of hypnotic suggestion on respiratory function. *International Journal of Clinical and Experimental Hypnosis*, **13**, 149–156.

Anderson, J. W., Story, L., Sieling, B., Chen, W. L., Petro, M. S., and Story, J. (1984). Hypocholesterolemic effects of oat-bran or bean intake for hypercholesterolemic men. *The American Journal of Clinical Nutrition*, **40**, 1146–1155.

Anderson, M. J., and Legendre, P. (1999). An empirical comparison of permuation methods for tests of partial regression coefficients in a linear model. *Journal of Statistical Computation and Simulation*, **62**, 271–303.

Anderson, M. J., and Robinson, J. (2001). Permutation tests for linear models. *Australian and New Zealand Journal of Statistics*, **43**, 75–88.

Atkinson, A. C. (1986). Comment: aspects of diagnostic regression analysis (discussion of paper by Chatterjee and Hadi). *Statistical Science*, **1**, 397–402.

Atkinson, A. C. (1988). Transformations unmasked, *Technometrics*, **30**, 311–318.

Baklizi, A. (2005). A continuously adaptive rank test for shift in location. *Australian and New Zealand Journal of Statistics*, **47**, 203–209.

Bandyopadhyay, U., and Dutta, D. (2007). Adaptive nonparametric tests for a single sample location problem. *Statistical Methodology*, **4**, 423–433.

Bartlett, M. S. (1947). The use of transformations. *Biometrics*, **3**, 39–52.

Blumberg, E. M., West, P. M., and Ellis, F. W. (1954). A possible relationship between psychological factors and human cancers, *Psychosomatic Medicine*, **16**, 277–286.

Box, G. E. P. and Cox, D. R. (1964). Analysis of transformations. *Journal of the Royal Statistical Society, Series B*, **26**, 211–243.

Brunner, E. and Munzel, U. (2000). The nonparametric Behrens–Fisher problem: asymptotic theory and a small-sample approximation. *Biometrical Journal*, **42**, 17–25.

Büning, H. (1996). Adaptive tests for the c-sample location problem-the case of two-sided alternatives. *Communications in Statistics–Theory and Methods*, **25**, 1569–1582.

Büning, H. (1999). Adaptive Johckheere-type tests for ordered alternatives. *Journal of Applied Statistics*, **26**, 541–551.

Büning, H. (2002). An adaptive distribution-free test for the general two-sample problem. *Computational Statistics*, **17**, 297–313.

Büning, H. and Kössler, W. (1998). Adaptive tests for umbrella alternatives. *Biometrical Journal*, **40**, 573–587.

Büning, H. and Thadewald, T. (2000). An adaptive two-sample location-scale test of Lepage type for symmetric distributions. *Journal of Statistical Computation and Simulation*, **65**, 287–310.

Carroll, R. J. and Rupert, D. (1988). *Transformation and Weighting in Regression*. Chapman and Hall, New York.

Chambers, J. M. and Hastie, T. J. (1992). *Statistical Models in S*. Wadsworth, Pacific Grove, California.

Chatterjee, S., Hadi, A. S., and Price, B. (2000). *Regression Analysis by Example*, 3rd edition. Wiley, New York.

Collins, M. F. (1987). A permutation test for planar regression. *Australia and New Zealand Journal of Statistics* **29**, 303–308.

Coleman, J. S., Cambell, E. Q., Hobson, C.J., McPartland, J., Mood, A. M., Weinfield, F.D., and York, R. L. (1966). *Equality of Educational Opportunity*, two volumes, Office of Education, U.S. Department of Health, Washington.

Conover, W. J. (1973). Rank tests for one sample, two samples and k samples without the assumption of a continuous distribution function. *The Annals of Statistics* **1**, 1105–1125.

Consumers Union of U.S. Inc. (2010). *Consumers Reports Best and Worst New Cars*, Consumers Union of U.S. Inc., Yonkers, New York.

Damaser, E. C., Shor, R. E., and Orne, M. T. (1963). Physiological effects during hypnotically requested emotions. *Psychosomatic Medicine* **25**, 334–343.

Davis C. S. (2002). *Statistical Methods for the Analysis of Repeated Measurements*. Springer-Verlag, New York.

Davison, A. C. and Hinkley, D. V. (1997). *Bootstrap Methods and Their Application*. Cambridge University Press.

DeGroot, M. H. (1986). *Probability and Statistics*, 2nd edition. Addison-Wesley, Reading, Massachusetts.

Draper, N. R. and Stoneman, D. M. (1966). Testing for the inclusion of variables in linear regression by a randomization technique. *Technometrics* **8**, 695–699.

Durstenfeld, R. (1964). Algorithm 235: Random permutation. *Communications of the ACM*, **7**, 420.

Farber, S. L. (1981). *Identical Twins Reared Apart.* Basic Books, New York.

Fenstad, G. U., Kjaernes, M., and Walløe, L. (1980). Robust estimation of standard deviation. *Journal of Statistical Computation and Simulation*, **10**, 113–132.

Fisher, R. A. and Yates, F. (1938). *Statistical Tables for Biological, Agricultural, and Medical Researchers.* Oliver and Boyd, London.

Fitzmaurice, G. M., Laird, N. M., and Ware, J. H. (2004). *Applied Longitudinal Analysis.* Wiley, New York.

Fleiss, J. L. (1986). *The Design and Analysis of Clinical Experiments.* Wiley, New York.

Fligner, M. A. and Policello, G. E. (1981). Robust Rank Procedures for the Behrens-Fisher Problem. *Journal of the American Statistical Association*, **76**, 162–168.

Freedman, D. and Lane, D. (1983). A nonstochastic interpretation of reported significance levels. *Journal of Business and Economic Statistics*, **1**, 292–298.

Freeman, P. (1989). The performance of the two-stage analysis of two treatment, two period crossover trials. *Statistics in Medicine*, **8**, 1421–1432.

Freidlin, B., Miao, W., and Gastwirth, J. L. (2003). On the use of the Shapiro-Wilk test in two-stage adaptive inference for paired data from moderate to very heavy tailed distributions. *Biometrical Journal*, **45**, 887–900.

Friedman, M. (1937). The use of ranks to avoid the assumption of normality implicit in the analysis of variance. *Journal of the American Statistical Association*, **11**, 86–92.

Garthwaite, P. (1996). Confidence intervals from randomization tests. *Biometrics*, **52**, 1387–1393.

Garthwaite, P. H. and Buckland, S. T. (1992). Generating Monte Carlo confidence intervals by the Robbins–Monro process. *Applied Statistics*, **41**, 159–171.

Gastwirth, J. L. (1965). Percentile modifications of two sample rank tests. *Journal of the American Statistical Association*, **60**, 1127–1141.

Gibbons, J. D. and Chakraborti, S.(2003). *Nonparametric Statistical Inference, 4th edition.* Marcel Dekker, New York.

Grizzle, J. E. (1965). The two period change-over design and its use in clinical trials. *Biometrics* **21**, 467–468.

Hájek, J. (1962). Asymptotically most powerful rank-order tests. *The Annals of Mathematical Statistics*, **33**, 1124–1147.

Haldane, J. B. S. (1948). Note on the median of a multivariate distribution. *Biometrika*, **35**, 414–415.

Hall, P. and Padmanabhan, A. (1997). Adaptive inference for the two-sample scale problem. *Technometrics*, **39**, 412–422.

Hampel, F. R., Ronchetti, E. M., Rousseeuw, P. J., Stahel, W. A. (1986). *Robust Statistics.* Wiley, New York.

Harrell, F. E. and Davis, C. E. (1982). A new distribution-free quantile estimator. *Biometrika*, **69**, 635–640.

Hernandez, F. and Johnson, R. A. (1980). The large-sample behavior of transformations to normality. *Journal of the American Statistical Association*, **75**, 855.

Hettmansperger, T. P. and Sheather, S. J. (1992). A cautionary note on the method of least median squares. *The American Statistician*, **46**, 79–83.

Hill, N. J. and Padmanabhan, A. R., (1991). Some adaptive robust estimators which work with real data. *Biometrical Journal*, **33**, 81–101.

Hill, N. J., Padmanabhan, A. R., and Puri, M. L. (1988). Adaptive nonparametric procedures and applications. *Applied Statistics*, **37**, 205–218.

Hills, M. and Armitage, P. (1979). The two-period cross-over clinical trial. *British Journal of Clinical Pharmacology*, **8**, 7–20.

Hodges, J. L., and Lehmann, E. L. (1963). Estimates of location based on rank tests. *Annals of Mathematical Statistics*, **34**, 598–611.

Hogg, R. V. (1967). Some observations on robust estimation. *Journal of the American Statistical Association*, **62**, 1179–1186.

Hogg, R. V. (1974). Adaptive robust procedures: A partial review and some suggestions for future applications and theory. *Journal of the American Statistical Association*, **69**, 909–923.

Hogg, R. V., Fisher, D. M., and Randles, R. H. (1975). A two-sample adaptive distribution-free test. *Journal of the American Statistical Association*, **70**, 656–661.

Hogg, R. V. and Lenth, R. V. (1984). A review of some adaptive statistical techniques, *Communications in Statistics–Theory and Methods*, **13**, 1551–1579.

Hüsler, J. (1987). On the two-sample adaptive distribution free test. *Communications in Statistics–Theory and Methods*, **16**, 55–68.

Ibrahim, J. G. and Molenberghs, G. (2009). Missing data methods in longitudinal studies: A review. *Test* **18**, 1–43.

Johnson, M. E. (1987). *Multivariate Statistical Simulation*. Wiley, New York.

Jones, B. and Kenward, M. G. (2003). *Design and Analysis of Cross-Over Trials*, 2nd edition. Chapman and Hall, Boca Raton.

Kaigh, W. D. and Lachenbruch, P. A. (1982). A generalized quantile estimator. *Communications in Statistics–Theory and Methods*, **11**, 2217–2238.

Kennedy, P. E. and Cade, B. S. (1996). Randomization tests for multiple regression. *Communications in Statistics–Simulation and Computation*, **25**, 923–936.

Kenward, M. G. and Jones, B. (1987). The analysis of data from 2×2 cross-over trials with baseline measurements. *Statistics in Medicine*, **6**, 911–926.

Kenward, M. G., and Roger, J. H. (1997). Small sample inference for fixed effects from restricted maximum likelihood. *Biometrics*, **53**, 983–997.

Kowalchuk, R. K., Keselman, H. J., Algina, J., and Wolfinger, R. D. (2004). The analysis of repeated measurements with mixed-model adjusted F tests. *Educational and Psychological Measurement*, **64**, 224–242.

Koziol, J. A., Maxwell, D. A., Fukushima, M., Colmerauer, M. E. M. , and Pilch (1981). A distribution-free test for tumor-growth curve analyses with application to an animal tumor immunotherapy experiment. *Biometrics*, **37**, 383–390.

Lehmann, E. L. (1975). *Nonparametrics: Statistical Methods Based on Ranks*. Holden-Day, San Francisco.

Lemmer, H. H. (1993). Adaptive tests for the median. *IEEE Transactions on Reliability*, **42**, 442–448.

Lepage, Y. (1971). A combination of Wilcoxon's and Ansari–Bradley's statistics. *Biometrika*, **58**, 213–217.

Leppik, I. E., Dreifuss, F. E., Bowman-Cloyd, T., Santilli, N., Jacobs, M., Crosby, C., Cloyd, J., Stockman, J., Graves, N., Sutula, T., Welty, T., Vickery, J., Brundage, R., Gumnit, R., and Gutierres, A. (1985). A double-blind crossover evaluation of progabide in partial seizures. *Neurology*, **35**, (Suppl.), 285.

Linton, L. R., Edgington, E. S., and Davies, R. W. (1989). A view of niche overlap amenable to statistical analysis. *Canadian Journal of Zoology*, **67**, 55–60.

Little, R. J. A. and Rubin, D. B. (1987). *Statistical Analysis with Missing Data*. Wiley, New York.

Manly, B. F. J. (1997). *Randomization, Bootstrap and Monte Carlo Methods in Biology*, 2nd edition. Chapman and Hall, London.

Mardia, K.V., Kent, J.T., and Bibby, J. M. (1979). *Multivariate Analysis*. Academic Press, London.

Miao, W. and Gastwirth, J. (2009). A new two stage adaptive nonparametric test for paired differences. *Statistics and its Interface*, **2**, 213–221.

Miller, R. G. (1980). *Simultaneous Statistical Inference*, 2nd edition. Springer-Verlag, New York.

Mitchell, H. H., Burroughs, W., and Beadles, J. R. (1936). The significance and accuracy of biological values of proteins computed from nitrogen metabolism data. *Journal of Nutrition*, **11**, 257–274.

Neuhäuser, M., Büning, and Hothorn, L. (2004). Maximum test versus adaptive tests for the two-sample location problem, *Journal of Applied Statistics*, **31**, 215–227.

O'Brien, P. C. (1988). Comparing two samples: extensions of the *t*, rank-sum, and log-rank tests. *Journal of the American Statistical Association*, **83**, 52–61.

O'Gorman, T. W. (1995). The effect of unequal variances on the power of several two-sample tests. *Communications in Statistics–Simulation and Computation*, **24**, 853–867.

O'Gorman, T. W. (1996). An adaptive two-sample test based on modified wilcoxon scores. *Communications in Statistics–Simulation and Computation*, **25**, 459–479.

O'Gorman, T. W. (1997). An adaptive test for the one-way layout. *The Canadian Journal of Statistics*, **25**, 269–279.

O'Gorman, T. W. (2001). An adaptive permutation test procedure for several common tests of significance. *Computational Statistics and Data Analysis*, **35**, 335–350.

O'Gorman, T. W. (2002). An adaptive test of significance for a subset of regression coefficients. *Statistics in Medicine*, **21**, 3527–3542.

O'Gorman, T. W. (2004). *Applied Adaptive Statistical Methods*. Society for Industrial and Applied Mathematics, Philadelphia.

O'Gorman, T. W. (2005). The performance of randomization tests that use permutations of independent variables. *Communications in Statistics–Simulation and Computation*, **34**, 895–908.

O'Gorman, T. W. (2006a). An adaptive test for a subset of coefficients in a multivariate regression model. *Biometrical Journal*, **48**, 849–859.

O'Gorman, T. W. (2006b). An adaptive test of a subset of regression coefficients using permutation of residuals. *Journal of Statistical Computation and Simulation*, **76**, 1095–1105.

O'Gorman, T. W. (2008a). Using adaptive tests for the analysis of repeated measurements, *Journal of Biopharmaceutical Statistics*, **18**, 595–610.

O'Gorman, T. W. (2008b). A comparison of several adaptive tests for paired data, *Journal of Statistical Computation and Simulation*, **79**, 1083–1093.

O'Gorman, T. W. (2010). An adaptive test for the analysis of repeated measurements having missing data, *Statistics in Biopharmaceutical Research*, **2**, 109–121.

Oja, H. (1987). On permutation tests in multiple regression and analysis of covariance problems. *Australian Journal of Statistics*, **29**, 91–100.

Ott, L. and Longnecker, M. (2001). *An Introduction to Statistical Methods and Data Analysis, 5th ed.*. Duxbury, Pacific Grove, California.

Overall J.E., Ahn C., Shivakumar C., and Kalburg, Y. (1999). Problematic formulations of SAS PROC. MIXED models for repeated measurements. *Journal of Biopharamaceutical Statistics*, **9**, 189–216.

Park T., Park J., and Davis C. S. (2001). Effects of covariance model assumptions on hypothesis tests for repeated measurements: analysis of ovarian hormone data and pituitary-pteryomaxillary distance data. *Statistics in Medicine*, **20**, 2441–2453.

Patel, H. I. (1983). Use of baseline measurements in the two-period crossover design. *Communications in Statistics–Theory and Methods*, **12**, 2693–2712.

Polansky, A. M. (1998). A smooth nonparametric approach to process capability. *Quality and Reliability Engineering International*, **14**, 43–48.

Potthoff, R. F. and Roy, S. N. (1964). A generalized multivariate analysis of variance model useful especially for growth curve problem. *Biometrika*, **51**, 313–326.

Powell, G. L., and Russell, A. P. (1984). The diet of the eastern short-horned lizard (Phrynosoma douglassi brevirostre) in Alberta and its relationship to sexual size dimorphism. *Canadian Journal of Zoology*, **62**, 428–440.

Ramberg, J.S., Dudewicz, E.J., Tadikamalla, P.R., and Mykytka, E.F. (1979). A probability distribution and its uses in fitting data. *Technometrics*, **21**, 201–214.

Randles, R. H. and Hogg, R. V. (1973). Adaptive distribution-free tests. *Communications in Statistics*, **2**, 337–356.

Randles, R.H. and Wolfe, D.A. (1979). *Introduction to the Theory of Nonparametric Statistics*. Wiley, New York.

Rawlings, J. O., Pantula, S. G., and Dickey, D. A. (1998). *Applied Regression Analysis*. Springer-Verlag, New York.

Robbins, H. and Monro, S. (1951). A stochastic approximation method. *The Annals of Mathematical Statistics*, **22**, 400–407.

Rousseeuw, P.J. (1984). Least median of squares regression. *Journal of the American Statistical Association*, **79**, 871–880.

Rousseeuw, P.J. and Leroy, A.M. (1987). *Robust Regression and Outlier Detection*. Wiley, New York.

Ruberg, S. J. (1986). A continuously adaptive nonparametric two-sample test. *Communications in Statistics–Theory and Methods*, **15**, 2899–2920.

SAS Institute. (2003). *SAS/STAT User's Guide, Version 9*. SAS Institute, Cary, North Carolina.

Sen, P. K. (1968). On a class of aligned rank order tests in two-way layouts. *The Annals of Mathematical Statistics*, **39**, 1115–1124.

Senn, S. (2007). *Statistical Issues in Drug Development*, 2nd edition. Wiley, West Sussex.

Silverman, B. W. (1986). *Density Estimation for Statistics and Data Analysis*. Chapman and Hall, London.

Simon, D. (2001). *A Nutritional Assessment of a Mature Adult Population in Santo Domingo, Dominican Republic*. Master's Thesis, Northern Illinois University. Raw data courtesy of the author.

Simpson, G. G., Roe, A. and Lewontin, R. C. (1960). *Quantitative Zoology*, Harcourt, Brace and World, New York.

Stromberg, A. J. (1993). Comment on A cautionary note on the method of least median squares. *The American Statistician*, **47**, 87.

Thall, P. F. and Vail, S. C. (1990). Some covariance models for longitudinal count data with overdispersion. *Biometrics*, **46**, 657–671.

Van der Waerden, B. L. (1952, 1953). Order Tests for the two-sample problem and their power. *Indagationes Mathematicae*, **14**, 453–458, **15**, 303–316.

Wichmann, B. A. and Hill, I. D. (1982). An efficient and portable pseudo-random number generator. *Applied Statistics*, **31**, 188–190.

Wilcox, R. (2003). Approximating Tukey's depth. *Communications in Statistics–Simulation and Computation*, **32**, 977–985.

Yuh, L. and Hogg, R. V. (1988). On adaptive M-regression. *Biometrics*, **44**, 433–445.

Zerbe, G. O. (1979). Randomization analysis of the completely randomized design extended to growth and response curves. *Journal of the American Statistical Association*, **74**, 215–221.

INDEX

Adaptive Tests of Significance Using Permutations of Residuals with R and SAS®.
By Thomas W. O'Gorman

Printed and bound by CPI Group (UK) Ltd, Croydon, CR0 4YY
16/09/2026
14963414-0002

Finite Element Methods for Flow Problems

Finite Element Methods for Flow Problems

Jean Donea and Antonio Huerta

WILEY

Telephone (+44) 1243 779777

Email (for orders and customer service enquiries): cs-books@wiley.co.uk

Visit our Home Page on www.wileyeurope.com or www.wiley.com

Reprinted April 2004

Other Wiley Editorial Offices

John Wiley & Sons Inc., 111 River Street, Hoboken, NJ 07030, USA

Jossey-Bass, 989 Market Street, San Francisco, CA 94103-1741, USA

Wiley-VCH Verlag GmbH, Boschstr. 12, D-69469 Weinheim, Germany

John Wiley & Sons Australia Ltd, 33 Park Road, Milton, Queensland 4064, Australia

John Wiley & Sons (Asia) Pte Ltd, 2 Clementi Loop #02-01, Jin Xing Distripark, Singapore 129809

John Wiley & Sons Canada Ltd, 22 Worcester Road, Etobicoke, Ontario, Canada M9W 1L1

Wiley also publishes its books in a variety of electronic formats. Some content that appears in print may not be available in electronic books.

British Library Cataloguing in Publication Data

A catalogue record for this book is available from the British Library

ISBN 0-471-49666-9

Typeset from pdf files supplied by the author
Printed and bound by CPI Antony Rowe, Eastbourne

Contents

Preface

This book is based on lectures regularly taught in the fifth year of engineering diplomas or basic graduate courses at the University of Liège and at the Universitat Politècnica de Catalunya. The goal of the book is not to provide an exhaustive account of finite elements in fluids, which is an extremely active area of research. The objective is to present the fundamentals of stabilized finite element methods for the analysis of steady and time-dependent convection–diffusion and fluid dynamics problems with an engineering rather than a mathematical bias. Organized into six chapters, it combines theoretical aspects and practical applications and covers some of the recent research in several areas of computational fluid dynamics.

The authors wish to thank their colleagues and friends at the Joint Research Centre of the European Commission, Ispra, Alenia, Torino, ENEL Research Hydraulics and Structures, Milano, Politecnico di Milano, Université de Liège and Universitat Politècnica de Catalunya, who contributed to the development of several of the concepts presented in this book and provided worked examples. We are especially grateful to Folco Casadei, Luigi Quartapelle and Vittorio Selmin who reviewed parts of this book, provided numerous suggestions and corrections, and contributed with several illustrative examples. We are particularly in debt to our colleagues at the Laboratori de Càlcul Numèric: Sonia Férnandez-Méndez, Esther Sala-Lardies, Antonio Rodríguez-Ferran, Pedro Díez, Josep Sarrate, Agustí Pérez-Foguet and Xevi Roca. Their reviews, suggestions and contributions have made this book a reality. The second author must also thank Gilles Pijaudier-Cabot and the École Centrale de Nantes for giving him sanctuary during the last stages of this project.

Above all, however, our thanks go to our wives, Marie-Paule and Sara, for their continuous and unfailing support and patience for many years and particularly during the writing of this book.

J. DONEA AND A. HUERTA

1

Introduction and preliminaries

1.1 FINITE ELEMENTS IN FLUID DYNAMICS

Introduced in the late 1950s in the aircraft industry, see, for instance, the historical outline by Felippa (2001), the finite element method (FEM) has emerged as one of the most powerful numerical methods so far devised. Among the basic attributes of the method which have led to its widespread acceptance and use are the ease in modeling complex geometries, the consistent treatment of differential-type boundary conditions and the possibility to be programmed in a flexible and general purpose format.

Standard finite element approximations are based upon the Galerkin formulation of the method of weighted residuals. This formulation has proven eminently successful in application to problems in solid/structural mechanics and in other situations, such as heat conduction, governed by diffusion-type equations. The reason for this success is that, when applied to problems governed by self-adjoint elliptic or parabolic partial differential equations, the Galerkin finite element method leads to symmetric stiffness matrices. In this case the difference between the finite element solution and the exact solution is minimized with respect to the energy norm, see, for instance, Strang and Fix (1973). In practice, the Galerkin formulation is optimal in problems governed by self-adjoint equations. In such cases, there exists a quadratic functional the minimum of which corresponds to satisfying the partial differential equation governing the problem at hand. For instance, in linear elasticity the equilibrium position of a structure corresponds to the minimum of the quadratic functional expressing the total potential energy of the system. Similarly, in steady heat conduction problems the thermal equilibrium resulting from satisfying the Laplace or Poisson equation corresponds to the minimum of a quadratic functional expressed in terms of the thermal flux, which physically represents the total thermal energy of the system.

The success of the Galerkin finite element method in structural mechanics, heat conduction and other problems of the potential type provided, in the early 1970s, a strong impetus for the utilization of the method in the simulation of problems in fluid dynamics. It was thought that the significant advantages gained in structural mechanics and diffusion-type problems would again be open to exploitation in the area of fluid flow simulation. Actually, this proved to be an optimistic point of view, especially with regard to modeling convection-dominated transport phenomena.

The main difficulty was due to the presence of convection operators in the formulation of flow problems based on kinematical descriptions other than Lagrangian. Convection operators are, in fact, non-symmetric and thus the best approximation property in the energy norm of the Galerkin method, which is the basis for success in symmetric cases, is lost when convection dominates the transport process.

In practice, solutions to convection-dominated transport problems by the Galerkin method are often corrupted by spurious node-to-node oscillations. These can only be removed by severe mesh (and time-step) refinement which clearly undermine the practical utility of the method. This has motivated the development of alternatives to the standard Galerkin formulation which preclude oscillations without requiring mesh or time-step refinement. Such alternatives are called stabilization techniques and have provided a major breakthrough in finite element modeling of problems in fluid dynamics.

In truly transient situations, another important issue is to ensure the proper coupling between spatial and temporal approximations. In fact, a stable and accurate spatial representation will be quickly spoiled if the algorithm used for transporting the solution in time is not of comparable accuracy. Space–time coupling is indeed particularly crucial when convection dominates the transport process, due to the directional character of propagation of information in hyperbolic problems. Significant progress has also been achieved in recent years in the development of algorithms for accurately tracing the transient solution of highly convective transport problems.

1.2 SUBJECTS COVERED

The purpose of this book is to describe methods of finite element analysis for steady and time-dependent convection–diffusion and fluid dynamics problems. The intent is to provide an introduction to a variety of modern methods, while preserving a pedagogical character through the presentation of simple worked examples.

The present chapter starts with a review of the basic kinematical descriptions used in fluid mechanics and recalls the conservation laws for mass, momentum and energy in differential and integral forms. It then provides an introduction to the basic ingredients of the finite element analysis of flow problems.

Chapter 2 introduces stabilized finite element methods for steady convection-dominated transport problems (elliptic equations). The difficulties of Galerkin finite elements are first recognized. This allows the design of possible cures for the node-to-node oscillations. The first alternative formulations proposed in the early 1970s to improve the standard Galerkin method tried to reproduce in the finite element context the effect of upwind differencing used in the finite difference context to stabilize the oscillatory results obtained by central difference approximations. These early, but not fully satisfactory, developments were quickly followed by more convincing finite element procedures, such as the Streamline-Upwind Petrov–Galerkin (SUPG) and the Galerkin/Least-squares (GLS) methods. Such methods do enjoy interesting stability and consistency properties and are nowadays widely used by the finite element community for solving convection-dominated transport problems.

After presenting the difficulties and remedies of Galerkin finite elements in steady convection-dominated problems, transient problems are introduced. In fact, Chapter 3 is devoted to pure convection. The scalar linear first-order hyperbolic equation allows discussion of time-dependent situations. In these problems, the objective has been to develop spatially stable and time-accurate finite element methods that take into account the role of the characteristics in the wave-like solution of hyperbolic equations. This has favored the development of solution algorithms in which attention is focused on achieving a proper coupling between the spatial discretization provided by the finite element method and the time discretization. Methods in this class include various characteristic Galerkin techniques, some classical time-stepping schemes and Taylor–Galerkin methods. The concept of accuracy and numerical stability is introduced. Moreover, spatial formulations especially suited for hyperbolic problems (least-squares and discontinuous Galerkin) are also introduced. A simple model problem also motivates a brief introduction to more recent techniques such as space–time formulations.

Engineering practice goes beyond linear scalar equations. Chapter 4 extends the concepts of the previous chapter to systems of nonlinear equations. In fact, it is concerned with a particular problem: the finite element modeling of inviscid compressible flows governed by the Euler equations of gas dynamics. Moreover, this extension allows discussion of the specificities of numerical methods to capture shocks. First, a brief review of the basic mathematical properties of nonlinear hyperbolic equations is presented. Second, a simple two-step procedure is introduced for the explicit integration of the governing conservation equations of mass, momentum and energy. It ensures second-order accuracy in the smooth part of the flow and, at the same time, allows easy incorporation of a modulated dissipation to avoid oscillatory results in the vicinity of shocks and other discontinuities in the flow. Then, various high-resolution shock-capturing techniques are described and their implementation in the finite element context is illustrated by several worked examples. The chapter closes with a discussion on the use of the Arbitrary Lagrangian–Eulerian (ALE) description for the finite element simulation of problems involving fluid–structure interaction. Both academic and industrial examples are proposed to illustrate the flexibility of the ALE technique in the modeling of coupled transient dynamic problems.

Once the basis of time integration (hyperbolic equations), Chapters 3 and 4, and spatial stabilization in steady problems (elliptic equations), Chapter 2, have been discussed, both methodologies converge in transient convection–diffusion problems (parabolic equations). Chapter 5 is still concerned with accurate time integration but has to deal with the second-order spatial operator introduced by the diffusion. This allows the incorporation in the transient schemes of the spatial stabilization introduced in Chapter 2, and, moreover, discussion of specific time integration techniques for convection–diffusion problems in order to obtain high-order accurate schemes.

The generalization to nonlinear systems of equations is done in Chapter 6. It provides an introduction to the finite element modeling of incompressible viscous flows governed by the Navier–Stokes equations. And consequently, apart from the numerical difficulties due to the presence of the nonlinear convective term, the incompressibility condition is also a major issue in this chapter. The problem is formulated in the primitive variables, namely velocity and pressure. Mixed and penalty methods in the framework of Stokes and Navier–Stokes equations are introduced. And a brief account is given of stabilization procedures capable of rendering convergent mixed finite element formulations which are unstable in the traditional Galerkin approach. To treat unsteady incompressible flows, a basic fractional-step projection method is introduced and some variants of it are discussed. Emphasis is placed on the treatment of the boundary conditions in each step of the time integration procedure and on the stable treatment of the pressure/incompressibility phase. The chapter closes with applications of the fractional-step method to forced and natural convection problems.

1.3 KINEMATICAL DESCRIPTIONS OF THE FLOW FIELD

In this section and the next one we summarize the continuum mechanics concepts that are needed for the mathematical description of flow problems. Classical references for the basic theory of fluid mechanics are Batchelor (1999), Landau and Lifshitz (1959) and Lamb (1993).

An important consideration when simulating fluid flow problems by any numerical method is the choice of an appropriate *kinematical description* of the flow field. The algorithms of continuum mechanics make use of three distinct types of description of motion: the Lagrangian description, the Eulerian description and the ALE description.

Lagrangian algorithms, in which each individual node of the computational mesh follows the associated material particle during motion, are mainly used in structural mechanics. Classical applications of the Lagrangian description in large deformation problems are the simulation of vehicle crash tests and the modeling of metal forming operations. In these applications, the Lagrangian algorithms are used in combination with both solid and structural (beam, plate, shell) elements. Numerical solutions are often characterized by large displacements and deformations and history-dependent constitutive relations are employed to describe elasto-plastic and visco-plastic material behavior. The Lagrangian description allows easy tracking of free surfaces and interfaces between different materials. Its weakness is its inability to follow large distortions of the computational domain without recourse to frequent remeshing operations.

Eulerian algorithms are widely used in fluid mechanics. Here, the computational mesh is fixed and the fluid moves with respect to the grid. The Eulerian formulation facilitates the treatment of large distortions in the fluid motion and is indispensable for

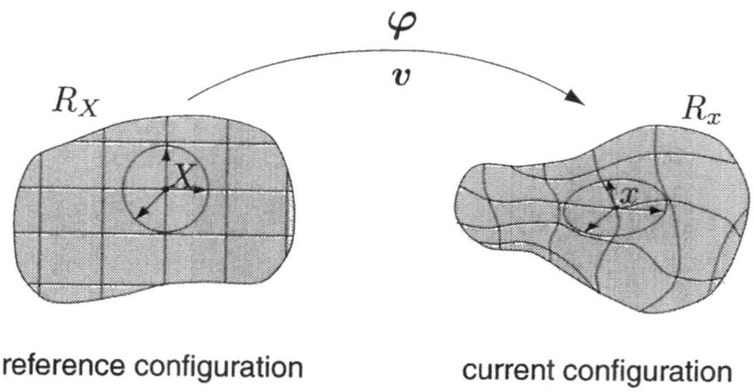

Fig. 1.1 Lagrangian description of motion.

the simulation of turbulent flows. Its handicap is the difficulty to follow free surfaces and interfaces between different materials or different media (e.g., fluid–fluid and fluid–solid interfaces).

ALE algorithms are particularly useful in flow problems involving large distortions in the presence of mobile and deforming boundaries. Typical examples are problems describing the interaction between a fluid and a flexible structure and the simulation of metal forming processes. The key idea in the ALE formulation is the introduction of a computational mesh which can move with a velocity independent of the velocity of the material particles. With this additional freedom with respect to the Eulerian and Lagrangian descriptions, the ALE method succeeds to a certain extend in minimizing the problems encountered in the classical kinematical descriptions, while combining at best their respective advantages.

1.3.1 Lagrangian and Eulerian descriptions

Two domains are commonly used in continuum mechanics: the material domain $R_{\boldsymbol{X}} \subset \mathbb{R}^{\mathrm{n_{sd}}}$, with $\mathrm{n_{sd}}$ spatial dimensions, made up of material particles $\boldsymbol{X}$, and the spatial domain $R_{\boldsymbol{x}}$, consisting of spatial points $\boldsymbol{x}$.

The Lagrangian viewpoint consists of following the material particles of the continuum in their motion. To this end, one introduces, as suggested in Figure 1.1, a computational grid which follows the continuum in its motion, the grid nodes being permanently connected to the same material points. The material coordinates, $\boldsymbol{X}$, allow us to identify the reference configuration, $R_{\boldsymbol{X}}$.

In the *total Lagrangian formulation*, $R_{\boldsymbol{X}}$ is considered fixed and it corresponds usually to the configuration of the continuum at the initial time. In the *updated Lagrangian formulation*, the reference configuration changes during the calculation and generally corresponds to the configuration relative to the previous time (or load) step.

The motion of the material points relates the material coordinates, $\boldsymbol{X}$, to the spatial ones, $\boldsymbol{x}$. It is defined by an application φ such that

$$\begin{aligned} \varphi : R_{\boldsymbol{X}} \times [t_0, t_{\text{final}}[&\longrightarrow R_{\boldsymbol{x}} \times [t_0, t_{\text{final}}[\\ (\boldsymbol{X}, t) &\longmapsto \varphi(\boldsymbol{X}, t) = (\boldsymbol{x}, t), \end{aligned} \tag{1.1}$$

which allows us to link $\boldsymbol{X}$ and $\boldsymbol{x}$ during time by the law of motion, namely

$$\boldsymbol{x} = \boldsymbol{x}(\boldsymbol{X}, t), \qquad t = t,$$

which explicitly states the particular nature of φ: first, the spatial coordinates $\boldsymbol{x}$ depend on both the material particle, $\boldsymbol{X}$, and time t, and, second, physical time is measured by the same variable t in both material and spatial domains. For every fixed instant t, the mapping φ defines a configuration in the spatial domain. It is convenient to employ a matrix representation for the gradient of φ,

$$\frac{\partial \varphi}{\partial(\boldsymbol{X}, t)} = \begin{pmatrix} \dfrac{\partial \boldsymbol{x}}{\partial \boldsymbol{X}} & \boldsymbol{v} \\ \boldsymbol{0}^T & 1 \end{pmatrix},$$

where $\boldsymbol{0}^T$ is a null row vector and the material velocity $\boldsymbol{v}$ is

$$\boldsymbol{v}(\boldsymbol{X}, t) = \left.\frac{\partial \boldsymbol{x}}{\partial t}\right|_{\boldsymbol{X}}, \tag{1.2}$$

with $\left.\right|_{\boldsymbol{X}}$ meaning "holding $\boldsymbol{X}$ fixed".

Obviously, the one-to-one mapping φ must verify $\det(\partial \boldsymbol{x}/\partial \boldsymbol{X}) > 0$ (non-zero to impose a one-to-one correspondence and positive to avoid change of orientation in the reference axes) at each point $\boldsymbol{X}$ and instant $t > t_0$. This allows us to keep track of the history of motion and, by the inverse transformation $(\boldsymbol{X}, t) = \varphi^{-1}(\boldsymbol{x}, t)$, to identify at any instant the initial position of the material particle occupying position $\boldsymbol{x}$ at time t.

Since the material points coincide with the same grid points during the whole motion, there are no convective effects in Lagrangian calculations: the material derivative reduces to a simple time derivative. The fact that each finite element of a Lagrangian mesh always contains the same material particles represents a significant advantage from the computational viewpoint, especially in problems involving materials with history-dependent behavior. These concepts are discussed in detail by Bonet and Wood (1997) in their excellent textbook on nonlinear continuum mechanics for finite element analysis. However, when large material deformations do occur, for instance vortices in fluids, Lagrangian algorithms undergo a loss in accuracy, and may even be unable to conclude a calculation, due to excessive distortions of the computational mesh linked to the material.

The difficulties caused by an excessive distortion of the finite element grid are overcome in the Eulerian formulation. The basic idea in the Eulerian formulation, which is very popular in fluid mechanics, consists in examining as time evolves the physical quantities associated with the fluid particles passing through a fixed region of space. In an Eulerian description the finite element mesh is thus fixed and the continuum moves and deforms with respect to the computational grid. The conservation equations are formulated in terms of the spatial coordinates $\boldsymbol{x}$ and the time t. Therefore, the Eulerian description of motion only involves variables and functions having an instantaneous significance in a fixed region of space. The material velocity $\boldsymbol{v}$ at a given mesh node corresponds to the velocity of the material point coincident at the considered time t with the considered node. The velocity $\boldsymbol{v}$ is consequently expressed with respect to the fixed element mesh without any reference to the initial configuration of the continuum and the material coordinates $\boldsymbol{X}$:

$$\boldsymbol{v} = \boldsymbol{v}(\boldsymbol{x}, t).$$

Since the Eulerian formulation dissociates the mesh nodes from the material particles, convective effects appear due to the relative motion between the deforming material and the computational grid, see Remark 1.1. As will be seen in Chapter 2, this presents numerical difficulties, but permits an easy treatment of complex material motion. By contrast with the Lagrangian description, serious difficulties are now found in following deforming material interfaces and mobile boundaries.

Remark 1.1 (Material and spatial time derivatives). In order to relate the time derivative in the material and spatial domain let a scalar physical quantity be described by $f(\boldsymbol{x}, t)$ and $f^{**}(\boldsymbol{X}, t)$ in the spatial and material domains, respectively. Asterisks are employed to emphasize that the functional forms are, in general, different. Since the particle motion φ is a mapping, $f(\boldsymbol{x}, t)$ and $f^{**}(\boldsymbol{X}, t)$ can be related as

$$f^{**}(\boldsymbol{X}, t) = f\big(\varphi(\boldsymbol{X}, t), t\big) \quad \text{or} \quad f^{**} = f \circ \varphi.$$

The gradient of this expression can be easily computed as

$$\frac{\partial f^{**}}{\partial(\boldsymbol{X}, t)}(\boldsymbol{X}, t) = \frac{\partial f}{\partial(\boldsymbol{x}, t)}(\boldsymbol{x}, t) \quad \frac{\partial \varphi}{\partial(\boldsymbol{X}, t)}(\boldsymbol{X}, t),$$

which is amenable to the matrix form

$$\begin{pmatrix} \dfrac{\partial f^{**}}{\partial \boldsymbol{X}} & \dfrac{\partial f^{**}}{\partial t} \end{pmatrix} = \begin{pmatrix} \dfrac{\partial f}{\partial \boldsymbol{x}} & \dfrac{\partial f}{\partial t} \end{pmatrix} \begin{pmatrix} \dfrac{\partial \boldsymbol{x}}{\partial \boldsymbol{X}} & \boldsymbol{v} \\ \boldsymbol{0}^T & 1 \end{pmatrix},$$

which renders, after block multiplication, an obvious first expression, that is $(\partial f^{**}/\partial \boldsymbol{X}) = (\partial f/\partial \boldsymbol{x})(\partial \boldsymbol{x}/\partial \boldsymbol{X})$; however, the second one is more interesting:

$$\frac{\partial f^{**}}{\partial t} = \frac{\partial f}{\partial t} + \frac{\partial f}{\partial \boldsymbol{x}} \boldsymbol{v}.$$

This is the well-known equation that relates the material and the spatial time derivatives. Dropping the asterisks to ease the notation, this relation is finally cast as

$$\left.\frac{\partial f}{\partial t}\right|_{\boldsymbol{X}} = \left.\frac{\partial f}{\partial t}\right|_{\boldsymbol{x}} + \boldsymbol{v} \cdot \boldsymbol{\nabla} f, \qquad \text{or} \qquad \frac{df}{dt} = \frac{\partial f}{\partial t} + \boldsymbol{v} \cdot \boldsymbol{\nabla} f, \tag{1.3}$$

which can be interpreted in the usual way: the variation of a physical quantity for a given particle $\boldsymbol{X}$ is the local variation plus a convective term taking into account the relative motion between the material and the spatial (laboratory) system. Moreover, in order not to overload the rest of the text with notation, except for specific sections, the material time derivative is denoted as

$$\frac{d\,\cdot}{dt} := \left.\frac{\partial\,\cdot}{\partial t}\right|_{\boldsymbol{X}},$$

and the spatial time derivative as

$$\frac{\partial\,\cdot}{\partial t} := \left.\frac{\partial\,\cdot}{\partial t}\right|_{\boldsymbol{x}}.$$

1.3.2 ALE description of motion

The above review of the classical Lagrangian and Eulerian descriptions has highlighted the advantages and drawbacks of each individual formulation. It has also shown the potential interest of a generalized description capable of combining at best the interesting aspects of the classical mesh descriptions, while minimizing as far as possible their drawbacks. Such a generalized description is termed an ALE description. ALE methods were first proposed in the finite difference context where original developments were made, among others, by Noh (1964), Trulio (1966) and Hirt, Amsden and Cook (1974); this last contribution was reprinted in 1997. The method was subsequently adopted in the finite element context and early applications are to be found in the work of Donea, Fasoli-Stella and Giuliani (1977), Belytschko, Kennedy and Schoeberle (1978), Belytschko and Kennedy (1978) and Hughes, Liu and Zimmermann (1978).

In the ALE description of motion, neither the material $R_{\boldsymbol{X}}$ nor the spatial $R_{\boldsymbol{x}}$ configuration is taken as the reference. Thus, a third domain is needed: the referential configuration $R_{\boldsymbol{\chi}}$ where reference coordinates $\boldsymbol{\chi}$ are introduced to identify the grid points. Figure 1.2 shows these domains and the one-to-one transformations relating the configurations. The referential domain $R_{\boldsymbol{\chi}}$ is mapped into the material and spatial domains by $\boldsymbol{\Psi}$ and $\boldsymbol{\Phi}$ respectively. The particle motion $\boldsymbol{\varphi}$ may then be expressed as $\boldsymbol{\varphi} = \boldsymbol{\Phi} \circ \boldsymbol{\Psi}^{-1}$, clearly showing that, of course, the three mappings $\boldsymbol{\Psi}$, $\boldsymbol{\Phi}$ and $\boldsymbol{\varphi}$ are not independent.

The mapping $\boldsymbol{\Phi}$ from the referential domain to the spatial domain, which can be understood as the motion of the grid points in the spatial domain, is represented by

$$\begin{aligned} \boldsymbol{\Phi} : R_{\boldsymbol{\chi}} \times [t_0, t_{\text{final}}[\; &\longrightarrow R_{\boldsymbol{x}} \times [t_0, t_{\text{final}}[\\ (\boldsymbol{\chi}, t) \; &\longmapsto \boldsymbol{\Phi}(\boldsymbol{\chi}, t) = (\boldsymbol{x}, t), \end{aligned} \tag{1.4}$$

and its gradient is

$$\frac{\partial \boldsymbol{\Phi}}{\partial(\boldsymbol{\chi}, t)} = \begin{pmatrix} \dfrac{\partial \boldsymbol{x}}{\partial \boldsymbol{\chi}} & \hat{\boldsymbol{v}} \\ \boldsymbol{0}^T & 1 \end{pmatrix},$$

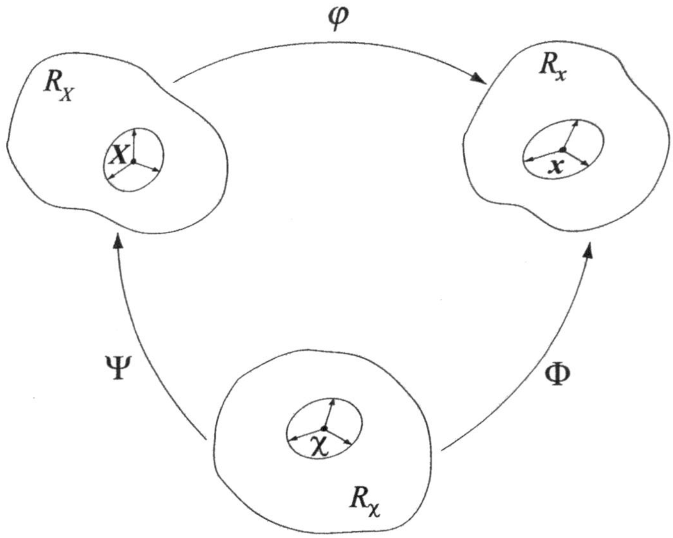

Fig. 1.2 The motion of the ALE computational mesh is independent of the material motion.

where now the mesh velocity $\hat{\boldsymbol{v}}$ is involved,

$$\hat{\boldsymbol{v}}(\boldsymbol{\chi}, t) = \left.\frac{\partial \boldsymbol{x}}{\partial t}\right|_{\boldsymbol{\chi}}. \tag{1.5}$$

Note that both the material and the mesh move with respect to the laboratory. Thus, the corresponding material and mesh velocities have been defined, deriving with respect to time the equations of material motion and mesh motion respectively, see equations (1.2) and (1.5).

Finally, regarding $\boldsymbol{\Psi}$, it is convenient to represent directly its inverse $\boldsymbol{\Psi}^{-1}$,

$$\begin{aligned} \boldsymbol{\Psi}^{-1} : R_{\boldsymbol{X}} \times [t_0, t_{\text{final}}[&\longrightarrow R_{\boldsymbol{\chi}} \times [t_0, t_{\text{final}}[\\ (\boldsymbol{X}, t) &\longmapsto \boldsymbol{\Psi}^{-1}(\boldsymbol{X}, t) = (\boldsymbol{\chi}, t), \end{aligned} \tag{1.6}$$

and its gradient is

$$\frac{\partial \boldsymbol{\Psi}^{-1}}{\partial(\boldsymbol{X}, t)} = \begin{pmatrix} \dfrac{\partial \boldsymbol{\chi}}{\partial \boldsymbol{X}} & \boldsymbol{w} \\ \boldsymbol{0}^T & 1 \end{pmatrix},$$

where velocity $\boldsymbol{w}$ is defined as

$$\boldsymbol{w} = \left.\frac{\partial \boldsymbol{\chi}}{\partial t}\right|_{\boldsymbol{X}} \tag{1.7}$$

and can be interpreted as the particle velocity in the referential domain, since it measures the time variation of the referential coordinate $\boldsymbol{\chi}$ holding the material particle

$\boldsymbol{X}$ fixed. The relation between velocities $\boldsymbol{v}$, $\hat{\boldsymbol{v}}$ and $\boldsymbol{w}$ can be obtained by differentiating $\boldsymbol{\varphi} = \boldsymbol{\Phi} \circ \boldsymbol{\Psi}^{-1}$,

$$\frac{\partial \boldsymbol{\varphi}}{\partial(\boldsymbol{X},t)}(\boldsymbol{X},t) = \frac{\partial \boldsymbol{\Phi}}{\partial(\boldsymbol{\chi},t)}\left(\boldsymbol{\Psi}^{-1}(\boldsymbol{X},t)\right) \quad \frac{\partial \boldsymbol{\Psi}^{-1}}{\partial(\boldsymbol{X},t)}(\boldsymbol{X},t)$$
$$= \frac{\partial \boldsymbol{\Phi}}{\partial(\boldsymbol{\chi},t)}(\boldsymbol{\chi},t) \quad \frac{\partial \boldsymbol{\Psi}^{-1}}{\partial(\boldsymbol{X},t)}(\boldsymbol{X},t)$$

or, in matrix format,

$$\begin{pmatrix} \dfrac{\partial \boldsymbol{x}}{\partial \boldsymbol{X}} & \boldsymbol{v} \\ \boldsymbol{0}^T & 1 \end{pmatrix} = \begin{pmatrix} \dfrac{\partial \boldsymbol{x}}{\partial \boldsymbol{\chi}} & \hat{\boldsymbol{v}} \\ \boldsymbol{0}^T & 1 \end{pmatrix} \begin{pmatrix} \dfrac{\partial \boldsymbol{\chi}}{\partial \boldsymbol{X}} & \boldsymbol{w} \\ \boldsymbol{0}^T & 1 \end{pmatrix},$$

which yields, after block multiplication,

$$\boldsymbol{v} = \hat{\boldsymbol{v}} + \frac{\partial \boldsymbol{x}}{\partial \boldsymbol{\chi}} \boldsymbol{w}.$$

This equation may be rewritten as

$$\boldsymbol{c} := \boldsymbol{v} - \hat{\boldsymbol{v}} = \frac{\partial \boldsymbol{x}}{\partial \boldsymbol{\chi}} \boldsymbol{w}, \tag{1.8}$$

thus defining the convective velocity $\boldsymbol{c}$, that is the relative velocity between the material and the mesh.

Remark 1.2. The convective velocity $\boldsymbol{c}$, see (1.8), should not be confused with $\boldsymbol{w}$, see (1.7). As stated before, $\boldsymbol{w}$ is the particle velocity as seen from the referential domain R_{χ}, whereas $\boldsymbol{c}$ is the particle velocity relative to the mesh as seen from the spatial domain $R_{\boldsymbol{x}}$ (both $\boldsymbol{v}$ and $\hat{\boldsymbol{v}}$ are variations of coordinate $\boldsymbol{x}$). In fact, equation (1.8) implies that $\boldsymbol{c} = \boldsymbol{w}$ if and only if $\partial \boldsymbol{x}/\partial \boldsymbol{\chi} = \boldsymbol{I}$ (where $\boldsymbol{I}$ is the identity tensor); that is, when the mesh motion is purely translational, without rotations or deformations of any kind.

Remark 1.3. After the fundamentals on ALE kinematics have been presented, it should be remarked that both Lagrangian and Eulerian formulations may be obtained as particular cases. With the choice $\boldsymbol{\Psi} = \boldsymbol{I}$, (1.6) reduces to $\boldsymbol{X} \equiv \boldsymbol{\chi}$ and a Lagrangian description results: the material and mesh velocities, equations (1.2) and (1.5), coincide, and the convective velocity $\boldsymbol{c}$, see (1.8), is null (there are no convective terms in the conservation laws).

If, on the other hand, $\boldsymbol{\Phi} = \boldsymbol{I}$, (1.4) simplifies into $\boldsymbol{x} \equiv \boldsymbol{\chi}$, thus implying an Eulerian description: a null mesh velocity is obtained from equation (1.5) and the convective velocity $\boldsymbol{c}$ is simply identical to the material velocity $\boldsymbol{v}$.

Remark 1.4 (Evaluation of the grid velocity). In the ALE formulation, the freedom of moving the mesh is very attractive. It helps to combine the respective advantages of the Lagrangian and Eulerian formulations. This could, however, be overshadowed by the burden of specifying grid velocities well suited to the particular problem under consideration. As a consequence, the practical implementation of the ALE description requires that an automatic mesh displacement prescription algorithm be supplied. In practice, such an algorithm selects the grid velocity to adapt the motion of the computational mesh to the peculiarities of the problem under investigation, trying to minimize both the squeeze and the distortion of the elements. See for instance the ALE mesh displacement algorithm developed by Giuliani (1982). When the ALE description is employed as a tool to adapt a finite element mesh to particular needs, for instance, to properly capture strong solution gradients, an error indicator or an error estimator is normally used to drive the algorithm in charge of computing the mesh velocities, see, for instance, Huerta et al. (1999).

1.3.3 The fundamental ALE equation

In order to express the conservation laws in an ALE framework, a relation between time derivatives is needed. In fact, the so-called *total* (or material) time derivatives, which are inherent to conservation laws, must be related to referential time derivatives. The relation presented in Remark 1.1 must be therefore extended in order to include the referential time derivative. Let a scalar physical quantity be described by $f(\boldsymbol{x}, t)$, $f^*(\boldsymbol{\chi}, t)$ and $f^{**}(\boldsymbol{X}, t)$ in the spatial, referential and material domains respectively. Asterisks are employed to emphasize that the functional forms are, in general, different.

With the help of mapping $\boldsymbol{\Psi}$, the transformation from $f^*(\boldsymbol{\chi}, t)$ to $f^{**}(\boldsymbol{X}, t)$ can be written as

$$f^{**} = f^* \circ \boldsymbol{\Psi}^{-1},$$

and its gradient can be easily computed as

$$\frac{\partial f^{**}}{\partial(\boldsymbol{X}, t)}(\boldsymbol{X}, t) = \frac{\partial f^*}{\partial(\boldsymbol{\chi}, t)}(\boldsymbol{\chi}, t) \quad \frac{\partial \boldsymbol{\Psi}^{-1}}{\partial(\boldsymbol{X}, t)}(\boldsymbol{X}, t),$$

or, in matrix form,

$$\begin{pmatrix} \dfrac{\partial f^{**}}{\partial \boldsymbol{X}} & \dfrac{\partial f^{**}}{\partial t} \end{pmatrix} = \begin{pmatrix} \dfrac{\partial f^*}{\partial \boldsymbol{\chi}} & \dfrac{\partial f^*}{\partial t} \end{pmatrix} \begin{pmatrix} \dfrac{\partial \boldsymbol{\chi}}{\partial \boldsymbol{X}} & \boldsymbol{w} \\ \boldsymbol{0}^T & 1 \end{pmatrix},$$

which renders, after block multiplication,

$$\frac{\partial f^{**}}{\partial t} = \frac{\partial f^*}{\partial t} + \frac{\partial f^*}{\partial \boldsymbol{\chi}} \boldsymbol{w}.$$

Note that this equation relates the material and the referential time derivatives. However, it also requires the evaluation of the gradient in the referential domain. This can be done, but in computational mechanics it is usually easier to work in the spatial (or material) domain. Moreover, in fluids, constitutive relations are naturally expressed in the spatial configuration and the Cauchy stress tensor, which will be introduced next, is the natural measure for stresses. Thus, using the definition of $\boldsymbol{w}$ given in (1.8), the previous equation may be rearranged into

$$\frac{\partial f^{**}}{\partial t} = \frac{\partial f^{*}}{\partial t} + \frac{\partial f}{\partial \boldsymbol{x}}\boldsymbol{c}.$$

Dropping the asterisks to ease the notation, the fundamental ALE relation between material time derivatives, referential time derivatives and spatial gradient is finally cast as

$$\left.\frac{\partial f}{\partial t}\right|_{\boldsymbol{X}} = \left.\frac{\partial f}{\partial t}\right|_{\boldsymbol{\chi}} + \frac{\partial f}{\partial \boldsymbol{x}}\boldsymbol{c} = \left.\frac{\partial f}{\partial t}\right|_{\boldsymbol{\chi}} + \boldsymbol{c}\cdot\boldsymbol{\nabla} f, \tag{1.9}$$

which can be interpreted in the usual way: the variation of the physical quantity f for a given particle $\boldsymbol{X}$ is the local variation (i.e., with respect to the reference $\boldsymbol{\chi}$) plus a convective term taking into account the relative motion between the material and the reference system. This equation is equivalent to (1.3) but in the ALE formulation; that is, when $(\boldsymbol{\chi}, t)$ is the reference.

1.3.4 Time derivative of integrals over moving volumes

To establish the integral form of the basic conservation laws for mass, momentum and energy, we also need to consider the rate of change of integrals of scalar and vector functions over a moving volume occupied by fluid.

Consider therefore a material volume V_t bounded by a smooth closed surface S_t whose points at time t move with the material velocity

$$\boldsymbol{v} = \boldsymbol{v}(\boldsymbol{x}, t),$$

where $\boldsymbol{x} \in S_t$. A material volume is a volume that permanently contains the same particles of the continuum under consideration. The material time derivative of the integral of a scalar function $f(\boldsymbol{x}, t)$ (note that f is defined in the spatial domain) over the time-varying material volume V_t is given by the following well-known expression, often referred to as the *Reynolds transport theorem*:

$$\frac{d}{dt}\int_{V_t} f(\boldsymbol{x}, t)\, dV = \int_{V_c \equiv V_t} \frac{\partial f(\boldsymbol{x}, t)}{\partial t}\, dV + \int_{S_c \equiv S_t} f(\boldsymbol{x}, t)\, \boldsymbol{v}\cdot\boldsymbol{n}\, dS, \tag{1.10}$$

which holds for smooth functions $f(\boldsymbol{x}, t)$. The volume integral on the r.h.s. is defined over a control volume V_c (fixed in space) which coincides with the moving material volume V_t at the considered instant, t, in time. Similarly, the fixed control surface S_c coincides at time t with the closed surface S_t bounding the material volume V_t. In the surface integral, $\boldsymbol{n}$ denotes the unit outward normal to the surface S_t at time t, and $\boldsymbol{v}$ is the material velocity of points of the boundary S_t. The first term on the r.h.s. of (1.10) is the *local time derivative* of the volume integral. The boundary integral represents the flux of the scalar quantity f across the fixed boundary of the control volume $V_c \equiv V_t$. A similar expression holds for the volume integral of a vector or tensor quantity.

An analogous formula can be deduced in the ALE context, that is with a referential time derivative. In this case, however, $\boldsymbol{v}$ is no longer the material velocity, as will be seen in Section 1.4.5, see also Huerta and Liu (1988).

Equation (1.10) constitutes the starting point for the derivation of the ALE integral form of the conservation equations of mass, momentum and energy, which are discussed next.

1.4 THE BASIC CONSERVATION EQUATIONS

To serve as an introduction to the discussion of finite element models for both incompressible and compressible flow problems, we recall in this section the differential and integral forms of the conservation equations for mass, momentum and energy.

1.4.1 Mass equation

A fundamental law of Newtonian mechanics is the conservation of the mass contained in a material volume. The law of mass conservation for a varying material volume V_t occupied by fluid is given by

$$0 = \frac{dM}{dt} = \frac{d}{dt}\int_{V_t} \rho \, dV,$$

where ρ is the fluid density. Applying to this integral expression the formula (1.10) for the rate of change of integrals over a moving volume and the divergence theorem, one obtains

$$0 = \frac{dM}{dt} = \int_{V_t} \frac{\partial \rho}{\partial t}\, dV + \int_{S_t} \rho \boldsymbol{v} \cdot \boldsymbol{n} \, dS = \int_{V_t} \left(\frac{\partial \rho}{\partial t} + \boldsymbol{\nabla} \cdot (\rho \boldsymbol{v}) \right) dV.$$

Since this relation is valid for all choices of the volume V_t, the integrand must be identically zero. Hence

$$\frac{\partial \rho}{\partial t} + \boldsymbol{\nabla} \cdot (\rho \boldsymbol{v}) = 0 \tag{1.11}$$

at all points in the fluid. Equation (1.11) is called the *mass-conservation equation*, or *continuity equation*. A different form of equation (1.11) is obtained by expanding the divergence term and noting that two of the terms together make up the material derivative of the density:

$$\frac{d\rho}{dt} + \rho \boldsymbol{\nabla} \cdot \boldsymbol{v} = 0. \tag{1.12}$$

1.4.2 Momentum equation

The *momentum equation*, also termed the *equation of motion*, is a relation equating the rate of change of momentum of a selected portion of fluid and the sum of all forces acting on that portion of fluid. For the portion of fluid of volume V_t enclosed by the *material surface* S_t, the momentum is $\int_{V_t} \rho \boldsymbol{v}\, dV$ and making use of the Reynolds transport theorem in vector form, that is (1.10), its rate of change is found to be

$$\begin{aligned} \frac{d}{dt}\int_{V_t} \rho \boldsymbol{v}\, dV &= \int_{V_t} \frac{\partial \rho \boldsymbol{v}}{\partial t}\, dV + \int_{S_t} (\rho \boldsymbol{v} \otimes \boldsymbol{v}) \cdot \boldsymbol{n}\, dS \\ &= \int_{V_t} \Big(\frac{\partial \rho \boldsymbol{v}}{\partial t} + \boldsymbol{\nabla} \cdot (\rho \boldsymbol{v} \otimes \boldsymbol{v}) \Big)\, dV, \end{aligned} \tag{1.13}$$

where the notation $\boldsymbol{v} \otimes \boldsymbol{v}$ denotes the tensor $[v_i v_j]$, $i, j = 1, \ldots, \mathrm{n_{sd}}$. Moreover, the i-th component of the divergence of a second-rank tensor $\mathbf{T}$ is

$$[\boldsymbol{\nabla} \cdot \mathbf{T}]_i := \sum_{j=1}^{\mathrm{n_{sd}}} \frac{\partial T_{ij}}{\partial x_j} \qquad \text{for } i = 1, \ldots, \mathrm{n_{sd}}.$$

Making use of the continuity equation (1.11) and expression (1.3) for the material time derivative, equation (1.13) can be transformed to

$$\frac{d}{dt}\int_{V_t} \rho \boldsymbol{v}\, dV = \int_{V_t} \rho \frac{d\boldsymbol{v}}{dt}\, dV, \tag{1.14}$$

which is simply the sum of the product of mass and acceleration for all the elements dV of the material volume V_t.

In general, a portion of fluid is acted upon by both volume and surface forces. We denote by $\boldsymbol{b}$ the volume force per unit mass of fluid, so that the total volume force on the selected portion of fluid is $\int_{V_t} \rho \boldsymbol{b}\, dV$. On the other hand, the i-component of the surface force exerted across a surface element of area dS and normal $\boldsymbol{n}$ is given by $\sigma_{ij}\, n_j\, dS$, so that the total force exerted on the selected portion of fluid by the surrounding matter can be expressed in terms of the Cauchy stress $\boldsymbol{\sigma}$ as

$$\int_{S_t} \sigma_{ij}\, n_j\, dS = \int_{V_t} \frac{\partial \sigma_{ij}}{\partial x_j} dV \qquad \text{or} \qquad \int_{S_t} \boldsymbol{\sigma} \cdot \boldsymbol{n}\, dS = \int_{V_t} \boldsymbol{\nabla} \cdot \boldsymbol{\sigma} dV.$$

In the previous expression use has been made of the summation convention on repeated indices. It will be used when the limits of the summation are clear as in this case (from 1 to $\mathrm{n_{sd}}$). However, in this introductory chapter the summation will be used more than strictly necessary in order to preclude misinterpretation.

The momentum balance for the selected material volume of fluid, which accounts for both previous actions, is expressed by

$$\int_{V_t} \rho \frac{d\boldsymbol{v}}{dt} dV = \int_{V_t} \rho \boldsymbol{b}\, dV + \int_{V_t} \boldsymbol{\nabla} \cdot \boldsymbol{\sigma} dV.$$

This integral relation holds for all choices of the material volume V_t. Thus,

$$\rho \frac{d\boldsymbol{v}}{dt} = \rho \boldsymbol{b} + \boldsymbol{\nabla} \cdot \boldsymbol{\sigma} \tag{1.15}$$

at all points of the fluid. This equation is termed the *equation of motion*; making use of (1.3) it becomes

$$\rho \frac{\partial \boldsymbol{v}}{\partial t} + \rho(\boldsymbol{v} \cdot \boldsymbol{\nabla})\boldsymbol{v} = \rho \boldsymbol{b} + \boldsymbol{\nabla} \cdot \boldsymbol{\sigma}, \quad \text{or} \quad \frac{\partial \rho \boldsymbol{v}}{\partial t} = \rho \boldsymbol{b} + \boldsymbol{\nabla} \cdot (\boldsymbol{\sigma} - \rho\, \boldsymbol{v} \otimes \boldsymbol{v}). \quad (1.16)$$

This last expression is the conservation form of the momentum equation.

Substituting equation (1.15) into (1.14) yields the following form of the rate of change of momentum for a material volume:

$$\frac{d}{dt} \int_{V_t} \rho \boldsymbol{v}\, dV = \int_{V_t} \rho \frac{d\boldsymbol{v}}{dt}\, dV = \int_{V_t} (\rho \boldsymbol{b} + \boldsymbol{\nabla} \cdot \boldsymbol{\sigma})\, dV.$$

For a fluid volume V (usually called control volume) whose position is *fixed* relative to the coordinate axes (i.e., fixed in the spatial domain), the rate of change of momentum is instead given from (1.16) by

$$\frac{\partial}{\partial t} \int_V \rho \boldsymbol{v}\, dV = \int_V \frac{\partial \rho \boldsymbol{v}}{\partial t} dV = \int_V (\rho \boldsymbol{b} + \boldsymbol{\nabla} \cdot \boldsymbol{\sigma})\, dV - \int_S \rho \boldsymbol{v}\, (\boldsymbol{v} \cdot \boldsymbol{n}) dS.$$

1.4.3 Internal energy equation

In incompressible flow problems, the energy equation is not required to solve the momentum and continuity equations. Its form is given here in view of its use in Chapter 4, where the finite element modeling of compressible flow problems is addressed.

Consider therefore the energy balance for the fluid of volume V_t contained within a *material surface* S_t. Work is being done on this mass of fluid by both volume and surface forces, and it may also be gaining energy by heat conduction across the boundary. Some of this total gain of energy is manifested as an increase in the kinetic energy of the fluid, and the remainder, according to the first law of thermodynamics, appears as an increase in the internal energy of the fluid.

The first law of thermodynamics states that the increase of the internal energy per unit mass of the fluid, Δe, is the sum

$$\Delta e = q + w,$$

where q denotes the gain of heat per unit mass and w represents the work performed on the fluid per unit mass. Herein, we shall assume that the fluid is thermally isolated, so that no exchange of heat can occur, $q = 0$.

The rate at which work is being done on the fluid in the material volume V_t is the sum of a contribution

$$\int_{V_t} \boldsymbol{v} \cdot \rho \boldsymbol{b}\, dV = \int_{V_t} v_i\, \rho b_i\, dV$$

from the applied body force, and a contribution

$$\int_{S_t} \boldsymbol{v} \cdot (\boldsymbol{\sigma} \cdot \boldsymbol{n}) dS = \int_{S_t} v_i\, \sigma_{ij}\, n_j\, dS = \int_{V_t} \frac{\partial (v_i\, \sigma_{ij})}{\partial x_j} dV = \int_{V_t} \boldsymbol{\nabla} \cdot (\boldsymbol{\sigma} \cdot \boldsymbol{v}) dV$$

from the surface forces exerted by the surrounding fluid. Thus from these two expressions the total rate of work on an arbitrary material element is

$$\boldsymbol{v} \cdot \boldsymbol{b} + \frac{\boldsymbol{v}}{\rho} \cdot (\boldsymbol{\nabla} \cdot \boldsymbol{\sigma}) + \frac{\boldsymbol{\sigma}}{\rho} : \boldsymbol{\nabla v} = v_i\, b_i + \frac{v_i}{\rho} \frac{\partial \sigma_{ij}}{\partial x_j} + \frac{\sigma_{ij}}{\rho} \frac{\partial v_i}{\partial x_j}$$

per unit mass of fluid. Using the equation of motion (1.15), we may rewrite the rate of work per unit mass of fluid as

$$\boldsymbol{v} \cdot \frac{d\boldsymbol{v}}{dt} + \frac{\boldsymbol{\sigma}}{\rho} : \boldsymbol{\nabla v} = v_i \frac{dv_i}{dt} + \frac{\sigma_{ij}}{\rho} \frac{\partial v_i}{\partial x_j}.$$

The first term is clearly related to the gain in kinetic energy, while the second term represents the rate of work w done in deforming the fluid element. In the absence of heat transfer effects, the rate of change of internal energy per unit mass of a material element of fluid is thus given by

$$\frac{de}{dt} = \frac{\sigma_{ij}}{\rho} \frac{\partial v_i}{\partial x_j} = \frac{\boldsymbol{\sigma}}{\rho} : \boldsymbol{\nabla v}. \tag{1.17}$$

For an *inviscid fluid* the stress reduces to the static pressure (constitutive law),

$$\sigma_{ij} = -p\, \delta_{ij}\, , \quad \text{or } \boldsymbol{\sigma} = -p\boldsymbol{I},$$

where $\boldsymbol{I}$ is the identity tensor (same dimensions as $\boldsymbol{\sigma}$ in this case), and the *internal energy equation* then takes the form

$$\rho \frac{de}{dt} + p\, \boldsymbol{\nabla} \cdot \boldsymbol{v} = 0, \tag{1.18}$$

or, making use of (1.12),

$$\frac{d\rho e}{dt} + (p + \rho e) \boldsymbol{\nabla} \cdot \boldsymbol{v} = 0.$$

In terms of the partial time derivative, using (1.3), the internal energy equation reads

$$\frac{\partial \rho e}{\partial t} + \boldsymbol{\nabla} \cdot (\rho e\, \boldsymbol{v}) + p \boldsymbol{\nabla} \cdot \boldsymbol{v} = 0. \tag{1.19}$$

In a spatial representation, the rate of change of the internal energy for a material volume of an inviscid fluid is then given from the Reynolds transport theorem, (1.10),

$$\frac{d}{dt} \int_{V_t} \rho e\, dV = \int_{V_t} \frac{\partial \rho e}{\partial t} dV + \int_{S_t} \rho e (\boldsymbol{v} \cdot \boldsymbol{n}) dS = \int_{V_t} \frac{\partial \rho e}{\partial t} dV + \int_{V_t} \boldsymbol{\nabla} \cdot (\rho e \boldsymbol{v}) dV.$$

Using (1.19) in the previous equation yields

$$\frac{d}{dt} \int_{V_t} \rho e\, dV = -\int_{V_t} p\, \boldsymbol{\nabla} \cdot \boldsymbol{v}\, dV.$$

For a control fluid volume whose position is *fixed* relative to the coordinate axes (in the spatial domain), the rate of change of internal energy is instead given in the form

$$\frac{\partial}{\partial t} \int_{V} \rho e\, dV = \int_{V} \frac{\partial \rho e}{\partial t} dV = -\int_{V} p\, \boldsymbol{\nabla} \cdot \boldsymbol{v}\, dV - \int_{S} \rho e (\boldsymbol{v} \cdot \boldsymbol{n}) dS.$$

1.4.4 Total energy equation

As shown by the structure of equation (1.19), the internal energy equation is not a conservation equation of the type

$$\frac{\partial q}{\partial t} + \nabla \cdot \boldsymbol{f} = 0.$$

In fact, the conserved quantity is the total energy. The total energy per unit mass of the fluid is the sum

$$E = e + \frac{1}{2}\|\boldsymbol{v}\|^2$$

of the internal energy, e, and the kinetic energy. The rate of change of the total energy per unit mass of a material element is thus

$$\frac{dE}{dt} = \frac{de}{dt} + \boldsymbol{v} \cdot \frac{d\boldsymbol{v}}{dt}. \tag{1.20}$$

For the particular case of an inviscid fluid, making use of the momentum equation (1.15) and of the internal energy equation (1.18), it becomes

$$\rho \frac{dE}{dt} = -\nabla \cdot (p\boldsymbol{v}) + \boldsymbol{v} \cdot \rho \boldsymbol{b}$$

or

$$\frac{d}{dt}(\rho E) = \rho \frac{dE}{dt} + E\frac{d\rho}{dt} = -(p + \rho E)\nabla \cdot \boldsymbol{v} + \boldsymbol{v} \cdot (\rho \boldsymbol{b} - \nabla p).$$

In terms of the partial time derivative, the total energy equation reads

$$\frac{\partial}{\partial t}(\rho E) = -\nabla \cdot \big((\rho E + p)\boldsymbol{v}\big) + \boldsymbol{v} \cdot \rho \boldsymbol{b}. \tag{1.21}$$

The rate of change of the total energy for a material volume is then given from (1.10) and (1.21) by

$$\begin{aligned}\frac{d}{dt}\int_{V_t} \rho E \, dV &= \int_{V_t} \frac{\partial}{\partial t}(\rho E)\, dV + \int_{S_t} \rho E\, \boldsymbol{v} \cdot \boldsymbol{n}\, dS \\ &= \int_{V_t} \boldsymbol{v} \cdot \rho \boldsymbol{b}\, dV - \int_{S_t} p\boldsymbol{v} \cdot \boldsymbol{n}\, dS.\end{aligned}$$

For a control fluid volume whose position is *fixed* relative to the coordinate axes, the rate of change of the total energy is also obtained using (1.21)

$$\frac{\partial}{\partial t}\int_V \rho E dV = \int_V \boldsymbol{v} \cdot \rho \boldsymbol{b}\, dV - \int_S (\rho E + p)\boldsymbol{v} \cdot \boldsymbol{n} dS = \int_V \boldsymbol{v} \cdot \rho \boldsymbol{b}\, dV - \int_S \rho H \boldsymbol{v} \cdot \boldsymbol{n} dS,$$

where

$$H = E + \frac{p}{\rho}$$

is the total specific enthalpy of the fluid.

1.4.5 ALE form of the conservation equations

The ALE differential form of the conservation equations is readily obtained from the corresponding material forms (1.12), (1.15), (1.17) and (1.20). The total derivatives are rewritten using (1.9) and the result is

$$\begin{aligned}
&\text{Mass:} && \left.\frac{\partial \rho}{\partial t}\right|_{\chi} + \boldsymbol{c}\cdot\nabla\rho = -\rho\,\nabla\cdot\boldsymbol{v},\\
&\text{Momentum:} && \rho\Big(\left.\frac{\partial \boldsymbol{v}}{\partial t}\right|_{\chi} + (\boldsymbol{c}\cdot\nabla)\boldsymbol{v}\Big) = \nabla\cdot\boldsymbol{\sigma} + \rho\boldsymbol{b},\\
&\text{Internal energy:} && \rho\Big(\left.\frac{\partial e}{\partial t}\right|_{\chi} + \boldsymbol{c}\cdot\nabla e\Big) = \boldsymbol{\sigma}:\nabla\cdot\boldsymbol{v},\\
&\text{Total energy:} && \rho\Big(\left.\frac{\partial E}{\partial t}\right|_{\chi} + \boldsymbol{c}\cdot\nabla E\Big) = \nabla\cdot(\boldsymbol{\sigma}\cdot\boldsymbol{v}) + \boldsymbol{v}\cdot\rho\boldsymbol{b}.
\end{aligned}\tag{1.22}$$

The starting point for deriving the ALE integral form of the conservation equations is expression (1.10) applied to an *arbitrary* volume V_t whose boundary $S_t = \partial V_t$ moves with the mesh velocity $\hat{\boldsymbol{v}}$:

$$\left.\frac{\partial}{\partial t}\right|_{\chi} \int_{V_t} f(\boldsymbol{x},t)\,dV = \int_{V_t} \left.\frac{\partial f(\boldsymbol{x},t)}{\partial t}\right|_{\boldsymbol{x}} dV + \int_{S_t} f(\boldsymbol{x},t)\,\hat{\boldsymbol{v}}\cdot\boldsymbol{n}\,dS,$$

where, in this case, we have explicitly indicated that the time derivative in the first term of the r.h.s. is a spatial time derivative, as in (1.10). We then successively replace the scalar $f(\boldsymbol{x},t)$ by the fluid density ρ, momentum $\rho\boldsymbol{v}$, internal energy ρe and total energy ρE. Similarly, the spatial time derivative $\partial f/\partial t$ is substituted by the correponding expressions for the mass equation, see (1.11), for the momentum equation, see (1.16), for the internal energy and for the total energy. The end result is the following set of ALE integral forms:

$$\begin{aligned}
&\left.\frac{\partial}{\partial t}\right|_{\chi} \int_{V_t} \rho\,dV + \int_{S_t} \rho\,\boldsymbol{c}\cdot\boldsymbol{n}\,dS = 0,\\
&\left.\frac{\partial}{\partial t}\right|_{\chi} \int_{V_t} \rho\boldsymbol{v}\,dV + \int_{S_t} \rho\boldsymbol{v}\,\boldsymbol{c}\cdot\boldsymbol{n}\,dS = \int_{V_t} (\nabla\cdot\boldsymbol{\sigma} + \rho\boldsymbol{b})\,dV,\\
&\left.\frac{\partial}{\partial t}\right|_{\chi} \int_{V_t} \rho e\,dV + \int_{S_t} \rho e\,\boldsymbol{c}\cdot\boldsymbol{n}\,dS = \int_{V_t} \boldsymbol{\sigma}:\nabla\cdot\boldsymbol{v}\,dV,\\
&\left.\frac{\partial}{\partial t}\right|_{\chi} \int_{V_t} \rho E\,dV + \int_{S_t} \rho E\,\boldsymbol{c}\cdot\boldsymbol{n}\,dS = \int_{V_t} \big(\nabla\cdot(\boldsymbol{\sigma}\cdot\boldsymbol{v}) + \boldsymbol{v}\cdot\rho\boldsymbol{b}\big)dV.
\end{aligned}\tag{1.23}$$

Note that the integral forms for the Lagrangian and Eulerian mesh descriptions are contained in the above ALE forms. The Lagrangian description corresponds to selecting $\hat{\boldsymbol{v}} = \boldsymbol{v}$ (i.e., $\boldsymbol{c} = \boldsymbol{0}$), while the Eulerian description corresponds to selecting $\hat{\boldsymbol{v}} = \boldsymbol{0}$ (i.e., $\boldsymbol{c} = \boldsymbol{v}$).

The ALE differential and integral forms of the conservation equations derived in the present section will be used in Chapter 4 in connection with the finite element modeling of coupled fluid–structure interaction problems.

1.4.6 Closure of the initial boundary value problem

Once the governing equations are defined, boundary and initial conditions (the latter only in transient problems) must be adequately prescribed in order to close the problem. Three types of boundary conditions are used in this text: Dirichlet, Neumann and Robin. The last are also called mixed boundary conditions.

Dirichlet boundary conditions prescribe the value of the unknown function. Neumann conditions impose the normal gradient of the unknown function along the boundary. Robin boundary conditions prescribe a combination of the unknown function and its gradient.

Not all boundary conditions can be applied arbitrarily everywhere along the boundary of the domain. For instance, in hyperbolic problems Dirichlet boundary conditions cannot be imposed on the whole boundary, see Chapters 3 and 4. In every chapter we shall present the model problem to be studied, and in each case boundary conditions shall be presented and discussed in more detail.

1.5 BASIC INGREDIENTS OF THE FINITE ELEMENT METHOD

The application of the finite element method for solving a boundary value problem requires a certain number of basic ingredients that we shall briefly recall in this section on the basis of a simple model problem. Before that, we wish to introduce some notations and mathematical terminology used throughout the book. Standard finite element texts, such as, for instance, Strang and Fix (1973), Oden and Reddy (1976), Carey and Oden (1983), Temam (2001), Girault and Raviart (1986), Hughes (2000), Gunzburger (1989), Pironneau (1989) and Quarteroni and Valli (1994) provide a detailed exposition of the mathematical concepts, which are the basis of the finite element method.

1.5.1 Mathematical preliminaries

The process of spatial discretization by the finite element method rests upon the discrete representation of a weak integral form of the partial differential equation to be solved. The formulation and subsequent discretization of such an integral form requires the definition of some function spaces and associated norms, as well as the introduction of compact forms involving the functions pertaining to those spaces.

Consider a spatial region (or domain) $\Omega \subset \mathbb{R}^{n_{sd}}$ with piecewise smooth boundary Γ. Here again, $n_{sd} = 1, 2$ or 3 denotes the number of space dimensions. We shall use the notation

$$f : \bar{\Omega} \to \mathbb{R}$$

to state that for each spatial point $x \in \bar{\Omega}$, $f(x) \in \mathbb{R}$. $\bar{\Omega}$ denotes the closure of Ω, that is the union of the domain Ω with its boundary Γ: $\bar{\Omega} = \Omega \cup \Gamma$.

A function $f : \bar{\Omega} \to \mathbb{R}$ is said to be of class $C^m(\Omega)$ if all its derivatives up to order m exist and are continuous functions. For instance, the notation $f(x) \in C^m(]a, b[)$ indicates that $f(x)$ possesses m continuous derivatives for $x \in]a, b[$.

In the finite element analysis we work with integral equations and, thus, we are interested in functions belonging to larger spaces than C^m. As we will see, instead of requiring the m-th derivative to be continuous, we will require that its square is integrable. In fact, finite element functions should possess generalized derivatives (i.e., derivatives in the sense of distributions) and some integrability properties. Such classes of functions are particular examples of Sobolev function spaces. A detailed account concerning Sobolev spaces may be found in the book by Adams (1975).

1.5.1.1 Some useful Sobolev spaces We shall denote by $\mathcal{L}_2(\Omega)$ the space of functions that are square integrable over the domain Ω. This space is equipped with the standard inner product

$$(u,v) = \int_\Omega uv\,d\Omega \quad \text{and norm} \quad \|v\|_0 = (v,v)^{1/2}.$$

Next we describe a particular class of Sobolev spaces, those of square integrable functions and derivatives. This restriction suffices for our purposes throughout the book. For any non-negative integer k, we define the Sobolev space $\mathcal{H}^k(\Omega)$ using multi-index notation: given the n-tuple $\alpha = (\alpha_1, \alpha_2, \ldots, \alpha_{n_{sd}}) \in \mathbb{N}^{n_{sd}}$ and the non-negative integer $|\alpha| := \alpha_1 + \alpha_2 + \cdots + \alpha_{n_{sd}}$,

$$\mathcal{H}^k(\Omega) = \left\{ u \in \mathcal{L}_2(\Omega) \middle| \frac{\partial^{|\alpha|} u}{\partial x_1^{\alpha_1} \partial x_2^{\alpha_2} \cdots \partial x_{n_{sd}}^{\alpha_{n_{sd}}}} \in \mathcal{L}_2(\Omega)\, \forall |\alpha| \le k \right\}.$$

Therefore, $\mathcal{H}^k(\Omega)$ consists of square integrable functions all of whose derivatives of order up to k are also square integrable. $\mathcal{H}^k(\Omega)$ is equipped with the norm

$$\|u\|_k = \left(\sum_{s=0}^{k} \sum_{|\alpha|=s} \left\| \frac{\partial^{|\alpha|} u}{\partial x_1^{\alpha_1} \partial x_2^{\alpha_2} \cdots \partial x_{n_{sd}}^{\alpha_{n_{sd}}}} \right\|_0^2 \right)^{1/2}.$$

Note that $\mathcal{L}_2$ is, in fact, a Sobolev space, $\mathcal{H}^0(\Omega) = \mathcal{L}_2(\Omega)$, while the Sobolev space for $k = 1$ is defined by

$$\mathcal{H}^1(\Omega) = \left\{ v \in \mathcal{L}_2(\Omega) \mid \frac{\partial v}{\partial x_i} \in \mathcal{L}_2(\Omega)\; i = 1, \ldots, n_{sd} \right\}.$$

This space is equipped with the inner product

$$(u,v)_1 = \int_\Omega \left(uv + \sum_{i=1}^{n_{sd}} \frac{\partial u}{\partial x_i} \frac{\partial v}{\partial x_i} \right) d\Omega$$

and its induced norm

$$\|u\|_1 = \sqrt{(u,u)_1}.$$

We shall also frequently use the subspace

$$\mathcal{H}_0^1(\Omega) = \left\{ v \in \mathcal{H}^1(\Omega) \mid v = 0 \text{ on } \Gamma \right\},$$

the elements of which possess a square integrable first derivative over the domain Ω and vanish on its boundary Γ. Moreover, its inner product and norm coincide with those of $\mathcal{H}^1(\Omega)$.

Remark 1.5. Note that the Sobolev spaces used in what follows, namely $\mathcal{H}^0 = \mathcal{L}_2, \mathcal{H}^1$ and $\mathcal{H}_0^1$, are Hilbert spaces with their corresponding inner product (recall that a Hilbert space is a linear space with an inner product in which all Cauchy sequences are convergent sequences).

Remark 1.6. $\mathcal{H}_0^1$ is usually defined as the closure of $C_0^\infty(\Omega)$ (the set of all continuous functions with continuous derivatives whose support is a bounded subset of Ω) with respect to the norm of $||\cdot||_1$. That is, $\mathcal{H}_0^1(\Omega)$ is the set of all functions u in $\mathcal{H}^1(\Omega)$ such that u is the limit in $\mathcal{H}^1(\Omega)$ of a sequence $\{u_s\}_{s=1}^\infty$ whose u_s are in $C_0^\infty(\Omega)$.

1.5.1.2 Extension to vector-valued functions In the finite element analysis of flow problems consideration will be given not only to scalar functions (such as temperature or pressure) but also to vector-valued functions (such as fluid velocity). For vector-valued functions with m components, that is $\boldsymbol{u}, \boldsymbol{v} : \Omega \to \mathbb{R}^m$, the procedure is in fact essentially the same as for scalar functions.

Consider again a domain $\Omega \subset \mathbb{R}^{\mathrm{n_{sd}}}$, $\mathrm{n_{sd}} \geq 1$, and denote by $\boldsymbol{\mathcal{H}}^k(\Omega)$ or $[\mathcal{H}^k(\Omega)]^m$ the space of vector functions with m components

$$\boldsymbol{u} = (u_1, u_2, \ldots, u_m)$$

for which each component $u_i \in \mathcal{H}^k(\Omega)$, $1 \leq i \leq m$. The space $\boldsymbol{\mathcal{H}}^k(\Omega)$ is equipped with an inner product inducing the following norm

$$||\boldsymbol{u}||_k = \left(\sum_{i=1}^{m} ||u_i||_k^2 \right)^{1/2} .$$

For the particular case of functions belonging to $\boldsymbol{\mathcal{L}}_2(\Omega) = \boldsymbol{\mathcal{H}}^0(\Omega)$, the inner product is given by

$$(\boldsymbol{u}, \boldsymbol{v}) = \int_\Omega \boldsymbol{u} \cdot \boldsymbol{v} \, d\Omega,$$

where there should be no ambiguity in using the same notation to represent the inner product of both scalar and vector-valued functions.

1.5.2 Trial solutions and weighting functions

To define the weak, or variational, form of the boundary value problems discussed in the present text, we need to define two classes, or collections, of functions: the *test* or weighting functions and the *trial* or admissible solutions. Here these spaces are defined in the context of the standard Galerkin formulation.

The first collection of functions, denoted by $\mathcal{V}$, is composed of *test* functions and consists of all functions which are square integrable, have square integrable first derivatives over the computational domain Ω, and vanish on the Dirichlet portion, Γ_D, of the boundary. It is defined as follows:

$$\mathcal{V} = \{w \in \mathcal{H}^1(\Omega) \mid w = 0 \quad \text{on } \Gamma_D\} \equiv \mathcal{H}^1_{\Gamma_D}(\Omega). \tag{1.24}$$

This is as previously noted a Sobolev space and its inner product and norm coincide with those of $\mathcal{H}^1(\Omega)$.

The second collection of functions is called the *trial solutions*. This collection is similar to the test functions except that these admissible functions are required to satisfy the Dirichlet conditions on Γ_D. This second collection is denoted by $\mathcal{S}$ and is defined by

$$\mathcal{S} = \{u \in \mathcal{H}^1(\Omega) \mid u = u_D \quad \text{on } \Gamma_D\} \equiv \mathcal{V} + \{\bar{u}_D\} \tag{1.25}$$

where $\bar{u}_D$ is any function in $\mathcal{H}^1(\Omega)$ such that $\bar{u}_D = u_D$ on Γ_D. Thus, $\mathcal{S}$ can be viewed as a translation of $\mathcal{V}$ and, consequently, it is an affine space. Note, for instance, that, for $u_D \neq 0$, the sum of two elements of $\mathcal{S}$ is not an element of $\mathcal{S}$. However, for homogeneous boundary conditions, $u_D = 0$, trial and test spaces coincide, $\mathcal{V} = \mathcal{S} = \mathcal{H}^1_0(\Omega)$.

The sets $\mathcal{S}$ and $\mathcal{V}$ clearly contain infinitely many functions. In the finite element method, $\mathcal{S}$ and $\mathcal{V}$ are approximated by convenient finite dimensional subsets of these collections which will be denoted by $\mathcal{S}^h$ and $\mathcal{V}^h$, respectively. These finite element spaces are characterized, among other things, by a partition of the domain.

At this point, we have to view the domain Ω as discretized into element domains. Let $\mathcal{T}^h(\Omega)$ be a regular partition, also called *triangulation*, of Ω into $\mathtt{n_{el}}$ convex subdomains $\Omega^e \neq \emptyset$, such that

$$\bar{\Omega} = \bigcup_{e=1}^{\mathtt{n_{el}}} \bar{\Omega}^e \qquad \text{and} \qquad \Omega^e \cap \Omega^f = \emptyset \qquad \text{for } e \neq f.$$

Each subdomain Ω^e has a piecewise smooth boundary $\Gamma_e = \partial\Omega^e$, and h is a characteristic mesh size (diam$(\Omega^e) \leq h$ for all elements).

The weighting functions $w^h \in \mathcal{V}^h$ vanish on Γ_D. The approximation u^h lies in $\mathcal{S}^h$ and satisfies, with the precision given by the characteristic mesh size h, the boundary condition u_D on Γ_D. In fact, along Γ_D we should have $u^h = u^h_D$; however, in order not to overload the notation u_D is also used instead of u^h_D unless it is necessary to explicitly show the precision of the boundary data. The interpolation spaces are defined as

$$\begin{aligned} \mathcal{V}^h &:= \{w \in \mathcal{H}^1(\Omega) \mid w|_{\Omega^e} \in \mathcal{P}_m(\Omega^e)\ \forall e \text{ and } w = 0 \text{ on } \Gamma_D\} \\ \mathcal{S}^h &:= \{u \in \mathcal{H}^1(\Omega) \mid u|_{\Omega^e} \in \mathcal{P}_m(\Omega^e)\ \forall e \text{ and } u = u_D \text{ on } \Gamma_D\}, \end{aligned} \tag{1.26}$$

where $\mathcal{P}_m$ is the *finite element interpolating space*. Note that $\mathcal{V}^h$ and $\mathcal{S}^h$ are finite dimensional subspaces of the spaces of test, $\mathcal{V}$, and trial, $\mathcal{S}$, functions. In several spatial dimensions three types of polynomial spaces are usually chosen: the set of polynomials, $\mathcal{P}_m$, of total degree $\leq m$ (usually defined over the reference triangle), the set of polynomials, $\mathcal{Q}_m$, of degree $\leq m$ in each variable (defined over the reference square), or the set of polynomials $\mathcal{S}_m := \mathcal{P}_m \oplus \text{span}\{x^m y, xy^m\}$ (known as serendipity or trunk spaces over the reference square). They all include a complete basis of the subspace of polynomials of degree m which characterizes the a priori convergence rate of the finite element approximation.

1.5.3 Compact integral forms

In establishing the weak, or variational, form of boundary value problems we shall make frequent use of the bilinear forms

$$a(\boldsymbol{u}, \boldsymbol{v}) = \int_\Omega \boldsymbol{\nabla u} : \boldsymbol{\nabla v} \, d\Omega \qquad \forall \boldsymbol{u}, \boldsymbol{v} \in \boldsymbol{\mathcal{H}}^1(\Omega), \tag{1.27a}$$

$$b(\boldsymbol{v}, q) = -\int_\Omega q \, \boldsymbol{\nabla} \cdot \boldsymbol{v} \, d\Omega \qquad \forall \boldsymbol{v} \in \boldsymbol{\mathcal{H}}^1(\Omega) \text{ and } q \in \mathcal{L}_2(\Omega), \tag{1.27b}$$

as well as of the trilinear form

$$c(\boldsymbol{v}; \boldsymbol{w}, \boldsymbol{u}) = \int_\Omega \boldsymbol{w} \cdot (\boldsymbol{v} \cdot \boldsymbol{\nabla}) \boldsymbol{u} \, d\Omega \qquad \forall \boldsymbol{u}, \boldsymbol{v}, \boldsymbol{w} \in \boldsymbol{\mathcal{H}}^1(\Omega), \tag{1.27c}$$

where the following notation has been introduced:

$$[\boldsymbol{\nabla u}]_{ij} = \frac{\partial u_i}{\partial x_j}, \qquad \text{for } i = 1, \dots, m \text{ and } j = 1, \dots, \mathrm{n_{sd}}$$

$$\boldsymbol{\nabla u} : \boldsymbol{\nabla v} = \sum_{i=1}^{m} \sum_{j=1}^{\mathrm{n_{sd}}} \frac{\partial u_i}{\partial x_j} \frac{\partial v_i}{\partial x_j} \quad \text{and} \quad \boldsymbol{w} \cdot (\boldsymbol{v} \cdot \boldsymbol{\nabla}) \boldsymbol{u} = \sum_{i=1}^{m} \sum_{j=1}^{\mathrm{n_{sd}}} w_i \, v_j \frac{\partial u_i}{\partial x_j}.$$

1.5.4 Strong and weak forms of a boundary value problem

The process of spatial discretization by the finite element method rests upon an integral form of the considered partial differential problem. This was the reason for introducing larger spaces than C^m based on integrability properties. Thus, the first task in a finite element analysis consists of formulating a (continuous) variational problem associated with the given partial differential equation and its boundary conditions.

Let us introduce a model boundary value problem that will illustrate the various steps in the practical implementation of the finite element method. Consider solving the Poisson equation

$$-\boldsymbol{\nabla}^2 u = s \quad \text{in } \Omega, \tag{1.28a}$$

where Ω is enclosed by a piecewise smooth boundary Γ, and $s \in C^0(\bar{\Omega})$ is a specified source term, which may depend on $\boldsymbol{x}$. The following notation has been introduced:

$$\boldsymbol{\nabla}^2 u := (\boldsymbol{\nabla} \cdot \boldsymbol{\nabla}) u = \sum_{i=1}^{\mathrm{n_{sd}}} \frac{\partial^2 u}{\partial x_i^2}.$$

We further assume that the value of the unknown u is prescribed on the Dirichlet portion Γ_D of the boundary,

$$u = u_D \quad \text{on } \Gamma_D, \tag{1.28b}$$

while the normal derivative of u is prescribed on the remaining Neumann portion Γ_N of the boundary Γ,

$$\frac{\partial u}{\partial n} := \boldsymbol{n} \cdot \boldsymbol{\nabla} u = h \quad \text{on } \Gamma_N. \tag{1.28c}$$

A function $u \in C^2(\Omega) \cap C^0(\bar{\Omega})$ that satisfies (1.28) is called a *classical solution* of the boundary value problem.

The first step in a *weighted residual formulation* leading to the finite element discretization of our model problem consists of formulating a *weak (or variational) form* of the boundary value problem. This is achieved by multiplying the governing equation (1.28a) by the weighting function w and integrating over the computational domain Ω:

$$-\int_\Omega w\, \boldsymbol{\nabla}^2 u \, d\Omega = \int_\Omega w\, s \, d\Omega.$$

Note that the continuity requirements on u still impose that it must be twice differentiable. Now, however, the second derivatives of u are not required to be continuous, they only need to be square integrable. Thus, it is sufficient that $u \in \mathcal{H}^2(\Omega)$. In any case, it is obvious that classical solutions of (1.28) will also verify this integral equation for all admissible functions w.

In order to derive the weak form and produce a natural Neumann boundary condition on Γ_N, we apply to the l.h.s. of the previous equation the Green–Gauss divergence theorem:

$$\begin{aligned} -\int_\Omega w\, \boldsymbol{\nabla}^2 u \, d\Omega &= -\int_\Omega \Big(\boldsymbol{\nabla} \cdot (w \boldsymbol{\nabla} u) - \boldsymbol{\nabla} w \cdot \boldsymbol{\nabla} u \Big) d\Omega \\ &= \int_\Omega \boldsymbol{\nabla} w \cdot \boldsymbol{\nabla} u \, d\Omega - \int_\Gamma w (\boldsymbol{n} \cdot \boldsymbol{\nabla} u) d\Gamma. \end{aligned}$$

Now the regularity requirements on w and u are modified; u is only differentiated once and w must be differentiable. In fact, its derivatives must be square integrable; thus, we should have $w \in \mathcal{H}^1(\Omega)$. If $w \in \mathcal{V}$, recall from (1.24) that $w = 0$ on Γ_D, and if we take into account the prescribed Neumann boundary condition (1.28c), we obtain the following weak form of our model problem:

$$\int_\Omega \boldsymbol{\nabla} w \cdot \boldsymbol{\nabla} u \, d\Omega = \int_\Omega w\, s \, d\Omega + \int_{\Gamma_N} w\, h \, d\Gamma.$$

Note that the application of the divergence theorem has allowed us to naturally introduce the Neumann boundary condition on Γ_N. At the same time, it has removed the second-derivative terms of the Laplacian operator from the volume integral. This reduces the continuity requirements on u and allows us to select this function in $\mathcal{H}^1(\Omega)$, since only first derivatives appear under the integral sign.

Classical solutions of (1.28) will also satisfy this integral equation. But it only accounts for (1.28a) and (1.28c), not Dirichlet boundary conditions (1.28b). This is done by means of the proper choice of the space where u belongs, namely $\mathcal{S}$. Recall from (1.25) that every member of $\mathcal{S}$ satisfies the Dirichlet boundary condition $u = u_D$ on Γ_D. Noting that provided $u \in \mathcal{S}$ the Dirichlet boundary condition on Γ_D is satisfied, the weak form of our model problem can then be formally stated as follows:

$$\text{find } u \in \mathcal{S} \text{ such that } \quad a(w,u) = (w,s) + (w,h)_{\Gamma_N} \quad \forall w \in \mathcal{V}. \tag{1.29}$$

Here use has been made of the scalar version of (1.27a),

$$a(w,u) = \int_\Omega \nabla w \cdot \nabla u \, d\Omega, \tag{1.30}$$

and the linear functionals

$$(w,s) = \int_\Omega w\, s\, d\Omega, \quad \text{and} \quad (w,h)_{\Gamma_N} = \int_{\Gamma_N} w\, h\, d\Gamma.$$

By construction a classical solution u of (1.28) is a solution of the weak form (1.29). Let us show that a weak solution $u \in \mathcal{S}$ of (1.29) is unique; this is done by means of the Lax–Milgram lemma. Therefore, if we assume that a classical solution of (1.28) exists the same function u is a solution of both problems.

Remark 1.7. Weak solutions can also be found in cases that do not have a classical solution. This is the case of many applications in which data do not have the required smoothness, for instance a discontinuous source term $s(\boldsymbol{x})$. In these circumstances the weak problem (1.29) can be formulated and, if the Lax–Milgram lemma can be applied, it admits a unique solution, but this solution is not a classical solution.

Theorem 1.1 (Lax–Milgram lemma). *Let $a(\cdot,\cdot)$ be a bilinear form on a Hilbert space $\mathcal{H}$ equipped with norm $\|\cdot\|_{\mathcal{H}}$. If $a(\cdot,\cdot)$ is continuous, that is*

$$\exists \gamma_1 > 0 \text{ such that } |a(w,v)| \leq \gamma_1 \|w\|_{\mathcal{H}} \|v\|_{\mathcal{H}} \qquad \forall w, v \in \mathcal{H},$$

and coercive (or $\mathcal{H}$-elliptic), that is

$$\exists \alpha > 0 \text{ such that } a(v,v) \geq \alpha \|v\|^2_{\mathcal{H}} \qquad \forall v \in \mathcal{H},$$

then for all $l(\cdot)$ bounded linear mappings on $\mathcal{H}$ (thus continuous), that is

$$\exists \gamma_2 > 0 \text{ such that } |l(w)| \leq \gamma_2 \|w\|_{\mathcal{H}} \qquad \forall w \in \mathcal{H},$$

there exists a unique $u \in \mathcal{H}$ such that

$$a(w,u) = l(w) \qquad \forall w \in \mathcal{H}$$

is satisfied.

A proof can be found elsewhere, see for instance Ciarlet (1978), Yosida (1995) or Quarteroni and Valli (1994).

In order to show that this theorem can be applied to (1.29) some further considerations are needed. First, note that both arguments of the bilinear form in Theorem 1.1 must belong to the the same Hilbert space. This is not the case in (1.29) unless, recalling the definition of $\mathcal{S}$ in (1.25), u_0 is defined as $u = u_0 + \bar{u}_D$ with $u_0 \in \mathcal{V}$. Replacing this decomposition in (1.29) the weak form becomes

$$\text{find } u_0 \in \mathcal{V} \text{ such that } a(w, u_0) = (w, s) + (w, h)_{\Gamma_N} - a(w, \bar{u}_D)\ \forall w \in \mathcal{V}. \quad (1.31)$$

Thus, in this case $\mathcal{H} = \mathcal{V} = \mathcal{H}^1_{\Gamma_D}(\Omega)$ and the associated norm is the $\mathcal{H}^1$ norm, $\|\cdot\|_{\mathcal{H}} = \|\cdot\|_1$.

Second, since s, h and $\bar{u}_D$ are data, we can define the linear operator

$$l(w) := (w, s) + (w, h)_{\Gamma_N} - a(w, \bar{u}_D).$$

And third, we need to check that the bilinear form defined in (1.30) is coercive and continuous, and that the linear form is bounded. The continuity of $a(\cdot, \cdot)$ and $l(\cdot)$ is easily verified assuming some regularity in the data and applying systematically the Cauchy–Schwartz inequality. However, coercivity of $a(\cdot, \cdot)$ is more involved in particular when Neumann or Robin boundary conditions are present. Quarteroni and Valli (1994, Sec. 6.1.2) give an excellent presentation of the conditions for existence and uniqueness of a solution, which is beyond the scope of this text.

Remark 1.8 (Symmetric bilinear form). If, under the assumptions of Theorem 1.1, the bilinear form $a(\cdot, \cdot)$ is symmetric, then it can define a scalar product on $\mathcal{H}$. Note that this is the case of (1.30) for functions in $\mathcal{V} = \mathcal{H}^1_{\Gamma_D}(\Omega)$. This inner product induces in $\mathcal{H}$ the so-called energy norm

$$\|u\| = \sqrt{a(u, u)},$$

which is a natural norm in the engineering finite element context. Moreover, in this case the weak problem can be viewed as a minimization problem as noted in the next remark.

Remark 1.9 (Variational principle). Under the assumptions of the previous remark, if we pose $w = \delta u$, where δu is a variation of the function u, then it can be shown that the weak form (1.29) emanates from the minimization of the quadratic functional

$$I(u) = \frac{1}{2} \int_\Omega (\nabla u)^2 \, d\Omega - \int_\Omega u\, s \, d\Omega - \int_{\Gamma_N} u\, h \, d\Gamma,$$

or, equivalently,

$$I(u) = \frac{1}{2} a(u, u) - (s, u) - (h, u)_{\Gamma_N}.$$

Indeed, $\delta I = 0$ implies that

$$\int_\Omega (\nabla \delta u) \cdot \nabla(u)\, d\Omega - \int_\Omega \delta u\, s\, d\Omega - \int_{\Gamma_N} \delta u\, h\, d\Gamma = 0.$$

It follows that the solution of the boundary value problem (1.28) coincides with the value of u which minimizes the potential $I(u) \in \mathcal{H}^1_{\Gamma_D}(\Omega)$.

This is the case here for the Poisson equation, but also, for instance, in linear elasticity where the displacement field satisfying the differential equations of equilibrium of a body corresponds to the displacement field which minimizes the quadratic functional expressing the total potential energy of the body. It must, however, be emphasized that the finite element method is in no way dependent upon the existence of such a potential or variational principle.

1.5.5 Finite element spatial discretization

We now have all the necessary ingredients to discretize the weak form (1.29) by means of the Galerkin finite element method. Assuming that the reader has some familiarity with this method, we shall limit ourselves to a brief review of the main steps in the process of finite element spatial discretization.

For a partition $\mathcal{T}^h$, see Section 1.5.2, the Galerkin formulation of our model problem is obtained by restricting the weak form (1.29) to the finite dimensional spaces $\mathcal{S}^h \subset \mathcal{S}$ and $\mathcal{V}^h \subset \mathcal{V}$ defined in (1.26),

$$\text{find } u^h \in \mathcal{S}^h \text{ such that} \quad a(w^h, u^h) = (w^h, s) + (w^h, h)_{\Gamma_N} \quad \forall w^h \in \mathcal{V}^h. \tag{1.32}$$

1.5.5.1 Convergence of Galerkin approximations

Convergence in finite elements rests upon a priori error bounds (this terminology is used because these bounds are stated before the approximation u^h is actually computed). Moreover, as previously done, the solution of (1.32) is decomposed as $u^h = u_0^h + \bar{u}_D^h$ with $u_0^h \in \mathcal{V}^h$ and $\bar{u}_D^h = u_D$ along Γ_D; then, (1.32) can be written as: find $u_0^h \in \mathcal{V}^h$ such that

$$a(w^h, u_0^h) = (w^h, s) + (w^h, h)_{\Gamma_N} - a(w, \bar{u}_D^h) \quad \forall w^h \in \mathcal{V}^h. \tag{1.33}$$

Theorem 1.2 (Céa's lemma). *Under the assumptions of Theorem 1.1 there exists a unique solution u_0^h to (1.33) and u_0 to (1.31); moreover, u_0^h is the near-best fit to u_0 in the $\|\cdot\|_1$ norm and the error is bounded as*

$$\|u_0 - u_0^h\|_1 \le \frac{\alpha}{\gamma_1} \min_{v^h \in \mathcal{V}^h} \|u_0 - v^h\|_1.$$

The proof, which can be found elsewhere, see for instance Quarteroni and Valli (1994) or Wait and Mitchell (1985), is characterized by the *Galerkin orthogonality* property. This property is obtained by subtracting (1.32) from (1.29), or subtracting (1.33) from (1.31), where w is particularized as $w^h \in \mathcal{V}^h \subset \mathcal{V}$, namely

$$a(w^h, u - u^h) = 0 \text{ or } a(w^h, u_0 - u_0^h) = 0 \qquad \forall w^h \in \mathcal{V}^h.$$

Remark 1.10 (Symmetric bilinear form). Note that if the bilinear form is also symmetric, and thus $a(\cdot,\cdot)$ induces the energy norm (see Remark 1.8), the Galerkin orthogonality states in fact that the approximation error $u - u^h$ is orthogonal to $\mathcal{V}^h$ under the energy norm. Thus u^h can be viewed as an orthogonal projection of u with respect to the natural (energy) scalar product. Moreover, Céa's lemma can be particularized in terms of the energy norm and the approximation u^h is said to be the best fit in terms of energy

$$\|u_0 - u_0^h\| = \min_{v^h \in \mathcal{V}^h} \|u_0 - v^h\|.$$

Note that this best fit, which is standard in self-adjoint problems, is rarely used in flow problems where convection induces non-symmetric bilinear forms.

These bounds are of interest because, if we intuitively accept that the interpolation of any function in $\mathcal{V}$ is improved as we refine the finite element mesh (consistency), that is a family of partitions $\mathcal{T}^h(\Omega)$ of Ω, see Section 1.5.2, can be defined such that

$$\forall v \in \mathcal{V}, \qquad \min_{v^h \in \mathcal{V}^h} \|v - v^h\|_1 \to 0 \quad \text{as} \quad h \to 0, \tag{1.34}$$

then Theorem 1.2 ensures that convergence is enforced as we decrease h.

In fact, piecewise polynomial interpolation analysis allows us to ensure that

$$\forall v \in \mathcal{V}, \qquad \min_{v^h \in \mathcal{V}^h} \|v - v^h\|_1 \leq C(v)\, h^p$$

where $C(v)$ is a positive constant, which depends on the smoothness of v and the distortion of the finite element mesh, and p is a positive number, which also depends on the smoothness of v and the degree m of the complete basis of polynomials used (see the end of Section 1.5.2).

This bound combined with Céa's lemma produces the standard a priori error bound

$$\|u_0 - u_0^h\|_1 \leq C(u_0)\frac{\alpha}{\gamma_1}\, h^p. \tag{1.35}$$

Note, however, that such a bound is only interesting from a theoretical point of view. Based on interpolation theory and the Lax–Milgram and Céa lemmas, this bound states that a sequence of finite element solutions converges to the exact solution of the weak problem. In engineering practice it is also important to quantify the error associated with an approximation u^h. This is done by means of a posteriori error bounds, see for instance Ainsworth and Oden (2000) or Ladevèze and Pelle (2001).

1.5.5.2 Computational aspects Due to the presence of Dirichlet boundary conditions, a distinction must be made between the number of nodal points, $\mathrm{n_{np}}$, of the discretized domain and the number of nodal unknowns, that is the number of equations $\mathrm{n_{eq}}$, of the system which will result from the Galerkin approximation. Following the terminology introduced by Hughes (2000), we denote by $\eta = \{1, 2, \ldots, \mathrm{n_{np}}\}$ the set of global node numbers in the finite element mesh. Furthermore, we denote by $\eta_D \subset \eta$ the subset of nodes belonging to the Dirichlet portion of the boundary. It follows, in scalar problems, that the cardinal of $\eta \setminus \eta_D$ is equal to $\mathrm{n_{eq}}$, the number of equations.

With this notation, the approximation u^h can be written as

$$u^h(\boldsymbol{x}) = \sum_{A \in \eta \setminus \eta_D} N_A(\boldsymbol{x})\, \mathrm{u}_A + \sum_{A \in \eta_D} N_A(\boldsymbol{x})\, u_D(\boldsymbol{x}_A) \tag{1.36}$$

where N_A is the shape function (see details below) associated with node number A in the finite element mesh and u_A is the nodal unknown. Moreover, in the Galerkin formulation the arbitrary test functions, w^h, are defined such that

$$w^h \in \mathcal{V}^h := \underset{A \in \eta \setminus \eta_D}{\text{span}} \{N_A\}. \tag{1.37}$$

Thus, using the definitions of (1.36) and (1.37) in (1.32), we obtain the following discrete weak form of our model problem:

$$\sum_{B \in \eta \setminus \eta_D} a(N_A, N_B)\mathrm{u}_B = (N_A, s) + (N_A, h)_{\Gamma_N}$$
$$- \sum_{B \in \eta_D} a(N_A, N_B) u_D(\boldsymbol{x}_B), \quad \forall A \in \eta \setminus \eta_D. \tag{1.38}$$

Remark 1.11. Upper-case letters, such as A and B, are used to represent global node numbers in the finite element mesh: $1 \leq A, B \leq \mathrm{n_{np}}$. On the other hand, lower-case letters, such as a and b, will be used to represent local node numbers in an element: $1 \leq a, b \leq \mathrm{n_{en}}$ ($\mathrm{n_{en}}$ is the number of element nodes).

Remark 1.12. A suitable mesh generator is used to subdivide the computational domain Ω into element domains Ω^e. In two dimensions, meshes generally consist of triangular and/or quadrilateral elements. An interesting feature of the finite element method is that it handles naturally unstructured meshes, which can concentrate the elements in regions such as internal or boundary layers where sharp solution gradients are expected. The term "unstructured mesh" means that the number of elements meeting at a node (the element vertices) may vary from node to node. A representative unstructured mesh in two dimensions is shown in Figure 1.3. In the present text, we shall assume most of the time that the computational domain Ω is two-dimensional and that its boundary Γ is polygonal.

In the practical implementation of the finite element method, attention is focused on the computations in an individual element. For every element $\Omega^e \in \mathcal{T}^h$, the shape functions N_a, $a = 1, \ldots, \mathrm{n_{en}}$, are defined on a master element in terms of normalized coordinates. They define the finite element interpolation space, which includes a complete basis of the subspace of polynomials of degree m. These subspaces are denoted as $\mathcal{P}_m$, $\mathcal{Q}_m$ or $\mathcal{S}_m$, see Section 1.5.2.

As representative of the meshes used in two dimensions, let us consider, as suggested in Figure 1.4, a subdivision of the computational domain Ω into four-node

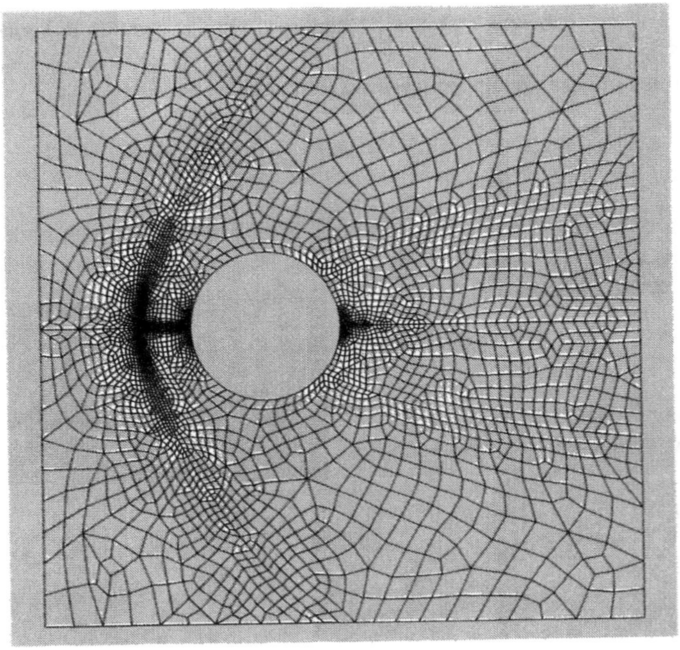

Fig. 1.3 Representative unstructured finite element mesh in two dimensions.

quadrilaterals. Each quadrilateral is mapped onto a canonical square with normalized local coordinates $(\xi, \eta) \in [-1, 1] \times [-1, 1]$ and the element shape functions are tensor products of those used in one dimension. This leads to a so-called isoparametric bilinear approximation on each element where both the global coordinates (x, y) and the unknown u are expressed by the same bilinear expansion parameterized by the nodal values:

$$\left\{\begin{matrix} x \\ y \end{matrix}\right\} = \sum_{a=1}^{4} N_a(\xi, \eta) \left\{\begin{matrix} x_a \\ y_a \end{matrix}\right\}, \quad \text{and} \quad u^h(x, y) \equiv u^h(\xi, \eta) = \sum_{a=1}^{4} N_a(\xi, \eta)\, \mathrm{u}_a,$$

where (x_a, y_a) are the coordinates of node a of the element.

All the element contributions to the discrete weak form (1.38) are computed in the local coordinates (ξ, η) using numerical quadrature, see Hughes (2000) or Zienkiewicz and Taylor (2000a) for details.

The assembly of the element contributions to the discrete weak form into the complete system results in a matrix equation of the form

$$\mathbf{K}\,\mathbf{u} = \mathbf{f}, \tag{1.39}$$

where $\mathbf{u}$ is the vector of the unknown nodal values. Its dimension is n_{eq}, in fact, $\mathbf{u}^T = [\ldots, \mathrm{u}_A, \ldots]$ with $A \in \eta \setminus \eta_D$.

In practice, the global matrix, $\mathbf{K}$, and nodal vector, $\mathbf{f}$, result from the topological assembly of element contributions. The addition of the element contributions to the

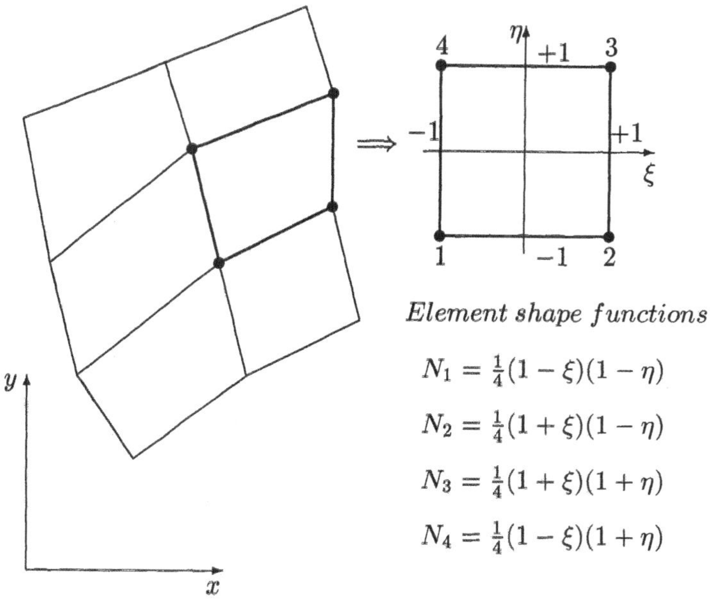

Fig. 1.4 Four-node quadrilateral elements and normalized reference element.

appropriate locations in the global matrix, $\mathbf{K}$, and nodal vector, $\mathbf{f}$, can be represented through the action of an assembly operator $\mathop{\mathbf{A}}^{e}$ acting on the local element matrix and nodal vector as follows:

$$\mathbf{K} = \mathop{\mathbf{A}}^{e}\mathbf{K}^e, \quad K^e_{ab} = \int_{\Omega^e} \boldsymbol{\nabla} N_a \cdot \boldsymbol{\nabla} N_b \, d\Omega = a(N_a, N_b)_{\Omega^e},$$

$$\mathbf{f} = \mathop{\mathbf{A}}^{e}\mathbf{f}^e, \qquad \mathrm{f}^e_a = (N_a, s)_{\Omega^e} + (N_a, h)_{\partial\Omega^e \cap \Gamma_N} - \sum_{b=1}^{\mathrm{n_{en}}} a(N_a, N_b)_{\Omega^e} u^e_{Db}.$$

Here, $u^e_{Db} = u_D(\boldsymbol{x}^e_b)$ if u_D is prescribed at node number b and equals zero otherwise. Hughes (2000) provides a detailed exposition on the topological assembly of the matrices and nodal vectors arising from the Galerkin finite element discretization.

Remark 1.13 (Alternative implementation of essential boundary conditions). The treatment of Dirichlet boundary conditions as indicated previously is not the only possible one. In fact, equation (1.39) is the matrix form of the weak form (1.33), but the blunt matrix form of (1.32) induces, in general, a singular system

$$\mathbf{K}^\star \mathbf{u}^\star = \mathbf{f}^\star$$

of $\mathrm{n_{np}}$ equations where some of the components (the cardinal of η_D in scalar problems) of $\mathbf{u}^\star = [\ldots, \mathrm{u}_A, \ldots]^T$, $A \in \eta$, are given by the Dirichlet boundary conditions.

Another popular method to implement essential boundary conditions is by means of Lagrange multipliers. If the Lagrange multiplier technique is chosen, the prescribed values on the Dirichlet portion of the boundary are introduced by means of linear constraints and the original system (singular in general) is enlarged by adding n_λ equations (the linear constraints) and n_λ unknowns (n_λ Lagrange multipliers λ, one per constraint, for instance the cardinal of η_D, i.e., $\mathrm{n_{np}} - \mathrm{n_{eq}}$, in scalar problems). The resulting system with the linear constraints is written as

$$\mathbf{r}(\mathbf{u}, \boldsymbol{\lambda}) = \begin{Bmatrix} \mathbf{K}^\star \mathbf{u}^\star + \mathbf{A}^T \boldsymbol{\lambda} - \mathbf{f}^\star \\ \mathbf{A}\mathbf{u}^\star - \mathbf{b} \end{Bmatrix} = \mathbf{0} \tag{1.40}$$

where $\mathbf{A}^T \boldsymbol{\lambda}$ is the vector of reaction forces that enforce the constraints, $\boldsymbol{\lambda}$ is a vector of n_λ Lagrange multipliers, $\mathbf{A}$ is an $n_\lambda \times \mathrm{n_{np}}$ rectangular matrix, and $\mathbf{b}$ a vector listing the prescribed values of the constraints.

Standard linear and nonlinear equation solvers have been adapted to handle linear constraints via the Lagrange-multiplier technique, see for instance Rodríguez-Ferran and Huerta (1999), and this procedure is very popular in object-oriented codes but it is not standard in other commercial programs. Note that apart from increasing the number of constraints the new matrix of the enlarged system,

$$\begin{pmatrix} \mathbf{K}^\star & \mathbf{A}^T \\ \mathbf{A} & \mathbf{0} \end{pmatrix},$$

is no longer positive definite as it is usually the case in (1.39). A practical example of how to incorporate the constraints represented by Dirichlet boundary conditions by use of Lagrange multipliers is given in Section 4.7.2.3. Ainsworth (2001) discusses some further possibilities for implementing linear constraints.

2

Steady transport problems

This chapter recalls first the deficiencies of the standard Galerkin formulation in convection-dominated problems. Then, it provides an introduction to generalized methods of the Petrov–Galerkin type designed to produce stable and accurate results in the presence of highly convective effects. This includes the Streamline-Upwind Petrov–Galerkin (SUPG) method, the Galerkin/Least-squares (GLS) method, as well as an introduction to other stabilization methods such as bubble function and wavelet-based methods, and, finally, introduces the variational multiscale method.

2.1 PROBLEM STATEMENT

As an introduction to the study of steady convection–diffusion problems by means of generalized Galerkin methods, we start by recalling the basic steps in the formulation of the standard Galerkin finite element method. This will serve as a basis to point out the deficiencies of the classical Galerkin approach in the solution of convection-dominated transport problems and thus introduce more adequate formulations.

2.1.1 Strong form

Let us consider the transport by convection and diffusion of a scalar quantity $u = u(\boldsymbol{x})$ in a domain $\Omega \subset \mathbb{R}^{n_{sd}}$, $n_{sd} = 2$ or 3, with smooth boundary Γ. The boundary is assumed to consist of a portion Γ_D on which the value of u is prescribed and of a complementary portion Γ_N on which the diffusive flux is prescribed. Conditions on Γ_D are Dirichlet (or essential) conditions, while conditions on Γ_N are known as Neumann (or natural) conditions. The boundary value problem associated with steady convective–diffusive transport is defined by the following equations:

$$\boldsymbol{a} \cdot \boldsymbol{\nabla} u - \boldsymbol{\nabla} \cdot (\nu \boldsymbol{\nabla} u) = s \qquad \text{in } \Omega, \tag{2.1a}$$

$$u = u_D \qquad \text{on } \Gamma_D, \tag{2.1b}$$

$$\boldsymbol{n} \cdot \nu \boldsymbol{\nabla} u = \nu \frac{\partial u}{\partial \boldsymbol{n}} = h \qquad \text{on } \Gamma_N, \tag{2.1c}$$

where u is the scalar unknown, $\boldsymbol{a}(\boldsymbol{x})$ is the convection velocity (also known as advection), $\nu > 0$ is the coefficient of diffusivity and $s(\boldsymbol{x})$ is a volumetric source term. The function u_D denotes prescribed values of u on the Dirichlet portion Γ_D of the boundary, while function h denotes prescribed values of the normal diffusive flux on the Neumann portion Γ_N. The unit outward normal vector to Γ is denoted by $\boldsymbol{n}$.

Remark 2.1 (Pure convection). In the absence of diffusion, $\nu = 0$, problem (2.1) is hyperbolic and boundary conditions can only be prescribed on the inflow portion of the boundary. In this case it is convenient to consider a partition of the boundary Γ such that

$$\Gamma = \overline{\Gamma^{in} \cup \Gamma^{out}}$$

$$\Gamma^{in} = \{\boldsymbol{x} \in \Gamma \mid \boldsymbol{a} \cdot \boldsymbol{n} < 0\} \qquad \text{(inflow boundary)}$$

$$\Gamma^{out} = \{\boldsymbol{x} \in \Gamma \mid \boldsymbol{a} \cdot \boldsymbol{n} > 0\} \qquad \text{(outflow boundary).}$$

In the case of Dirichlet inlet conditions, the equations governing purely convective transport are

$$\boldsymbol{a} \cdot \boldsymbol{\nabla} u = s \qquad \text{in } \Omega,$$

$$u = u_D \qquad \text{on } \Gamma^{in}.$$

Note that the solution of linear pure convection problems can be discontinuous in the cross-flow direction (a jump may exist orthogonal to a characteristic curve; Chapter 3 presents in detail the concept of characteristic curves). This occurs when the data u_D, that are prescribed on the inflow boundary, are discontinuous. By contrast, in the case of a convection–diffusion problem (2.1), the solution is continuous over the whole computational domain, any possible jump being spread over a layer of width proportional to the square root of the diffusivity, $\sqrt{\nu}$, around the near characteristic.

Remark 2.2. Instead of prescribing the normal diffusive flux on Γ_N, one may consider prescribing the total normal flux, h_T, consisting of the sum of a diffusive component and a convective component:

$$(\nu \boldsymbol{\nabla} u - u\boldsymbol{a}) \cdot \boldsymbol{n} = h_T. \tag{2.2}$$

This is standard in conservation equations. Two options are available to deal with a total flux boundary condition. The first one consists in rewriting (2.2) as a standard diffusive flux boundary condition, but with a value of the flux depending on the solution u itself:

$$\nu \frac{\partial u}{\partial n} = h_T + u(\boldsymbol{a} \cdot \boldsymbol{n}). \tag{2.3}$$

In most applications the previous equation appears as

$$\nu \frac{\partial u}{\partial n} = (\boldsymbol{a} \cdot \boldsymbol{n})(u - u_{\text{ext}}),$$

when the total flux, h_T, is replaced by $-(\boldsymbol{a} \cdot \boldsymbol{n})u_{\text{ext}}$, where u_{ext} is the value of u outside the domain. Note that, in the numerical implementation, such a boundary condition will contribute to the construction of the l.h.s. matrix. This, as will be seen next, does not occur with the standard Neumann boundary condition, equation (2.1c).

The second option consists in expressing the convective term in the form $\boldsymbol{a} \cdot \boldsymbol{\nabla} u = \boldsymbol{\nabla} \cdot (u\,\boldsymbol{a}) - u\,\boldsymbol{\nabla} \cdot \boldsymbol{a}$. This allows us to rewrite the convection–diffusion equation, see (2.1a), in the form

$$\boldsymbol{\nabla} \cdot (u\,\boldsymbol{a} - \nu \boldsymbol{\nabla} u) - u\,\boldsymbol{\nabla} \cdot \boldsymbol{a} = s. \tag{2.4}$$

This option incorporates the divergence of the total flux appearing in the boundary term (2.2), and, as is shown next, will simplify the treatment of the boundary integral in the weak form.

2.1.2 Weak form

The first step towards the finite element spatial discretization of the convection–diffusion problem is to associate an equivalent weak (or variational) form to the strong form (2.1) of the boundary value problem. As seen in Section 1.5.2, the trial solution space $\mathcal{S}$ consists of real-valued functions, u, defined on Ω such that all members of $\mathcal{S}$ satisfy the Dirichlet condition in (2.1b). Similarly, the space $\mathcal{V}$ of the weighting functions, w, is chosen such that $w = 0$ on Γ_D. Namely, $\mathcal{S} := \{u \in \mathcal{H}^1(\Omega) \mid u = u_D \text{ on } \Gamma_D\}$ and $\mathcal{V} := \mathcal{H}^1_{\Gamma_D}(\Omega) = \{w \in \mathcal{H}^1(\Omega) \mid w = 0 \text{ on } \Gamma_D\}$.

The weak formulation of the convection–diffusion problem (2.1) then takes the following form: find $u \in \mathcal{S}$ such that

$$\int_\Omega w(\boldsymbol{a} \cdot \boldsymbol{\nabla} u)\, d\Omega - \int_\Omega w \boldsymbol{\nabla} \cdot (\nu \boldsymbol{\nabla} u)\, d\Omega = \int_\Omega w\, s\, d\Omega \quad \text{for all } w \in \mathcal{V},$$

or, using the divergence theorem on the diffusion term and noting that $w = 0$ on Γ_D,

$$\int_\Omega w(\boldsymbol{a} \cdot \boldsymbol{\nabla} u)\, d\Omega + \int_\Omega \boldsymbol{\nabla} w \cdot (\nu \boldsymbol{\nabla} u)\, d\Omega = \int_\Omega w\, s\, d\Omega + \int_{\Gamma_N} w\, h\, d\Gamma \quad \text{for all } w \in \mathcal{V}. \tag{2.5}$$

This weak form is at the basis of the finite element spatial discretization of the convection–diffusion problem. Note that the integration by parts of the diffusion term allows us to naturally introduce the prescribed flux condition on portion Γ_N of the boundary.

At this point, it is convenient to introduce a compact version of (2.5) based upon the integral forms defined in (1.27), namely

$$\begin{aligned} a(w,u) &= \int_\Omega \nabla w \cdot (\nu \nabla u) d\Omega, & (w,s) &= \int_\Omega w\, s\, d\Omega, \\ c(\boldsymbol{a};w,u) &= \int_\Omega w(\boldsymbol{a}\cdot\nabla u) d\Omega, & (w,h)_{\Gamma_N} &= \int_{\Gamma_N} w\, h\, d\Gamma. \end{aligned} \tag{2.6}$$

This allows us to rewrite the weak form (2.5) in the following compact form:

$$a(w,u) + c(\boldsymbol{a};w,u) = (w,s) + (w,h)_{\Gamma_N}. \tag{2.7}$$

Remark 2.3. When the total flux boundary condition (2.2) is prescribed, see Remark 2.2, an alternative weak form of the convection–diffusion problem may be constructed from the differential equation in conservation form (2.4). After integration by parts the weak form in this case becomes

$$\int_\Omega \nabla w \cdot (\nu \nabla u - u\boldsymbol{a}) d\Omega - \int_\Omega w\, u (\nabla \cdot \boldsymbol{a}) d\Omega = \int_\Omega w\, s\, d\Omega + \int_{\Gamma_N} w\, h_T\, d\Gamma \quad \text{for all } w \in \mathcal{V}. \tag{2.8}$$

Note that an extra term (due to the integration by parts of the convection term) appears on the l.h.s. This term cancels if the convection velocity is divergence free, which is often the case in engineering applications.

2.2 GALERKIN APPROXIMATION

We now have all the necessary ingredients to perform the spatial discretization of the convection–diffusion problem (2.1) by means of the Galerkin finite element method. Let $\mathcal{S}^h$ and $\mathcal{V}^h$ be finite dimensional spaces' subsets of $\mathcal{S}$ and $\mathcal{V}$, respectively. These interpolation spaces were defined in Section 1.5.2. The weighting functions $w^h \in \mathcal{V}^h$ vanish on Γ_D. The approximation u^h lies in $\mathcal{S}^h$ and satisfies, with the precision given by the characteristic mesh size h, the boundary condition u_D on Γ_D. The Galerkin formulation is obtained by restricting the weak form (2.7) to the finite dimensional spaces, namely, find $u^h \in \mathcal{S}^h$ such that

$$a(w^h,u^h) + c(\boldsymbol{a};w^h,u^h) = (w^h,s) + (w^h,h)_{\Gamma_N} \quad \text{for all } w^h \in \mathcal{V}^h. \tag{2.9}$$

At this point, we have to view Ω as discretized into elements Ω^e, $1 \le e \le \mathrm{n_{el}}$. As shown by equation (1.36), the approximation, u^h, can be written as

$$u^h(\boldsymbol{x}) = \sum_{A\in\eta\setminus\eta_D} N_A(\boldsymbol{x})\,\mathrm{u}_A + \sum_{A\in\eta_D} N_A(\boldsymbol{x})\,u_D(\boldsymbol{x}_A) \tag{2.10}$$

where N_A is the shape function associated with node number A and u_A is the nodal unknown. Moreover, the test functions, w^h, are defined such that

$$w^h \in \mathcal{V}^h := \operatorname*{span}_{A \in \eta \setminus \eta_D} \{N_A\}. \tag{2.11}$$

Thus, after substitution of (2.10) into (2.9), and using (2.11), we obtain the discrete weak form

$$\begin{aligned}\sum_{B\in\eta\setminus\eta_D} \left[a(N_A, N_B) + c(\boldsymbol{a}; N_A, N_B)\right] \mathrm{u}_B &= (N_A, s) + (N_A, h)_{\Gamma_N} \\ - \sum_{B\in\eta_D} \left[a(N_A, N_B) + c(\boldsymbol{a}; N_A, N_B)\right] u_D(\boldsymbol{x}_B), &\text{ for all } A \in \eta \setminus \eta_D. \end{aligned} \tag{2.12}$$

Assembling the element contributions to this weak form, we obtain the algebraic system governing the nodal values of the discrete solution of the convection–diffusion problem. After inclusion of the Dirichlet boundary conditions as explained in Chapter 1, this system takes the matrix form

$$(\mathbf{C} + \mathbf{K})\mathbf{u} = \mathbf{f}, \tag{2.13}$$

where $\mathbf{u}$ is the vector of the unknown nodal values with dimension n_{eq}, while $\mathbf{C}$ and $\mathbf{K}$ are, respectively, the convection matrix and the diffusion matrix. Both matrices are obtained by topological assembly of the element contributions evaluated, as explained in Section 1.5.5. In terms of the local node numbers $1 \leq a, b \leq \mathrm{n}_{\mathrm{en}}$, the assembly process takes the following form:

$$\begin{aligned} \mathbf{C} &= \mathop{\mathsf{A}}^{e} \mathbf{C}^e \quad C^e_{ab} = \int_{\Omega^e} N_a(\boldsymbol{a} \cdot \boldsymbol{\nabla} N_b) d\Omega \quad \text{(convection matrix)} \\ \mathbf{K} &= \mathop{\mathsf{A}}^{e} \mathbf{K}^e \quad K^e_{ab} = \int_{\Omega^e} \boldsymbol{\nabla} N_a \cdot \nu \boldsymbol{\nabla} N_b \, d\Omega \quad \text{(diffusion matrix).} \end{aligned} \tag{2.14}$$

The r.h.s. vector in (2.13) considers the contribution of the source term, s, the prescribed flux, h, and the Dirichlet data u_D. It results from the assembly of elemental contributions of the form $\mathbf{f} = \mathop{\mathsf{A}}^{e} \mathbf{f}^e$ with

$$\begin{aligned} f^e_a = (N_a, s)_{\Omega^e} &+ (N_a, h)_{\partial\Omega^e \cap \Gamma_N} \\ &- \sum_{b=1}^{\mathrm{n}_{\mathrm{en}}} \left[a(N_a, N_b)_{\Omega^e} + c(\boldsymbol{a}; N_a, N_b)_{\Omega^e}\right] u^e_{Db}, \end{aligned} \tag{2.15}$$

where $u^e_{Db} = u_D(\boldsymbol{x}^e_b)$ if u_D is prescribed at node number b and equals zero otherwise.

2.2.1 Piecewise linear approximation in 1D

As a model problem and for illustration purposes we consider a 1D scalar equation. Moreover, in order to determine the element matrices the convection and diffusion coefficients, a and ν, are assumed constants. However, the source term s is variable with x. The treatment of the boundary conditions is simplified by imposing homogeneous Dirichlet conditions on each side. Note, however, that other boundary conditions could easily be implemented. Thus the model problem is written as

$$au_x - \nu u_{xx} = s(x) \qquad \text{in }]0, L[\tag{2.16a}$$

$$u = 0 \qquad \text{at } x = 0 \text{ and } x = L. \tag{2.16b}$$

The weak form associated with this model problem is, after integration by parts of the diffusion term, given by

$$\int_0^L (w\, a\, u_x + w_x\, \nu\, u_x)\, dx = \int_0^L w\, s\, dx, \tag{2.17}$$

or in the compact form, see (2.6),

$$a(w, u) + c(a; w, u) = (w, s).$$

The weak form will now be discretized using a uniform mesh of linear elements of size h. Without loss of generality the node numbering is assumed to be consecutive; $\mathrm{n_{eq}}$ is, in this case, the number of interior nodes of the spatial discretization, that is $\eta_D = \{1, \mathrm{n_{np}}\}$ and $\eta = \{1, 2, \ldots, \mathrm{n_{eq}}, \mathrm{n_{eq}} + 1, \mathrm{n_{np}}\}$. The trial and weighting functions are defined as before, see (2.10) and (2.11). Then they are introduced in (2.17) to yield the following discrete equation at an interior node A, $A = 2, \ldots, \mathrm{n_{eq}}+1$:

$$\int_0^L \sum_{B=2}^{\mathrm{n_{eq}}+1} \left(aN_A \frac{\partial N_B}{\partial x} + \nu \frac{\partial N_A}{\partial x} \frac{\partial N_B}{\partial x} \right) \mathrm{u}_B\, dx = \int_0^L N_A\, s\, dx. \tag{2.18}$$

A linear element in 1D is defined by two nodes, $\mathrm{n_{en}} = 2$, locally denoted as 1 and 2. The shape functions of a linear element are given by

$$N_1(\xi) = \frac{1}{2}(1 - \xi) \qquad N_2(\xi) = \frac{1}{2}(1 + \xi),$$

where ξ is the normalized coordinate, $-1 \leq \xi \leq +1$. As usual, at any interior point of the element one has $u(\xi) = N_1(\xi)\, \mathrm{u}_1 + N_2(\xi)\, \mathrm{u}_2$, and $x(\xi) = N_1(\xi)\, x_1 + N_2(\xi)\, x_2$. Note that for a uniform mesh of size h

$$dx = \frac{1}{2}(x_2 - x_1) d\xi = \frac{h}{2} d\xi,$$

and thus

$$\frac{\partial N_b}{\partial x} = \frac{\partial N_b}{\partial \xi} \frac{\partial \xi}{\partial x} = \frac{2}{h} \frac{\partial N_b}{\partial \xi} \qquad \text{for } b = 1, 2.$$

With these preliminaries, the integrals in expression (2.18) are readily evaluated in closed form to obtain the element convection matrix, $\mathbf{C}^e$, and the element diffusion matrix, $\mathbf{K}^e$. In terms of the local node numbers 1 and 2 the result is

$$\mathbf{C}^e = a \int_{\Omega^e} \begin{pmatrix} N_1 \dfrac{\partial N_1}{\partial x} & N_1 \dfrac{\partial N_2}{\partial x} \\ N_2 \dfrac{\partial N_1}{\partial x} & N_2 \dfrac{\partial N_2}{\partial x} \end{pmatrix} dx = \frac{a}{2} \begin{pmatrix} -1 & +1 \\ -1 & +1 \end{pmatrix} \tag{2.19a}$$

$$\mathbf{K}^e = \nu \int_{\Omega^e} \begin{pmatrix} \dfrac{\partial N_1}{\partial x} \dfrac{\partial N_1}{\partial x} & \dfrac{\partial N_1}{\partial x} \dfrac{\partial N_2}{\partial x} \\ \dfrac{\partial N_2}{\partial x} \dfrac{\partial N_1}{\partial x} & \dfrac{\partial N_2}{\partial x} \dfrac{\partial N_2}{\partial x} \end{pmatrix} dx = \frac{\nu}{h} \begin{pmatrix} +1 & -1 \\ -1 & +1 \end{pmatrix}. \tag{2.19b}$$

Here $\Omega^e = [x_e, x_{e+1}]$ with $e = 1, \ldots, \mathrm{n_{el}}$ ($\mathrm{n_{el}} = \mathrm{n_{eq}} + 1 = \mathrm{n_{np}} - 1$ in this case). Furthermore, interpolating the source term s by means of the element shape functions, namely $s(\xi) = N_1(\xi)\, s_1 + N_2(\xi)\, s_2$, one finds that the components of the load vector $\mathbf{f}$ in (2.13) are obtained from element contributions of the form

$$\mathbf{f}^e = \int_{\Omega^e} \{N_1\,(N_1 s_1 + N_2 s_2),\; N_2\,(N_1 s_1 + N_2 s_2)\}^T\, dx.$$

With these results, and assembling in the usual finite element manner the contributions emanating from both elements to which a given node belongs, one finds that the Galerkin method delivers the following discrete equation at an interior node j:

$$a \left(\frac{u_{j+1} - u_{j-1}}{2h} \right) - \nu \left(\frac{u_{j+1} - 2u_j + u_{j-1}}{h^2} \right) = \frac{1}{6}(s_{j-1} + 4s_j + s_{j+1}). \tag{2.20}$$

Remark 2.4. Notice that the l.h.s. of the discrete equation produced with linear elements coincides with that of second-order central differences. In this respect, the Galerkin method based on linear elements and the central difference method appear to be closely related. Nevertheless, a significant difference between finite element and finite difference approximations is observed in the treatment of the source term. On one hand, the Galerkin method generates a weighted average on the r.h.s. of (2.20). That is, the finite element method uses the so-called consistent mass matrix to weigh the nodal values of the source term s_j. This consistent mass matrix is defined by

$$\mathbf{M} = \mathop{\mathbf{A}}^{e} \mathbf{M}^e$$

where the element mass matrix can be written in terms of the element shape functions as

$$M^e_{ab} = \int_{\Omega^e} \begin{pmatrix} N_1 N_1 & N_1 N_2 \\ N_2 N_1 & N_2 N_2 \end{pmatrix} dx = \frac{h}{6} \begin{pmatrix} 2 & 1 \\ 1 & 2 \end{pmatrix}.$$

On the other hand, the difference method simply uses a local value of the source, namely s_j at node j. That is, the difference method employs a so-called lumped (or diagonal) mass matrix, $\mathbf{M}^L$, defined at the element level by

$$[M^L]^e_{ab} = \int_{\Omega^e} \begin{pmatrix} N_1 & 0 \\ 0 & N_2 \end{pmatrix} dx = \frac{h}{2} \begin{pmatrix} 1 & 0 \\ 0 & 1 \end{pmatrix}.$$

2.2.2 Analysis of the discrete equation

In this section we show that the Galerkin finite element method as described in the previous section is not ideally suited to solve convection-dominated problems. To characterize the relative importance of convective and diffusive effects in a given flow problem, it is useful to introduce the mesh Péclet number

$$Pe = \frac{ah}{2\nu} \tag{2.21}$$

which expresses the ratio of convective to diffusive transport. This allows us to rewrite the discrete equation (2.20) in the form

$$\frac{a}{2h}\left(\frac{Pe-1}{Pe}u_{j+1} + \frac{2}{Pe}u_j - \frac{Pe+1}{Pe}u_{j-1}\right) = \frac{1}{6}(s_{j-1} + 4s_j + s_{j+1}). \tag{2.22}$$

The importance of the Péclet number will become more and more clear as we move along. However, the previous equation is already a good example. Note that the l.h.s. of (2.22), which corresponds (for linear elements and a Galerkin formulation) to the discrete counterpart of the l.h.s. of the weak form (2.9) or the differential operator in (2.1a), is characterized by the Péclet number.

To illustrate the deficiencies of the Galerkin finite element method in the solution of convection-dominated problems, we shall first present a simple numerical example based on linear elements and then use this example to identify a possible remedy for improving the finite element solution.

This problem simply consists of solving the 1D boundary value problem (2.16) with a constant source term, $s = 1$, in a dimensionless domain, $L = 1$. A uniform source has been chosen on purpose in order to avoid truncation errors due to the spatial discretization of the source term. In this manner, the truncation error, which will arise from the Galerkin discretization of our model problem, must be attributed fully to the discrete representation of the convection and diffusion operators.

The exact solution to the above model problem is given by

$$u(x) = \frac{1}{a}\left(x - \frac{1-\exp(\gamma x)}{1-\exp\gamma}\right), \tag{2.23}$$

where $\gamma = a/\nu$. The numerical approximation to (2.23) has been computed with a mesh of 10 uniform elements using the Galerkin scheme (2.20) or (2.22) for several values of the mesh Péclet number, see (2.21), namely, $Pe = 0.25, 0.9$ and 5 (the convection velocity a is taken as one). The results are displayed in Figure 2.1 in comparison with the exact solution. One notes that the Galerkin solution is corrupted by non-physical oscillations when the Péclet number is larger than one. The Galerkin method loses its best approximation property when the non-symmetric convection operator dominates the diffusion operator in the transport equation, and consequently spurious node-to-node oscillations appear.

The Galerkin equation (2.20) has a truncation error. This is standard in discrete equations. We shall now analyze it and, as a result, discover why the Galerkin method

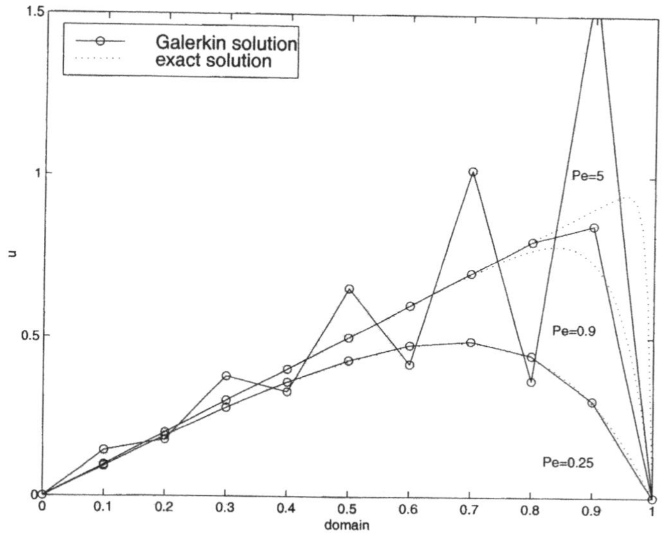

Fig. 2.1 Galerkin solution (solid lines) of the convection–diffusion problem (2.16) with $L = 1, s = 1$ using a uniform mesh of 10 linear elements. Dotted lines show the exact solution.

(and the intimately related central difference method) are not optimal methods for solving convection-dominated problems.

To reach our objective, we need to find a discrete scheme similar in structure to (2.22), but giving the exact solution at each node of a uniform mesh of linear elements for any mesh size h and all values of the Péclet number. The comparison between this exact scheme and the Galerkin scheme will then allow us to identify which modifications are necessary, in the Galerkin method, in order to improve its response in highly convective situations.

To obtain an exact scheme, we must identify the value of three coefficients, say α_1, α_2 and α_3, such that

$$\alpha_1 \, u_{j-1} + \alpha_2 \, u_j + \alpha_3 \, u_{j+1} = 1 \tag{2.24}$$

for all nodal coordinates x_j, mesh dimensions h and Péclet numbers Pe. From the exact solution (2.23) we have

$$\begin{cases} u_{j-1} = \dfrac{1}{a}\left(x_j - h - \dfrac{1 - \exp(\gamma x_j)\exp(-2Pe)}{1 - \exp\gamma}\right), \\ u_j = \dfrac{1}{a}\left(x_j - \dfrac{1 - \exp(\gamma x_j)}{1 - \exp\gamma}\right), \\ u_{j+1} = \dfrac{1}{a}\left(x_j + h - \dfrac{1 - \exp(\gamma x_j)\exp(2Pe)}{1 - \exp\gamma}\right). \end{cases}$$

When these expressions are introduced in (2.24), the following three conditions are obtained:

$$\begin{cases} \alpha_1 + \alpha_2 + \alpha_3 = 0 \\ -\alpha_1 + \alpha_3 = a/h \\ \alpha_1 \exp(-2Pe) + \alpha_2 + \alpha_3 \exp(2Pe) = 0. \end{cases}$$

Solving for α_1, α_2 and α_3, we obtain

$$\begin{cases} \alpha_1 = -a(1 + \coth Pe)/(2h) \\ \alpha_2 = a(\coth Pe)/h \\ \alpha_3 = a(1 - \coth Pe)/(2h). \end{cases}$$

When these expressions are substituted in (2.24), the desired exact scheme becomes

$$\frac{a}{2h}\left[(1 - \coth Pe)u_{j+1} + (2\coth Pe)u_j - (1 + \coth Pe)u_{j-1}\right] = 1. \tag{2.25}$$

As expected, this equation presents similarities but it is not identical to the one obtained with the Galerkin formulation (2.22). To highlight the similarities and the differences between both schemes, equation (2.25) may be rearranged in two interesting ways.

First, in a form similar to the original Galerkin scheme (2.20)

$$a\frac{u_{j+1} - u_{j-1}}{2h} - (\nu + \bar{\nu})\frac{u_{j+1} - 2u_j + u_{j-1}}{h^2} = 1, \tag{2.26}$$

where $\bar{\nu}$ is an added numerical diffusion defined as

$$\bar{\nu} = \beta\frac{ah}{2} = \beta\nu Pe \qquad \text{with } \beta = \coth Pe - \frac{1}{Pe}. \tag{2.27}$$

Note that this added numerical diffusion only depends on the parameters of the governing differential equation and the element size h.

Second, the same scheme (2.25) can also be rewritten as

$$\frac{1-\beta}{2}\left(a\frac{u_{j+1} - u_j}{h}\right) + \frac{1+\beta}{2}\left(a\frac{u_j - u_{j-1}}{h}\right) - \nu\frac{u_{j+1} - 2u_j + u_{j-1}}{h^2} = 1, \tag{2.28}$$

where the discretization of the convective term appears as a weighted average of the fluxes (convection) of the solution to the left and to the right of node j. That is, the convection term is not discretized using a centered scheme.

This analysis is at the origin of two families of early techniques developed for improving the standard Galerkin method: those based on adding an artificial diffusion and those concerned with non-centered discretizations of the convection operator, also called upwind schemes. These early methods are presented in Section 2.3 and the current stabilization techniques are discussed in Section 2.4.

Remark 2.5. To reduce the computational costs (repeated evaluations of hyperbolic cotangents), it is common practice to replace the optimal formula (2.27) by its doubly asymptotic approximation, see Figure 2.2, given by

$$\beta \approx \beta_{\text{approx}} = \begin{cases} Pe/3 & \text{if } -3 \leq Pe \leq 3, \\ \text{sign}(Pe) & \text{if } |Pe| > 3. \end{cases} \tag{2.29}$$

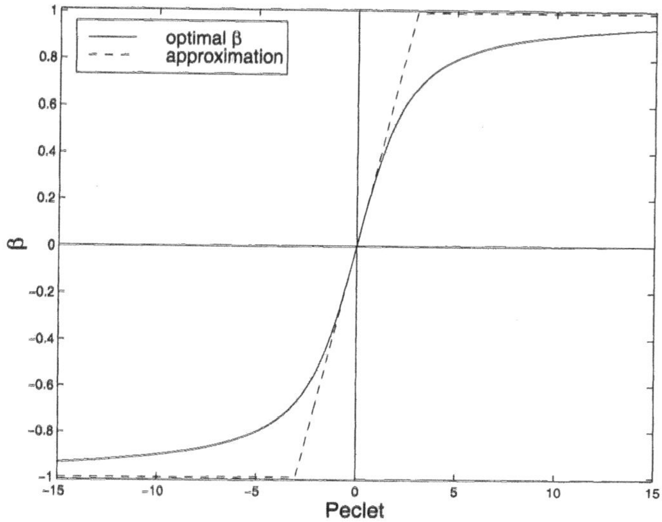

Fig. 2.2 Optimal value of the parameter β, see equation (2.27), and its doubly asymptotic approximation, see equation (2.29).

Remark 2.6 (The negative diffusion of the Galerkin method). A simple comparison between the Galerkin discrete scheme (2.20) and the exact difference scheme (2.26) indicates that the Galerkin method introduces a truncation error in the form of a diffusion operator. That is, the Galerkin scheme lacks $-\bar{\nu}(u_{j+1} - 2u_j + u_{j-1})/h^2$ to represent the exact solution. The diffusion coefficient $\bar{\nu}$, see (2.27), gives a measure of the magnitude of the truncation error as a function of the Péclet number. This truncation error is systematically negative for all values of P_e. Due to this negative truncation error, a *modified equation* is actually solved by the Galerkin method, which possesses a reduced diffusion coefficient. Such a modified equation is the differential equation that is solved exactly (exact nodal values) by the Galerkin method, namely

$$a\, u_x - \left[\nu - \bar{\nu}\frac{\sinh^2(P_e)}{P_e^2}\right] u_{xx} = 1. \tag{2.30}$$

Notice that as the Péclet number increases the diffusion coefficient in the modified equation (2.30) may become negative. In this case, no stable solution is guaranteed. Figure 2.3 shows the actual diffusion coefficient of the partial differential equation associated with the Galerkin scheme, namely the modified equation (2.30). Note that it becomes negative for values of the grid Péclet number larger than one.

The negative numerical diffusion inherent in the Galerkin finite element method (and the second-order central difference method) is indeed the cause of the numerical difficulties encountered in the simulation of highly convective transport problems.

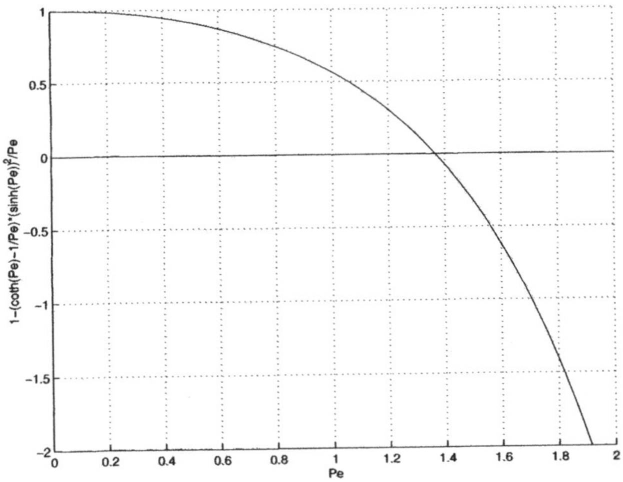

Fig. 2.3 Normalized diffusivity of the modified equation associated with a Galerkin formulation, see equation (2.30), namely $1-(\bar{\nu}/\nu)(\sinh^2(Pe)/Pe^2)$, as a function of $Pe = ah/(2\nu)$.

Remark 2.7 (Linear difference equation). To further highlight the appearance of non-physical oscillations when the grid Péclet number exceeds unity, we note that the homogeneous form of the discrete equation (2.22) can be rewritten as

$$(1-Pe)(u_{j+1}-u_j) = (1+Pe)(u_j-u_{j-1}),$$

which shows that the slopes of the solution to the left and to the right of node j are of opposite sign when the Péclet number is larger than one.

Moreover, this linear difference equation can be solved exactly, see the details in Isaacson and Keller (1994, Chap. 8, Sec. 4). The characteristic equation is

$$(1-Pe)\lambda^2 - 2\lambda + (1+Pe) = 0.$$

Its roots are $\lambda_1 = 1$ and $\lambda_2 = (1+Pe)/(1-Pe)$, and thus the solution of the homogeneous form of equation (2.22) is given by

$$u_j = C_1 + C_2 \left(\frac{1+Pe}{1-Pe}\right)^j,$$

where C_1 and C_2 are constants fixed by the boundary conditions. This expression clearly shows that the approximate solution delivered by the Galerkin method is oscillatory when $Pe > 1$.

By contrast, the homogeneous form of the discrete equation corresponding to the exact scheme (2.26) or (2.28) reads

$$[1+(\beta-1)Pe]\,(u_{j+1}-u_j) = [1+(\beta+1)Pe]\,(u_j-u_{j-1}), \tag{2.31}$$

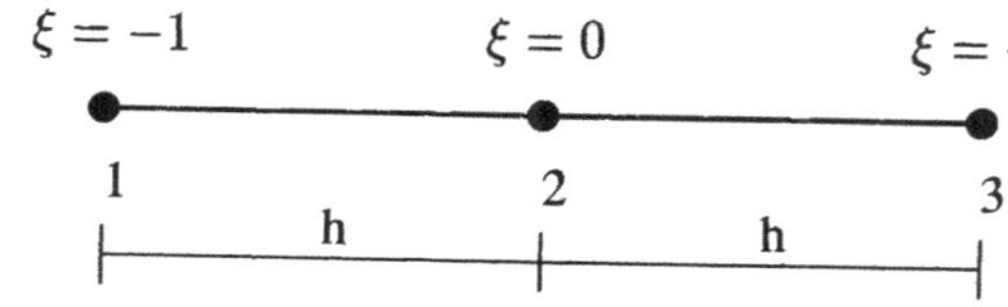

Fig. 2.4 Quadratic element in 1D.

where β is given in (2.27). The analytical solution of this linear difference equation can also be determined, and is given by

$$u_j = C_1 + C_2 \left(\frac{1 + (1+\beta)Pe}{1 - (1-\beta)Pe} \right)^j .$$

To avoid spurious oscillations of the Galerkin method, the absolute value of β should be larger than the critical value

$$|\beta| \geq \beta_{\text{crit}} = 1 - \frac{1}{|Pe|}. \tag{2.32}$$

Note that this condition is verified by both the optimal value of beta, (2.27), and its doubly asymptotic approximation defined in (2.29).

2.2.3 Piecewise quadratic approximation in 1D

To further illustrate the deficiencies of the Galerkin finite element formulation in convection-dominated problems, let us now examine the numerical solution of a 1D convection–diffusion problem discretized with quadratic shape function elements. First, we shall establish the form of the convection and diffusion matrices. Then the discrete equations for the model problem (2.16) are analyzed.

One of the objectives of this section is to show that quadratic elements and, in general, high-order finite elements present serious difficulties in convection-dominated problems. This is mainly due to the fact that the behavior of interior and corner nodes is different. This will be illustrated in detail in the next section.

As shown in Figure 2.4, we consider a generic element with end nodes 1 and 3, and a mid-side node 2. With reference to the normalized coordinate $-1 \leq \xi \leq +1$, the shape functions of the element are

$$N_1(\xi) = \frac{1}{2}\xi(\xi - 1), \quad N_2(\xi) = 1 - \xi^2, \quad N_3(\xi) = \frac{1}{2}\xi(\xi + 1). \tag{2.33}$$

It follows that at any interior point of an element, one has $u(\xi) = N_1(\xi)\,\mathrm{u}_1 + N_2(\xi)\,\mathrm{u}_2 + N_3(\xi)\,\mathrm{u}_3$ and $x(\xi) = N_1(\xi)\,x_1 + N_2(\xi)\,x_2 + N_3(\xi)\,x_3$. If a uniform mesh is used, the middle node is located at $x_2 = (x_1 + x_3)/2$. Then, if the characteristic size is h (h is the distance between nodes *not the element size*), the following relations hold between the normalized and physical coordinates: $dx = h\,d\xi$ and

$$\frac{\partial N_b}{\partial x} = \frac{\partial N_b}{\partial \xi}\frac{\partial \xi}{\partial x} = \frac{1}{h}\frac{\partial N_b}{\partial \xi}, \qquad b = 1, 2, 3.$$

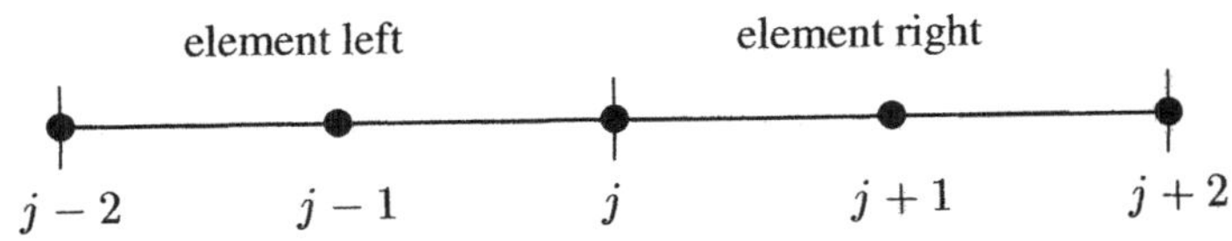

Fig. 2.5 Node numbering for a uniform mesh of quadratic elements.

With these interpolation functions the integrals in expression (2.18) are readily evaluated in closed form to obtain the convection, $\mathbf{C}^e$, and diffusion, $\mathbf{K}^e$, matrices of the quadratic element:

$$\mathbf{C}^e = a\int_{\Omega^e}\begin{pmatrix} N_1\frac{\partial N_1}{\partial x} & N_1\frac{\partial N_2}{\partial x} & N_1\frac{\partial N_3}{\partial x} \\ N_2\frac{\partial N_1}{\partial x} & N_2\frac{\partial N_2}{\partial x} & N_2\frac{\partial N_3}{\partial x} \\ N_3\frac{\partial N_1}{\partial x} & N_3\frac{\partial N_2}{\partial x} & N_3\frac{\partial N_3}{\partial x} \end{pmatrix} dx$$

$$= \frac{a}{2}\begin{pmatrix} -1 & 4/3 & -1/3 \\ -4/3 & 0 & 4/3 \\ 1/3 & -4/3 & 1 \end{pmatrix} \qquad (2.34a)$$

$$\mathbf{K}^e = \nu\int_{\Omega^e}\begin{pmatrix} \frac{\partial N_1}{\partial x}\frac{\partial N_1}{\partial x} & \frac{\partial N_1}{\partial x}\frac{\partial N_2}{\partial x} & \frac{\partial N_1}{\partial x}\frac{\partial N_3}{\partial x} \\ \frac{\partial N_2}{\partial x}\frac{\partial N_1}{\partial x} & \frac{\partial N_2}{\partial x}\frac{\partial N_2}{\partial x} & \frac{\partial N_2}{\partial x}\frac{\partial N_3}{\partial x} \\ \frac{\partial N_3}{\partial x}\frac{\partial N_1}{\partial x} & \frac{\partial N_3}{\partial x}\frac{\partial N_2}{\partial x} & \frac{\partial N_3}{\partial x}\frac{\partial N_3}{\partial x} \end{pmatrix} dx$$

$$= \frac{\nu}{6h}\begin{pmatrix} 7 & -8 & 1 \\ -8 & 16 & -8 \\ 1 & -8 & 7 \end{pmatrix}. \qquad (2.34b)$$

Here $\Omega^e = [x_{2e-1}, x_{2e+1}]$ with $e = 1, \ldots, \mathrm{n_{el}}$ ($\mathrm{n_{el}} = (\mathrm{n_{eq}} + 1)/2 = (\mathrm{n_{np}} - 1)/2$ in this case).

As with the linear element, we interpolate the source term $s(x)$ in (2.16) by means of the element shape functions,

$$s(\xi) = N_1(\xi)s_1 + N_2(\xi)s_2 + N_3(\xi)s_3, \qquad (2.35)$$

and obtain the components of the load vector $\mathbf{f}$ in (2.13) from element contributions of the form

$$\mathrm{f}_b^e = \int_{\Omega^e} N_b\,(N_1\,s_1 + N_2\,s_2 + N_3\,s_3)\,dx, \qquad b = 1, 2, 3.$$

In order to obtain in finite difference format the discrete equation at each node, the numbering sequence shown in Figure 2.5 is employed. With the element matrices presented in (2.34a) and (2.34b), and assembling in the usual finite element manner, one finds that the Galerkin method delivers two types of nodal equations representing the discrete counterpart of the convection–diffusion equation (2.16a):

1. *At a mid-side node* i (in Figure 2.5, $i = j+1$ or $i = j-1$) the discrete equation is obtained in the form

$$a\left(\frac{u_{i+1}-u_{i-1}}{2h}\right) - \nu\left(\frac{u_{i-1}-2u_i+u_{i+1}}{h^2}\right) = \frac{1}{10}\left(s_{i-1}+8s_i+s_{i+1}\right) \quad (2.36)$$

its l.h.s. is identical to equation (2.20) obtained using piecewise linear approximations.

2. *At the corner (inter-element) node* j (see Figure 2.5) the discrete equation involves a stencil of five nodes and reads

$$\begin{aligned} a &\left[2\left(\frac{u_{j+1}-u_{j-1}}{2h}\right) - \left(\frac{u_{j+2}-u_{j-2}}{4h}\right)\right] \\ &- \nu\left[2\left(\frac{u_{j-1}-2u_j+u_{j+1}}{h^2}\right) - \left(\frac{u_{j-2}-2u_j+u_{j+2}}{4h^2}\right)\right] \\ &= \frac{1}{10}\left(-s_{j-2}+2s_{j-1}+8s_j+2s_{j+1}-s_{j+2}\right). \end{aligned} \quad (2.37)$$

As already done for the linear element, a first test for quadratic elements is performed by solving the linear convection–diffusion problem (2.16) with $L = 1$ and $s = 1$. The results obtained with the Galerkin formulation on a uniform mesh of five quadratic elements are displayed in Figure 2.6 and compared with the exact solution given, as previously, by (2.23). Again the Galerkin solution is characterized by spurious node-to-node oscillations when convective effects become important.

Remark 2.8. Similar conclusions can be drawn with another well-known academic problem, see Section 2.6.2. The same linear convection–diffusion equation (2.16a) is solved over a unit domain, $L = 1$, with no source term, $s = 0$, and the Dirichlet boundary conditions $u(0) = 0$ and $u(1) = 1$. The exact solution is, in this case,

$$u(x) = \frac{1-\exp(\gamma x)}{1-\exp(\gamma)},$$

with $\gamma = a/\nu$.

2.2.4 Analysis of the discrete equations

We shall now analyze the discrete equations (2.36) and (2.37) delivered by the Galerkin method on a uniform mesh of quadratic elements. As previously done for the piecewise linear approximations, see Section 2.2.2, an exact scheme is identified for both the mid-side and corner nodes.

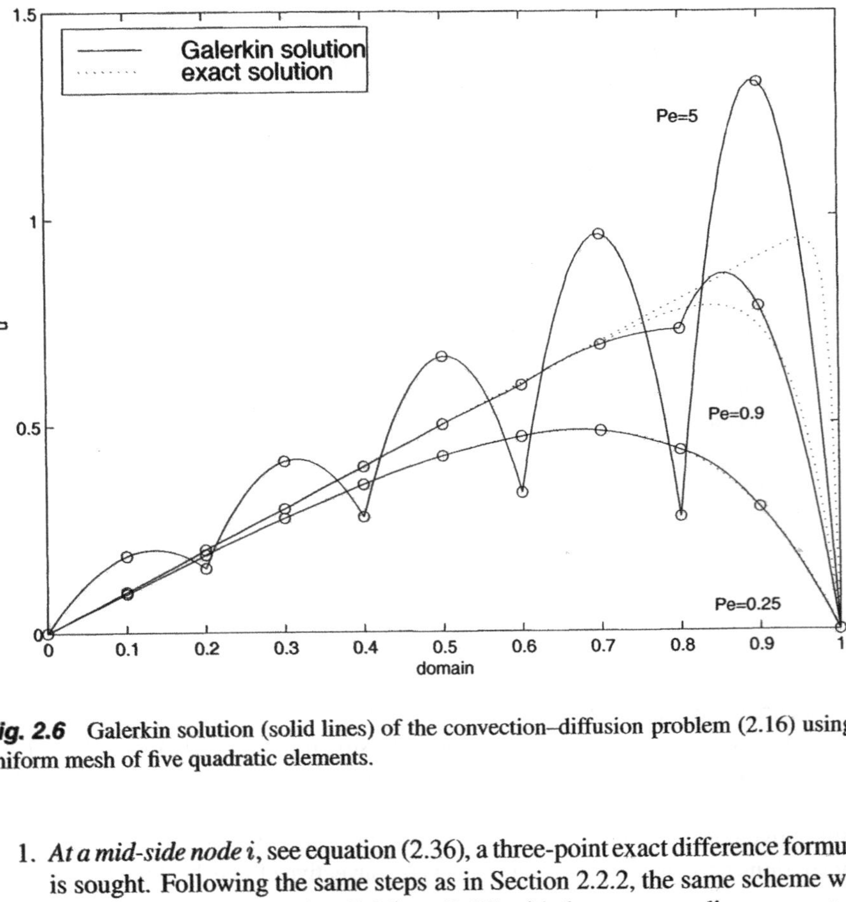

Fig. 2.6 Galerkin solution (solid lines) of the convection–diffusion problem (2.16) using a uniform mesh of five quadratic elements.

1. *At a mid-side node* i, see equation (2.36), a three-point exact difference formula is sought. Following the same steps as in Section 2.2.2, the same scheme will be obtained, that is equation (2.26) or (2.28) with the corresponding parameters given by (2.27). Thus as previously discussed, two interpretations (upwinding or artificial diffusion) are again possible. Moreover, it follows that the mid-side node equation delivered by the Galerkin method corresponds to a modified equation with a reduced diffusion coefficient.

2. *At the corner (inter-element) node* j (see Figure 2.5) a new exact discrete equation is needed. Each difference equation now involves five nodes, see (2.37). The same procedure as in Section 2.2.2 is used. Notice, however, that now five coefficients must be determined, which need to satisfy the condition

$$\alpha_1 \, u_{j-2} + \alpha_2 \, u_{j-1} + \alpha_3 \, u_j + \alpha_4 \, u_{j+1} + \alpha_5 \, u_{j+2} = 1, \tag{2.38}$$

but only three equations are available. This is reasonable because an exact scheme can be obtained with three nodes. If more nodes are employed, families of schemes are possible. For instance, the following two-parameter (c_1 and c_2)

family will induce exact nodal schemes:

$$\begin{cases} \alpha_1 = \dfrac{a}{4h}(1+\coth Pe)\, c_1, \\ \alpha_2 = \dfrac{a}{4h}\left[-2(1+\coth Pe) - (1+3\coth Pe)\, c_1 - (1+\coth Pe)\, c_2\right], \\ \alpha_3 = \dfrac{a}{4h}\left[4\coth Pe - (1-3\coth Pe)\, c_1 + (1+3\coth Pe)\, c_2\right], \\ \alpha_4 = \dfrac{a}{4h}\left[2(1-\coth Pe) + (1-\coth Pe)\, c_1 + 2(1-3\coth Pe)\, c_2\right], \\ \alpha_5 = \dfrac{-a}{4h}(1-\coth Pe)\, c_2. \end{cases} \tag{2.39}$$

However, as previously done, we prefer to rewrite equation (2.38) as a discrete equation delivered by the Galerkin formulation with an added numerical diffusion. To this end, we rewrite equation (2.37) in the form

$$a\left[2\left(\frac{u_{j+1}-u_{j-1}}{2h}\right) - \left(\frac{u_{j+2}-u_{j-2}}{4h}\right)\right] - (\nu + \bar{\nu}_{\mathtt{corner}})\left[2\left(\frac{u_{j-1}-2u_j+u_{j+1}}{h^2}\right) - \left(\frac{u_{j-2}-2u_j+u_{j+2}}{4h^2}\right)\right] = 1, \tag{2.40}$$

where a new artificial diffusion coefficient $\bar{\nu}_{\mathtt{corner}}$ must be defined. In fact, after substitution of (2.39) into (2.38) and imposing its equivalence to (2.40), one obtains

$$\begin{cases} c_1 = 1 - \dfrac{1}{2}\coth Pe\, \dfrac{1-\coth Pe}{2-\coth Pe} \\ c_2 = 1 + \dfrac{1}{2}\coth Pe\, \dfrac{1+\coth Pe}{2-\coth Pe} \end{cases}$$

and

$$\bar{\nu}_{\mathtt{corner}} = \frac{ah}{2}\,\frac{(\coth Pe - 1/Pe) - (\cosh Pe)^2(\coth 2Pe - 1/(2Pe))}{1-(\cosh Pe)^2/2}. \tag{2.41}$$

Comparing the discrete equation (2.37) delivered by the Galerkin method and the exact scheme (2.40), one notes that the truncation error introduced at corner nodes by the Galerkin method is again in the form of a spurious diffusion.

This analysis confirms that any standard discrete formulation will induce different responses for interior and corner nodes. Moreover, an exact nodal formulation must treat, as we have seen, differently interior and corner nodes.

We shall now look at ways to remedy the lack of stability of the Galerkin finite element method and thereby obtain stable and accurate approximations to convection-dominated problems.

2.3 EARLY PETROV–GALERKIN METHODS

The Galerkin formulation presents serious deficiencies in convection-dominated problems. These deficiencies can be interpreted and cured in two ways, see Section 2.2.2. First, diffusion was added in order to counterbalance the negative numerical diffusion introduced by the Galerkin approximation based on linear elements. Second, an upwind approximation of the convective term is used because the centered scheme employed is not ideal in convection-dominated problems. Precisely, the early remedies were based on these two philosophies. In fact, as will be seen later, both methodologies are equivalent. That is, an upwind approximation induces numerical diffusion and vice versa.

2.3.1 Upwind approximation of the convective term

In the framework of the *finite difference method*, numerical diffusion can be introduced by replacing the second-order accurate central approximation to the convective term by a first-order upwind approximation defined, for $a > 0$, by

$$u_x(x_j) \approx \frac{u_j - u_{j-1}}{h}. \tag{2.42}$$

If upwind differencing is used on the convective term instead of central differencing, the discrete convection–diffusion equation at node j becomes, in the absence of source term,

$$a\,\frac{u_j - u_{j-1}}{h} - \nu\,\frac{u_{j+1} - 2u_j + u_{j-1}}{h^2} = 0. \tag{2.43}$$

The upwind derivative of the convective term introduces a numerical dissipation which comes in addition to the physical diffusion ν, as can be seen from the Taylor series development of the convective term around x_j:

$$a\frac{u_j - u_{j-1}}{h} = a\,u_x(x_j) - \frac{ah}{2}\,u_{xx}(x_j) + \mathcal{O}(h^2).$$

An added diffusion of magnitude $ah/2$ has thus been introduced by the upwind approximation of the convective term. As a matter of fact, the discrete equation (2.43) also results from the central difference approximation of the equation

$$a\,u_x - \left(\nu + \frac{ah}{2}\right) u_{xx} = 0,$$

which includes an added numerical diffusion of magnitude $ah/2$. It has, however, been noted that in most cases the upwind treatment of the convective term as given in (2.43) leads to excessively dissipative results and this has given rise to many criticisms against the upwind technique, see for instance Davis and Mallinson (1976), Gresho and Lee (1979) and Leonard (1979). That a full upwind treatment of the convective term leads to stable, but overly diffusive, results can be appreciated from Figure 2.7 which shows the upwind solution of the model problem (2.16) in comparison with the

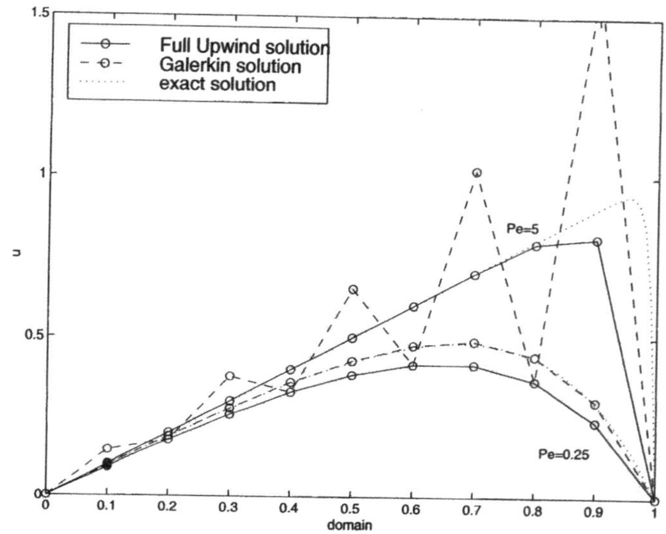

Fig. 2.7 Full upwind solution (solid lines) of the model convection–diffusion problem (2.16) with $L = 1$ and $s = 1$. Dashed lines show the Galerkin approximation and dotted lines the exact solution (2.23).

Galerkin approximation and the exact solution. Note that for large Péclet numbers the solution is stable and close to the exact one. However, for low values of the Péclet number, when the Galerkin approximation is accurate, the full upwind solution is clearly over diffusive.

2.3.2 First finite elements of upwind type

In a *finite element* framework, several different techniques can be utilized to achieve the upwind effect. The key idea in practically all the proposed finite element formulations of upwind type has been to replace the standard Galerkin formulation with a so-called *Petrov–Galerkin* weighted residual formulation in which the weighting function may be selected from a different class of functions than the approximate solution. The basic idea behind such an approach has been that the optimal approximation property, the basis of the success of the standard Galerkin finite element method in self-adjoint problems, could be carried over to convection–diffusion problems by means of the Petrov–Galerkin formulation.

The first upwind finite element formulations were presented in the 1970s by the Dundee and Swansea research groups (see, e.g., Christie et al., 1976; Heinrich et al., 1977; Heinrich and Zienkiewicz, 1979; Griffiths and Mitchell, 1979). These were based on *modified weighting functions* such that the element upstream of a node is weighted more heavily than the element downstream of a node. An example of such a weighting function is shown in Figure 2.8. In this way, the finite element equivalent

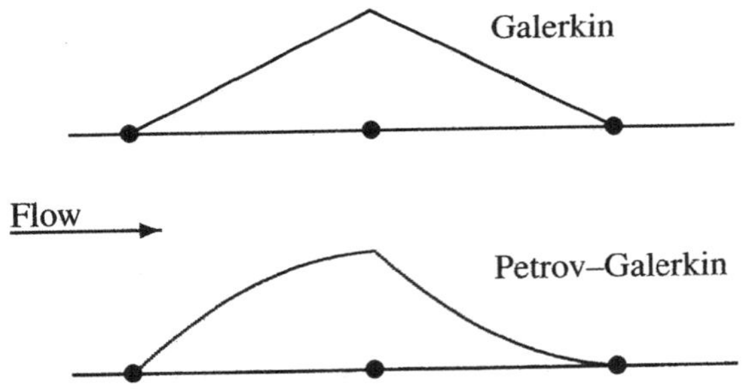

Fig. 2.8 Upwind-type weighting function.

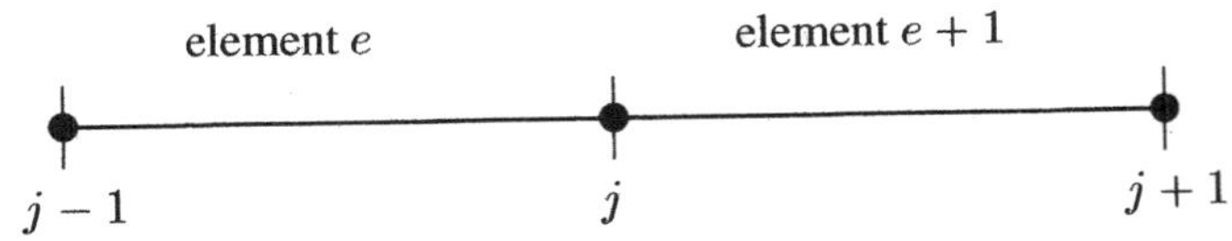

Fig. 2.9 Node numbering for two consecutive linear elements.

of upwind differencing was invented in the form of a Petrov–Galerkin method based upon an upwind distortion of the weighting function.

To give an example of upwind test functions, we consider again the solution of the 1D model problem (2.16) with piecewise linear approximations. Let j be an internal node belonging to elements e and $e+1$, see Figure 2.9. The shape functions of node j are given by

$$N_j = \begin{cases} N_2 = \frac{1}{2}(1+\xi) & \text{in element } e, \\ N_1 = \frac{1}{2}(1-\xi) & \text{in element } e+1, \end{cases} \quad \text{for } \xi \in [-1, 1].$$

Upwind test functions, $\tilde{w}_j$, giving more weight to the upwind element e than to the downwind element $e+1$, can be constructed by adding and subtracting from the Galerkin weights a bubble function of amplitude proportional to a free parameter β as follows:

$$\tilde{w}_j = \begin{cases} \tilde{w}_2 = \frac{1}{2}(1+\xi) + \frac{3}{4}\beta(1-\xi^2) & \text{in element } e, \\ \tilde{w}_1 = \frac{1}{2}(1-\xi) - \frac{3}{4}\beta(1-\xi^2) & \text{in element } e+1, \end{cases} \quad \text{for } \xi \in [-1, 1].$$

If we replace the Galerkin weighting function w_j with the above Petrov–Galerkin test functions in the weak form (2.17) of the convection–diffusion problem (2.16) with $s = 0$, the resulting modified weak form reads

$$\int_0^L (\tilde{w}\, a\, u_x + \tilde{w}_x\, \nu\, u_x)\, dx = 0.$$

Discretization of this weak form yields the following equation at the internal node j:

$$\big[1 + (1+\beta)Pe\big](u_j - u_{j-1}) - \big[1 - (1-\beta)Pe\big](u_{j+1} - u_j) = 0.$$

The parameter β controls the amount of numerical diffusion (or upwinding). Note that the previous equation is in fact equation (2.31). Thus, as noted previously, choosing the parameter $|\beta| \geq \beta_{\text{crit}} = 1 - 1/|Pe|$, see equation (2.32), will avoid an oscillatory solution. Moreover, the exact solution at the nodes is obtained with β defined by (2.27), that is $\beta = \coth Pe - 1/Pe$.

Though giving exact solutions for the 1D steady convection–diffusion equation, this initial upwind finite element formulation suffered from severe shortcomings in application to more complicated situations. These early Petrov–Galerkin methods were able to deliver stable numerical results; however, they produce in general an excessive numerical dissipation and were thus subject to the same criticism as upwind differences. Moreover, the formulation requires the use of higher-order weighting functions, which make the computer implementation of this Petrov–Galerkin method more difficult and more costly than the classical Galerkin method as regards numerical integration.

As suggested by Hughes (1978), the upwind effect can also be achieved through a modification of the numerical quadrature rule for the convective term. In the 1D case, in which piecewise linear elements are employed, a single quadrature point, $\bar{\xi}$, is positioned within the element according to

$$\bar{\xi} = \coth Pe - \frac{1}{Pe}.$$

In the case of a bilinear quadrilateral, the location of the quadrature point is defined by

$$\bar{\boldsymbol{\xi}} = \left\{ \begin{array}{c} \bar{\xi} \\ \bar{\eta} \end{array} \right\}, \tag{2.44a}$$

$$\bar{\xi} = \coth Pe_\xi - 1/Pe_\xi, \qquad \bar{\eta} = \coth Pe_\eta - 1/Pe_\eta, \tag{2.44b}$$

$$Pe_\xi = a_\xi h_\xi/(2\nu), \qquad Pe_\eta = a_\eta h_\eta/(2\nu), \tag{2.44c}$$

$$a_\xi = \boldsymbol{e}_\xi \cdot \boldsymbol{a}, \qquad a_\eta = \boldsymbol{e}_\eta \cdot \boldsymbol{a}, \tag{2.44d}$$

where the unit vectors, $\boldsymbol{e}_\xi$ and $\boldsymbol{e}_\eta$, and the element lengths, h_ξ and h_η, are defined as shown in Figure 2.10, and $\boldsymbol{a}$ and ν are evaluated at the origin of the $\{\xi, \eta\}$ coordinate system in the element. Unfortunately, this early two-dimensional scheme was found to exhibit the shortcoming of excessive crosswind diffusion.

Another early approach to the development of finite element schemes of the upwind type was suggested by Belytschko and Eldib (1979). Their method is based on amplifying certain terms in the convection matrix and for linear elements the behavior of the amplification scheme was shown to be similar to other upwind finite element schemes.

2.3.3 The concept of balancing diffusion

In the previous section non-centered discretizations of the convection term (upwind schemes) were presented. However, as noted previously, see Section 2.2.2, another

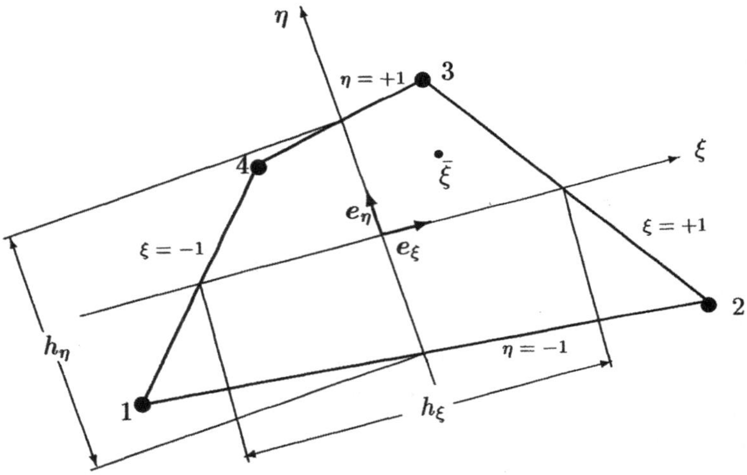

Fig. 2.10 Geometry of four-node quadrilateral element.

alternative to improve the behavior of the Galerkin method (with linear elements) in convection-dominated problems is to add artificial diffusion to counteract the negative dissipation introduced by the Galerkin formulation. This alternative is used next and will be shown to coincide with an upwind scheme.

2.3.3.1 *Linear elements in 1D* For the simple case of the linear convection–diffusion problem (2.16) with a constant source term, it is possible to formulate an optimal upwind technique producing the exact solution at the nodes of a uniform mesh of linear elements for all values of the Péclet number. Exploiting the results in Section 2.2.2, an optimal method can be constructed by considering a Galerkin approximation of the modified equation

$$a\,u_x - (\nu + \bar{\nu})\,u_{xx} = 0, \text{ with } \bar{\nu} = \beta\frac{ah}{2}$$

containing a free parameter β which governs the amplitude of the added numerical diffusion. Note that the source is taken as zero, $s = 0$, in order to simplify the developments.

For the convection–diffusion problem (2.16) with Dirichlet boundary conditions at $x = 0$ and $x = L$, Hughes and Brooks (1979) replace the usual weak formulation

$$\int_0^L (w\,a\,u_x + w_x\,\nu\,u_x)\;dx = 0,$$

where w denotes the weighting function, by the following weak statement

$$\int_0^L (w\,a\,u_x + w_x\,(\nu + \bar{\nu})\,u_x)\;dx = 0, \tag{2.45}$$

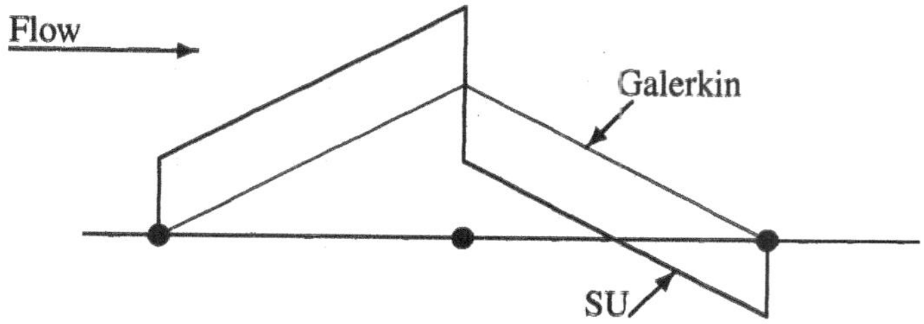

Fig. 2.11 Weighting function of the streamline-upwind (SU) method for linear elements.

where the magnitude of the added diffusion, $\bar{\nu}$, is governed by the free parameter β $(0 \leq \beta \leq 1)$. The optimal value of parameter β is given by (2.27), that is $\beta = \coth Pe - 1/Pe$, and the value $\beta = 1$ corresponds to full upwind differencing as in (2.42).

Since the added numerical diffusion counterbalances the negative diffusion introduced by the Galerkin method, it is also termed *balancing diffusion* (Kelly et al., 1980). In view of the definition of the added diffusion, $\bar{\nu} = \beta a h/2$, we note that the weak statement (2.45) can be rewritten in the form

$$\int_0^L \left[\left(w + \beta \frac{h}{2} w_x \right) a u_x + w_x \nu \, u_x \right] dx = 0, \tag{2.46}$$

which shows that the balancing diffusion method uses a *modified weighting function*, given by $\tilde{w} = w + \beta(h/2)w_x$ for the convective term only. Note that equation (2.46) does not correspond to a consistent Petrov–Galerkin formulation because the modified weighting function is only applied to the convective term. Moreover, the modified function is discontinuous at the inter-element boundaries, as shown in Figure 2.11 for the case of a linear element. Since it gives more weight to the element upstream of a node, the modified function is clearly an upwind-type weighting function. We shall refer to this scheme as the *streamline-upwind* (SU) method in view of its generalization to multiple dimensions. In fact, in several spatial dimensions diffusion is added in the flow direction only and not transversely.

To perform the spatial discretization of the weak form (2.46) using linear finite elements, we proceed as in Section 2.2.1 for the Galerkin formulation. The integrals in (2.46) are easily evaluated in closed form. They lead to the following SU form of the convection matrix:

$$\mathbf{C}^e_{SU} = \frac{a}{2} \begin{pmatrix} -(1-\beta) & 1-\beta \\ -(1+\beta) & 1+\beta \end{pmatrix}.$$

As shown by the weak form (2.46), the element diffusion matrix, $\mathbf{K}^e_{SU}$, possesses the same structure as the Galerkin diffusion matrix, $\mathbf{K}^e$, defined in (2.19b).

2.3.3.2 Quadratic elements in 1D Our objective is now to develop modified weighting functions giving exact results for all values of the Péclet number when approximating the 1D linear convection–diffusion equation by means of a uniform mesh of quadratic finite elements.

As shown in Section 2.2.3, the piecewise quadratic approximation generates two types of nodal equations which possess distinct truncation errors. This somehow complicates the formulation of an added diffusion method with respect to linear finite elements, since different values of the balancing diffusion should be incorporated into the corner node and the mid-side node equations given by the Galerkin formulation.

At mid-side nodes, the optimal value of the added diffusivity is given by the same formula obtained for linear 1D elements, that is equation (2.27). At corner nodes, the Galerkin method introduces another numerical diffusion and the optimal value of the diffusivity is given by equation (2.41). Thus, the artificial viscosity that needs to be added at the mid-side and corner nodes is, respectively, $\bar{\nu} = \beta\, ah/2$ with

$$\beta = \coth Pe - 1/Pe \tag{2.47a}$$

and $\bar{\nu}_{\text{corner}} = \beta_{\text{corner}}\, ah/2$, where

$$\beta_{\text{corner}} = \frac{(\coth Pe - 1/Pe) - (\cosh Pe)^2(\coth 2Pe - 1/(2Pe))}{1 - (\cosh Pe)^2/2}. \tag{2.47b}$$

We now have all the ingredients to formulate the balancing diffusion method for 1D quadratic elements. As with linear elements the weak form reads

$$\int_0^L (\tilde{w}\, a\, u_x + w_x\, \nu\, u_x)\, dx = 0,$$

but now the modified weighting function $\tilde{w}$ depends on the type of nodal equation to be weighted:

$$\tilde{w}_j = \begin{cases} N_j + \beta\,(h/2)\,(dN_j/dx) & \text{with } j \text{ mid-side node,} \\ N_j + \beta_{\text{corner}}\,(h/2)\,(dN_j/dx) & \text{with } j \text{ corner node,} \end{cases}$$

where β and β_{corner} are defined in (2.47).

On this basis, we can perform the spatial discretization of the 1D linear convection–diffusion equation using the quadratic shape functions defined in (2.33). Then, the SU form of the element convection matrix is found to be given by

$$\mathbf{C}^e_{SU} = \frac{a}{2}\begin{pmatrix} -1 + 7/6\,\beta_{\text{corner}} & 4/3\,(1-\beta_{\text{corner}}) & -1/6\,(2-\beta_{\text{corner}}) \\ -4/3\,(1+\beta) & 8/3\,\beta & 4/3\,(1-\beta) \\ 1/6\,(2+\beta_{\text{corner}}) & -4/3\,(1+\beta_{\text{corner}}) & 1+7/6\,\beta_{\text{corner}} \end{pmatrix},$$

while the element diffusion matrix, $\mathbf{K}^e_{SU}$, possesses the same structure as the Galerkin diffusion matrix, $\mathbf{K}^e$, defined in (2.34b).

The use of the above balancing diffusion methods for linear and quadratic elements in the solution of convection-dominated problems is illustrated in Figure 2.12 where exact solutions are obtained at the nodes. Compare these results with the corresponding Galerkin approximations shown in Figures 2.1 and 2.6.

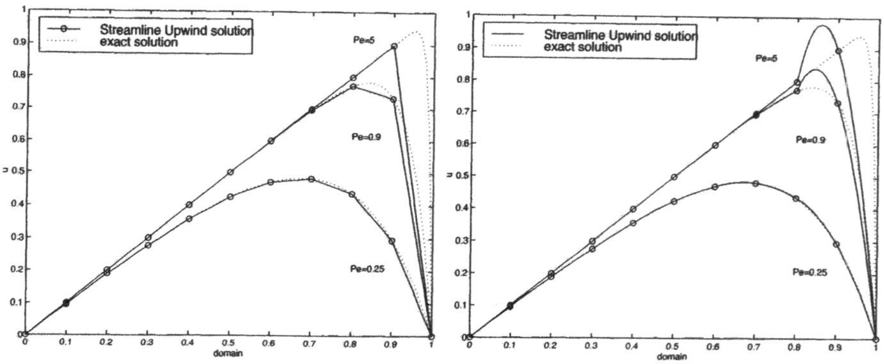

Fig. 2.12 Streamline-upwind (artificial diffusion) solution (solid lines) of the convection–diffusion problem (2.16) with $L = 1, s = 1$ using a uniform mesh of 10 linear elements (left) and 5 quadratic elements (right). Dotted lines show the exact solution.

Moreover, it is important to note that the exact nodal solution for quadratic elements presents, in convection-dominated cases, an important overshoot (recall that a uniform mesh is employed). This behavior, which is inherent to the quadratic approximation, will be an important drawback in nonlinear problems such as the ones discussed in Chapters 4 and 6.

2.3.3.3 Multidimensional case: the concept of SU schemes The extension to multidimensional domains of the concept of modified weighting functions discussed above is not trivial. The crucial issue is that the balancing diffusion should be added in the flow direction only, and not transversely. The reason is that convective transport takes place along the streamlines and adding diffusion transversely to the flow leads to overly diffusive results, due to an excess of crosswind diffusion.

In order to extend the added diffusion method to deal with multidimensional problems, Hughes and Brooks proposed in a series of papers (Hughes and Brooks, 1979; Brooks and Hughes, 1982; Hughes and Brooks, 1982) to construct the artificial diffusion operator in tensorial form to act only in the flow direction and not transversely. This leads to the concept of SU schemes, which account for the directional character of the convective term. The idea of adding diffusion along the streamlines was also exploited by Kelly et al. (1980) and described as *anisotropic balancing dissipation*. This is achieved by replacing the scalar artificial diffusion coefficient $\bar{\nu}$ in (2.45) by the tensor diffusivity

$$\tilde{\nu}_{ij} = \bar{\nu}\, a_i a_j / \|\boldsymbol{a}\|^2, \tag{2.48}$$

where a_i is the component of flow velocity, $\boldsymbol{a}$, along the coordinate direction x_i.

The development of a general theory to optimally select $\bar{\nu}$ is still an area of current research. For the basic isoparametric elements a simple generalization of the 1D definition is usually adopted. For example, in the case of the bilinear quadrilateral one takes

$$\bar{\nu} = \left(\bar{\xi}\, a_\xi h_\xi + \bar{\eta}\, a_\eta h_\eta\right)/2, \tag{2.49}$$

where $\bar{\xi}$ and $\bar{\eta}$ are defined in (2.44), and the standard notation for the normalized coordinates defined in Figure 2.10 is employed.

Remark 2.9. The tensor diffusivity, $\tilde{\boldsymbol{\nu}}$, represents a diffusion acting only in the direction of the flow and not transversely. In two dimensions this can be verified assuming that locally the x_1 direction is aligned with the streamlines and the x_2 direction is perpendicular. That is, the components of the convection velocity are $\boldsymbol{a} = (1, 0)^T$. In this coordinate system the artificial diffusivity matrix defined by (2.48) is

$$\tilde{\boldsymbol{\nu}} = \bar{\nu}\begin{pmatrix} 1 & 0 \\ 0 & 0 \end{pmatrix}$$

which clearly manifests the absence of crosswind diffusion.

In order to determine the *streamline-upwind test function*, that is the multidimensional modified weighting function, the weak form (2.45) is rewritten for $\mathrm{n_{sd}}$ spatial dimensions, namely: find $u \in \mathcal{S}$ such that for all $w \in \mathcal{V}$

$$\int_\Omega \Big[w(\boldsymbol{a}\cdot\boldsymbol{\nabla}u) + \boldsymbol{\nabla}w\cdot(\nu\mathbf{I}_{\mathrm{n_{sd}}} + \tilde{\boldsymbol{\nu}})\cdot\boldsymbol{\nabla}u \Big]\, d\Omega = 0, \tag{2.50}$$

where $\mathbf{I}_{\mathrm{n_{sd}}}$ is an identity matrix of dimension $\mathrm{n_{sd}}$ and the added diffusion $\tilde{\boldsymbol{\nu}}$ has been incorporated. In fact, given the definition of this added tensor diffusivity, see (2.48), the added term can be modified as follows:

$$\int_\Omega \boldsymbol{\nabla}w\cdot\tilde{\boldsymbol{\nu}}\cdot\boldsymbol{\nabla}u\, d\Omega = \int_\Omega \frac{\bar{\nu}}{\|\boldsymbol{a}\|^2}(\boldsymbol{a}\cdot\boldsymbol{\nabla}w)(\boldsymbol{a}\cdot\boldsymbol{\nabla}u)\, d\Omega.$$

Consequently, equation (2.50) can be rewritten as

$$\int_\Omega \Big\{ \Big[w + \frac{\bar{\nu}}{\|\boldsymbol{a}\|^2}(\boldsymbol{a}\cdot\boldsymbol{\nabla}w) \Big](\boldsymbol{a}\cdot\boldsymbol{\nabla}u) + \nu\boldsymbol{\nabla}w\cdot\boldsymbol{\nabla}u \Big\}\, d\Omega = 0.$$

Therefore, the SU test function, which only affects the convective term, is defined as

$$\tilde{w} = w + \frac{\bar{\nu}}{\|\boldsymbol{a}\|^2}(\boldsymbol{a}\cdot\boldsymbol{\nabla}w). \tag{2.51}$$

Finally, the SU method can be interpreted as the Galerkin method plus an extra term introducing the SU added numerical diffusivity:

$$\underbrace{\int_\Omega \Big[w(\boldsymbol{a}\cdot\boldsymbol{\nabla}u) + \nu\boldsymbol{\nabla}w\cdot\boldsymbol{\nabla}u \Big]\, d\Omega}_{\text{Standard Galerkin}} + \underbrace{\int_\Omega \frac{\bar{\nu}}{\|\boldsymbol{a}\|^2}(\boldsymbol{a}\cdot\boldsymbol{\nabla}w)(\boldsymbol{a}\cdot\boldsymbol{\nabla}u)\, d\Omega}_{\text{Added SU term}} = 0. \tag{2.52}$$

As noted previously, the perturbation added to the test function is discontinuous along the element edges or interfaces. Thus, the extra term in (2.52) is only computed in the element interiors. When this methodology is generalized for the original steady convection–diffusion problem, equations (2.1), the weak form, (2.7), associated with the SU stabilized technique becomes

$$
a(w,u) + c(\boldsymbol{a};w,u) \\
+ \underbrace{\sum_e \int_{\Omega^e} \frac{\bar{\nu}}{\|\boldsymbol{a}\|^2} (\boldsymbol{a}\cdot\boldsymbol{\nabla} w)(\boldsymbol{a}\cdot\boldsymbol{\nabla} u)\, d\Omega}_{\text{SU stabilization term}} = (w,s) + (w,h)_{\Gamma_N} \qquad (2.53)
$$

where $\bar{\nu}$ is given by (2.49) and the constants therein by equations (2.44).

This approach, called the SU formulation, makes use of upwind test functions only for the convective term. As was shown, it is equivalent to the use of an artificial diffusion coefficient acting only along the streamline direction. This method produces smooth solutions for high Péclet numbers. Moreover, it is able to deliver exact results on a uniform 1D grid for the linear, constant coefficient, convection–diffusion equation. However, accuracy problems still remain in more complicated cases, such as spatially variable source term or convection velocity, as well as time-dependent problems. In fact, modifying the test function only for the convective term produces a non-residual formulation. Note that the true solution of the differential equation is no longer a solution to the weak problem (2.53). A discussion of these accuracy problems (in a finite difference context) and numerical remedies to overcome them can be found in Christie (1985), Leonard (1979) and Morton (1996). Herein, we shall limit ourselves to indicate that when the governing equation includes a source term, this term must, for consistency, be discretized using the same SU weighting function as the convective term. Failing to do this would lead to erroneous results, as will be seen in the next section, see also the numerical results in Figure 2.13.

Hughes and Brooks (1982) subsequently proposed to apply the modified weighting function to all terms in the equation in order to obtain a consistent formulation. Moreover, they noted that for linear elements the perturbation to the standard test function can be neglected in the diffusion term. The concept of adding diffusion along the streamlines in a consistent manner has been successfully exploited in the Streamline-Upwind Petrov–Galerkin (SUPG) method that will be described next.

2.4 STABILIZATION TECHNIQUES

In order to stabilize the convective term in a consistent manner (*consistent stabilization*), ensuring that the solution of the differential equation is also a solution of the weak form, Hughes and co-workers have proposed several techniques. They follow a structure similar to (2.53). That is, an extra term over the element interiors is added to the Galerkin weak form. This term is a function of the residual of the differential equation to ensure consistency. Note that this is not the case in (2.53). These methods are especially designed for the steady convection–diffusion equation, see (2.1a), or more generally for the steady convection–diffusion–reaction equation, namely

$$
\boldsymbol{a}\cdot\boldsymbol{\nabla} u - \boldsymbol{\nabla}\cdot(\nu\boldsymbol{\nabla} u) + \sigma u = s \quad \text{in } \Omega, \qquad (2.54)
$$

with the usual Dirichlet and Neumann boundary conditions shown in (2.1). The residual of the differential equation (2.54) is

$$\mathcal{R}(u) = \underbrace{\boldsymbol{a}\cdot\boldsymbol{\nabla} u - \boldsymbol{\nabla}\cdot(\nu\boldsymbol{\nabla} u) + \sigma u}_{\mathcal{L}(u)} - s = \mathcal{L}(u) - s, \tag{2.55}$$

where $\mathcal{L}$ is the differential operator associated with the differential equation. Note that as soon as we restrict ourselves to the finite dimensional spaces standard in finite elements (and u is replaced by u^h) $\mathcal{R}(u)$ is computed only for each element interior Ω^e. The general form of these (consistent) *stabilization techniques* is

$$a(w,u) + c(\boldsymbol{a}; w,u) + (w, \sigma\, u) + \underbrace{\sum_e \int_{\Omega^e} \mathcal{P}(w)\, \tau\, \mathcal{R}(u)\, d\Omega}_{\text{stabilization term}} = (w,s) + (w,h)_{\Gamma_N} \tag{2.56}$$

where $\mathcal{P}(w)$ is a certain operator applied to the test function, τ is the stabilization parameter (also called intrinsic time), and $\mathcal{R}(u)$ is the residual of the differential equation, see (2.55). The stabilization techniques are characterized by the definition of $\mathcal{P}(w)$. Here the Streamline-Upwind Petrov–Galerkin (SUPG) and the Galerkin/Least-squares (GLS) methods will be presented, see for instance Codina (1998) for a general presentation.

Remark 2.10. Note that in this section we have introduced the reaction, or production, term in the original convection–diffusion problem. This is done because now stabilization techniques usually treat both equations, (2.1a) and (2.54), in a uniform manner. Specific analyses of stabilization techniques for the convection–diffusion–reaction equation can be found in Tezduyar and Park (1986), Harari and Hughes (1994), or Idelsohn et al. (1996) among others.

2.4.1 The SUPG method

This stabilization technique is defined by taking

$$\mathcal{P}(w) = \boldsymbol{a}\cdot\boldsymbol{\nabla} w, \tag{2.57}$$

which, in fact, corresponds to the perturbation of the test function introduced in the SU method, see (2.53). Note that this method corresponds to the standard weak forms (2.5) or (2.7) with the SU test function (2.51) consistently applied to all terms of the equation. Thus, since the space of the test functions does not coincide with the space of the interpolation functions, this is in fact a Petrov–Galerkin formulation.

The restriction of the weak form (2.56) for the SUPG method, that is with the perturbation defined in (2.57), to the usual finite dimensional subspaces leads to the discrete problem that must be solved: find $u^h \in \mathcal{S}^h$ such that

$$a(w^h, u^h) + c(\boldsymbol{a}; w^h, u^h) + (w^h, \sigma u^h)$$
$$+ \sum_e \int_{\Omega^e} (\boldsymbol{a} \cdot \boldsymbol{\nabla} w^h) \tau \big[\boldsymbol{a} \cdot \boldsymbol{\nabla} u^h - \boldsymbol{\nabla} \cdot (\nu \boldsymbol{\nabla} u^h) + \sigma u^h - s\big] \, d\Omega$$
$$= (w^h, s) + (w^h, h)_{\Gamma_N} \qquad \text{for all } w^h \in \mathcal{V}^h, \tag{2.58}$$

where the stabilization parameter τ can be defined as

$$\tau = \bar{\nu}/\|\boldsymbol{a}\|^2 \tag{2.59}$$

with $\bar{\nu}$ given in (2.49) for 2D and $\bar{\nu} = \beta a h/2$ in 1D. Section 2.4.3 discusses in more detail the definition of τ.

Remark 2.11 (Exact nodal solutions in 1D for convection–diffusion). For the convection–diffusion equation discretized with linear elements, the stabilization term reduces to

$$\sum_e \int_{\Omega^e} (\boldsymbol{a} \cdot \boldsymbol{\nabla} w^h) \tau (\boldsymbol{a} \cdot \boldsymbol{\nabla} u^h - s) \, d\Omega$$

because second derivatives of u^h cancel out. The difference operator acting on u^h is identical to the one obtained with the non-residual formulation presented in Section 2.3.3.1. Thus, $\beta = \coth Pe - 1/Pe$ or $\tau = (h/2a)(\coth Pe - 1/Pe)$ produce the exact nodal solution on a uniform mesh in 1D.

For quadratic elements, however, the stabilization term becomes

$$\sum_e \int_{\Omega^e} (\boldsymbol{a} \cdot \boldsymbol{\nabla} w^h) \tau \big[\boldsymbol{a} \cdot \boldsymbol{\nabla} u^h - \boldsymbol{\nabla} \cdot (\nu \boldsymbol{\nabla} u^h) - s\big] \, d\Omega,$$

and the second derivatives must be accounted for in the element interiors. Moreover, as previously observed in Section 2.3.3.2, the intrinsic time τ must be different for mid-side, τ_{m}, and corner, τ_{c}, nodes. The parameters computed previously, see (2.47), will not produce the nodally exact solution on a uniform mesh when used in (2.58) because the difference operator has changed. In fact, taking into account these two parameters, one for the mid-side nodes and the other for the corner nodes, the difference equations associated with the weak form (2.58) become:

1. At a *mid-side node*,

$$a\Big(\frac{u_{j+1} - u_{j-1}}{2h}\Big) - (\nu + \tau_{\text{m}} a^2)\Big(\frac{u_{j+1} - 2u_j + u_{j-1}}{h^2}\Big) = 0.$$

2. At a *corner node*,

$$2\Big(a - \tau_{\text{c}} \frac{3a\nu}{h^2}\Big)\Big(\frac{u_{j+1} - u_{j-1}}{2h}\Big) - \Big(a - \tau_{\text{c}} \frac{6a\nu}{h^2}\Big)\Big(\frac{u_{j+2} - u_{j-2}}{4h}\Big)$$
$$-2(\nu + \tau_{\text{c}} a^2)\Big(\frac{u_{j+1} - 2u_j + u_{j-1}}{h^2}\Big) + (\nu + \tau_{\text{c}} a^2)\Big(\frac{u_{j+2} - 2u_j + u_{j-2}}{4h^2}\Big) = 0.$$

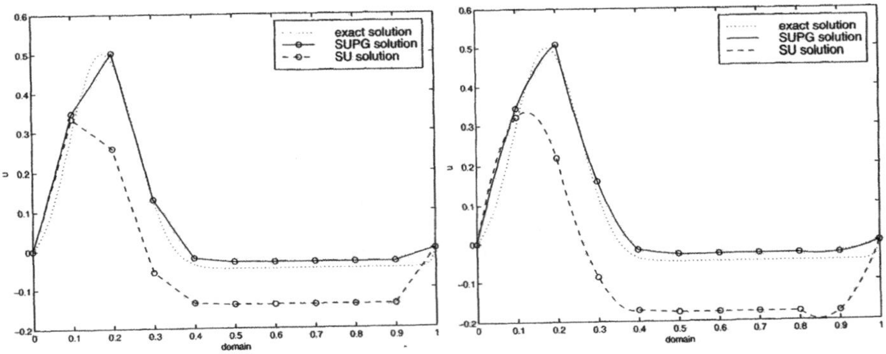

Fig. 2.13 SUPG (solid line) and SU (dashed line) solutions of the convection–diffusion problem (2.60) at $Pe = 5$ using a uniform mesh of 10 linear elements (left) and five quadratic elements (right). The dotted line shows the exact solution.

Note that the source term is taken equal to zero in order to simplify the expressions. However, this analysis can be extended to constant source terms as previously done in Section 2.3.3.2. Imposing now that the coefficients multiplying the nodal values are equal to the coefficients in the exact difference scheme, see Section 2.2.4, the optimal parameters are obtained: $\tau_m = \beta\, h/(2a)$ with, as always, $\beta = \coth Pe - 1/Pe$; and $\tau_c = \beta_c\, h/(2a)$ with

$$\beta_c = \frac{(2Pe - 1) + (-6Pe + 7)e^{-2Pe} + (-6Pe - 7)e^{-4Pe} + (2Pe + 1)e^{-6Pe}}{(Pe + 3) + (-7Pe - 3)e^{-2Pe} + (7Pe - 3)e^{-4Pe} - (Pe + 3)e^{-6Pe}}.$$

With these parameters, exact nodal solutions are obtained. That is, Figure 2.12 is reproduced exactly with the SUPG formulation. The above parameters were originally proposed by Codina (1993b) who used the concept of linear difference equation associated with the homogeneous linear convection–diffusion equation, see Remark 2.7, in order to determine them.

As noted earlier, the SUPG formulation performs better than the original SU technique. The influence of the consistent formulation can be noticed with a simple modification of the source term. Numerical solutions of the convection–diffusion problem,

$$\begin{cases} au_x - \nu u_{xx} = 5\, e^{-100(x-\frac{1}{8})^2} - 5\, e^{-100(x-\frac{1}{4})^2} & \text{in }]0,1[\\ u = 0 & \text{at } x = 0 \text{ and } x = 1, \end{cases} \tag{2.60}$$

are compared with the exact one in Figure 2.13.

One important issue in the SUPG method is the definition of the stabilization parameter τ. The stability and convergence analysis of this method (see Johnson, Nävert and Pitkäranta, 1984) allows us to determine the behavior of τ. The fact that the added stabilization term, see equation (2.58), is not symmetric introduces technical difficulties in establishing the stability of SUPG. This is avoided in the Galerkin/Least-squares stabilization technique because it introduces a symmetric stabilization term in a consistent manner.

2.4.2 The Galerkin/Least-squares method

The GLS technique is defined by imposing that the stabilization term in (2.56) is an element-by-element weighted least-squares formulation of the original differential equation. This corresponds to the following choice for the operator applied to the test function:

$$\mathcal{P}(w) = \mathcal{L}(w) = \boldsymbol{a} \cdot \boldsymbol{\nabla} w - \boldsymbol{\nabla} \cdot (\nu \boldsymbol{\nabla} w) + \sigma w. \tag{2.61}$$

With such a definition of operator $\mathcal{P}$, the weak form that must be solved is: find $u^h \in \mathcal{S}^h$ such that

$$a(w^h, u^h) + c(\boldsymbol{a}; w^h, u^h) + (w^h, \sigma\, u^h) + \sum_e \int_{\Omega^e} \mathcal{L}(w^h)\, \tau \left[\mathcal{L}(u^h) - s\right] d\Omega$$
$$= (w^h, s) + (w^h, h)_{\Gamma_N} \qquad \text{for all } w^h \in \mathcal{V}^h. \tag{2.62}$$

This equation can also be written, for the steady convection–diffusion–reaction equation (2.54), as

$$a(w^h, u^h) + c(\boldsymbol{a}; w^h, u^h) + (w^h, \sigma\, u^h)$$
$$+ \sum_e \int_{\Omega^e} \left[\boldsymbol{a} \cdot \boldsymbol{\nabla} w^h - \boldsymbol{\nabla} \cdot (\nu \boldsymbol{\nabla} w^h) + \sigma w^h\right] \tau \left[\boldsymbol{a} \cdot \boldsymbol{\nabla} u^h - \boldsymbol{\nabla} \cdot (\nu \boldsymbol{\nabla} u^h) + \sigma u^h\right] d\Omega$$
$$= (w^h, s) + (w^h, h)_{\Gamma_N} + \sum_e \int_{\Omega^e} \left[\boldsymbol{a} \cdot \boldsymbol{\nabla} w^h - \boldsymbol{\nabla} \cdot (\nu \boldsymbol{\nabla} w^h) + \sigma w^h\right] \tau s, \tag{2.63}$$

where it is important to notice that the stabilization term that affects the l.h.s. is symmetric. This symmetry is a major advantage in establishing stability.

From a practical point of view there is no major difference between SUPG and GLS methods. In fact, both methods are identical for convection–diffusion (no reaction) and with linear elements (the second-order derivatives are zero in the element interiors). Moreover, the qualitative influence of each term in the definition of $\mathcal{P}$, equation (2.61), may be interpreted as follows:

$$\mathcal{P}(w) = \mathcal{L}(w) = \underbrace{\boldsymbol{a} \cdot \boldsymbol{\nabla} w}_{\text{SUPG}} - \underbrace{\boldsymbol{\nabla} \cdot (\nu \boldsymbol{\nabla} w)}_{0} + \underbrace{\sigma w}_{\text{Galerkin}}.$$

The first term corresponds to the SUPG stabilization, the second term is zero for linear elements, and the third term is a Galerkin weighting. Thus, for linear elements and a constant positive reaction, GLS is SUPG with the Galerkin term weighted $1 + \sigma\,\tau$ times more. This implies that the instabilities introduced by Galerkin are a little more amplified in GLS compared with SUPG. See Figure 2.14 for a comparison of these methods.

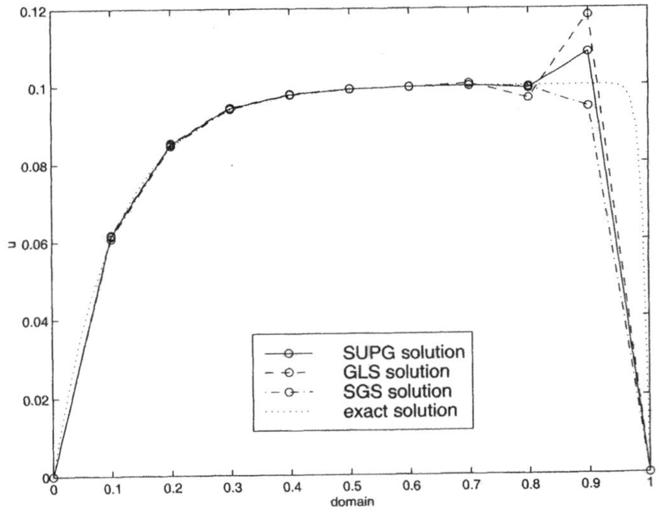

Fig. 2.14 Comparison between SUPG (solid line), GLS (dashed line) and SGS (dash–dot line) solutions of the convection–diffusion–reaction problem (2.54) for $a = 1$, $\nu = 10^{-2}$, $\sigma = 10$ and a uniform mesh of 10 linear elements. The dotted line shows the exact solution.

Remark 2.12. This minor problem of the GLS stabilization technique is overcome in the simplest version of the sub-grid scale (SGS) method, see Section 2.5.3, because in this case the stabilization operator involves the adjoint operator, $\mathcal{L}^*$, of $\mathcal{L}$, namely $\mathcal{P}(w) = -\mathcal{L}^*(w) = \boldsymbol{a} \cdot \boldsymbol{\nabla} w + \boldsymbol{\nabla} \cdot (\nu \boldsymbol{\nabla} w) - \sigma w$. Consequently, following the same rationale as before, in this case the Galerkin term is weighted by $1 - \sigma\,\tau$ and thus has less influence than in SUPG.

2.4.3 The stabilization parameter

The parameter τ plays a major role in stabilization techniques. Note first that τ is in fact a stabilization coefficient matrix, as noted earlier for quadratic elements or more typically for systems of differential equations, see for instance the original paper of Hughes and Mallet (1986a) or more recently Codina (2000). Here the scalar 1D convection–diffusion–reaction equation (2.54) is used for illustration purposes.

For scalar equations a number of definitions of parameter τ have been proposed and tested (see for instance Tezduyar and Ganjoo, 1986; Tezduyar and Osawa, 2000). Here, however, only three are recalled.

First, *superconvergence* (in the form of nodally exact results) is obtained for linear elements in 1D for convection–diffusion, see Remark 2.11 and Section 2.2.2, if

$$\tau = \frac{h}{2a}\Big(\coth(Pe) - 1/Pe\Big).$$

This property can be generalized for quadratic elements (Remark 2.11) but not for higher dimensions or general convection–diffusion systems.

It is obvious that τ must vanish when the mesh is refined (no stabilization is necessary for a fine enough mesh). In fact, convergence is affected by the asymptotic behavior of τ. An error analysis allows us to determine the structure of the stabilization coefficient as a function of the discretization (mesh size h) and the parameters in the differential equation (convection velocity a, diffusivity ν and reaction σ).

Second, an algebraic analysis is the basis of Codina's (2000) definition of τ,

$$\tau = \left(\frac{2a}{h} + \frac{4\nu}{h^2} + \sigma\right)^{-1} = \frac{h}{2a}\left(1 + \frac{1}{Pe} + \frac{h}{2a}\sigma\right)^{-1}, \tag{2.64}$$

which corresponds with the second-order accurate formula presented by Shakib, Hughes and Johan (1991) based on a local truncation error analysis.

Third, Shakib and co-workers propose the fourth-order accurate formula

$$\tau = \left(\left(\frac{2a}{h}\right)^2 + 9\left(\frac{4\nu}{h^2}\right)^2 + \sigma^2\right)^{-1/2} = \frac{h}{2a}\left(1 + \frac{9}{Pe^2} + \left(\frac{h}{2a}\sigma\right)^2\right)^{-1/2}. \tag{2.65}$$

An asymptotic analysis near the different limits (i.e., Taylor series expansions for $Pe = 0$, $Pe = \infty$, $\sigma = 0$ and $\sigma = \infty$) shows the superior convergence of (2.65) to the theoretical values compared with (2.64). However, the fourth-order accuracy does not extend to higher dimensions or general convection–diffusion systems. Nevertheless, numerical experiments indicate that (2.65) presents slightly lower errors (in the $\mathcal{L}_2$ norm) compared with (2.64).

It is important to note that for higher-order finite elements, apart from the results discussed in Section 2.2.4 and Remark 2.11 (see also Franca, Frey and Hughes, 1992; Codina, 1993b), no optimal definition of τ exits. Numerical experiments seem to indicate that for finite elements of order p the value of the stabilization parameter should be approximately τ/p.

Finally, note that this chapter has focused on convection–diffusion. Reaction has been added in some cases but diffusion–reaction problems are not discussed in detail, see Harari, Frey and Franca (2002) for the definition of stability parameters in such problems. The proper definition of the stabilization parameter is important in practical computations. Thus it must be adapted to the equation under consideration and to non-uniform meshes (element-by-element evaluation).

2.5 OTHER STABILIZATION TECHNIQUES AND NEW TRENDS

The subject of stabilization methods for convection-dominated transport problem is still an area of active research. Alternative approaches to the stabilization techniques discussed so far have been proposed in recent years and new trends are emerging. In this section we shall present a brief overview of three alternative stabilization techniques to SUPG and GLS and introduce a new promising technique based upon the so-called variational multiscale method.

2.5.1 Finite increment calculus

An alternative approach for deriving stabilized formulations for convection–diffusion and fluid flow problems has been suggested by Oñate (1998) under the name *finite increment calculus*. The basic idea in this method is that most stabilized numerical schemes can be derived by applying the standard Galerkin formulation to a so-called stabilized form of the governing differential equations of the problem. These equations are obtained by invoking higher-order balance (equilibrium) statements on a finite (as opposed to infinitesimal) domain.

We shall illustrate the process for a 1D convection–diffusion problem. The classical balance equation for this problem is given in (2.16) and reads

$$au_x - \nu u_{xx} = s(x).$$

For simplicity, we shall assume that the convection velocity a is constant. The idea is to replace the above balance equation by a higher-order one derived on a space slab $(x-\delta, x)$ of finite dimension δ. Over such a finite subdomain, the unknown u, the convective and diffusive fluxes and the source term may experience finite variations. As a consequence, the higher-order balance equation takes the following form in terms of the fluxes entering and leaving the subdomain:

$$\left(F^{out}_{conv} - F^{in}_{conv}\right) + \left(F^{out}_{diff} - F^{in}_{diff}\right) = \int_{x-\delta}^{x} s(x)\,dx,$$

where

$$F^{out}_{conv} = a\,u, \qquad F^{in}_{conv} = a\left(u - \delta\frac{\partial u}{\partial x} + \frac{1}{2}\delta^2\frac{\partial^2 u}{\partial x^2}\right) + \mathcal{O}(\delta^3),$$

$$F^{out}_{diff} = -\nu\,\frac{\partial u}{\partial x}, \qquad F^{in}_{diff} = -\nu\,\frac{\partial}{\partial x}\left(u - \delta\frac{\partial u}{\partial x} + \frac{1}{2}\delta^2\frac{\partial^2 u}{\partial x^2}\right) + \mathcal{O}(\delta^3),$$

$$\int_{x-\delta}^{x} s(x)\,dx = \frac{\delta}{2}\left(s(x-\delta) + s(x)\right) + \mathcal{O}(\delta^3) = \delta\,s(x) - \frac{\delta^2}{2}\frac{\partial s(x)}{\partial x} + \mathcal{O}(\delta^3).$$

This statement of equilibrium is easily seen to produce the higher-order balance equation, which is approximated to first order by

$$\left(1 - \frac{\delta}{2}\frac{\partial}{\partial x}\right)\left(a\,u_x - \nu\,u_{xx} - s(x)\right) = 0.$$

Note that the modified balance equation reduces to the standard form by assuming that the dimension of the balance space slab is infinitesimal. In the presence of Dirichlet conditions at both ends of the domain $[0, L]$, application of the standard Galerkin method to the higher-order balance equation produces the following consistent variational form:

$$\int_0^L \left(w + \frac{\delta}{2}\frac{\partial w}{\partial x}\right)\left(a\,u_x - \nu\,u_{xx} - s(x)\right)dx = 0,$$

in which the weighting function is identical in structure to an SU weighting function. Therefore, for suitable values of parameter δ, application of the standard Galerkin method to the higher-order balance equation produces stabilized finite element schemes for convection–diffusion similar to those obtained with the Petrov–Galerkin approaches discussed in the previous sections of this chapter. A similar approach had been invoked earlier by Donea, Belytschko and Smolinski (1985) with application to quadratic shape functions.

As shown by Oñate and co-workers, the modified differential equation can be used to derive a numerical scheme for iteratively computing the stabilization parameters in a sort of model adaptivity procedure. Further details on the finite increment calculus approach to stabilization and its application to convection–diffusion and fluid flow problems can be found in Oñate (1998) and Oñate and Manzán (1999) and references therein. A similar approach may also be found in the work of Ilinca, H étu and Peletier (2000).

2.5.2 Bubble functions and wavelet approximations

As amply discussed in the present chapter, the Galerkin method employing standard finite elements is not a robust approach for the numerical modeling of transport problems exhibiting multiscale behavior in the form of localized phenomena such as interior and boundary layers. This has motivated the development of new methods capable of dealing with phenomena involving both coarse-scale and fine-scale aspects. In this context, p-type finite elements typically used in adaptive refinement procedures, as well as other non-classical finite element procedures, such as elements enriched with bubble or wavelet functions, have proven to provide an interesting alternative to the previously described stabilized methods. In this section, we shall illustrate the construction of finite element models enriched with bubbles or wavelet functions to improve the resolution of localized phenomena and thereby enhance the stability of solutions to convection-dominated problems.

2.5.2.1 Bubble functions Bubbles are functions defined on the interiors of finite elements which vanish on the element boundaries. Baiocchi, Brezzi and Franca (1993) were the first to point out that the enrichment of the finite element space by summation of polynomial bubble functions results in a stabilization procedure for convection–diffusion problems that is formally similar to SUPG and GLS. Unfortunately, standard (low-order) polynomial bubbles are not able to adequately resolve thin internal layers with steep gradients or other small-scale phenomena. Then, the concept of enriching the Galerkin finite element method with so-called residual-free bubbles, see for instance Brezzi, Franca and Russo (1998) or Franca, Nesliturk and Stynes (1998), or with element Green's functions, see Hughes (1995) or Hughes et al. (1998), was introduced to provide a more general framework for the discretization of problems involving multiscale phenomena.

The basic idea in bubble function methods is to decompose the solution of a given boundary value problem into the sum of a coarse-scale solution and a fine-scale one. The classical Galerkin finite element method is used to represent the coarse-scale response, that is the resolvable part of the solution for the given finite element mesh. Bubble functions then take care of the fine-scale aspects of the solution which cannot be resolved by the finite element mesh. Bubbles have a stabilizing effect for convection–diffusion problems similar to that of the added diffusion method. A simple application of the bubble function method as a stabilizing formulation for convection–diffusion problems is described in Section 2.6.1.

2.5.2.2 Wavelet functions Recently, the finite element application of wavelets has become an active research area, in particular, to model localization phenomena. Like bubbles, wavelets are functions with local support. Wavelet elements have the potential of capturing localized phenomena and computing multiscale solutions to partial differential equations with higher convergence rates than conventional finite element methods. Dahmen et al. (1997) provide an overview of recent progress in the development and use of wavelet methods in the fluid mechanics area.

In the finite element solution of convection-dominated problems exhibiting internal or boundary layers, advantage can be taken of the properties of wavelet-based approximations to achieve an accurate localization of such layers. When wavelet approximations are combined with the Galerkin finite element method, the enhanced accuracy is obtained at the cost of some computational complexity. In particular, the computation of the numerical values of the wavelet functions and of their derivatives at the quadrature points is not an easy task. Another delicate aspect of the method is the prescription of the boundary conditions which is more complicated than for the standard finite element method. Several methods have been suggested to implement them, including Lagrange multipliers and penalty methods.

2.5.3 The variational multiscale method

The variational multiscale method introduced by Hughes (1995) provides the necessary mathematical framework for the construction of so-called sub-grid scale models. In its simplest form it is very similar to the *sub-grid viscosity* methods presented in a series of papers by Guermond (1999a; 1999b). The idea here is again that standard finite element approximations can only resolve the coarse-scale aspects of problems involving multiscale behavior. The variational multiscale method is based on the additive decomposition of the solution, u, on a coarse-scale component, $\bar{u}$, which can be resolved by the considered finite element mesh, and a fine-scale component, u', which one attempts to determine analytically. Notice that the fine-scale solution, u', actually represents the error, $u - \bar{u}$, of the coarse-scale component, while $\bar{u}$ is the resolvable scale approximated by the finite element method.

In order to simplify the exposition, the variational multiscale method is illustrated in the context of the convection–diffusion–reaction equation, see (2.54), with homogeneous Dirichlet boundary conditions (and with the usual assumption of a divergence-free convection velocity, $\boldsymbol{a}$), see Hughes et al. (1998) for a complete exposition. The standard variational form of this boundary value problem is, as usual, find $u \in \mathcal{V}$ such that

$$a(w,u) + c(\boldsymbol{a};w,u) + \sigma\ (w,u) = (w,s) \qquad \text{for all } w \in \mathcal{V}.$$

The additive decomposition of the solution, $u = \bar{u} + u'$, and test function, $w = \bar{w} + w'$, is accompanied by the corresponding splitting of the functional space $\mathcal{V} = \bar{\mathcal{V}} \oplus \mathcal{V}'$ into a finite dimensional coarse-scale subspace and a necessarily infinite dimensional fine-scale subspace. Note that all the previously presented functions verify the homogeneous Dirichlet boundary conditions. Under these conditions the weak form becomes

$$a(\bar{w}+w',\bar{u}+u') + c(\boldsymbol{a};\bar{w}+w',\bar{u}+u') + \sigma\ (\bar{w}+w',\bar{u}+u') = (\bar{w}+w',s),$$

which, by virtue of the linear independence of $\bar{w}$ and w', splits into two problems:

$$\begin{aligned} &a(\bar{w},\bar{u}) + c(\boldsymbol{a};\bar{w},\bar{u}) + \sigma\ (\bar{w},\bar{u}) \\ &\qquad + a(\bar{w},u') + c(\boldsymbol{a};\bar{w},u') + \sigma\ (\bar{w},u') = (\bar{w},s) \quad \forall \bar{w} \in \bar{\mathcal{V}}, \end{aligned} \tag{2.66a}$$

$$\begin{aligned} &a(w',u') + c(\boldsymbol{a};w',u') + \sigma\ (w',u') \\ &\qquad + a(w',\bar{u}) + c(\boldsymbol{a};w',\bar{u}) + \sigma\ (w',\bar{u}) = (w',s) \quad \forall w' \in \mathcal{V}'. \end{aligned} \tag{2.66b}$$

The first problem governs the resolved scales and the second one the unresolvable scales.

The objective now is to solve analytically the fine-scale problem (2.66b) as a function of the coarse-scale solution $\bar{u}$. Then, this u', as a function of $\bar{u}$, is substituted in (2.66a) to obtain an equation for $\bar{u}$. A Green's function technique would be appropriate in some problems to obtain the fine-scale solution. However, the fine-scale Green's function belongs to an infinite dimensional subspace and is non-local. It is then standard to impose $u' = 0$ along the finite element edges in order to localize the fine-scale problem in the interior of each finite element.

Then the Euler–Lagrange equations associated with the fine-scale problem (2.66b) become

$$\begin{cases} \Pi\,\mathcal{L}(u') = -\Pi\big[\mathcal{L}(\bar{u}) - s\big] & \text{in } \Omega^e \\ u' = 0 & \text{on } \Gamma^e \end{cases} \tag{2.67}$$

where $\mathcal{L}(u) = \boldsymbol{a}\cdot\boldsymbol{\nabla}u - \boldsymbol{\nabla}\cdot(\nu\boldsymbol{\nabla}u) + \sigma u$ is the differential operator associated with the original convection–diffusion–reaction problem, and Π denotes the $\mathcal{L}_2$ projection onto $\mathcal{V}'$. Note that this problem for u' is driven by the residual of the coarse (resolvable) finite element solution. The simplest way to approximate the solution of problem (2.67) is to assume that

$$u' = -\tau\big[\mathcal{L}(\bar{u}) - s\big], \tag{2.68}$$

which obviously does not even verify the boundary conditions.

At this point we note, in view of the above assumption, that

$$\begin{aligned} &a(\bar{w},u') + c(\boldsymbol{a};\bar{w},u') + \sigma\ (\bar{w},u') \\ &\quad = \sum_e \int_{\Omega^e} \big[-\boldsymbol{a}\cdot\boldsymbol{\nabla}\bar{w} - \boldsymbol{\nabla}\cdot(\nu\boldsymbol{\nabla}\bar{w}) + \sigma\bar{w}\big]\, u'\, d\Omega = \sum_e \big(\mathcal{L}^*(\bar{w}), u'\big)_{\Omega^e}, \end{aligned} \tag{2.69}$$

where the adjoint operator $\mathcal{L}^*(u) = -\boldsymbol{a} \cdot \boldsymbol{\nabla} u - \boldsymbol{\nabla} \cdot (\nu \boldsymbol{\nabla} u) + \sigma u$ has been introduced. Now, the approximation (2.68) is substituted in (2.69), which is used to write the coarse-scale problem (2.66a) in a standard stabilization form

$$a(\bar{w}, \bar{u}) + c(\boldsymbol{a}; \bar{w}, \bar{u}) + \sigma\ (\bar{w}, \bar{u}) + \sum_e \tau\big(-\mathcal{L}^*(\bar{w}), \mathcal{L}(\bar{u}) - s\big)_{\Omega^e} = (\bar{w}, s).$$

Note that this equation corresponds to the standard stabilized form (2.56) with $\mathcal{P}(w) = -\mathcal{L}^*(w)$. This method is called *sub-grid scale* (SGS); it delivers similar results to the SUPG and GLS techniques unless the reaction term is dominant, see Remark 2.12. Figure 2.14 presents a comparison of these methods. The results for quadratic elements present a similar qualitative behavior and thus are not shown.

This method was proposed by Hughes (1995). A detailed presentation can be found in Hughes et al. (1998) where better approximations to the solution of the fine-scale problem (2.67) than the simplest algebraic approximation (2.68) are also discussed. A rigorous development of this technique for systems of differential equations without the restriction that $u' = 0$ along the element edges can be found in Codina (2000). Note, however, that the potentiality of this techniques extends far beyond stabilization techniques for convection–diffusion–reaction equations. See, for instance, Hughes, Mazzei and Jansen (2000) or Hughes et al. (2001) where this technique is used in turbulent flows to introduce modeling (Reynolds stresses) at small scales.

2.5.4 Complements

The idea of locally enriching the approximation basis in order to better reflect the local character of the solution has also been exploited with success in other classes of methods. For instance, the so-called generalized finite element method, see Strouboulis, Copps and Babuška (2000), which results from a combination of the standard finite element method and the partition of unity method introduced by Melenk and Babuška (1996), also provides an efficient way to model localization phenomena (internal and boundary layers) with higher convergence rates than conventional finite element methods. Similarly, modern mesh-free (or particle) methods pioneered by Nayroles, Touzot and Villon (1992) may be used to enrich a finite element mesh where needed without any need for remeshing, see Huerta and Fern ández-Méndez (2000). Methods in this class include the Element-Free Galerkin (EFG) method developed by Belytschko et al. (1994; 1996) and the Reproducing Kernel Particle Method (RKPM) introduced by Liu et al. (1995). Also noteworthy is the h-p clouds method introduced by Duarte and Oden (1996).

2.6 APPLICATIONS AND SOLVED EXERCISES

2.6.1 Construction of a bubble function method

In order to illustrate the use of the bubble function method introduced in Section 2.5.2.1, we consider the convection–diffusion equation with homogeneous Dirichlet boundary conditions. The standard Galerkin formulation of this problem consists of the following: find $u^h \in \mathcal{V}^h$ such that $\forall w^h \in \mathcal{V}^h$

$$a(w^h, u^h) + c(a; w^h, u^h) = (w^h, s).$$

In the bubble function method, each interpolation and test function, u^h and w^h, is decomposed in the standard piecewise continuous polynomials of Galerkin finite elements plus bubble functions:

$$u^h = u^h_G + u^h_B, \quad \text{and } w^h = w^h_G + w^h_B,$$

where subscript G refers to the standard polynomial approximating function of the Galerkin method and subscript B to the bubble part of the approximation, which is required to take on zero value at element boundaries.

This last property of the bubble functions allows us to use the classical static condensation procedure to extract the bubble part u^h_B of the solution in each individual element as a function of the piecewise polynomial part u_G.

We shall illustrate this idea on the homogeneous version of the 1D boundary value problem (2.16). In 1D the bubble function for a linear element is given by $(1 - \xi^2)$, so that the decomposition becomes

$$u^h = \overbrace{\frac{1}{2}(1-\xi)\,u_1 + \frac{1}{2}(1+\xi)\,u_2}^{u^h_G} + \overbrace{(1-\xi^2)\,u_3}^{u^h_B} \tag{2.70a}$$

$$w^h = \underbrace{\frac{1}{2}(1-\xi)\,w_1 + \frac{1}{2}(1+\xi)\,w_2}_{w^h_G} + \underbrace{(1-\xi^2)\,w_3}_{w^h_B} \tag{2.70b}$$

where subscripts 1 and 2 characterize the end nodes of the linear element and subscript 3 refers to the bubble function. The weak form of the homogeneous ($s = 0$) problem is given by

$$\int_0^L \left(w^h\, a\, u^h_x + w^h_x\, \nu\, u^h_x\right) dx = 0. \tag{2.71}$$

By setting $w^h = w^h_B$, see (2.70b), one finds

$$\int_{-1}^{+1} \left[a(1-\xi^2) - 4\frac{\nu}{h}\xi\right] \left[\frac{u_2 - u_1}{2} - 2\xi u_3\right] d\xi = 0,$$

from which we extract the value of parameter u_3 associated with the bubble function in terms of the element nodal parameters:

$$u_3 = -\frac{ah}{8\nu}(u_2 - u_1) = -\frac{Pe}{4}(u_2 - u_1).$$

This allows us to rewrite (2.70a) in terms of the nodal values as

$$\begin{aligned} u^h &= \frac{1}{2}(1-\xi)\Big(1 + \frac{Pe}{2}(1+\xi)\Big)\,u_1 + \frac{1}{2}(1+\xi)\Big(1 - \frac{Pe}{2}(1-\xi)\Big)\,u_2 \\ &= P_1(\xi)\,u_1 + P_2(\xi)\,u_2. \end{aligned}$$

Introducing this expression into the weak form (2.71) and now imposing $w^h = w^h_G$, one finds that the bubble function method delivers the following element matrices:

$$\mathbf{C}^e = a\int_0^h \begin{pmatrix} N_1\dfrac{\partial P_1}{\partial x} & N_1\dfrac{\partial P_2}{\partial x} \\ N_2\dfrac{\partial P_1}{\partial x} & N_2\dfrac{\partial P_2}{\partial x} \end{pmatrix} dx = \frac{a}{2}\begin{pmatrix} -1+\dfrac{Pe}{3} & 1-\dfrac{Pe}{3} \\ -1-\dfrac{Pe}{3} & 1+\dfrac{Pe}{3} \end{pmatrix}$$

$$\mathbf{K}^e = \nu\int_0^h \begin{pmatrix} \dfrac{\partial N_1}{\partial x}\dfrac{\partial P_1}{\partial x} & \dfrac{\partial N_1}{\partial x}\dfrac{\partial P_2}{\partial x} \\ \dfrac{\partial N_2}{\partial x}\dfrac{\partial P_1}{\partial x} & \dfrac{\partial N_2}{\partial x}\dfrac{\partial P_2}{\partial x} \end{pmatrix} dx = \frac{\nu}{h}\begin{pmatrix} 1 & -1 \\ -1 & 1 \end{pmatrix}.$$

Comparing these element matrices with those derived in Section 2.2 for the Galerkin formulation, one notes that the bubble function method only modifies the convection matrix $\mathbf{C}^e$ of the linear element without affecting the diffusion matrix $\mathbf{K}^e$. With these results, and assembling in the usual finite element manner the contributions emanating from both elements to which a given node belongs, one finds that the bubble function method delivers the following discrete equation at an interior node j of a mesh of uniform linear elements:

$$a\frac{u_{j+1}-u_{j-1}}{2h} - \left(\nu + \frac{Pe}{3}\frac{ah}{2}\right)\frac{u_{j+1}-2u_j+u_{j-1}}{h^2} = 0.$$

This shows that bubbles have a stabilizing effect for convection–diffusion problems similar to that of the added diffusion method. In fact, the added diffusion in the previous equation corresponds to that given in (2.26) with the coefficient β given by the doubly asymptotic approximation in (2.29).

2.6.2 One-dimensional transport

This example is concerned with the solution of the 1D homogeneous convection–diffusion equation with prescribed Dirichlet conditions at both ends of the domain $[0,1]$, namely $u(0) = 0$ and $u(1) = 1$. The problem is first solved using a mesh of 10 uniform linear elements and then repeated with five uniform quadratic elements. A unit convection velocity is assumed and three values of the mesh Péclet number are considered: $Pe = 0.25, 0.9$ and 5.0. Figure 2.15 reports the results obtained with linear elements using the Galerkin method, a full upwind approach, and the SUPG method. The results obtained with quadratic elements are reported in Figure 2.16. For small to moderate Pe, the exact solution varies rather smoothly over the entire domain. However, as Pe is increased beyond unity, the solution has a boundary layer, which cannot be resolved by the standard Galerkin method. This is in contrast with the stable and accurate results obtained with SUPG. As anticipated, the full upwind approximation delivers stable, but inaccurate, results due to an excess of numerically added dissipation.

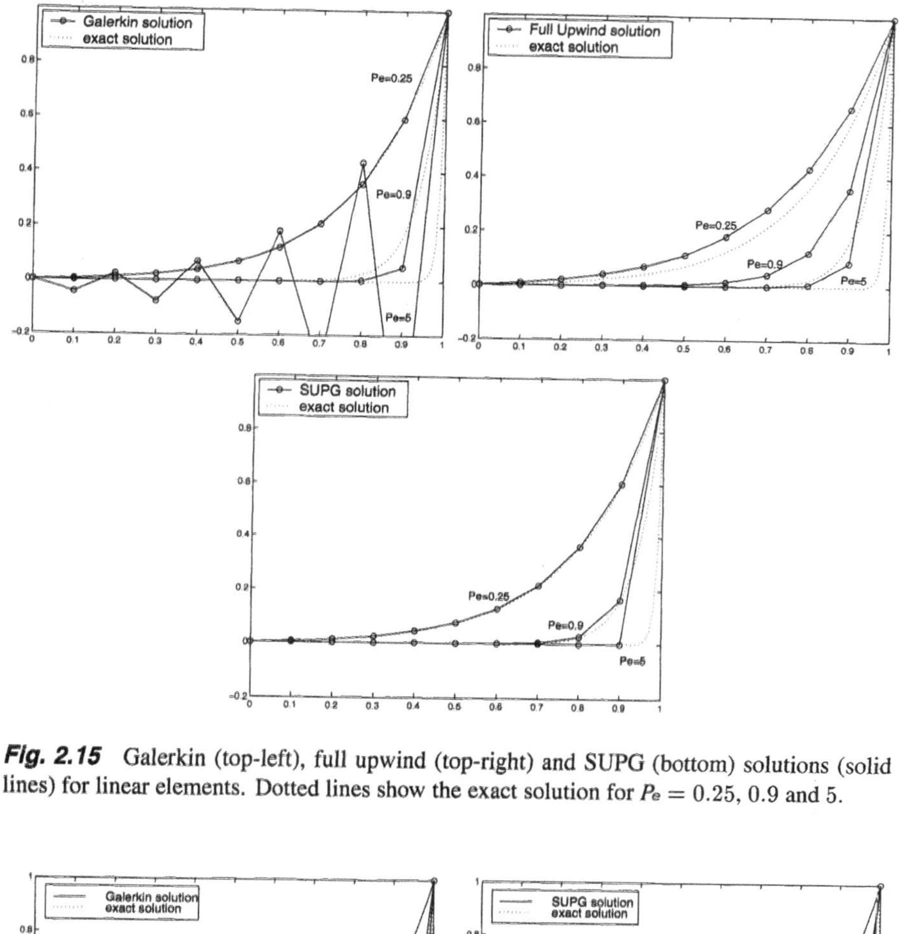

Fig. 2.15 Galerkin (top-left), full upwind (top-right) and SUPG (bottom) solutions (solid lines) for linear elements. Dotted lines show the exact solution for $P_e = 0.25$, 0.9 and 5.

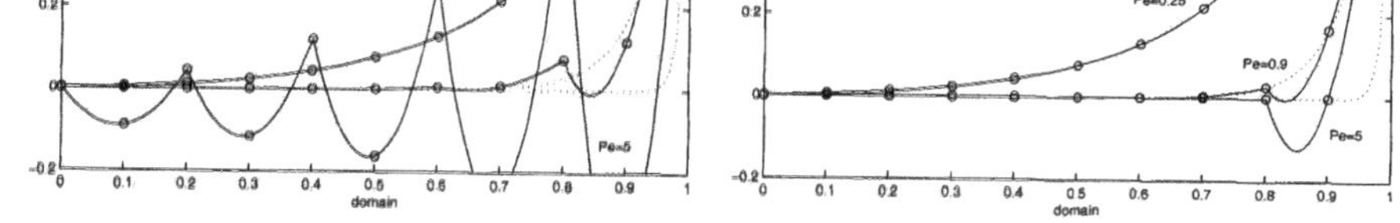

Fig. 2.16 Galerkin (left) and SUPG (right) solutions (solid lines) for quadratic elements. Dotted lines show the exact solution for $P_e = 0.25$, 0.9 and 5.

2.6.3 Convection–diffusion across a source term

The previous example is further studied introducing a source term:

$$\begin{cases} a\,u_x - \nu\,u_{xx} = 10e^{-5x} - 4e^{-x} & \text{for } x \in]0,1[\\ u(0) = 0 \text{ and } u(1) = 1. \end{cases}$$

Again, a unit convection velocity is assumed and the Péclet number is chosen as 0.25 or 5. Galerkin, SU (the non-consistent formulation) and SUPG are compared in Figure 2.17 for linear and quadratic elements. GLS and SGS results are not shown because they are almost identical to those of SUPG.

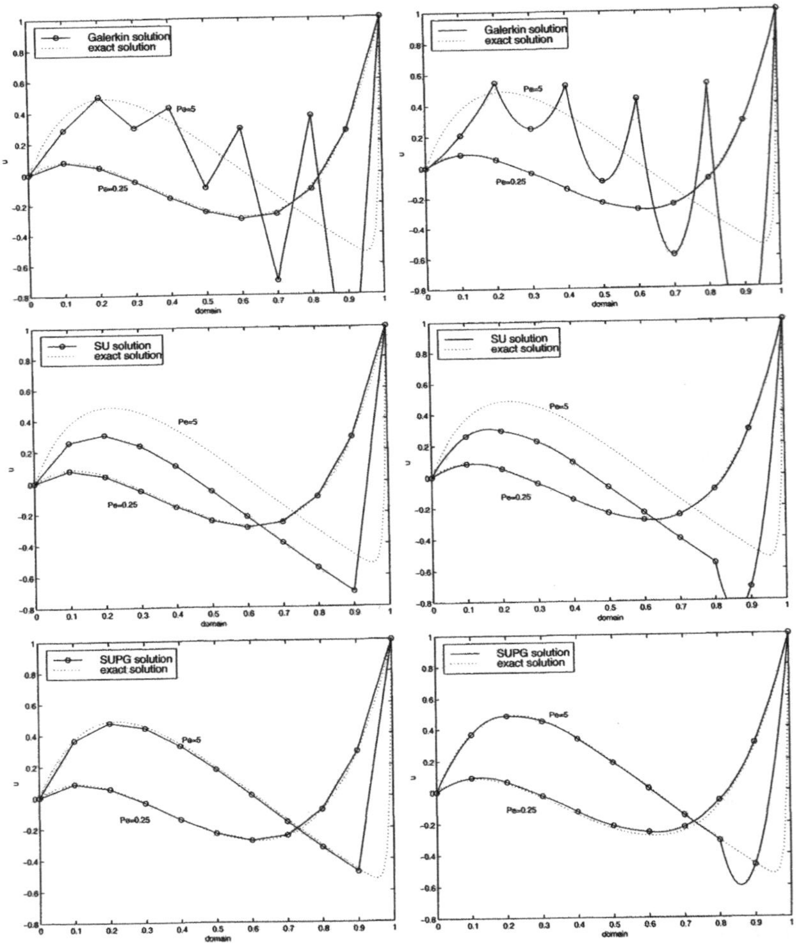

Fig. 2.17 Galerkin (top), SU (center) and SUPG (bottom) solutions (solid lines) for linear (left) and quadratic (right) elements. Dotted lines show the exact solution for $Pe = 0.25$ and 5.

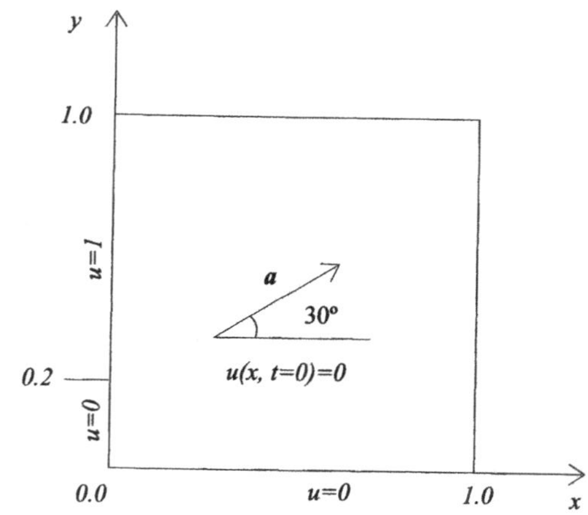

Fig. 2.18 Convection of discontinuous inlet data skew to the mesh: problem statement.

2.6.4 Convection–diffusion skew to the mesh

One of the major contributions in the stabilization of steady convection–diffusion problems is the concept of *anisotropic balancing diffusion* or *streamline upwind test functions* discussed in Section 2.3.3.3. The equivalence between added diffusion and upwinding is simple in 1D but its blunt extension to multiple dimensions leads to excessive crosswind diffusion. This is observed in the present example. Galerkin and SUPG formulations are compared with an artificial diffusion method; that is, adding a scalar artificial diffusivity defined by equation (2.59), namely $\bar{\nu} = \tau \, \|\boldsymbol{a}\|^2$, with τ determined by any of the definitions in Section 2.4.3. Note that the present artificial diffusion method is based on a scalar diffusivity. Thus, it should not be confused with the SU method which possesses the tensorial structure indicated in (2.48) and is further illustrated in Remark 2.9.

The problem statement is depicted in Figure 2.18 where the unit square is taken as the computational domain, $\bar{\Omega} = [0, 1] \times [0, 1]$. This 2D test case has been widely used to illustrate the effectiveness of stabilized finite element methods in the modeling of convection-dominated flows. A mesh of 10 by 10 equal-sized bilinear elements is considered.

The flow is unidirectional and constant, $\|\boldsymbol{a}\| = 1$, but the convective velocity is skew to the mesh with an angle of 30°. The diffusivity coefficient is taken to be 10^{-4}, corresponding to a mesh Péclet number of 10^4. The inlet boundary data are discontinuous and two types of boundary conditions are considered at the outlet:

- *Downwind homogeneous natural boundary conditions.* The results for this case are displayed in Figure 2.19. Given the elevated value of the Péclet number, the solution is practically one of pure convection. The Galerkin method is not able to satisfactorily resolve the discontinuity and produces spurious oscillations. The artificial diffusion method and the SUPG method yield better results, but SUPG introduces less crosswind diffusion.

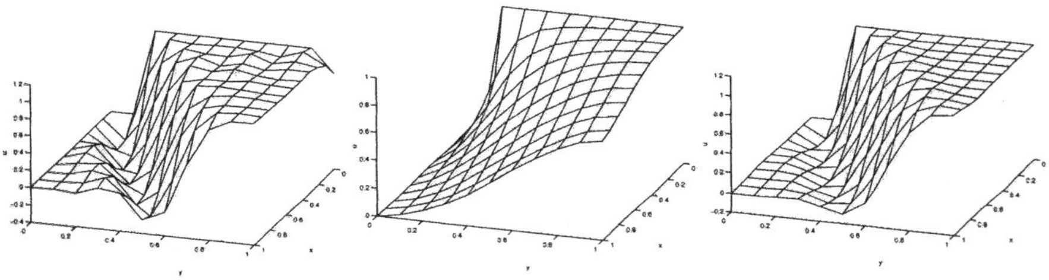

Fig. 2.19 Galerkin (left), artificial diffusion (center) and SUPG (right) solutions for the 2D convection–diffusion problem with downwind natural conditions.

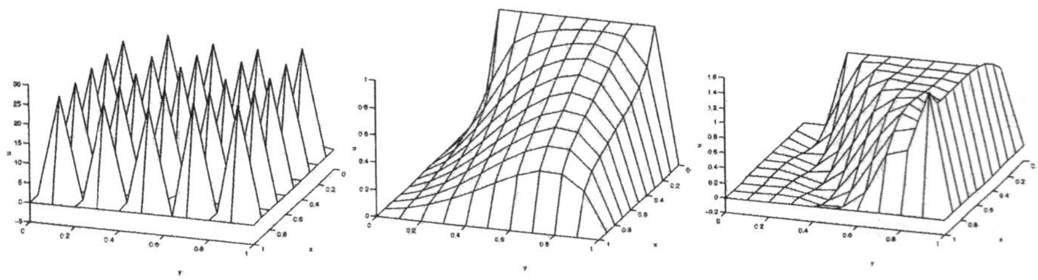

Fig. 2.20 Galerkin (left), artificial diffusion (center) and SUPG (right) solutions for the 2D convection–diffusion problem with downwind essential boundary conditions.

- *Downwind homogeneous essential boundary conditions.* Here, we impose $u = 0$ on the outlet portion of the boundary. The solution now involves a thin boundary layer at the outlet. As shown in Figure 2.20, the crude 10 by 10 mesh has difficulty capturing the details of the solution. The Galerkin results are wildly oscillatory and bear no resemblance to the exact results. Better results are obtained with the stabilized formulations. The artificial diffusion technique introduces excessive numerical diffusion.

2.6.5 Convection–diffusion–reaction in 2D

On the same unit square domain as in the previous example, $\bar{\Omega} = [0, 1] \times [0, 1]$, the influence of the convection and reaction terms are compared for the Galerkin, SUPG, GLS and SGS formulations. As in the previous example, convection is taken skew to the mesh with an angle of 30° and the diffusion is small, $\nu = 10^{-4}$. Two cases are studied: first the convection–reaction-dominated case, $||\boldsymbol{a}|| = 1/2$ and $\sigma = 1$, and then the reaction-dominated case, $||\boldsymbol{a}|| = 10^{-3}$ and $\sigma = 1$. The results are shown in Figures 2.21 and 2.22. The stabilized methods produce similar results. Nevertheless, as noted in Remark 2.12 and Section 2.5.3 in the presence of reaction (in particular for the reaction-dominated example, see Figure 2.22) SGS reduces the oscillations near the boundary layers.

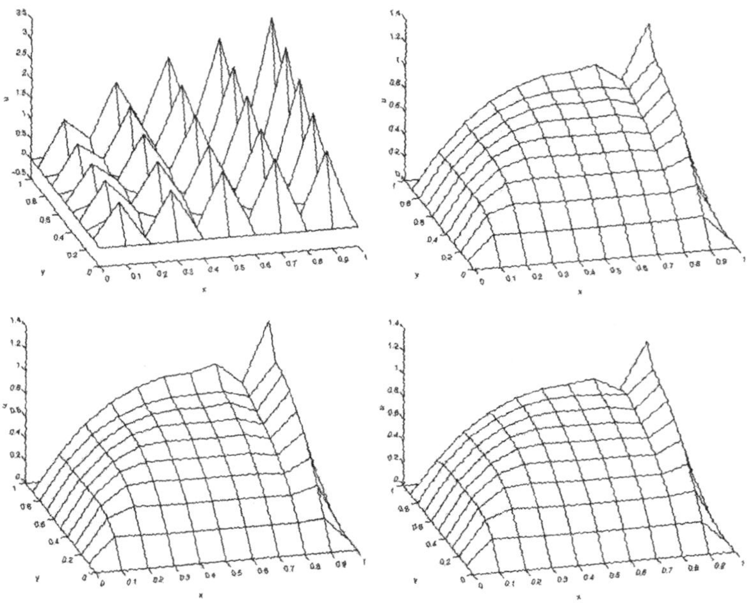

Fig. 2.21 Galerkin, SUPG, GLS and SGS (from left to right and top to bottom) solutions for the 2D convection–reaction-dominated problem.

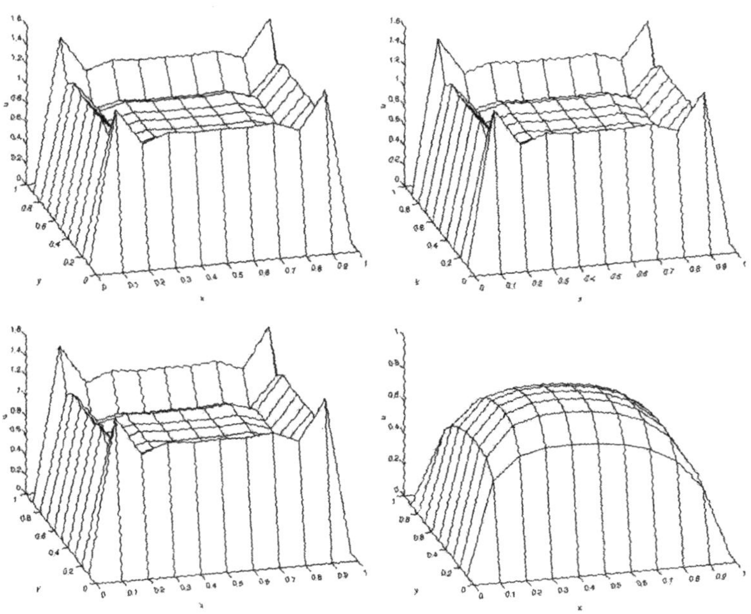

Fig. 2.22 Galerkin, SUPG, GLS and SGS (from left to right and top to bottom) solutions for the 2D reaction-dominated problem.

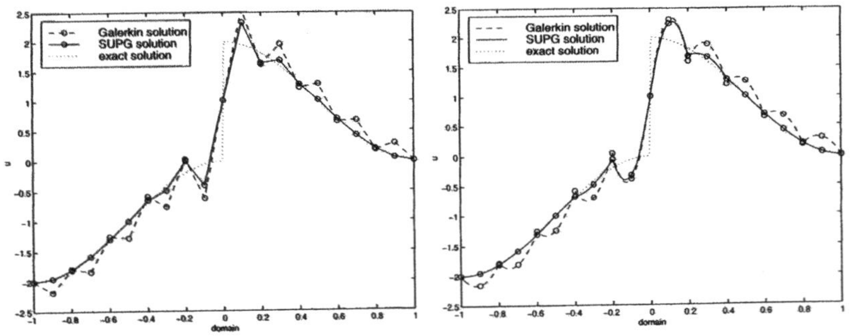

Fig. 2.23 Galerkin and SUPG solutions of Hemker problem for linear (left) and quadratic (right) elements.

2.6.6 The Hemker problem

The so-called Hemker problem is studied, which consists of solving the convection–diffusion equation

$$x\frac{\partial u}{\partial x} + \nu\frac{\partial^2 u}{\partial x^2} = -\nu\pi^2\cos(\pi x) - \pi x\sin(\pi x) \qquad \text{on } [-1,1],$$

with space-dependent convection velocity and source term. The boundary conditions are specified as

$$u(-1) = -2; \qquad u(1) = 0.$$

The exact solution is

$$u(x) = \cos(\pi x) + \text{erf}(x/\sqrt{(}2\nu))/\text{erf}(1/\sqrt{(}2\nu))$$

and has a turning point in the middle of the domain.

We take $\nu = 10^{-10}$ and use a mesh of 20 uniform linear elements ($h = 0.1$). The problem is first solved in the Galerkin formulation. The numerical solution is depicted in Figure 2.23 in comparison with the exact solution. Node-to-node oscillations extending over the whole computational domain characterize the Galerkin approach. The solution is then repeated using the SUPG formulation. Since the convection velocity $a(x) = x$ is negative on $]-1,0[$ and positive on $]0,1[$, care must be taken to add a positive diffusion everywhere in the mesh. The SUPG results are displayed in Figure 2.23. Though the SUPG method succeeds in removing the oscillations in the smooth part of the solution, residual oscillations remain near the turning point where a near-discontinuity is present. Additional nonlinear viscosity should be added locally to suppress these residual oscillations. This aspect will be treated in detail in Chapter 4.

3

Unsteady convective transport

This chapter addresses the development of time-accurate methods for solving transient convection problems using finite elements. First, a brief review is presented of the method of characteristics and of its use in the finite element context. The chapter then proceeds with the presentation and analysis of classical, first- and second-order accurate, time-stepping algorithms. This is followed by the description of higher-order methods which indirectly account for the role of the characteristics in convection problems. The last part of the chapter considers the coupling of a finite-element-based spatial discretization and high-order accurate time-stepping algorithms. Emphasis is placed on the use of least-squares-type finite elements to ensure stability when non-dissipative schemes are used for marching in time.

3.1 INTRODUCTION

Time-dependent problems describing convective transport are governed by hyperbolic equations and the characteristics play a dominant role in their solution. This has important consequences as regards the development of accurate numerical methods for solving evolutionary problems in this class. So, space and time being linked through the characteristics, the discretization of one certainly has an influence on the discretization of the other. In fact, an accurate spatial representation can be quickly eroded when it is transported in time if the time integration algorithm is not able to propagate the information along the directions prescribed by the convection problem.

In principle, one could circumvent the issue of convecting the information along the streamlines by resorting to a Lagrangian formulation. In such a moving coordinate system, convective terms disappear from the governing equations. Unfortunately, a purely Lagrangian description is in general not practicable in fluid flow problems because of excessive distortions of the computational mesh. However, a number of numerical methods based on the concept of convecting the information along the streamlines have been developed and applied in the finite element modeling of convective transport using a fixed (Eulerian) element mesh. The chapter is organized as follows.

In Section 3.2, we define the strong form of the initial boundary value problem for convective transport. Then, Section 3.3 illustrates the role of the characteristics in the solution of these problems. Several methods exploiting the characteristics are described. This includes semi-Lagrangian, Lagrange–Galerkin and characteristic-based methods. Their common feature is to exploit the fact that the solution is constant along a particle path or a characteristic. Unfortunately, finite element methods exclusively based on the characteristics are rather difficult to implement and expensive to use. This is the reason why direct time integration of the convection equations is often preferred. However, it must be based on suitable time-stepping schemes. In Section 3.4, we review the properties of second-order time-stepping algorithms classically used to produce the transient response of unsteady problems. Their accuracy analysis in Section 3.5 reveals that such algorithms are not optimal for convection problems. This is because second-order methods are unable to take into account the directional character of propagation of information in convective transport. In particular, a rapid fall-off in accuracy occurs as the time step is increased. Moreover, in explicit finite element schemes the stability range is reduced with respect to the corresponding finite difference schemes. Attention is therefore focused on higher-order time-stepping schemes capable of indirectly taking into account the propagation of information along the characteristics.

One of the indirect methods which exploit the characteristics in the numerical solution of convective transport problems, the Taylor–Galerkin method, is described in Section 3.6. It represents an attempt to simulate by a Taylor series in time the fact that the solution of convective transport problems is constant along the characteristics. This Taylor series is usually extended to third or fourth order. The discussion of time-stepping algorithms closes with an introductory presentation in Section 3.7 of monotonicity-preserving schemes. Such schemes are used near sharp solution gradients to suppress the residual oscillations that linear stabilization techniques cannot remove.

We then address the problem of coupling highly accurate time-stepping algorithms and finite element spatial discretization techniques based on the classical $\mathcal{C}^0$-continuous spatial representation. In Chapter 2 we have shown that the Galerkin finite element method is not optimal to model steady transport problems in which convective effects are dominant. Stabilization techniques capable of introducing sufficient numerical dissipation are needed to remedy the lack of numerical stability of the Galerkin formulation. In the present context of transient problems, the required amount of numerical damping can be provided either by the use of dissipative time-stepping methods or, as in Chapter 2, by the finite element spatial discretization. In Section 3.8 least-squares-based finite elements are introduced to provide the numerical dissipation needed to achieve stable results with non-dissipative time-stepping algorithms. To conclude the discussion of discretization procedures for convection problems, we briefly introduce in Section 3.9 the discontinuous Galerkin method and then discuss space–time formulations in Section 3.10. Finally, representative test problems in 1D and 2D are solved in Section 3.11 to illustrate the use of time-accurate finite element methods in the numerical solution of convective transport problems.

3.2 PROBLEM STATEMENT

Before discussing numerical algorithms, we shall introduce the strong form of the linear convection problems treated in this chapter. In the so-called *conservation law form*, the equations governing unsteady convective transport are

$$u_t + \nabla \cdot \boldsymbol{f}(u) = s(\boldsymbol{x}, t) \qquad \text{in } \Omega \times]0, T[, \tag{3.1a}$$
$$u(\boldsymbol{x}, 0) = u_0(\boldsymbol{x}) \qquad \text{on } \Omega \text{ at } t = 0, \tag{3.1b}$$
$$u = u_D \qquad \text{on } \Gamma_D^{in} \times]0, T[, \tag{3.1c}$$
$$-\boldsymbol{f} \cdot \boldsymbol{n} = h \qquad \text{on } \Gamma_N^{in} \times]0, T[, \tag{3.1d}$$

where now the solution u and the source term s are a function of time. Note that in the hyperbolic case we cannot specify boundary conditions on the outflow portion of the boundary $\Gamma = \partial\Omega$ of the domain Ω. As in Chapter 2, Γ^{in} denotes the inflow portion of the boundary (see Remark 2.1). We employ the partition $\Gamma^{in} = \Gamma_D^{in} \cup \Gamma_N^{in}$ to identify the Dirichlet and the Neumann portions of the inflow boundary.

The value of u is specified on the Dirichlet portion of the inflow boundary, while the inlet normal flux is prescribed on the Neumann portion. The vector-valued flux function $\boldsymbol{f}(u)$ can be defined in terms of the convection velocity $\boldsymbol{a}$ by

$$\boldsymbol{f}(u) = \boldsymbol{a}\, u. \tag{3.2}$$

If the convection velocity $\boldsymbol{a}$ is independent of u, the problem is linear. Nonlinear problems are studied in Chapter 4.

When the convection velocity $\boldsymbol{a}$ is divergence free, use can be made of the identity

$$\nabla \cdot (\boldsymbol{a}\, u) = u\, \nabla \cdot \boldsymbol{a} + \boldsymbol{a} \cdot \nabla u \tag{3.3}$$

to recast problem (3.1) in the *convective* (also termed *advective*) form:

$$u_t + (\boldsymbol{a} \cdot \nabla) u = s \qquad \text{in } \Omega \times]0, T[, \tag{3.4a}$$
$$u(\boldsymbol{x}, 0) = u_0(\boldsymbol{x}) \qquad \text{on } \Omega \text{ at } t = 0, \tag{3.4b}$$
$$u = u_D \qquad \text{on } \Gamma_D^{in} \times]0, T[\tag{3.4c}$$
$$-\boldsymbol{a} u \cdot \boldsymbol{n} = h \qquad \text{on } \Gamma_N^{in} \times]0, T[,. \tag{3.4d}$$

Equations (3.1a) and (3.4a) describe the convective transport of a quantity u whose speed of propagation is given by the vector function $\boldsymbol{a}(\boldsymbol{x}, t)$.

The numerical solution of the above initial boundary value problems clearly involves a double discretization process, namely the spatial and temporal discretizations. While the former will be performed here by the finite element method, two broad classes of methods will be employed to trace the temporal evolution of the solution of convection problems. The first class is based on the characteristics and exploits them to transport the information in time. The second class of methods is based on a standard coordinate system and makes use of time-stepping algorithms to advance the solution from one time level to the next until the final time for the problem is reached.

3.3 THE METHOD OF CHARACTERISTICS

In the present section, we recall the basic properties of convection equations with an emphasis on the role of the characteristics in their solution. We then review finite-element-based algorithms capable of producing time-accurate solutions of convective transport problems through a direct use of the characteristics. See also the introductory paper by Donea and Quartapelle (1992).

3.3.1 The concept of characteristic lines

For first-order equations, such as the convection equations studied in this chapter, the prototype linear partial differential equation (PDE) is of the form

$$u_t + a\,u_x = s, \tag{3.5}$$

where a and s may depend on x and t. We may interpret this equation as follows: for a given solution $u(x,t)$ and at a given space–time point (x,t), the total derivative of u with respect to time in the direction defined by the slope

$$dx/dt = a$$

is equal to s. In this way, we have introduced the concept of propagation of the information u, with a speed a, as a function of the considered space–time point (x,t). For linear PDEs with constant coefficients the speed of propagation is constant.

When the PDE involves the physical concept of propagation it is said to be *hyperbolic* and the direction $dx/dt = a$ is termed the *characteristic direction*. We thus have the following result: *any first-order PDE is hyperbolic*. If the PDE is linear, the characteristic curves are fixed in the (x,t) plane, independent of the solution $u(x,t)$. Furthermore, if the linear PDE has constant coefficients, the characteristics are straight lines. Nonlinear first-order PDEs in which the speed a depends on the solution u are studied in Chapter 4. To further illustrate the concept of transport typical of first-order PDEs, let us consider the homogeneous form of equation (3.5) with a constant. The change of variables

$$\xi = x - at, \qquad \eta = x + at$$

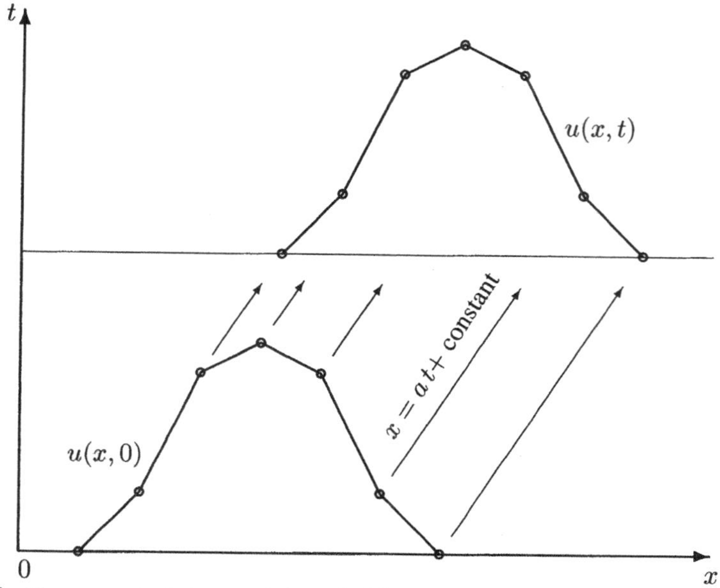

Fig. 3.1 The concept of characteristics: $du/dt = 0$ and consequently u is constant along the lines $x = a\,t +$ constant.

induces the transformation

$$\begin{Bmatrix} u_t \\ u_x \end{Bmatrix} = \begin{pmatrix} \dfrac{\partial \xi}{\partial t} & \dfrac{\partial \eta}{\partial t} \\ \dfrac{\partial \xi}{\partial x} & \dfrac{\partial \eta}{\partial x} \end{pmatrix} \begin{Bmatrix} u_\xi \\ u_\eta \end{Bmatrix} = \begin{pmatrix} -a & a \\ 1 & 1 \end{pmatrix} \begin{Bmatrix} u_\xi \\ u_\eta \end{Bmatrix}.$$

Thus, the PDE $u_t + au_x = 0$ becomes

$$2a\,u_\eta = 0,$$

whose general solution is

$$u = f(\xi) = f(x - at),$$

where f is an arbitrary function. It follows that the solution $u(x, t)$ at point x and time t is equal to the solution at time $t - \Delta t$ at the point $x - a\Delta t$:

$$u(x, t) = u(x - a\Delta t, t - \Delta t). \tag{3.6}$$

In other words, the solution is a rigid transport in time of the spatial profile of u. For this reason, the name *transport equation* is given to equation (3.5). The concept of transport along the characteristics is illustrated in Figure 3.1 which shows how an initial distribution $u(x, 0) = u_0(x)$ of the unknown undergoes a uniform translation along the x coordinate axis as time goes on.

3.3.2 Properties of the linear convection equation

To provide a basis for the discussion of characteristic-based finite element algorithms, we recall in this section the basic mathematical properties of the unsteady convection equation. For this purpose, we consider first the case of a scalar quantity u transported by a prescribed convection velocity field $\boldsymbol{a}(\boldsymbol{x}, t)$ in the presence of a known source term $s(\boldsymbol{x}, t)$. The initial boundary value problem for the quantity u is assumed linear and defined as in (3.4).

As seen previously, the analytical solution of this linear initial boundary value problem at a given point in space and time can be determined with the aid of the concept of characteristic lines associated with the velocity field $\boldsymbol{a}(\boldsymbol{x}, t)$. The idea is to replace the operator $\partial/\partial t + \boldsymbol{a} \cdot \boldsymbol{\nabla}$ in the l.h.s. of equation (3.4a), which represents the material derivative in the Eulerian description, by a simple time derivative using the Lagrangian viewpoint.

Accordingly, for a given space–time point $(\boldsymbol{x}, \tau)$, where $\boldsymbol{x} \in \Omega$ and $\tau \in]0, T[$, one determines the characteristic line $\boldsymbol{X} = \boldsymbol{X}(\boldsymbol{x}, \tau; t)$ passing through the considered space–time point by solving the ordinary (vector) differential equation

$$\frac{d\boldsymbol{X}}{dt}(\boldsymbol{x}, \tau; t) = \boldsymbol{a}(\boldsymbol{X}(\boldsymbol{x}, \tau; t)), \qquad t \in]0, T[, \tag{3.7a}$$

subject to the condition

$$\boldsymbol{X}(\boldsymbol{x}, \tau; \tau) = \boldsymbol{x}. \tag{3.7b}$$

With reference to the Lagrangian point of view, $\boldsymbol{X}(\boldsymbol{x}, \tau; t)$ can be interpreted as providing the position at time t of a fluid particle transported by the convection velocity field $\boldsymbol{a}$, which occupies the spatial position $\boldsymbol{x}$ at time τ. In other words, (3.7a) defines the particle trajectory. Along this trajectory, the *material (or total) derivative*

$$\frac{du}{dt} = \frac{\partial u}{\partial t} + \boldsymbol{a} \cdot \boldsymbol{\nabla} u,$$

which is the l.h.s. of the convection equation (3.4a), reduces to a simple time derivative. In fact, by definition, the material derivative is the time rate of change felt by an observer moving with the material particles.

It is worth noting that equation (3.7a) is generally nonlinear, except when the convection field has a specific spatial dependence of the form $\boldsymbol{a}(\boldsymbol{x}, t) = \boldsymbol{\alpha}(t) + \boldsymbol{\beta}(t)\, \boldsymbol{x}$, where $\boldsymbol{\alpha}(t)$ and $\boldsymbol{\beta}(t)$ are arbitrary functions. Excluding such a situation, the presence of the nonlinearity is the price to pay for the reduction of the linear unsteady convection equation (3.4a) to an ordinary differential equation.

The integration of equation (3.7a) proceeds from $t = \tau$ backwards, until the characteristic curve either intersects the boundary Γ_{in} or reaches the initial time $t = 0$, see Pironneau (1989) and Douglas and Russell (1982) for details:

- The intersection of the characteristic with Γ_{in} occurs at a time $t_\Gamma(\boldsymbol{x}, \tau)$ that depends on the considered characteristic curve. If the intersection point is denoted by $\boldsymbol{X}_\Gamma = \boldsymbol{X}_\Gamma(\boldsymbol{x}, \tau)$, one has

$$\boldsymbol{X}_\Gamma(\boldsymbol{x}, \tau) = \boldsymbol{X}(\boldsymbol{x}, \tau; t_\Gamma).$$

It follows that the solution at the considered space–time position $(\boldsymbol{x}, \tau)$ depends on the value $u_D(\boldsymbol{X}_\Gamma, t_\Gamma)$ prescribed at the boundary point $\boldsymbol{X}_\Gamma$. To determine $u(\boldsymbol{x}, \tau)$, the governing equation (3.4a) is written in the so-called characteristic form. We shall denote the value of the unknown along the characteristic curve $\boldsymbol{X}(\boldsymbol{x}, \tau; t)$ by $U(t) := u(\boldsymbol{X}(\boldsymbol{x}, \tau; t), t)$, or, omitting the explicit indication of the dependence on $(\boldsymbol{x}, \tau)$ to simplify the notation, by $U(t) = u(\boldsymbol{X}(t), t)$. In characteristic form equation (3.4a) along $\boldsymbol{X}(t)$ reduces to the following linear ordinary differential equation:

$$\frac{dU}{dt} = S(t), \tag{3.8}$$

where $S(t) = s(\boldsymbol{X}(t), t)$. Equation (3.8) must then be integrated up to $t = \tau$ subject to the *initial condition*

$$U(t_\Gamma) = u_D(\boldsymbol{X}_\Gamma, t_\Gamma)$$

at the intersection time with Γ_{in}. Thus, for characteristic lines intersecting Γ^{in} the solution reads

$$u(\boldsymbol{x}, \tau) = U(\tau) = u_D(\boldsymbol{X}_\Gamma, t_\Gamma) + \int_{t_\Gamma}^{\tau} S(t)dt. \tag{3.9}$$

- On the contrary, when the characteristic curve reaches the plane $t = 0$, one has $\boldsymbol{X}(\boldsymbol{x}, \tau; 0) \in \Omega$. Then, the solution $u(\boldsymbol{x}, t)$ is determined uniquely by integrating the characteristic equation (3.8) with the initial condition $U(0) = u_0(\boldsymbol{X}(0))$ stemming from the original initial condition (3.1b). This gives

$$u(\boldsymbol{x}, \tau) = U(\tau) = u_0(\boldsymbol{X}(0)) + \int_0^{\tau} S(t)\, dt. \tag{3.10}$$

Remark 3.1. In this chapter we shall assume differentiability of $u(x, t)$. However, if the characteristics transport the initial condition, one could easily construct a "solution" transporting initial non-smooth data. In fact, any singularity in the initial data is transported along the characteristics. This is a fundamental property of linear hyperbolic equations. The concept of non-smooth solutions of hyperbolic equations, which is intimately related to the concept of *weak solutions*, will be discussed in the next chapter. Nonlinear hyperbolic equations may have non-smooth solutions even if initial data are smooth.

Remark 3.2 (Equation in conservation form). The analysis just described also applies when the linear hyperbolic equation is formulated in the conservation form (3.1a), that is

$$u_t + \boldsymbol{\nabla} \cdot (\boldsymbol{a}\, u) = s, \tag{3.11}$$

where the velocity field $\boldsymbol{a}(\boldsymbol{x}, t)$ is still assumed to be a known quantity. In view of the identity (3.3), and introducing the quantities

$$b(\boldsymbol{x}, t) = \boldsymbol{\nabla} \cdot \boldsymbol{a}(\boldsymbol{x}, t) \quad \text{and} \quad B(t) = b(\boldsymbol{X}(t), t),$$

equation (3.11) becomes along the characteristic $\boldsymbol{X}(t)$

$$\frac{dU}{dt} + B(t)\,U = S(t), \tag{3.12}$$

where again $S(t) = s(\boldsymbol{X}(t), t)$. Due to the additional variable coefficient term, $B(t)$, the solution for the case where the characteristic line crosses the domain Ω at $t = 0$ is

$$u(\boldsymbol{x}, \tau) = e^{-\int_0^\tau B(t)\,dt}\left[u_0(\boldsymbol{X}(0)) + \int_0^\tau S(t')\,e^{\int_0^{t'} B(t)\,dt}\,dt'\right]. \tag{3.13}$$

A similar expression is obtained when the characteristic intersects the boundary Γ_{in}, see Pironneau (1989, pp. 75–76).

Remark 3.3 (System of hyperbolic equations). In the case of a system of coupled hyperbolic equations, such as the Euler equations of gas dynamics to be discussed in Chapter 4, the problem cannot be reduced to a standard ordinary differential problem along the characteristics, unless the coefficients of the equations are constants. It is actually still possible to introduce characteristic curves passing through a given space–time point $(\boldsymbol{x}, \tau)$, the number of such characteristics being in general equal to the number of components in the governing system. However, the original system of coupled equations written along these characteristic curves takes the form of an ordinary differential system of a rather peculiar nature. Its solution is in fact defined by a system of integral equations in which each component of the vector unknown involves integration along a different curve, see John (1991, pp. 46–52) for a detailed description of the 1D case.

Remark 3.4 (Nonlinear convection equation). Let us briefly extend the above concepts to the case of a nonlinear convection equation of the form

$$u_t + \boldsymbol{a}(u) \cdot \boldsymbol{\nabla} u = s, \tag{3.14a}$$

where $\boldsymbol{a}(u) = \partial \boldsymbol{f}(u)/\partial u$. This is the quasi-linear form of a nonlinear scalar conservation law. Here, the source term s may also depend on the unknown. If the solution remains smooth and the equation holds in the classical sense, the characteristic curve $\boldsymbol{X}(t)$ reaching the space–time point $(\boldsymbol{x}, \tau)$ satisfies the following ordinary differential equation with final condition:

$$\frac{d\boldsymbol{X}}{dt} = \boldsymbol{a}(u(\boldsymbol{X}(t), t)), \qquad \boldsymbol{X}(\tau) = \boldsymbol{x}. \tag{3.14b}$$

Due to the dependence of the convection field on the unknown u, both equations (3.14) must be solved in a coupled manner. It is not possible to solve equation (3.14b) first in order to obtain the characteristic curves separately from the solution of the convection equation (3.14a) itself.

Assume that the foot of the considered characteristic lies inside Ω at $t = 0$, so that the initial condition for u has to be imposed. If one introduces the unknown $U(t) := u(\boldsymbol{X}(t), t)$, with $0 \leq t \leq \tau$, the equation and condition (3.14b) become

$$\frac{d\boldsymbol{X}}{dt} = \boldsymbol{a}(U(t)), \qquad \boldsymbol{X}(\tau) = \boldsymbol{x}, \tag{3.15a}$$

whereas the nonlinear convection equation (3.14a) expressed in characteristic form and the associated initial condition take the form

$$\frac{dU}{dt} = s(U(\boldsymbol{X}(t), t); t), \qquad U(0) = u_0(\boldsymbol{X}(0)). \tag{3.15b}$$

Equations (3.15) represent a coupled system of first-order ordinary differential equations to be solved over the interval $0 \leq t \leq \tau$ and supplemented by conditions at both ends of the integration interval. System (3.15) therefore represents a two-point boundary value problem for a first-order system with separated end-conditions. For solving such a problem, use can be made of the numerical technique proposed by Quartapelle and Rebay (1990) in combination with Newton iterations. The initial condition consists of a homogeneous relationship between the initial values of U and $\boldsymbol{X}$ in the form

$$U(0) \equiv u_0(\boldsymbol{X}(0)),$$

which is a nonlinear equation with respect to the unknown $\boldsymbol{X}(0)$ whenever $u_0(\boldsymbol{x})$ is a nonlinear function of its argument. Summarizing, the nonlinearity of the convection equation does not preclude a complete reduction of the partial differential problem to an ordinary differential one. However, the nonlinear convection equation introduces two extra difficulties:

1. The equation defining the characteristic curve and the governing equation in characteristic form, which are in general both nonlinear, must be solved as a coupled set of equations;

2. The initial condition is a coupled and nonlinear condition for the unknowns $(\boldsymbol{X}, U)$ of the system.

3.3.3 Methods based on the characteristics

In this section we briefly present finite element methods exploiting the characteristics for solving the convection equation (3.4a) with a space–time variable convection field $\boldsymbol{a}(\boldsymbol{x}, t)$. A more complete review of characteristic-based finite element methods can be found in a paper by Donea and Quartapelle (1992). The reader interested in a detailed exposition of such methods should consult the books by Pironneau (1989) and Morton (1996).

3.3.3.1 Semi-Lagrangian method A popular method of characteristic type specifically adapted to treat unsteady convection problems is the semi-Lagrangian method introduced by Purnell (1976), Robert (1981; 1982), Pudykiewicz and Staniforth (1984) and Staniforth and Pudykiewicz (1985).

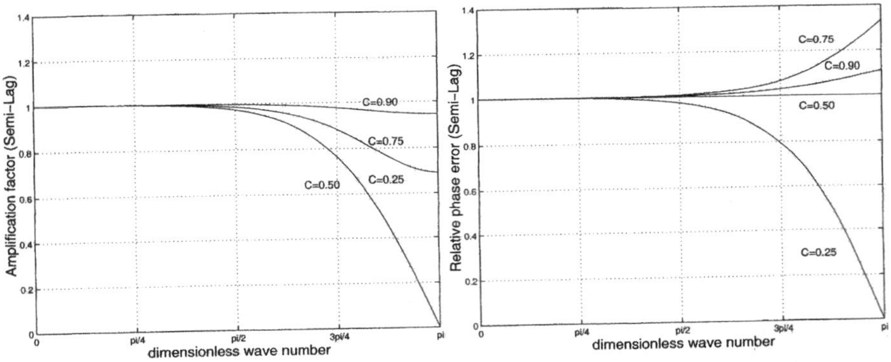

Fig. 3.2 Accuracy properties in pure convection of the semi-Lagrangian scheme with cubic interpolation: amplification factor modulus $|G|$ (left) and relative phase error $\phi_{numer}/\phi_{exact}$ (right). $C = \|\boldsymbol{a}\|\Delta t/h$ is the Courant number, see Section 3.5.1

With reference to the linear convection equation (3.4a), let $\boldsymbol{x}$ be a mesh point and $u^n(\boldsymbol{x})$ the finite element solution obtained at time $t^n = n\Delta t$, Δt being the time increment. The solution $u^{n+1}(\boldsymbol{x})$ at the next time level $t^{n+1} = t^n + \Delta t$ is obtained by first approximating equation (3.7a) defining the characteristic over the interval $]t^n, t^{n+1}[$ with second-order accuracy using the mid-point rule. This gives

$$\Delta \boldsymbol{X}(\boldsymbol{x}) = \boldsymbol{a}\Big(\boldsymbol{x} - \frac{1}{2}\Delta \boldsymbol{X},\, t^{n+1/2}\Big)\,\Delta t.$$

This nonlinear equation for the displacement vector $\Delta \boldsymbol{X}$ can be solved iteratively using the recursive scheme

$$\Delta \boldsymbol{X}_{(k+1)}(\boldsymbol{x}) = \boldsymbol{a}\Big(\boldsymbol{x} - \frac{1}{2}\Delta \boldsymbol{X}_{(k)},\, t^{n+1/2}\Big)\,\Delta t \tag{3.16}$$

and an interpolation formula to evaluate $\boldsymbol{a}$ between mesh points. Then, the characteristic equation (3.8) is approximated with second-order accuracy in time, according to the relationship

$$\frac{u^{n+1}(\boldsymbol{x}) - u^n(\boldsymbol{x} - \Delta \boldsymbol{X})}{\Delta t} = \frac{1}{2}\left[s^{n+1}(\boldsymbol{x}) + s^n(\boldsymbol{x} - \Delta \boldsymbol{X})\right].$$

Here $s^n(\boldsymbol{x}) = s(\boldsymbol{x}, t^n)$ and the quantities $u^n(\boldsymbol{x} - \Delta \boldsymbol{X})$ and $s^n(\boldsymbol{x} - \Delta \boldsymbol{X})$ are evaluated from $u^n(\boldsymbol{x})$ and $s^n(\boldsymbol{x})$ by interpolation. In practice, bi-cubic splines have been considered for two- and three-dimensional calculations. The value of $\Delta \boldsymbol{X}(\boldsymbol{x})$ at the previous time step can be used as a first guess to solve the nonlinear system (3.16) at each node. The convergence of the iterative procedure (3.16) is guaranteed by the fixed-point theorem provided that $\boldsymbol{a}(\boldsymbol{x}, t^{n+1/2})$ and its first partial derivatives are continuous, the time step Δt is sufficiently small and the first guess sufficiently close to the actual solution.

The accuracy properties of the semi-Lagrangian method with cubic interpolation are reported in Figure 3.2. They were derived, as explained in detail in Section 3.5, using Fourier mode analysis. The accuracy of a time integration method is usually characterized by its amplification factor and its relative phase error. These quantities are expressed in terms of the dimensionless wave number $\xi = k\,h$, where k is the wave number of the considered Fourier mode and h the mesh size.

In the present case of pure convection, the amplification factor for the exact solution is $G_{\text{exact}} = 1$ (no damping). As can be seen from Figure 3.2, the semi-Lagrangian scheme exhibits excellent amplitude response for dimensionless wave numbers in the range $0 \le \xi \le \pi/4$, which can be accurately resolved by the numerical method. Also the phase response of the semi-Lagrangian method appears to be excellent at small and intermediate wave numbers.

The same properties apply to the characteristic Galerkin method of Morton mentioned below. For its accuracy and efficiency the semi-Lagrangian method is widely used in meteorological forecasting as well as in the modeling of environmental flows using non-uniform Cartesian meshes, see for instance Staniforth and C ôté (1991) and Tanguay, Simard and Staniforth (1989) for practical applications of the method.

3.3.3.2 Lagrange–Galerkin methods Methods in this class also exploit the property of constant solution along a fluid particle trajectory to approximate the total derivative in the l.h.s. of (3.4a). The particle paths coincide with the associated characteristic curves and are defined by equations (3.7). Lagrange–Galerkin methods bear similarity with semi-Lagrangian methods, but are specifically adapted to the use of a finite element spatial discretization based on the Galerkin projection. Two variants of the basic Lagrange–Galerkin method are described next.

Direct integration along the characteristics: If the convection velocity $\boldsymbol{a}(\boldsymbol{x}, t)$ is known in advance over the entire time interval $]0, T[$, the complete solution of the linear convection equation (3.4a) can be determined as follows.

For a given triangulation of the domain Ω, one constructs the characteristic curves passing through all the nodes in the finite element mesh by approximating them as continuous lines consisting of straight segments. Recall that the characteristics are defined by equations (3.7). The nodal values of the unknown u at time t are then obtained by evaluating integrals in equations (3.9) or (3.10) and using numerical quadrature. A detailed description of a method of this kind is provided by Pironneau (1981/82).

In nonlinear situations, the field $\boldsymbol{a}(u, \boldsymbol{x}, t)$ depends on the solution itself, so that it is necessary to advance the solution by increments from a given time t^n to time $t^{n+1} = t^n + \Delta t$, where Δt is a convenient time step. As suggested by Pironneau (1981/82), equation (3.4a) is integrated along trajectories drawn backwards from $(\boldsymbol{x}, t^{n+1})$ to $\boldsymbol{X}(\boldsymbol{x}, t^{n+1}; t^n)$ using a suitable approximation for the characteristic velocity $\boldsymbol{a}$. Then, the standard Galerkin projection is employed to obtain the nodal values of the solution at time t^{n+1}. Both first-order methods, the conditionally stable explicit Euler scheme and the unconditionally stable backward Euler scheme, can be used in the Galerkin formulation which reads

$$\int_\Omega w\, u^{n+1}\, d\Omega = \int_\Omega w\, u^n \left(\boldsymbol{X}(\cdot, t^{n+1}; t^n)\right) d\Omega + \Delta t \int_\Omega w\, s^{n+\theta}\, d\Omega.$$

The choice $\theta = 0$ corresponds to the explicit method, and $\theta = 1$ corresponds to the implicit one.

Two critical issues arise in the implementation of the Lagrange–Galerkin method. First, the numerical approximation of the first term on the r.h.s. of the previous equation can give rise to instability phenomena due to errors introduced by the numerical quadrature. Second, at each integration point, $\boldsymbol{\xi}$, the interpolation of $u^n(\boldsymbol{X}(\boldsymbol{\xi}, t^{n+1}; t^n))$ must be evaluated and this may not be a trivial task. Note that, when the mapping $\boldsymbol{\xi} \to \boldsymbol{X}(\boldsymbol{\xi}, t^{n+1}; t^n)$ is used, the new position $\boldsymbol{X}(\boldsymbol{\xi}, t^{n+1}; t^n)$ may fall in a different element. See also the interesting discussion by Bermejo (1995) and Allievi and Bermejo (2000)

Characteristic variational formulation: The fact that the solution of convection problems remains constant along fluid particle paths can also be exploited in a closer connection with a weak variational formulation, as exemplified by the characteristic-based finite element method proposed by Benqué et al. (1980; 1982).

In this method, the convection equation (3.4a) is first recast into weak form by multiplying it by a suitable test function $\psi(\boldsymbol{x}, t)$ in space–time and then integrating over the space–time domain $Q_n := \Omega \times]t^n, t^{n+1}[$. This produces the weak form

$$\iint_{Q_n} \psi \,(u_t + \boldsymbol{a} \cdot \boldsymbol{\nabla} u)\, d\Omega\, dt = \iint_{Q_n} \psi\, s\, d\Omega\, dt.$$

After integration by parts in both space and time, and assuming that the function $\psi(\boldsymbol{x}, t)$ vanishes on the spatial boundary, the previous integral equation becomes

$$\int_\Omega (\psi^{n+1}\, u^{n+1} - \psi^n\, u^n) d\Omega = \iint_{Q_n} \big((\psi_t + \boldsymbol{a} \cdot \boldsymbol{\nabla}\psi)\, u + \psi\, s\big) d\Omega\, dt, \tag{3.17}$$

where we have assumed that the convection velocity $\boldsymbol{a}$ is divergence free.

In order to simplify the integral on the r.h.s. of (3.17) the arbitrary test function $\psi(\boldsymbol{x}, t)$ is required to satisfy the homogeneous initial boundary value problem

$$\psi_t + \boldsymbol{a} \cdot \boldsymbol{\nabla}\psi = 0 \qquad \text{on } Q_n, \tag{3.18a}$$
$$\psi(\boldsymbol{x}, t) = 0 \qquad \text{on } \Gamma, \tag{3.18b}$$
$$\psi(\boldsymbol{x}, t^{n+1}) = w(\boldsymbol{x}), \tag{3.18c}$$

where $w(\boldsymbol{x})$ is the standard weighting function associated with the considered finite element approximation over the domain Ω.

Under these circumstances, the weak form (3.17) can be rewritten as

$$\int_\Omega w\, u^{n+1}\, d\Omega = \int_\Omega \psi^n\, u^n\, d\Omega + \iint_{Q_n} \psi\, s\, d\Omega\, dt,$$

where the unknown u^{n+1} appears explicitly on the l.h.s. Note that the finite element discretization of this equation yields a linear system of algebraic equations governed by a symmetric positive definite matrix in the form of a consistent mass matrix.

However, this simplicity is only apparent. The function $\psi(\boldsymbol{x}, t)$ must be known at t^n and where needed for the evaluation of the last integral on the r.h.s. of the previous equation. The difficulty now lies in the resolution of the problem (3.18). The homogeneous pure convection equation (3.18a) can be solved using the characteristics as shown in Section 3.3.2. In fact, the function $\psi(\boldsymbol{x}, t)$ remains constant along the characteristics because the source term is zero, see (3.9) and (3.10). Thus,

$$\psi(\boldsymbol{x}, t) = w(\boldsymbol{X}(\boldsymbol{x}, t^{n+1}; t)) \quad \text{with } t \in [t^n, t^{n+1}].$$

Two issues are relevant for this method: the evaluation of the characteristics and the computational burden due to the interpolation of $w(\boldsymbol{X}(\boldsymbol{x}, t^{n+1}; t))$. Note that this was also the case in the previous method based on a direct integration along the characteristics.

The method is nonetheless very accurate as confirmed by Morton (1996, p. 329) who demonstrates that "*on a uniform mesh in one dimension and for linear constant coefficient advection, the Lagrange–Galerkin method based on continuous piecewise linear basis functions is equivalent to a semi-Lagrangian method using cubic spline interpolation*".

In addition to the original works previously mentioned, the idea of using a Lagrangian (or characteristic-based) approach to integrate the convective terms in transport equations has been exploited by many researchers. In particular, mention should be made of the characteristic Galerkin method introduced by Morton (1982; 1983; 1985), which, with linear elements, possesses accuracy properties similar to the semi-Lagrangian method with cubic interpolation. Also noteworthy is the least-squares characteristic method due to Li (1990). The starting consideration for this method is that the solution obtained by convection of the initial data along the characteristics does not belong to the interpolation space spanned by the shape functions of the base finite element mesh. To cure this situation, the idea of Li (1990) was to employ the least-squares approach to minimize the difference between the rigidly translated solution and the desired one expressed in terms of shape functions referred to the base element mesh.

As we have seen, finite element methods directly exploiting the characteristics present technical difficulties in their practical implementation. For this reason, instead of discretizing convection problems along the particle trajectories in a Lagrangian fashion, we prefer to discretize them along the Cartesian coordinates, a choice which in our view simplifies the numerical developments. The fact nevertheless remains that time discretization schemes for convection problems should be able to mimic the role of the characteristics, or, in other words, to take into account the directional character of propagation of the information.

3.4 CLASSICAL TIME AND SPACE DISCRETIZATION TECHNIQUES

As seen in Chapter 2, achieving a stable and accurate spatial representation is the main objective in the finite element solution of steady transport problems. This is not the only issue in time-dependent problems. In fact, the spatial representation provided by finite elements needs to be accurately transported in time to trace the transient response. This implies that one must select an appropriate algorithm for numerical time integration. A proper balance between the spatial and the temporal approximations must be considered. In other words, phase accuracy now becomes an important consideration in addition to spatial stability.

A usual practice in the finite analysis of time-dependent problems consists of discretizing first with respect to the spatial variables, thus obtaining a system of coupled first-order ordinary differential equations (with respect to time), a procedure called *semi-discrete method*. Then, to complete the discretization of the PDE, it remains to integrate the first-order differential system forward in time to trace the temporal evolution of the solution starting from the initial data $u_0(\boldsymbol{x})$. The complete apparatus of numerical methods for ordinary differential equations can be used; this procedure is called in the numerical analysis literature the *method of lines*. Which discretization is performed first is not an issue for linear spatial operators with constant coefficients and a Galerkin formulation. However, for future cases (nonlinear problems, stabilization techniques for transient analysis, etc.) it is preferable that the time discretization be performed before the spatial discretization.

In this section we recall some of the most usual algorithms for solving first-order differential equations. Their accuracy and stability properties when used in conjunction with linear finite elements are presented in Section 3.5.

3.4.1 Time discretization

3.4.1.1 The θ family of methods This family is widely used for integrating first-order differential equations. It is a single step method, that is the value u^{n+1} of the problem unknown at time $t^{n+1} = t^n + \Delta t$ is determined from the value u^n at time t^n. In this case, this is done by a weighted average of u_t^n and u_t^{n+1} at the end points of the integration step:

$$\frac{u(t^{n+1}) - u(t^n)}{\Delta t} = \theta\, u_t(t^{n+1}) + (1-\theta)\, u_t(t^n) + \mathcal{O}\big((1/2-\theta)\Delta t, \Delta t^2\big).$$

The time step Δt is assumed constant for the time being and θ is a parameter taken to be in the interval $[0, 1]$. Note that u_t will be replaced using equation (3.4a).

Several well-known methods are obtained for different values of the θ parameter. For values of $\theta < 1/2$ the schemes are conditionally stable. The *Euler method*, that is $\theta = 0$, is the best known. For values of $\theta \geq 1/2$ the methods are unconditionally stable. *Backward Euler*, $\theta = 1$, *Galerkin*, $\theta = 2/3$, and *Crank–Nicolson*, $\theta = 1/2$, are the most usual ones. As the truncation error in the previous equation indicates, the only method with second-order accuracy is Crank–Nicolson (see the details in Ames, 1992; Mitchell and Griffiths, 1980; Wait and Mitchell, 1985; Johnson, 1987; Lambert, 1991). The stability properties of these methods are studied in Section 3.5.

In practice, it is usual and sometimes preferable to solve for the incremental unknown $\Delta u = u^{n+1} - u^n$ rather than for u^{n+1}:

$$\frac{\Delta u}{\Delta t} - \theta\, \Delta u_t = u_t^n, \tag{3.19}$$

where $\Delta u_t = u_t^{n+1} - u_t^n$, u^n is the approximation of $u(t^n)$, $t^n = t^0 + n\Delta t$ (t^0 is the initial time), and u_t is again replaced using the original PDE (3.4a). This results in the following semi-discrete equation:

$$\frac{\Delta u}{\Delta t} + \theta(\boldsymbol{a} \cdot \boldsymbol{\nabla})\Delta u = \theta s^{n+1} + (1-\theta)s^n - \boldsymbol{a} \cdot \boldsymbol{\nabla} u^n. \tag{3.20}$$

3.4.1.2 The Lax–Wendroff method In addition to Crank–Nicolson, second-order accurate, explicit, time-stepping algorithms are also widely used. The Lax–Wendroff method, which is based upon a truncated Taylor series expansion, is one of the most popular. In the Taylor series expansion

$$u(t^{n+1}) = u(t^n) + \Delta t\, u_t(t^n) + \frac{1}{2}\Delta t^2 u_{tt}(t^n) + \mathcal{O}(\Delta t^3),$$

the first and second time derivatives are substituted by spatial derivatives using the governing equation (3.4a). Thus, the following relationships are used:

$$u_t^n = s^n - \boldsymbol{a} \cdot \boldsymbol{\nabla} u^n,$$
$$u_{tt}^n = s_t^n - \boldsymbol{a} \cdot \boldsymbol{\nabla} u_t^n = s_t^n - \boldsymbol{a} \cdot \boldsymbol{\nabla} s^n + (\boldsymbol{a} \cdot \boldsymbol{\nabla})^2 u^n,$$

and the second-order explicit method is obtained from the truncation of the previous Taylor series expansion, that is $(u^{n+1} - u^n)/\Delta t = u_t^n + \Delta t\, u_{tt}^n/2$. The resulting time-stepping algorithm reads

$$\frac{\Delta u}{\Delta t} = -\boldsymbol{a} \cdot \boldsymbol{\nabla} u^n + \frac{\Delta t}{2}(\boldsymbol{a} \cdot \boldsymbol{\nabla})^2 u^n + s^n + \frac{\Delta t}{2}\left(s_t^n - \boldsymbol{a} \cdot \boldsymbol{\nabla} s^n\right). \tag{3.21}$$

One of the main distinguishing features of the Lax–Wendroff method is that, despite its appearance, the second term on the r.h.s. of (3.21) is not to be thought of as an added artificial diffusion term. In fact, as far as time-dependent solutions are concerned, the second-order spatial derivatives are simply a consequence of the second-order accurate temporal approximation. The tensorial structure of these terms indicates that the correction introduced by the second time derivative acts only in the direction of the streamlines, in complete analogy with the SUPG method discussed in Chapter 2.

3.4.1.3 The leap-frog method Another widely used explicit time-stepping method is the three-level leap-frog method

$$u(t^{n+1}) = u(t^{n-1}) + 2\Delta t\, u_t(t^n) + \mathcal{O}(\Delta t^3),$$

which is second-order accurate in the time step Δt because the rate of change u_t^n is evaluated at the mid-point between the time stations t^{n-1} and t^{n+1}. When applied to the convection equation (3.4a) the leap-frog method reads

$$\frac{u^{n+1} - u^{n-1}}{2\Delta t} = u_t^n = s^n - \boldsymbol{a} \cdot \boldsymbol{\nabla} u^n. \tag{3.22}$$

3.4.2 Galerkin spatial discretization

3.4.2.1 The integral equation First, the standard Galerkin method of weighted residuals will be employed for the spatial discretization of the convection problem (3.4). This can be done directly on the differential equation (3.4a), or as previously indicated, on the time discretized equations (3.19), (3.21) or (3.22). For completeness, we present the *semi-discrete* scheme first. By means of the weighted residuals method applied directly to (3.4a) the following integral equation is obtained:

$$\int_\Omega w\, u_t\, d\Omega + \int_\Omega w(\boldsymbol{a}\cdot\boldsymbol{\nabla} u) d\Omega = \int_\Omega w\, s\, d\Omega, \tag{3.23}$$

where, as in previous chapters, the test functions w belong to the space $\mathcal{V}$ and satisfy the homogeneous boundary conditions on Γ^{in}_D:

$$\mathcal{V} = \left\{ w(\boldsymbol{x}) \in \mathcal{H}^1(\Omega) \mid w(\boldsymbol{x}) = 0 \text{ for } \boldsymbol{x} \in \Gamma^{in}_D \right\}.$$

Note that the functions in $\mathcal{V}$ do not depend on time. By contrast, the solution u of (3.4) lies in $\mathcal{L}_2(0, T; \mathcal{H}^1(\Omega))$. In fact, the time dependence of u can be translated to the trial space $\mathcal{S}_t$, which varies as a function of time,

$$\mathcal{S}_t := \left\{ u \mid u(\cdot, t) \in \mathcal{H}^1(\Omega), t \in [0, T] \text{ and } u(\boldsymbol{x}, t) = u_D \text{ for } \boldsymbol{x} \in \Gamma^{in}_D \right\}.$$

Dirichlet boundary conditions are taken into account by the definition of trial space. But, to account for the Neumann boundary conditions, the convective term, the second term on the l.h.s. of (3.23), is integrated by parts. Then, the variational problem associated with the initial boundary value problem defined in (3.4) becomes: for any $t \in [0, T]$ find $u \in \mathcal{S}_t$ such that for all weighting functions $w \in \mathcal{V}$

$$\begin{cases} (w, u_t) - (\boldsymbol{\nabla} w, \boldsymbol{a}\, u) + \big(w, u(\boldsymbol{a}\cdot\boldsymbol{n})\big)_{\Gamma^{out}} = (w, h)_{\Gamma^{in}_N} + (w, s), \\ (w, u(\boldsymbol{x}, 0)) = (w, u_0(\boldsymbol{x})), \end{cases} \tag{3.24}$$

where the standard notation presented in Section 1.5 is used for the $\mathcal{L}_2$ scalar product, and h is the prescribed normal flux on Γ^{in}_N. Note that the surface integral on the l.h.s. is limited to Γ^{out} because $w = 0$ on Γ^{in}.

> **Remark 3.5 (Only Dirichlet boundary conditions).** If no Neumann conditions are prescribed on the inflow portion of the boundary the variational form presented in (3.23) can be used directly. Then, given the source term s, the Dirichlet boundary condition u_D and the initial condition u_0, one has to find $u(\boldsymbol{x}, t) \in \mathcal{S}_t$ and $t \in]0, T[$ such that for all $w \in \mathcal{V}$,
>
> $$\begin{cases} (w, u_t) + c(\boldsymbol{a}; w, u) = (w, s), \\ (w, u(\boldsymbol{x}, 0)) = (w, u_0(\boldsymbol{x})), \end{cases}$$
>
> where the trilinear form, already introduced in (2.6), is employed, namely
>
> $$c(\boldsymbol{a}; w, u) = \int_\Omega w(\boldsymbol{a}\cdot\boldsymbol{\nabla} u) d\Omega.$$

3.4.2.2 Galerkin formulation of the semi-discrete scheme The spatial discretization of the unsteady transport equation by means of the Galerkin formulation consists of defining two finite dimensional spaces $\mathcal{S}^h$ and $\mathcal{V}^h$ as subsets of $\mathcal{S}$ and $\mathcal{V}$,

$$\mathcal{V}^h := \{w \in \mathcal{H}^1(\Omega) \mid w|_{\Omega^e} \in \mathcal{P}_p(\Omega^e)\ \forall e \text{ and } w = 0 \text{ on } \Gamma_D\}$$
$$\mathcal{S}_t^h := \{u \mid u(\cdot,t) \in \mathcal{H}^1(\Omega), u(\cdot,t)|_{\Omega^e} \in \mathcal{P}_p(\Omega^e)\ t \in [0,T]\ \forall e \text{ and } u = u_D \text{ on } \Gamma_D\}$$

where $\mathcal{P}_p$ is the finite element interpolating space consisting of polynomials of order p. The *semi-discrete* Galerkin formulation is obtained by restricting forms (3.24) to the above finite dimensional spaces, namely, for any $t \in [0,T]$ find $u^h \in \mathcal{S}_t^h$ such that for all $w^h \in \mathcal{V}^h$,

$$\begin{cases} (w^h, u_t^h) - (\boldsymbol{\nabla} w^h, \boldsymbol{a}\, u^h) + (w^h, u^h(\boldsymbol{a}\cdot\boldsymbol{n}))_{\Gamma^{out}} = (w^h, h)_{\Gamma_N^{in}} + (w^h, s), \\ (w^h, u^h(\boldsymbol{x},0)) = (w^h, u_0(\boldsymbol{x})). \end{cases}$$

Once the Galerkin forms are defined, we follow the same rationale as already discussed in Section 2.2. Note, however, that now the time dependence of the solution, $u^h(\boldsymbol{x},t)$, is taken into account in the following manner: the shape functions $N_A(\boldsymbol{x})$ do not depend on time and the time dependence is accounted for by the nodal values of the unknown. Thus, (2.10) becomes

$$u^h(\boldsymbol{x},t) = \sum_{A\in\eta\setminus\eta_D} N_A(\boldsymbol{x})\, \mathrm{u}_A(t) + \sum_{A\in\eta_D} N_A(\boldsymbol{x})\, u_D(\boldsymbol{x}_A,t),$$

where, as before, η is the set of global node numbers in the finite element mesh and $\eta_D \subset \eta$ the subset of nodes belonging to the Dirichlet portion of the boundary, Γ_D^{in}. The test functions are defined as before, see (2.11), $w^h \in \mathcal{V}^h = \text{span}_{B\in\eta\setminus\eta_D}\{N_B\}$.

Finally, the usual assembly process delivers the *semi-discrete* system

$$\mathbf{M}\dot{\mathbf{u}} - (\mathbf{C}^T - \mathbf{B}^{out})\mathbf{u} = \mathbf{f}, \tag{3.25}$$

which governs the transient response of the convection problem. Note that vectors $\mathbf{u}$ and $\dot{\mathbf{u}}$ contain, respectively, the nodal values of the unknown u and of its time derivative, while $\mathbf{M}$ and $\mathbf{C}$ are, respectively, the consistent mass matrix and the convection matrix. Matrix $\mathbf{B}^{out}$ is related to the outflow boundary, Γ^{out}. These matrices are obtained by topological assembly of element contributions as follows:

$$\begin{aligned} \mathbf{M} &= \mathbf{A}^e \mathbf{M}^e & M_{ab}^e &= \int_{\Omega^e} N_a N_b \, d\Omega \\ \mathbf{C} &= \mathbf{A}^e \mathbf{C}^e & C_{ab}^e &= \int_{\Omega^e} N_a (\boldsymbol{a}\cdot\boldsymbol{\nabla} N_b) d\Omega \\ \mathbf{B}^{out} &= \mathbf{A}^e [\mathbf{B}^{out}]^e & [B^{out}]_{ab}^e &= \int_{\partial\Omega^e\cap\Gamma^{out}} N_a N_b (\boldsymbol{a}\cdot\boldsymbol{n})\, d\Omega \end{aligned} \tag{3.26}$$

where $\mathbf{A}$ denotes the assembly operator, and $1 \le a, b \le \mathrm{n_{en}}$. The r.h.s. vector, $\mathbf{f}$, considers the contribution of the source term, s, the prescribed flux, h, and the

Dirichlet data u_D. It results from the assembly of nodal contributions of the form $\mathbf{f} = \mathbf{A}^e \mathbf{f}^e$ and

$$\mathrm{f}_a^e = (N_a, s)_{\Omega^e} + (N_a, h)_{\partial\Omega^e \cap \Gamma_N^{in}}$$
$$- \sum_{b=1}^{n_{en}} \left[(N_a, N_b)_{\Omega^e} \frac{\partial u_{Db}^e(t)}{\partial t} - (\boldsymbol{\nabla} N_a, \boldsymbol{a} N_b)_{\Omega^e} u_{Db}^e(t) \right.$$
$$\left. + (N_a, N_b(\boldsymbol{a}\cdot\boldsymbol{n}))_{\partial\Omega^e \cap \Gamma^{out}} u_{Db}^e(t) \right],$$

where n_{en} is the number of element nodes, and $u_{Db}^e(t) = u_D(\boldsymbol{x}_b, t)$ if u_D is prescribed at node number b and equals zero otherwise.

Remark 3.6. It is worth noting that a particularly high spatial accuracy is obtained by the Galerkin approach when a uniform mesh of linear finite elements is employed to discretize a pure convection problem. In this simple case,

$$u_t + a u_x = 0 \tag{3.27}$$

produces the following semi-discrete equation at an interior node j:

$$\dot{\mathrm{u}}_j + \frac{1}{6}\left(\dot{\mathrm{u}}_{j-1} - 2\dot{\mathrm{u}}_j + \dot{\mathrm{u}}_{j+1}\right) + a\left(\frac{u_{j+1} - u_{j-1}}{2h}\right) = 0,$$

where h is the size of the elements. It is easy to show, by means of Taylor series developments, that the previous scheme is fourth-order accurate. This is indeed a significant gain with respect to the second-order accuracy obtained with the central difference method. Such a gain is due to the presence of the consistent mass matrix; note the second term on the l.h.s. in addition to the term $\dot{\mathrm{u}}_j$ present in the central finite difference method. However, this high accuracy is lost on a non-uniform mesh or when data are not smooth enough (recall that the proof requires Taylor series developments).

Note also that the differential system (3.25) must be integrated in time to produce the transient response of the convection problem. It is easily conceived that the fourth-order spatial accuracy would be quickly eroded if the numerical time integration procedure were not of comparable accuracy. The issue of an adequate coupling between space and time discretizations is particularly important in convection problems due to the role of the characteristics in their solution.

3.4.2.3 Galerkin formulation of the θ family methods As previously noted, which discretization is performed first is not an issue for linear spatial operators with constant coefficients and a Galerkin formulation. However, for future cases (nonlinear problems, stabilization techniques for transient analysis, etc.) it is preferable that time discretization precedes the spatial one. The time discretized equations, (3.19), (3.21) and (3.22), do have a truncation error. However, if the temporal truncation error is neglected, these equations can be interpreted as a spatial differential operator applied to u, or in some cases to the unknown increment Δu. In fact, they represent a strong form that must be solved at each time step. Under this rationale it is easy to determine from (3.19) the variational form associated with the θ family methods:

$$\left(w, \frac{\Delta u}{\Delta t}\right) - \theta\left(w, \Delta u_t\right) = \left(w, u_t^n\right).$$

When u_t is replaced using the original PDE, (3.4a), or if a weighted residual formulation is applied to (3.20), we obtain an equation for the unknown Δu at each time-step:

$$\begin{aligned}\left(w, \frac{\Delta u}{\Delta t}\right) &- \theta(\boldsymbol{\nabla} w, \boldsymbol{a}\,\Delta u) + \theta\big((\boldsymbol{a}\cdot\boldsymbol{n})w, \Delta u\big)_{\Gamma^{out}} \\ &= (\boldsymbol{\nabla} w, \boldsymbol{a}\,u^n) - \big((\boldsymbol{a}\cdot\boldsymbol{n})w, u^n\big)_{\Gamma^{out}} \\ &\quad + \big(w, \theta h^{n+1} + (1-\theta)h^n\big)_{\Gamma_N^{in}} + \big(w, \theta s^{n+1} + (1-\theta)s^n\big). \qquad (3.28)\end{aligned}$$

Note that in the previous equation the unknown Δu appears in the same three terms as in the l.h.s. of (3.24). Thus, after spatial discretization, the unknown Δu will be governed by the same mass matrix, convection matrix and outflow boundary matrix as presented in (3.25) and (3.26), but scaled by $1/\Delta t$ and θ, respectively. The r.h.s. terms include also the influence of the inflow Neumann boundary condition, h, and the source term, s. These terms are linearly interpolated, using θ, between time t^n and t^{n+1}. Moreover, since the unknown is the increment, $\Delta u = u^{n+1} - u^n$, the r.h.s. also includes the convection part at t^n.

It is important to note that in equation (3.28) time is already discretized. Therefore, its solutions are approximations of order 1 or 2 depending on the value of θ. Moreover, as previously noted, the test functions belong to the space $\mathcal{V}$ presented earlier (no variation with time) and the unknown Δu belongs to a solution space $\mathcal{S}_t$ which varies as a function of time:

$$\mathcal{S}_t := \left\{\Delta u \mid \Delta u(\cdot, t) \in \mathcal{H}^1(\Omega),\ t \in [0, T] \text{ and } \Delta u(\boldsymbol{x}, t) \text{ is given for } \boldsymbol{x} \in \Gamma_D^{in}\right\}.$$

3.4.2.4 Galerkin formulation of the Lax–Wendroff method Similar variational forms can be derived for other time-stepping schemes. From (3.21), which characterizes the Lax–Wendroff method, the following variational form is obtained:

$$\left(w, \frac{\Delta u}{\Delta t}\right) = -\left(w, \boldsymbol{a}\cdot\boldsymbol{\nabla} u^n - \frac{\Delta t}{2}(\boldsymbol{a}\cdot\boldsymbol{\nabla})^2 u^n\right) + \left(w, s^n + \frac{\Delta t}{2}\left(s_t^n - \boldsymbol{a}\cdot\boldsymbol{\nabla} s^n\right)\right).$$

After integration by parts, this equation becomes

$$\begin{aligned}\left(w, \frac{\Delta u}{\Delta t}\right) &= \left(\boldsymbol{a}\cdot\boldsymbol{\nabla} w, u^n + \frac{\Delta t}{2}[s^n - (\boldsymbol{a}\cdot\boldsymbol{\nabla})u^n]\right) \\ &\quad - \left((\boldsymbol{a}\cdot\boldsymbol{n})w, u^n + \frac{\Delta t}{2}[s^n - (\boldsymbol{a}\cdot\boldsymbol{\nabla})u^n]\right)_{\Gamma^{out}} \\ &\quad + \left(w, h^{n+1/2}\right)_{\Gamma_N^{in}} + \left(w, s^n + \frac{\Delta t}{2} s_t^n\right), \qquad (3.29)\end{aligned}$$

where the following expression has been used for the prescribed normal flux:

$$h^{n+1/2} = -\Big(u^n + \frac{\Delta t}{2}[s^n - (\boldsymbol{a}\cdot\boldsymbol{\nabla})u^n]\Big)(\boldsymbol{a}\cdot\boldsymbol{n}) + \mathcal{O}(\Delta t^2),$$

because it introduces a truncation error of the same order as the time-stepping scheme. Note that, as expected, this one step Lax–Wendroff method requires, at each time step, the resolution of an algebraic system. This system has a constant matrix, the consistent mass matrix. Note also that some authors call this method *second-order Taylor–Galerkin*. Thus, we shall refer to this scheme as TG2.

3.4.2.5 *Galerkin formulation of the leap-frog method* This method induces a variational equation obtained from (3.22) in the form

$$\left(w, \frac{u^{n+1}}{2\Delta t}\right) = \left(w, \frac{u^{n-1}}{2\Delta t}\right) + \left(w, s^n - \boldsymbol{a}\cdot\boldsymbol{\nabla}u^n\right), \tag{3.30}$$

which, after integration by parts of the convective term, becomes

$$\left(w, \frac{u^{n+1}}{2\Delta t}\right) = \left(w, \frac{u^{n-1}}{2\Delta t} + s^n\right) + (\boldsymbol{a}\cdot\boldsymbol{\nabla}w, u^n) - (w, u^n(\boldsymbol{a}\cdot\boldsymbol{n}))_{\Gamma^{\text{out}}} + (w, h^n)_{\Gamma_N^{\text{in}}}.$$

In this case, as for the previous method, the matrix governing the system of discrete equations is the consistent mass matrix.

3.5 STABILITY AND ACCURACY ANALYSIS

A fundamental result, which is classical in the finite difference literature, is the *Lax equivalence theorem* (see Richtmyer and Morton, 1967). It relates the concepts of convergence and consistency to the concept of *stability* of a numerical scheme. Section 1.5.5.1 in Chapter 1 already introduces consistency, see (1.34), and convergence, see (1.35), in the context of elliptic problems. They can be extended to initial value problems, see Quarteroni and Valli (1994, Sec. 14.3) for a detailed presentation. More precisely, the Lax equivalence theorem states that for a *well-posed* linear initial value problem, a consistent scheme is convergent if and only if it is stable. In order to comprehend this theorem some basic concepts such as *well-posed* or *stable* problem shall be introduced. A problem is considered *well-posed* when its solution depends continuously on its initial value and is uniformly bounded in any compact interval. For linear problems such as the ones considered here this concept can be formalized as follows: for a homogeneous problem (source term equal to zero) with homogeneous boundary conditions for every $T > 0$ there exists $0 < C(T) < \infty$ such that $\|u(\boldsymbol{x}, t)\| \leq C(T)$ uniformly for all $t \in [0, T]$ (the constant C depends on the initial condition $\|u(\boldsymbol{x}, 0)\|$). For instance, the convection problem described by (3.4) with constant coefficients is well-posed because the norm of the solution is constant in time. The transient convection–diffusion–reaction problem (i.e., the transient counterpart of (2.54) that will be studied in Chapter 5) is also well-posed because diffusion will induce a decay in the norm of the solution. Thus the Lax equivalence theorem ensures that, for the complete analysis of the numerical scheme, it suffices to study its stability. Consequently, we concentrate our efforts on the stability analysis.

Stability of numerical schemes for PDEs is not a trivial issue. Here we will not study stability in detail. There are excellent references that study stability in a more formal manner for pure convection or convection–diffusion–reaction problems. The classical one is by Richtmyer and Morton (1967). Mitchell and Griffiths (1980) present a more accessible textbook for finite differences and the monograph by Quarteroni and Valli (1994), which is more finite element oriented, is also worth reading; other textbooks can also be very helpful, see for instance Hughes (2000) or Morton (1996). Different techniques (energy methods, eigenvalue techniques) can be employed to perform a stability analysis, but here we will use the Fourier method. One of the advantages of using such a technique is that apart from stability another crucial issue is studied: *accuracy*. In fact, at the end of this section, the *modified equation method* of Warming and Hyett (1974) is also introduced to investigate the accuracy properties of the various time integration schemes.

3.5.1 Analysis of stability by Fourier techniques

The classical Fourier analysis, also called *Von Neumann stability analysis* (see for instance Ames, 1992; Mitchell and Griffiths, 1980; Wait and Mitchell, 1985), considers a homogeneous linear differential equation with constant coefficients, namely equation (3.4a) with $s = 0$. But it is important to note that such a technique is designed to analyze *Cauchy problems*. That is, in a 1D situation, problems defined over the whole real axis ($-\infty < x < \infty$), and thus with no boundary conditions (there is, however, the requirement that the solution is square integrable). Hence, it is applied automatically to problems with periodic boundary conditions and it can also be extended to problems with Dirichlet conditions over finite domains for the PDEs studied here. Perhaps more important is the necessity that for each point the fully discretized equation should be identical, thus restricting treatment to uniform discretizations. In fact, we are actually going to deal with linear elements and the fully discrete equation at one interior node; this is standard procedure in finite differences where the stability literature is vast and accessible to practitioners.

Instead of restricting ourselves to the pure convection problem, we present this technique for the homogeneous transient convection–diffusion–reaction equation with constant coefficients, namely

$$u_t + \boldsymbol{a} \cdot \boldsymbol{\nabla} u - \boldsymbol{\nabla} \cdot (\nu \boldsymbol{\nabla} u) + \sigma u = 0. \tag{3.31}$$

Furthermore, we assume that the Fourier series for the initial condition, $u(\boldsymbol{x}, 0)$, is absolutely convergent (see Richtmyer and Morton, 1967, Chap. 4). This equation is studied in detail in Chapter 5.

The linearity of the problem is crucial. The evolution of the initial condition is the sum of the evolutions for each single mode in which the initial condition is decomposed. The analytical solution of (3.31) for one Fourier mode in n_{sd} spatial

Table 3.1 Spatial, discrete and mode transformation operators for 1D.

Spatial operator	Discrete	Mode transf.	Expression
$(N_i, N_j)/h^2$	$1 + \frac{1}{6}\delta^2$	$\mathcal{M}(\xi)$	$1 - (2/3)\sin^2(\xi/2)$
$\Delta t\,(N_i, \boldsymbol{a}\cdot\boldsymbol{\nabla} N_j)/h^2$	$\frac{1}{2}C\,\delta$	$\mathcal{A}(\xi, C)$	$i\,C\,\sin\xi$
$-\Delta t\,\nu\,(\boldsymbol{\nabla} N_i, \boldsymbol{\nabla} N_j)/h^2$	$d\,\delta^2$	$\mathcal{K}(\xi, d)$	$-4\,d\,\sin^2(\xi/2)$
$\Delta t\,(\boldsymbol{a}\cdot\boldsymbol{\nabla} N_i, N_j)/h^2$	$-\frac{1}{2}C\,\delta$	$-\mathcal{A}(\xi, C)$	$-i\,C\,\sin\xi$
$-\Delta t^2\,(\boldsymbol{a}\cdot\boldsymbol{\nabla} N_i, \boldsymbol{a}\cdot\boldsymbol{\nabla} N_j)/h^2$	$C^2\,\delta^2$	$\mathcal{D}(\xi, C)$	$-4\,C^2\,\sin^2(\xi/2)$

dimensions reads

$$u(\boldsymbol{x}, t) = \prod_{j=1,\dots,\mathrm{n_{sd}}} e^{-(\nu k_j^2+\sigma)t}\, e^{ik_j(x_j - a_j t)}, \tag{3.32}$$

where a_j is the j component of the convective velocity $\boldsymbol{a} = (a_1, a_2, \dots, a_{\mathrm{n_{sd}}})^T$ and k_j is the j component of the wave vector $\mathbf{k} = (k_1, k_2, \dots, k_{\mathrm{n_{sd}}})^T$, which defines the generalized Fourier component

$$e^{i\mathbf{k}\cdot\boldsymbol{x}} = e^{ik_1x_1}e^{ik_2x_2}\dots e^{ik_{\mathrm{n_{sd}}}x_{\mathrm{n_{sd}}}}.$$

The *exact amplification factor*, G_{ex}, which relates the value of the unknown u at two consecutive instants t^n and t^{n+1} and at each point in Ω, is then determined:

$$u^{n+1} = G_{\text{ex}}u^n, \qquad G_{\text{ex}} = e^{-(\delta_{\text{ex}}+i\omega_{\text{ex}})\Delta t} \tag{3.33}$$

where $\delta_{\text{ex}} = \nu\,\mathbf{k}\cdot\mathbf{k} + \sigma$ and $\omega_{\text{ex}} = \boldsymbol{a}\cdot\mathbf{k}$.

To determine the numerical amplification factor for a given time-stepping scheme, taking into account the spatial discretization, it is necessary to determine the influence of each spatial operator on a Fourier mode. Thus discrete operators associated with any of the weak continuous spatial operators defined in Section 3.4.2 must be determined. In Table 3.1 the relations between spatial operators, finite difference discrete operators and the mode transformations are shown. This table is particularized for uniform linear 1D elements of size h and consequently it requires the well-known difference operators (see Wait and Mitchell, 1985):

- Second-order centered first difference

$$\delta u(x, t) = u(x + h, t) - u(x - h, t).$$

- Second-order centered second difference

$$\delta^2 u(x, t) = u(x - h, t) - 2u(x, t) + u(x + h, t).$$

To determine the influence of the numerical scheme on a Fourier component one introduces the dimensionless wave vector ("number" in the 1D case), $\boldsymbol{\xi} = h\,\mathbf{k}$. Moreover, using Δt and h the parameters in the differential equation, (3.31), are scaled. The

Courant number is $C = \|\boldsymbol{c}\|$ with $\boldsymbol{c} = \boldsymbol{a}\Delta t/h$, the diffusion number is $d = (\nu\Delta t)/h^2$, and the dimensionless reaction is $r = \sigma\Delta t$. The exact amplification factor, (3.33), is rewritten as

$$G_{\text{ex}}(\boldsymbol{\xi}; \boldsymbol{c}, d, r) = e^{-(\delta_{\text{ex}} + i\omega_{\text{ex}})\Delta t} = e^{-(d\,\boldsymbol{\xi}\cdot\boldsymbol{\xi} + r + i\boldsymbol{c}\cdot\boldsymbol{\xi})}. \tag{3.34}$$

As noted previously we will restrict ourselves to 1D. Thus, the exact amplification factor (3.34) will be compared with the numerical one obtained for a particular numerical scheme at an arbitrary interior node j,

$$u_j^{n+1} = G(\xi, C, d, r)\; u_j^n, \qquad G(\xi, C, d, r) = e^{-(\delta + i\omega)\Delta t}. \tag{3.35}$$

Here, $u^n(x)$ denotes the piecewise linear finite element solution obtained at time $t^n = n\Delta t$ on a uniform mesh and the corresponding nodal values are denoted $u_j^n = u^n(jh)$.

Then, if the method is stable, the amplification factor G must verify

$$|\,G(\xi, C, d, r)| \leq 1$$

for all values of the dimensionless wave number $\xi \in [0, \pi]$ (note that $\xi = \pi$ corresponds to two elements per wavelength) and the dimensionless numbers C, d and r. Moreover, the accuracy of the numerical scheme can be assessed by comparing numerical and exact damping and phase values. These relative errors are defined respectively as

$$\frac{\delta}{\delta_{\text{ex}}} = \frac{|\,G(\xi, C, d, r)|}{|\,G_{\text{ex}}(\xi, C, d, r)|} \quad \text{and} \quad \frac{\omega}{\omega_{\text{ex}}} = \frac{\arg(G(\xi, C, d, r))}{\arg(G_{\text{ex}}(\xi, C, d, r))}. \tag{3.36}$$

From the point of view of accuracy only the range $0 \leq \xi \leq \pi/4$ is of interest because $\xi = \pi/4$ corresponds to eight elements per wavelength.

3.5.2 Analysis of classical time-stepping schemes

We shall now apply the Fourier technique to investigate the stability and accuracy properties of the previously presented time-stepping techniques combined with the Galerkin finite element formulation, see Section 3.4.2, in pure convection. Unfortunately, disadvantages of the standard Galerkin finite element method will become apparent when it is combined with the classical time-stepping schemes described in Section 3.4 (see also Morton and Parrott, 1980; Donea, Quartapelle and Selmin, 1987; Donea and Quartapelle, 1992; Morton, 1996). In particular, we shall see that second-order time schemes do not properly combine with linear finite elements in convection problems, because large values of the time step imply large phase errors.

The θ family of methods, the Lax–Wendroff method and the leap-frog method are considered for time discretization. The discrete equation obtained at an interior node j with the **θ family method** is determined from (3.28) (recall that there is no source term, $s = 0$, and we consider an interior node). The fully discrete equation can be written from Table 3.1 as

$$\left(1 + \frac{1}{6}\delta^2 + \frac{1}{2}\theta\, C\, \delta\right)(u_j^{n+1} - u_j^n) = -\frac{1}{2}\theta\, C\, \delta\, u_j^n,$$

which allows us to find the equation that modifies each Fourier component using also Table 3.1,

$$\big(\mathcal{M}(\xi) + \theta\mathcal{A}(\xi, C)\big)(u_j^{n+1} - u_j^n) = -\mathcal{A}(\xi, C)\, u_j^n.$$

Thus the numerical amplification factor is

$$G_\theta = \frac{\mathcal{M}(\xi) - (1-\theta)\mathcal{A}(\xi, C)}{\mathcal{M}(\xi) + \theta\mathcal{A}(\xi, C)} = \frac{1 - \frac{2}{3}\sin^2\frac{\xi}{2} - i(1-\theta)C\,\sin\xi}{1 - \frac{2}{3}\sin^2\frac{\xi}{2} + i\,\theta\, C\,\sin\xi}$$

The stability can now be studied for each value of θ verifying the condition $|G_\theta| \leq 1$. In particular for $\theta = 1/2$, that is Crank–Nicolson, one can verify that the amplification factor is always (for $0 \leq \xi \leq \pi$ and $C \geq 0$) equal to one. Thus, Crank–Nicolson is ***unconditionally stable*** and non dissipative (with a Galerkin formulation). Moreover, the accuracy of this method can be evaluated by means of (3.36).

A similar analysis can be performed for the **Lax–Wendroff method** departing from (3.29) (with no source term and at an interior node). The equation that modifies each Fourier component is in this case

$$\mathcal{M}(\xi)\,(u_j^{n+1} - u_j^n) = \left(\frac{1}{2}\mathcal{D}(\xi, C) - \mathcal{A}(\xi, C)\right) u_j^n,$$

and the amplification factor is

$$G_{TG2} = \frac{\mathcal{M}(\xi) + \frac{1}{2}\mathcal{D}(\xi, C) - \mathcal{A}(\xi, C)}{\mathcal{M}(\xi)} = \frac{1 - (\frac{2}{3} + 2C^2)\sin^2\frac{\xi}{2} - i\,C\,\sin\xi}{1 - \frac{2}{3}\sin^2\frac{\xi}{2}}$$

which induces the condition of numerical stability: $C^2 < 1/3$.

Finally, from (3.30) the analysis of the **leap-frog method** can be performed, namely

$$\mathcal{M}(\xi)\, u_j^{n+1} = \mathcal{M}(\xi)\, u_j^{n-1} - \mathcal{A}(\xi, C)\, u_j^n,$$

where the amplification factor is determined by solving a quadratic equation (hint: use $u_j^{n+1} = G_{LF}^2 u_j^{n-1}$)

$$G_{LF} = \frac{-\mathcal{A}(\xi, C) \pm \sqrt{\mathcal{M}(\xi)\mathcal{M}(\xi) + \mathcal{A}(\xi, C)\mathcal{A}(\xi, C)}}{\mathcal{M}(\xi)}$$

$$= \frac{-i\,C\,\sin\xi \pm \sqrt{(1 - \frac{2}{3}\sin^2\frac{\xi}{2}) - C^2\sin^2\frac{\xi}{2}}}{1 - \frac{2}{3}\sin^2\frac{\xi}{2}}.$$

Genuine finite element schemes are based on a so-called consistent mass representation in which the mass matrix is defined as in (3.26). Another option consists of

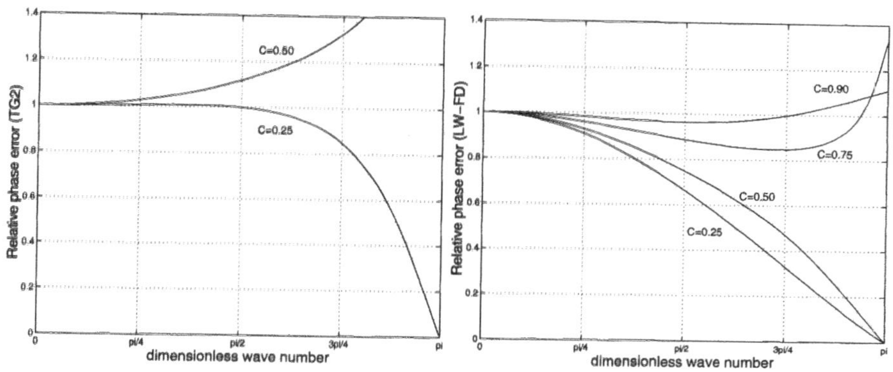

Fig. 3.3 Relative phase error in pure convection of the Lax–Wendroff method combined with linear finite elements using a consistent (left) and a diagonal (right) mass representation.

using a diagonal mass matrix. In this case, the discrete equations obtained in 1D on a uniform mesh of linear elements are identical to those obtained with second-order central differences. The finite element scheme corresponding to (3.29) with a diagonal mass representation instead of the consistent mass matrix yields the following scheme for an interior node:

$$u_j^{n+1} = u_j^n - \frac{1}{2}C\big(u_{j+1}^n - u_{j-1}^n\big) + \frac{1}{2}C^2\big(u_{j-1}^n - 2u_j^n + u_{j+1}^n\big).$$

This scheme corresponds to the central finite difference Lax–Wendroff discretization (LW-FD). Its amplification factor is

$$G_{\text{LW-FD}} = 1 - 2C^2 \sin^2\frac{\xi}{2} - i\,C\,\sin\xi.$$

It is stable up to $C^2 = 1$ and possesses the so-called *unit CFL property*; that is, the exact nodal solution is obtained on a uniform mesh when $C^2 = 1$. Recall that CFL stands for the initials of Courant, Friedrichs and Lewy authors of the celebrated paper originally published in 1928 and reprinted in english in 1967. Note that the finite difference method is far more economical from a computational point of view than its finite element analogue (a diagonal instead of a consistent matrix is employed) and it also has a larger domain of stability. In order to further compare the Lax–Wendroff finite element scheme (TG2) and its finite difference analogue the relative phase errors are depicted in Figure 3.3.

One notes that for small values of the time step Δt, that is small Courant numbers, the consistent finite element scheme has a higher-order phase accuracy than the finite element scheme using a diagonal mass representation. This is because semi-discrete (time continuous) consistent finite element schemes are fourth-order accurate on a uniform mesh, while methods based on a diagonal mass matrix only produce a second-order accurate spatial discretization. At intermediate and short wavelengths the phase error of the TG2 scheme becomes positive as the time step is increased and deteriorates

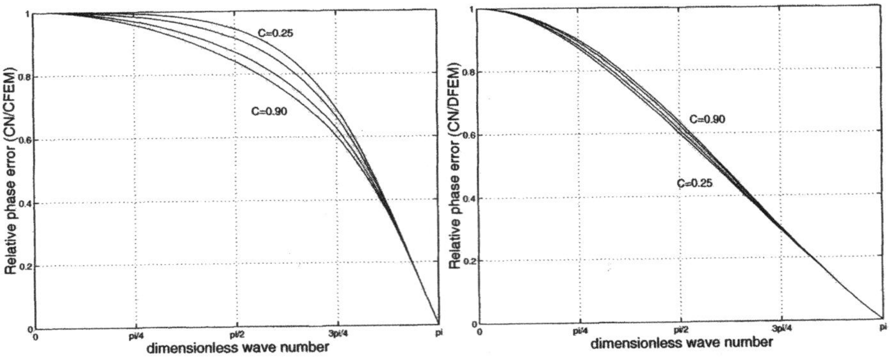

Fig. 3.4 Relative phase error in pure convection of the Crank–Nicolson method combined with linear finite elements using a consistent (left) and a diagonal (right) mass representation.

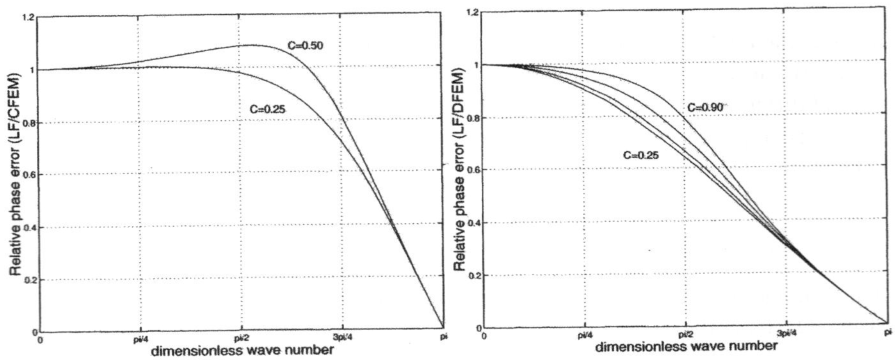

Fig. 3.5 Relative phase error in pure convection of the leap-frog method combined with linear finite elements using a consistent (left) and a diagonal (right) mass representation.

seriously as the stability limit $C^2 \to \frac{1}{3}$ is approached. By contrast, the scheme with diagonal mass matrix has a predominantly lagging phase error, except for large wave numbers when $C > \frac{1}{2}$. Here, the phase error is seen to decrease as the Courant number approaches the stability limit $C = 1$.

A similar analysis can be performed for the Crank–Nicolson and leap-frog methods. The relative phase errors of the Crank–Nicolson and leap-frog schemes combined with linear finite elements using either a consistent or a diagonal mass representation are reported in Figures 3.4 and 3.5, respectively. One notes that the phase response of the consistent finite element schemes deteriorates as the Courant number increases, while the phase accuracy in the case of a lumped-mass matrix improves as C approaches the stability limit of the explicit scheme. Furthermore, in the case of the leap-frog method, the combination with finite elements using a consistent mass matrix leads to a reduced stability range, namely $C^2 \leq 1/3$, while the scheme based on a diagonal mass representation is stable up to $C^2 \leq 1$.

In summary, the superior phase accuracy of the finite element schemes based on a consistent mass matrix, which is due to their fourth-order spatial accuracy on a uniform mesh, cannot be exploited as the time step is increased. Moreover, the stability interval in explicit methods is reduced when compared with the corresponding lumped-mass (or central difference) methods. These facts will be confirmed in the next section by an analysis based on the modified equation method.

Further evidence of the difficulties in coupling linear finite elements and second-order time-stepping algorithms is provided in Section 3.11 with the solved exercises. Here we have considered the pure convection equation, namely the first-order wave equation. Christon (1991) presents a complete analysis of the influence of the mass matrix on the dispersive nature of the semi-discrete second-order wave equation.

3.5.3 The modified equation method

Numerical methods for PDEs inevitably introduce truncation errors. Thus, numerical schemes do not solve the original PDE, but instead what Warming and Hyett (1974) call a *modified equation*. The modified equation is the PDE which is actually solved, apart from round-off errors, when a given numerical scheme is applied to solve an initial value problem. For example, the numerical solution of the linear convection equation in 1D, see (3.27), induces a modified equation of the generic form

$$\frac{\partial u}{\partial t} + a\frac{\partial u}{\partial x} + \sum_{p=1}^{\infty} \mu_{2p+1} \frac{\partial^{2p+1} u}{\partial x^{2p+1}} + \sum_{p=1}^{\infty} \mu_{2p} \frac{\partial^{2p} u}{\partial x^{2p}} = 0, \tag{3.37}$$

where $\mu_p = \mu_p(h, \Delta t)$, and h is the element size.

The two summation terms that appear in the modified equation and do not belong to the original equation represent the truncation error of the numerical scheme. These terms provide immediate information on the dissipation and dispersion properties of the numerical scheme. In fact, even and odd derivatives are associated, respectively, with amplitude and phase errors. Moreover, the correct sign of the coefficient of the lowest-order even-derivative term in the modified equation is, in pure convection, a necessary, but not always sufficient, condition for numerical stability.

The procedure to determine the modified equation requires some tedious algebra and goes as follows:

1. The difference equation resulting from time and space discretization of the governing equation is first expanded in a double Taylor series in time and space around the space–time point $jh, n\Delta t$.

2. This produces a PDE including an infinite number of terms and also spatial, temporal and mixed derivatives of every order.

3. One then eliminates the temporal derivatives of order higher than one, as well as the mixed derivatives in favor of purely spatial derivatives using the PDE obtained in step 2. This produces the modified equation in the form indicated by equation (3.37).

Table 3.2 Modified equation in pure convection for second-order Lax–Wendroff, leap-frog and Crank–Nicolson schemes combined with linear finite elements.

TG2 + Linear elements

$$u_t + au_x =$$
$$\tfrac{ah^2}{6}C^2\tfrac{\partial^3 u}{\partial x^3} - \tfrac{ah^3}{24}C(1-3C^2)\tfrac{\partial^4 u}{\partial x^4} + \tfrac{ah^4}{180}(1-\tfrac{15}{2}C^2+9C^4)\tfrac{\partial^5 u}{\partial x^5} + \cdots$$

Leap-frog + Linear elements

$$u_t + au_x =$$
$$\tfrac{ah^2}{6}C^2\tfrac{\partial^3 u}{\partial x^3} + \tfrac{ah^4}{360}(2-27C^4)\tfrac{\partial^5 u}{\partial x^5} + \cdots$$

Crank–Nicolson + Linear elements

$$u_t + au_x =$$
$$-\tfrac{ah^2}{12}C^2\tfrac{\partial^3 u}{\partial x^3} + \tfrac{ah^4}{720}(4-9C^4)\tfrac{\partial^5 u}{\partial x^5} + \cdots$$

The modified equations associated with the classical second-order schemes presented so far for the 1D convection equation will now be presented. They offer an alternative approach to the Fourier techniques to investigate the accuracy properties of the discrete convection schemes. The modified equations associated with the various second-order time schemes are obtained in the form indicated in Table 3.2. They are derived assuming uniform linear elements and a Galerkin spatial discretization.

Note that fourth-order spatial accuracy of finite element schemes is clearly apparent from the modified equations (the mesh size h does not affect the coefficient of the third-derivative terms, which only depends on the square of the time step: $h^2C^2 = a^2\Delta t^2$). These schemes exhibit a dominating phase error controlled by third-derivative terms which is due to the time discretization only. Unfortunately, the dominating phase error increases with the square of the time step. This implies a rapid accuracy fall-off as the time step is increased. Furthermore, the stability range of these explicit schemes (TG2 and LF) is drastically reduced compared with the corresponding lumped-mass finite element (or central difference) schemes, usually by a factor $\sqrt{3}$. See in Table 3.2 the coefficient of $\partial^4 u/\partial x^4$ in the modified equation for the Lax–Wendroff method (recall that this coefficient must be negative for stability).

By contrast, the modified equations in Table 3.3 show that when finite elements with a diagonal mass representation (or second-order central differences) are used for space discretization, the phase error of explicit schemes decreases as the Courant number increases. Note, however, that the implicit Crank–Nicolson method exhibits the same drawback as its consistent-mass counterpart. The phase error in both schemes increases with the square of the time-step size.

The behavior of second-order time schemes in the finite element solution of convection problems is further illustrated by the solved exercises in Section 3.11. There, they are compared with the higher-order time-stepping methods described in the next sections.

Table 3.3 Modified equation in pure convection for second-order Lax–Wendroff, leap-frog and Crank–Nicolson schemes combined with second-order central differences or lumped-mass linear finite elements.

Lax–Wendroff + Finite differences
$u_t + au_x =$
$-\frac{ah^2}{6}(1-C^2)\frac{\partial^3 u}{\partial x^3} - \frac{ah^3}{8}C(1-C^2)\frac{\partial^4 u}{\partial x^4} - \frac{ah^4}{120}(1+5C^2-6C^4)\frac{\partial^5 u}{\partial x^5} + \cdots$
Leap-frog + Finite differences
$u_t + au_x =$
$-\frac{ah^2}{6}(1-C^2)\frac{\partial^3 u}{\partial x^3} + \frac{ah^4}{120}(1-10C^2+9C^4)\frac{\partial^5 u}{\partial x^5} + \cdots$
Crank–Nicolson + Finite differences
$u_t + au_x =$
$-\frac{ah^2}{6}(1+\frac{1}{2}C^2)\frac{\partial^3 u}{\partial x^3} - \frac{ah^4}{120}(1+5C^2+\frac{3}{2}C^4)\frac{\partial^5 u}{\partial x^5} + \cdots$

3.6 TAYLOR–GALERKIN METHODS

3.6.1 The need for higher-order time schemes

We have underlined in Section 3.5 the difficulties in coupling second-order time schemes with linear finite elements in convection problems. These difficulties can be overcome by extending the time accuracy beyond second order.

To highlight the need for accurate time-stepping methods in the finite element solution of time-dependent convection problems we will restrict ourselves to the homogeneous, $s = 0$, 1D case, see equation (3.5).

Section 3.3 showed that the solution of this equation remains constant along the characteristic lines $dx/dt = a$, see equation (3.6). Therefore, provided the solution is regular enough, the value of the unknown u at two consecutive time levels $t^n = n\Delta t$ and $t^{n+1} = t^n + \Delta t$ satisfies

$$\begin{aligned} u(x, t^{n+1}) &= u(x - a\Delta t, t^n) \\ &= u(x, t^n) - a\Delta t \frac{\partial u}{\partial x}(x, t^n) + \frac{(a\Delta t)^2}{2}\frac{\partial^2 u}{\partial x^2}(x, t^n) + \cdots \\ &= \exp\left(-a\Delta t \frac{\partial}{\partial x}\right) u(x, t^n). \end{aligned} \tag{3.38}$$

Moreover, for an infinitely differentiable function a Taylor series of $u(x, t^{n+1})$ indicates that

$$u(x, t^{n+1}) = \exp\left(\Delta t \frac{\partial}{\partial t}\right) u(x, t^n). \tag{3.39}$$

Relations (3.38) and (3.39) highlight a key aspect of the numerical approximation of convection problems: space and time are linked by the characteristics and the discretization of one certainly influences the other. Good methods are thus needed for both numerical time integration and spatial representation; that is, the exponential function in (3.39) and the spatial operator in (3.38) must be properly approximated. The fact is that second-order time schemes do not allow a sufficiently accurate approximation of the exponential operator in (3.39), or, stated in other words, they do not properly account for the directional character of propagation of information in hyperbolic problems. Higher-order time-stepping schemes provide a better approximation to the exponential function in (3.39), and consequently allow a better account of the propagation of information along the characteristics. Higher-order implicit time-stepping methods are discussed in Chapter 5 where the finite element solution of transient convection–diffusion problems is addressed. In the present chapter devoted to pure convection problems, we shall present higher-order explicit methods known as Taylor–Galerkin (TG) methods (Donea, 1984; Donea et al., 1987; Donea and Quartapelle, 1992). Such methods represent an attempt to simulate, by a Taylor series in time extended to third or fourth order, the concept that in convective transport the solution remains constant along the characteristics.

3.6.2 Third-order explicit Taylor–Galerkin method

3.6.2.1 Time discretization Explicit Taylor–Galerkin methods represent a generalization of the Lax–Wendroff method discussed in Section 3.4.1.2. They are based on a Taylor series expansion up to the desired order. Thus, the solution u must be sufficiently smooth. In order to obtain a third-order method the Taylor series is taken as

$$\frac{u(t^{n+1}) - u(t^n)}{\Delta t} = u_t(t^n) + \frac{1}{2}\Delta t\, u_{tt}(t^n) + \frac{1}{6}\Delta t^2 u_{ttt}(t^n) + \mathcal{O}(\Delta t^3). \qquad (3.40)$$

Then, time derivatives of u ($u_t, u_{tt}, \dots$) are replaced using the original differential equation, (3.4a) in our case. For pure convection equations the order of the time derivatives does correspond with that of the space derivatives. In this case, in order to substitute the time derivatives in (3.40), they are expressed as

$$u_t = s - \boldsymbol{a} \cdot \boldsymbol{\nabla} u, \qquad (3.41a)$$

$$u_{tt} = s_t - \boldsymbol{a} \cdot \boldsymbol{\nabla} u_t = s_t - \boldsymbol{a} \cdot \boldsymbol{\nabla} s + (\boldsymbol{a} \cdot \boldsymbol{\nabla})^2 u, \qquad (3.41b)$$

$$u_{ttt} = s_{tt} - \boldsymbol{a} \cdot \boldsymbol{\nabla} s_t + (\boldsymbol{a} \cdot \boldsymbol{\nabla})^2 u_t. \qquad (3.41c)$$

Note, however, that in (3.41c) u_t has not been substituted by (3.41a). This is done on purpose to have at most second spatial derivatives and preserve the use of standard $\mathcal{C}^0$-continuous finite element approximations. In practice, u_t in (3.41c) is substituted by $(u^{n+1} - u^n)/\Delta t$, and the third-order approximation of the Taylor series (3.40) becomes

$$\left[1 - \frac{\Delta t^2}{6}(\boldsymbol{a} \cdot \boldsymbol{\nabla})^2\right] \frac{u^{n+1} - u^n}{\Delta t} = -(\boldsymbol{a} \cdot \boldsymbol{\nabla}) u^n + \frac{\Delta t}{2}(\boldsymbol{a} \cdot \boldsymbol{\nabla})^2 u^n + s^n$$
$$+ \frac{\Delta t}{2}(s_t^n - \boldsymbol{a} \cdot \boldsymbol{\nabla} s^n) + \frac{\Delta t^2}{6}(s_{tt}^n - \boldsymbol{a} \cdot \boldsymbol{\nabla} s_t^n). \qquad (3.42)$$

This time-stepping scheme only involves first and second time derivatives and is the basis of the one-step, third-order accurate, explicit Taylor–Galerkin method.

3.6.2.2 Spatial discretization Let us now illustrate the construction of an explicit third-order Taylor–Galerkin method for the convection problem (3.4a). Mixed Dirichlet and Neumann conditions are prescribed on the inlet portion of the boundary, as specified in equations (3.4c) and (3.4d). The convection velocity $\boldsymbol{a}$ is assumed to be divergence free, so that $\boldsymbol{a} \cdot \boldsymbol{\nabla} u = \boldsymbol{\nabla} \cdot (u\, \boldsymbol{a})$.

To produce the Galerkin variational form associated with scheme (3.42), we multiply it by the test function w and integrate over the spatial domain Ω. After integration by parts, the following variational problem is obtained where $\Delta u = u^{n+1} - u^n$ denotes the incremental unknown.

Given u^n, find $\Delta u \in \mathcal{S}$, such that for all $w \in \mathcal{V}$,

$$\begin{aligned}\left(w, \frac{\Delta u}{\Delta t}\right) &+ \frac{\Delta t^2}{6}\left(\boldsymbol{a} \cdot \boldsymbol{\nabla} w, \boldsymbol{a} \cdot \boldsymbol{\nabla} \frac{\Delta u}{\Delta t}\right) - \frac{\Delta t^2}{6}\left((\boldsymbol{a} \cdot \boldsymbol{n})w, \boldsymbol{a} \cdot \boldsymbol{\nabla} \frac{\Delta u}{\Delta t}\right)_{\Gamma^{out}} \\ &= \left(\boldsymbol{a} \cdot \boldsymbol{\nabla} w, u^n - \frac{\Delta t}{2}(\boldsymbol{a} \cdot \boldsymbol{\nabla} u^n)\right) - \left((\boldsymbol{a} \cdot \boldsymbol{n})w, u^n - \frac{\Delta t}{2}(\boldsymbol{a} \cdot \boldsymbol{\nabla} u^n)\right)_{\Gamma^{out}} \\ &\quad + \frac{\Delta t}{2}\left(\boldsymbol{a} \cdot \boldsymbol{\nabla} w, s^{n+1/3}\right) - \frac{\Delta t}{2}\left((\boldsymbol{a} \cdot \boldsymbol{n})w, s^{n+1/3}\right) \\ &\quad + \left(w, \frac{3}{4}s^{n+2/3} + \frac{1}{4}s^n\right) + \left(w, \frac{3}{4}h^{n+2/3} + \frac{1}{4}h^n\right)_{\Gamma_N^{in}}. \end{aligned} \tag{3.43}$$

Note that the term responsible for the third-order accuracy of the explicit Taylor–Galerkin method, which we call TG3, introduces a modification to the standard mass matrix which nevertheless remains symmetric if only Dirichlet boundary conditions are present.

Notice also that the integration by parts produces boundary terms, which on a portion Γ_N^{in} of the inlet boundary generate a natural condition for the prescribed normal inlet flux h in (3.4d). As shown by equation (3.43), the boundary term on the outlet boundary contributes partly to the matrix governing the incremental unknown Δu, the remainder being part of the known independent term. The boundary terms on Γ^{out} are necessary to obtain the correct form of the discrete equation at nodes on the outlet boundary of the computational domain. Neglecting these terms generates spurious reflections at outflow boundaries (Donea and Quartapelle, 1992).

The linear system to be solved at each time step has characteristics similar to the one obtained by applying the second-order Lax–Wendroff scheme with finite elements. The advantage here is that, with similar computational cost, we obtain third-order accuracy. The generalized mass matrix of TG3 is tridiagonal in 1D and has the typical profile of a stiffness matrix in 2D and 3D. Generally, the matrix is characterized by a diagonal dominance that allows an approximate but accurate solution of the corresponding linear system with a few Jacobi iterations.

Remark 3.7. When the velocity field $\boldsymbol{a}$ and the source term s are functions of space and time, as originally stated for the initial boundary value problem (3.4), the third-order explicit method (3.42) becomes

$$\left[1 - \frac{\Delta t^2}{6}(\boldsymbol{a}^{n+1}\cdot\boldsymbol{\nabla})^2 + \frac{\Delta t^2}{6}(\boldsymbol{a}_t^{n+1}\cdot\boldsymbol{\nabla})\right] u^{n+1}$$
$$= \left[1 + \frac{\Delta t^2}{3}(\boldsymbol{a}^{n}\cdot\boldsymbol{\nabla})^2 - \frac{\Delta t^2}{3}(\boldsymbol{a}_t^{n}\cdot\boldsymbol{\nabla}) - \Delta t(\boldsymbol{a}^{n}\cdot\boldsymbol{\nabla})\right] u^{n} + \Delta t\, s^n$$
$$+ \frac{\Delta t^2}{6}\left[s_t^{n+1} - (\boldsymbol{a}^{n+1}\cdot\boldsymbol{\nabla})s^{n+1}\right] + \frac{\Delta t^2}{3}\left[s_t^{n} - (\boldsymbol{a}^{n}\cdot\boldsymbol{\nabla})s^{n}\right], \quad (3.44)$$

where $\boldsymbol{a}^{n+1} = \boldsymbol{a}(\boldsymbol{x}, t^{n+1})$, etc. Here, the time dependence of the characteristic velocity causes the modified mass matrix to lose its symmetric character.

Remark 3.8. The Taylor–Galerkin method can be applied as well in the solution of nonlinear conservation law equations of the form

$$\frac{\partial u}{\partial t} + \boldsymbol{\nabla}\cdot\boldsymbol{f}(u) = s.$$

Examples will be given in Chapter 4 where the inviscid Burgers' equation and the Euler equations of gas dynamics are considered.

3.6.2.3 Stability analysis For the 1D convection equation (3.27), the TG3 scheme (3.43) produces the following discrete equation at an interior node j of a uniform mesh of linear elements of size h:

$$\left(1 + \frac{1}{6}(1 - C^2)\delta^2\right)(u_j^{n+1} - u_j^n) = -\frac{1}{2}C\delta u_j^n + \frac{1}{2}C^2\delta^2 u_j^n, \quad (3.45)$$

where use has been made of the central difference operators, see Table 3.1. The amplification factor of the TG3 scheme is given by

$$G_{\mathrm{TG3}}(\xi,\, C) = \frac{1 - \frac{2}{3}(1 + 2C^2)\sin^2\frac{\xi}{2} - iC\sin\xi}{1 - \frac{2}{3}(1 - C^2)\sin^2\frac{\xi}{2}}, \quad (3.46)$$

and the stability condition is $|C| \leq 1$. The TG3 scheme possesses the unit CFL property and its accuracy characteristics are illustrated in Figure 3.6 which shows the diagrams of its phase and amplitude errors. The comparison with the same data in Figure 3.2 for the semi-Lagrangian method with cubic interpolation and the characteristic Galerkin method of Morton reveals that the phase is exact in the three methods for $C = 1/2$ and $C = 1$, and that the phase error is almost identical, the TG3 scheme being slightly inferior in the range $1/2 < C < 1$.

Unfortunately, the TG3 scheme experiences a drastic reduction of its stability range in multidimensional situations. As shown in Figure 3.7, the limit curve for stability in 2D strongly depends on the orientation of the Courant vector $\boldsymbol{c}$ whose components are (c_x, c_y). It is given by the equation

$$c_x^{2/3} + c_y^{2/3} = 1,$$

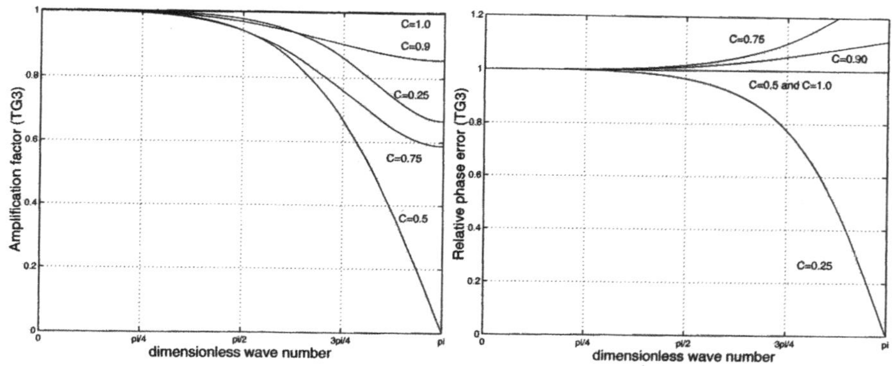

Fig. 3.6 Accuracy properties in pure convection of the third-order explicit Taylor–Galerkin scheme TG3: amplification factor modulus $|G|$ (left) and relative phase error $\phi_{numer}/\phi_{exact}$ (right).

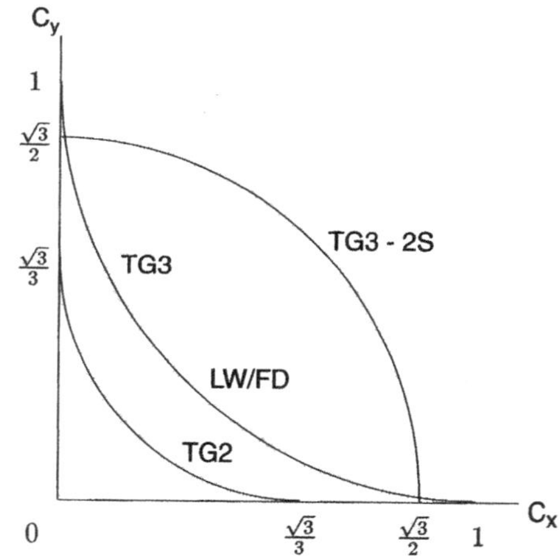

Fig. 3.7 Stability range in two-dimensional convection of one-step Taylor–Galerkin method (TG3), two-step Taylor–Galerkin method (TG3-2S), and Lax–Wendroff finite element (TG2) and finite difference (LW/FD) methods.

where the components of the Courant vector are defined as follows:

$$c_x = a_x \Delta t / h_x, \quad c_y = a_y \Delta t / h_y,$$

with a_x and a_y being the velocity components and h_x and h_y the mesh sizes along the Cartesian coordinate directions. The 1D stability condition $|C| \leq 1$ becomes in the most restrictive 2D case, that is $|c_x| = |c_y| := |C|$, $|C| < 1/(2\sqrt{2}) = 0.353$ and $|C| < 1/(3\sqrt{3}) = 0.192$ in 3D. As shown below, a two-step version of the method can be developed which possesses a more extended stability domain in multidimensional situations.

Table 3.4 Modified equation in pure convection for two second-order schemes (TG2 and leap-frog) and two higher-order methods (3rd-order TG3 and 4th-order LFTG).

TG2 + Linear elements

$$u_t + au_x =$$
$$\frac{ah^2}{6}C^2\frac{\partial^3 u}{\partial x^3} - \frac{ah^3}{24}C(1-3C^2)\frac{\partial^4 u}{\partial x^4} + \frac{ah^4}{180}(1-\tfrac{15}{2}C^2+9C^4)\frac{\partial^5 u}{\partial x^5} + \cdots$$

TG3 + Linear elements

$$u_t + au_x =$$
$$-\frac{ah^3}{24}C(1-C^2)\frac{\partial^4 u}{\partial x^4} + \frac{ah^4}{180}(1-5C^2+4C^4)\frac{\partial^5 u}{\partial x^5} + \cdots$$

Leap-frog + Linear elements

$$u_t + au_x =$$
$$\frac{ah^2}{6}C^2\frac{\partial^3 u}{\partial x^3} + \frac{ah^4}{360}(2-27C^4)\frac{\partial^5 u}{\partial x^5} + \cdots$$

LFTG + Linear elements

$$u_t + au_x =$$
$$\frac{ah^4}{360}(2+5C^2-7C^4)\frac{\partial^5 u}{\partial x^5} + \cdots$$

Figure 3.7 also shows that the situation is even worse for the Lax–Wendroff finite element scheme. The stability limit for Lax–Wendroff in 2D, called TG2, is in fact given by

$$c_x^{2/3} + c_y^{2/3} = (1/3)^{1/3}.$$

3.6.2.4 *Properties of the explicit Taylor–Galerkin method* The basic distinguishing features of the third-order explicit Taylor–Galerkin method TG3 can be underlined as follows:

1. The modified equation associated with TG3 is reported in Table 3.4 where it can be compared with that of the second-order Lax–Wendroff finite element method (TG2). Note that the leading dispersion error due to the time discretization has moved from the third- to the fifth-order derivative because of the increased time accuracy. The leading term of the dissipation error is a fourth-order derivative in both the second-order TG2 scheme and the third-order TG3 scheme. The necessary condition for numerical stability in 1D is $C^2 \leq 1$ for TG3 as compared with $C^2 \leq 1/3$ for the Lax–Wendroff finite element method (TG2). Moreover, the lowest-order terms of both dispersion and dissipation errors for TG3 are found to be zero for $C^2 = 1$. The optimal stability limit and the unit CFL property of TG3 are confirmed by the amplification factor (3.46) of the scheme.

2. As already pointed out for the Lax–Wendroff method, the terms in the TG3 scheme, see (3.43), involving $(\boldsymbol{a}\cdot\nabla)^2$ are not to be thought of as an artificial

numerical diffusion inherent to the scheme. In fact, as far as time-dependent solutions are concerned, the second-order spatial terms are only an element of the improved temporal approximation. The tensorial structure of these terms indicates that the correction introduced by the second time derivative acts only in the direction of the streamlines, in complete analogy with the SUPG method discussed in Chapter 2.

3.6.3 Fourth-order explicit leap-frog method

The concept in the Taylor–Galerkin method of improving the time accuracy of finite element schemes for unsteady convection can be extended to other time-stepping algorithms. As an illustrative example, we mention the fourth-order accurate extension of the standard second-order leap-frog method

$$\frac{u^{n+1}-u^{n-1}}{2\Delta t}=u_t^n.$$

The improved method is based upon the forward and backward Taylor series

$$\begin{aligned} u(t^{n+1}) &= u(t^n)+\Delta t\, u_t^n(t^n)+\frac{\Delta t^2}{2}u_{tt}^n(t^n)+\frac{\Delta t^3}{6}u_{ttt}^n(t^n)+\mathcal{O}(\Delta t^4),\\ u(t^{n-1}) &= u(t^n)-\Delta t\, u_t^n(t^n)+\frac{\Delta t^2}{2}u_{tt}^n(t^n)-\frac{\Delta t^3}{6}u_{ttt}^n(t^n)+\mathcal{O}(\Delta t^4), \end{aligned} \tag{3.47}$$

which by subtraction produce the fourth-order accurate leap-frog method

$$\frac{u^{n+1}-u^{n-1}}{2\Delta t}=u_t^n+\frac{1}{6}\Delta t^2 u_{ttt}^n. \tag{3.48}$$

The construction of the fourth-order leap-frog Taylor–Galerkin (LFTG) method then proceeds as for the TG3 method. The LFTG scheme is fourth-order accurate in time and also fourth-order accurate in space on a mesh of uniform linear elements in 1D. Its stability condition in 1D is $C^2 \leq 1$, while the second-order leap-frog finite element scheme has the reduced stability limit $C^2 \leq 1/3$.

Because the leap-frog time discretization is a centered discretization, the method is non-dissipative, that is $|G| = 1$, in pure convection. The relative phase error of the LFTG scheme for $C = 0.25, 0.50, 0.75$ and 0.90 is illustrated in Figure 3.8 in comparison with its second-order counterpart which cannot be operated at the last two values of the Courant number. To further appraise the gain in accuracy brought about by the fourth-order leap-frog Taylor–Galerkin scheme LFTG, we compare in Table 3.4 its modified equation for pure convection with that for the corresponding second-order leap-frog method. As can be seen, the LFTG scheme is characterized by a fifth-order dominating phase error, as compared with the third-order one for the classical second-order leap-frog method.

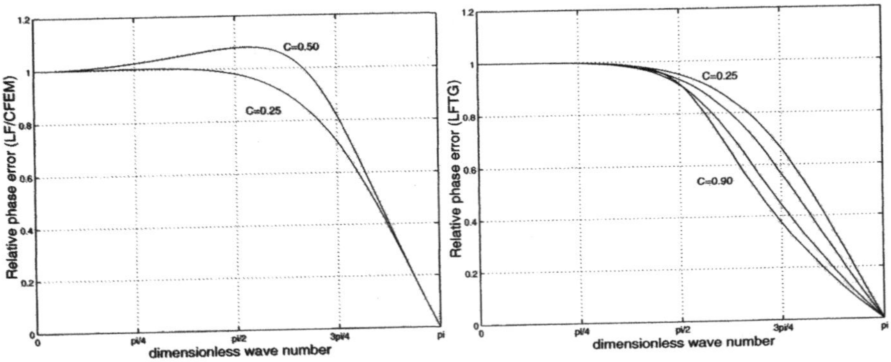

Fig. 3.8 Relative phase error $\phi_{numer}/\phi_{exact}$ in pure convection of the second-order (left) and fourth-order (right) leap-frog schemes combined with consistent linear finite elements.

3.6.4 Two-step explicit Taylor–Galerkin methods

We shall now describe two-step versions of the explicit Taylor–Galerkin method TG3 which include second time derivatives only and are thus easier to implement than the one-step method TG3, especially for solving nonlinear multidimensional hyperbolic problems, see Selmin, Donéa and Quartapelle (1985) for details. A further advantage of the two-step Taylor–Galerkin methods is their extended stability range in multidimensional situations as compared with the one-step TG3 method. A two-step third-order method is discussed first. Then, fourth-order ones are briefly presented.

3.6.4.1 Two-step third-order method A two-step version of the third-order TG3 scheme has been proposed by Selmin (1987). The method, called TG3-2S, is based on the idea of achieving a third-order temporal accuracy by means of the following two-step procedure:

$$\begin{aligned} \tilde{u}^n &= u^n + \frac{1}{3}\Delta t\, u_t^n + \alpha\, \Delta t^2\, u_{tt}^n, \\ u^{n+1} &= u^n + \Delta t\, u_t^n + \frac{1}{2}\Delta t^2\, \tilde{u}_{tt}^n, \end{aligned} \tag{3.49}$$

where the value of the parameter α is left unspecified for the time being. Third-order accuracy is achieved when combining both steps; the parameter α only influences the coefficient of the fourth-order term in the overall time series. As a consequence, its value will only affect the modulus of the amplification factor of the resulting scheme but not its phase. A convenient way to determine the available degree of freedom, α, consists in selecting it in order to have, in the 1D linear case, the same phase speed as that of the one-step TG3 scheme, equation (3.45). For the 1D linear convection equation, see (3.27), the fully discrete version of the two-step scheme (3.49) becomes

$$\begin{aligned}\left(1+\frac{1}{6}\delta^2\right)(\tilde{u}^n-u^n) &= -\frac{1}{6}\,C\,\delta u^n+\alpha\,C^2\,\delta^2 u^n,\\ \left(1+\frac{1}{6}\delta^2\right)(u^{n+1}-u^n) &= -\frac{1}{2}\,C\,\delta u^n+\frac{1}{2}\,C^2\,\delta^2\tilde{u}^n,\end{aligned} \tag{3.50}$$

and the amplification factor $G_{\text{TG3-2S}}$ of the parameterized two-step scheme is derived in the form

$$G_{\text{TG3-2S}}(\xi,C,\alpha)=\frac{1-\frac{2}{3}\sin^2\frac{\xi}{2}-2C^2\,\tilde{G}(\xi,C,\alpha)\,\sin^2\frac{\xi}{2}-iC\sin\xi}{1-\frac{2}{3}\sin^2\frac{\xi}{2}}$$

where

$$\tilde{G}(\xi,C,\alpha)=1+\frac{-\frac{1}{3}iC\sin\xi-4\alpha C^2\sin^2\frac{\xi}{2}}{1-\frac{2}{3}\sin^2\frac{\xi}{2}}.$$

It is immediately verified that

$$G_{\text{TG3-2S}}(\xi,C,1/9)=G_{\text{TG3}}(\xi,C)\,R(\xi,C), \tag{3.51}$$

where $R(\xi,C)$ is the real function

$$R(\xi,C)=\frac{\left(1-\frac{2}{3}(1-C^2)\sin^2\frac{\xi}{2}\right)\left(1-\frac{2}{3}(1+C^2)\sin^2\frac{\xi}{2}\right)}{\left(1-\frac{2}{3}\sin^2\frac{\xi}{2}\right)^2}. \tag{3.52}$$

Therefore, for the choice $\alpha=1/9$ the two-step procedure (3.49) reproduces exactly the phase–speed characteristics of the single-step TG3 scheme. The condition of numerical stability for the two-step method is $|C|\le\sqrt{3}/2\simeq0.866$. Figure 3.9 illustrates the amplitude response of the two-step scheme which appears to be slightly more dissipative than its one-step counterpart, especially for Courant numbers close to the stability limit.

It is important to note that the excellent phase properties of the one-step third-order TG3 scheme can be reproduced exactly by the two-step procedure in 2D and 3D, see Selmin (1987) and Quartapelle (1993). Moreover, as shown in Figure 3.7, the stability range of the two-step scheme in 2D remains practically unaltered with respect to that in 1D. In fact, the stability limit in 2D is defined by

$$c_x^2+c_y^2=3/4.$$

This is in sharp contrast with the one-step Taylor–Galerkin scheme (TG3), which experiences a drastic reduction of its stability limit in multidimensional situations. Thus, the two-step formulation of the explicit Taylor–Galerkin method, besides making high-order accuracy accessible for truly nonlinear problems, offers the additional advantage of giving an isotropic stability domain in multidimensional problems.

Observe that the effect of the modified mass matrix in TG3 is achieved here, for the two-step scheme, through a double application of the standard consistent mass

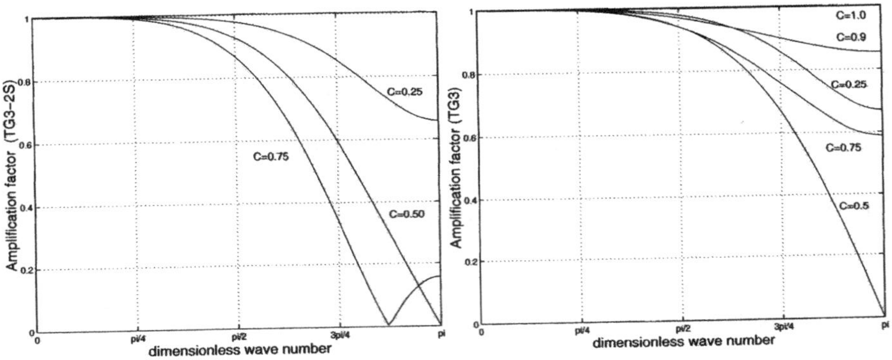

Fig. 3.9 Amplification factor modulus $|G|$ of Selmin's two-step third-order explicit scheme TG3-2S (left) compared with the one-step third-order method TG3 (right).

matrix. This alleviates the computational effort with respect to the one-step method (3.43) in which the modified mass matrix depends on the problem unknown.

Remark 3.9 (Vector convection equation). When a system of nonlinear convection equations is considered, the application of the two-step method becomes much more involved than with a single equation. Examples will be given in Chapter 4 where the method is applied for solving the Euler equations. We refer to the articles by Laval (1988) and Laval and Quartapelle (1990), or to the book by Quartapelle (1993), for the implementation details of the two-step method in application to a vector convection equation.

3.6.4.2 Two-step fourth-order methods As a last example of high-order explicit schemes for convective transport, we shall illustrate the construction of two fourth-order methods suggested by Quartapelle (1993).

Consider the fourth-order explicit temporal approximation

$$u(t^{n+1}) = u(t^n) + \Delta t\, u_t^n(t^n) + \frac{\Delta t^2}{2} u_{tt}^n(t^n) + \frac{\Delta t^3}{6} u_{ttt}^n(t^n) + \frac{\Delta t^4}{24} u_{tttt}^n(t^n) + \mathcal{O}(\Delta t^5).$$

It can be transformed into a two-stage method by means of the following factorization:

$$1 + \Delta t\frac{\partial}{\partial t} + \frac{\Delta t^2}{2}\frac{\partial^2}{\partial t^2} + \frac{\Delta t^3}{6}\frac{\partial^3}{\partial t^3} + \frac{\Delta t^4}{24}\frac{\partial t}{\partial t^4} = 1 + \Delta t\frac{\partial}{\partial t} + \frac{\Delta t^2}{2}\frac{\partial^2}{\partial t^2}\left(1 + \frac{\Delta t}{3}\frac{\partial}{\partial t} + \frac{\Delta t^2}{12}\frac{\partial^2}{\partial t^2}\right),$$

which produces the two-stage explicit method

$$\begin{aligned} \tilde{u} &= u^n + \frac{1}{3}\Delta t\, u_t^n + \frac{1}{12}\Delta t^2\, u_{tt}^n, \\ u^{n+1} &= u^n + \Delta t\, u_t^n + \frac{1}{2}\Delta t^2\, \tilde{u}_{tt}. \end{aligned} \tag{3.53}$$

This two-step method has the same stability and accuracy properties as the classical fourth-order explicit Runge–Kutta method. For the linear convection equation (3.4a) the method is stable up to $C^2 = 1$.

As shown by Quartapelle (1993), there is another possibility of achieving a fourth-order temporal accuracy with a two-step strategy. It is based on two second-order expansions

$$\begin{aligned} \tilde{u} &= u^n + \alpha\Delta t\, u_t^n + \beta\Delta t^2 u_{tt}^n, \\ u^{n+1} &= u^n + \Delta t\, \tilde{u}_t + \gamma\Delta t^2 \tilde{u}_{tt}, \end{aligned} \tag{3.54}$$

where the parameters are defined as

$$\alpha = 0.141, \quad \beta = 0.116, \quad \gamma = 0.359.$$

The stability limit of this alternative two-step fourth-order scheme is $|C| \leq 0.847$ and its phase response properties are found to be superior to those of the two-step method (3.53). The book by Quartapelle (1993) contains a detailed presentation of the above two-step fourth-order explicit methods.

3.7 AN INTRODUCTION TO MONOTONICITY-PRESERVING SCHEMES

Up to now, the emphasis has been placed on achieving time-accurate solutions to unsteady convection problems. Here, we wish to recall that discontinuities may appear in the solution of convection problems, even if they are governed by linear equations. This will indeed be the case when discontinuous data are prescribed on the inflow portion of the boundary.

In such a situation, the numerical solution delivered by stabilized finite element methods, see Chapter 2, may still present localized oscillations in the vicinity of sharp solution gradients. The consequence of such undesirable overshoots may be that some physical constraint, such as for instance positivity of the unknown, might be violated, thus leading to non-physical results.

There is therefore a need to ensure non-oscillatory, or *monotone*, numerical behavior near strong solution gradients to obtain physically correct results. This aspect will be discussed in a certain depth in Chapter 4 in connection with the solution of the Euler equations of gas dynamics. In this section, our aim is simply to introduce the concept of high-order *monotonicity-preserving schemes* along the lines introduced by Harten (1983) (or the recent reprint Harten, 1997). For simplicity, we shall make reference to the 1D convection equation.

A scheme is said to be *monotonicity preserving* if monotonicity of the solution u^{n+1} at time t^{n+1} follows from monotonicity of the solution u^n at time t^n. By monotonicity of a solution, one means that the following inequalities hold for all of its components:

$$\min(u_{j-1}, u_{j+1}) \leq u_j \leq \max(u_{j-1}, u_{j+1}).$$

An important theorem concerning monotonicity-preserving schemes is due to Godunov (1959) who showed that any linear monotonicity-preserving scheme is at most first-order accurate in space. Linear monotonicity-preserving schemes add the same artificial dissipation all over the computational domain and, consequently, are too dissipative in the smooth part of the solution.

In response to the deficiency of monotone schemes, so-called *high-resolution schemes* were developed in the 1980s. Such schemes are at least second-order accurate in the smooth part of the solution, but include a nonlinear (solution-dependent) damping mechanism that permits a sharp resolution of strong solution gradients.

Central to the development of high-resolution schemes is the concept of *total variation diminishing* (TVD) schemes introduced by Harten (1983). The total variation of a smooth function u is defined as

$$\text{TV}(\mathbf{u}) = \sum_{\forall j} |u_j - u_{j-1}|,$$

and a scheme is said to be TVD if

$$\text{TV}(\mathbf{u}^{n+1}) \leq \text{TV}(\mathbf{u}^n).$$

The TVD requirement is less stringent than the monotonicity preservation requirement and allows a significant improvement in accuracy.

The guiding idea behind the design of TVD schemes is that physical solutions to scalar hyperbolic equations do not allow the appearance of any new extremum in the evolution of the unknown. Accordingly, in a TVD scheme the total variation of the numerical solution is controlled in a nonlinear way to prevent the appearance of any spurious new extremum.

Most high-order TVD schemes can be viewed as centered-difference schemes with an appropriate numerical dissipation calibrated so as to preserve monotonicity without compromising the accuracy. Modern methods use nonlinear numerical dissipation with diffusion coefficients dependent upon the local behavior of the solution, being larger near discontinuities than in smooth regions.

Basic references for the construction of TVD schemes in the finite difference context are, in addition to the pioneering work of Harten, Yee (1987), Hirsch (1990), LeVeque (1992) and Quarteroni and Valli (1994). Notice that the development of TVD schemes for use in connection with finite elements is still the object of active research.

To give a simple example of a TVD scheme, suppose we wish to solve the 1D convection equation (3.27) with $a > 0$ using the Crank–Nicolson method and linear finite elements. Using a diagonal mass representation, we obtain the following discrete equation at an interior node j:

$$u_j^{n+1} + \frac{\Delta t}{2h}\left(h_{j+1/2}^{n+1} - h_{j-1/2}^{n+1}\right) = u_j^n - \frac{\Delta t}{2h}\left(h_{j+1/2}^n - h_{j-1/2}^n\right),$$

where

$$h_{j+1/2} = \frac{1}{2}a(u_{j+1} + u_j) \qquad \text{and} \qquad h_{j-1/2} = \frac{1}{2}a(u_j + u_{j-1}) \tag{3.55}$$

are the numerical fluxes for the considered centered scheme. Note that a first-order upwind scheme would use

$$h_{j+1/2} = a\,u_j \qquad \text{and} \qquad h_{j-1/2} = a\,u_{j-1}.$$

As seen in Chapter 2, full upwinding generally introduces an excessive amount of numerical damping. Following Yee (1987), the above second-order scheme can be rendered into a TVD scheme by transforming the fluxes (3.55) as follows:

$$\begin{aligned} h_{j+1/2} &= \frac{1}{2}\Big(a\,(u_{j+1} + u_j) - |a|(1 - Q_{j+1/2})\Delta_{j+1/2}\,u\Big), \\ h_{j-1/2} &= \frac{1}{2}\Big(a\,(u_j + u_{j-1}) - |a|(1 - Q_{j-1/2})\Delta_{j-1/2}\,u\Big), \end{aligned} \tag{3.56}$$

where $Q_{j+1/2}$ and $Q_{j-1/2}$ are limiting functions depending on the solution gradient. The variations $\Delta_{j+1/2}\,u$ and $\Delta_{j-1/2}\,u$ are defined as follows:

$$\Delta_{j+1/2}\,u = u_{j+1} - u_j, \qquad \Delta_{j-1/2}\,u = u_j - u_{j-1}.$$

The limiting function $Q_{j+1/2}$ depends on three consecutive element gradients

$$\Delta_{j-1/2}u, \quad \Delta_{j+1/2}u, \quad \Delta_{j+3/2}u$$

and is of the form

$$Q_{j+1/2} = Q(r_{j+1/2}^-\,,\, r_{j+1/2}^+),$$

where

$$r_{j+1/2}^- = \frac{\Delta_{j-1/2}\,u}{\Delta_{j+1/2}\,u} \qquad \text{and} \qquad r_{j+1/2}^+ = \frac{\Delta_{j+3/2}\,u}{\Delta_{j+1/2}\,u}.$$

Two examples for the limiting function Q are

$$\begin{aligned} Q(r^-\,,\, r^+) &= \text{minmod}(1, r^-, r^+), \\ Q(r^-\,,\, r^+) &= \text{minmod}\big(2, 2r^-, 2r^+, 0.5(r^- + r^+)\big). \end{aligned}$$

The "minmod" function of a list of arguments is equal to the smallest number in absolute value if all arguments are of the same sign, or is equal to zero if any argument is of the opposite sign.

After substitution of the modified fluxes (3.56) in equation (3.55), one gets

$$\begin{aligned}
u_j^{n+1} &+ \frac{\Delta t}{4h}\Big(a\,(u_{j+1}^{n+1} + u_j^{n+1}) - |a|\,(1 - Q_{j+1/2}^{n+1})\,\Delta_{j+1/2}\,u^{n+1}\Big) \\
&- \frac{\Delta t}{4h}\Big(a\,(u_j^{n+1} + u_{j-1}^{n+1}) - |a|\,(1 - Q_{j-1/2}^{n+1})\,\Delta_{j-1/2}\,u^{n+1}\Big) \\
&\qquad = u_j^n - \frac{\Delta t}{4h}\Big(a\,(u_{j+1}^n + u_j^n) - |a|\,(1 - Q_{j+1/2}^n)\,\Delta_{j+1/2}\,u^n\Big) \\
&\qquad\quad + \frac{\Delta t}{4h}\Big(a\,(u_j^n + u_{j-1}^n) - |a|\,(1 - Q_{j-1/2}^n)\,\Delta_{j-1/2}\,u^n\Big).
\end{aligned}$$

Notice that the above TVD algorithm is nonlinear even when it is applied to the linear convection equation (3.27). To solve this set of nonlinear equations non-iteratively, a linearized version is usually considered. It consists in the replacement of Q^{n+1} by Q^n in the l.h.s., see Yee (1987). Another benefit of this linearization is that the scheme, though involving five points, now leads to a tridiagonal system of linear equations. This is because, at the $(n+1)$-th time level, only three points are involved, namely u_{j-1}^{n+1}, u_j^{n+1} and u_{j+1}^{n+1}, the other two points being at the n-th time level.

After linearization the implicit TVD scheme reads

$$\begin{aligned}
u_j^{n+1} &+ \frac{a\Delta t}{4h}(u_{j+1}^{n+1} - u_{j-1}^{n+1}) \\
&- \frac{|a|\Delta t}{4h}\Big((1 - Q_{j+1/2}^n)(u_{j+1}^{n+1} - u_j^{n+1}) - (1 - Q_{j-1/2}^n)(u_j^{n+1} - u_{j-1}^{n+1})\Big) \\
&= u_j^n - \frac{a\Delta t}{4h}(u_{j+1}^n - u_{j-1}^n) \\
&\quad + \frac{|a|\Delta t}{4h}\Big((1 - Q_{j+1/2}^n)(u_{j+1}^n - u_j^n) - (1 - Q_{j-1/2}^n)(u_j^n - u_{j-1}^n)\Big).
\end{aligned}$$

Note that the TVD method selectively adds a numerical dissipation, the maximum value of which is

$$\nu_{\max} = \frac{|a|h}{2}.$$

The above 1D strategy can be adapted to deal with 2D situations. In this case, as suggested by Donea, Selmin and Quartapelle (1988), one works with segments connecting adjacent nodes, for instance the element sides in a mesh of triangular elements. Other examples of high-order monotonicity-preserving schemes will be given in Chapter 4 where shock-capturing techniques are discussed in detail.

3.8 LEAST-SQUARES-BASED SPATIAL DISCRETIZATION

The implicit Crank–Nicolson scheme discussed in Section 3.4 is not dissipative (pure convection is considered here). It must therefore be combined with a finite element spatial representation capable of introducing the amount of numerical dissipation required to produce stable results in the presence of steep solution gradients.

In the case of purely convective transport, a stabilization technique can be developed on the basis of a pure least-squares minimization. Two variants of such an approach will be described in this section. They require that time discretization is performed before the spatial discretization and rely on a quadratic functional associated with the semi-discrete version of the governing equation. Further extensions of least squares to convection problems using space–time formulations are discussed in Section 3.10. The book by Jiang (1998) can be consulted for a detailed account of least-squares finite element methods.

3.8.1 Least-squares approach for the θ family of methods

An interesting least-squares method for pure convection was originally proposed by Carey and Jiang (1988) in connection with the θ family of methods presented in Section 3.4.1. Consider the linear convection equation (3.4a) discretized with respect to time by means of this family of methods, and recall (3.20):

$$\frac{\Delta u}{\Delta t} + \theta(\boldsymbol{a} \cdot \boldsymbol{\nabla})\Delta u = \theta s^{n+1} + (1-\theta)s^n - (\boldsymbol{a} \cdot \boldsymbol{\nabla})u^n.$$

This equation can be viewed as a spatial strong form that must be solved at each time step, namely

$$\mathcal{L}(\Delta u) - f = 0,$$

where $\mathcal{L} = 1/\Delta t + \theta \boldsymbol{a} \cdot \boldsymbol{\nabla}$ is the spatial differential operator, and the known source term is $f = \theta s^{n+1} + (1-\theta)s^n - \boldsymbol{a} \cdot \boldsymbol{\nabla} u^n$. Minimization of the least-squares functional, $\big(\mathcal{L}(\Delta u) - f, \mathcal{L}(\Delta u) - f\big)$, produces the least-squares equation (still in the spatial continuum)

$$\big(\mathcal{L}(w), \mathcal{L}(\Delta u) - f\big) = 0,$$

which takes the following explicit form:

$$\begin{aligned}\Big(\frac{w}{\Delta t} + \theta \boldsymbol{a} \cdot \boldsymbol{\nabla} w, \frac{\Delta u}{\Delta t} + \theta \boldsymbol{a} \cdot \boldsymbol{\nabla}\Delta u\Big)&\\ = \Big(\frac{w}{\Delta t} + \theta \boldsymbol{a} \cdot \boldsymbol{\nabla} w, \theta s^{n+1} + (1-\theta)s^n - \boldsymbol{a} \cdot \boldsymbol{\nabla} u^n\Big).&\qquad (3.57)\end{aligned}$$

This equation highlights the symmetric character of the implicit operator of the least-squares method. On a mesh of uniform linear elements in 1D, this scheme provides the fully discrete equation

$$\Big(1 + \Big(\frac{1}{6} - \theta^2 C^2\Big)\delta^2\Big)\,\Delta u = -\frac{1}{2} C\,\delta u_j^n + \theta\, C^2 \delta^2 u_j^n,$$

and consequently the following amplification factor:

$$G_{\mathrm{CJ}} = \frac{1 - 2\big(\frac{1}{3} + 2(1-\theta)\theta C^2\big)\sin^2\frac{\xi}{2} - ic\sin\xi}{1 - 2\big(\frac{1}{3} - 2\theta^2 C^2\big)\sin^2\frac{\xi}{2}}.$$

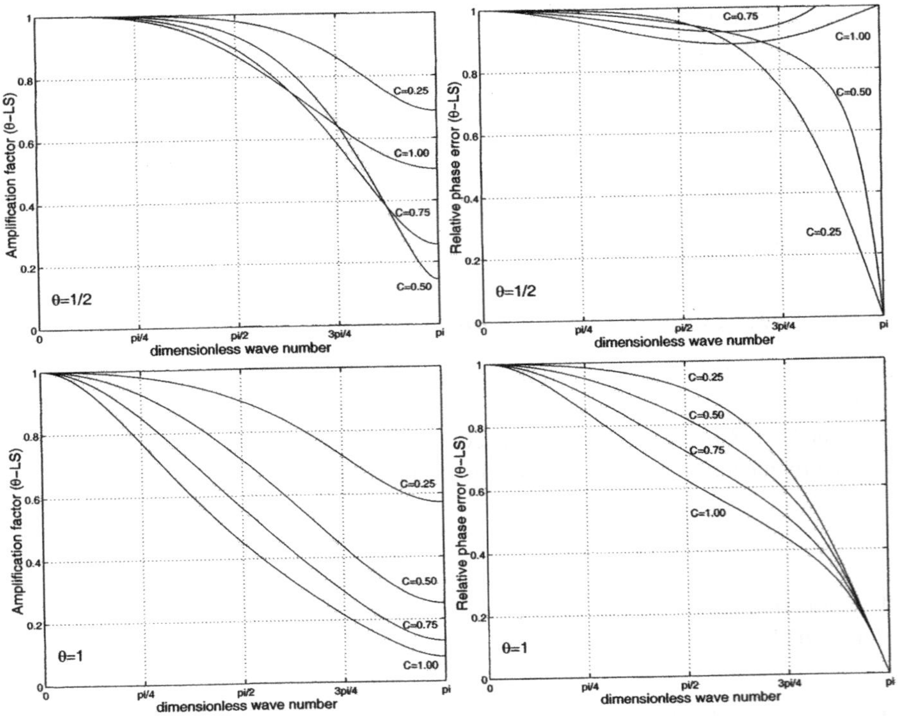

Fig. 3.10 Numerical properties of the least-squares method of Carey and Jiang for Crank–Nicolson $\theta = 1/2$ (top) and backward Euler $\theta = 1$ (bottom) for several values of the Courant number.

From the previous equations one can see that the least-squares scheme is unconditionally stable for $\frac{1}{2} \leq \theta \leq 1$ (Carey and Jiang, 1988). Figure 3.10 displays its numerical properties for $\theta = 1/2$ and $\theta = 1$. For $\theta = 1/2$ the scheme has a rather accurate phase response provided it is operated with $|C| \leq 1$. But, the fully implicit scheme corresponding to $\theta = 1$ has a poor phase accuracy, except for small values of the Courant number. This scheme can nevertheless be of interest for computing steady-state solutions by means of a false transient.

3.8.2 Taylor least-squares method

An extremely accurate least-squares method for the spatial discretization of transient convection problems was introduced by Park and Liggett (1990). Their method combines Taylor–Galerkin and least-squares concepts. It starts from the time discretized version of the linear convection equation (3.4a) provided by the following fourth-order accurate implicit scheme originally used by Harten and Tal-Ezer (1981):

$$\frac{\Delta u}{\Delta t} = \frac{1}{2}\left(u_t^{n+1} + u_t^n\right) - \frac{\Delta t}{12}\left(u_{tt}^{n+1} - u_{tt}^n\right). \tag{3.58}$$

Assuming a time-independent convection velocity, the scheme reads

$$\mathcal{L}\Big(\frac{\Delta u}{\Delta t}\Big) = -\boldsymbol{a}\cdot\boldsymbol{\nabla}u^n + f^*, \tag{3.59}$$

where

$$\mathcal{L} = 1 + \frac{\Delta t}{2}\boldsymbol{a}\cdot\boldsymbol{\nabla} + \frac{\Delta t^2}{12}(\boldsymbol{a}\cdot\boldsymbol{\nabla})^2 \quad \text{and}$$
$$f^* = \frac{1}{2}(s^{n+1}+s^n) - \frac{\Delta t}{12}(s_t^{n+1}-s_t^n) + \frac{\Delta t}{12}(\boldsymbol{a}\cdot\boldsymbol{\nabla})(s^{n+1}-s^n).$$

The quadratic functional associated with this semi-discrete equation is

$$\left(\mathcal{L}\Big(\frac{\Delta u}{\Delta t}\Big) + \boldsymbol{a}\cdot\boldsymbol{\nabla}u^n - f^*, \mathcal{L}\Big(\frac{\Delta u}{\Delta t}\Big) + \boldsymbol{a}\cdot\boldsymbol{\nabla}u^n - f^*\right),$$

and the minimization procedure leads to the following least-squares weighted residual formulation:

$$\left(\mathcal{L}(w), \mathcal{L}\Big(\frac{\Delta u}{\Delta t}\Big)\right) = \big(\mathcal{L}(w), -\boldsymbol{a}\cdot\boldsymbol{\nabla}u^n + f^*\big).$$

This equation can also be derived if one recalls the strong form (Euler–Lagrange equation) associated with a least-squares functional induced by $\mathcal{L}(u) = s$. It corresponds to the higher-order problem $\mathcal{L}^*\mathcal{L}(u) = \mathcal{L}^*(s)$, where $\mathcal{L}^*$ is the formal adjoint of operator $\mathcal{L}$. In the present case the differential equation is defined in (3.59), and thus the adjoint operator is

$$\mathcal{L}^* = 1 - \frac{\Delta t}{2}(\boldsymbol{a}\cdot\boldsymbol{\nabla}) + \frac{\Delta t^2}{12}(\boldsymbol{a}\cdot\boldsymbol{\nabla})^2.$$

It follows that the Taylor least-squares weak formulation for the semi-discrete equation (3.59) could be obtained by applying the standard Galerkin projection to the time integration scheme:

$$\left(1 - \frac{\Delta t^2}{12}(\boldsymbol{a}\cdot\boldsymbol{\nabla})^2 + \Big(\frac{\Delta t^2}{12}\Big)^2(\boldsymbol{a}\cdot\boldsymbol{\nabla})^4\right)\frac{\Delta u}{\Delta t}$$
$$= \left(1 - \frac{\Delta t}{2}(\boldsymbol{a}\cdot\boldsymbol{\nabla}) + \frac{\Delta t^2}{12}(\boldsymbol{a}\cdot\boldsymbol{\nabla})^2\right)\big(-\boldsymbol{a}\cdot\boldsymbol{\nabla}u^n + f^*\big).$$

Due to the presence of third- and fourth-order derivatives in this least-squares equation, Park and Liggett (1990) employed $\mathcal{C}^1$ finite elements (cubic Hermitian polynomials) instead of the standard $\mathcal{C}^0$ finite elements.

$\mathcal{C}^1$ finite elements require, as nodal unknowns, the value of the interpolating function and also its derivatives, namely u and u_x in 1D. This implies an important increase in nodal unknowns. For instance, a 2D four-noded element constructed by the tensor product of the 1D Hermite cubics has 16 unknowns, four per node: u, u_x, u_y and u_{xy}, see for instance Carey and Oden (1983). This increase is more dramatic in 3D.

The study of the amplification factor and accuracy properties of the resulting scheme is given in the original paper. It also compares the Taylor/Least-squares scheme with the third-order explicit Taylor–Galerkin method and other least-squares methods. Numerical results for 2D calculations indicate that the Taylor/Least-squares scheme is extremely accurate. Park and Liggett (1991) also implemented this method in 3D using serendipity-type Hermitian elements, which have less nodal unknowns compared with the complete Hermite interpolation and thus less accuracy.

3.9 THE DISCONTINUOUS GALERKIN METHOD

Due to the compactness of its formulation, the discontinuous Galerkin method initially introduced by Lasaint and Raviart (1974) is being increasingly used in the solution of convection and convection–diffusion problems, see for instance the work of Cockburn (1998) and Baumann and Oden (1999). We briefly introduce the discontinuous Galerkin method in application to the linear first-order hyperbolic problem

$$u_t + \nabla \cdot (\boldsymbol{a}\, u) = s(\boldsymbol{x}, t) \qquad \text{in } \Omega \times]0, T[, \tag{3.60a}$$

$$u(\boldsymbol{x}, 0) = u_0(\boldsymbol{x}) \qquad \text{on } \Omega \text{ at } t = 0, \tag{3.60b}$$

$$u = u_D \qquad \text{on } \Gamma_D^{in} \times]0, T[. \tag{3.60c}$$

The discontinuous approximations belong to a so-called *broken space*. Given a regular partition, $\mathcal{T}^h$, of the computational domain Ω into subdomains Ω^e, the test functions belonging to the broken space $\mathcal{V}(\mathcal{T}^h)$ are continuous and smooth in every element $\Omega^e \in \mathcal{T}^h(\Omega)$, but discontinuous across inter-element boundaries. Moreover, as previously done in Section 3.4.2, the trial space varies with time, namely

$$\mathcal{V}(\mathcal{T}^h) := \{ w \in \mathcal{L}_2(\Omega) \mid w|_{\Omega^e} \in \mathcal{H}^2(\Omega^e) \, \forall \, \Omega^e \in \mathcal{T}^h(\Omega) \}$$

$$\mathcal{S}_t(\mathcal{T}^h) = \{ u \mid u(\cdot, t) \in \mathcal{L}_2(\Omega),\; u(\cdot, t)|_{\Omega^e} \in \mathcal{H}^2(\Omega^e) \, t \in [0, T] \, \forall \, \Omega^e \in \mathcal{T}^h(\Omega) \} .$$

Typical elements of $\mathcal{S}_t(\mathcal{T}^h)$ and $\mathcal{V}(\mathcal{T}^h)$ are non-zero functions on one element Ω^e and zero everywhere else on the mesh.

The discontinuous Galerkin formulation of problem (3.60) can then be stated as follows: given $u_0(\boldsymbol{x})$, for any $t \in]0, T[$ find $u \in \mathcal{S}_t(\mathcal{T}^h)$, such that $u(\boldsymbol{x}, 0) = u_0(\boldsymbol{x})$ and

$$\big(w, u_t + \nabla \cdot (\boldsymbol{a}\, u) - s\big)_{\Omega^e} - \big(w, (u^+ - u^-)(\boldsymbol{a} \cdot \boldsymbol{n}_e)\big)_{\partial\Omega^{e,in} \setminus \Gamma^{in}} - \big(w, (u^+ - u_D)(\boldsymbol{a} \cdot \boldsymbol{n}_e)\big)_{\partial\Omega^{e,in} \cap \Gamma^{in}} = 0$$

for all $w \in \mathcal{V}(\mathcal{T}^h)$ and $\Omega^e \in \mathcal{T}^h(\Omega)$. Here $\boldsymbol{n}_e$ is the outward normal to element e. Note that this weak form is local, that is it is defined over one element, not global, over the complete spatial domain Ω. This is possible because the test functions are discontinuous along inter-element boundaries. Note that this does not imply that each element can be solved independently. In fact, the boundary integral introduces

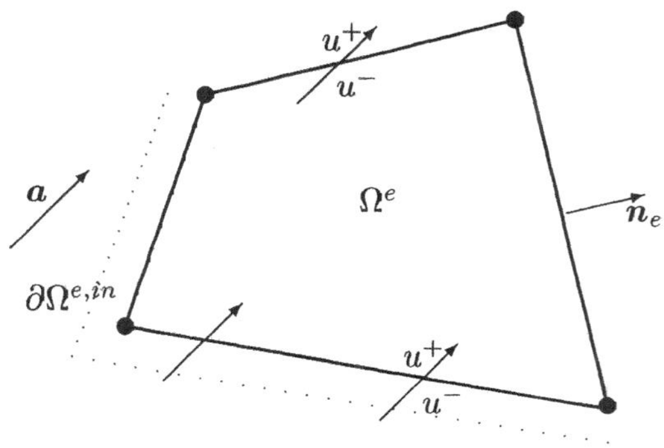

Fig. 3.11 Notation for discontinuous Galerkin method.

a weakly enforced continuity condition across the boundaries, which couples the unknowns of adjacent elements. As shown in Figure 3.11, to cope with the discontinuity of the field variable across inter-element boundaries, one defines

$$u^{\pm} = \lim_{\epsilon \to 0^+} u(\boldsymbol{x} \pm \epsilon \boldsymbol{a}) \qquad \text{for } \boldsymbol{x} \in \partial\Omega^e.$$

Moreover, the inlet portion of the element boundary, $\partial\Omega^e$, is defined by

$$\partial\Omega^{e,in} = \{\boldsymbol{x} \in \partial\Omega^e \,|\, \boldsymbol{a} \cdot \boldsymbol{n}_e(\boldsymbol{x}) < 0\}.$$

An alternative discontinuous Galerkin formulation is obtained by integration by parts of the divergence term. It provides a natural boundary condition for the convective flux and reads as follows:

$$\begin{aligned}(w, u_t - s)_{\Omega^e} &- (\boldsymbol{\nabla} w \cdot \boldsymbol{a}, u)_{\Omega^e} \\ &+ \big(w, u^-(\boldsymbol{a} \cdot \boldsymbol{n}_e)\big)_{\partial\Omega^e \setminus \Gamma^{in}} + \big(w, u_D(\boldsymbol{a} \cdot \boldsymbol{n}_e)\big)_{\partial\Omega^e \cap \Gamma^{in}} = 0. \quad (3.61)\end{aligned}$$

Despite their conditional stability, the use of explicit time-stepping schemes is very convenient here, because the global mass matrix is block diagonal with uncoupled blocks. The problem can then be solved element by element in one sweep by inverting at very low cost the element mass matrix. In Section 3.11.4 we shall illustrate the application of the discontinuous Galerkin method in the solution of a simple 1D problem of propagation of a steep front. Another attractive aspect of this method is its easy combination with adaptive refinement procedures. This is because in each element the polynomial order can be adapted to the local smoothness of the solution. A typical example is the adaptive-order discontinuous Galerkin method developed by Baumann and Oden (2000). In Chapter 4 this method is discussed in more detail and used to solve the Euler equations of gas dynamics.

3.10 SPACE–TIME FORMULATIONS

Up to here, time discretization of the convection equation has been performed using finite difference formula while finite elements are employed for spatial discretization only. As a matter of fact, finite element interpolations can also be used in the time domain. Methods in this class include the time-discontinuous Galerkin method (Jamet, 1978; Johnson et al., 1984), the space–time least-squares method proposed by Nguyen and Reynen (1984), as well as the space–time Galerkin/Least-squares algorithms developed by Shakib and Hughes (1991) and the space–time integrated least-squares approach proposed by Perrochet and Azérad (1995). In fact, the weighted residual formulation is now extended over the space–time domain. Usually, low-order approximations, such as piecewise constant or piecewise linear ones, are employed to describe time dependency. If a linear interpolation over a time slab $]t^n, t^{n+1}[$ is assumed, the local space–time interpolation is written in product form

$$u^h(\boldsymbol{x}, t) = \sum_{A=1}^{\mathrm{n_{np}}} N_A(\boldsymbol{x})\big((1-\theta)\mathrm{u}_A^n + \theta\,\mathrm{u}_A^{n+1}\big),$$

where $\theta = (t - t^n)/(t^{n+1} - t^n)$.

Let us now illustrate the extension to the space–time domain of finite element methods for transient convection problems. Note that the method has also been applied to the Navier–Stokes equations to be discussed in Chapter 6, see for instance Masud and Hughes (1997).

We consider piecewise continuous approximations in space and discontinuous approximations in time. Discontinuous approximations in time allow us to solve independently for each time slab instead of solving a global problem over the whole time domain. Recall that such a procedure was already used in space when we discussed the discontinuous Galerkin method in Section 3.9.

The time domain is partitioned in $\mathrm{n_{st}}$ sub-intervals, where each sub-interval is defined as $I^n =]t^n, t^{n+1}[, n = 0, 1, \ldots, \mathrm{n_{st}} - 1$. Space–time slabs are then obtained in the form

$$Q^n = \Omega \times I^n.$$

For the considered space–time slab Q^n, the spatial domain Ω is subdivided into $\mathrm{n_{el}}$ elements, Ω^e, $e = 1, \ldots, \mathrm{n_{el}}$, giving space–time element domains

$$Q_e^n = \Omega^e \times I^n, \quad e = 1, \ldots, \mathrm{n_{el}}.$$

3.10.1 Time-discontinuous Galerkin formulation

To introduce time-discontinuous approximations in a simple context, let us first consider the time-discontinuous Galerkin formulation (Johnson et al., 1984; Shakib, 1989; Shakib and Hughes, 1991). Since the finite element interpolation is discontinuous at the space–time slab interfaces, it is useful to employ the notation

$$u^h(t_\pm^n) = \lim_{\epsilon \to 0+} u^h(t^n \pm \epsilon).$$

The finite element spaces for the trial and weighting functions are defined as follows:

$$\begin{aligned}
\mathcal{S}^h &= \bigcup_{n=0}^{\mathrm{n_{st}}-1} \mathcal{S}_n^h,\ \mathcal{S}_n^h = \{u^h \mid u^h \in C^0(Q^n),\ u^h|_{Q_e^n} \in \mathcal{P}_k(Q_e^n),\ u^h|_{\Gamma^{in}} = u_D\} \\
\mathcal{V}^h &= \bigcup_{n=0}^{\mathrm{n_{st}}-1} \mathcal{V}_n^h,\ \mathcal{V}_n^h = \{w^h \mid w^h \in C^0(Q^n),\ w^h|_{Q_e^n} \in \mathcal{P}_k(Q_e^n),\ w^h|_{\Gamma^{in}} = 0\}.
\end{aligned} \tag{3.62}$$

Note that continuity over Q^n implies continuity in space but does not require a continuous interpolation between time slabs. Although $\mathcal{P}_k$ indicates the space of polynomials of total degree $\leq k$ (a complete basis of degree k, see Section 1.5.2) the degrees of the polynomials in space and time can be chosen independently.

The weighted residual formulation of the homogeneous linear convection equation (3.4a) with Dirichlet inlet conditions is: for $n = 0, 1, \ldots, \mathrm{n_{st}} - 1$, find $u^h \in \mathcal{S}_n^h$, such that for all $w^h \in \mathcal{V}_n^h$,

$$\iint_{Q^n} w^h \left(u_t^h + \boldsymbol{a} \cdot \boldsymbol{\nabla} u^h\right) d\Omega\, dt + \int_{\Omega} w^h(t_+^n) \left(u^h(t_+^n) - u^h(t_-^n)\right) d\Omega = 0, \tag{3.63}$$

with the initial condition $u^h(\boldsymbol{x}, t_-^0) = u_0(\boldsymbol{x})$. The last integral in (3.63) is a jump condition which imposes a weakly enforced continuity condition across the slab interfaces and is the mechanism by which information is propagated from one slab to another.

We shall use finite element approximations over a space–time slab which are piecewise polynomials in space and linear in time; that is, for $(\boldsymbol{x}, t) \in Q^n = \Omega \times I^n$,

$$u^h(\boldsymbol{x}, t) = \sum_{A=1}^{\mathrm{n_{np}}} N_A(\boldsymbol{x}) \left(\Theta_1(t)\, \tilde{\mathrm{u}}_A^n + \Theta_2(t)\, \mathrm{u}_A^{n+1}\right).$$

$N_A(\boldsymbol{x})$ is the spatial shape function at node A; $\Theta_1(t)$ and $\Theta_2(t)$ are the time interpolation functions defined for the linear case as

$$\begin{aligned}
\Theta_1(t) &= \frac{t^{n+1} - t}{t^{n+1} - t^n} = \frac{t^{n+1} - t}{\Delta t}, \quad \text{and} \\
\Theta_2(t) &= \frac{t - t^n}{t^{n+1} - t^n} = \frac{t - t^n}{\Delta t};
\end{aligned}$$

and the nodal values of u^h for node A at t_+^n and t_-^{n+1} are, respectively, $\tilde{\mathrm{u}}_A^n$ and u_A^{n+1}. The test functions w^h for each time slab (recall: piecewise polynomials in space and linear in time) are similarly defined, $N_A(\boldsymbol{x})\,\Theta_1(t)$ and $N_A(\boldsymbol{x})\,\Theta_2(t)$ for $A = 1, \ldots, \mathrm{n_{np}}$. With these definitions the weighted residual equation (3.63) yields the following couple of equations for each node A:

$$\sum_{B=1}^{\mathrm{n_{np}}} \left\{ \iint_{Q^n} N_A\, \Theta_1 \left[N_B \frac{\mathrm{u}_B^{n+1} - \tilde{\mathrm{u}}_B^n}{\Delta t} + \left(\Theta_1\, \tilde{\mathrm{u}}_B^n + \Theta_2\, \mathrm{u}_B^{n+1}\right)(\boldsymbol{a} \cdot \boldsymbol{\nabla}) N_B \right] d\Omega\, dt \right\} = 0$$

$$\sum_{B=1}^{\mathrm{n_{np}}} \left\{ \iint_{Q^n} N_A \,\Theta_2 \Big[N_B \frac{\mathrm{u}_B^{n+1} - \tilde{\mathrm{u}}_B^n}{\Delta t} + (\Theta_1 \,\tilde{\mathrm{u}}_B^n + \Theta_2 \,\mathrm{u}_B^{n+1})(\boldsymbol{a} \cdot \boldsymbol{\nabla}) N_B \Big] d\Omega \, dt \right\}$$
$$+ \int_\Omega N_A \sum_{B=1}^{\mathrm{n_{np}}} N_B (\tilde{\mathrm{u}}_B^n - \mathrm{u}_B^n) d\Omega = 0.$$

The time-discontinuous formulation employing linear approximations in time is third-order accurate with respect to Δt and unconditionally stable, see Shakib and Hughes (1991). Like the standard Galerkin method, the time-discontinuous Galerkin method must be stabilized in space. For instance, a least-squares approach can be employed to improve stability without compromising the accuracy. Two variants of a least-squares space–time methodology are described in the next sections.

3.10.2 Time-discontinuous least-squares formulation

Consider again the homogeneous form of the linear convection equation (3.4a). As before, the finite element approximations are assumed to be discontinuous in time and continuous in space. However, instead of the Galerkin formulation, a least-squares approach is considered, based upon the following space–time quadratic functional:

$$\iint_{Q^n} (u_t + \boldsymbol{a} \cdot \boldsymbol{\nabla} u)^2 \; d\Omega \, dt.$$

This functional is minimized with respect to a variation of the time-dependent unknown $u(\boldsymbol{x}, t)$. The jump condition enforcing continuity across the slab interfaces must also be added. Finally, using the trial and test spaces defined by (3.62), we obtain the following least-squares weighted residual formulation: for $n = 0, 1, \ldots, \mathrm{n_{st}} - 1$, find $u^h \in \mathcal{S}_n^h$, such that for all $w^h \in \mathcal{V}_n^h$,

$$\iint_{Q^n} (w_t^h + \boldsymbol{a} \cdot \boldsymbol{\nabla} w^h)(u_t^h + \boldsymbol{a} \cdot \boldsymbol{\nabla} u^h) \, d\Omega \, dt$$
$$+ \frac{1}{\Delta t} \int_\Omega w^h(t_+^n) \big(u^h(t_+^n) - u^h(t_-^n)\big) \, d\Omega = 0. \quad (3.64)$$

This time-discontinuous least-squares formulation is also third-order accurate with respect to Δt when a linear-in-time approximation is employed.

3.10.3 Space–time Galerkin/Least-squares formulation

To conclude the presentation of space–time finite element methods for convection problems, we shall describe the space–time Galerkin/Least-squares method proposed by Shakib and Hughes (1991). For the homogeneous convection equation (3.4a), the space–time Galerkin/Least-squares weighted residual formulation becomes: for

$n = 0, 1, \ldots, n_{\text{st}} - 1$, find $u^h \in \mathcal{S}_n^h$, such that for all $w^h \in \mathcal{V}_n^h$,

$$\iint_{Q^n} w^h \left(u_t^h + \boldsymbol{a} \cdot \boldsymbol{\nabla} u^h\right) d\Omega\, dt + \iint_{Q^n} \left(w_t^h + \boldsymbol{a} \cdot \boldsymbol{\nabla} w^h\right) \tau \left(u_t^h + \boldsymbol{a} \cdot \boldsymbol{\nabla} u^h\right) d\Omega\, dt + \int_\Omega w^h(t_+^n) \left(u^h(t_+^n) - u^h(t_-^n)\right) d\Omega = 0, \quad (3.65)$$

where $u^h(\boldsymbol{x}, t_-^0) = u(\boldsymbol{x}, 0)$. The first and the last integrals are the same as in the time-discontinuous Galerkin formulation (3.63). As before, the last integral is a jump condition. The second integral is the least-squares operator already encountered in (3.64). Parameter τ was already discussed in Section 2.4.3, but here it is extended to transient problems and particularized for pure convection following Shakib (1989)

$$\tau = \left(\left(\frac{2}{\Delta t}\right)^2 + \left(\frac{2a}{h}\right)^2\right)^{-1/2}. \quad (3.66)$$

The stability and accuracy analysis performed by Shakib and Hughes (1991) indicates that for linear-in-time approximations the method is third-order accurate with respect to Δt and unconditionally stable.

3.11 APPLICATIONS AND SOLVED EXERCISES

Several test problems describing purely convective transport are presented in this section to confirm the accuracy characteristics of the finite element schemes discussed in the present chapter.

Tables 3.5 and 3.6 summarize the amplification factors and the stability properties of the most relevant classical schemes discussed here for a uniform linear/bilinear mesh.

3.11.1 Propagation of a cosine profile

A simple 1D problem is proposed to illustrate and compare the performance of explicit and implicit schemes. The convection equation (3.27) is solved over the spatial interval $]0, 1[$ considering the following initial:

$$u(x, 0) = \begin{cases} \frac{1}{2}\left(1 + \cos(\pi(x - x_0)/\sigma)\right) & \text{if } |x - x_0| \leq \sigma, \\ 0 & \text{otherwise,} \end{cases}$$

and boundary condition: $u(0, t) = 0$ for $t \geq 0$, where $x_0 = 0.2$ and $\sigma = 0.12$.

Explicit methods. The exact solution of equation (3.27) with $a = 1$ corresponds to the translation to the right of the initial profile at unit speed. Figure 3.12 compares

Table 3.5 Amplification factor for different schemes.

Scheme	Amplification factor
CN	$G_{\mathrm{CN}}(\xi, C) = \dfrac{1 - \frac{2}{3}\sin^2\frac{\xi}{2} - i\frac{1}{2}C\sin\xi}{1 - \frac{2}{3}\sin^2\frac{\xi}{2} + i\frac{1}{2}C\sin\xi}$
TG2	$G_{\mathrm{TG2}}(\xi, C) = \dfrac{1 - (\frac{2}{3} + 2C^2)\sin^2\frac{\xi}{2} - iC\sin\xi}{1 - \frac{2}{3}\sin^2\frac{\xi}{2}}$
TG3	$G_{\mathrm{TG3}}(\xi,\ C) = \dfrac{1 - \frac{2}{3}(1 + 2C^2)\sin^2\frac{\xi}{2} - iC\sin\xi}{1 - \frac{2}{3}(1 - C^2)\sin^2\frac{\xi}{2}}$
TG3-2S	$G_{\mathrm{TG3\text{-}2S}}(\xi, C, \alpha) = \dfrac{1 - \frac{2}{3}\sin^2\frac{\xi}{2} - 2C^2\ \tilde{G}(\xi, C, \alpha)\ \sin^2\frac{\xi}{2} - iC\sin\xi}{1 - \frac{2}{3}\sin^2\frac{\xi}{2}}$
CJ	$G_{\mathrm{CJ}}(\xi, C, \theta) = \dfrac{1 - 2(\frac{1}{3} + 2(1-\theta)\theta C^2)\sin^2\frac{\xi}{2} - iC\sin\xi}{1 - 2(\frac{1}{3} - 2\theta^2 C^2)\sin^2\frac{\xi}{2}}$

Table 3.6 Stability limits for different schemes.

Scheme	Stability limits	
	1D	2D
CN	unconditional stability	
TG2	$C^2 \leq 1/3$	$c_x^{2/3} + c_y^{2/3} \leq (1/3)^{1/3}$
TG3	$C^2 \leq 1$	$c_x^{2/3} + c_y^{2/3} \leq 1$
TG3-2S	$C^2 \leq 3/4$	$c_x^2 + c_y^2 \leq 3/4$
CJ	unconditional stability for $1/2 \leq \theta \leq 1$	

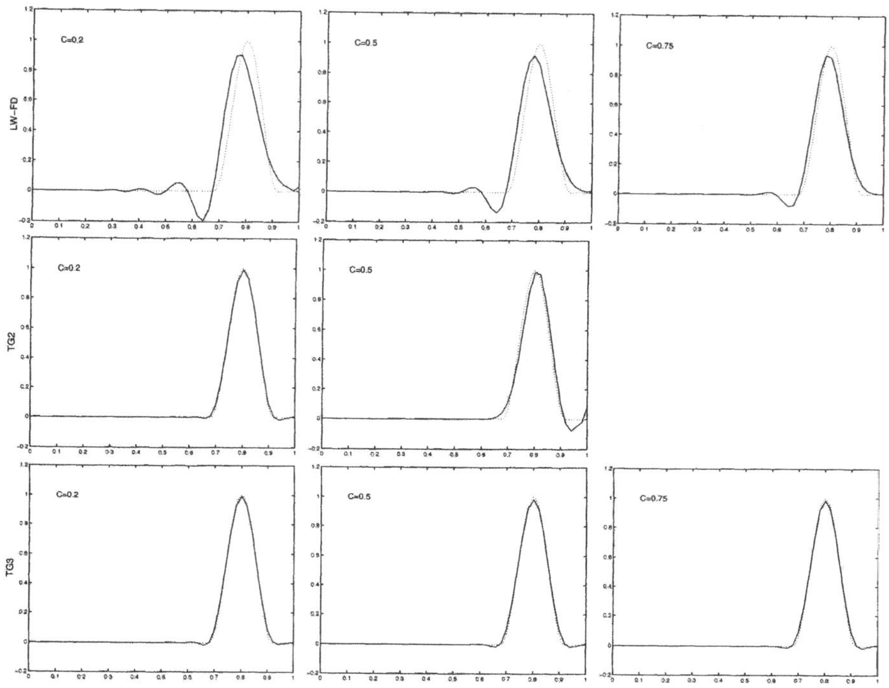

Fig. 3.12 Propagation of a cosine profile (explicit methods): comparison between the exact solution (dotted line) and the Lax–Wendroff finite element solution with diagonal mass (LW-FD), consistent mass (TG2) and the third-order Taylor–Galerkin solution (TG3).

the numerical solutions obtained at time $t = 0.6$ using a mesh of 50 uniform linear elements and different values of the Courant number C.

The problem is solved using in succession with:

- the second-order Lax–Wendroff finite element method is combined with a diagonal mass representation (LW-FD);
- the Lax–Wendroff finite element method is combined with a consistent mass representation (TG2);
- the third-order explicit Taylor–Galerkin scheme TG3.

We refer to Section 3.4.2.4 for the Galerkin formulation of the Lax–Wendroff method and to Section 3.6.2 for the Taylor–Galerkin method.

Figure 3.12 shows that both schemes using a consistent mass matrix exhibit a better phase accuracy than the Lax–Wendroff scheme combined with a diagonal mass representation (LW-FD). Note that the Lax–Wendroff scheme with consistent mass representation (TG2) cannot be operated with $C^2 > 1/3$. Moreover, it shows a phase lead at $C = 1/2$.

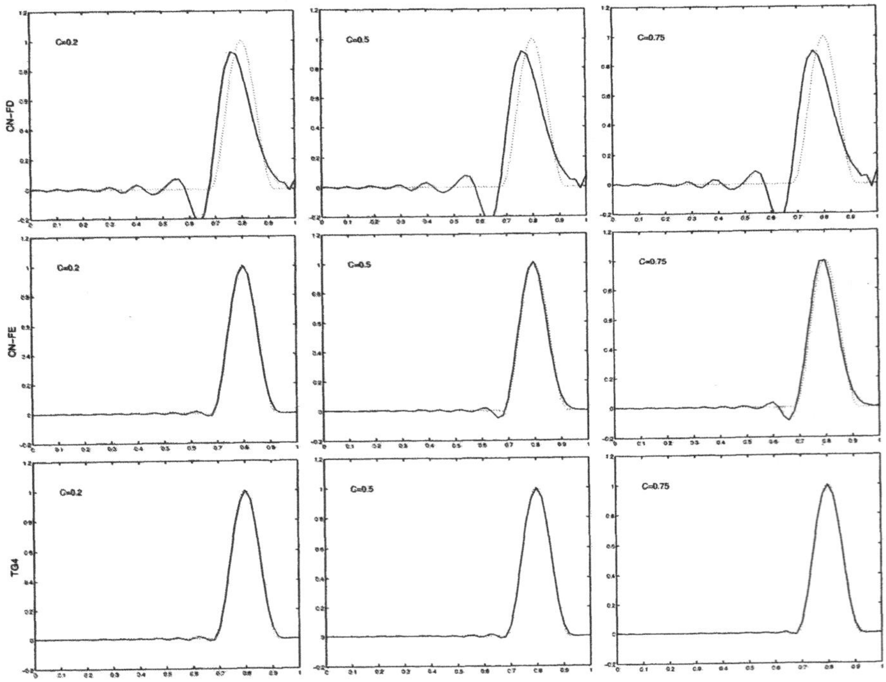

Fig. 3.13 Propagation of a cosine profile (implicit methods): comparison between the exact solution (dotted line) and the Crank–Nicolson solution with diagonal mass (CN-FD), consistent mass (CN-FE) and the fourth-order Taylor–Galerkin solution (TG4).

These findings are fully consistent with the phase error diagrams in Figure 3.3. As regards the third-order accurate TG3 scheme, we note that it has a rather uniform phase accuracy over the entire stable interval $0 < C < 1$ of the Courant number. This is in agreement with the phase–speed characteristics reported in Figure 3.6.

Implicit methods. Figure 3.13 shows the results of the same convection problem obtained with Galerkin and three implicit schemes, namely:

- the second-order Crank–Nicolson method combined with linear elements and a diagonal mass (CN-FD);
- the same method but with a consistent mass representation (CN-FE);
- the fourth-order Taylor–Galerkin method (TG4) resulting from the combination of the Harten and Tal-Ezer time scheme in equation (3.58) with linear elements and a consistent mass representation.

The Galerkin formulation of the Crank–Nicolson method is described in Section 3.4.2.3. For the TG4 method, the following variational equation is obtained from equations (3.27) and (3.58):

$$\int_0^L \left(w\left(\Delta u + \frac{a\Delta t}{2}\frac{\partial \Delta u}{\partial x}\right) - \frac{a^2\Delta t^2}{12}\frac{\partial w}{\partial x}\frac{\partial \Delta u}{\partial x}\right) dx = -\int_0^L w a\Delta t \frac{\partial u^n}{\partial x} dx,$$

where $\Delta u = u^{n+1} - u^n$. Similarly to the explicit methods, the schemes using finite elements with a consistent mass matrix show a superior phase accuracy. This is in agreement with the phase–speed characteristics reported in Figure 3.4. Observe that the phase error for the Crank–Nicolson/consistent finite element scheme increases with the time-step size, while the fourth-order accurate implicit method shows an excellent phase response at all values of the Courant number.

3.11.2 Travelling wave package

This example tries to visualize the influence of numerical damping and phase lag for different explicit and implicit methods. The conclusions drawn from the graphical representations of equations (3.36) (accuracy analysis) are recovered in this practical example. An initial condition is convected at a unitary speed over a uniform mesh of linear elements. It is defined as the product of a square wave an a sinusoidal wave,

$$u(x,0) = \exp(-(\alpha_1(x+1-\beta_1)^{n_1}))\exp(-(\alpha_2(1-x-\beta_2)^{n_2}))\sin(\kappa x),$$

with $\alpha_1 = \alpha_2 = 7$, $n_1 = n_2 = 30$, $\beta_1 = 1/\sqrt{2}$, $\beta_2 = 0$, and $\kappa = 20\pi$. The element size is chosen such that the sinusoidal wave induces a dimensionless wave number $\xi = kh = \pi/4$. This value corresponds to the minimum number of elements (eight) per wave length to accurately represent the sinus. On one hand, this high frequency at $\xi = \pi/4$ will be affected by the phase errors. On the other, the amplitude of this wave (controlled by the square function induced by the exponentials) will be affected by the amplitude error.

The time increment for the explicit methods has been chosen such that the Courant number is 90% of the stability limit, Section 3.5.2 and Table 3.6. This criterion is usually employed in practice. Implicit methods are stable for every time increment. Their precision degrades for large values of the Courant number, C, and, usually, at $C = 1$ they present their maximum accuracy. Thus, for comparison purposes, the time increment is chosen such that $C = 0.9$.

Figure 3.14 shows the results for different methods. Phase errors clearly affect the second-order, explicit, methods: Lax–Wendroff with a diagonal mass matrix representation (LW-FD), that is finite differences, Lax–Wendroff with the consistent mass matrix (TG2) and leap-frog. As noted earlier, the diagonal matrix representation increases the stability range (LW-FD uses a larger time step) but accuracy is compromised: important amplitude errors. The third-order explicit Taylor–Galerkin scheme (TG3) shows its superior behavior, in particular with respect to phase errors, due to its higher-order time accuracy.

There is not a clear improvement with second-order implicit methods: Crank–Nicolson with diagonal matrix representation (CN-FD), Crank–Nicolson with a consistent mass matrix (CN-FE) and the least-squares Crank–Nicolson (CJ), see Section

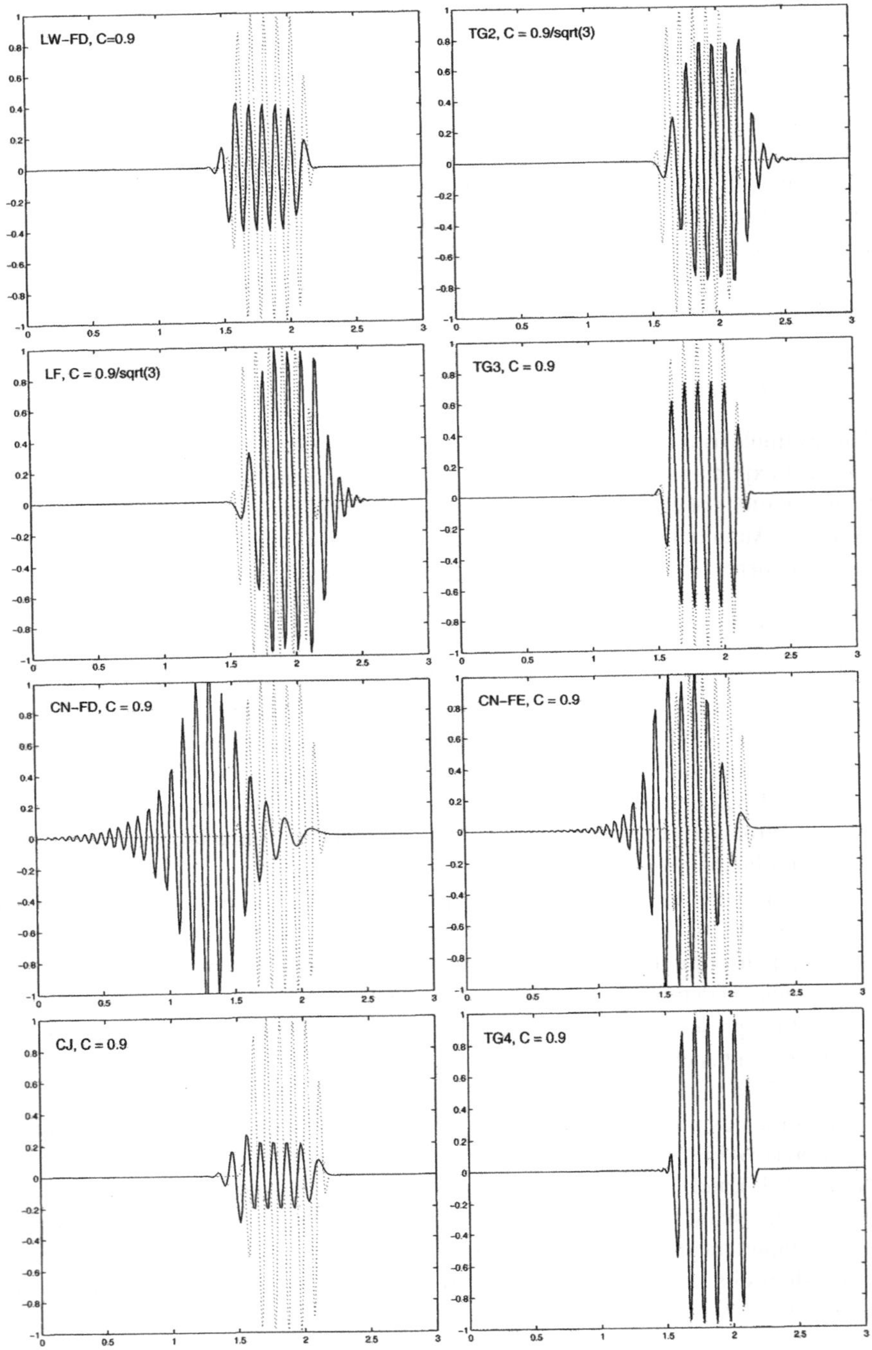

Fig. 3.14 Comparison between different explicit and implicit methods for $\xi = \pi/4$.

3.8.1. They all show large phase errors and the least-squares formulation, as expected, is over-diffusive. The one-step fourth-order Taylor–Galerkin method (TG4) introduced in the previous example, shows very accurate results because of its high-order time accuracy.

This example, however, represents a limit case: $\xi = kh = \pi/4$. Figure 3.15 shows results of the same problem (in particular, same mesh) with a different initial condition: the parameter κ is modified, $\kappa = 10\pi$, in order to have $\xi = kh = \pi/8$. All methods show a clear improvement, except for Crank–Nicolson with diagonal matrix representation (CN-FD), which still presents large phase errors, all the other results seem acceptable. The accuracy of high-order schemes, TG3 and TG4, is still remarkable.

3.11.3 The rotating cone problem

This classical test problem for 2D convection schemes considers the convection of a product-cosine hill in a pure rotation velocity field. The initial data is

$$u(\boldsymbol{x},0) = \begin{cases} \frac{1}{4}(1+\cos\pi X_1)(1+\cos\pi X_2) & \text{if } X_1^2 + X_2^2 \leq 1, \\ 0 & \text{otherwise,} \end{cases}$$

where $\boldsymbol{X} = (\boldsymbol{x} - \boldsymbol{x}_0)/\sigma$, and the boundary condition is $u = 0$ on Γ^{in}. The initial position of the center and the radius of the cosine hill are $\boldsymbol{x}_0$ and σ, respectively. In the examples they are chosen as $\boldsymbol{x}_0 = (\frac{1}{6}, \frac{1}{6})$ and $\sigma = 0.2$. The convection field is a pure rotation one with unit angular velocity, namely $\boldsymbol{a}(\boldsymbol{x}) = (-x_2, x_1)$. A uniform mesh of 30×30 four-node elements over the unit square $[-\frac{1}{2}, \frac{1}{2}] \times [-\frac{1}{2}, \frac{1}{2}]$ is employed in the calculations.

Explicit methods. The numerical solutions are shown in Figure 3.16 after a full revolution completed in 200 time steps ($\Delta t = 2\pi/200$). They have been computed using three explicit finite element schemes in the Galerkin formulation, namely:

- the second-order Lax–Wendroff method combined with bilinear elements and a diagonal mass representation;
- the Lax–Wendroff method combined with bilinear elements and a consistent mass representation (TG2);
- the third-order explicit Taylor–Galerkin scheme (TG3).

The weak form for the Lax–Wendroff method is the particularization of equation (3.29) for the case of $s = 0$ and $h = 0$, namely

$$\left(w, \frac{\Delta u}{\Delta t}\right) = \left(\boldsymbol{a}\cdot\boldsymbol{\nabla}w, u^n - \frac{\Delta t}{2}(\boldsymbol{a}\cdot\boldsymbol{\nabla})u^n\right) - \left((\boldsymbol{a}\cdot\boldsymbol{n})w, u^n - \frac{\Delta t}{2}(\boldsymbol{a}\cdot\boldsymbol{\nabla})u^n\right)_{\Gamma^{out}}.$$

Similarly, from (3.43) we determine the weak form for the TG3 method

$$\left(w, \frac{\Delta u}{\Delta t}\right) + \frac{\Delta t^2}{6}\left(\boldsymbol{a}\cdot\boldsymbol{\nabla}w, \boldsymbol{a}\cdot\boldsymbol{\nabla}\frac{\Delta u}{\Delta t}\right) - \frac{\Delta t^2}{6}\left((\boldsymbol{a}\cdot\boldsymbol{n})w, \boldsymbol{a}\cdot\boldsymbol{\nabla}\frac{\Delta u}{\Delta t}\right)_{\Gamma^{out}}$$

$$= \left(\boldsymbol{a}\cdot\boldsymbol{\nabla}w, u^n - \frac{\Delta t}{2}(\boldsymbol{a}\cdot\boldsymbol{\nabla}u^n)\right) - \left((\boldsymbol{a}\cdot\boldsymbol{n})w, u^n - \frac{\Delta t}{2}(\boldsymbol{a}\cdot\boldsymbol{\nabla}u^n)\right)_{\Gamma^{out}}.$$

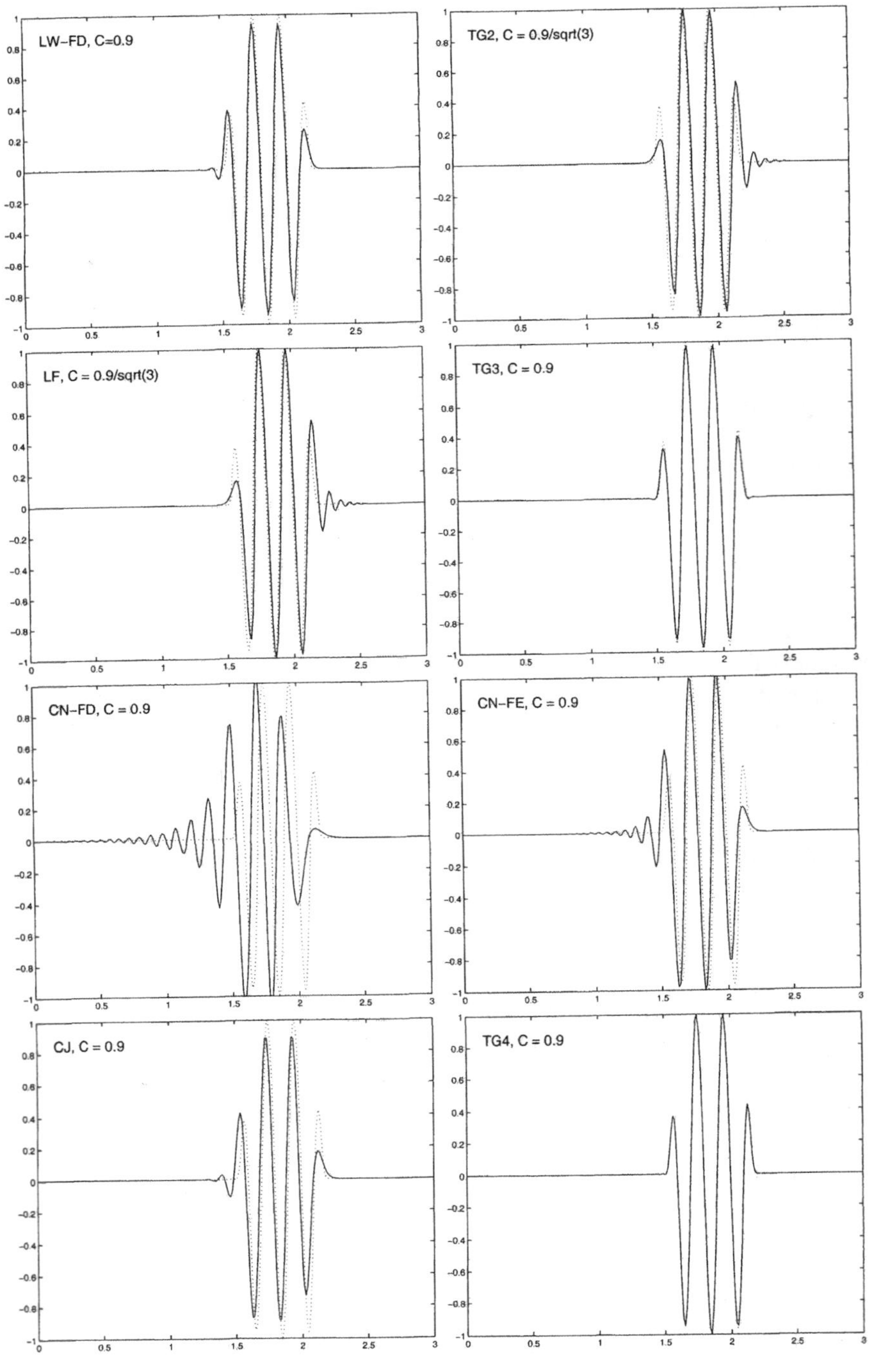

Fig. 3.15 Comparison between different explicit and implicit methods for $\xi = \pi/8$.

To compare the accuracy of the various explicit methods, the maximum and minimum values of the computed solutions are provided in Figure 3.16. The greater accuracy of the finite element schemes employing a consistent (TG2) or generalized (TG3) mass matrix is clearly apparent. Admittedly, the consistent finite element schemes are computationally more expensive than the Lax–Wendroff scheme using a diagonal mass matrix, because the solution of a banded (symmetric) linear system is required at each time step.

Implicit methods. We repeat the same problem using now the implicit Crank–Nicolson method. The first tests are performed using the Galerkin finite element formulation. The associated weak form is given from (3.28) for s and h equal to zero, namely

$$\left(w, \frac{\Delta u}{\Delta t}\right) - \frac{1}{2}(\boldsymbol{\nabla} w, \boldsymbol{a}\,\Delta u) + \frac{1}{2}\big((\boldsymbol{a}\cdot\boldsymbol{n})w, \Delta u\big)_{\Gamma^{out}}$$

$$= \big(\boldsymbol{\nabla} w, \boldsymbol{a}\,u^n\big) - \big((\boldsymbol{a}\cdot\boldsymbol{n})w, u^n\big)_{\Gamma^{out}}.$$

Three increasing values of the time step are employed, which correspond to a complete revolution of the cone in, respectively, 120, 60 and 30 time steps. In this way, we shall appraise the behavior of the Crank–Nicolson/Galerkin method beyond the stability limit of the explicit TG3 scheme. The numerical results after one complete revolution of the cone are displayed in Figure 3.17. Note that the Crank–Nicolson scheme with the Galerkin formulation represents a non-dissipative method in pure convection. Moreover, the phase accuracy of the method decreases when the time step is increased. As a result, significant non-physical oscillations develop as soon as the time-step size exceeds the stability limit of the explicit schemes.

These oscillations can be attenuated using a dissipative spatial formulation, such as the least-squares FEM of Carey and Jiang described in Section 3.8.1. From (3.57) we obtain the weak form for this method

$$\left(\frac{w}{\Delta t} + \frac{1}{2}\boldsymbol{a}\cdot\boldsymbol{\nabla} w, \frac{\Delta u}{\Delta t} + \frac{1}{2}\boldsymbol{a}\cdot\boldsymbol{\nabla}\Delta u\right) = -\left(\frac{w}{\Delta t} + \frac{1}{2}\boldsymbol{a}\cdot\boldsymbol{\nabla} w, \boldsymbol{a}\cdot\boldsymbol{\nabla} u^n\right).$$

Note that the dissipative effect of the least-squares approach increases with the square of the time step. Again, the rotating cone problem is solved using 120, 60 and 30 time increments for a full rotation. Figure 3.18 reports the results which can be compared with those in Figure 3.17 for the Galerkin approach. The dissipative nature of the least-squares formulation results in lower values of the cone height with respect to the Galerkin results in Figure 3.17. Also, reduced minimum values of the computed solutions are obtained with the least-squares approach.

3.11.4 Propagation of a steep front

Crank–Nicolson methods. This 1D problem considers the convection at unit speed of discontinuous initial data. The discontinuity occurs over one element and is initially located at position $x = 0.2$ of the computational domain $]0, 1[$. The inlet condition $u(0, t) = 1$ is imposed. A mesh of uniform linear elements of size $h = 1/50$

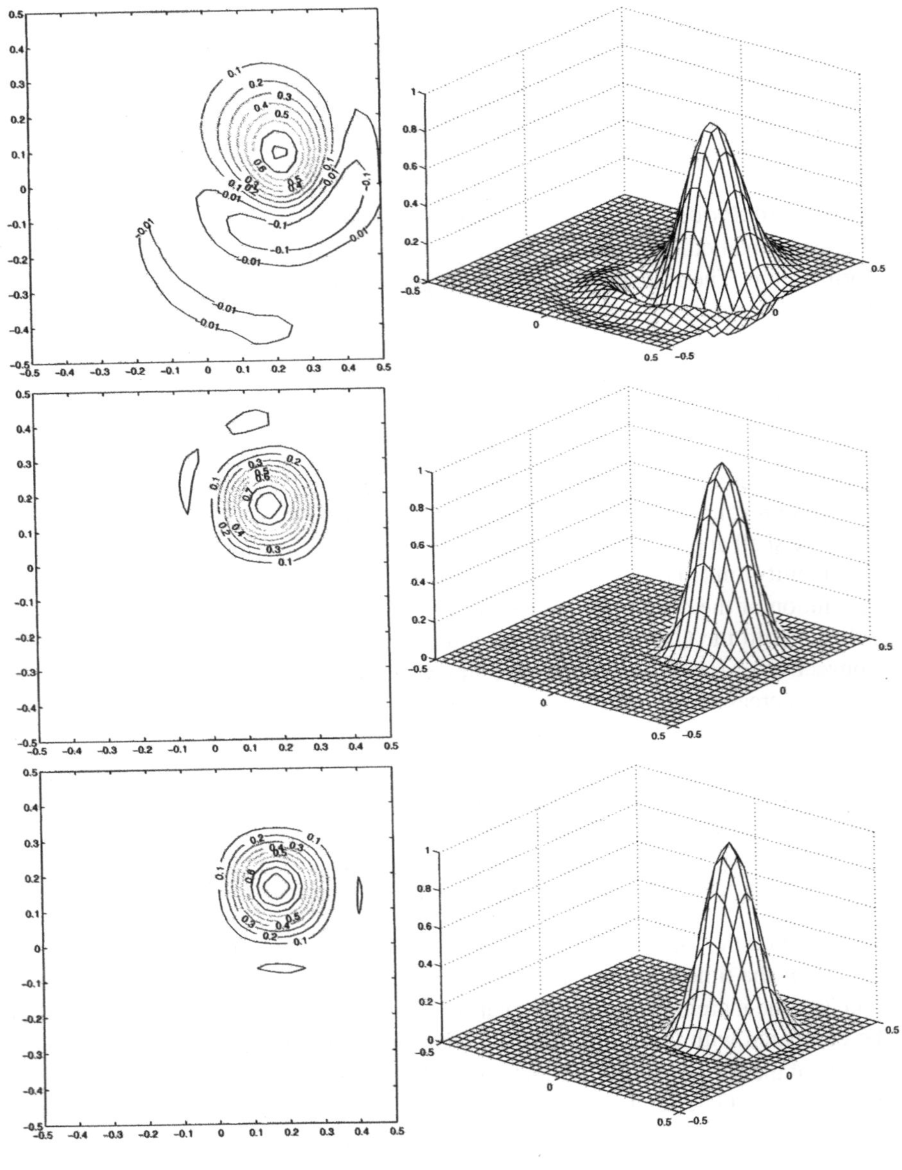

Fig. 3.16 Convection of a cosine hill in a pure rotation velocity field: comparison of the numerical solutions after a complete revolution calculated with $\Delta t = 2\pi/200$ by means of (Top): Lax–Wendroff/diagonal mass scheme ($u_{\max} = 0.8186, u_{\min} = -0.1774$); (Middle): TG2 ($u_{\max} = 0.9830, u_{\min} = -0.0186$); and (Bottom): TG3 ($u_{\max} = 0.9835, u_{\min} = -0.0148$).

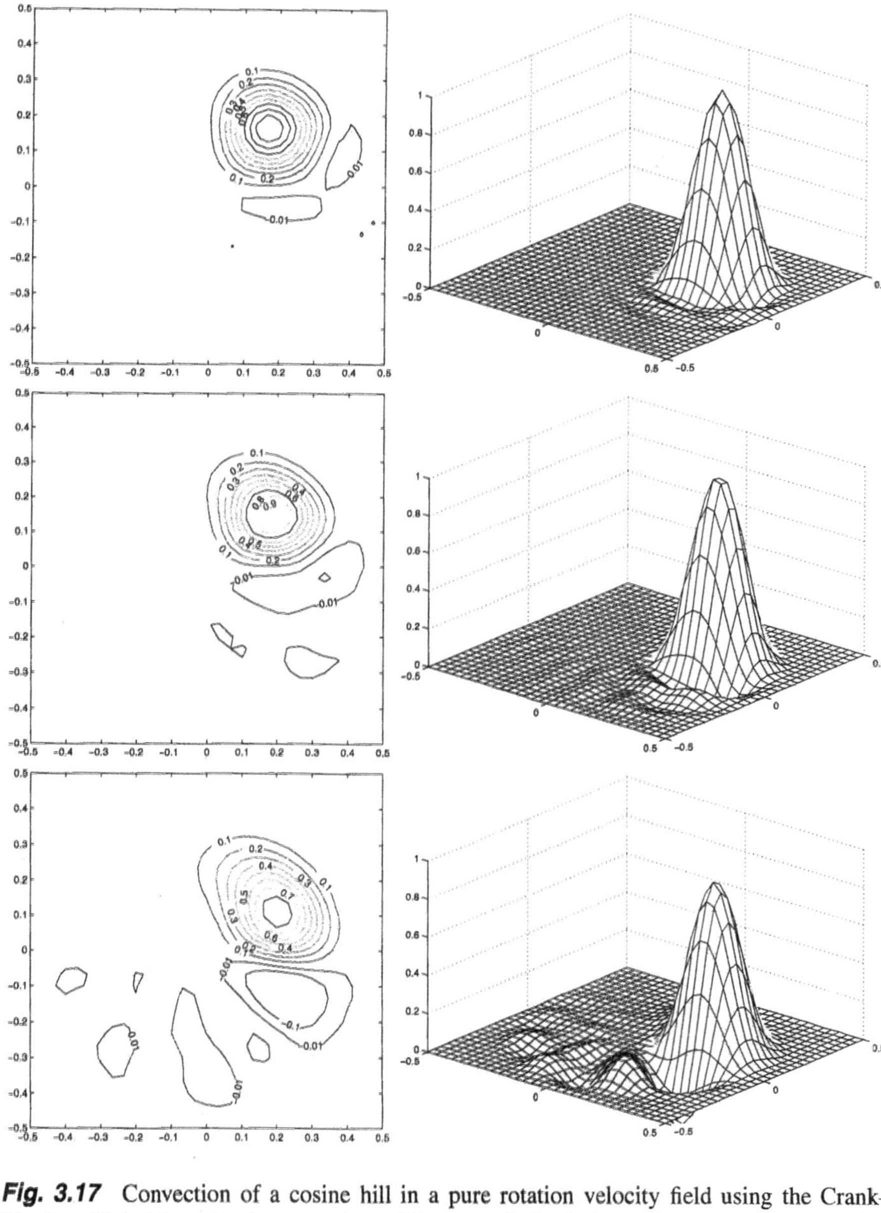

Fig. 3.17 Convection of a cosine hill in a pure rotation velocity field using the Crank–Nicolson/Galerkin method: comparison of the numerical solutions after a complete revolution computed with (Top): $\Delta t = 2\pi/120$, $u_{\max} = 0.9969$, $u_{\min} = -0.0454$; (Middle): $\Delta t = 2\pi/60$, $u_{\max} = 0.9691$, $u_{\min} = -0.1096$; and (Bottom): $\Delta t = 2\pi/30$, $u_{\max} = 0.8931$, $u_{\min} = -0.2694$.

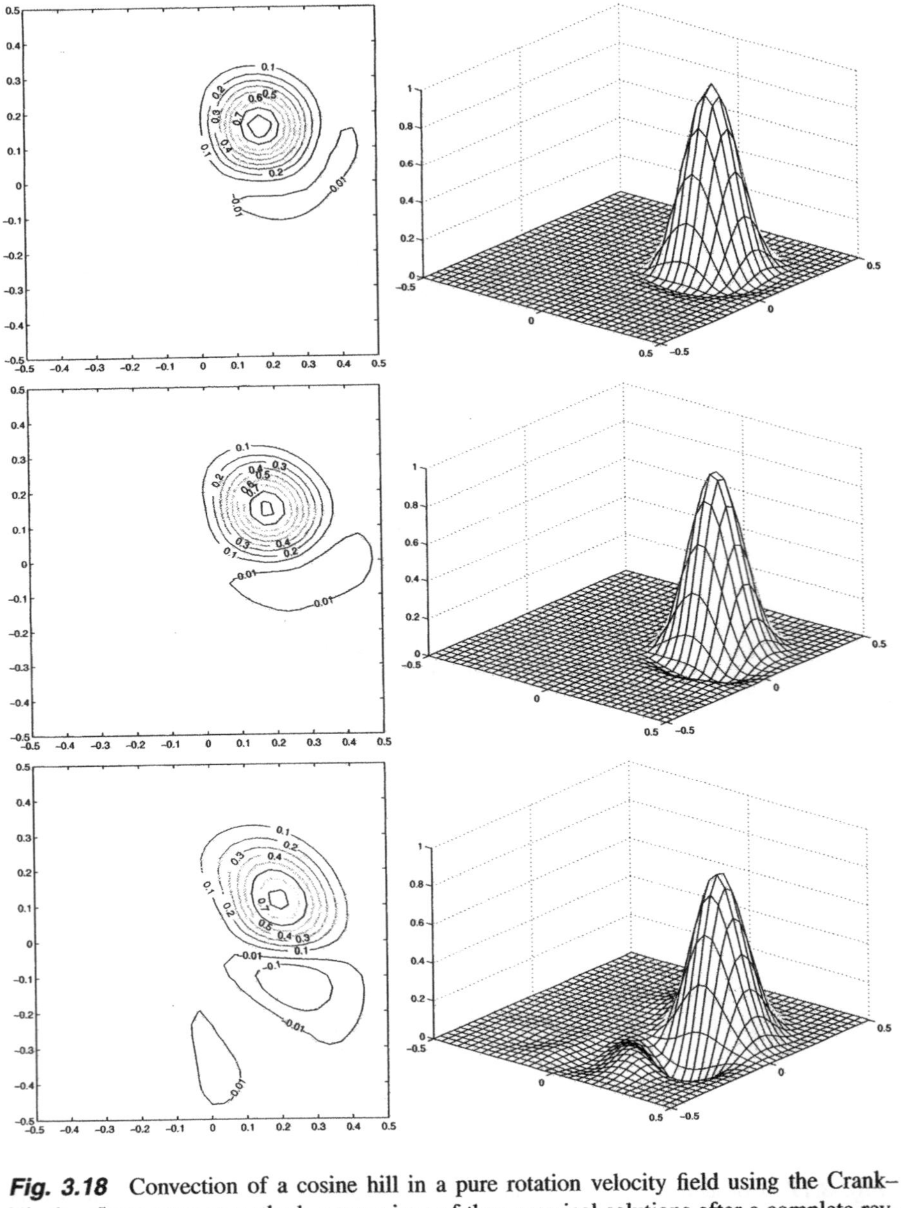

Fig. 3.18 Convection of a cosine hill in a pure rotation velocity field using the Crank–Nicolson/least-squares method: comparison of the numerical solutions after a complete revolution computed with (Top): $\Delta t = 2\pi/120$, $u_{\max} = 0.9691$, $u_{\min} = -0.0266$; (Middle): $\Delta t = 2\pi/60$, $u_{\max} = 0.9165$, $u_{\min} = -0.0616$; and (Bottom): $\Delta t = 2\pi/30$, $u_{\max} = 0.8370$, $u_{\min} = -0.2009$.

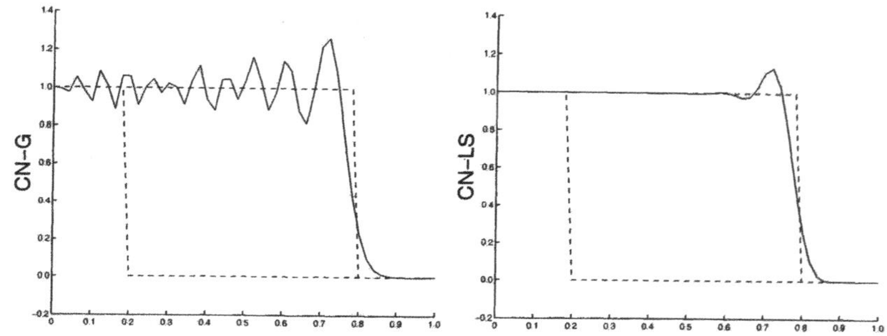

Fig. 3.19 Propagation of a steep front using the Crank–Nicolson scheme with the Galerkin method (left) and with the least-squares method (right). The Courant number is $C = 0.75$ in both cases. The graphs show the computed solutions at time $t = 0.60$, together with the initial profile and the exact solution.

is employed. The results at time $t = 0.6$ are displayed in Figure 3.19 together with the initial data. They were obtained (for a Courant number $C = 0.75$) by combining the Crank–Nicolson scheme (with linear elements) and (1) the Galerkin formulation, and (2) the least-squares formulation of Carey and Jiang (1988), see Section 3.8.1.

Note that Crank–Nicolson with least-squares succeeds in removing the spurious oscillations induced by the Galerkin formulation over the whole computational domain. Since Crank–Nicolson is not a monotone scheme, residual oscillations remain at the front. These could be removed using nonlinear viscosity, which is added at the front to render the scheme locally first-order accurate, see Section 3.7.

Discontinuous Galerkin in space and the second-order two-step Lax–Wendroff method in time. The steep-front problem is again solved with a uniform mesh of 50 linear elements. The two-step TG2 method integrates in time the semi-discrete equations resulting from the discontinuous Galerkin method. The method follows the two-step rationale presented in Section 3.6.4, see also Section 4.2.3.2:

$$
\begin{aligned}
u^{n+1/2} &= u^n + \frac{\Delta t}{2}\, u_t^n & &= u^n - \frac{\Delta t}{2}\, a\, u_x^n, \\
u^{n+1} &= u^n + \Delta t\, u_t^{n+1/2} & &= u^n - \Delta t\, a\, u_x^{n+1/2}.
\end{aligned}
$$

Mid-interval values of the unknown are computed within each element using a diagonal mass representation. Denoting by indices i and j the nodal values of the unknown u in a typical linear element, we have

$$
\left(u_i^+\right)^{n+1/2} = \left(u_i^+\right)^n - \frac{a\Delta t}{2h}\left(u_j^- - u_i^+\right)^n, \; \left(u_j^-\right)^{n+1/2} = \left(u_j^-\right)^n - \frac{a\Delta t}{2h}\left(u_j^- - u_i^+\right)^n.
$$

To compute the end-of-step values, we use a consistent mass representation and obtain the following equations from expression (3.61):

$$
\begin{aligned}
\frac{h}{6}\left(2\dot{u}_i^+ + \dot{u}_j^-\right) &= -\frac{a}{2}\left(u_i^+ + u_j^-\right)^{n+1/2} + a\left(u_i^-\right)^{n+1/2},\\
\frac{h}{6}\left(\dot{u}_i^+ + 2\dot{u}_j^-\right) &= \frac{a}{2}\left(u_i^+ + u_j^-\right)^{n+1/2} - a\left(u_j^-\right)^{n+1/2},
\end{aligned}
$$

from which we obtain upon inversion of the element mass matrix:

$$
\begin{aligned}
\left(u_i^+\right)^{n+1} &= \left(u_i^+\right)^n + \frac{a\Delta t}{h}\left(4u_i^- - 3u_i^+ - u_j^-\right)^{n+1/2},\\
\left(u_j^-\right)^{n+1} &= \left(u_j^-\right)^n + \frac{a\Delta t}{h}\left(3u_i^+ - 2u_i^- - u_j^-\right)^{n+1/2}.
\end{aligned}
$$

As explained in detail in Section 4.4.2, artificial diffusion is needed close to zones of sharp gradients to avoid spurious oscillations in the numerical solution. Here, artificial diffusion consists in adding the following terms to the r.h.s. of the second-step equations:

$$
\begin{aligned}
&\frac{\nu_e\Delta t}{h^2}\left((u_j^- - u_i^+)^n - (u_i^- - u_{i-1}^+)^n\right) \quad \text{at node } i,\\
&\frac{\nu_e\Delta t}{h^2}\left((u_{j+1}^- - u_j^+)^n - (u_j^- - u_i^+)^n\right) \quad \text{at node } j.
\end{aligned}
$$

Note that these terms emanate from the usual Galerkin term for the diffusion operator:

$$
\sum_{\Omega^e}\int_{\Omega^e}\nu_e\frac{\partial w}{\partial x}\frac{\partial u}{\partial x}dx.
$$

Zero diffusive flux is imposed at both extremes of the computational domain. The coefficient of numerical diffusion ν_e is computed in each element Ω^e as a function of the local value of the second derivative of the unknown: $\nu_e = \frac{ah}{4}\max(d_i, d_j)$, where

$$
d_i = \frac{u_j^- - u_i^+ - u_i^- + u_{i-1}^+}{u_j^- + u_i^+ + u_i^- + u_{i-1}^+}, \qquad d_j = \frac{u_{j+1}^- - u_j^+ - u_j^- + u_i^+}{u_{j+1}^- + u_j^+ + u_j^- + u_i^+}.
$$

Note that we are actually solving a convection–diffusion problem, instead of a pure convection one. The stability limit of the two-step Lax–Wendroff method applied to the steep-front problem was found to be $C = 0.30$. This reduced stability is compensated by the extremely low cost per time step of the discontinuous Galerkin method. The results obtained with $C = 0.30$ are displayed in Figure 3.20. There are no spurious oscillations, but the discontinuity is spread over a few elements through the action of the added diffusivity.

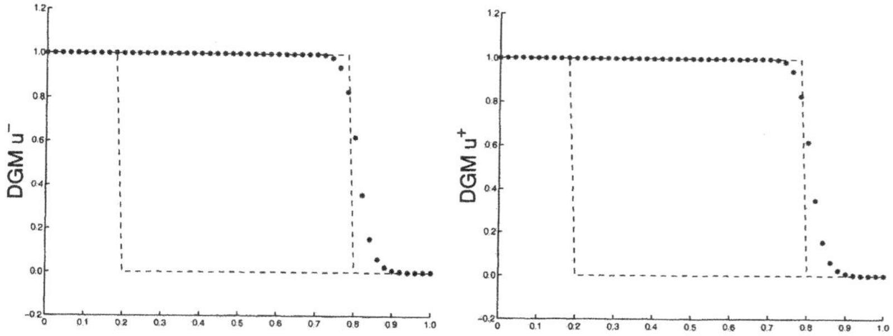

Fig. 3.20 Propagation of a steep front using the discontinuous Galerkin method and the two-step Lax–Wendroff method in time. The Courant number is $C = 0.30$. The graphs show the computed solutions u^- (left) and u^+ (right) at time $t = 0.60$, together with the initial profile and the exact solution.

Space–time methods. We repeat the solution of the problem using three space–time methods, namely:

- the time-discontinuous Galerkin formulation described in Section 3.10.1;
- the time-discontinuous least-squares formulation described in Section 3.10.2;
- the space–time Galerkin/Least-squares formulation described in Section 3.10.3.

Linear finite element approximations are employed in both space and time. Two values of the Courant number are used with these unconditionally stable methods, namely, $C = 1$, and $C = 2$.

Time-discontinuous Galerkin: the developments of Section 3.10.1 induce the following partitioned matrix system for the nodal unknowns $\mathbf{u}^{n+1}$ and $\mathbf{u}^{n^+}$:

$$\left(\mathbf{M} + \frac{2}{3}\Delta t\mathbf{C}\right)\mathbf{u}^{n+1} - \left(\mathbf{M} - \frac{1}{3}\Delta t\mathbf{C}\right)\mathbf{u}^{n^+} = 0$$
$$\left(\mathbf{M} + \frac{1}{3}\Delta t\mathbf{C}\right)\mathbf{u}^{n+1} + \left(\mathbf{M} + \frac{2}{3}\Delta t\mathbf{C}\right)\mathbf{u}^{n^+} = 2\,\mathbf{M}\,\mathbf{u}^{n^-},$$

where $\mathbf{M}$ denotes the mass matrix and $\mathbf{C}$ the convection matrix. The condition $u^{n+1}(1) = u^{n^+}(1) = 1$ is then enforced to satisfy the inlet condition. Note that this third-order accurate and unconditionally stable method requires the solution of an algebraic system double the size of usual time-stepping algorithms. Results are displayed in Figure 3.21. Once again, oscillations extending over the whole computational domain characterize the Galerkin formulation. Note that the level of the spurious oscillations appears to decrease when the Courant number is increased.

Time-discontinuous least-squares: here, we use the pure least-squares approach described in Section 3.10.2. The partitioned matrix system for the nodal unknowns $\mathbf{u}^{n+1}$ and $\mathbf{u}^{n^+}$ is now obtained in the form

$$\left(\mathbf{M} + \frac{1}{3}a^2\Delta t^2\mathbf{K}\right)\mathbf{u}^{n+1} - \left(\mathbf{M} - \Delta t\mathbf{C} + \frac{1}{6}a^2\Delta t^2\mathbf{K}\right)\mathbf{u}^{n^+} = 0$$
$$\left(-\mathbf{M} - \Delta t\mathbf{C} + \frac{1}{6}a^2\Delta t^2\mathbf{K}\right)\mathbf{u}^{n+1} + \left(2\mathbf{M} + \frac{1}{3}a^2\Delta t^2\mathbf{K}\right)\mathbf{u}^{n^+} = \mathbf{M}\,\mathbf{u}^{n^-}.$$

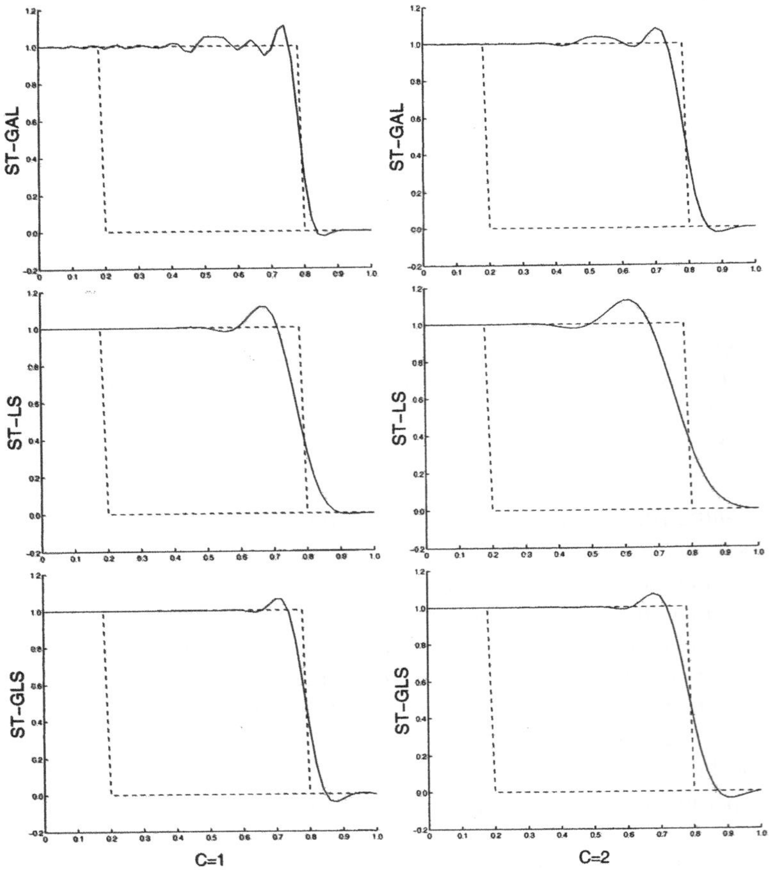

Fig. 3.21 Propagation of a steep front using space–time methods with Courant numbers $C = 1$ (left) and $C = 2$ (right). Different methods are employed: the space–time Galerkin (top), the space–time least-squares (middle), and the space–time Galerkin/Least-squares (bottom). The curves show the computed solutions at time $t = 0.60$, together with the initial profile and the exact solution.

Here, matrix $\mathbf{K}$ is a diffusivity matrix accounting for the numerical dissipation introduced by the least-squares method. Results displayed in Figure 3.21 show that pure least-squares is over diffusive and smears out excessively the steep front. In fact, from the previous equations, we see that the added diffusivity is proportional to the square of the Courant number.

Time-discontinuous Galerkin/Least-squares: we now follow the GLS approach for bilinear space–time elements described in Section 3.10.3. The partitioned matrix system for the nodal unknowns $\mathbf{u}^{n+1}$ and $\mathbf{u}^{n^+}$ now depends on the stabilization parameter τ. We have selected the value of τ as given in expression (3.66). In terms of the Courant number, this gives

$$\tau = \frac{\Delta t}{2\sqrt{1+C^2}}.$$

The partitioned matrix system for the nodal unknowns $\mathbf{u}^{n+1}$ and $\mathbf{u}^{n^+}$ is now obtained in the form:

$$\left(\left(1+\frac{2\tau}{\Delta t}\right)\mathbf{M}+\frac{2}{3}\Delta t\mathbf{C}+\frac{2}{3}\tau a^2\Delta t\mathbf{K}\right)\mathbf{u}^{n+1} - \left(\left(1+\frac{2\tau}{\Delta t}\right)\mathbf{M}-\left(\frac{1}{3}\Delta t+2\tau\right)\mathbf{C}-\frac{1}{3}\tau a^2\Delta t\mathbf{K}\right)\mathbf{u}^{n^+} = 0$$

$$\left(\left(1-\frac{2\tau}{\Delta t}\right)\mathbf{M}+\left(\frac{1}{3}\Delta t-2\tau\right)\mathbf{C}+\frac{1}{3}\tau a^2\Delta t\mathbf{K}\right)\mathbf{u}^{n+1} + \left(\left(1+\frac{2\tau}{\Delta t}\right)\mathbf{M}+\frac{2}{3}\Delta t\mathbf{C}+\frac{2}{3}\tau a^2\Delta t\mathbf{K}\right)\mathbf{u}^{n^+} = 2\,\mathbf{M}\,\mathbf{u}^{n^-}.$$

Figure 3.21 depicts the results. Time-discontinuous GLS clearly delivers the best solution for this steep-front problem. Note that the dissipation introduced by the GLS operator also increases with the value of the Courant number. In fact, from the previous equations and the definition of the stabilization parameter τ, we see that the added diffusivity of the GLS method is proportional to $C^2/\sqrt{1+C^2}$. The GLS method is nevertheless significantly less diffusive than pure least-squares.

4

Compressible Flow Problems

Engineering practice goes beyond scalar linear hyperbolic equations. This chapter is concerned with nonlinear systems of hyperbolic equations. An important and well-known problem of this class is the modeling of inviscid compressible flows governed by the Euler equations of gas dynamics. The nonlinear nature of the Euler equations typically leads to non-smooth solutions characterized by the presence of shocks. Various finite-element-based strategies are discussed for tracing transient solutions in the presence of flow discontinuities. The Eulerian description is used throughout the chapter, except for the finite element treatment of coupled fluid–structure problems where the Arbitrary Lagrangian–Eulerian (ALE) description is employed because it is more convenient.

4.1 INTRODUCTION

The development of numerical methods for simulating complex flow problems is of major importance in view of the numerous applications of fluid dynamics in many different areas of applied science and engineering. Numerical modeling of fluid dynamics problems is indeed of particular relevance to the aerospace and automotive industries, to the oil industry, in meteorology, hydrology, oceanography, protection of the environment, as well as in the safety assessment of industrial plants. In recent years, numerical simulation of flow problems has been extended to several new application areas and disciplines. These include coupled problems which combine fluid mechanics and electromagnetic theory (Maxwell equations), or fluid mechanics and biomechanics, such as the modeling of arterial blood flow with the aim of providing guidance for surgery planning.

The equations governing fluid motion are generally quite complex and of a form strongly dependent upon the particular problem under investigation. Consider, for instance, the motion of a fluid under the effect of an explosion: the transient phenomenon is of very short duration and viscosity effects can be neglected. Moreover, a realistic modeling of such fast-transient dynamics problems requires the compressibility of the fluid to be taken into account. By contrast, if we are to analyze the flow of a visco-plastic fluid, such as, for instance, in the simulation of forming processes for fiber-reinforced thermo-plastics, viscosity will be the important factor, while the fluid can generally be considered as incompressible and its flow as steady. Given the great variety of fluid mechanics problems, it is advisable to focus attention on a particular class of problems at a time, rather than attempting to consider the problem of fluid motion under a general form.

The present chapter is devoted to the finite element treatment of compressible flow problems governed by the Euler equations of gas dynamics. It starts with a review in Section 4.2 of the properties of nonlinear hyperbolic equations. We underline the numerical difficulties introduced by the directional character of propagation of information in hyperbolic problems and by the possible generation of discontinuities in the solution, even starting from continuous initial data. Emphasis is then placed on the numerical treatment of non-smooth solutions by a weak formulation in which spatial derivatives are no longer acting on the problem variables but only on smooth weighting functions.

Systems of hyperbolic equations are then treated in Section 4.3 where the Euler equations governing compressible high-speed gas flow are introduced. We first discuss the basic properties of the Euler equations and the possible form of the associated boundary conditions according to the flow regime under consideration. Then, starting from the strong form of the conservation equations for mass, momentum and energy, we construct the associated weak variational forms which are the basis of the spatial discretization by the finite element method. Various options for the spatial discretization are considered. This includes continuous or discontinuous interpolation of all the conservation variables, as well as mixed representations where some variables are continuous and other are discontinuous across inter-element boundaries.

As already seen in Chapter 3, all second- and higher-order time discretization schemes are not monotone methods and generate oscillations in the vicinity of sharp solution gradients. These have to be damped by the addition of artificial dissipation terms. We discuss in Section 4.4 various spatial discretization techniques of the upwind type able to cope with the directional character of propagation of information in hyperbolic problems. Then, the finite element implementation of specific artificial viscosity techniques for the accurate representation of shocks and other flow discontinuities is discussed in Section 4.5. This includes the construction of so-called high-resolution schemes.

Most numerical methods designed for compressible flow problems present difficulties when applied to low-speed (low Mach number) flow situations. This is due to the fast propagation of pressure waves as flow conditions approach the incompressible limit. Finite element algorithms can be developed that work for both the compressible and the nearly incompressible regime. A summarized account is given in Section 4.6.

We then describe in Section 4.7 finite element models for coupled fluid–structure problems. The response of a linear elastic structure interacting with an acoustic fluid is treated first. Then, ALE finite element models are discussed for fluid–structure interaction in the nonlinear regime. Illustrative examples indicate the effectiveness of the ALE method for problems with moving boundaries and deforming fluid–structure interfaces. The chapter closes with the presentation in Section 4.8 of solved exercises illustrating in the simple context of one spatial dimension the finite element solution of Burgers' and Euler equations.

4.2 NONLINEAR HYPERBOLIC EQUATIONS

4.2.1 Scalar equations

In nonlinear hyperbolic problems discontinuities can be generated from continuous initial conditions. Such solutions cannot verify the partial differential equation in the classical sense, but they may satisfy a weak form of this equation. This leads us to introduce the concept of *generalized* or *weak solutions* which admit the presence of discontinuities provided a condition, called the *jump condition*, is verified. Weak solutions are not necessarily unique; an extra condition, the *entropy condition*, allows us to determine the physically correct solution. We shall briefly recall these concepts which are the basis of the numerical approximation of nonlinear hyperbolic equations. Specialized texts, such as for instance Hirsch (1990), LeVeque (1992), Quarteroni and Valli (1994) or Godlewski and Raviart (1996), should be consulted for a more complete exposition of the properties of nonlinear hyperbolic equations.

Let us start the discussion of nonlinear scalar equations by considering the 1D Cauchy problem

$$\begin{cases} u_t + f_x(u) = 0, \\ u(x,0) = u_0(x), \end{cases} \tag{4.1}$$

where $f(u)$ is a nonlinear function of the unknown u. As in previous chapters u_t represents the partial time derivative of the unknown u, while f_x indicates the spatial derivative of $f(u)$. A classical example of this class of equations is the inviscid *Burgers' equation*, where $f(u) = u^2/2$, so that equation (4.1) becomes

$$\begin{cases} u_t + \left(\dfrac{u^2}{2}\right)_x = 0, \\ u(x,0) = u_0(x), \end{cases} \quad \text{or in convective form:} \quad \begin{cases} u_t + u\, u_x = 0, \\ u(x,0) = u_0(x). \end{cases} \tag{4.2}$$

This is a nonlinear transport equation where the convection velocity is the solution u itself. Thus, the characteristics, see Section 3.3.1, satisfy the equation

$$\frac{dx}{dt} = u(x,t). \tag{4.3}$$

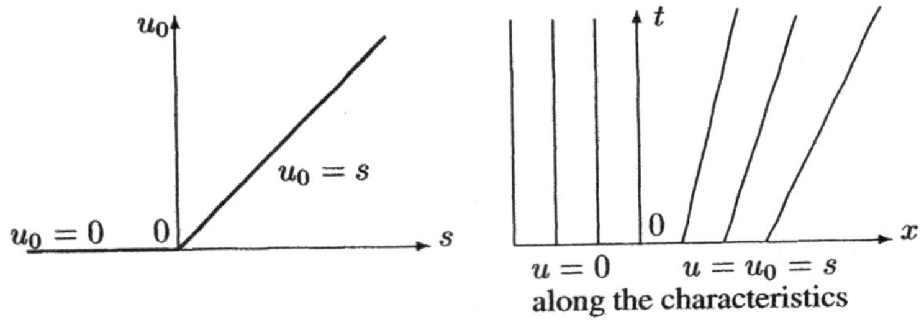

Fig. 4.1 Increasing continuous initial data produce a continuous solution to the inviscid Burgers' equation.

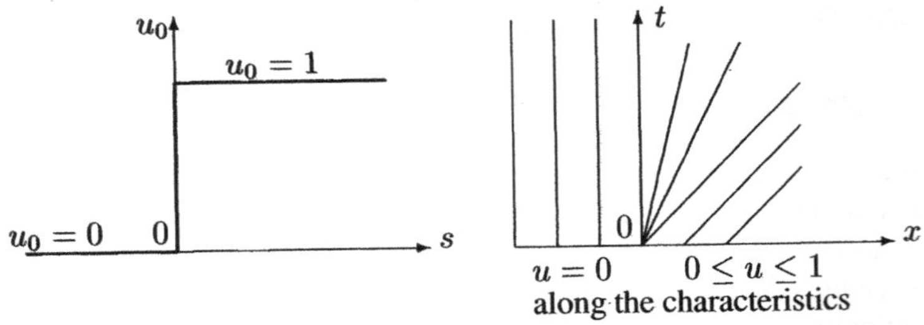

Fig. 4.2 Increasing discontinuous initial data produce a continuous solution to Burgers' equation.

The material derivative of u along a characteristic can be evaluated from Burgers' equation (4.2) and the characteristic (4.3)

$$\frac{du}{dt} = \frac{\partial u}{\partial t} + \frac{\partial u}{\partial x}\frac{dx}{dt} = u_t + u\,u_x = 0.$$

This shows that the solution u is constant along each characteristic. Thus, equation (4.3) implies that the slope dx/dt of the characteristics is constant. Therefore, the characteristics are straight lines. When the initial data are continuous this property implies that the characteristic equation is

$$x = s + u_0(s)\,t, \tag{4.4a}$$

for any typical space–time point $(s, 0)$ along the x-axis and with a slope $u_0(s)$ defined by the initial data. Moreover, along this characteristic

$$u = \text{constant} = u_0(s). \tag{4.4b}$$

Equations (4.4) provide a parametric representation of the solution $u(x, t)$ of Burgers' equation. This equation possesses a unique solution as long as the characteristics do not intersect. As shown in Figures 4.1 and 4.2, non-intersecting characteristics (and thus a unique solution) are obtained for increasing continuous initial data, as well as for increasing discontinuous initial data. By contrast, see Figure 4.3, if the

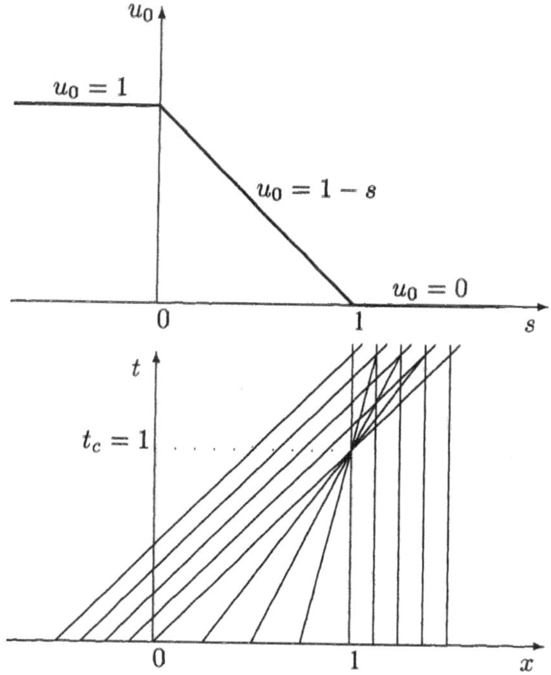

Fig. 4.3 With decreasing initial data the characteristics $x = s + u_0(s)\,t$ intersect causing the solution to become discontinuous.

initial data are such that the derivative $du_0(s)/ds$ is negative at some point of the x-axis, then there is a critical time t_c at which characteristics first cross, causing the solution to become discontinuous. Let us illustrate the possible situations by means of three examples.

First, for continuous initial data, reproduced in Figure 4.1,

$$u_0(s) = \begin{cases} 0 & \text{for } s \leq 0, \\ s & \text{for } s \geq 0, \end{cases}$$

relations (4.4) indicate that the solution of Burgers' problem (4.2) is given by

$$\begin{aligned} x &= s, & u &= 0 & &\text{for } s \leq 0,\ t > 0, \\ x &= s\,(1+t), & u &= s & &\text{for } s \geq 0,\ t > 0. \end{aligned}$$

Elimination of parameter s between these equations yields the solution u:

$$u(x,t) = \begin{cases} 0 & \text{for } x \leq 0,\ t > 0, \\ x/(1+t) & \text{for } x \geq 0,\ t > 0. \end{cases}$$

The corresponding characteristics are drawn in Figure 4.1.

Second, for the discontinuous initial data shown in Figure 4.2,

$$u_0(s) = \begin{cases} 0 & \text{for } s < 0, \\ 1 & \text{for } s > 0, \end{cases}$$

the solution is also determined from (4.4) and consists of three branches:

$$u(x,t) = \begin{cases} 0 & \text{for } x \leq 0,\ t > 0, \\ x/t & \text{for } 0 \leq x \leq t,\ t > 0, \\ 1 & \text{for } x \geq t,\ t > 0. \end{cases}$$

These expressions represent a continuous function $u(x,t)$ in the semi-plane $t > 0$. This continuous solution is called an *expansion fan* and the corresponding characteristics are drawn in Figure 4.2. This situation is representative of a supersonic expansion ramp. The important point to note here is that, due to the nonlinearity of Burgers' equation, discontinuous initial data may generate a continuous solution. This is, however, not always the case.

And third, in the case of decreasing initial data, even if continuously distributed, the situation is in fact quite different. Characteristics eventually cross and a discontinuity of the solution forms. Consider the following decreasing, and continuous, initial data:

$$u_0(s) = \begin{cases} 1 & \text{for } s \leq 0, \\ 1-s & \text{for } 0 \leq s \leq 1, \\ 0 & \text{for } s \geq 1. \end{cases}$$

As can be seen from the characteristics in Figure 4.3, the solution is unique for $t < t_c = 1$, but not beyond t_c. For $t \geq 1$ the solution is discontinuous because signals from the left portion of the considered domain travel faster than those from the right. This causes signals to pile up, thereby creating a discontinuity. This situation is typical for a compression profile and the analogy with a supersonic compression ramp should be noted.

In the more general case of the conservation law (4.1), it can be shown that if $u_0(s_1) > u_0(s_2)$ with $s_1 < s_2$, the two characteristics $x = s_1 + u_0(s_1)t$ and $x = s_2 + u_0(s_2)t$ intersect at time $t_c = [s_2 - s_1]/[u_0(s_1) - u_0(s_2)]$. Beyond time t_c a classical solution of the conservation law (4.1) no longer exists; however, a generalized (or weak) solution (which is discontinuous) can be defined.

4.2.2 Weak solutions and entropy condition

Weak solutions were already introduced in Chapter 1, see Remark 1.7, for steady problems. Here the same concept is extended to transient problems, in particular, for equation (4.1). The objective is, as in Section 1.5.4, to weaken the continuity requirements on the solution $u(x,t)$. In this case, since we are confronted with a first-order hyperbolic equation, the objective is to avoid differentiating $u(x,t)$. Thus, discontinuous solutions will be admissible. Denote by $C_0^1(\mathbb{R} \times [0,\infty[)$ the space of test functions $w(x,t)$ that are continuously differentiable and with compact support in the space–time domain $\mathbb{R} \times [0,\infty[$. That is, $w(x,t) = 0$ outside of some bounded set. Multiplying the conservation law (4.1) by $w(x,t)$, integrating over space and time, and integrating by parts (using Green's formula), yields the integral relationship

$$\int_0^\infty \int_{-\infty}^\infty \Big[w_t(x,t)\,u(x,t) \;+ w_x(x,t)\,f\big(u(x,t)\big)\Big]\,dx\,dt$$
$$= -\int_{-\infty}^\infty w(x,0)\,u_0(x)\,dx. \quad (4.5)$$

Note that boundary terms disappear because $w(x,t)$ has compact support and thus vanishes at infinity (only the initial condition remains). A function $u(x,t)$ is called a *weak solution* of the conservation law (4.1) provided it is measurable and the integral relation (4.5) holds for all functions $w \in C_0^1(\mathbb{R} \times [0,\infty[)$.

Note that equation (4.5) does not include derivatives of the unknown u, nor of the flux function $f(u)$. Weak solutions are not necessarily unique in the sense that different solutions can be obtained with the same initial condition. The physically correct solution is the one that satisfies the so-called *entropy condition*, see for instance Oleĭnik (1957) for an in-depth analysis.

The correct solution can be determined by the *vanishing viscosity* approach. For instance, the physically correct weak solution of the inviscid Burgers' equation (4.2) corresponds to the solution of Burgers' equation when viscosity goes to zero. That is, the inviscid case (4.2) is seen as a model of

$$u_t^\epsilon + u^\epsilon u_x^\epsilon = \epsilon\, u_{xx}^\epsilon$$

with $u^\epsilon(x,0) = u_0(x)$, valid only for very small ϵ and smooth u^ϵ. In order to avoid working with the viscous equation there are other conditions easier to check and that will also allow us to determine the correct solution. They are called *entropy conditions*. This is also the reason for calling the viscous solution the entropy solution.

In order to establish the generic form of the entropy condition, we consider the conservation equation (4.1) with piecewise constant initial data with a single discontinuity. This is known as the *Riemann problem*. As an example, let us take Burgers' equation (4.2) with the following piecewise constant initial data:

$$u_0(s) = \begin{cases} u_l & \text{for } s < 0, \\ u_r & \text{for } s > 0. \end{cases}$$

When $u_l > u_r$, the unique weak solution is given by

$$u(x,t) = \begin{cases} u_l & \text{for } x < \sigma t \\ u_r & \text{for } x > \sigma t \end{cases} \quad (4.6)$$

where

$$\sigma = \frac{u_l + u_r}{2} \quad (4.7)$$

is the speed at which the discontinuity propagates. At each instant $t \geq 0$ the solution has two constant states and, as illustrated in Figure 4.4, the characteristics penetrate into the discontinuity from both regions. This kind of discontinuity is called a *jump*, or a *shock* in application to gas dynamics, and is physically admissible. Note that the smooth solution u^ϵ, namely

$$u^\epsilon(x,t) = u_r + \frac{1}{2}(u_l - u_r)\left[1 - \tanh\left(\frac{(u_l - u_r)(x - \sigma t)}{2\epsilon}\right)\right],$$

of Burgers' equation converges to (4.6) as ϵ goes to zero. Therefore, this unique solution is the desired vanishing viscosity solution.

If $u_l < u_r$ there are infinitely many weak solutions. One of them is again (4.6), but now, as indicated in Figure 4.5, the characteristics emanate from the discontinuity. It can be shown that this solution is unstable in the sense that small perturbations of the data produce large changes of the solution. Moreover, if a small amount of viscosity is introduced in the equation the solution changes completely. Thus solution (4.6) is not physically desirable. Among the other weak solutions, one that is stable to perturbations is the *rarefaction wave* given by

$$u(x,t) = \begin{cases} u_l, & x < u_l\, t \\ x/t, & u_l\, t \leq x \leq u_r\, t \\ u_r, & x > u_r\, t. \end{cases}$$

This is in fact the vanishing viscosity solution to Burgers' equation, see Figure 4.6.

In order to generalize these concepts for a generic flux function f we need first to define the propagation speed of the discontinuity, that is the *Rankine–Hugoniot jump condition*, and second to establish if this discontinuity is physically acceptable using the previous observation that shock should have characteristic lines going *into* and *not out of* the discontinuity, as time advances.

Still in the context of the Riemann problem, that is piecewise constant data with a single discontinuity, we want to obtain for a generic flux function f the propagation speed of the discontinuity, namely the generalization of (4.7) which is also denoted as σ. We integrate equation (4.1) with respect to the spatial variable x from left, $\sigma t - 1$, to right, $\sigma t + 1$, of the discontinuity. This yields

$$\int_{\sigma t-1}^{\sigma t+1} \big(u_t(x,t) + f_x(u)\big) dx = 0 \Rightarrow \int_{\sigma t-1}^{\sigma t+1} u_t(x,t)\, dx = -f(u_r) + f(u_l) \qquad (4.8)$$

for each $t > 0$. On the other hand, since data are piecewise constant we may write

$$\begin{aligned} 0 = \frac{d}{dt}(u_l + u_r) &= \frac{d}{dt}\left[\int_{\sigma t-1}^{\sigma t} u(x,t)\, dx + \int_{\sigma t}^{\sigma t+1} u(x,t)\, dx\right] \\ &= \int_{\sigma t-1}^{\sigma t+1} u_t(x,t)\, dx - \sigma\,(u_l - u_r), \end{aligned} \qquad (4.9)$$

where the left, $u_l = u(\sigma t - 1, t)$, and right, $u_l = u(\sigma t + 1, t)$, values of the solution are used on each side of the discontinuity, $x = \sigma t$. Then, from (4.8) and (4.9) we

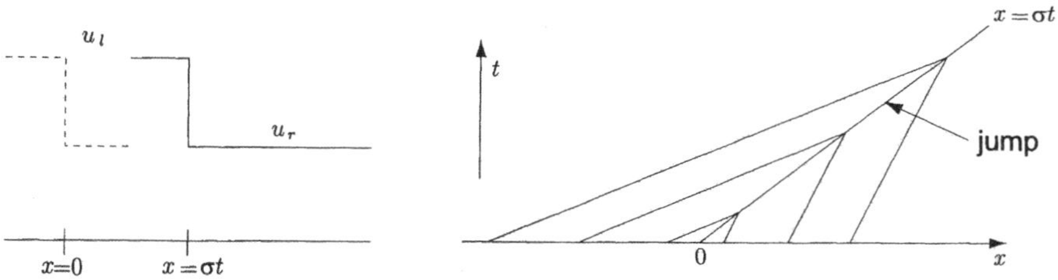

Fig. 4.4 Physically acceptable (entropy-compliant) weak solution: shock wave.

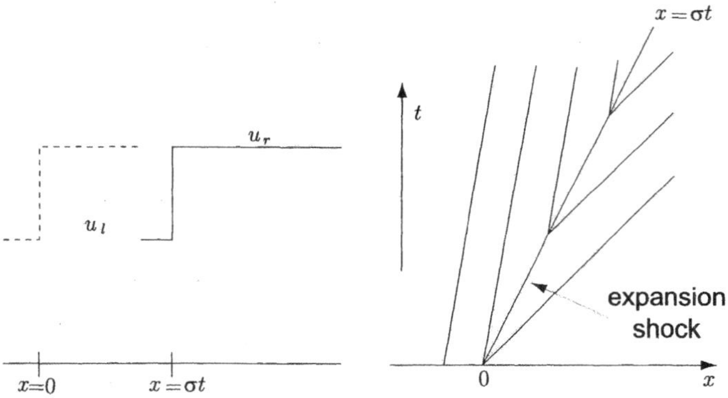

Fig. 4.5 Unstable (entropy-violating) weak solution.

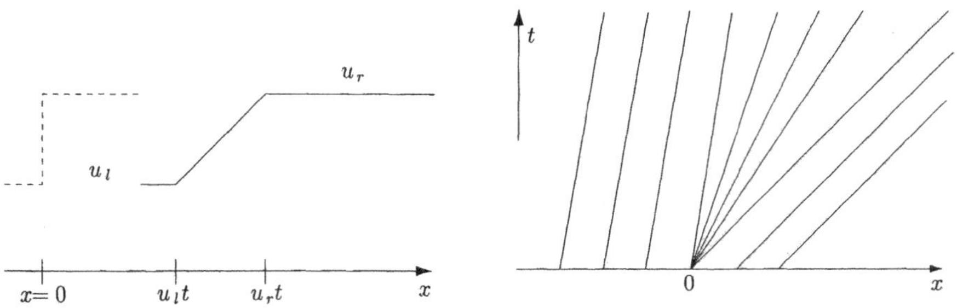

Fig. 4.6 Entropy-compliant weak solution: rarefaction wave.

obtain the propagation speed of the discontinuity for a scalar problem and for any flux function f,

$$\sigma = \frac{f(u_l) - f(u_r)}{u_l - u_r}. \tag{4.10}$$

As expected, the particularization to Burgers' equation recovers expression (4.7).

The previous analysis was carried out for the Riemann problem, that is for piecewise constant data. It can be generalized to more typical problems where the solution to the left and right of the discontinuity is varying smoothly. In general relation (4.10) across any propagating shock where u_l and u_r denote the values immediately to the left and right of the discontinuity is called the *Rankine–Hugoniot jump condition.*

Now that the speed of the discontinuity is defined we need to determine if the discontinuity corresponding to the weak solution is physically acceptable (i.e., it corresponds to the vanishing viscosity solution). Previous observations showed that characteristics must penetrate the discontinuity as time advances in order to have a physical solution. When the characteristics emanate from a discontinuity, small perturbations of the initial data, or of the equation itself (for instance, adding a small amount of viscosity), may result in a strong change of the solution (and typically this will cause a rarefaction of characteristics). From this analysis, the first entropy condition is derived:

- A discontinuity propagating with speed σ given by (4.10) satisfies the entropy condition if

$$a(u_l) = \frac{df}{du}(u_l) > \sigma > \frac{df}{du}(u_r) = a(u_r),$$

where the advection velocity is used, see Remark 3.4. For a convex flux function f, the speed of the discontinuity σ, namely the Rankine–Hugoniot condition, see (4.10), must lie between $a(u_l)$ and $a(u_r)$. Thus, the previous condition simply requires that $a(u_l) > a(u_r)$, which, in view of the convexity of the flux function f, amounts to imposing $u_l > u_r$.

A more general statement for the entropy condition, which applies to non-convex flux functions f, is:

- A discontinuity propagating with speed σ given by (4.10) satisfies the entropy condition if

$$\frac{f(u) - f(u_l)}{u - u_l} \geq \sigma \geq \frac{f(u) - f(u_r)}{u - u_r}$$

for all u between u_l and u_r.

4.2.3 Time and space discretization

Classical time integration methods, such as the θ family of methods and the Taylor–Galerkin schemes discussed in Chapter 3, can be applied in the solution of nonlinear hyperbolic problems. Note, however, that implicit methods generate systems of nonlinear equations and thus require iterative solution techniques. When the relevant time scales of the problem are close enough to justify the use of explicit time-marching schemes, this type of formulation should be favored. In fact, most explicit schemes keep their simple algorithmic structure in application to nonlinear problems. Multi-step explicit algorithms only involving first time derivatives are particularly suited, as they do not require evaluation of the spatial derivatives of the flux function. To illustrate the algorithmic simplicity of explicit methods, we shall consider the one-step and two-step Taylor–Galerkin methods, but note that other explicit methods can be implemented just as easily.

4.2.3.1 *One-step Taylor–Galerkin method* Consider the multidimensional scalar conservation law

$$u_t + \boldsymbol{\nabla} \cdot \boldsymbol{f}(u) = 0. \tag{4.11}$$

Proceeding as in Section 3.6.2, we replace the first and second time derivatives in the Taylor expansion with spatial derivatives using equation (4.11). This gives

$$u_t = -\boldsymbol{\nabla} \cdot \boldsymbol{f},$$

$$u_{tt} = -\boldsymbol{\nabla} \cdot \boldsymbol{f}_t = -\boldsymbol{\nabla} \cdot \left(\frac{\partial \boldsymbol{f}}{\partial u} u_t \right) = \boldsymbol{\nabla} \cdot \left(\boldsymbol{a}(u) \boldsymbol{\nabla} \cdot \boldsymbol{f} \right),$$

where the convection/advection velocity $\boldsymbol{a}(u) = \partial \boldsymbol{f}(u)/\partial u$ is again used, see Remark 3.4 or equation (3.2). The resulting second-order accurate time-stepping method reads

$$\frac{u^{n+1} - u^n}{\Delta t} = -\boldsymbol{\nabla} \cdot \boldsymbol{f}^n + \frac{\Delta t}{2} \boldsymbol{\nabla} \cdot \left(\boldsymbol{a}(u^n) \boldsymbol{\nabla} \cdot \boldsymbol{f}^n \right),$$

where as usual $\boldsymbol{f}^n = \boldsymbol{f}(u^n)$. A weighted residual formulation is obtained with the usual procedure: multiply by a test function belonging to

$$\mathcal{V} = \left\{ w(\boldsymbol{x}) \in \mathcal{H}^1(\Omega) \mid w(\boldsymbol{x}) = 0 \text{ for } \boldsymbol{x} \in \Gamma_D^{in} \right\},$$

integrate over space, and use Green's formula to weaken the smoothness requirement on the trial solution. The problem is then: find $u \in \mathcal{S}_t$, such that for all $w \in \mathcal{V}$,

$$\left(w, \frac{u^{n+1} - u^n}{\Delta t} \right) = \left(\boldsymbol{\nabla} w, \boldsymbol{f}^n \right) - \frac{\Delta t}{2} \left(\boldsymbol{\nabla} w, \boldsymbol{a}(u^n)\, \boldsymbol{\nabla} \cdot \boldsymbol{f}^n \right)$$

$$- \left(w\, \boldsymbol{n}, \boldsymbol{f}^n - \frac{\Delta t}{2} \boldsymbol{a}(u^n)\, \boldsymbol{\nabla} \cdot \boldsymbol{f}^n \right)_{\Gamma^{out}},$$

where a Dirichlet inlet condition is assumed and the trial space is now

$$\mathcal{S}_t := \left\{ u \mid u(\cdot, t) \in \mathcal{L}_2(\Omega), t \in [0, T] \text{ and } u(\boldsymbol{x}, t) = u_D \text{ for } \boldsymbol{x} \in \Gamma^{in} \right\}.$$

Note that discontinuous solutions are admissible. Nevertheless, in finite element practice continuous spatial approximations are usually chosen and shocks are modeled via adaptive mesh refinement and smearing the discontinuity through the introduction of artificial viscosity.

From a computational point of view at each step a system of algebraic equations must be solved. The matrix is the mass matrix, which in an Eulerian formulation does not vary with time and possesses nice properties for iterative solvers. The flux $\boldsymbol{f}(u)$ within an element is usually linearly interpolated in terms of its nodal values and the advection velocity $\boldsymbol{a}(u)$ is evaluated at the element integration points.

A possible disadvantage of the one-step TG method in application to nonlinear equations is the need to evaluate the second time derivative of the unknown in terms of the flux and its derivatives. Though perfectly feasible for scalar equations (Burgers), such evaluation becomes much more involved for systems of hyperbolic equations (Euler). Therefore, a two-step version of the method involving only first time derivatives is generally preferred in the modeling of nonlinear systems. Moreover, the divergence of the flux in the boundary integral of the previous weak form (last term on the r.h.s.) is also avoided in the two-step TG method.

4.2.3.2 Two-step Taylor–Galerkin method Consider now solving the multidimensional scalar conservation law (4.11) by means of the two-step time scheme

$$u^{n+1/2} = u^n + \frac{\Delta t}{2} u_t^n \qquad = u^n - \frac{\Delta t}{2} \boldsymbol{\nabla} \cdot \boldsymbol{f}^n,$$
$$u^{n+1} = u^n + \Delta t\, u_t^{n+1/2} \qquad = u^n - \Delta t\, \boldsymbol{\nabla} \cdot \boldsymbol{f}^{n+1/2},$$

where $\boldsymbol{f}^n = \boldsymbol{f}(u^n)$ and $\boldsymbol{f}^{n+1/2} = \boldsymbol{f}(u^{n+1/2})$.

In weak form, the problem in the second integration step is defined as

$$\left(w, \frac{u^{n+1} - u^n}{\Delta t}\right) = (\boldsymbol{\nabla} w, \boldsymbol{f}^{n+1/2}) - (w\boldsymbol{n}, \boldsymbol{f}^{n+1/2})_{\Gamma_{out}}. \tag{4.12}$$

Note that the integrals on the r.h.s. require the evaluation of the intermediate flux, $\boldsymbol{f}^{n+1/2}$, at the Gauss points only. As a result, the two-step TG method is implemented as follows:

Step 1: Compute at the integration points of the elements the intermediate value $u^{n+1/2}$ of the unknown using

$$u^{n+1/2} = u^n - \frac{1}{2}\Delta t \boldsymbol{\nabla} \cdot \boldsymbol{f}(u^n).$$

The divergence of the flux within an element is evaluated using the so-called group representation of the flux components (see Remark 4.1 below):

$$f_i = \sum_{a=1}^{\mathrm{n_{en}}} N_a(\boldsymbol{x})\, f_{ia}, \quad i = 1, \dots, \mathrm{n_{sd}}.$$

Once the intermediate value $u^{n+1/2}$ is obtained, the intermediate flux $\boldsymbol{f}^{n+1/2}$ at the considered element integration point is readily evaluated. Notice that Step 1 only involves element-based computations and thus avoids completely the need for the assembly of element contributions.

Step 2: It remains to compute the end-of-step values u^{n+1} by spatial discretization of the weak form (4.12). This results in an algebraic system with a mass matrix.

Remark 4.1 (Flux representation). An issue in the finite element discretization of nonlinear hyperbolic problems is the choice of a local approximation for the nonlinear flux function:

- A first possible option consists of using an elementwise constant representation of the flux:

$$\boldsymbol{f}(u) = \boldsymbol{f}(\bar{u}),$$

where $\bar{u}$ is a mean value of the unknown within the considered element.

- A second option is to interpolate first u at the desired point in the element and then evaluate the flux:

$$\boldsymbol{f}(u) = \boldsymbol{f}\left(\sum_{a=1}^{\mathrm{n_{en}}} N_a(\boldsymbol{x})\, u_a\right).$$

This is used in the one-step Lax–Wendroff method: the flux is evaluated at the Gauss points once the value of the unknown u is interpolated.

- Finally, another possible choice is to interpolate the flux directly by means of the same shape functions used for u:

$$\boldsymbol{f}(u) = \sum_{a=1}^{\mathrm{n_{en}}} N_a(\boldsymbol{x})\boldsymbol{f}(u_a).$$

This option is called *group representation* and is often used in Step 1 of the two-step Lax–Wendroff method to compute the divergence of the flux within an element. In this case, the group representation of the nonlinear flux function presents computational advantages over the other interpolation methods. It simplifies the evaluation of the flux divergence because the spatial dependency of the flux function is described directly by the element shape functions. Note also that on a uniform mesh of linear elements, the group representation provides a fourth-order accurate representation of the spatial derivative of the flux function (Donea and Giuliani, 1981).

To appraise the performance of the various flux representations, the reader is referred to the solved exercises in Section 4.8.

4.3 THE EULER EQUATIONS

4.3.1 Strong form of the conservation equations

The Euler equations of gas dynamics considered here express conservation of mass, momentum and energy in a compressible, inviscid and non-conducting fluid. They were developed in Section 1.4 both in differential and integral form. Recall the strong form of mass (1.11), momentum (1.16) and energy (1.21) conservation

$$\begin{aligned}\frac{\partial \rho}{\partial t} + \boldsymbol{\nabla}\cdot(\rho\boldsymbol{v}) &= 0\\ \frac{\partial \rho\boldsymbol{v}}{\partial t} + \boldsymbol{\nabla}\cdot(\rho\,\boldsymbol{v}\otimes\boldsymbol{v} + p\,\mathbf{I}) &= \rho\boldsymbol{b}\\ \frac{\partial \rho E}{\partial t} + \boldsymbol{\nabla}\cdot\big((\rho E + p)\boldsymbol{v}\big) &= \boldsymbol{v}\cdot\rho\boldsymbol{b}\end{aligned} \tag{4.13}$$

where the Cauchy stress $\boldsymbol{\sigma}$ has been replaced by $-p\,\mathbf{I}$ because we assume an inviscid fluid (constitutive law), and the rest of the variables were already defined in Chapter 1: the density ρ, the momentum $\rho\boldsymbol{v}$, the total energy ρE per unit volume of the fluid, the fluid pressure p and the external volume force per unit volume $\rho\boldsymbol{b}$. In vector form and in 3D ($\mathrm{n_{sd}} = 3$), these equations can be rewritten as

$$\mathbf{U}_t + \frac{\partial \mathbf{F}_1}{\partial x_1} + \frac{\partial \mathbf{F}_2}{\partial x_2} + \frac{\partial \mathbf{F}_3}{\partial x_3} = \mathbf{B}, \tag{4.14}$$

where $\mathbf{U}$ is the vector of conservation variables, $\mathbf{F}_i$ are the associated flux vectors for each spatial dimension and $\mathbf{B}$ is a source term. They are defined as follows:

$$\mathbf{U} = \begin{pmatrix} \rho \\ \rho \boldsymbol{v} \\ \rho E \end{pmatrix}, \quad \mathbf{F}_i = \begin{pmatrix} \rho\, v_i \\ \rho \boldsymbol{v}\, v_i + p \\ (\rho E + p)\, v_i \end{pmatrix}, \quad i = 1, \ldots, \mathrm{n_{sd}}, \text{ and } \mathbf{B} = \begin{pmatrix} 0 \\ \rho \boldsymbol{b} \\ \boldsymbol{v} \cdot \rho \boldsymbol{b} \end{pmatrix}. \quad (4.15)$$

The previous differential equation can be compacted further using the divergence of the flux vector

$$\mathbf{F} = \begin{pmatrix} \mathbf{F}_1 \\ \mathbf{F}_2 \\ \mathbf{F}_3 \end{pmatrix},$$

and finally, the Euler conservation equations, boundary conditions and initial condition induce the strong form of the problem:

$$\mathbf{U}_t + \boldsymbol{\nabla} \cdot \mathbf{F} = \mathbf{B} \qquad \text{in } \Omega \times]0, T[, \qquad (4.16a)$$

$$\mathbf{F}^{in} \cdot \boldsymbol{n} = \mathbf{G} \qquad \text{on } \Gamma \times]0, T[, \qquad (4.16b)$$

$$\mathbf{U}(\boldsymbol{x}, 0) = \mathbf{U}_0(\boldsymbol{x}) \qquad \text{on } \Omega \text{ at } t = 0. \qquad (4.16c)$$

In the previous chapter we have already seen that boundary conditions for scalar hyperbolic problems are only imposed on the inflow portion of the boundary. For systems of hyperbolic equations only inflow *components* of the flux vector can be prescribed. This is discussed in detail in Section 4.3.4.

Boundary conditions for this problem will be discussed in detail in Section 4.3.4. Before then, we need to define the equation of state to complete the previous system of conservation equations. The equation of state relates the internal energy, $e = E - (1/2)\|\boldsymbol{v}\|^2$, see Section 1.4.4, to pressure, p, and density, ρ. For an ideal gas, internal energy is only a function of temperature. In fact, for a perfect polytropic gas the internal energy per unit mass is proportional to temperature: $e = c_v T$ where c_v is constant and known as the specific heat at constant volume. Thus the state equation relating e to p and ρ is obtained using the ideal gas law

$$p = R\rho T = \frac{R}{c_v}\rho\, e = \frac{R}{c_v}\rho\Big(E - \frac{1}{2}\|\boldsymbol{v}\|^2\Big)$$

where R is the gas constant per unit mass, which is equal to the universal gas constant $\mathcal{R}$ divided by the molecular mass of the fluid.

Other expressions of the equation of state are also common. For instance, for a polytropic gas the total enthalpy of the fluid $H = E + p/\rho$ is also proportional to temperature: $H = c_p T$ where c_p is constant and known as the specific heat at constant pressure. The ratio $\gamma = c_p/c_v$ of the specific heat coefficients can relate the specific heat constants and the gas constant per unit mass, R,

$$c_p = \frac{\gamma}{\gamma - 1} R \qquad \text{and} \qquad c_v = \frac{R}{\gamma - 1}.$$

With the help of this new constant γ the usual form of the equation of state is obtained

$$E = \frac{p}{\rho(\gamma - 1)} + \frac{1}{2}\|\boldsymbol{v}\|^2 \tag{4.17a}$$

or in terms of the total enthalpy of the fluid

$$H = E + \frac{p}{\rho} = \frac{\gamma p}{\rho(\gamma - 1)} + \frac{1}{2}\|\boldsymbol{v}\|^2. \tag{4.17b}$$

The speed of sound, c, is also common in this formulation. It enters in the definition of the Mach number

$$M = \frac{\|\boldsymbol{v}\|}{c},$$

and it is given by

$$c = \sqrt{\frac{\gamma p}{\rho}}.$$

Thus using the state equations (4.17) the speed of sound can also be written as

$$c^2 = \gamma(\gamma - 1)\Big(E - \frac{1}{2}\|\boldsymbol{v}\|^2\Big) \quad \text{or} \quad c^2 = (\gamma - 1)\Big(H - \frac{1}{2}\|\boldsymbol{v}\|^2\Big).$$

Remark 4.2 (Non-conservative form). The Euler equations can be written in various equivalent forms, depending on the choice of the dependent flow variables. For instance, use can be made of the primitive variables, ρ, $\boldsymbol{v}$, p. In this case, equations (4.13) and (4.16) transform to the following non-conservative form, see for example Hirsch (1990, Chap. 16) for details:

$$\begin{aligned}
\rho_t + \boldsymbol{v}\cdot\boldsymbol{\nabla}\rho + \rho\,\boldsymbol{\nabla}\cdot\boldsymbol{v} &= 0,\\
\boldsymbol{v}_t + (\boldsymbol{v}\cdot\boldsymbol{\nabla})\boldsymbol{v} + \frac{1}{\rho}\,\boldsymbol{\nabla}p &= \boldsymbol{b},\\
p_t + \boldsymbol{v}\cdot\boldsymbol{\nabla}p + \rho\, c^2\,\boldsymbol{\nabla}\cdot\boldsymbol{v} &= 0,
\end{aligned}$$

where c is the speed of sound. It is important to note that the conservation form of the Euler equations is preferred in order to correctly compute the propagation speed and the intensity of flow discontinuities.

4.3.2 The quasi-linear form of the Euler equations

In order to investigate the basic properties of the Euler equations (see for instance Hirsch, 1990) it is necessary to write these equations in a quasi-linear form. This is similar in structure to the convection equations studied in Chapter 3. In the case of a perfect polytropic gas or, more generally, for fluids satisfying the relation $p = \rho\, f(e)$, the components $\mathbf{F}_i$ of the inviscid flux vector $\mathbf{F}$, see (4.15), are *homogeneous functions of degree 1*, see Remark 4.3, in the conservation variables $\mathbf{U}$. Therefore, the flux components can be written as

$$\mathbf{F}_i(\mathbf{U}) = \mathbf{A}_i(\mathbf{U})\,\mathbf{U}, \tag{4.18}$$

where $\mathbf{A}_i(\mathbf{U}) = \partial \mathbf{F}_i / \partial \mathbf{U}$, with $i = 1, \ldots, \mathrm{n_{sd}}$, are the Jacobian matrices. Thus, from (4.14), the quasi-linear form of the Euler equations is given by

$$\mathbf{U}_t + \mathbf{A}_1 \frac{\partial \mathbf{U}}{\partial x_1} + \mathbf{A}_2 \frac{\partial \mathbf{U}}{\partial x_2} + \mathbf{A}_3 \frac{\partial \mathbf{U}}{\partial x_3} = \mathbf{B},$$

or in compact form

$$\mathbf{U}_t + (\mathbf{A} \cdot \boldsymbol{\nabla})\, \mathbf{U} = \mathbf{B}. \tag{4.19}$$

Remark 4.3 (Homogeneous functions). Let f be a scalar function differentiable in an open set $S \subset \mathbb{R}^n$. f is a homogeneous function of degree p in S if

$$f(t\boldsymbol{x}) = t^p f(\boldsymbol{x}) \qquad \forall t > 0 \text{ and } \forall \boldsymbol{x} \in S.$$

Moreover, the so-called Euler theorem for homogeneous functions claims that a homogeneous function of degree p verifies

$$p\, f(\boldsymbol{x}) = \boldsymbol{x} \cdot \boldsymbol{\nabla} f(\boldsymbol{x}).$$

Remark 4.4 (Euler equations in 2D). In 2D, the vector $\mathbf{U}$ of conservative variables and the associated flux vector $\mathbf{F}$ reduce to

$$\mathbf{U} = \begin{pmatrix} \rho \\ \rho v_1 \\ \rho v_2 \\ \rho E \end{pmatrix}, \quad \mathbf{F}_1 = \begin{pmatrix} \rho v_1 \\ \rho v_1^2 + p \\ \rho v_1 v_2 + p \\ H \rho v_1 \end{pmatrix} \quad \text{and} \quad \mathbf{F}_2 = \begin{pmatrix} \rho v_2 \\ \rho v_1 v_2 + p \\ \rho v_2^2 + p \\ H \rho v_2 \end{pmatrix}.$$

In 2D the Jacobian matrices for a polytropic ideal gas are

$$\mathbf{A}_1 = \begin{pmatrix} 0 & 1 & 0 & 0 \\ \frac{\gamma-1}{2}\|\boldsymbol{v}\|^2 - v_1^2 & (3-\gamma) v_1 & (1-\gamma) v_2 & \gamma - 1 \\ -v_1 v_2 & v_2 & v_1 & 0 \\ \frac{\gamma-1}{2} v_1 \|\boldsymbol{v}\|^2 - v_1 H & -v_1^2(\gamma-1) + H & (1-\gamma) v_1 v_2 & \gamma v_1 \end{pmatrix},$$

$$\mathbf{A}_2 = \begin{pmatrix} 0 & 0 & 1 & 0 \\ -v_1 v_2 & v_2 & v_1 & 0 \\ \frac{\gamma-1}{2}\|\boldsymbol{v}\|^2 - v_2^2 & (1-\gamma) v_1 & (3-\gamma) v_2 & \gamma - 1 \\ \frac{\gamma-1}{2} v_2 \|\boldsymbol{v}\|^2 - v_2 H & (1-\gamma) v_1 v_2 & -v_2^2(\gamma-1) + H & \gamma v_1 \end{pmatrix}.$$

Remark 4.5 (Euler equations in 1D). In 1D, the vector $\mathbf{U}$ of conservative variables and the associated flux vector $\mathbf{F}$ reduce to

$$\mathbf{U} = \begin{pmatrix} \rho \\ \rho v \\ \rho E \end{pmatrix} \quad \text{and} \quad \mathbf{F} = \begin{pmatrix} \rho v \\ \rho v^2 + p \\ \rho v H \end{pmatrix}.$$

The corresponding Jacobian matrix $\mathbf{A} = \partial \mathbf{F} / \partial \mathbf{U}$ takes the form

$$\mathbf{A} = \begin{pmatrix} 0 & 1 & 0 \\ -\frac{1}{2}(3-\gamma)\, v^2 & (3-\gamma)\, v & \gamma - 1 \\ (\gamma-1)\, v^3 - \gamma\, v\, E & \gamma E - \frac{3}{2}(\gamma-1)\, v^2 & \gamma v \end{pmatrix}.$$

4.3.3 Basic properties of the Euler equations

The Euler equations form a hyperbolic system in time and the solution of the homogeneous quasi-linear system (4.19),

$$\mathbf{U}_t + (\mathbf{A} \cdot \boldsymbol{\nabla})\mathbf{U} = 0,$$

can be expressed in Fourier series. A typical term (mode) of the series has the wave-like form

$$\mathbf{U} = \hat{\mathbf{U}} e^{i(\mathbf{k} \cdot \boldsymbol{x} - \omega t)}.$$

Here, $\mathbf{k} = (k_1, \ldots, k_{\mathrm{n_{sd}}})^T$ is the wave-number vector and ω represents (up to a factor 2π) the frequency of the propagating wave. Note that $(\mathbf{k} \cdot \boldsymbol{x} - \omega t)$ represents the phase of the wave propagating in the direction $\mathbf{k}$ with frequency ω. Introducing the previous single mode solution into the homogeneous equation leads to the following condition for the existence of a wave-like solution:

$$\det\big(\omega \mathbf{I} - \mathbf{A} \cdot \mathbf{k}\big) = 0,$$

where $\mathbf{I}$ is the identity matrix of order $\mathrm{n_{sd}}$. Thus, wave-like solutions will exist if the eigenvalues λ_j, $j = 1, \ldots, \mathrm{n_{sd}} + 2$, of the projection matrix $\mathbf{A_k} = \mathbf{A} \cdot \mathbf{k}$ are real with linear independence of the corresponding right eigenvectors. These eigenvalues represent the frequency ω. Let $\mathbf{R}$ be the matrix whose columns are the right eigenvectors of $\mathbf{A_k}$, that is

$$\mathbf{A_k R} = \mathbf{R \Lambda}, \tag{4.20}$$

where $\mathbf{\Lambda} := \mathrm{diag}(\lambda_j), j = 1, \ldots, \mathrm{n_{sd}}+2$, is a diagonal matrix, such that the diagonal elements are the eigenvalues λ_j of $\mathbf{A_k}$. These eigenvalues are associated with an arbitrary direction of propagation $\mathbf{k}$, $\lambda_j(\mathbf{k})$, and can be evaluated in terms of the fluid velocity $\boldsymbol{v}$ and the sound speed c as

$$\begin{aligned} \lambda_1 = \cdots = \lambda_{\mathrm{n_{sd}}} &= \boldsymbol{v} \cdot \mathbf{k}, \\ \lambda_{\mathrm{n_{sd}}+1} &= \boldsymbol{v} \cdot \mathbf{k} + c, \\ \lambda_{\mathrm{n_{sd}}+2} &= \boldsymbol{v} \cdot \mathbf{k} - c. \end{aligned}$$

Note that the characteristics associated with the eigenvalue $\boldsymbol{v} \cdot \mathbf{k} = \sum_{s=1}^{\mathrm{n_{sd}}} v_s k_s$ of multiplicity $\mathrm{n_{sd}}$ represent the trajectories of the particles. Finally, we can consider the decomposition of the projection Jacobian matrices along the direction of the propagation $\mathbf{k}$

$$\mathbf{A_k} = \sum_{s=1}^{\mathrm{n_{sd}}} k_s \mathbf{A}_s = \mathbf{R \Lambda R}^{-1}. \tag{4.21}$$

It is important to note that this equation shows that a projection of the Jacobian matrix $\mathbf{A}$ along an arbitrary direction is diagonalizable, not the Jacobian matrix itself. This has a major consequence in the diagonalization of the Euler equations. In fact, the Euler equations can no longer be diagonalized in 2D and 3D. This is because

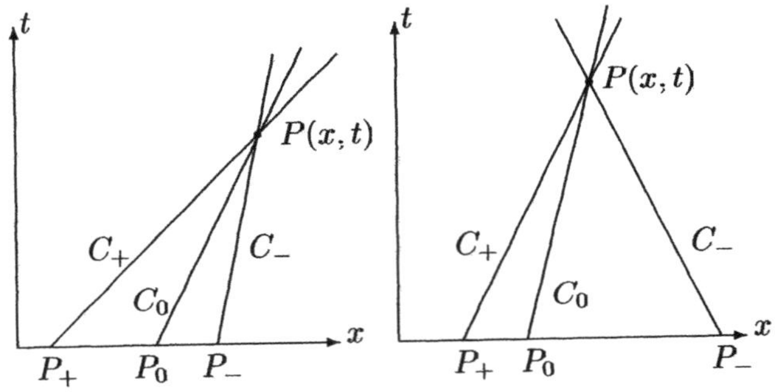

Fig. 4.7 Illustration of characteristics at point $P(x,t)$ for supersonic (left) and subsonic (right) flow conditions.

waves can travel in an infinite number of directions and the decomposition into scalar waves is no longer unique. In *one dimension*, see Remark 4.5, we may multiply the system (4.19) in quasi-linear form by matrix $\mathbf{R}^{-1}$ to obtain

$$\mathbf{R}^{-1}\Big(\mathbf{U}_t + \mathbf{A} \cdot \mathbf{U}_x\Big) = \mathbf{R}^{-1}\mathbf{B}.$$

Then, introducing the *characteristic variables* $\mathbf{W} = \mathbf{R}^{-1}\mathbf{U}$ and under the assumption that the coefficients of matrix $\mathbf{R}$ are constant, the previous equation becomes:

$$\mathbf{W}_t + \mathbf{\Lambda}\mathbf{W}_x = \mathbf{R}^{-1}\mathbf{B}. \tag{4.22}$$

Unfortunately, the characteristic variables can only be defined for the 1D Euler equations and not for the more general case of multidimensional flows. The characteristic variables, also called *Riemann variables*, are (for isentropic flows of a polytropic ideal gas)

$$\mathbf{W} = \left(s\,,\, v + \frac{2}{\gamma - 1}c\,,\, v - \frac{2}{\gamma - 1}c\right)^T, \tag{4.23}$$

where s is the entropy per unit mass of fluid, see Remark 4.6. The associated eigenvalues are

$$\mathbf{\Lambda} = \text{diag}(v, v + c, v - c), \tag{4.24}$$

and they define the three characteristics, see Figure 4.7:

$$\begin{aligned} \frac{dx}{dt} &= v && \text{on } C_0; \\ \frac{dx}{dt} &= v + c && \text{on } C_+; \\ \frac{dx}{dt} &= v - c && \text{on } C_-. \end{aligned} \tag{4.25}$$

From (4.22) the 1D homogeneous Euler equations are

$$\left(\frac{\partial}{\partial t} + v\frac{\partial}{\partial x}\right)s = 0,$$
$$\left(\frac{\partial}{\partial t} + (v+c)\frac{\partial}{\partial x}\right)\left(v + \frac{2c}{\gamma - 1}\right) = 0,$$
$$\left(\frac{\partial}{\partial t} + (v-c)\frac{\partial}{\partial x}\right)\left(v - \frac{2c}{\gamma - 1}\right) = 0.$$

This allows us to observe that the material derivative of s is zero, that is the specific entropy is constant along the particle trajectories in the smooth part of the flow (away from discontinuities). In fact, the entropy s is transported at speed v along the straight characteristic C_0 of equation $dx/dt = v$, which coincides with the particle path. The quantities $v \pm 2c/(\gamma - 1)$ are transported at speed $v \pm c$ along the characteristic lines $C_\pm$ illustrated in Figure 4.7.

It is important to note that, in contrast to the scalar case, the characteristics associated with the Euler equations are not in general straight lines and do not transport constant values of the unknowns. The above 1D system is an exception, because it is homogeneous (the r.h.s. is zero), and the diagonalization of the system induces that each l.h.s. expresses the vanishing of the material derivative of the transported quantity.

These properties, which are restricted to the 1D problem, are useful for the specification of suitable boundary conditions in multiple dimensions. Moreover, information on the sign of the eigenvalues $\mathbf{\Lambda}$, which represent the propagation speeds, will be exploited in Section 4.4.2 to derive directional (upwind-type) spatial discretization models for the Euler equations.

Remark 4.6 (Entropy per unit mass). This is another fundamental thermodynamic variable, which can be interpreted as a measure of the disorder in the system. The entropy per unit mass, s, is defined from a reference state (up to a constant) as

$$s - s_{\text{ref}} = c_v \ln \frac{p/p_{\text{ref}}}{(\rho/\rho_{\text{ref}})^\gamma} = c_p \ln \frac{T}{T_{\text{ref}}} - R \ln \frac{p}{p_{\text{ref}}}.$$

4.3.4 Boundary conditions

The issue of deciding which boundary conditions to impose on the Euler equations is not a trivial one; an inadequate choice can affect the existence and uniqueness of solutions.

For instance, at a *solid boundary*, only the normal component of the velocity must be specified equal to the solid velocity normal to the fluid–solid interface. Consequently, in the case of a fixed solid boundary, the normal component v_n of the fluid velocity is set to zero. Moreover, in the absence of thermal conduction, the energy flux across the boundary vanishes. The same applies to the flux of mass. Thus, neither the density ρ nor the total energy ρE must be specified at a solid boundary. The conditions of zero normal flux for both mass and energy are automatically satisfied in the finite element context, because of the natural boundary conditions on prescribed normal fluxes.

In general, the computational domain is delimited by an *external boundary*. A linearized Riemann analysis in the direction of the outward normal to the contour is required to determine the speed of wave propagation. As indicated in the previous section, the Euler equations can be diagonalized in 1D; the outward normal to the boundary is this dimension. Then, three distinct values of the propagation speed are obtained from the linearization of the Euler equations, namely

$$\begin{aligned} \lambda_1 = \cdots = \lambda_{\mathrm{n_{sd}}} &= v_n; \\ \lambda_{\mathrm{n_{sd}}+1} &= v_n + c, \\ \lambda_{\mathrm{n_{sd}}+2} &= v_n - c, \end{aligned} \tag{4.26}$$

where $v_n = \boldsymbol{v} \cdot \boldsymbol{n}$ is the velocity along the outward normal. Then, at each point of the boundary of the computational domain one should prescribe as many conditions as there are negative eigenvalues (incoming information) in (4.26).

Two distinct possibilities must be considered in the case of *supersonic flow*, that is when the local Mach number is such that $M = (|v_n|/c) \geq 1$.

1. *Supersonic inflow boundaries,* where $v_n < 0$ and $|v_n| > c$: for such boundaries all components of vector $\mathbf{U}$ defined in (4.15) must be specified because all eigenvalues (4.26) are negative.

2. *Supersonic outflow boundaries* where $v_n > 0$ and $v_n > c$: here, all eigenvalues are positive and no component of vector $\mathbf{U}$ can be prescribed.

For *subsonic boundaries*, $M = (|v_n|/c) < 1$, the situation is more complex and the components of $\mathbf{U}$ that can be specified are those of the incoming Riemann variables. Here we also distinguish two cases:

1. *Subsonic inflow boundaries*, where $v_n < 0$ and $|v_n| < c$: only the eigenvalue $\lambda_{\mathrm{n_{sd}}+1}$ is not negative, and $\mathrm{n_{sd}} + 1$ appropriate conditions can be imposed.

2. *Subsonic outflow boundaries*, where $v_n > 0$ and $v_n < c$: only one condition (for instance, the pressure) can be prescribed in this case, because only the eigenvalue $\lambda_{\mathrm{n_{sd}}+2}$ is negative.

4.4 SPATIAL DISCRETIZATION TECHNIQUES

Various options are available for the finite element spatial discretization of the Euler equations. This comprises several possible choices for the weighted residual formulation, as well as various options for the interpolation of the field variables. As an alternative to the standard Galerkin formulation, formulations of upwind-type are available which possess good stability properties and produce a directional discretization in accordance with the physical behavior of inviscid flows. Finite element models of upwind type include stabilized formulations, such as SUPG and GLS, as well as discontinuous Galerkin formulations combined with flux vector splitting or other devices such as approximate Riemann solvers. We start by introducing the standard (continuous) Galerkin formulation of the Euler equations, and then discuss ways of producing more stable upwind-type discretizations.

4.4.1 Galerkin formulation

To perform the classical Galerkin spatial discretization of the Euler equations (4.16) we need to generalize the functional spaces defined in Section 3.4.2 for the scalar case. For clarity of exposition we assume that no Dirichlet boundary conditions are prescribed at the inflow boundaries, and as indicated by equation (4.16b), the boundary conditions are specified in terms of prescribed normal flux. The test functions, $\mathbf{W}$, belong to $\boldsymbol{\mathcal{V}} := \boldsymbol{\mathcal{H}}^1(\Omega)$ (space of vector functions, see Section 1.5.1.2) and do not depend on time, and the space of trial functions, $\boldsymbol{\mathcal{S}}_t$, varies as a function of t

$$\boldsymbol{\mathcal{S}}_t := \left\{\mathbf{U} \mid \mathbf{U}(\cdot,t) \in \boldsymbol{\mathcal{H}}^1(\Omega), t \in [0,T]\right\}.$$

The Galerkin spatial discretization of the Euler equations (4.16) can then be stated as follows: for $t \in]0,T[$, find $\mathbf{U}(\boldsymbol{x},t) \in \boldsymbol{\mathcal{S}}_t$ such that $\mathbf{U}(\boldsymbol{x},0) = \mathbf{U}_0(\boldsymbol{x})$ and

$$\int_\Omega \mathbf{W}^T\,\mathbf{U}_t\,d\Omega - \int_\Omega \boldsymbol{\nabla}\mathbf{W}^T\,\mathbf{F}(\mathbf{U})\,d\Omega = \int_\Omega \mathbf{W}^T\,\mathbf{B}\,d\Omega - \int_\Gamma \mathbf{W}^T\,\mathbf{G}\,d\Gamma$$

for all test functions $\mathbf{W} \in \boldsymbol{\mathcal{V}}$. In compact form, this equation reads

$$(\mathbf{W},\mathbf{U}_t) - (\boldsymbol{\nabla}\mathbf{W},\mathbf{F}) = (\mathbf{W},\mathbf{B}) - (\mathbf{W},\mathbf{G})_\Gamma. \tag{4.27}$$

The conservation variables in vector $\mathbf{U}$ are approximated spatially within an element using standard finite element shape functions:

$$\rho_e = \mathbf{N}_\rho^T\,\widetilde{\boldsymbol{\rho}}, \qquad (\rho\boldsymbol{v})_e = \mathbf{N}_{\rho v}^T\,\widetilde{\boldsymbol{\rho v}}, \qquad (\rho E)_e = \mathbf{N}_{\rho E}^T\,\widetilde{\boldsymbol{\rho E}},$$

where vectors $\widetilde{\boldsymbol{\rho}}$, $\widetilde{\boldsymbol{\rho v}}$ and $\widetilde{\boldsymbol{\rho E}}$ list the values of the conservation variables at their respective nodal points and $\mathbf{N}$ denotes the associated shape functions. Note that each conservation variable can be interpolated differently.

A common practice in compressible flow problems is to use low-order approximations. In fact, high-order polynomial approximations are not optimal in zones where the solution presents steep gradients. Moreover, the polynomial approximation for the density and the energy is often taken at one order less than for the momentum components. Note, however, that, in contrast to the incompressible case discussed in Chapter 6, this is not indispensable for compressible flows. For instance, when using primitive variables a common choice is constant element-by-element approximations for the density and the specific internal energy (and hence the pressure), and piecewise linear/multilinear representations for the velocity.

As seen in Chapters 2 and 3, the Galerkin approach to the spatial discretization of highly convective problems is often characterized by a lack of sufficient stability.

This is also the case for the Euler system of hyperbolic equations. Finite element formulations of the upwind type are therefore better adapted for the spatial discretization of the Euler equations. Such formulations are presented in the next section.

4.4.2 Upwind-type discretizations

Centered spatial discretization techniques, such as the Galerkin finite element method, do not account for the physical propagation properties of the solutions of the Euler equations. That is, the propagation of perturbations along the characteristics is not taken into account. For this reason, finite element formulations of upwind type are generally preferred for the Euler equations. As already seen in Chapters 2 and 3 for scalar convection problems, spatial discretization schemes of streamline-upwind type are able to produce directional discretizations in accordance with the physical behavior of the solution of convective transport problems. In the case of the Euler equations, the situation is more complex than in the scalar case, because there are generally mixed sign eigenvalues in the coupled system of conservation equations.

Three basic classes of upwind-type methods have been developed to produce a directional spatial discretization of the Euler equations in accordance with the physical behavior of inviscid compressible flows. Note, however, that standard upwind schemes are only first-order accurate in space and therefore generally too dissipative. They have to be corrected in the form of high-resolution methods to achieve, as indicated in Section 4.5.3, higher-order accuracy away from discontinuities.

A first family of upwind schemes is inspired by Godunov's (1959) method which solves the locally 1D Euler equations for discontinuous neighboring states (the Riemann problem in Figure 4.4). This approach introduces in the discretization scheme information from the exact local solution of the nonlinear Euler equations. It has generated many variants which are based on *approximate Riemann solvers*. A popular method in this class is Roe's (1981) approximate Riemann solver in which a constant coefficient linear system of conservation laws is considered instead of the original nonlinear system. The reader interested in an in-depth discussion of approximate Riemann solvers is urged to consult specialized textbooks, such as, for instance, Hirsch (1990) and LeVeque (1992).

A second family of upwind-type methods are the so-called *flux vector splitting* methods introduced in the finite difference context by Steger and Warming (1981) and van Leer (1982). These methods use information on the sign of the eigenvalues of the Jacobian matrices to split the flux terms in the Euler equations and discretize them directionally according to the sign of the associated propagation speeds. We shall describe the flux vector splitting technique in the next section and then illustrate its use in conjunction with the discontinuous Galerkin method.

A third family of upwind-type finite element methods for the Euler equations includes the SUPG and GLS methods introduced in Chapter 2. Sections 4.4.2.3 and 4.4.2.4 provide an introduction to such stabilization methods in the context of the Euler equations.

4.4.2.1 *Flux vector splitting technique* The rationale of flux vector splitting techniques is presented for a 1D problem, and then extended to multidimensional situations. The Euler equations have in general positive and negative eigenvalues, which correspond to the inflow and outflow components. To produce an upwind-type discretization, the idea is to decompose the flux vector $\mathbf{F}$ into the sum

$$\mathbf{F} = \mathbf{F}^+ + \mathbf{F}^-,$$

such that $\partial\mathbf{F}^+/\partial\mathbf{U}$ has only positive eigenvalues and $\partial\mathbf{F}^-/\partial\mathbf{U}$ has only negative eigenvalues. Hence, the spatial terms in a 1D problem

$$\mathbf{U}_t + \mathbf{F}^+_x + \mathbf{F}^-_x = 0$$

can be discretized directionally according to the sign of the eigenvalues by expressions such as

$$\frac{\partial\mathbf{F}^+}{\partial x}(x_i) \approx \frac{\mathbf{F}^+_i - \mathbf{F}^+_{i-1}}{h} \qquad \text{and} \qquad \frac{\partial\mathbf{F}^-}{\partial x}(x_i) \approx \frac{\mathbf{F}^-_{i+1} - \mathbf{F}^-_i}{h}.$$

The splitting of the flux vector is defined using (4.18), that is $\mathbf{F} = \mathbf{A}\,\mathbf{U}$, as

$$\mathbf{F} = \mathbf{F}^+ + \mathbf{F}^-, \qquad \mathbf{F}^+ = \mathbf{A}^+\,\mathbf{U}, \qquad \mathbf{F}^- = \mathbf{A}^-\,\mathbf{U}.$$

Therefore we need to split the Jacobian matrix $\mathbf{A}$ into $\mathbf{A} = \mathbf{A}^+ + \mathbf{A}^-$. To define these matrices we recall the diagonalization of the 1D projection of the Jacobian matrix $\mathbf{A}$, see Section 4.3.3 and equation (4.21), that is $\mathbf{A} = \mathbf{R}\boldsymbol{\Lambda}\mathbf{R}^{-1}$, where, for the 1D Euler equations, $\boldsymbol{\Lambda}$ is defined in (4.24) and $\mathbf{R}$ is the matrix whose columns are the right eigenvectors of $\mathbf{A}$, see (4.20). The splitting of the Jacobian matrix becomes obvious from the splitting of the diagonal matrix of eigenvalues $\boldsymbol{\Lambda}$, that is

$$\boldsymbol{\Lambda} = \boldsymbol{\Lambda}^+ + \boldsymbol{\Lambda}^- \text{ induces: } \mathbf{A}^+ = \mathbf{R}\boldsymbol{\Lambda}^+\mathbf{R}^{-1}, \text{ and } \mathbf{A}^- = \mathbf{R}\boldsymbol{\Lambda}^-\mathbf{R}^{-1},$$

where $\boldsymbol{\Lambda}^+$ and $\boldsymbol{\Lambda}^-$ have the positive and negative eigenvalues, respectively,

$$\begin{aligned}\boldsymbol{\Lambda}^+ &= \text{diag}\left(\frac{(v-c)+|v-c|}{2}, \frac{v+|v|}{2}, \frac{(v+c)+|v+c|}{2}\right),\\ \boldsymbol{\Lambda}^- &= \text{diag}\left(\frac{(v-c)-|v-c|}{2}, \frac{v-|v|}{2}, \frac{(v+c)-|v+c|}{2}\right).\end{aligned} \tag{4.28}$$

Remark 4.7. In this 1D case, each of the three characteristic variables (4.23) can be integrated separately, and a scalar upwind scheme can then be used for each one:

$$\frac{u_i^{n+1} - u_i^n}{\Delta t} = -\frac{1}{h}\left(a^+(u_i^n - u_{i-1}^n) + a^-(u_{i+1}^n - u_i^n)\right),$$

with $a^+ = \max(0,a) = (a+|a|)/2$, $a^- = \min(a,0) = (a-|a|)/2$, and where it suffices to replace the convection speeds a^+ and a^- with the positive and negative projections of the eigenvalues defined in (4.28). Note that u in the previous equation must be replaced by the Riemann variables.

This final decomposition allows us to compute $\mathbf{A}^+$ and $\mathbf{A}^-$ and consequently $\mathbf{F}^+$ and $\mathbf{F}^-$. Then, as noted previously, the numerical scheme can account for inflow or outflow fluxes. This, however, is only valid for a 1D problem. The generalization to higher spatial dimension is not trivial. As seen in Section 4.3.3 the Euler equations cannot be diagonalized in 2D or 3D. Consequently, in higher dimensions the 1D concept is used restricting the splitting of the flux to the normal direction. The normal flux to an element side (face in 3D) with outward normal $\boldsymbol{n}$ is

$$\mathbf{F}_{\boldsymbol{n}}(\mathbf{U}) = \sum_{i=1}^{n_{sd}} \mathbf{F}_i(\mathbf{U})\, n_i.$$

Consequently, the upwind characterization leading to the splitting of the normal flux is based on the signs of the eigenvalues of the projected Jacobian matrix

$$\mathbf{A}_{\boldsymbol{n}}(\mathbf{U}) = \frac{\partial \mathbf{F}_{\boldsymbol{n}}(\mathbf{U})}{\partial \mathbf{U}} = \sum_{i=1}^{n_{sd}} \mathbf{A}_i(\mathbf{U})\, n_i = (\mathbf{A}\cdot\boldsymbol{n}).$$

The normal flux $\mathbf{F}_{\boldsymbol{n}}(\mathbf{U})$ can then be split, see for instance Figure 4.8, into inflow and outflow components $\mathbf{F}_{\boldsymbol{n}}^-$ and $\mathbf{F}_{\boldsymbol{n}}^+$, respectively, which are defined by

$$\mathbf{F}_{\boldsymbol{n}}^{\pm}(\mathbf{U}) = \mathbf{A}_{\boldsymbol{n}}^{\pm}\,\mathbf{U} = \left(\mathbf{R}\,\boldsymbol{\Lambda}^{\pm}\,\mathbf{R}^{-1}\right)\mathbf{U}, \text{ with } \boldsymbol{\Lambda}^{\pm} = \frac{1}{2}\left(\boldsymbol{\Lambda}\pm|\boldsymbol{\Lambda}|\right).$$

The splittings $\mathbf{F}_{\boldsymbol{n}}^{\pm}$ are known in closed form for several flux vector splittings, see Hirsch (1990, Chap. 20) for specific cases. In the next section, we show how flux vector splitting techniques ideally combine with the discontinuous Galerkin method for the upwind discretization of the Euler equations. Examples of applications of the flux vector splitting technique in the finite element context are described in Section 4.8.2.

4.4.2.2 Discontinuous Galerkin method with flux vector splitting The solution of the Euler equations is often characterized by discontinuities. An alternative to the classical Galerkin method (continuous in space) is a finite element method where the approximation $\mathbf{U}^h$ and the fluxes $\mathbf{F}(\mathbf{U}^h)$ are allowed to be discontinuous across element boundaries. The *discontinuous Galerkin method* was introduced in Section 3.9. Here we follow the seminal work of Baumann and Oden (2000).

The discontinuous Galerkin method is globally conservative and also elementwise conservative (i.e., the conservation equations are approximately satisfied at the element level) if a conservative formulation is employed, see (4.16). Moreover, the method involves a weak imposition of the Rankine–Hugoniot jump conditions across inter-element and domain boundaries. Thus, in principle, one could easily capture shocks (discontinuities) with such a method. This requires, however, adaptive remeshing in order to align element boundaries and discontinuities.

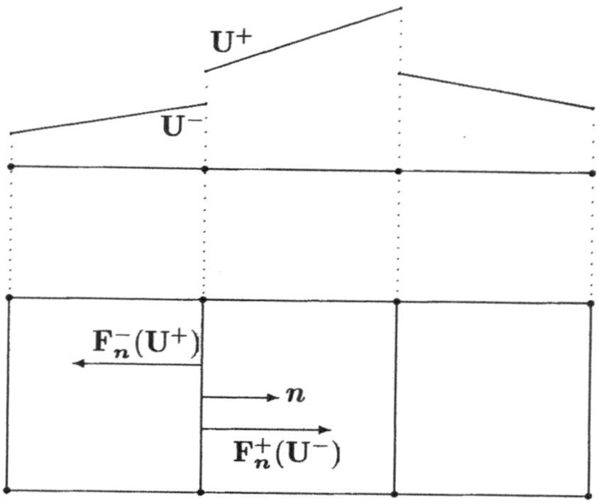

Fig. 4.8 Schematic representation of discontinuous Galerkin with flux vector splitting.

For clarity of exposition we drop the superscript h that denotes the approximation to the solution. Thus, in this section, $\mathbf{U}$ represents the finite element approximation to the conservative variables. The functional spaces needed for the resolution of the Euler equations with the discontinuous Galerkin method are the vector analogues, $\mathcal{V}(\mathcal{T}^h)$ and $\mathcal{S}_t(\mathcal{T}^h)$, of the *broken spaces* introduced in Section 3.9, namely

$$\mathcal{V}(\mathcal{T}^h) = \{\mathbf{W} \in \mathcal{L}_2(\Omega) \mid \mathbf{W}|_{\Omega^e} \in \mathcal{H}^2(\Omega^e)\ \forall \Omega^e \in \mathcal{T}^h(\Omega)\},$$
$$\mathcal{S}_t(\mathcal{T}^h) = \{\mathbf{U} \mid \mathbf{U}(\cdot,t) \in \mathcal{L}_2(\Omega),\ \mathbf{U}(\cdot,t)|_{\Omega^e} \in \mathcal{H}^2(\Omega^e)\, t \in [0,T]\,\forall \Omega^e \in \mathcal{T}^h(\Omega)\}.$$

Note that a partition $\mathcal{T}^h$ of the domain Ω is also required.

As shown in Figure 4.8, the approximation of the conservation variables in the state vector $\mathbf{U}$ is discontinuous across inter-element boundaries. To deal with discontinuous approximations, it is useful to employ the following notation:

$$\mathbf{U}^{\pm} = \lim_{\epsilon \to 0^+} \mathbf{U}(\boldsymbol{x} \pm \epsilon \boldsymbol{n})$$

where $\boldsymbol{x}$ is a point on the boundary and $\boldsymbol{n}$ the outward normal. Likewise, as indicated in Figure 4.9, the normal flux $\mathbf{F}_{\boldsymbol{n}}(\mathbf{U})$ at any point on a boundary with outward normal $\boldsymbol{n}$ is split into an inflow component $\mathbf{F}_{\boldsymbol{n}}^-$ and an outflow component $\mathbf{F}_{\boldsymbol{n}}^+$. With this notation, $\mathbf{F}_{\boldsymbol{n}}^+(\mathbf{U}^-)$ represents the flux of mass, momentum and energy in the direction $\boldsymbol{n}$, while $\mathbf{F}_{\boldsymbol{n}}^-(\mathbf{U}^+)$ is the flux in the opposite direction. Using this notation, the influx boundary condition of the Euler problem, (4.16b), can be rewritten as

$$\mathbf{F}_{\boldsymbol{n}}^-(\mathbf{U}^-) = \mathbf{G} \qquad \text{on } \Gamma \times]0, T[.$$

The space discretization of Euler equations is based on the following weak formulation: given $\mathbf{U}_0(\boldsymbol{x})$ and the normal flux $\mathbf{G}(\boldsymbol{x},t)$ on Γ, find $\mathbf{U}(\boldsymbol{x},t) \in \mathcal{S}_t(\mathcal{T}^h)$, such

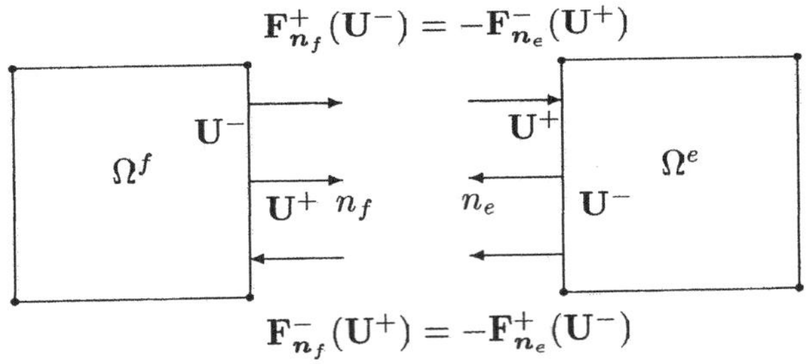

Fig. 4.9 Fluxes at interface between contiguous elements.

that for all weighting functions $\mathbf{W} \in \mathcal{V}(\mathcal{T}^h)$ and every element Ω^e

$$\begin{aligned}(\mathbf{W}, \mathbf{U}_t)_{\Omega^e} - (\nabla \mathbf{W}, \mathbf{F}(\mathbf{U}))_{\Omega^e} + (\mathbf{W}, \mathbf{F}^+_{\boldsymbol{n}_e}(\mathbf{U}^-))_{\partial\Omega^e} \\ + (\mathbf{W}, \mathbf{F}^-_{\boldsymbol{n}_e}(\mathbf{U}^+))_{\partial\Omega^e/\Gamma} = (\mathbf{W}, \mathbf{B})_{\Omega^e} - (\mathbf{W}, \mathbf{G})_{\partial\Omega^e\cap\Gamma};\end{aligned}$$

where

$$\mathbf{F}_{\boldsymbol{n}_e}(\mathbf{U}) = \sum_{i=1}^{\mathrm{n_{sd}}} \mathbf{F}_i(\mathbf{U})\,[\boldsymbol{n}_e]_i.$$

For each flux vector splitting a closed form of $\mathbf{F}^{\pm}_{\boldsymbol{n}}$ is known, see Hirsch (1990, Chap. 20) for specific cases. The first two terms on the l.h.s. of the previous equation and those on the r.h.s. have the same structure as in a standard Galerkin formulation. The last two terms on the l.h.s. represent a jump condition. The jump condition enforces a weak continuity of the normal flux across the inter-element boundaries.

Explicit time-stepping algorithms are very convenient for the time discretization of the semi-discrete equations resulting from the discontinuous Galerkin method. Baumann and Oden (2000) suggest using a discontinuous space–time discretization of the Euler equations with piecewise constant approximation in time. Space–time methods of this type were discussed in Section 3.10 of Chapter 3.

Remark 4.8. The discontinuous Galerkin method makes it very easy to use different spatial approximations in adjacent elements to maximize the accuracy. Note, however, that only low-order approximations (piecewise linear or piecewise constant) should be used near sharp solution gradients. A good resolution of discontinuities can only be obtained through the use of a locally fine enough mesh of low-order elements and, as shown in the worked examples of Section 4.8, through the injection of a suitable amount of artificial viscosity. Higher-order polynomial approximations can be used in the smooth part of the flow to maximize the accuracy.

4.4.2.3 Extension of the SUPG stabilization technique Hughes and coworkers (see Hughes, Tezduyar and Brooks, 1982; Hughes and Tezduyar, 1984) present the first attempts to extend SUPG methods, see Section 2.4.1, to systems of nonlinear hyperbolic conservation laws. A breakthrough in the formulation of stabilized finite element methods for compressible flow problems came with the idea of working with symmetrized conservation laws (Hughes, Franca and Mallet, 1987). The symmetrization is accomplished by a change of variables leading to the formulation of the Euler equations in the entropy variables described in Section 4.3.2.

For practical reasons, however, it is desirable to develop stabilized finite element methods for the Euler equations which instead of entropy variables employ conservation variables. With this choice, the state equations for general gas laws are simpler to establish, the numerical implementation is easier, and a closer similarity exists with traditional finite element formulations. Hansbo (1993) was the first to introduce a formulation of the SUPG method for the Euler equations expressed in conservation variables. Let us briefly illustrate the derivation of the SUPG formulation of the Euler equations and underline the difficulty of constructing the stabilization matrix which plays the role of the stabilization parameter in the scalar convection–diffusion case discussed in Chapter 2.

Consider the initial boundary value problem associated with the Euler equations (4.16). Generalizing the formulation given in Section 2.4.1 of Chapter 2, the SUPG method for the Euler equations consists of adding to the Galerkin variational formulation, see (4.27), an element-by-element contribution depending on the local residual of equation (4.16a). This results in the following problem: for $t \in \,]0,T[$, find $\mathbf{U}(\boldsymbol{x},t) \in \boldsymbol{\mathcal{S}}_t$ for all $\mathbf{W} \in \boldsymbol{\mathcal{V}}$, such that $\mathbf{U}(\boldsymbol{x},0) = \mathbf{U}_0(\boldsymbol{x})$ and

$$(\mathbf{W},\mathbf{U}_t) - (\boldsymbol{\nabla}\mathbf{W},\mathbf{F}) + \sum_e \big((\mathbf{A}\cdot\boldsymbol{\nabla})\mathbf{W}, \boldsymbol{\tau}\mathcal{R}(\mathbf{U})\big)_{\Omega^e} = (\mathbf{W},\mathbf{B}) - (\mathbf{W},\mathbf{G})_\Gamma$$

where $\mathcal{R}(\mathbf{U}) = \mathbf{U}_t + \boldsymbol{\nabla}\cdot\mathbf{F} - \mathbf{B}$ is the local residual of the governing equation. Note that in complete analogy with the scalar convection–diffusion case discussed in Chapter 2, the present stabilized formulation results from the introduction of a stabilizing term governed by the matrix $\boldsymbol{\tau}$. Since

$$\boldsymbol{\nabla}\cdot\mathbf{F} = \sum_{j=1}^{n_{sd}} \frac{\partial \mathbf{F}_j}{\partial x_j} = \sum_{j=1}^{n_{sd}} \frac{\partial \mathbf{F}_j}{\partial \mathbf{U}}\frac{\partial \mathbf{U}}{\partial x_j} = \sum_{j=1}^{n_{sd}} \mathbf{A}_j \frac{\partial \mathbf{U}}{\partial x_j} = (\mathbf{A}\cdot\boldsymbol{\nabla})\mathbf{U},$$

we note that the stabilization term includes the integral

$$\big((\mathbf{A}\cdot\boldsymbol{\nabla})\mathbf{W}, \boldsymbol{\tau}(\mathbf{A}\cdot\boldsymbol{\nabla})\mathbf{U}\big)_{\Omega^e},$$

which is responsible for adding diffusion along the characteristic directions, provided the matrix $\boldsymbol{\tau}$ is properly designed.

The structure of the stabilization matrix $\boldsymbol{\tau}$ is therefore essential in SUPG methods for compressible flow problems. Only in simple cases, such as 1D systems and/or systems amenable to diagonal form, can one find optimal definitions of the coefficients of matrix $\boldsymbol{\tau}$. An appropriate stabilization matrix $\boldsymbol{\tau}$ should be symmetric, positive definite, have dimensions of time, and scale linearly with the element size (no stabilization is needed for a fine enough mesh). In multiple dimensions, matrix $\boldsymbol{\tau}$ should be designed to introduce numerical dissipation along the characteristic directions only and not transversely. Its construction is, in general, not a trivial task, because the Jacobian matrices $\mathbf{A}_i$, $i = 1, \ldots, \mathrm{n_{sd}}$, are not simultaneously diagonalizable in multidimensional Euler equations.

In 1D, the system of Euler equations can be diagonalized as explained in Section 4.3.3. Then, the scalar stabilization parameter τ, see Section 2.4.3, can be used for each individual equation. In pure convection, $\tau = h/(2a)$, see both (2.64) and (2.65). For Euler equations in 1D this definition can be generalized, producing a stabilization matrix $\boldsymbol{\tau}$, if we recall the diagonalization of $\mathbf{A} = \mathbf{R}\boldsymbol{\Lambda}\mathbf{R}^{-1}$, see Section 4.3.3, and basic algebra definitions, such as $\mathbf{A}^p = \mathbf{R}\boldsymbol{\Lambda}^p\mathbf{R}^{-1}$ and $|\mathbf{A}| = \mathbf{R}\,|\boldsymbol{\Lambda}|\,\mathbf{R}^{-1}$,

$$\boldsymbol{\tau} = \frac{h}{2}\left(\mathbf{A}^2\right)^{-\frac{1}{2}} = \frac{h}{2}|\mathbf{A}|^{-1}.$$

We have assumed that all the eigenvalues, see (4.24), are non-zero. If there is any zero eigenvalue it is suppressed, as well as its corresponding eigenvector. This produces a formula similar to the previous one, namely

$$\boldsymbol{\tau} = \frac{h}{2}\,\tilde{\mathbf{R}}\,|\tilde{\boldsymbol{\Lambda}}|^{-1}\tilde{\mathbf{R}}^{-1},$$

where $\tilde{\boldsymbol{\Lambda}}$ only includes the non-zero eigenvalues and $\tilde{\mathbf{R}}$ their corresponding right eigenvectors.

Such a definition of $\boldsymbol{\tau}$ induces, in general, a non-symmetric positive definite matrix ($\mathbf{A}$ is non-symmetric). Hughes and Mallet (1986a) generalize the previous expression to multidimensional situations in the framework of a symmetric system of equations,

$$\boldsymbol{\tau} = \frac{h}{2}\Big(\sum_{j=1}^{\mathrm{n_{sd}}}\mathbf{A}_j^2\Big)^{-\frac{1}{2}}.$$

Furthermore, they extend it to non-regular meshes using

$$\boldsymbol{\tau} = \Big(\sum_{j=1}^{\mathrm{n_{sd}}}\mathbf{B}_j^2\Big)^{-1/2}, \text{ where } \quad \mathbf{B}_j = \sum_{j=1}^{\mathrm{n_{sd}}}\frac{\partial \xi_j}{\partial x_i}\mathbf{A}_i$$

takes into account the parametric mapping $\boldsymbol{x} = \boldsymbol{x}(\boldsymbol{\xi})$ between the actual coordinates, $\boldsymbol{x}$, and the normalized local coordinates, $\boldsymbol{\xi} = (\xi_1, \ldots, \xi_{\mathrm{n_{sd}}})^T$. In practice, the stabilization matrix, $\boldsymbol{\tau}$, is assumed constant in each individual element. If there is any zero eigenvalue the same procedure as before is applied.

Finally, Soulaïmani and Fortin (1994) show that the previous expression is also valid for non-symmetric Euler formulations such as the conservation form used in this chapter and employ the equivalence between norms to produce a simpler formula

$$\boldsymbol{\tau} = \Big(\sum_{j=1}^{\mathrm{n_{sd}}}|\mathbf{B}_j|\Big)^{-1}.$$

In practice, it is possible to obtain an algebraic expression for $\boldsymbol{\tau}^{-1}$ to avoid the solution of an eigenvalue problem, see the cited references.

4.4.2.4 Stabilized space–time formulation Here we illustrate the use of stabilized formulations in the context of a space–time formulation. As in Section 3.10, to which we refer for the notation, we consider piecewise continuous approximations in space and discontinuous approximations in time. Accordingly, we introduce space–time slabs $Q^n = \Omega \times I^n$, where $I^n =]t^n, t^{n+1}[$, and space–time element domains $Q_e^n = \Omega^e \times I^n$, $e = 1, \dots, \mathrm{n_{el}}$. The finite element space for the trial and weighting functions is then defined as follows:

$$\mathcal{S}^h = \bigcup_{n=0}^{\mathrm{n_{st}}-1} \mathcal{S}_n^h, \quad \mathcal{S}_n^h = \{\mathbf{U}^h \mid \mathbf{U}^h \in [\mathcal{C}(Q^n)]^{\mathrm{n_{sd}}+2},\ \mathbf{U}^h|_{Q_e^n} \in [\mathcal{P}_k(Q_e^n)]^{\mathrm{n_{sd}}+2},\ \forall e\}.$$

It is the vector analogue of (3.62) with no Dirichlet boundary conditions.

With reference to the Euler equations in conservative variables, system (4.16), the GLS method can now be formulated as follows: for $n = 0, 1, \dots, \mathrm{n_{st}} - 1$, find $\mathbf{U}^h \in \mathcal{S}_n^h$ such that for all $\mathbf{W} \in \mathcal{S}_n^h$

$$\begin{aligned}
&\iint_{Q^n} (\mathbf{W}^h)^T \left(\mathbf{U}_t^h + \boldsymbol{\nabla}\cdot\mathbf{F}^h - \mathbf{B}^h\right) d\Omega\, dt \\
&\quad + \sum_e \iint_{Q_e^n} \left(\mathbf{W}_t^h + (\mathbf{A}\cdot\boldsymbol{\nabla})\mathbf{W}^h\right)^T \boldsymbol{\tau} \left(\mathbf{U}_t^h + \boldsymbol{\nabla}\cdot\mathbf{F}^h - \mathbf{B}^h\right) d\Omega\, dt \\
&\qquad + \int_\Omega \left(\mathbf{W}^h(t_+^n)\right)^T \left(\mathbf{U}^h(t_+^n) - \mathbf{U}^h(t_-^n)\right) d\Omega = 0,
\end{aligned}$$

where $\mathbf{U}^h(t_-^0) = \mathbf{U}_0$. The last integral in this equation imposes a weakly enforced continuity condition across the slab interfaces at t^n, see Section 3.10, and it is the mechanism by which information is propagated from one slab to another.

For instance, in the case of a (discontinuous) piecewise constant approximation in time, all time derivatives in the previous equation vanish, the GLS formulation coincides with SUPG, and an implicit approximation over the time interval $\Delta t = t^{n+1} - t^n$ is obtained

$$\begin{aligned}
&\Delta t \int_\Omega (\mathbf{W}^h)^T \left((\mathbf{A}\cdot\boldsymbol{\nabla})\mathbf{U}_{n+1}^h - \mathbf{B}_{n+1}^h\right) d\Omega\, dt \\
&\quad + \Delta t \sum_e \int_{\Omega^e} \left((\mathbf{A}\cdot\boldsymbol{\nabla})\mathbf{W}^h\right)^T \boldsymbol{\tau} \left((\mathbf{A}\cdot\boldsymbol{\nabla})\mathbf{U}_{n+1}^h - \mathbf{B}_{n+1}^h\right) d\Omega\, dt \\
&\qquad + \int_\Omega (\mathbf{W}^h)^T \left(\mathbf{U}_{n+1}^h - \mathbf{U}_n^h\right) d\Omega = 0.
\end{aligned}$$

In deriving this expression we have used the quasi-linear form (4.19) of the Euler equations to replace the divergence of the flux, $\boldsymbol{\nabla}\cdot\mathbf{F}^h = (\mathbf{A}\cdot\boldsymbol{\nabla})\mathbf{U}^h$. The following convention has been used: $U_{n+1}^h = U^h(t^{n+1}) := U^h(t)$ for $t \in]t^n, t^{n+1}[$. If the values at t^n are known, the only unknown in the previous equation is $\mathbf{U}_{n+1}^h$. An implicit scheme is thus obtained that is reminiscent of the first-order backward Euler method.

4.4.2.5 Residual decomposition technique Alternative ways of extending the SUPG formulation to the Euler equations have been suggested using the so-called multidimensional residual decomposition technique. This technique was originated by Roe and Deconinck and their co-workers and a detailed account of it may, for instance, be found in the thesis of Carette (1997) and the references therein. The main idea consists of decomposing the elementwise residual of the system of Euler equations into a set of scalar components which can be treated separately with stabilized convection schemes. This formulation allows a natural equation-by-equation definition of the stabilization parameter.

In 1D, the Euler equations can be diagonalized and, as seen in Section 4.3.3, they can be decomposed into three scalar convection equations associated with the characteristics defined in (4.25). In 2D and 3D, the Euler equations can no longer be diagonalized. Waves can travel in an infinite number of directions and the decomposition of the flux divergence into scalar waves is no longer unique. However, in the framework of the multidimensional residual decomposition approach, several techniques have been developed to decompose the system of Euler equations into scalar contributions just as in the 1D case. The basic ingredient of the method is an *approximate diagonalization* along the lines suggested by Roe (1986) and Deconinck, Hirsch and Peuteman (1986) in their characteristic-based approach. Various techniques are actually available to decompose the multidimensional Euler equations into a set of scalar convection equations. They are described in detail in the theses of Paill ère (1995) and Carette (1997) and the references therein to the original works of Roe, Deconinck, Hirsch, and their collaborators.

4.5 NUMERICAL TREATMENT OF SHOCKS

4.5.1 Introduction

One of the most striking features of compressible fluid flow is the presence of shock waves. A shock wave is produced in a fluid when a succession of compression waves, each propagating faster than its predecessor, pile up, determining an abrupt transition of the field variables.

In the solution of inviscid compressible equations, shocks are seen as discontinuities of the solution, that is surfaces across which the fluid variables are discontinuous. In nature, shocks have a finite thickness, usually very small, associated with the mean free path of the particles of the fluid. Shock waves produce entropy; thus over the thickness of the shock the kinetic energy of the short wavelengths is transferred into internal energy.

From a mathematical point of view, the Rankine–Hugoniot jump conditions, which relate fluid variables across the shock, guarantee the solution of the inviscid equations in the case of a shock discontinuity. However, the numerical computation of shock flows via a direct application of the Rankine–Hugoniot jump conditions would simply be prohibitive. Fortunately, it can be shown that, provided appropriate *conservative discretization schemes* are employed, the Rankine–Hugoniot jump conditions are satisfied by numerical solutions.

The key idea for the numerical treatment of shocks and other flow discontinuities is the introduction into the considered conservation law equation of an artificial (numerical) viscosity term. This term has the effect of broadening the thickness of a shock over a few elements in the mesh, thereby smearing the discontinuity and removing spurious oscillations at its front. Discontinuity surfaces are thus replaced by thin transition layers over which the flow variables (density, pressure, velocity) vary rapidly, but continuously. Provided the structure of the artificial dissipation terms is correctly chosen, the Rankine–Hugoniot conditions are satisfied without degrading the results over the remainder of the flow field. Recall the concept of vanishing viscosity solution introduced in Section 4.2.2.

There are several ways of introducing terms representing an added dissipation in the numerical schemes resulting from the finite element method. In Chapter 2 added numerical diffusion was already introduced in the context of stabilization of convection-dominated problems. Among early *shock-capturing techniques* for the Euler equations, mention must be made of the method of pseudo-viscous pressure due to von Neumann and Richtmyer (1950). This method is discussed in Section 4.5.2, together with another early method for treating shocks due to Lapidus (1967). These early methods introduce free parameters whose calibration is not trivial, because they must introduce just enough numerical viscosity to preserve monotonicity without spreading shocks over too many grid points.

Higher-order shock-capturing schemes have been developed for the approximation of nonlinear hyperbolic conservation laws. They produce non-oscillatory discontinuities by a different approach than explicitly adding an artificial viscosity term to the considered hyperbolic conservation law. The damping mechanism is implicitly introduced by choosing an appropriate form of the discrete equations. Here, mention must be made, among others, of the works of van Leer (1974), Harten (1983), Roe (1984), Sweby (1984) and Woodward and Colella (1984). These modern schemes, called *high-resolution schemes*, were briefly introduced in Section 3.7 and are further discussed in Section 4.5.3. They are based on conditions less severe than monotonicity and have the following properties:

- They are at least of second-order accuracy in smooth parts of the flow.
- They sharply resolve discontinuities without generating spurious oscillations and can be modified to produce solutions satisfying the entropy condition.
- They do not need explicit artificial viscosity such as in the pseudo-viscous pressure method; damping is implicitly included in the form of the discrete equations through the definition of appropriate flux functions incorporating an upwind effect.

We close the discussion of shock-capturing techniques by mentioning in Section 4.5.3.3 that locally monotone solutions can also be obtained by a suitable choice of the stabilization matrix in the SUPG and GLS methods.

4.5.2 Early artificial diffusion methods

Early methods for the numerical treatment of shocks and other flow discontinuities consisted of introducing a suitable amount of artificial diffusion near sharp solution gradients. Among the various artificial diffusion methods proposed to produce a local smoothing of the solution to the Euler equations, we have selected the widely used method of pseudo-viscous pressure and the Lapidus viscosity.

4.5.2.1 Method of pseudo-viscous pressure The treatment of shocks by the method of pseudo-viscous pressure due to von Neumann and Richtmyer (1950) consists of rewriting the equations in Section 4.3.1 of conservation of momentum and total energy in the form

$$\frac{\partial \rho \boldsymbol{v}}{\partial t} + \boldsymbol{\nabla} \cdot (\rho \boldsymbol{v} \otimes \boldsymbol{v} + (p+q)\mathbf{I}) = \rho \boldsymbol{b}$$
$$\frac{\partial \rho E}{\partial t} + \boldsymbol{\nabla} \cdot \big((\rho E + p + q)\boldsymbol{v}\big) = \boldsymbol{v} \cdot \rho \boldsymbol{b},$$

where p is the fluid static pressure and the scalar q added to the fluid pressure is called *pseudo-viscous pressure*. This additional term is defined below and represents the dissipative mechanism. Its value should be negligible everywhere in the flow domain, except in the vicinity of a shock.

Since dissipation is being introduced for purely numerical reasons, the pseudo-viscous pressure q can be defined through any appropriate function of p, ρ, $\boldsymbol{v}$, etc., and their derivatives. Nevertheless, the following conditions must be satisfied by any artificial viscosity technique:

1. The conservation equations must have solutions free of discontinuities (i.e., the conservation variables must be continuous across the shock).
2. The shock thickness must be of the order of magnitude of the mesh size h.
3. The effect of the terms containing q must be negligible outside the shocks.
4. The Rankine–Hugoniot jump conditions must be verified when the dimensions characterizing the flow are large compared with the shock thickness.

von Neumann and Richtmyer (1950) show that in 2D the choice

$$q = \begin{cases} c_Q\, \rho\, A_e\, (\boldsymbol{\nabla} \cdot \boldsymbol{v})^2 & \text{if } \boldsymbol{\nabla} \cdot \boldsymbol{v} < 0, \\ 0 & \text{if } \boldsymbol{\nabla} \cdot \boldsymbol{v} \geq 0, \end{cases}$$

allows all the above conditions to be satisfied. In this expression A_e is the element area and c_Q a dimensionless constant (close to unity). The constant c_Q controls the spread of the shock and it is determined empirically. Note from the previous definition that the dissipation mechanism introduced by q is a nonlinear viscosity.

Frequently, and in particular for modeling liquids, a so-called linear pseudo-viscosity is required as suggested by Wilkins (1969). It is added to the previous definition of q,

$$q_L = \begin{cases} c_L\, c\, \rho\, A_e^{1/2}\, |\boldsymbol{\nabla} \cdot \boldsymbol{v}| & \text{if } \boldsymbol{\nabla} \cdot \boldsymbol{v} < 0, \\ 0 & \text{if } \boldsymbol{\nabla} \cdot \boldsymbol{v} \geq 0. \end{cases}$$

This linear pseudo-viscosity is added to the previously defined nonlinear pseudo-viscosity to reduce spurious oscillations; c_L must be small to avoid excessive diffusion of the shock wave front. A value of c_L of the order of 0.05 is standard.

The implementation of the pseudo-viscous pressure in the framework of the finite element method is simple. The pseudo-viscous pressure q is added to the static fluid pressure p in the discrete form of the momentum and energy equations.

4.5.2.2 Lapidus viscosity Another widely used artificial viscosity method to capture shocks was proposed by Lapidus (1967) for finite difference context. In 1D, the added viscosity normally acts as a diffusion operator

$$\nu_{\text{Lap}} = c_{\text{Lap}}\, h^2 |\partial v / \partial x|,$$

where c_{Lap} is a user-specified coefficient, which normally varies between 0.0 and 2.0. In order to extend this concept to multiple dimensions Löhner, Morgan and Peraire (1985) and Peraire (1986) constructed a local coordinate system oriented in the direction of $\boldsymbol{\ell}$, the unit vector in the direction of maximum change in the absolute value of the velocity $||\boldsymbol{v}||$, that is

$$\ell_i = \left[\frac{\partial ||\boldsymbol{v}||}{\partial x_i}\right] \Bigg/ \left[\left(\sum_{j=1}^{n_{\text{sd}}} \frac{\partial ||\boldsymbol{v}||}{\partial x_j}\frac{\partial ||\boldsymbol{v}||}{\partial x_j}\right)^{1/2}\right], \quad i = 1, \ldots, n_{\text{sd}},$$

or

$$\boldsymbol{\ell} = \frac{\boldsymbol{\nabla} ||\boldsymbol{v}||}{||(\boldsymbol{\nabla} ||\boldsymbol{v}||)||}.$$

Then, with the added viscosity

$$\nu_{\text{Lap}} = c_{\text{Lap}}\, h^2 \left|\frac{\partial(\boldsymbol{v}\cdot\boldsymbol{\ell})}{\partial \boldsymbol{\ell}}\right| = c_{\text{Lap}}\, h^2 |\boldsymbol{\ell}\cdot\boldsymbol{\nabla}(\boldsymbol{v}\cdot\boldsymbol{\ell})|,$$

one can finally define the added diffusion term used to smooth the velocities

$$\frac{\partial}{\partial \boldsymbol{\ell}}\left(\nu_{\text{Lap}} \frac{\partial \boldsymbol{v}}{\partial \boldsymbol{\ell}}\right) = (\boldsymbol{\ell}\cdot\boldsymbol{\nabla})\big(\nu_{\text{Lap}}\,(\boldsymbol{\ell}\cdot\boldsymbol{\nabla})\boldsymbol{u}\big).$$

This artificial diffusion method has the essential features required in these techniques. The sensing function, which here is $|\boldsymbol{\ell}\cdot\boldsymbol{\nabla}(\boldsymbol{v}\cdot\boldsymbol{\ell})|$, induces a zero added viscosity when velocity is smooth. It is invariant under coordinate rotation; thus it is independent of the mesh orientation. It does not smear shear layers (contact discontinuities) because velocity is continuous across them and neither boundary layers because $\boldsymbol{v}$ and $\boldsymbol{\ell}$ are orthogonal in these regions.

To limit the addition of artificial viscosity close to zones of sharp gradients, other sensors can be employed which are activated, for instance, by the variation of the pressure gradient (see the exercises in Section 4.8). Moreover, instead of explicitly adding artificial viscosity terms, a damping mechanism may be introduced implicitly by choosing an appropriate form of the discrete equations. This leads to so-called *high-resolution schemes*, some of which are illustrated in the next section.

4.5.3 High-resolution methods

As already mentioned, monotone schemes are generally too dissipative and cannot produce accurate solutions for complex flow problems. Higher-order shock-capturing schemes are introduced for the purpose of adding minimal numerical dissipation and to give non-oscillatory solutions in the presence of steep solution gradients.

As explained in Section 3.7, the basic idea underlying high-order shock-capturing methods is to produce a high-order method in the smooth part of the flow and to modify it by adding numerical dissipation only in the neighborhood of a discontinuity. Modern methods use high-order TVD schemes which introduce a controlled amount of nonlinear dissipation to give non-oscillatory shocks and ensure convergence to the entropy solution. The term nonlinear dissipation means that the diffusion coefficient depends on the local behavior of the solution. Thus, diffusion should be larger near discontinuities than in smooth regions of the flow.

The guiding idea behind the design of TVD schemes for inviscid flow problems is that physical solutions to the scalar hyperbolic equations do not allow the appearance of any new extremum in the evolution of the flow variables. In a TVD scheme, the total variation of the numerical solution is controlled in a nonlinear way, such as to prevent the appearance of any new extremum. High-resolution TVD schemes in the finite difference context have been developed by van Leer (1974), Harten (1983), Roe (1984), Sweby (1984) and Yee (1987). A detailed account of the theory underlying modern shock-capturing schemes can be found in the textbooks by Hirsch (1990), LeVeque (1992), Quarteroni and Valli (1994) and Morton (1996). Note that the adaptation of TVD schemes for use in connection with finite elements is still an area of active development, see Donea et al. (1988) for an early application.

The critical issue in the design of high-order TVD schemes is to introduce enough dissipation to preserve monotonicity without affecting the level of accuracy away from the flow discontinuities. To this end, most schemes incorporate a mechanism which automatically controls the amount of added numerical dissipation. Such mechanisms are often in the form of *limiters* which impose constraints on the gradient of the considered dependent variable (*slope limiters*) or on the flux function (*flux limiters*). Some of the approaches to obtain a better resolution of flow discontinuities are briefly presented in the next paragraphs.

4.5.3.1 Flux-limiter method Boris and Book (1997) and van Leer (1974) introduced high-resolution schemes using nonlinear flux limiters. They generally require a first-order upwind scheme and an anti-diffusion mechanism. The latter compensates the large amount of diffusion introduced by the first-order scheme and thus induces sharper resolution of discontinuities. Moreover, high accuracy in the smooth parts of the flow and desirable properties, in particular satisfaction of the entropy condition, are preserved. Such schemes satisfy the basic requirements of a good convection scheme: they are TVD, conservative and less diffusive than the simple upwind schemes.

We illustrate this concept of flux limiter on a 1D scalar convection equation:

$$u_t + f_x = 0, \qquad \text{with} \quad f = a\,u.$$

The forward Euler time discretization of this equation can be expressed at node i in terms of the convective flux as

$$u_i^{n+1} = u_i^n - \frac{\Delta t}{h}\left(f_{i+1/2}^n - f_{i-1/2}^n\right).$$

In the case of a centered (second-order) approximation of the flux term, one has

$$\left[f_{i+1/2}^n\right]_{\text{cent}} = \frac{1}{2}\left(f_i^n + f_{i+1}^n\right), \qquad \left[f_{i-1/2}^n\right]_{\text{cent}} = \frac{1}{2}\left(f_i^n + f_{i-1}^n\right).$$

These approximations give rise to instabilities. But, a first-order upwind approximation based upon

$$\left[f_{i+1/2}^n\right]_{\text{up}} = \frac{1}{2}\Big(\left(a_{i+1/2}^n + |a_{i+1/2}^n|\right) u_i^n + \left(a_{i+1/2}^n - |a_{i+1/2}^n|\right) u_{i+1}^n\Big)$$

generally introduces an excessive numerical diffusion. Then, an optimal strategy consists in using the second-order numerical flux in smooth regions of the flow, and the monotonic upwind method in the vicinity of discontinuities. This procedure is best implemented by introducing *limiter schemes* based on the local gradient of the solution u (slope limiters), or of the flux function f (flux limiters). In practice, a flux-limiter method uses a flux function of the form

$$\left[f_{i+1/2}^n\right]_{\text{lim}} = \left[f_{i+1/2}^n\right]_{\text{up}} + \Phi(r_{i+1/2})\Big(\left[f_{i+1/2}^n\right]_{\text{cent}} - \left[f_{i+1/2}^n\right]_{\text{up}}\Big), \qquad (4.29)$$

where

$$r_{i+1/2} = \begin{cases} (u_i^n - u_{i-1}^n)/(u_{i+1}^n - u_i^n) & \text{if } a_{i+1/2} \geq 0, \\ (u_{i+2}^n - u_{i+1}^n)/(u_{i+1}^n - u_i^n) & \text{if } a_{i+1/2} < 0. \end{cases}$$

The term in equation (4.29) depending on the function $\Phi(r)$ is called anti-diffusive flux. It compensates the high numerical diffusion introduced by the upwind approximation. For stability, flux-limiter schemes must be TVD. This imposes the following restrictions on the function $\Phi(r)$:

$$0 \leq \frac{\Phi(r)}{r} \leq 2 \qquad \text{and} \qquad 0 \leq \Phi(r) \leq 2.$$

Note, however, that the function $\Phi(r)$ is normally selected so that $0 \leq \Phi(r) \leq 1$. Two classical examples are the superbee limiter of Roe (1985) for which

$$\Phi(r) = \max\big(0, \min(1, 2r), \min(2, r)\big),$$

and the limiter suggested by van Leer (1974):

$$\Phi(r) = \frac{r + |r|}{1 + |r|}.$$

4.5.3.2 *Practical implementation of high-resolution schemes* As an example of the previous discussion, we describe a method (see Donea et al., 1988), which introduces a locally controlled dissipation, in the two-step Taylor–Galerkin method (TG2). This locally controlled dissipation induces a non-oscillatory scheme for the representation of shocks. In complete analogy with the idea behind the flux-limiter method discussed in the previous section, the proposed modification combines the good resolution offered by a second-order time scheme in the regular part of the flow and the capability to damp out the non-physical oscillations in the vicinity of flow discontinuities.

Suppose we want to apply the two-step TG2 method, see Section 4.2.3.2, to the Euler equations, (4.16a). After Galerkin spatial discretization of the second step, the following algebraic system is obtained:

$$\mathbf{M}\big(\mathbf{U}^{n+1} - \mathbf{U}^{n}\big) = \Delta t\, \mathbf{r}^{n+1/2} \tag{4.30}$$

where the mass matrix $\mathbf{M}$ has dimensions $\mathrm{n_{eq}} \times (\mathrm{n_{sd}}+2)$ and vector $\mathbf{r}^{n+1/2}$ accounts for the mid-step value of the non-inertia terms, see (4.12). Note that, in (4.30), $\mathbf{M}$ could, in principle, be replaced by the diagonal matrix $\mathbf{M}^L$ obtained by row sum; a second-order scheme would also be obtained but, as seen in 3.5.2, with poor phase properties.

To reduce the second-order scheme (TG2) to first-order near shocks, we replace the consistent mass matrix $\mathbf{M}$ acting on $\mathbf{U}^{n+1}$ with the diagonal matrix $\mathbf{M}^L$. This substitution produces the first-order accurate scheme

$$\mathbf{M}^L\mathbf{U}^{n+1} = \mathbf{M}\mathbf{U}^{n} + \Delta t\, \mathbf{r}^{n+1/2},$$

which can be rewritten as

$$\mathbf{M}^L\mathbf{U}^{n+1} = \mathbf{M}^L\mathbf{U}^{n} + \big(\mathbf{M} - \mathbf{M}^L\big)\mathbf{U}^{n} + \Delta t\, \mathbf{r}^{n+1/2}. \tag{4.31}$$

Remark 4.9. The second term on the r.h.s. of (4.31), which contains the difference between the consistent mass matrix and its diagonal counterpart, represents an added dissipation. For instance, the scalar 1D linear convection equation scheme associated with (4.31) delivers the following discrete equation at an interior node j:

$$u_j^{n+1} = u_j^n - \frac{1}{2}C\,\big(u_{j+1}^n - u_{j-1}^n\big) + \big(\frac{1}{2}C^2 + \frac{1}{6}\big)\big(u_{j-1}^n - 2u_j^n + u_{j+1}^n\big),$$

where C is the Courant number. This discrete equation is clearly of the Lax–Wendroff type with an added dissipative term. It is stable for $|C| \leq \sqrt{2/3}$ and monotone over this interval, as shown by Harten, Hyman and Lax (1976).

In order to modulate the added dissipation and thereby better mimic the flux-limiter method, in the sense of equation (4.29), we introduce a parameter d in the diffusion term of the first-order scheme (4.31). That is,

$$\mathbf{M}^L\mathbf{U}^{n+1} = \mathbf{M}^L\mathbf{U}^{n} + d\big(\mathbf{M} - \mathbf{M}^L\big)\mathbf{U}^{n} + \Delta t\, \mathbf{r}^{n+1/2}. \tag{4.32}$$

Note the conceptual relation between d and the anti-diffusive flux: $d = 1 - \Phi$. A second-order scheme is obtained for $d = 0$ and a first-order method with maximum dissipative effect is obtained when $d = 1$.

Furthermore, in order to separate the effect of convective transport from that associated with the added dissipation (and to recover the benefits of the consistent mass matrix), it is appropriate to implement the first-order scheme (4.32) according to the following two-stage procedure:

$$\mathbf{M}(\mathbf{U}^* - \mathbf{U}^n) = \Delta t\, \mathbf{r}^{n+1/2}, \tag{4.33a}$$

$$\mathbf{M}^L(\mathbf{U}^{n+1} - \mathbf{U}^*) = d(\mathbf{M} - \mathbf{M}^L)\mathbf{U}^*. \tag{4.33b}$$

The first stage, (4.33a), corresponds to the second-order Taylor–Galerkin scheme (4.30) and is characterized by the same (complex) amplification factor. The second stage, (4.33b), only introduces a multiplicative coefficient into the amplification factor of the first stage. The advantage of the two-stage procedure is to preserve the second-order phase accuracy of the TG2 method in the smooth part of the flow, while facilitating the introduction of a modulated dissipation around shocks.

The dissipative term on the r.h.s. of (4.33b) must be further modified in order to produce a *local modulation* of the added numerical dissipation. This is achieved by replacing the global parameter d with a local parameter defined in terms of a sensor which recognizes the discontinuities in the flow. At a given finite element node A the dissipative term of the first-order scheme (4.33b) takes the form

$$d \sum_B M_{AB} U_B^* - M_{AA}^L U_A^*, \tag{4.34a}$$

where the summation extends to all nodes B topologically connected to node A. Since the lumped-mass matrix is obtained by row sum $M_{AA}^L = \sum_B M_{AB}$, we may rewrite the dissipative term (4.34a) as

$$\sum_B d_{AB}\, M_{AB}(U_B^* - U_A^*), \tag{4.34b}$$

where $0 \le d_{AB} \le 1$ is the required local modulation coefficient. Observe that the dissipation operator consists of segment contributions (the sides of the elements for a triangular mesh); moreover, the 1D character of segments allows an easy adaptation to a finite element context of the procedures developed in 1D to limit the dissipative effect.

To illustrate the proposed limitation procedure, we consider its application in connection with the artificial viscosity method. As already mentioned, the local parameter d_{AB} must be expressed in terms of a sensor which recognizes the discontinuities in the flow. As suggested by Jameson (1985), an effective sensor to detect the presence of shocks can be constructed by considering the second derivative of the pressure. To this end, for a given segment A–B one introduces the quantities

$$d_A = \left| \frac{p_B - 2p_A + p_{A_-}}{p_B + 2p_A + p_{A_-}} \right| \quad \text{and} \quad d_B = \left| \frac{p_{B_+} - 2p_B + p_A}{p_{B_+} + 2p_B + p_A} \right|, \tag{4.35a}$$

where

$$p_{A_-} = p_B - 2(\boldsymbol{x}_B - \boldsymbol{x}_A)\cdot[\boldsymbol{\nabla} p]_A \quad \text{and} \quad p_{B_+} = p_A + 2(\boldsymbol{x}_B - \boldsymbol{x}_A)\cdot[\boldsymbol{\nabla} p]_B. \tag{4.35b}$$

Then, the artificial viscosity coefficient for segment A–B is evaluated from the relation

$$d_{AB} = \min(\chi \max(d_A, d_B), 1), \tag{4.35c}$$

where χ is an adjustable parameter. In this way d_{AB} is maximum on both sides of a shock and zero inside it. Note the presence of a free parameter in this last expression. This is avoided in more elaborate TVD schemes, such as, for instance, those developed by Harten (1983) and Yee (1987).

4.5.3.3 Nonlinear artificial dissipation for stabilized methods

Stabilized methods such as SUPG, see Chapter 2, guarantee that spurious oscillations created in the neighborhood of sharp solution gradients do not propagate all over the computational domain. Recall that this is not the case with the standard Galerkin method. Nevertheless, stabilization techniques such as SUPG are linear high-order methods and consequently some oscillations will remain close to near-discontinuities (recall Godunov's theorem). This is due to the absence of control of the solution gradient in directions other than along the streamlines. This has motivated the development of nonlinear discontinuity-capturing versions of stabilized methods, with the objective of obtaining monotone solutions in the presence of sharp layers. For simplicity, we shall illustrate the discontinuity-capturing version of the SUPG method with reference to a scalar convection–diffusion problem.

To suppress the residual oscillations near strong solution gradients an additional nonlinear diffusion term is added. That is, an extra consistent (residual-based) term is added apart from the standard stabilization term presented in Section 4.4.2.3. Recall that in scalar form this stabilization term is $\sum_e \big(\mathcal{P}(w), \tau\, \mathcal{R}(u)\big)_{\Omega^e}$ where $\mathcal{R}(u)$ is the residual of the differential equation. The added diffusion can be isotropic or anisotropic. The former has the same structure as standard SUPG, see equation (2.57), namely $\mathcal{P}(w) = \boldsymbol{a}\cdot\boldsymbol{\nabla} w$. Thus, the shock-capturing term takes the form

$$\sum_e \big(\hat{\boldsymbol{a}}\cdot\boldsymbol{\nabla} w, \hat{\tau}\,\mathcal{R}(u)\big)_{\Omega^e}; \tag{4.36}$$

where $\hat{\boldsymbol{a}}$ is a newly defined vector field. For instance, Hughes and Mallet (1986b) propose using the projection of $\boldsymbol{a}$ onto $\boldsymbol{\nabla} u$. That is, in the case $\|\boldsymbol{\nabla} u\| \neq 0$, the projected velocity is given by

$$\hat{\boldsymbol{a}} = \frac{\boldsymbol{a}\cdot\boldsymbol{\nabla} u}{\|\boldsymbol{\nabla} u\|^2}\boldsymbol{\nabla} u.$$

The new stabilization parameter $\hat{\tau}$ depends on the standard τ, see Section 2.4.3. Recall that for each element

$$\tau = \beta\,\frac{h}{2\|\boldsymbol{a}\|},$$

where h is a measure of the element size, whereas β is a function of the mesh Péclet number defined by

$$Pe = \frac{h\|\boldsymbol{a}\|}{2\nu},$$

where ν is the physical diffusivity. As seen in Chapter 2, in 1D convection–diffusion the optimal choice for β is given by

$$\beta = \coth Pe - \frac{1}{Pe}$$

and the choice $\beta = 1$ corresponds to the full upwind case, which maximizes the stabilization parameter τ.

The stability parameter $\hat{\tau}$ controlling the discontinuity-capturing term can be defined in a number of ways. Hughes and Mallet (1986b) initially suggest using $\hat{\tau} = \tau$, but this choice is overly diffusive when the convection velocity and the gradient of u are parallel vectors, that is when $\hat{\boldsymbol{a}} = \boldsymbol{a}$. It amounts to double the SUPG stabilization term. They subsequently propose using

$$\hat{\tau} = \max(0, \tau_{\hat{\mathbf{a}}} - \tau), \qquad \text{with} \quad \tau_{\hat{\mathbf{a}}} = \beta \frac{h}{2\|\hat{\boldsymbol{a}}\|}. \tag{4.37}$$

This method produces an added diffusion of form, see equation (4.36),

$$\sum_e \left(\hat{\boldsymbol{a}} \cdot \boldsymbol{\nabla} w, \hat{\tau}\, \mathcal{R}(u)\right)_{\Omega^e} = \sum_e \int_{\Omega^e} \left[\hat{\tau} \frac{\boldsymbol{a} \cdot \boldsymbol{\nabla} u}{\|\boldsymbol{\nabla} u\|^2} \mathcal{R}(u)\right] \boldsymbol{\nabla} u \cdot \boldsymbol{\nabla} w \, d\Omega, \tag{4.38}$$

where the bracketed term, which acts as nonlinear diffusion, may be negative.

An alternative approach is to change the definition of $\hat{\boldsymbol{a}}$ and still use a residual-based approach. Johnson, Szepessy and Hansbo (1990), Hansbo and Johnson (1991) and Galeão and do Carmo (1988) suggest using the following definition of $\hat{\boldsymbol{a}}$:

$$\hat{\boldsymbol{a}} = \frac{\mathcal{R}(u)}{\|\boldsymbol{\nabla} u\|^2} \boldsymbol{\nabla} u,$$

where $\mathcal{R}(u)$ is the residual of the governing equation. The nonlinear artificial viscosity introduced by this method is given by

$$\nu_{\text{shock}} = \begin{cases} \hat{\tau}\, \mathcal{R}(u)^2 / \|\boldsymbol{\nabla} u\|^2 & \text{if } \|\boldsymbol{\nabla} u\| \neq 0, \\ 0 & \text{otherwise.} \end{cases}$$

Here, parameter $\hat{\tau}$ is computed as in equation (4.37) but making use of the new definition of $\hat{\boldsymbol{a}}$. Finally, note that the nonlinearity of the discontinuity-capturing term always originates from the definition of $\hat{\boldsymbol{a}}$, which depends on $\boldsymbol{\nabla} u$.

Codina (1993a) suggests, based on the discrete maximum principle, an anisotropic nonlinear viscosity that should act in the crosswind direction only. The objective is to add diffusion without affecting the streamline dissipation introduced by the linear form of the SUPG method. This requires, first, a nonlinear added diffusion,

$$\nu_{\text{shock}} = \begin{cases} \frac{1}{2}\alpha h |\mathcal{R}(u)| / \|\boldsymbol{\nabla} u\| & \text{if } \|\boldsymbol{\nabla} u\| \neq 0, \\ 0 & \text{otherwise,} \end{cases}$$

where $\alpha = \max\big(0, \chi - \frac{2\nu}{h\|a\|}\big)$. The recommended value of the constant χ for linear elements is $\chi = 0.7$. Second, it also requires a tensorial structure in the added term

$$\sum_{e=1}^{n_{el}} \int_{\Omega^e} \nu_{\text{shock}} \, \boldsymbol{\nabla} w \cdot \left(\mathbf{I} - \frac{\boldsymbol{a} \otimes \boldsymbol{a}}{\|\boldsymbol{a}\|^2} \right) \cdot \boldsymbol{\nabla} u \, d\Omega,$$

where $\mathbf{I}$ is the unit tensor.

Finally, it is important to note that both isotropic and anisotropic added dissipations are only active when $\|\boldsymbol{\nabla} u\| \neq 0$ and that numerical results are sensitive to the numerical tolerance ($\|\boldsymbol{\nabla} u\| \leq$ tolerance $\Leftrightarrow \|\boldsymbol{\nabla} u\| = 0$) implemented in the code.

4.6 NEARLY INCOMPRESSIBLE FLOWS

Most numerical methods for the analysis of compressible flow problems present numerical difficulties when applied to low-speed, nearly incompressible flows. This is clearly due to the fast propagation of pressure signals as flow conditions approach the incompressible limit. In the case of low-speed flow, the accurate representation of pressure transients with explicit finite element schemes designed for the compressible regime would require the use of extremely small time steps, clearly undermining the practical utility of such schemes. In more mathematical terms, pressure behavior changes from hyperbolic (wave propagation) to elliptic character when passing from the compressible to the incompressible regime: the mass-conservation equation and the equation of state are replaced by the incompressibility constraint requiring that the velocity field be divergence-free.

Different approaches have been pursued to develop computational strategies for approximating flows at all speeds, from highly compressible to nearly incompressible. Methods for all flow speeds require an implicit treatment of the mass-conservation equation and of the pressure terms in the momentum and energy equations, in order to introduce the required elliptic behavior into the governing equations. A fractional-step approach to the time integration of the conservation equations is usually adopted, as it allows the separation of the necessarily implicit pressure terms from the other terms in the conservation equations.

An interesting splitting-up scheme for 2D flows consisting of three distinct phases was originally developed in finite difference format by Hirt et al. (1974) and subsequently extended to 3D problems by Stein, Gentry and Hirt (1977). The key idea in deriving such a fractional-step approach has been to note that the convective terms in the conservation equations vanish in the Lagrangian formulation; that is, when the computational mesh moves with the fluid. It is this reduction that suggests to first solve the Lagrangian form of the equations in the first phase of the calculation cycle and then to add the convective contributions as a separate step in the last phase. The middle phase adds an implicit pressure calculation that permits solutions to be obtained at all flow speeds.

This strategy has subsequently been adapted to a finite element context by Donea, Giuliani and Halleux (1982), see also Donea (1983), and has been widely used in the simulation of transient dynamic problems involving fluid–structure interaction. An application of this splitting-up technique is described in Section 4.7.2.

More recently, the splitting-up approach to flows at all speeds has been further exploited in the finite element context. Particularly noteworthy is the characteristic-based split (CBS) algorithm introduced by Zienkiewicz and Codina (1995), see also Zienkiewicz et al. (1995), Codina, Vázquez and Zienkiewicz (1998) and Zienkiewicz and Taylor (2000b). A related modeling approach of nearly incompressible flows has been proposed by Sampaio and Moreira (2000).

The central feature of the above methods is the derivation of an evolution equation for the pressure field through the elimination of the density from the mass balance equation. The pressure, momentum and energy equations are then discretized in time by a splitting-up procedure which ensures an implicit treatment of the equation for the pressure field and of the pressure terms in the momentum and energy equations. In this way, nearly incompressible flow situations can be treated without numerical stability problems. A further advantage of the splitting-up procedure is the possibility to isolate the convective terms and treat them by means of algorithms, such as the Taylor–Galerkin (or characteristic Galerkin) methods, designed for pure convection problems.

Stabilized finite element algorithms of the GLS or SUPG type can also be adapted to deal with nearly incompressible flows. In the low Mach number limit, the classical choice for the stabilization matrix τ given in Section 4.4.2.3 fails to provide adequate stabilization. This is essentially due to a mismatch which occurs for low Mach numbers between the magnitude of the fluxes in the original equations and the corresponding terms in the numerically added viscosity. An alternative definition of τ has been proposed by Wong, Darmofal and Peraire (2001) which leads to a formulation which can handle very low Mach number flows accurately.

4.7 FLUID–STRUCTURE INTERACTION

Fluid–structure interaction phenomena are an important consideration in several engineering fields. This is manifestly the case in the design of automotive and aerospace structures, as well as in modeling the response of offshore structures, long-span bridges and high-rise buildings. Fluid–structure interaction also plays an important role in the safety assessment of power generation plants and many other industrial installations.

The general topic of fluid–structure interaction is indeed a particularly broad subject in that it simultaneously brings together all the aspects associated with both structural mechanics and fluid mechanics. Each of these two areas are complex by themselves; however, when considered together, the situation becomes even more complex. In fact, the interaction (or coupling) between the fluid and solid response can be viewed as a feedback loop of the type illustrated in Figure 4.10: the structure surface loading is not known a priori but depends on the interface pressures in the fluid; the fluid response is in turn a function of the structure's surface motion.

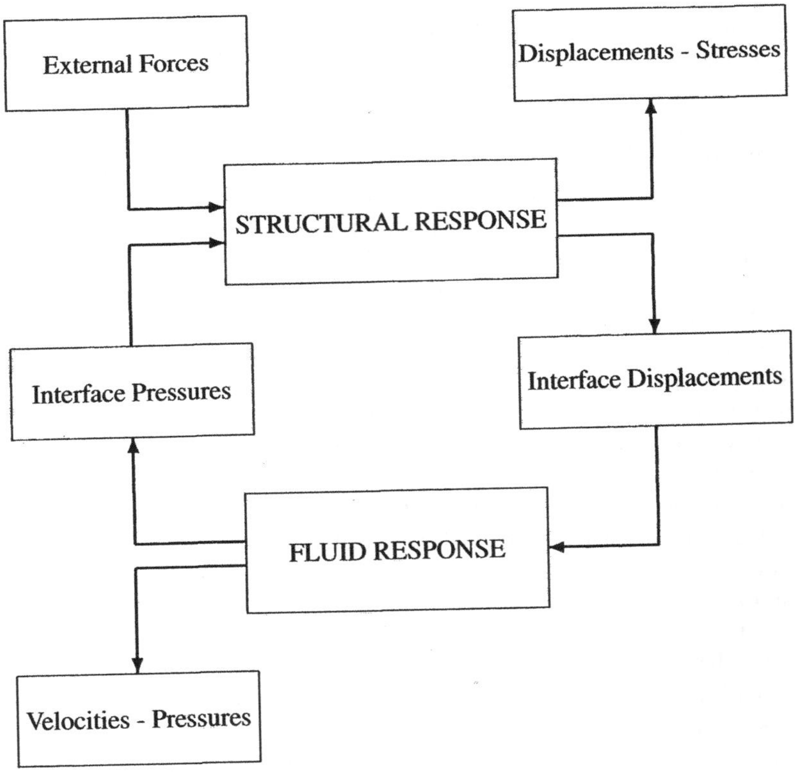

Fig. 4.10 Feedback loop in fluid–structure interaction.

In this section we restrict the fluid under consideration to be compressible and inviscid. There are indeed well-developed fluid–structure formulations for viscous fluids, in both the compressible and the incompressible regime. The reader interested in fluid–structure algorithms for viscous fluids should consult the relevant literature, such as, for instance, the original works of Liu and co-workers (Liu and Chang, 1985; Liu and Chang, 1986; Liu and Gvildys, 1986)

Here, we shall consider two classes of fluid–structure problems. The first is concerned with situations in which the displacements of the fluid are small, so that the so-called *acoustic approximation* for the fluid is valid. Modeling the response of a linear elastic structure interacting with an acoustic fluid is of particular interest in the framework of vibration studies of aeronautical structures containing an inviscid fluid. The second class of fluid–structure problems is concerned with situations involving a large-displacement response of the fluid–structure system. In such cases, the use of the Arbitrary Lagrangian–Eulerian (ALE) formulation introduced in Section 1.4 is recommended to model the nonlinear fluid domain and ease its coupling with the nonlinear structural domain usually treated in the Lagrangian description.

4.7.1 Acoustic approximation

When displacements are small, the acoustic approximation leads to simplified formulations of the fluid response, either in terms of fluid displacements or in terms of pressure. The displacement formulation of the fluid is attractive in that it is directly compatible with the standard formulation used for structures, thus facilitating the coupling between fluid and structural domains. Nevertheless, a formulation based upon the discretization of the fluid pressure is generally preferred. The reason is that only one dependent variable is involved in the pressure formulation, while the displacement formulation includes as many unknowns as the number of space dimensions. In this section subscript F will be used to refer to the fluid domain and subscript S to the structure.

The derivation of the pressure equation governing the fluid response in the acoustic approximation starts from the linearized form (convective terms neglected) of the conservation equation for momentum introduced in Section 1.4. When specialized to an inviscid compressible fluid, the linearized form of the momentum equation reads

$$\rho \frac{\partial^2 \boldsymbol{u}_F}{\partial t^2} = -\boldsymbol{\nabla} p, \tag{4.39}$$

where $\boldsymbol{u}_F$ is the fluid displacement vector and the pressure p is assumed to be given by the barotropic (independent of temperature) equation of state

$$p = -\kappa \boldsymbol{\nabla} \cdot \boldsymbol{u}_F, \tag{4.40}$$

where κ is the fluid bulk modulus. Note that pressure is positive in compression. Differentiating this equation twice with respect to time and using the linearized form (4.39) of the momentum equation yields the wave equation

$$\frac{1}{c^2}\frac{\partial^2 p}{\partial t^2} = \boldsymbol{\nabla}^2 p \tag{4.41}$$

for the pressure field, where $c = \sqrt{\kappa/\rho}$ is the speed of wave propagation.

The boundary conditions associated with the wave equation (4.41) usually consist of both Dirichlet and Neumann conditions. On portion Γ_D^F of the boundary of the fluid domain Ω_F the value of pressure is prescribed as $p = p_D$, where p_D is a given function. The remaining part of the boundary consist of a portion Γ_N^F where the Neumann condition $\partial p/\partial n = h_F$ is prescribed (h_F is given), and a portion Γ_I, the interface between the fluid and the structure, on which the fluid–structure interaction relations must be enforced.

The key point in coupling the fluid and structural domains lies in the imposition of the interface conditions. The relations on the fluid–solid interface Γ_I are:

- *Displacement/velocity compatibility.* An inviscid fluid is free to slip along the structural interface, but the displacement/velocity of the fluid along the normal to the interface must be equal to the displacement/velocity of the structure along the same direction, namely

$$\boldsymbol{n}_F \cdot \boldsymbol{u}_F = \boldsymbol{n}_F \cdot \boldsymbol{u}_S \qquad \text{(continuity of normal displacements)}, \tag{4.42a}$$

$$\boldsymbol{n}_F \cdot \boldsymbol{v}_F = \boldsymbol{n}_F \cdot \boldsymbol{v}_S \qquad \text{(continuity of normal velocities)}. \tag{4.42b}$$

This condition ensures that the fluid and structural domains will not detach or overlap during the motion. Both equations are equivalent and we use one or the other depending on the formulation employed (displacements or velocities).

- *Traction equilibrium*. That is, the stresses in the fluid must be equal to the stresses in the structure. For inviscid fluids this condition reduces to

$$\boldsymbol{n}_F\, p = \boldsymbol{n}_S \cdot \boldsymbol{\sigma} \qquad \text{(continuity of interface traction vector)}, \tag{4.42c}$$

where $\boldsymbol{\sigma}$ is the stress tensor on the structure.

Then, denoting by q the weighting function for the fluid, the weak form of the pressure formulation is given by

$$\left(q, \frac{1}{c^2}\ddot{p}\right)_{\Omega_F} + (\boldsymbol{\nabla} q, \boldsymbol{\nabla} p)_{\Omega_F} - (q, h_F)_{\Gamma_N^F} + (q, \rho_F(\boldsymbol{n}_F \cdot \ddot{\boldsymbol{u}}_S))_{\Gamma_I} = 0. \tag{4.43}$$

The last integral represents the coupling term with the structure and is obtained using

$$\frac{\partial p}{\partial n} = -\rho_F\,(\ddot{u}_n)_F = -\rho_F\,(\ddot{u}_n)_S$$

from the linearized momentum equation (4.39) and the continuity condition (4.42a).

The governing equation for the structural domain is the linearized momentum equation

$$\rho_S \frac{\partial^2 \boldsymbol{u}_S}{\partial t^2} = \boldsymbol{\nabla} \cdot \boldsymbol{\sigma} + \boldsymbol{b}, \tag{4.44}$$

where $\boldsymbol{u}_S$ is the structural displacement vector, $\boldsymbol{\sigma}$ the Cauchy stress and $\boldsymbol{b}$ the prescribed body force per unit volume. The boundary conditions consist of the Dirichlet condition $\boldsymbol{u}_S = \boldsymbol{u}_D$ on Γ_D^S, the Neumann condition $\boldsymbol{n} \cdot \boldsymbol{\sigma} = \mathbf{h}_S$ on Γ_N^S and relations (4.42) on the interface Γ_I with the fluid.

Denoting by $\boldsymbol{w}_S$ the vector of weighting functions for the momentum equation (4.44) and noting that the normal traction applied by the fluid on Γ_I is $p\,\boldsymbol{n}_F$, the weak form of the structural problem is

$$\begin{aligned}(\boldsymbol{w}_S, \rho_S\, \ddot{\boldsymbol{u}}_S)_{\Omega_S} + (\boldsymbol{\nabla} \boldsymbol{w}_S, \boldsymbol{\sigma})_{\Omega_S} - (\boldsymbol{w}_S, \boldsymbol{b})_{\Omega_S}& \\ - (\boldsymbol{w}_S, \mathbf{h}_S)_{\Gamma_N} - (\boldsymbol{w}_S, p\,\boldsymbol{n}_F)_{\Gamma_I} &= 0.\end{aligned} \tag{4.45}$$

For a linear elastic structure interacting with an acoustic fluid, the system of semi-discrete equations resulting from the finite element spatial discretization of the weak forms (4.43) and (4.45) takes the following partitioned matrix form in the absence of damping, see Belytschko (1983) for details:

$$\begin{pmatrix} \mathbf{M}_S & \mathbf{0} \\ -\rho_F\, \mathbf{R} & \mathbf{M}_F \end{pmatrix} \begin{pmatrix} \ddot{\mathbf{u}}_S \\ \ddot{\mathbf{p}} \end{pmatrix} + \begin{pmatrix} \mathbf{K}_S & \mathbf{R}^T \\ \mathbf{0} & \mathbf{K}_F \end{pmatrix} \begin{pmatrix} \mathbf{u}_S \\ \mathbf{p} \end{pmatrix} = \begin{pmatrix} \mathbf{f}_S^{\text{ext}} \\ \mathbf{f}_F^{\text{ext}} \end{pmatrix}. \tag{4.46}$$

Unfortunately, the partitioned matrices in this system are not symmetric. This lack of symmetry represents a drawback in the case of implicit time integration of the equations and requires non-standard algorithms to extract eigenvalues for vibration analysis. This has motivated the development of symmetric formulations, such as the ones proposed by Moran and Ohayon (1979) and Kehr-Candille and Ohayon (1992), see also the references therein. In the framework of vibro-acoustic studies of aeronautical structures, such as liquid-propelled launch vehicles, they consider the response of an elastic structure containing an inviscid compressible fluid. The structure is again described by the displacement field $\boldsymbol{u}_S$, but the fluid is now described by two scalar fields, the pressure p and the displacement potential ϕ, such that $u_F = \boldsymbol{\nabla}\phi$ and $\int_{\Omega_F} \phi = 0$. This choice of flow variables leads to symmetric formulations and allows the incorporation of a damping mechanism on the fluid–structure interface.

Remark 4.10 (Added mass concept). If the fluid can be considered incompressible, the wave equation (4.41) for the pressure field becomes the Laplace equation $\boldsymbol{\nabla}^2 p = 0$ and thus, there is no pressure-mass term in the second equation of system (4.46). The pressure is then obtained from the second equation as

$$\mathbf{p} = \mathbf{K}_F^{-1}\left(\mathbf{f}_F^{\text{ext}} + \rho_F\,\mathbf{R}\,\ddot{\mathbf{u}}_S\right)$$

and can be substituted in the first equation of system (4.46) to obtain the modified structural equation

$$\left(\mathbf{M}_S + \rho_F\,\mathbf{R}^T\,\mathbf{K}_F^{-1}\,\mathbf{R}\right)\ddot{\mathbf{u}}_S + \mathbf{K}_S\,\mathbf{u}_S = \mathbf{f}_S^{\text{ext}} - \mathbf{R}^T\,\mathbf{K}_F^{-1}\,\mathbf{f}_F^{\text{ext}}.$$

The need for a coupled solution has thus been eliminated and the effect of the incompressible fluid on the structure is described by a so-called *added mass* term that modifies the standard mass matrix $\mathbf{M}_S$. Note that, in contrast to the usual sparse and banded finite element matrices, the added mass matrix is full for the structural nodes in contact with the fluid.

4.7.2 Nonlinear transient dynamic problems

The second class of fluid–structure problems we wish to consider is concerned with situations involving a large-displacement response of the fluid, possibly accompanied by a nonlinear response of the structure. In such cases, the use of the ALE formulation introduced in Section 1.4 is recommended to model the fluid domain and ease its coupling with the structural domain usually treated in the Lagrangian description. We shall illustrate a particular problem in this class, namely the nonlinear transient dynamic response of fluid–structure systems. Fluid–structure interaction phenomena of this type are often encountered in the safety studies of industrial plants or components, where transient events such as explosions, impacts or crashes must be simulated. The particular algorithms discussed below are implemented in PLEXIS-3C, a finite element computer code for fast transient analyses (Bung et al., 1989).

4.7.2.1 Structural modeling A Lagrangian mesh in which material points and nodal points remain coincident throughout the calculation is usually adopted for the structural domain. If equilibrium is expressed in the current configuration in terms of the Cauchy stress, $\boldsymbol{\sigma}$, the weak form of the momentum equation is identical to that in equation (4.45) for the linear theory. Then, the following set of differential equations in time results from its finite element spatial discretization:

$$\mathbf{M}_S\, \mathbf{a}_S = \mathbf{f}_{ext} - \mathbf{f}_{int}, \tag{4.47}$$

where $\mathbf{M}_S$ is the mass matrix of the structure, $\mathbf{a}_S$ the vector of nodal accelerations, $\mathbf{f}_{ext}$ is a nodal vector of externally applied loads and $\mathbf{f}_{int}$ the nodal vector of internal forces arising from the discretization of the stress-divergence term in the momentum equation. See for instance Belytschko (1983) or Belytschko, Liu and Moran (2000) for a more detailed account of the Lagrangian mesh/Cauchy stress formulation. In fast transient problems, the time integration of the differential system (4.47) is usually performed using the explicit central difference scheme

$$\mathbf{v}_S^{n+1} = \mathbf{v}_S^n + \frac{\Delta t}{2}\left(\mathbf{a}_S^n + \mathbf{a}_S^{n+1}\right) \quad \text{and} \quad \mathbf{u}_S^{n+1} = \mathbf{u}_S^n + \Delta t\left(\mathbf{v}_S^n + \frac{\Delta t}{2}\mathbf{a}_S^n\right),$$

where $\mathbf{v}_S$ is the vector of nodal velocities and $\mathbf{u}_S$ the vector of nodal displacements. The last equation gives the new mesh node displacements and thus the updated configuration of the structural domain. In this new configuration, the stresses $\boldsymbol{\sigma}^{n+1}$ are then evaluated by application of constitutive models relating an objective rate of Cauchy stress to the strain rate (or stretching). The latter is evaluated from the half-step velocity $\mathbf{v}_S^{n+1/2} = \mathbf{v}_S^n + \frac{\Delta t}{2}\mathbf{a}_S^n$.

Note that the compatibility conditions (4.42) must be enforced along the interface between the structure and the fluid. Their implementation in the context of an ALE description of the fluid domain is discussed in Section 4.7.2.4.

4.7.2.2 ALE algorithms for compressible fluids The fluid is again assumed inviscid and compressible and its flow governed by the Euler equations. To model the hydrodynamic domain in the ALE description, a common practice is to use simple finite elements with linear or multilinear local approximations for the fluid velocity, $\boldsymbol{v}$, and the mesh velocity, $\hat{\boldsymbol{v}}$. Furthermore, the density, ρ, the specific internal energy, e, and hence the pressure, p, will be assumed elementwise constant.

According to the developments in Section 1.4, where the relative velocity $\boldsymbol{c} := \boldsymbol{v} - \hat{\boldsymbol{v}}$ between the material and the mesh has been introduced, the following integral form of the mass and internal energy equations is used for updating the element mass and internal energy:

$$\begin{aligned}
\left.\frac{\partial M^e}{\partial t}\right|_{\chi} &= \left.\frac{\partial}{\partial t}\right|_{\chi} \int_{\Omega^e} \rho\, d\Omega = -\int_{\Gamma^e} \rho\, \boldsymbol{c}\cdot\boldsymbol{n}\, d\Gamma, \\
\left.\frac{\partial I^e}{\partial t}\right|_{\chi} &= \left.\frac{\partial}{\partial t}\right|_{\chi} \int_{\Omega^e} \rho\, e\, d\Omega = -\int_{\Omega^e} p\, \boldsymbol{\nabla}\cdot\boldsymbol{v}\, d\Omega - \int_{\Gamma^e} \rho\, e\, \boldsymbol{c}\cdot\boldsymbol{n}\, d\Gamma.
\end{aligned}$$

In addition, use is made of the rate of change of the element volume V^e given by

$$\left.\frac{\partial V^e}{\partial t}\right|_{\chi} = \int_{\Gamma^e} \hat{\boldsymbol{v}}\cdot\boldsymbol{n}\, d\Gamma$$

to update the element density and specific internal energy via

$$\rho^e = \frac{M^e}{V^e} \qquad \text{and} \qquad e^e = \frac{I^e}{M^e}.$$

On the other hand, the weak form of the ALE momentum equation (1.22) is given by

$$(\boldsymbol{w}_F, \rho \boldsymbol{v}_t) + (\boldsymbol{w}_F, \rho(\boldsymbol{c}\cdot\boldsymbol{\nabla})\boldsymbol{v}) = (\boldsymbol{\nabla}\boldsymbol{w}_F, p\mathbf{I}) + (\boldsymbol{w}_F, \rho\boldsymbol{b}) + (\boldsymbol{w}_F, \mathbf{t}_F)_{\Gamma_N},$$

where $\boldsymbol{w}_F$ represents the vector of weighting functions, $\mathbf{t}_F = p\boldsymbol{n}$ is the normal traction vector prescribed on portion Γ_N of the fluid domain boundary. The finite element discretization of the previous momentum equation results in the matrix system

$$\mathbf{M}_F \left.\frac{\partial \boldsymbol{v}}{\partial t}\right|_{\boldsymbol{\chi}} = \mathbf{f}_t + \mathbf{f}_b + \mathbf{f}_p + \bar{\mathbf{f}}, \tag{4.48}$$

where $\mathbf{M}_F$ is the mass matrix, $\mathbf{f}_t$ represents the nodal loads induced by the transport of momentum components, $\mathbf{f}_b$ denotes the nodal loads due to the applied body forces $\rho\boldsymbol{b}$, $\mathbf{f}_p$ represents the nodal loads due to the fluid pressure p, and $\bar{\mathbf{f}}$ accounts for the prescribed boundary pressure. This last vector includes the reaction induced by prescribed displacements and velocities, as well as the compatibility conditions at the fluid–structure interface

Because of the presence of transport terms in the above ALE equations for mass, momentum and internal energy, special treatment is suggested for time integration of the equations in the fluid domain. As already mentioned in Section 4.6, a time integration scheme valid for flows at all speeds can be derived by adapting the fractional-step method introduced in the finite difference context by Hirt et al. (1974) and Stein et al. (1977). The method is organized in three steps. Since the transport terms in the semi-discrete conservation equations vanish in the Lagrangian formulation, when $\boldsymbol{c} = \boldsymbol{v} - \hat{\boldsymbol{v}} = 0$, the Lagrangian form of the equations is solved in the first step using an explicit time scheme. This provides intermediate values of the density, velocity and internal energy. Note that the element mass does not change in the Lagrangian phase. In the second step, which is also assumed Lagrangian, an implicit pressure calculation is introduced. This step eliminates the usual stability condition of explicit schemes that limits sound waves to travel no further than one element per time step. This allows solutions to be obtained for flows at all speeds, from highly compressible to nearly incompressible. The transport (convection) contributions are then added in the last step of the time integration procedure. The evaluation of the convective terms requires an updating of the mesh velocities $\hat{\boldsymbol{v}}$. An automatic algorithm, such as the one proposed by Giuliani (1982), should be employed for this purpose. Note that the transport phase is hyperbolic and must be stabilized. Donea, Giuliani and Halleux (1982) discuss additional details on the above fractional-step integration of the conservation equations.

4.7.2.3 Boundary conditions Having established the general computational framework for the treatment of the structural and fluid domains, we now consider how boundary conditions are enforced. As usual, they are subdivided into natural and essential conditions. Natural conditions are prescribed external loads, such as concentrated forces or distributed pressures. Their treatment does not pose any problem, as they fit as natural boundary conditions in the relevant variational formulation. The essential boundary conditions include prescribed displacements and velocities, as well as the compatibility conditions at the fluid–structure interface. Essential boundary conditions on the fluid domain are best implemented by means of the Lagrange multiplier method, see Remark 1.13.

Consider for instance that the essential conditions are imposed on the half-step value $\mathbf{v}^{n+1/2}$ of the velocity at the fluid–structure interface (thus, $\mathbf{f}_t$ is zero in this step). Such conditions can be expressed as a linear set of constraints of the form

$$\mathbf{A}\mathbf{v}^{n+1/2} = \mathbf{b}^{n+1/2}, \tag{4.49}$$

where $\mathbf{A}$ is a sparse (most of its coefficients are zero, in particular those related to degrees of freedom with no essential boundary condition) rectangular matrix (number of degrees of freedom for which essential boundary conditions are imposed times total number of degrees of freedom), $\mathbf{v}$ is the vector of nodal velocities and $\mathbf{b}$ a vector of (possibly time-dependent) prescribed values. The equilibrium equations for the half-step are written in the form

$$\mathbf{M}_F \left(\frac{\mathbf{v}^{n+1/2} - \mathbf{v}^{n-1/2}}{\Delta t} \right) = \mathbf{f}_b^n + \mathbf{f}_p^n - \mathbf{f}_{\text{reac}}^n, \tag{4.50}$$

where $\mathbf{M}_F$ is the mass matrix, $\mathbf{f}_b$ and $\mathbf{f}_p$ were already defined in (4.48), and finally, $\mathbf{f}_{\text{reac}}$ stands for the unknown reaction forces produced by the essential boundary conditions. We assume, for simplicity, that all the prescribed boundary pressure is due to Dirichlet boundary conditions, that is $\bar{\mathbf{f}}^n \equiv -\mathbf{f}_{\text{reac}}$. In order to introduce the constraints defined by (4.49) in the equilibrium equations (4.50), Lagrange multipliers are introduced. The unknown reactions are expressed as

$$\Delta t\, \mathbf{f}_{\text{reac}}^n = \mathbf{A}^T \boldsymbol{\lambda}^n, \tag{4.51}$$

where $\boldsymbol{\lambda}$ is the vector of Lagrange multipliers. Substituting this expression into the equilibrium equations (4.50) and rewriting (4.49) yields

$$\begin{pmatrix} \mathbf{M}_F & \mathbf{A}^T \\ \mathbf{A} & \mathbf{0} \end{pmatrix} \begin{pmatrix} \mathbf{v}^{n+1/2} \\ \boldsymbol{\lambda}^n \end{pmatrix} = \begin{pmatrix} \mathbf{M}_F\, \mathbf{v}^{n-1/2} + \Delta t(\mathbf{f}_b^n + \mathbf{f}_p^n) \\ \mathbf{b}^{n+1/2} \end{pmatrix}.$$

Since the mass matrix is regular (in fact, it is symmetric and positive definite), we can compute its inverse and the previous system can be solved in two steps. The first equation can be written as

$$\mathbf{v}^{n+1/2} = \mathbf{v}^{n-1/2} + \mathbf{M}_F^{-1}\left(\Delta t(\mathbf{f}_b^n + \mathbf{f}_p^n) - \mathbf{A}^T \boldsymbol{\lambda}^n\right), \tag{4.52a}$$

and after substitution in the second one, we solve for $\boldsymbol{\lambda}^n$

$$\mathbf{A}\mathbf{M}_F^{-1}\mathbf{A}^T \boldsymbol{\lambda}^n = \mathbf{A}\mathbf{v}^{n-1/2} + \Delta t \mathbf{A}\mathbf{M}_F^{-1}(\mathbf{f}_b^n + \mathbf{f}_p^n) - \mathbf{b}^{n+1/2}. \tag{4.52b}$$

Note that $\mathbf{A}\mathbf{M}_F^{-1}\mathbf{A}^T$ is regular if and only if the rank of $\mathbf{A}$ is maximum, namely rank $\mathbf{A}$ is equal to the number of degrees of freedom for which essential boundary conditions are imposed. Once the Lagrange multipliers are known, $\mathbf{v}^{n+1/2}$ is determined from the first equation. Finally, if required the reaction forces $\mathbf{f}^n_{\text{reac}}$ are then obtained from expression (4.51).

This approach as stated has an important drawback (specially in the context of an explicit code): the inverse of the mass matrix is full! Although the system of equations in (4.52b) is not large (its size is only the number of degrees of freedom for which essential boundary conditions are imposed), the evaluation of the inverse matrix precludes the use of the algorithm as presented. Note however, that in transient dynamics it is standard to use the lumped (or diagonal) mass matrix, $\mathbf{M}^L_F$. If this matrix is used in (4.52) things change drastically: its inverse is diagonal (and trivial) and thus $\mathbf{A}[\mathbf{M}^L_F]^{-1}\mathbf{A}^T$ can be constructed easily. Moreover, the lumped-mass matrix uncouples each degree of freedom of $\mathbf{v}^{n+1/2}$ and consequently equations (4.52) can be particularized for the subset of degrees of freedom for which essential boundary conditions are imposed, that is each matrix in the product $\mathbf{A}[\mathbf{M}^L_F]^{-1}\mathbf{A}^T$ can be square. Under these circumstances the previous algorithm is efficient and can be used in the context of fast transient dynamics computations.

4.7.2.4 *Fluid–structure coupling* As already noticed for the linear acoustic approximation, Section 4.7.1, a key point in the fluid–structure coupling lies in the prescription of the interface conditions. These conditions are already described by equations (4.42).

To illustrate the ALE coupling procedure along fluid–solid interfaces, consider a structural member in permanent contact with an inviscid fluid which is allowed to slide along the structure. Two nodes are placed at each point of the interface: one fluid node and one structural node. Since the fluid is treated in the ALE formulation, the movement of the fluid mesh may be chosen completely independent of the movement of the fluid itself. In particular, we may constrain the fluid nodes to remain contiguous to the structural nodes, so that all nodes on the sliding interface remain permanently aligned. This is achieved by imposing that the grid velocity $\hat{\boldsymbol{v}}_F$ of the fluid nodes at the interface be equal to the material velocity $\boldsymbol{v}_S$ of the adjacent structural nodes.

The permanent alignment of nodes at ALE interfaces greatly facilitates the flow of information between the fluid and structural domains and permits fluid–structure coupling to be effected in the simplest and most elegant manner; that is, the imposition of equations (4.42) is simple because of the node alignment.

The equilibrium condition, see equation (4.42c), as in the linear acoustic regime, states that at each point of the interface between structure and inviscid fluid a contact pressure is transmitted. Moreover, the interaction pressure is directed along the normal to the interface. In the finite element representation, the continuous interface is replaced with a discrete approximation and instead of a distributed interaction pressure, consideration is given to its resultant at each interface node, that is to the interaction force $\mathbf{r}$.

The compatibility condition, stated in equation (4.42b), imposes continuity of the velocity components normal to the interface. Recall that the tangential velocities at the fluid and solid nodes on the interface are unconstrained. This condition can be expressed in the form (4.49). For instance, in 2D and for one point we have

$$\mathbf{A} = \left(n_x,\ n_y,\ -n_x,\ -n_y\right); \quad \mathbf{v} = \left(v_{Fx},\ v_{Fy},\ v_{Sx},\ v_{Sy}\right)^T; \quad \mathbf{b} = (0).$$

The interaction force $\mathbf{r}$, see (4.51), at this point is expressed in terms of one Lagrange multiplier (only one condition is presented):

$$\Delta t\left(r_{Fx},\ r_{Fy},\ r_{Sx},\ r_{Sy}\right) = \left(n_x,\ n_y,\ -n_x,\ -n_y\right)\lambda.$$

This is repeated for each interaction node to construct the vectors and matrices in (4.49) and (4.51). This allows us to solve (4.52b) and from (4.51) obtain the interaction force at each node on the ALE interface. In this manner, the fluid and structural responses can be solved separately, though achieving a tight coupling between the two subdomains.

Remark 4.11 (Normal to a discrete interface). In practice, especially in complex 3D configurations, one major difficulty is to determine the normal vector at each node of the fluid–structure interface. Various algorithms have been developed to deal with this issue, Casadei and Halleux (1995) and Casadei and Sala (1999) present detailed solutions. In 2D the tangent to the interface at a given node is usually defined as parallel to the line connecting the nodes at the ends of the interface segments meeting at that node.

Remark 4.12. The above treatment of the coupling problem only applies to those portions of the structure which are always submerged during the calculation. As a matter of fact, there may exist portions of the structure which only come into contact with the fluid some time after the calculation begins. This is, for instance, the case for structural parts above a fluid-free surface. For such portions of the structural domain some sort of sliding treatment is necessary, as for standard Lagrangian methods.

4.7.3 Illustrative examples

Two numerical simulations are presented that illustrate ALE finite elements in transient dynamic fluid–structure interaction as outlined in the previous sections. Calculations were performed with PLEXIS-3C finite element code (Bung et al., 1989).

4.7.3.1 Flexible vessel experiment A schematic of the configuration is depicted in Figure 4.11. A thin cylindrical vessel with a hemispherical bottom is nearly completely filled with liquid and impulsively loaded by the detonation of an explosive charge located on the vessel axis. The vessel contains a flexible inner shield hinged at its base. The top of the vessel is clamped to a rigid cover. Due to symmetry, only half of the configuration is actually modeled. First, a purely Lagrangian solution was attempted using the finite element mesh shown in Figure 4.11. Conical shell elements are used to model the structural parts, while triangular and quadrilateral fluid elements

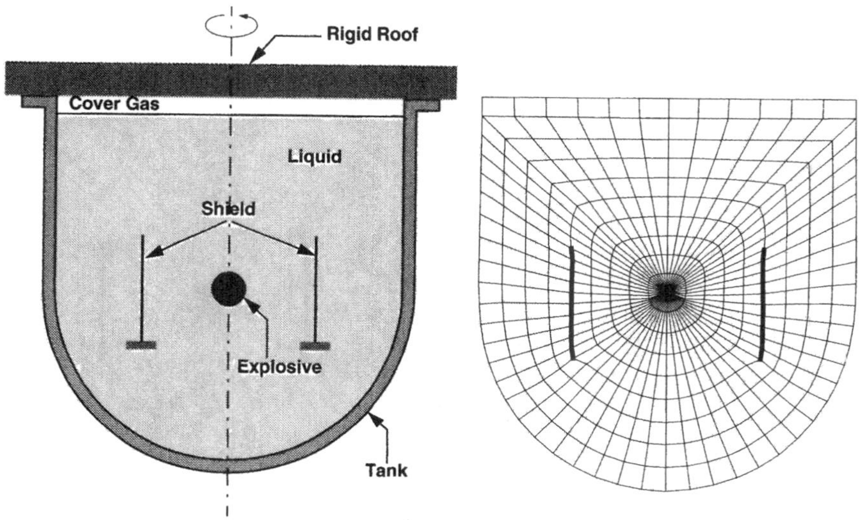

Fig. 4.11 Model of containment vessel and element mesh for Lagrangian and multiphase, multicomponent ALE solutions.

are employed to model the explosive charge, the liquid and the cover gas at the top of the vessel. The fluid–structure interfaces are not properly treated in this Lagrangian example because fluid and structural nodes coincide, causing the fluid to stick to the structural walls. As can be observed from Figure 4.12, large distortions of the fluid mesh do occur and due to mesh entanglement the calculations failed after 95 ms.

The experiment was then repeated using the ALE description in the hydrodynamic domain and treating fluid–structure interaction at the ALE interfaces as explained in the previous sections. The fluid is now able to slide along the structural walls and, as shown in Figure 4.12, the automatic mesh rezoning algorithm succeeds in keeping the hydrodynamic mesh reasonably regular.

Another experiment was then performed in which the ALE algorithm was used together with a heterogeneous (multiphase, multicomponent) fluid formulation. Hence, there are no Lagrangian fluid–fluid interfaces in this calculation. The finite element mesh is depicted in Figure 4.13. As can be appreciated from this same figure, the mesh distortions are now minimal. The position of physical fluid–fluid interfaces may be estimated from Figure 4.13 where the mass fractions of the various components are plotted. The present results are found to be in substantial agreement with those reported in Figure 4.12 obtained with the ALE multi-fluid formulation.

4.7.3.2 An industrial application As an example of an industrial application of the ALE methodology, we consider the simulation of an explosion in a power transformer cell which is part of an underground power plant, see Figure 4.14a. This problem is discussed in detail by Casadei et al. (2001) and the results are reproduced

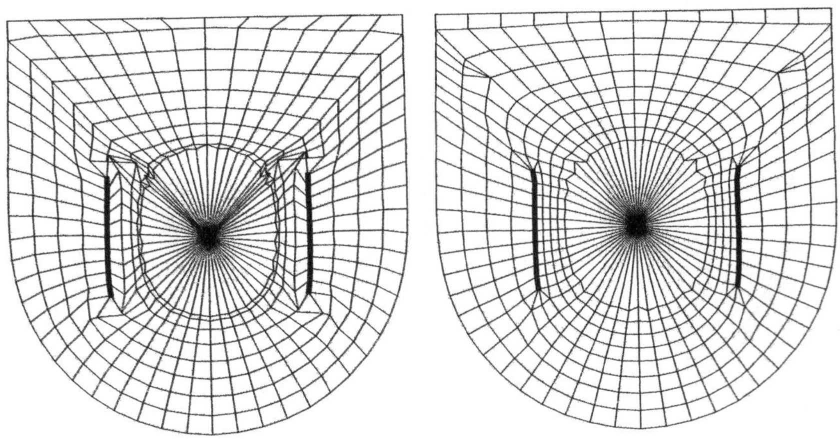

Fig. 4.12 Purely Lagrangian solution of containment vessel (left) and ALE solution using Lagrangian fluid–fluid interfaces (right).

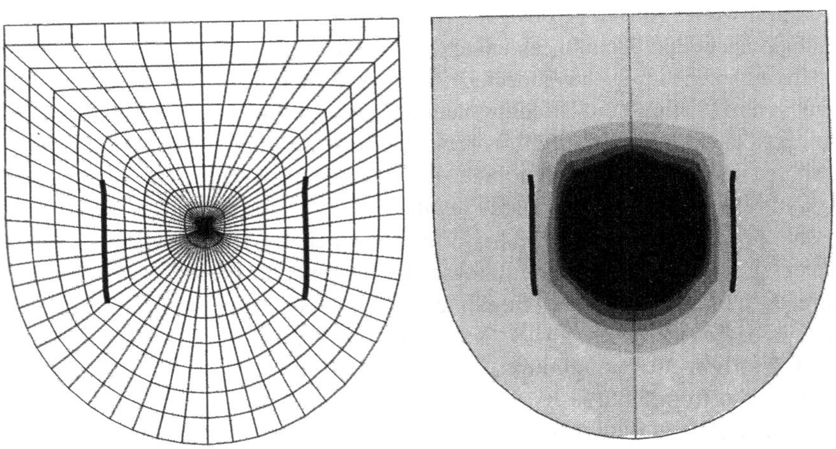

Fig. 4.13 ALE solution of containment vessel using multicomponent fluid formulation: final mesh configuration (left) and mass fractions of the various components (right).

here with their kind permission. If an electrical fault occurs inside the oil-insulated equipment, an arc is likely to occur causing pyrolysis of the mineral oil. As a result, a large quantity of gaseous hydrocarbons is produced containing over 70% hydrogen which may ultimately break the transformer tank and propagate inside the cell.

Under critical circumstances, the air–hydrocarbon mixture may ignite giving rise to a strong blast. The aim of the study is to assess the structural response in case of explosion of the aluminum tubes (1 m diameter) that contain and insulate by means of SF_6 gas the high-voltage bars running out from the power transformer. Suitable assumptions allow the determination of the initial conditions of a high-pressure and temperature bubble that simulates the energy associated with the explosion. The explosive bubble is surrounded by air at normal conditions, see Figure 4.14b. An elasto-plastic material describes the behavior of the aluminum tubes represented in Figure 4.14c.

Figure 4.14d shows the structural portion of the computational domain. The aluminum tubes are discretized by means of shell elements, while the transformer and cell walls are assumed rigid and are thus not discretized. Figure 4.14e shows a portion of the fluid domain discretized by means of tetrahedral elements and the embedded shell structural elements for the aluminum tubes.

Figures 4.14f and 4.14g show the computed mass fractions of air at half the time and at the end of the transient. A pressure plot at half the time of the transient is shown in Figure 4.14h, while Figure 4.14i illustrates the deformed shape of the aluminum tubes at 50 ms. Deformation is amplified 10 times.

As shown by the above example, ALE finite element computer codes, such as PLEXIS-3C, represent very useful modeling tools for the simulation of postulated accident situations in industrial plants.

4.8 SOLVED EXERCISES

4.8.1 One-step Taylor–Galerkin solution of Burgers' equation

The inviscid Burgers' equation (4.2) is solved over the 1D domain $]0, L[$. The inlet boundary condition $u(0,t) = 1$ is prescribed and the following initial data are used:

$$u(x,0) = \begin{cases} 1 & 0 \le x \le 0.64, \\ 1 - (x - 0.64)/(0.20) & 0.64 \le x \le 0.84, \\ 0 & 0.84 \le x \le 1.0. \end{cases}$$

This skew initial profile straightens as time goes on until a discontinuity is formed which propagates at the dimensionless speed of 0.5. Time integration is performed by means of the one-step second-order scheme

$$\frac{u^{n+1} - u^n}{\Delta t} = u_t^n + \frac{\Delta t}{2} u_{tt}^n,$$

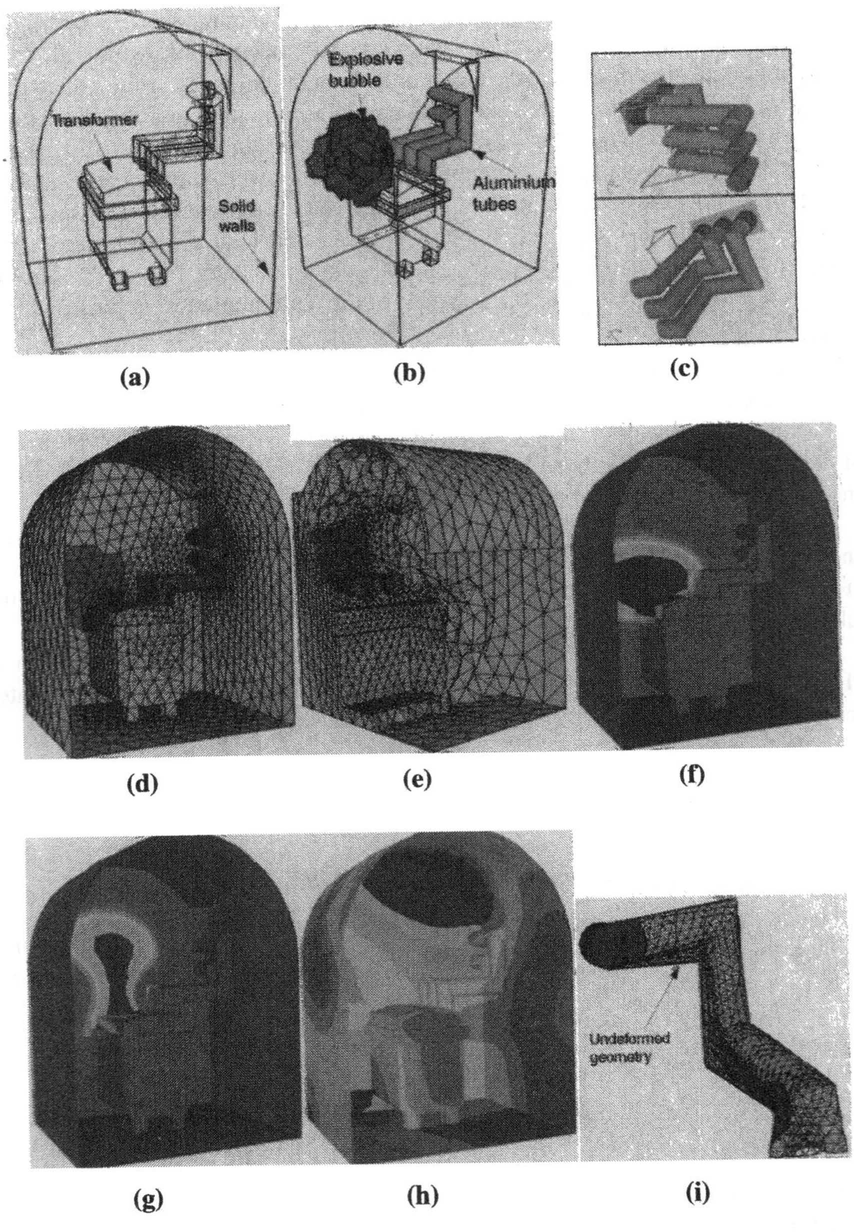

Fig. 4.14 Simulation of an explosion in a power transformer cell (Courtesy of ENEL Research Hydraulics and Structures, Milano).

where in view of the differential equation, see (4.2),

$$u_t = -f_x, \quad \text{and } u_{tt} = -f_{xt} = -f_{tx} = -\left(f_u\, u_t\right)_x = \left(u^2\, u_x\right)_x.$$

The time discretized version of Burgers' equation therefore reads

$$\frac{u^{n+1} - u^n}{\Delta t} = -f_x^n + \frac{\Delta t}{2}\left((u^n)^2\, u_x^n\right)_x.$$

After integration by parts of both spatial terms, the second-order Taylor–Galerkin formulation is given as follows:

$$\int_0^L w\, \frac{u^{n+1} - u^n}{\Delta t}\, dx = \int_0^L w_x\, f^n\, dx - \frac{\Delta t}{2} \int_0^L w_x\, (u^n)^2\, u_x^n\, dx$$
$$- \left[w \left(f^n - \frac{\Delta t}{2} (u^n)^2\, u_x^n \right) \right]_{x=0}^{x=L}.$$

The boundary term can be rewritten using $f_t = f_u\, u_t = -u^2\, u_x$, as

$$-\left[w \left(f^n - \frac{\Delta t}{2} (u^n)^2\, u_x^n \right) \right]_{x=0}^{x=L} = -\left[w \left(f^n + \frac{\Delta t}{2}\, f_t^n \right) \right]_{x=0}^{x=L},$$

which provides a natural condition for prescribed boundary flux. We therefore replace the inlet boundary condition $u(0,t) = 1$ by the condition $f(0,t) = \left(\frac{1}{2}u^2\right)_{x=0} = 0.5$.

The problem unknown u is linearly interpolated over an element and three different representations of the flux f in the convective term are tested:

1. f is elementwise constant and determined from the mean value of u in the considered element (*constant representation*);

2. f is determined from the value of u at the two element Gauss points and a two-point Gaussian quadrature is used to evaluate the convective term (*classical representation*);

3. f is linearly interpolated using its evaluation at the element nodes (*group representation*).

The two-point quadrature rule is also used on the diffusion term.

The scope of the exercise is essentially to compare the above three representations of the flux. The results obtained at dimensionless times $t = 0.16$ and $t = 0.30$ using a mesh of 50 uniform linear elements and a Courant number $C = u_{max}\Delta t/h = 0.5$ are shown in Figure 4.15. As long as the initial profile straightens, the solutions obtained with the three flux representations are quite similar. Nevertheless, one may already note a more pronounced dissipative effect when the linear interpolation of the flux is employed. The differences are definitely more marked when the discontinuity propagates, the group representation of the flux giving the best solution.

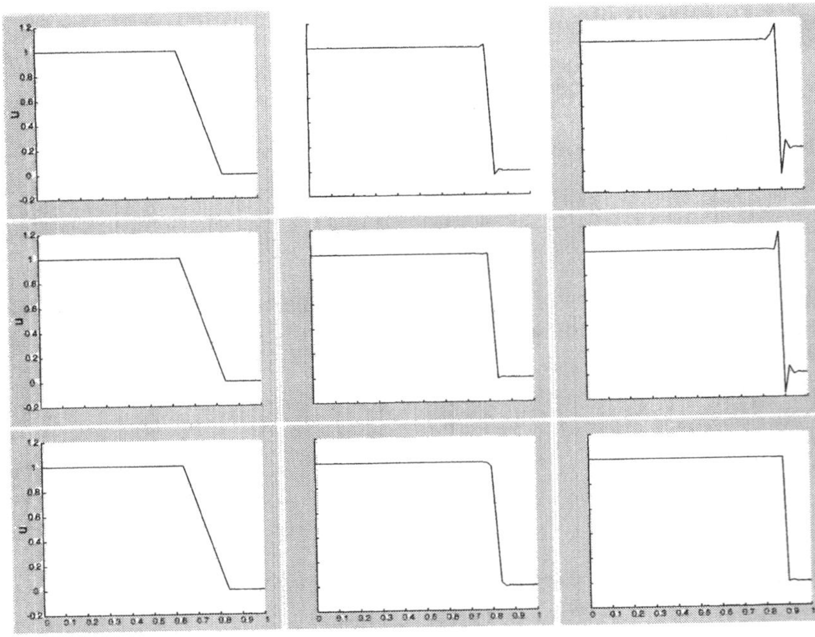

Fig. 4.15 Simulation by the one-step second-order Taylor–Galerkin method of the formation and propagation of a discontinuity. Comparison of three local representations of the flux in Burgers' equation: constant flux interpolation (top); classical interpolation (middle); linear flux interpolation (bottom). Solution times are from left to right: $t = 0, t = 0.16$ and $t = 0.30$.

4.8.2 The shock tube problem

We now solve the classical shock tube problem proposed by Sod (1978) for which there exists an exact solution to the 1D Euler equations. The problem involves a shock wave, a contact discontinuity and an expansion fan. A contact discontinuity is an interface between two fluid regions of different densities, but equal pressure. The shock tube problem is thus representative of the numerical difficulties encountered in the solution of the Euler equations of gas dynamics.

On the spatial domain $]0, 1[$, we solve the 1D system of Euler equations, see (4.16a)

$$\mathbf{U}_t + \mathbf{F}(\mathbf{U})_x = 0,$$

where $\mathbf{U}$, $\mathbf{F}$ and the Jacobian matrix $\mathbf{A}$ are described in Remark 4.5, the eigenvalues of $\mathbf{A}$ are explicitly determined in (4.24), and they define the three characteristics described by (4.25) and plotted in Figure 4.7. Under the assumption of a perfect gas, the pressure is given by (4.17a), namely

$$p = (\gamma - 1)\rho\big(E - \frac{v^2}{2}\big),$$

and we take $\gamma = 1.4$. The following initial data are used:

$0 \leq x \leq 1/2$	$1/2 < x \leq 1$
$\rho = 1.0$	$\rho = 0.125$
$\rho v = 0.0$	$\rho v = 0.0$
$\rho E = 2.5$	$\rho E = 0.25$

The initial density and pressure difference is maintained by a diaphragm which is ruptured at $t = 0$. The objective is to compute the evolution of the conservation variables as a shock and a contact discontinuity propagate along the tube.

One-step Taylor–Galerkin. Time integration of the Euler equations is performed by means of the one-step second-order Taylor–Galerkin scheme

$$\frac{\mathbf{U}^{n+1} - \mathbf{U}^n}{\Delta t} = \mathbf{U}_t^n + \frac{\Delta t}{2}\mathbf{U}_{tt}^n.$$

The associated variational form is given as follows after integration by parts of the spatial terms:

$$\int_0^L \mathbf{W} \cdot \frac{\mathbf{U}^{n+1} - \mathbf{U}^n}{\Delta t}\,dx = \int_0^L \mathbf{W}_x \cdot \mathbf{F}^n\,dx$$
$$- \frac{\Delta t}{2}\int_0^L \mathbf{W}_x \cdot (\mathbf{A}^n)^2\,\mathbf{U}_x^n\,dx - \left[\mathbf{W} \cdot \left(\mathbf{F}^n + \frac{\Delta t}{2}\mathbf{F}_t^n\right)\right]_{x=0}^{x=L},$$

where, as in the previous example, the differential equation, $\mathbf{U}_t = -\mathbf{F}_x$, is used to determine

$$\mathbf{U}_{tt} = -\mathbf{F}_{xt} = -\mathbf{F}_{tx} = -\left(\mathbf{A}\,\mathbf{U}_t\right)_x = \left(\mathbf{A}\,\mathbf{F}_x\right)_x = \left(\mathbf{A}^2\,\mathbf{U}_x\right)_x.$$

Note that the boundary term allows us to introduce prescribed flux components. In the present example, the definitions of $\mathbf{U}$, $\mathbf{F}$, see Remark 4.5, indicate that the boundary value of the density and energy flux is zero, while the boundary value of the momentum flux is given by the left and right pressures associated with the initial data.

The problem was solved using a mesh of 100 uniform linear elements. Time integration has been performed until $t = 0.2$ using a fixed time step $\Delta t = 1.5 \times 10^{-3}$. This corresponds to a Courant number

$$C = (v_{\max} + c)\frac{\Delta t}{h} = 0.45.$$

The convective and diffusive terms were integrated using a two-point Gaussian quadrature. A linear approximation of the flux was used in compression regions ($\partial v/\partial x < 0$), and the classical flux representation otherwise. No artificial viscosity was introduced into the Galerkin formulation.

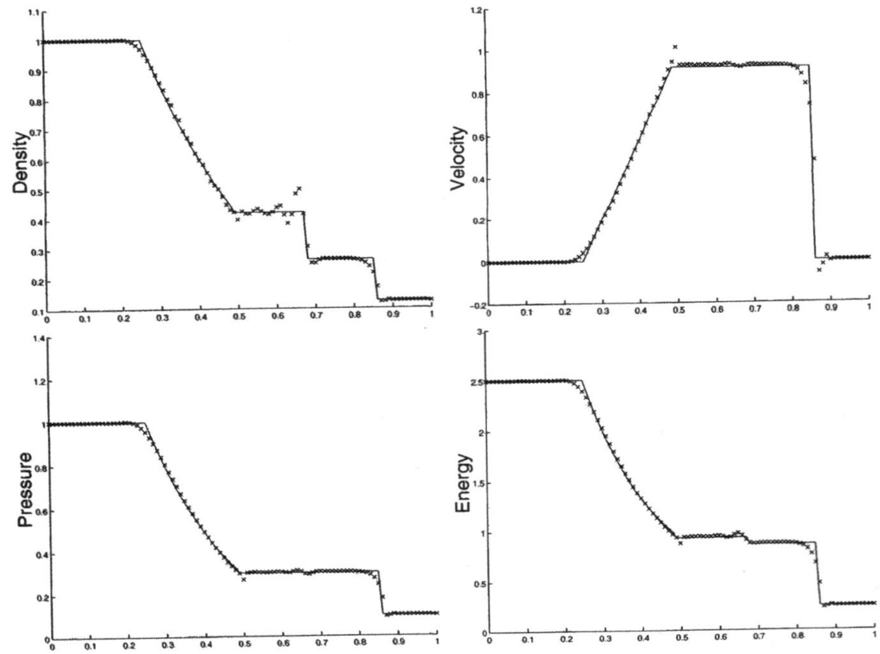

Fig. 4.16 Sod's shock tube: one-step Taylor–Galerkin without artificial viscosity.

Figure 4.16 shows a comparison with the exact solution. One notes that the position of the flow discontinuities is well predicted. However, in the absence of shock-capturing terms, oscillations are generated around sharp solution gradients. They are reduced in the next example by the selective addition of artificial viscosity.

Stabilization of the one-step Taylor–Galerkin solution. We repeat the solution of Sod's shock tube problem with the selective addition of artificial viscosity, as explained in Section 4.5.3.2. The idea is to first compute the solution to the Euler equations using the one-step second-order method as in the previous test case. Then, the second-order scheme is locally reduced to first order as indicated in equation (4.32).

The local modulation of the added viscosity is performed according to expressions (4.34) using the artificial viscosity method in equations (4.35) with parameter $\chi = 3.0$. The results obtained at time $t = 0.2$ are displayed in Figure 4.17 in comparison with the exact solution. The effect of the shock-capturing terms is to smooth out the flow discontinuities and thereby suppress the oscillations near sharp gradients typical of the second-order method.

Two-step Taylor–Galerkin. This method is described in Section 4.2.3.2. In the first step of the time integration procedure, we compute the half-step value $\mathbf{U}^{n+1/2}$

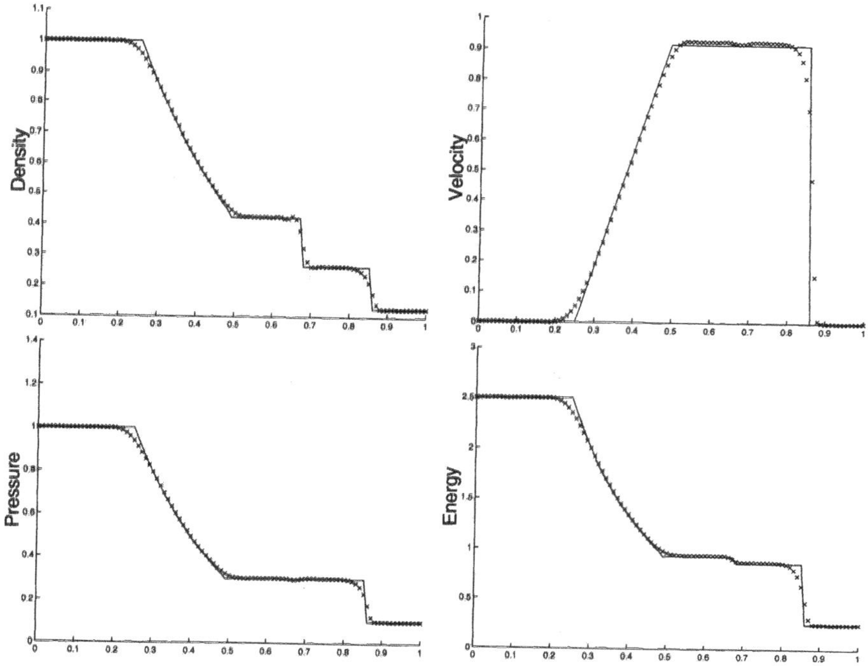

Fig. 4.17 Sod's shock tube: one-step Taylor–Galerkin with artificial viscosity.

of the flow variables and associated flux components at the center of the elements. Once the flux vector $\mathbf{F}^{n+1/2}$ is obtained, the end-of-step values are computed solving the variational equation

$$\int_0^L \mathbf{W} \cdot \frac{\mathbf{U}^{n+1} - \mathbf{U}^n}{\Delta t}\, dx = \int_0^L \mathbf{W}_x \cdot \mathbf{F}^{n+1/2}\, dx - \Big[\mathbf{W} \cdot \mathbf{F}^{n+1/2}\Big]_{x=0}^{x=L},$$

where the flux term has again been integrated by parts to produce a natural condition of prescribed boundary flux.

Once a solution is obtained with the second-order method in time, it is smoothed, locally, as in the previous test to reduce the non-physical oscillations near the discontinuities. The results obtained at time $t = 0.2$ with a time-step size $\Delta t = 0.002$ are displayed in Figure 4.18 in comparison with the exact solution. One notes that the simple two-step Taylor–Galerkin method gives results of the same quality as the one-step method.

Flux vector splitting. In order to produce an upwind-type spatial discretization of the 1D Euler equations we split the flux vector in the conservation equation

$$\mathbf{U}_t + \mathbf{F}_x = 0$$

into the form

$$\mathbf{U}_t + \mathbf{F}_x^+ + \mathbf{F}_x^- = 0$$

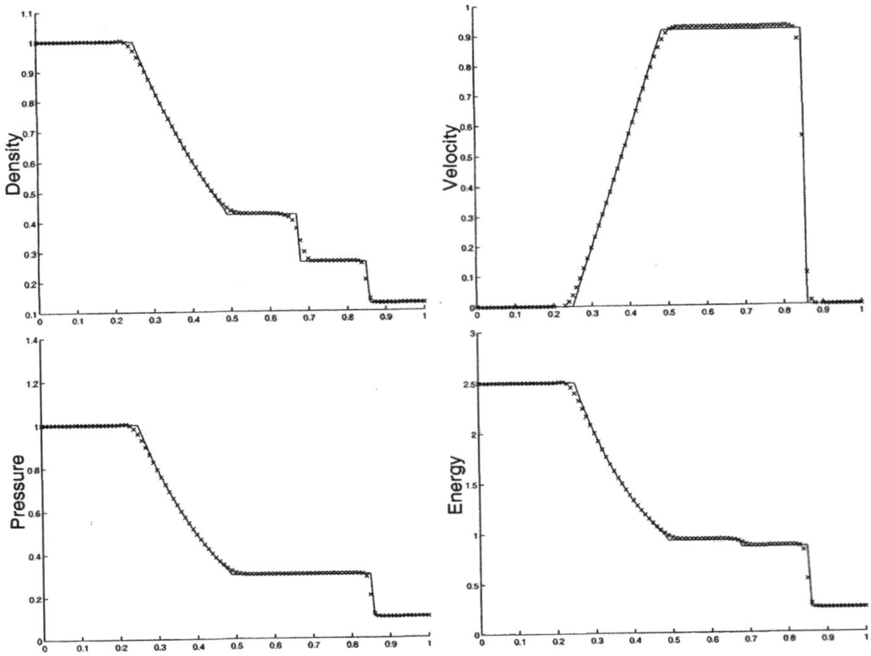

Fig. 4.18 Sod's shock tube: two-step Taylor–Galerkin with artificial viscosity.

as explained in Section 4.4.2.1. Moreover, we generalize the standard first-order upwind scheme as suggested by Steger and Warming (1981):

$$\frac{\mathbf{U}_i^{n+1} - \mathbf{U}_i^n}{\Delta t} = -\frac{\left(\mathbf{F}_i^+ - \mathbf{F}_{i-1}^+\right)^n}{h} - \frac{\left(\mathbf{F}_{i+1}^- - \mathbf{F}_i^-\right)^n}{h}.$$

Recall that $\mathbf{F}^+$ is always positive (or zero), while $\mathbf{F}^-$ is always negative (or zero).

To reproduce the flux vector splitting technique in the finite element context, we make use of the construction of upwind finite element schemes discussed in Chapter 2 and consider the weak form

$$\int_0^L \mathbf{W} \cdot \mathbf{U}_t \, dx = -\int_0^L \left(\mathbf{W} + \frac{h}{2}\mathbf{W}_x\right) \cdot \mathbf{F}_x^+ \, dx - \int_0^L \left(\mathbf{W} - \frac{h}{2}\mathbf{W}_x\right) \cdot \mathbf{F}_x^- \, dx,$$

or, after integration by parts of the convective terms

$$\int_0^L \mathbf{W} \cdot \mathbf{U}_t \, dx = \int_0^L \mathbf{W}_x \cdot \left(\mathbf{F}^+ - \frac{h}{2}\mathbf{F}_x^+\right) dx$$
$$+ \int_0^L \mathbf{W}_x \cdot \left(\mathbf{F}^- + \frac{h}{2}\mathbf{F}_x^-\right) dx - \Big[\mathbf{W} \cdot (\mathbf{F} \cdot \boldsymbol{n})\Big]_{x=0}^{x=L}.$$

We use a uniform mesh of 100 elements with piecewise linear interpolation of the conservation variables and associated flux components. Figure 4.19 shows the results.

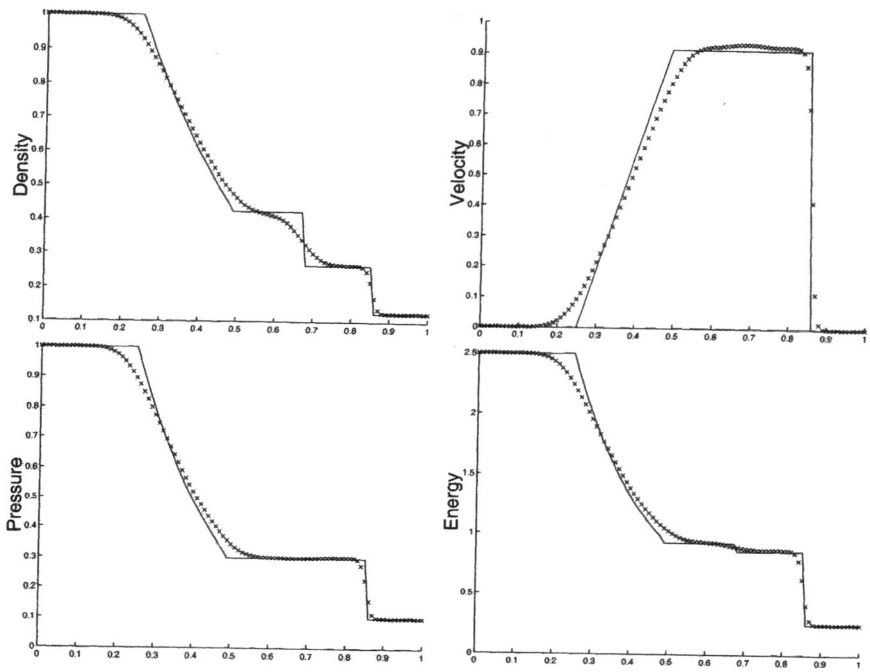

Fig. 4.19 Sod's shock tube: first-order in time and space flux vector splitting technique. Note the excessive diffusion introduced by standard monotone methods.

A time step $\Delta t = 0.004$, corresponding to a Courant number of 0.8, was employed with the Euler explicit method. The following flux vector splitting introduced by Steger and Warming (1981) was used (see also Hirsch, 1990, Chap. 20):

$$\mathbf{F}^{+} = \frac{1}{2\gamma}\begin{pmatrix} (2\gamma-1)\rho v + \rho c \\ 2(\gamma-1)\rho v^2 + \rho(v+c)^2 \\ (\gamma-1)\rho v^3 + \frac{1}{2}\rho(v+c)^3 + \frac{1}{2}\frac{3-\gamma}{\gamma-1}\rho c^2(v+c) \end{pmatrix}; \tag{4.53a}$$

and

$$\mathbf{F}^{-} = \frac{1}{2\gamma}\begin{pmatrix} \rho v - \rho c \\ \rho(v-\rho c)^2 \\ \frac{1}{2}\rho(v-c)^3 + \frac{1}{2}\frac{3-\gamma}{\gamma-1}\rho c^2(v-c) \end{pmatrix}. \tag{4.53b}$$

Here, $c = \sqrt{\gamma p/\rho}$ is the speed of sound, and p the pressure defined by the equation of state of a perfect gas.

At this point, it is important to recall the definitions of $\mathbf{F}_{\boldsymbol{n}}^{+}$ and $\mathbf{F}_{\boldsymbol{n}}^{-}$ adopted in Section 4.4.2.1, see Figure 4.9. It follows from such definitions that the splitting in (4.53) directly applies in the present example only at element ends where the unit outward normal has components $\boldsymbol{n} = (1, 0)$. At element ends with outward normal

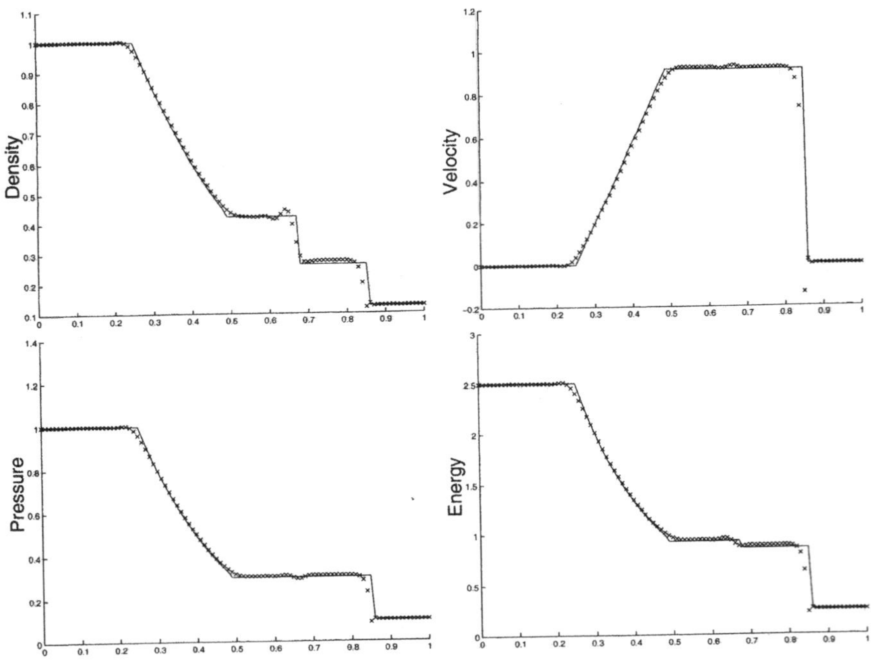

Fig. 4.20 Sod's shock tube: discontinuous Galerkin with artificial diffusion and forward Euler for time integration.

with components $\boldsymbol{n} = (-1, 0)$, the positive flux component obtained from (4.53a) is actually $\mathbf{F}^-_{\boldsymbol{n}}$ and the negative component is $\mathbf{F}^+_{\boldsymbol{n}}$.

As shown by the results in Figure 4.19, the use of a first-order flux vector splitting technique gives stable, but overly diffusive, results, especially as regards the representation of the contact discontinuity. Flux-limiter techniques in Section 4.5.3.1 should be applied to concentrate the added numerical diffusion around sharp gradients only.

Discontinuous Galerkin. Here, the shock tube problem is solved using the discontinuous Galerkin method described in Section 4.4.2.2. The flux vector $\mathbf{F}_{\boldsymbol{n}}(\mathbf{U})$ is split into inflow and outflow components using Steger and Warming's (1981) flux splitting, see equations (4.53). Figure 4.20 shows the results. Conservative variables and the associated flux components are linearly interpolated over an element. The Euler first-order method was used for time integration. Artificial viscosity was added to the momentum and energy equations, but only on elements close to sharp gradients. The coefficient of artificial diffusion was taken as $\nu^e = h^e \lambda^e / 16$, where h^e is the size of the element and $\lambda^e = \lambda^e(\mathbf{U}^e)$ is the maximum characteristic speed, taken as $c + v$, where $c = \sqrt{\gamma p/\rho}$ is the speed of sound and v the modulus of the velocity vector.

5

Unsteady convection–diffusion problems

After studying elliptic problems in Chapter 2 and hyperbolic problems in Chapters 3 and 4, this chapter is concerned with parabolic problems. It combines the spatial discretization techniques for steady convection–diffusion problems discussed in Chapter 2 and the numerical schemes for time integration presented in Chapter 3. The objective is to produce time-accurate finite element methods for unsteady problems describing transport by convection and diffusion. Only linear and scalar problems are considered. Emphasis is placed on the accuracy properties of the Galerkin and spatially stabilized formulations.

5.1 INTRODUCTION

Time-accurate numerical methods for solving unsteady convection–diffusion problems using finite elements are studied here. Problems in this class are parabolic. In contrast to the pure convection problems discussed in Chapters 3 and 4, there are no discontinuous solutions in the presence of physical diffusivity and boundary conditions must be imposed everywhere on the boundary of the domain. However, Dirichlet boundary conditions may produce internal and boundary layers with steep solution gradients. Thus, on one hand, the stability problems of the Galerkin formulation studied in Chapter 2 are also present here. And, on the other hand, in convection-dominated problems, the need stressed in Chapter 3 to use time-stepping algorithms capable of simulating the role of characteristics is also present here. However, because of the presence in convection–diffusion of a second-order spatial operator, the finite element techniques discussed so far for achieving spatial stabilization and time accuracy need to be suitably adapted to deal with parabolic transport problems.

The chapter begins with the presentation in Section 5.2 of the initial boundary value problem for convection–diffusion–reaction. This is followed in Section 5.3 by the description of time-stepping algorithms specifically adapted to trace the transient solution of mixed transport problems. Various classes of methods are considered. After a brief review of classical algorithms, such as the θ family of methods, we discuss operator-splitting methods separating convection and diffusion operators, as well as high-order accurate multistep methods. This includes time-stepping techniques based on Runge–Kutta methods and multistage schemes based on the factorization of Padé approximations of the exponential function.

Spatial discretization procedures for unsteady convection–diffusion problems are then introduced in Section 5.4. The classical Galerkin formulation is presented first and its lack of sufficient stability in convection-dominated cases is underlined. The stabilization procedures for steady problems discussed in Chapter 2, such as SUPG, GLS and SGS, are then extended to the transient case. Moreover, pure least-squares formulations are introduced along the lines discussed in Chapter 3 for convective transport problems. The accuracy properties of fully discrete schemes are also illustrated. To conclude the discussion of spatial discretization procedures, we briefly consider in Section 5.5 the extension of stabilized formulations to the space–time domain. Finally, Section 5.6 presents solved numerical examples.

5.2 PROBLEM STATEMENT

The strong form of the convection–diffusion–reaction initial/boundary value problem is stated as follows: given the velocity field $\boldsymbol{a}(\boldsymbol{x},t)$, the diffusion coefficient $\nu(\boldsymbol{x},t)$, the reaction $\sigma(\boldsymbol{x},t)$, the source term $s(\boldsymbol{x},t)$, and the necessary initial and boundary conditions, find $u(\boldsymbol{x},t)$ such that

$$u_t + \boldsymbol{a}\cdot\boldsymbol{\nabla}u - \boldsymbol{\nabla}\cdot(\nu\boldsymbol{\nabla}u) + \sigma u = s \qquad \text{in } \Omega\times]0,T[, \tag{5.1a}$$

$$u(\boldsymbol{x},0) = u_0(\boldsymbol{x}) \qquad \text{on } \Omega, \tag{5.1b}$$

$$u = u_D \qquad \text{on } \Gamma_D, \tag{5.1c}$$

$$\nu(\boldsymbol{n}\cdot\boldsymbol{\nabla})u = h \qquad \text{on } \Gamma_N. \tag{5.1d}$$

The numerical solution of convection–diffusion–reaction problems clearly involves a double discretization process, that is space discretization and time discretization. The former will be performed by the finite element method and will be discussed in Section 5.4. Various classes of methods can be employed to trace the temporal evolution of the solution of convection–diffusion–reaction problems. They are presented in the necessary detail in the next section.

Remark 5.1 (Boundary conditions). Equation (5.1a) is parabolic. Thus boundary conditions are imposed on Γ which is the smooth boundary of $\Omega \subset \mathbb{R}^{n_{sd}}$. Moreover, the boundary is assumed to consist of a portion Γ_D on which the value of u is prescribed, see (5.1c), and of a complementary portion Γ_N on which the diffusive flux is prescribed, see (5.1d), where h is given. Conditions on Γ_D are Dirichlet (or essential) conditions, while conditions on Γ_N are known as Neumann (or natural) conditions. Note that this last condition could be generalized in the form

$$\nu(\boldsymbol{n}\cdot\boldsymbol{\nabla})u = \varrho(\boldsymbol{n}\cdot\boldsymbol{a})(u-u_e)+\eta \qquad \text{on } \Gamma_N$$

where u_e is also given. This condition is either Neumann or Robin depending on the given parameters ϱ and η (they may vary with $\boldsymbol{x}$ and time). For simplicity, we shall assume that the Neumann condition (5.1d) is prescribed on Γ_N.

Remark 5.2 (Existence and uniqueness). Quarteroni and Valli (1994, Chap. 12) and Morton (1996, Thm 2.5.1) present the exact conditions for existence and uniqueness of the solution. In convection-dominated problems, it is sufficient for existence and uniqueness to impose that $\nu \geq \nu_0 > 0$ and $0 < \mu_0 \leq \sigma - \frac{1}{2}\boldsymbol{\nabla}\cdot\boldsymbol{a} \leq \mu_1$ in Ω. With these conditions, the bilinear form associated with the spatial operator (in a Galerkin formulation) is continuous and coercive, with a coercivity constant $\alpha = \min\{\nu_0, (\nu_0 + C_\Omega\mu_0)/(1+C_\Omega)\}$, where C_Ω is the Poincaré inequality constant ($\int_\Omega v^2 d\Omega \leq C_\Omega \int_\Omega |\boldsymbol{\nabla} v|^2 d\Omega$). That is, the Lax–Milgram lemma, see Section 1.5.4, can be applied.

5.3 TIME DISCRETIZATION PROCEDURES

5.3.1 Classical methods

Finite differences are usually employed for time discretization. The time-stepping schemes presented in Section 3.4.1 for purely convective transport can be directly employed for convection–diffusion–reaction problems. A noticeable exception is the one-step Lax–Wendroff method introduced in Section 3.4.1.2, which cannot be used in connection with C^0-continuous finite elements due to the presence of the second-order diffusion operator. Thus, if standard finite elements must be used, the time-marching schemes should only involve first time derivatives of the unknown.

The most popular methods for parabolic problems are the *θ family of methods* already discussed in Section 3.4.1.1. The θ scheme, see (3.19), applied to the convection–diffusion–reaction equation (5.1a) for constant coefficients yields

$$\frac{\Delta u}{\Delta t} + \theta\big[\boldsymbol{a}\cdot\boldsymbol{\nabla} - \boldsymbol{\nabla}\cdot(\nu\boldsymbol{\nabla}) + \sigma\big]\Delta u$$
$$= \theta s^{n+1} + (1-\theta)s^n - \big[\boldsymbol{a}\cdot\boldsymbol{\nabla} - \boldsymbol{\nabla}\cdot(\nu\boldsymbol{\nabla}) + \sigma\big]u^n. \quad (5.2)$$

Note that now second derivatives of u are present due to the diffusion operator. As previously observed in Section 3.4.1.1, Crank–Nicolson, $\theta = 1/2$, is the only second-order accurate method. $\theta = 1$ corresponds to the implicit backward Euler method, and $\theta = 0$ is the explicit forward Euler method. For $\theta \geq 1/2$, the scheme is A-stable. Usually, the Crank–Nicolson scheme is used for true transient problems where time accuracy is important. The backward Euler or the so-called Galerkin scheme, $\theta = 2/3$, may be used to obtain steady-state solutions by means of a time-marching approach with large time steps (usually called relaxation methods).

Another classical method for transient problems is the *Adams–Bashforth method*. This is a second-order accurate explicit method which requires evaluation of the first time derivative at two consecutive time levels. Its derivation rests upon the forward Taylor series

$$u(t^{n+1}) = u(t^n) + \Delta t\, u_t(t^n) + \frac{\Delta t^2}{2} u_{tt}(t^n) + \mathcal{O}(\Delta t^3),$$

in which the second time derivative is approximated by

$$u_{tt}(t^n) = \frac{u_t(t^n) - u_t(t^{n-1})}{\Delta t} + \mathcal{O}(\Delta t),$$

thus giving the two-level second-order explicit method

$$u^{n+1} = u^n + \frac{\Delta t}{2}\left(3u_t^n - u_t^{n-1}\right), \tag{5.3}$$

where the time derivative of the unknown u_t at a given instant must be replaced using equation (5.1a). Note that the Adams–Bashforth method is not self-starting. The first step is usually performed using the forward Euler method, preferably with a small time step.

Note that all of these methods only involve first time derivatives. Thus, they are easily implemented in conjunction with standard $\mathcal{C}^0$-continuous finite elements. The reader interested in an in-depth discussion of classical time-stepping methods for convection–diffusion problems may consult the books by Mitchell and Griffiths (1980) and Gresho and Sani (2000).

Remark 5.3 (Linear multistep methods). Linear multistep methods are popular because they make use of previously computed values of u and u_t at previous steps. A k-step linear multistep (LMS) needs u^{n+1-i} and u_t^{n+1-i}, $i = 1, \ldots, k$, to compute u^{n+1}. They are defined by the following expression, see for instance Gear (1971) or Hughes (2000),

$$\sum_{i=0}^{k} \alpha_i\, u^{n+1-i} = \Delta t \sum_{i=0}^{k} \beta_i\, u_t(t^{n+1-i}, u^{n+1-i}),$$

where α_i and β_i are parameters defining a specific method. The normalization $\alpha_0 = 1$ is standard and the method is explicit if $\beta_0 = 0$; otherwise, it is implicit. For instance, taking

$$\alpha_0 = -\alpha_1 = 1 \quad \text{and} \quad \beta_0 = -\theta,\ \beta_1 = \theta - 1,$$

gives the θ family of methods. Similarly, taking

$$\alpha_0 = 1,\ \alpha_1 = -1,\ \alpha_2 = 0 \quad \text{and} \quad \beta_0 = 0,\ \beta_1 = -3/2,\ \beta_2 = 1/2$$

reproduces the explicit Adams–Bashforth method, see equation (5.3). Usually, explicit methods are combined with implicit ones, of the Adams–Moulton family for instance, to produce predictor–corrector schemes.

Unfortunately, the second Dahlquist barrier (see, for instance, Hairer and Wanner, 1996) states that A-stable LMS methods are at most second-order accurate. Among the second-order methods, the Crank–Nicolson method has the smallest truncation error.

5.3.2 Fractional-step methods

Fractional-step (or operator-splitting) methods are widely used for time integration of unsteady problems. They can take different forms according to the way they split a complex problem into two or more simpler ones. For instance, in Chapter 4 a fractional-step approach was suggested, in the context of fluid–structure interaction, to separate material (Lagrangian) and convective terms. Another application of fractional-step methods will be discussed in Chapter 6 to separate the viscous terms and the pressure/incompressibility terms to properly account for the incompressibility constraint in the unsteady Navier–Stokes equations.

Here fractional-step techniques are applied to the solution of unsteady convection–diffusion–reaction problems. In this context, the original problem is transformed into the sum of a pure convection problem and a diffusion–reaction problem to be solved in sequence at each station of the time integration procedure. Methods in this class have been thoroughly studied by the Russian mathematicians Yanenko (1971) and Marchuk (1982; 1990). In view of the different characteristics of convection and diffusion operators, operator-splitting techniques allow us to select the most appropriate numerical algorithms to solve the convection and the diffusion phases.

Consider again equation (5.1a) and assume for simplicity that the convection velocity $\boldsymbol{a}$, the diffusivity ν and the reaction σ are constant. A class of fractional-step methods operate at the level of the semi-discrete form, that is, the system of ordinary differential equations obtained after spatial discretization of (5.1a), namely

$$\frac{d\mathbf{u}}{dt} + \mathbf{L}\mathbf{u} = \mathbf{f},$$

where matrix $\mathbf{L}$ arises from the spatial discretization of the convection, diffusion and reaction operators and vector $\mathbf{f}$ accounts for the source term. In this case, the splitting is of the form

$$\frac{d\mathbf{u}}{dt} + (\mathbf{L}_1 + \mathbf{L}_2)\mathbf{u} = \mathbf{f}$$

and is called *algebraic splitting*, see Quarteroni and Valli (1994).

For instance, Yanenko (1971) proposed a first-order accurate implicit method in the form of the following algebraic splitting:

$$\frac{\mathbf{u}^{n+1/2} - \mathbf{u}^n}{\Delta t} + \mathbf{L}_1\, \mathbf{u}^{n+1/2} = 0,$$

$$\frac{\mathbf{u}^{n+1} - \mathbf{u}^{n+1/2}}{\Delta t} + \mathbf{L}_2\, \mathbf{u}^{n+1} = \mathbf{f}^n.$$

This method produces a couple of implicit problems, but unconditional stability can only be guaranteed for cases where both $\mathbf{L}_1$ and $\mathbf{L}_2$ are positive definite matrices.

Other widely used splitting methods are described in the technical literature (e.g., Ames, 1992; Mitchell and Griffiths, 1980; Quarteroni and Valli, 1994). These include the method of Peaceman and Rachford

$$\frac{\mathbf{u}^{n+1/2}-\mathbf{u}^n}{\Delta t/2}+\mathbf{L}_1\mathbf{u}^{n+1/2}=\mathbf{f}^{n+1/2}-\mathbf{L}_2\mathbf{u}^n,$$
$$\frac{\mathbf{u}^{n+1}-\mathbf{u}^{n+1/2}}{\Delta t/2}+\mathbf{L}_2\mathbf{u}^{n+1}=\mathbf{f}^{n+1/2}-\mathbf{L}_1\mathbf{u}^{n+1/2},$$

and a similar one by Douglas and Rachford:

$$\frac{\mathbf{u}^{n+1/2}-\mathbf{u}^n}{\Delta t}+\mathbf{L}_1\mathbf{u}^{n+1/2}=\mathbf{f}^n-\mathbf{L}_2\mathbf{u}^n,$$
$$\frac{\mathbf{u}^{n+1}-\mathbf{u}^{n+1/2}}{\Delta t}+\mathbf{L}_2\mathbf{u}^{n+1}=\mathbf{L}_2\mathbf{u}^n.$$

Other splitting methods operate at the level of the differential operators. They are best explained by rewriting the governing equation (5.1a) in the symbolic form

$$u_t+\mathcal{L}u=s, \qquad \text{with} \quad \mathcal{L}=\boldsymbol{a}\cdot\boldsymbol{\nabla}-\boldsymbol{\nabla}\cdot(\nu\boldsymbol{\nabla})+\sigma,$$

and then the convection–diffusion–reaction operator $\mathcal{L}$ splits into a sum of two components as follows:

$$\mathcal{L}=\mathcal{L}_1+\mathcal{L}_2, \qquad \text{with} \quad \mathcal{L}_1=\boldsymbol{a}\cdot\boldsymbol{\nabla}, \quad \text{and} \quad \mathcal{L}_2=-\boldsymbol{\nabla}\cdot(\nu\boldsymbol{\nabla})+\sigma.$$

An example of splitting at operator level is described by Donea, Giuliani, Laval and Quartapelle (1982) who follow Marchuk (1982). They consider a convection–diffusion problem ($\sigma=0$) and split the operator as in the previous equation on a typical time interval Δt. Moreover, the source term is associated with the second step. This induces the following algorithm for each time step:

$$\text{First step:} \quad \begin{cases} v_t+\mathcal{L}_1 v=0 & \text{in } \Omega\times[t^n,t^{n+1}[\\ v(t^n)=u^n \end{cases}$$

$$\text{Second step:} \quad \begin{cases} w_t+\mathcal{L}_2 w=s & \text{in } \Omega\times[t^n,t^{n+1}[\\ w(t^n)=v^{n+1} \end{cases}$$

$$\text{Final update:} \quad u^{n+1}=w^{n+1}.$$

Since the first step represents a pure convection problem, methods for hyperbolic equations discussed in Chapter 3 can be employed. For instance, the third-order explicit Taylor–Galerkin scheme (TG3), see (3.42), would produce the scheme

$$\Big(1-\frac{\Delta t^2}{6}\mathcal{L}_1^2\Big)\frac{v^{n+1}-v^n}{\Delta t}=-\Big(\mathcal{L}_1-\frac{\Delta t}{2}\mathcal{L}_1^2\Big)v^n.$$

The spatial discretization of this semi-discrete equation can then be performed with the standard Galerkin finite element method.

The Crank–Nicolson method can be used for the second step, that is the diffusion equation. Thus,

$$\Big(1 + \frac{\Delta t}{2}\mathcal{L}_2\Big)\frac{w^{n+1} - w^n}{\Delta t} = -\mathcal{L}_2\, w^n + \frac{1}{2}\big(s^n + s^{n+1}\big).$$

Here again, the spatial discretization would be performed by means of the standard Galerkin method. Donea, Giuliani, Laval and Quartapelle (1982) present a detailed discussion of the above fractional-step method which confirms its interest for approximating the solution of unsteady convection–diffusion problems.

Note that in the case of operator splitting, the sequential treatment of the convection and diffusion phases requires a splitting between the boundary conditions to endow each individual phase of a fractional-step algorithm with consistent boundary data. For instance, in the convection phase, it is clear that data can only be prescribed on the inflow portion of the boundary. Fractional-step methods are also sensitive to the treatment of the source term, see for instance Peraire (1986) and Peraire, Zienkiewicz and Morgan (1986). Another critical issue with operator-splitting methods is the overall accuracy of the procedure. Most methods are in fact only first-order accurate. The reader interested in a detailed exposition of fractional-step and operator-splitting methods is urged to consult the specialized textbooks by Marchuk (1982; 1990), Yanenko (1971), as well as Glowinski and Le Tallec (1989) and Quarteroni and Valli (1994).

5.3.3 High-order time-stepping schemes

Unsteady convection–diffusion–reaction problems are more difficult to solve using high-order time-stepping methods than the pure convection problems discussed in Chapter 3. The reason is the presence in the governing equation (5.1a) of the second-order diffusion operator. This operator limits one-step time integration algorithms to second-order temporal accuracy when they are combined with $\mathcal{C}^0$-continuous finite element approximations. In order to use a standard implementation of $\mathcal{C}^0$ finite elements, time-stepping schemes for convection–diffusion should only involve first time derivatives. This leads to the two-step schemes presented in Section 3.6.4, or more generally, to Runge–Kutta methods (Lambert, 1991; Hairer, Nørsett and Wanner, 1993) or to multistage schemes emanating from the factorization of Padé approximations to the exponential function (Donea, Roig and Huerta, 1998).

5.3.3.1 Runge–Kutta methods The technical literature contains many references to the use of Runge–Kutta time-stepping schemes in connection with finite element and finite volume algorithms. Only a few representative examples are mentioned here. For instance, they can be used to reach a steady solution (the concern is efficiency and not time accuracy), see Jameson, Schmidt and Turkel (1981) or Liu and Jameson (1993). Iannelli and Baker (1991) use a stiffly stable, second-order accurate, implicit Runge–Kutta method for Euler and Navier–Stokes aerodynamic applications. Tezduyar, Mittal and Shih (1991) and Jiang and Kawahara (1993) employ a third-order explicit Runge–Kutta method for the finite element analysis of unsteady incompressible flows. Giles (1997) presents an application of modern stability analysis to the Galerkin formulation of Runge–Kutta methods applied to the Navier–Stokes equations. In particular, examples of predictor–corrector and explicit Pad é approximations are shown. A fourth-order method is used by Pereira et al. (2001) for a finite volume solution of the incompressible Navier–Stokes equations. In the context of nonlinear convection-dominated problems, Cockburn and Shu (2001) review the development of the Runge–Kutta discontinuous Galerkin methods.

Runge–Kutta methods are multistage methods that only make use of the solution u^n at time t^n to compute the solution u^{n+1} at t^{n+1}. This is achieved by computing a number $\mathrm{n_{tg}}$ of intermediate values of the time derivative of the unknown u, within the interval $\Delta t = t^{n+1} - t^n$. The differential equation (5.1a) is rewritten as

$$u_t + \mathcal{L}(u) = s, \quad \text{or} \quad u_t = \mathcal{F}(t, u) \tag{5.4}$$

where the spatial differential operators are defined as

$$\mathcal{L} := \boldsymbol{a} \cdot \boldsymbol{\nabla} + \sigma - \boldsymbol{\nabla} \cdot (\nu \boldsymbol{\nabla}), \quad \text{and} \quad \mathcal{F}(t, u) := s - \mathcal{L}(u). \tag{5.5}$$

Then, the standard form (see Lambert, 1991; Hairer et al., 1993; Hairer and Wanner, 1996) of an $\mathrm{n_{tg}}$*-stage Runge–Kutta method* is

$$u^{n+1} = u^n + \Delta t \sum_{i=1}^{\mathrm{n_{tg}}} b_i \, \ell_i, \tag{5.6a}$$

$$\ell_i = \mathcal{F}\Big(t^n + c_i \Delta t\,,\, u^n + \Delta t \sum_{j=1}^{\mathrm{n_{tg}}} a_{i,j} \, \ell_j\Big), \quad i = 1, 2, \ldots, \mathrm{n_{tg}}. \tag{5.6b}$$

It can also be written in another convenient and usual form,

$$u^{n+1} = u^n + \Delta t \sum_{i=1}^{\mathrm{n_{tg}}} b_i \, \mathcal{F}(t^n + c_i \Delta t\,,\, u^{n+\beta_i}), \tag{5.7a}$$

$$u^{n+\beta_i} = u^n + \Delta t \sum_{j=1}^{\mathrm{n_{tg}}} a_{i,j} \, \mathcal{F}(t^n + c_j \Delta t\,,\, u^{n+\beta_j}), \quad i = 1, 2, \ldots, \mathrm{n_{tg}} \tag{5.7b}$$

using the interpretation

$$\ell_i = \mathcal{F}(t^n + c_i \Delta t\,,\, u^{n+\beta_i}), \quad i = 1, 2, \ldots, \mathrm{n_{tg}}.$$

In the above relations the step increment is the time increment Δt in view of the time-dependent problem (5.1) under consideration. The parameters β_i can be substituted by a stage counter, although each β_i can be interpreted as $u^{n+\beta_i} \simeq u(t^{n+\beta_i})$ and thus equation (5.7b) gives an approximation of u at an intermediate instant. The coefficients in (5.6), or (5.7), must verify the consistency conditions

$$c_i = \sum_{j=1}^{\mathrm{n_{tg}}} a_{i,j} \quad \text{and} \quad \sum_{i=1}^{\mathrm{n_{tg}}} b_i = 1,$$

and they are conveniently displayed as a *Butcher array*:

$$
\begin{array}{c|cccc}
c_1 & a_{1,1} & a_{1,2} & \cdots & a_{1,\mathrm{n_{tg}}} \\
c_2 & a_{2,1} & a_{2,2} & \cdots & a_{2,\mathrm{n_{tg}}} \\
\vdots & \vdots & \vdots & & \vdots \\
c_{\mathrm{n_{tg}}} & a_{\mathrm{n_{tg}},1} & a_{\mathrm{n_{tg}},2} & \cdots & a_{\mathrm{n_{tg}},\mathrm{n_{tg}}} \\
\hline
 & b_1 & b_2 & \cdots & b_{\mathrm{n_{tg}}}
\end{array}
$$

Implicit Runge–Kutta methods are those where at least one $a_{i,j} \neq 0$ for $j \geq i$. If this is not the case, the method is explicit.

5.3.3.2 *Explicit Runge–Kutta methods* It is well-known that $\mathrm{n_{tg}}$-stage explicit Runge–Kutta methods are of order $\mathrm{n_{tg}}$ for $\mathrm{n_{tg}} \leq 4$. For $\mathrm{n_{tg}} > 4$, $\mathrm{n_{tg}}$-stage methods systematically have order lower than $\mathrm{n_{tg}}$. For this reason, Runge–Kutta methods of fourth-order are very popular among the explicit methods. The classical fourth-order Runge–Kutta method is defined by the following Butcher array:

$$
\begin{array}{c|cccc}
0 & & & & \\
1/2 & 1/2 & & & \\
1/2 & 0 & 1/2 & & \\
1 & 0 & 0 & 1 & \\
\hline
 & 1/6 & 1/3 & 1/3 & 1/6
\end{array}
$$

The resulting time-stepping algorithm follows from (5.7) and reads

$$
\begin{aligned}
u^{n+\beta_1} &= u^n, & u^{n+\beta_2} &= u^n + \tfrac{\Delta t}{2}\mathcal{F}(t^n + \tfrac{\Delta t}{2}, u^{n+\beta_1}), \\
u^{n+\beta_3} &= u^n + \tfrac{\Delta t}{2}\mathcal{F}(t^n + \tfrac{\Delta t}{2}, u^{n+\beta_2}), & u^{n+\beta_4} &= u^n + \Delta t\mathcal{F}(t^n + dt, u^{n+\beta_3}), \\
u^{n+1} &= u^n + \tfrac{\Delta t}{6}\big(\mathcal{F}(t^n, u^{n+\beta_1}) + 2\mathcal{F}(t^n + \tfrac{\Delta t}{2}, u^{n+\beta_2}) \\
&\qquad + 2\mathcal{F}(t^n + \tfrac{\Delta t}{2}, u^{n+\beta_3}) + \mathcal{F}(t^n + \Delta t, u^{n+\beta_4})\big).
\end{aligned}
$$

Remark 5.4 (Stability properties). Explicit Runge–Kutta methods are only conditionally stable. In application to the modal equation

$$u_t + \lambda u = 0,$$

where λ denotes a typical eigenvalue, their amplification factor has the same structure as the corresponding Padé approximation (first row in Table 5.1)

$$G_{(RK,\mathrm{n_{tg}})}(\lambda\Delta t) = \mathsf{R}_{\mathrm{n_{tg}},0}(-\lambda\Delta t).$$

Their domain of numerical stability in the complex $\lambda\Delta t$ plane is depicted in Figure 5.1. The fourth-order method presented previously has interesting stability properties. The associated absolute stability curve, shown in Figure 5.1, is the same as that of Padé approximation $\mathsf{R}_{4,0}$. It cuts the real and imaginary axes at -2.78 and $\pm 2\sqrt{2}$, respectively. Since the absolute stability region

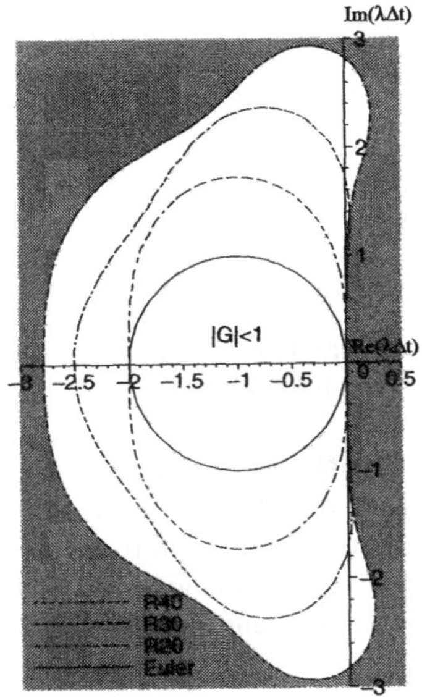

Fig. 5.1 Stability domain of explicit Runge–Kutta and Padé methods.

contains a finite portion of the imaginary axis, the fourth-order Runge–Kutta method can be used in convection-dominated situations where the problem eigenvalues are distributed close to the imaginary axis of the complex $\lambda\Delta t$ plane.

5.3.3.3 Explicit Padé approximations As an alternative to explicit Runge–Kutta methods, one may consider using a multistage factorization of the explicit Padé approximations $\mathsf{R}_{n,0}$ to the exponential function, see the first row in Padé Table 5.1 and set $z = \Delta t \partial/\partial t$, see Donea, Roig and Huerta (1998; 2000). They are easier to implement in conjunction with finite elements due to their simple algorithmic structure and possess the same properties as their corresponding Runge–Kutta methods, for instance the same stability domain, if the coefficients in (5.1a) do not depend on time.

For the second-order Padé approximation $\mathsf{R}_{2,0}$, a two-stage method can be derived from the factorization

$$u(t^{n+1}) = u(t^n) + \Delta t \frac{\partial}{\partial t}\Big(u + \frac{\Delta t}{2}\frac{\partial u}{\partial t}\Big)\Big|_{t=t^n} + \mathcal{O}(\Delta t^3),$$

which yields the two-stage ($\mathtt{n_{tg}} = 2$) Lax–Wendroff method

$$\begin{aligned} u^{n+1/2} &= u^n + \frac{\Delta t}{2} u_t^n; \\ u^{n+1} &= u^n + \Delta t\, u_t^{n+1/2}. \end{aligned}$$

Table 5.1 Padé approximations of the exponential function e^z.

$R_{n,m}(z)$	$n=0$	$n=1$	$n=2$	$n=3$
$m=0$	1	$1+z$	$1+z+\frac{z^2}{2}$	$1+z+\frac{z^2}{2}+\frac{z^3}{6}$
$m=1$	$\frac{1}{1-z}$	$\frac{1+\frac{z}{2}}{1-\frac{z}{2}}$	$\frac{1+\frac{2z}{3}+\frac{z^2}{6}}{1-\frac{z}{3}}$	$\frac{1+\frac{3z}{4}+\frac{z^2}{4}+\frac{z^3}{24}}{1-\frac{z}{4}}$
$m=2$	$\frac{1}{1-z+\frac{z^2}{2}}$	$\frac{1+\frac{z}{3}}{1-\frac{2z}{3}+\frac{z^2}{6}}$	$\frac{1+\frac{z}{2}+\frac{z^2}{12}}{1-\frac{z}{2}+\frac{z^2}{12}}$	$\frac{1+\frac{3z}{5}+\frac{3z^2}{20}+\frac{z^3}{60}}{1-\frac{2z}{5}+\frac{z^2}{20}}$
$m=3$	$\frac{1}{1-z+\frac{z^2}{2}-\frac{z^3}{6}}$	$\frac{1+\frac{z}{4}}{1-\frac{3z}{4}+\frac{z^2}{4}-\frac{z^3}{24}}$	$\frac{1+\frac{2z}{5}+\frac{z^2}{20}}{1-\frac{3z}{5}+\frac{3z^2}{20}-\frac{z^3}{60}}$	$\frac{1+\frac{z}{2}+\frac{z^2}{10}+\frac{z^3}{120}}{1-\frac{z}{2}+\frac{z^2}{10}-\frac{z^3}{120}}$

The third-order Padé approximation $R_{3,0}$ can also be factorized in the form

$$u(t^{n+1}) = u(t^n) + \Delta t \frac{\partial}{\partial t}\left(u + \frac{\Delta t}{2}\frac{\partial}{\partial t}\left(u + \frac{\Delta t}{3}\frac{\partial u}{\partial t}\right)\right)\bigg|_{t=t^n} + \mathcal{O}(\Delta t^4),$$

which yields the three-stage ($n_{tg} = 3$) method

$$\begin{aligned} u^{n+1/3} &= u^n + \frac{\Delta t}{3} u_t^n, \\ u^{n+1/2} &= u^n + \frac{\Delta t}{2} u_t^{n+1/3}, \\ u^{n+1} &= u^n + \Delta t\, u_t^{n+1/2}. \end{aligned}$$

This third-order scheme has been used by Jiang and Kawahara (1993) and by Tezduyar et al. (1991) in the finite element analysis of unsteady incompressible flows.

Similarly higher-order methods can be developed (for constant-in-time coefficients). In fact, multistage explicit methods of the Padé family are easily expressed in the incremental form

$$\begin{aligned} u^{n+\beta_1} &= u^n, \\ u^{n+\beta_i} &= u^n + \beta_i\, \Delta t\, u_t^{n+\beta_{i-1}}, \qquad i = 2, \ldots, n_{tg}+1, \end{aligned} \tag{5.8}$$

where $\beta_1 = 0$, and $\beta_i = 1/(n_{tg}+2-i), i = 2, \ldots, n_{tg}+1$. Note that these methods can be expressed using a Butcher array:

$$\begin{array}{c|ccccc} 0 & 0 & & & & \\ 1/n_{tg} & 1/n_{tg} & 0 & & & \\ \frac{1}{n_{tg}-1} & 0 & \frac{1}{n_{tg}-1} & 0 & & \\ \vdots & \vdots & \ddots & \ddots & \ddots & \\ 1/2 & 0 & \cdots & 0 & 1/2 & 0 \\ \hline & 0 & \cdots & \cdots & 0 & 1 \end{array}$$

Then, the time derivative is replaced using the convection–diffusion–reaction equation (5.1a). Finally, as shown in Section 5.4, a weak form of the resulting semi-discrete system is introduced to serve as a basis for the finite element spatial discretization.

Remark 5.5. Note that although explicit multistage Padé schemes can be written using a Butcher array, they are not Runge–Kutta methods. For instance, one can easily check that the coefficients of the three-stage method do not verify the required conditions (see for instance Hairer et al., 1993, p. 144). The factorization of the Taylor series and the subsequent substitution of u_t can be performed if the operators in (5.5) are linear (not dependent on u) and constants ($\boldsymbol{a}$, σ, ν and s are independent of t).

5.3.3.4 Implicit Runge–Kutta methods Implicit Runge–Kutta (IRK) methods are popular in the ordinary differential equations community (Lambert, 1991; Hairer and Wanner, 1996). Their elevated cost, when implemented in the context of convection–diffusion equations, has deterred their use. For each time step the IRK method requires the simultaneous solution of (5.7b) to determine $u^{n+\beta_i}$ for $i = 1, 2, \ldots, \mathtt{n_{tg}}$, or, equivalently, solution of (5.6b) to determine every ℓ_i. Then, equation (5.7a), or (5.6a), can be used to compute u^{n+1} explicitly.

The IRK methods based on Lobatto quadrature are of particular interest for the time integration of unsteady transport problems, because they include the end points of the integration interval, namely $u^{n+\beta_1} = u^n$ and $u^{n+\beta_{\mathtt{n_{tg}}}} = u^{n+1}$. This reduces the computational cost: the system of equations has one equation less and equation (5.7a) is verified automatically. They have a maximum order of $2\mathtt{n_{tg}} - 2$ where $\mathtt{n_{tg}}$ is the number of stages. The first family of such methods is called Lobatto IIIA and it is based on the following Butcher arrays:

$$\mathtt{n_{tg}} = 2 \text{ (thus order 2)} \qquad \begin{array}{c|cc} 0 & 0 & 0 \\ 1 & 1/2 & 1/2 \\ \hline & 1/2 & 1/2 \end{array}$$

$$\mathtt{n_{tg}} = 3 \text{ (thus order 4)} \qquad \begin{array}{c|ccc} 0 & 0 & 0 & 0 \\ 1/2 & 5/24 & 1/3 & -1/24 \\ 1 & 1/6 & 2/3 & 1/6 \\ \hline & 1/6 & 2/3 & 1/6 \end{array}$$

Note that the first row of the matrix in the Butcher arrays is, for Lobatto IIIA methods, always zero as well as c_1, that is $a_{1,1} = a_{1_2} = \cdots = a_{1,\mathtt{n_{tg}}} = 0$. Thus, as expected, the first stage that corresponds to u^n is not evaluated. Therefore, the system of simultaneous equations that must be solved at each time step has dimension $\mathtt{n_{tg}} - 1$. Moreover, the last explicit computation in (5.7a) corresponding to the evaluation of u^{n+1} is also unnecessary because the last row of that matrix is identical to $[b_1, b_2, \ldots, b_{\mathtt{n_{tg}}}]$.

5.3.3.5 *Implicit Padé approximations* Lobatto IIIA IRK methods in the previous section are closely related to the Padé approximations $\mathsf{R}_{n,n}$ on the diagonal of Table 5.1. As shown by Donea et al. (1998; 2000), multistage schemes can be derived directly from the diagonal Padé approximations. They are obtained in the following compact form:

$$\frac{\Delta \boldsymbol{u}}{\Delta t} - \mathbf{W}\, \Delta \boldsymbol{u}_t = \mathbf{w}\, u_t^n, \tag{5.9}$$

where the unknown $\Delta \boldsymbol{u} \in \mathbb{R}^{\mathtt{n_{tg}}-1}$ is a vector whose dimension varies with the number of stages, $\mathtt{n_{tg}}$, and the vector $\Delta \boldsymbol{u}_t$ is the partial derivative of $\Delta \boldsymbol{u}$ with respect to time, namely

$$\Delta \boldsymbol{u} = \begin{Bmatrix} u^{n+\beta_2} - u^n \\ u^{n+\beta_3} - u^{n+\beta_2} \\ \vdots \\ u^{n+1} - u^{n+\beta_{\mathtt{n_{tg}}-1}} \end{Bmatrix}, \quad \text{and } \Delta \boldsymbol{u}_t = \begin{Bmatrix} u_t^{n+\beta_2} - u_t^n \\ u_t^{n+\beta_3} - u_t^{n+\beta_2} \\ \vdots \\ u_t^{n+1} - u_t^{n+\beta_{\mathtt{n_{tg}}-1}} \end{Bmatrix}. \tag{5.10}$$

$\mathbf{W}$ is an $(\mathtt{n_{tg}}-1)\times(\mathtt{n_{tg}}-1)$ matrix and $\mathbf{w}$ an $(\mathtt{n_{tg}}-1)$ vector. They are defined below for two specific cases. As with IRK methods, the time derivatives in (5.9) are replaced by spatial derivatives using the original differential equation, see equations (5.4) and (5.5). For instance, in the case of a linear operator $\mathcal{L}$ with constant coefficients and a source term s, equation (5.9) becomes

$$\frac{\Delta \boldsymbol{u}}{\Delta t} + \mathbf{W}\mathcal{L}(\Delta \boldsymbol{u}) = \mathbf{w}\big[s^n - \mathcal{L}(u^n)\big] + \mathbf{W}\Delta \boldsymbol{s}. \tag{5.11}$$

The precise definition of $\Delta \boldsymbol{u}$ (recall that $\Delta \boldsymbol{u}_t$ is simply $\partial \Delta \boldsymbol{u}/\partial t$), $\Delta \boldsymbol{s}$, $\mathbf{w}$ and $\mathbf{W}$ depends on each particular method.

As noted earlier, these methods are closely related to Lobatto IIIA IRK methods. The noticeable difference is that, while the present scheme is based on non-overlapping step increments, the corresponding Runge–Kutta method employs total increments. This implies that a linear correspondence exits between both techniques, namely

$$\begin{Bmatrix} u^{n+\beta_2} - u^n \\ u^{n+\beta_3} - u^n \\ \vdots \\ u^{n+1} - u^n \end{Bmatrix} = \begin{pmatrix} 1 & 0 & \cdots & \cdots & 0 \\ 1 & 1 & \ddots & & \vdots \\ \vdots & \ddots & \ddots & \ddots & \vdots \\ \vdots & & \ddots & 1 & 0 \\ 1 & \cdots & \cdots & 1 & 1 \end{pmatrix} \begin{Bmatrix} u^{n+\beta_2} - u^n \\ u^{n+\beta_3} - u^{n+\beta_2} \\ \vdots \\ u^{n+1} - u^{n+\beta_{\mathtt{n_{tg}}-1}} \end{Bmatrix}.$$

This minor difference does not affect linear problems discretized with a Galerkin formulation. However, as shown in Section 5.4.6.5, it has major consequences as regards numerical stability when implemented in connection with stabilized spatial formulations.

We shall illustrate the construction of multistage Padé schemes with reference to the second-order approximation $\mathsf{R}_{1,1}$ and the fourth-order approximation $\mathsf{R}_{2,2}$, for which equation (5.9) takes the following particular forms:

Second-order Padé approximation: $\mathsf{R}_{1,1}$ **(Crank–Nicolson)**

$$\Delta \boldsymbol{u} = \boldsymbol{u}^{n+1} - \boldsymbol{u}^{n}, \qquad \Delta \boldsymbol{s} = \boldsymbol{s}^{n+1} - \boldsymbol{s}^{n},$$

$$\mathbf{W} = \frac{1}{2}, \qquad \mathbf{w} = 1.$$

Note that in this case $\mathtt{n_{tg}} = 1$ and, consequently, the vectors and matrix in (5.9) become scalars.

Fourth-order Padé approximation: $\mathsf{R}_{2,2}$

$$\Delta \boldsymbol{u} = \begin{Bmatrix} \boldsymbol{u}^{n+1/2} - \boldsymbol{u}^{n} \\ \boldsymbol{u}^{n+1} - \boldsymbol{u}^{n+1/2} \end{Bmatrix}, \qquad \Delta \boldsymbol{s} = \begin{Bmatrix} \boldsymbol{s}^{n+1/2} - \boldsymbol{s}^{n} \\ \boldsymbol{s}^{n+1} - \boldsymbol{s}^{n+1/2} \end{Bmatrix};$$

$$\mathbf{W} = \frac{1}{24}\begin{pmatrix} 7 & -1 \\ 13 & 5 \end{pmatrix}, \qquad \mathbf{w} = \frac{1}{2}\begin{Bmatrix} 1 \\ 1 \end{Bmatrix}.$$

Both equations (5.9) and (5.11) with the corresponding initial and boundary conditions define a problem in *strong form*, which must be solved at each time step. For simplicity, the truncation errors are not explicitly shown in (5.9) and (5.11). Note, however, that if these truncation errors are accounted for, equations (5.1a), (5.9) and (5.11) are equivalent, and the exact solution verifies the new strong form. The spatially continuous differential equations, (5.9) or (5.11), are the basis of the finite element discretization.

5.4 SPATIAL DISCRETIZATION PROCEDURES

5.4.1 Galerkin formulation of the semi-discrete scheme

Here, we present the Galerkin formulation associated with the strong form (5.1) of the unsteady convection–diffusion–reaction problem. As in Section 3.4.2, Galerkin can be applied directly on the differential equation (5.1a), or on the time discretized equations (5.2), (5.3), (5.8), (5.7) or (5.9). For completeness, we present the *semi-discrete* scheme first.

The space of weighting functions denoted by $\mathcal{V}$ satisfies the homogeneous boundary conditions on Γ_D. The functions, w, in $\mathcal{V}$ do not depend on time. As in Section 3.4.2.1, the time dependency of the approximate solution u can be translated to the trial space $\mathcal{S}_t$, which varies as a function of time,

$$\mathcal{S}_t := \left\{u \mid u(\cdot, t) \in \mathcal{H}^1(\Omega), t \in [0, T] \text{ and } u(\boldsymbol{x}, t) = u_D \text{ for } \boldsymbol{x} \in \Gamma_D\right\}.$$

As usual, trial solutions a priori verify Dirichlet boundary conditions. The weak form of the initial boundary value problem (5.1) is defined as follows: given s, u_D, h, u_e

and u_0, for any $t \in [0,T]$ find $u(\boldsymbol{x},t) \in \mathcal{S}_t$, such that for all $w \in \mathcal{V}$

$$(w, u_t) + c(\boldsymbol{a}; w, u) + a(w, u) + (w, \sigma u) = (w, s) + (w, h)_{\Gamma_N}. \tag{5.12}$$

We recall the definitions already presented in (2.6), namely

$$a(w,u) = \int_\Omega \boldsymbol{\nabla} w \cdot (\nu \boldsymbol{\nabla} u) d\Omega, \qquad (w,s) = \int_\Omega w\, s\, d\Omega,$$
$$c(\boldsymbol{a}; w, u) = \int_\Omega w(\boldsymbol{a} \cdot \boldsymbol{\nabla} u) d\Omega, \qquad (w,h)_{\Gamma_N} = \int_{\Gamma_N} w\, h\, d\Gamma.$$

The spatial discretization by means of the Galerkin formulation consists of defining two finite dimensional spaces $\mathcal{S}^h$ and $\mathcal{V}^h$, subsets of $\mathcal{S}$ and $\mathcal{V}$, see Section 3.4.2.2. Then, the semi-discrete Galerkin formulation is obtained by restricting the previous weak form to these finite dimensional spaces.

The system of ordinary differential equations is obtained following the same rationale as in Section 3.4.2.2. Recall that the time dependence of the solution, $u^h(\boldsymbol{x},t)$, is taken into account by the nodal values of the unknown. The shape functions $N_A(\boldsymbol{x})$ do not depend on time. Thus, (2.10) becomes

$$u^h(\boldsymbol{x},t) = \sum_{A \in \eta \setminus \eta_D} N_A(\boldsymbol{x})\, \mathrm{u}_A(t) + \sum_{A \in \eta_D} N_A(\boldsymbol{x})\, u_D(\boldsymbol{x}_A, t),$$

where, as before, η is the set of global node numbers in the finite element mesh and $\eta_D \subset \eta$ the subset of nodes belonging to the Dirichlet portion of the boundary, Γ_D. The test functions are defined as before, see (2.11), $w^h \in \mathcal{V}^h = \mathrm{span}_{B \in \eta \setminus \eta_D}\{N_B\}$.

Finally, the usual assembly process delivers the semi-discrete system of ordinary differential equations

$$\mathbf{M}\dot{\mathbf{u}} + (\mathbf{C} + \mathbf{K} + \sigma \mathbf{M})\mathbf{u} = \mathbf{f}, \tag{5.13}$$

where, for simplicity, we have assumed that the reaction, σ, is uniform and constant. Note that vectors $\mathbf{u}$ and $\dot{\mathbf{u}}$ contain, respectively, the nodal values of the unknown u and of its time derivative, while $\mathbf{M}$, $\mathbf{C}$ and $\mathbf{K}$ are, respectively, the consistent mass matrix, the convection matrix and the diffusion matrix. These matrices are obtained by topological assembly of element contributions as follows:

$$\begin{aligned}
\mathbf{M} &= \mathbf{A}^e \mathbf{M}^e & M^e_{ab} &= \int_{\Omega^e} N_a N_b \, d\Omega \\
\mathbf{C} &= \mathbf{A}^e \mathbf{C}^e & C^e_{ab} &= \int_{\Omega^e} N_a(\boldsymbol{a} \cdot \boldsymbol{\nabla} N_b) d\Omega \\
\mathbf{K} &= \mathbf{A}^e \mathbf{K}^e & K^e_{ab} &= \int_{\Omega^e} \boldsymbol{\nabla} N_a \cdot (\nu \boldsymbol{\nabla} N_b)\, d\Omega
\end{aligned} \tag{5.14}$$

where $\mathbf{A}$ denotes the assembly operator, $1 \le a, b \le \mathrm{n_{en}}$ and $\mathrm{n_{en}}$ is the number of element nodes. The r.h.s. vector, $\mathbf{f}$, considers the contribution of the source term,

s, the prescribed flux, h, and the Dirichlet data u_D. It results from the assembly of nodal contributions of the form $\mathbf{f} = \mathop{\mathbf{A}}^{e} \mathbf{f}^e$ and

$$\mathrm{f}^e_a = (N_a, s)_{\Omega^e} + (N_a, h)_{\partial\Omega^e \cap \Gamma_N} - \sum_{b=1}^{n_{en}} \Big[(N_a, N_b)_{\Omega^e} \frac{\partial u^e_{Db}(t)}{\partial t} + \Big(c(\boldsymbol{a}; N_a, N_b)_{\Omega^e} + a(N_a, N_b)_{\Omega^e} + (N_a, \sigma N_b)_{\Omega^e} \Big) u^e_{Db}(t) \Big],$$

where $u^e_{Db}(t) = u_D(\boldsymbol{x}_b, t)$ if u_D is prescribed at node number b and zero otherwise.

5.4.2 Galerkin formulation of θ family methods

Which discretization is performed first is not an issue for linear spatial operators with constant coefficients and a Galerkin formulation. The time discretized equations, (5.2), (5.3), (5.8), (5.7) or (5.9), do have a truncation error. However, if the temporal truncation error is neglected, these equations can be interpreted as a spatial differential operator. In fact, they represent a strong form that must be solved at each time step.

Under this rationale it is easy to determine from (5.2) the variational form associated with the θ family methods:

$$\begin{aligned} \left(w, \frac{\Delta u}{\Delta t}\right) + \theta\big[c(\boldsymbol{a}; w, \Delta u) + a(w, \Delta u) + (w, \sigma \Delta u)\big] \\ = -\big[c(\boldsymbol{a}; w, u^n) + a(w, u^n) + (w, \sigma u^n)\big] \\ + \big(w, \theta s^{n+1} + (1-\theta) s^n\big) + \big(w, \theta h^{n+1} + (1-\theta) h^n\big). \end{aligned} \tag{5.15}$$

Note that in the previous equation the unknown Δu appears in the same three terms as in the l.h.s. of (5.12). Thus, after spatial discretization, we obtain the same mass, convection and diffusion matrices (scaled by $1/\Delta t$ and θ) as in (5.13) and (5.14). The r.h.s. terms also include the influence of the Neumann boundary condition, h, and the source term, s. These terms are linearly interpolated between time t^n and t^{n+1}. It is important to note that time is already discretized in equation (5.15). Therefore, the solutions of (5.15) are first or second-order approximations in time depending on the value of θ.

In order to analyze this family of methods we follow the methodology proposed in Section 3.5.1, which already included diffusion and reaction. Thus to determine the influence of the numerical scheme on a Fourier mode of wavelength $1/k$ one introduces the dimensionless wave number, $\xi = h\,k$. Recall the dimensionless scalars: Courant number $C = a\Delta t/h$, diffusion number $d = (\nu \Delta t)/h^2$, and dimensionless reaction $r = \sigma \Delta t$. Then, we follow the same steps of Section 3.5.2; the discrete equation obtained at an interior node j with the θ family method is determined from (5.15) (recall that there is no source term, $s = 0$, and we consider an interior node).

The fully discrete equation can be written from Table 3.1 as

$$\left(\left(1+\frac{\delta^2}{6}\right)+\theta\left(\frac{C\,\delta}{2}-d\,\delta^2+r\left(1+\frac{\delta^2}{6}\right)\right)\right)(u_j^{n+1}-u_j^n)$$
$$=-\theta\left(\frac{C\,\delta}{2}-d\,\delta^2+r\left(1+\frac{\delta^2}{6}\right)\right)u_j^n,$$

which allows us to find the equation that modifies each Fourier component,

$$\Big(\mathcal{M}(\xi)-\theta\big(\mathcal{K}(\xi,d)-\mathcal{A}(\xi,C)-r\mathcal{M}(\xi)\big)\Big)(u_j^{n+1}-u_j^n)$$
$$=\big(\mathcal{K}(\xi,d)-\mathcal{A}(\xi,C)-r\mathcal{M}(\xi)\big)u_j^n.$$

Thus the numerical amplification factor for the convection–diffusion–reaction equation and a Galerkin formulation is

$$G_\theta=\frac{\mathcal{M}(\xi)-(1-\theta)(\mathcal{K}(\xi,d)-\mathcal{A}(\xi,C)-r\mathcal{M}(\xi))}{\mathcal{M}(\xi)-\theta(\mathcal{K}(\xi,d)-\mathcal{A}(\xi,C)-r\mathcal{M}(\xi))}. \tag{5.16}$$

Using the last column in Table 3.1 one can explicitly determine the dependence of G_θ on C, d, r, θ and ξ. Then stability can be studied for each value of θ verifying the condition $|G_\theta|\leq 1$. In particular for $\theta\geq 1/2$ one can verify that the amplification factor is always (for $0\leq\xi\leq\pi$ and $C\geq 0$) less than one. Thus, Crank–Nicolson and backward Euler are *unconditionally stable.* Figure 5.2 plots $|G_\theta|$ as a function of ξ for different values of the Péclet, $Pe=C/(2d)$, and Courant, C, numbers. The curves are plotted in polar coordinates, the angle is ξ and the radius is $|G_\theta|$. The unit radius circle is also plotted. Thus when a method produces a radius larger than one it is unstable. Note that as expected the Euler method is *conditionally stable,* see Remark 5.6. The stability results for the pure diffusion case are well-known (see Wait and Mitchell, 1985) and can be determined from (5.16),

$$d\leq\frac{1}{6(1-2\theta)}\leq\frac{1}{6(1-2\theta)}\,\frac{2\big(2+\cos(\xi)\big)}{1-\cos(\xi)}.$$

Remark 5.6 (Stability range of explicit methods). In pure diffusion in 1D, the forward Euler method is stable for

$$\Delta t\leq\frac{h^2}{6\nu}\quad\text{or}\quad d\leq\frac{1}{6}.$$

The method is unstable in pure convection if a centered (Galerkin) approximation is used for the convective term. When an upwind approximation is employed, the method is stable provided $\Delta t\leq h/a$ (Courant condition). In convection–diffusion, the stability condition depends on the value of Pe:

$$\begin{cases}\text{if } Pe\leq\sqrt{3}, & C\leq Pe/3,\\ \text{if } Pe>\sqrt{3}, & C\leq 1/Pe.\end{cases}$$

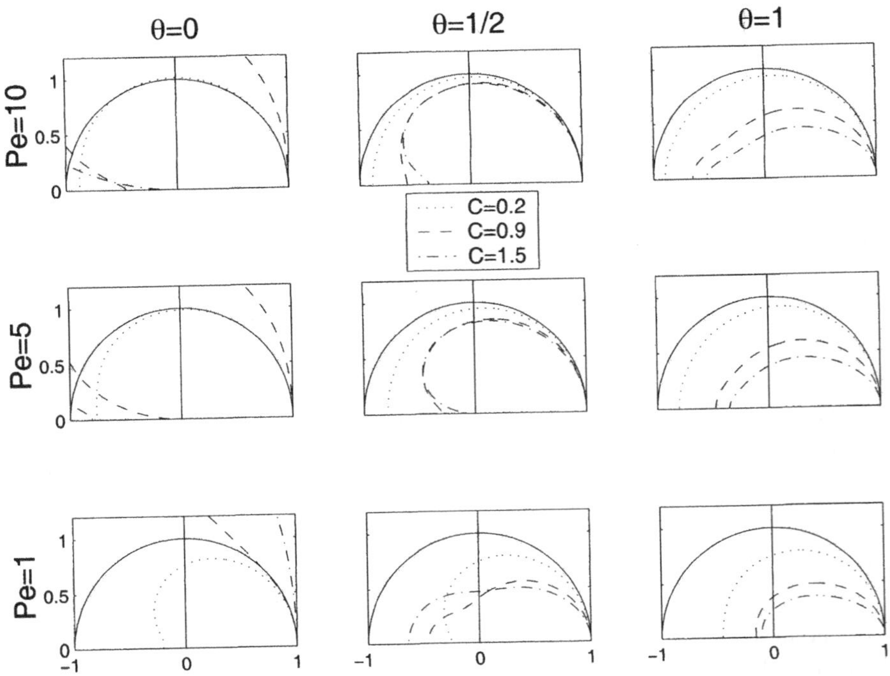

Fig. 5.2 Stability of the θ family methods for different values of Péclet and Courant numbers.

Stability of the second-order Adams–Bashforth method is governed by the same type of criteria, but its stability range is only half that for the Euler method. The leap-frog (LF) method is unstable in pure diffusion, and it is also unstable in pure convection if an upwind approximation is used for the convective term. When a centered approximation is employed, the LF method is stable under the Courant condition $\Delta t \leq h/a$. See, for instance, Gresho and Sani (2000) for additional information on the stability of explicit methods in convection–diffusion.

The accuracy of these methods can be evaluated by means of (3.36). Figure 5.3 shows relative errors at different Péclet numbers for numerical damping, $|G|/|G_{\text{ex}}|$, and phase error, $\arg(G)/\arg(G_{\text{ex}})$. From a precision point of view only the range $0 \leq \xi \leq \pi/4$ is of interest. Note that backward Euler is as expected overly diffusive, $|G|/|G_{\text{ex}}| \leq 1$. Moreover, reasonable phase errors require small Courant numbers.

Crank–Nicolson shows better accuracy due to its second-order accuracy, see Figure 5.4. However, its excellent behavior in damping for high values of Pe degrades as the diffusion coefficient and the Courant number increase. Moreover, its phase error is always important for moderately high values of the Courant number. For both methods the introduction of a reaction does not drastically affect these conclusions.

Apart from stability and accuracy (in time) we should recall the intrinsic deficiencies of the Galerkin formulation in the presence of boundary layers. As shown in the

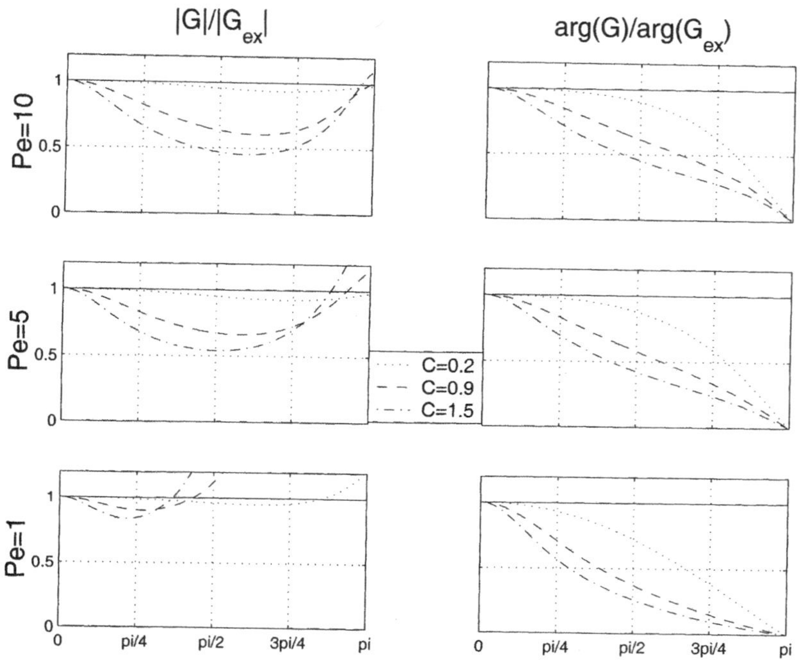

Fig. 5.3 Accuracy properties of backward Euler at different values of the Péclet number.

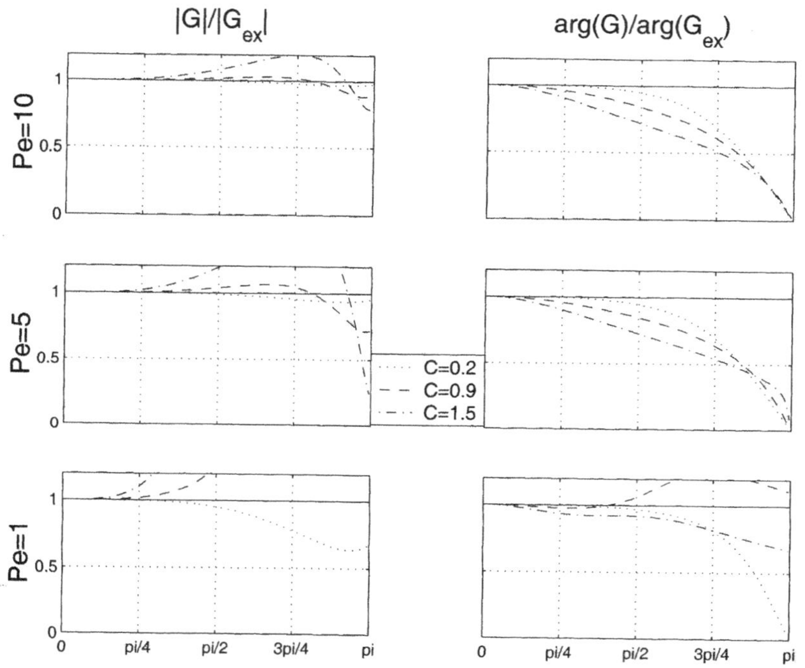

Fig. 5.4 Accuracy properties of Crank–Nicolson at different values of the Péclet number.

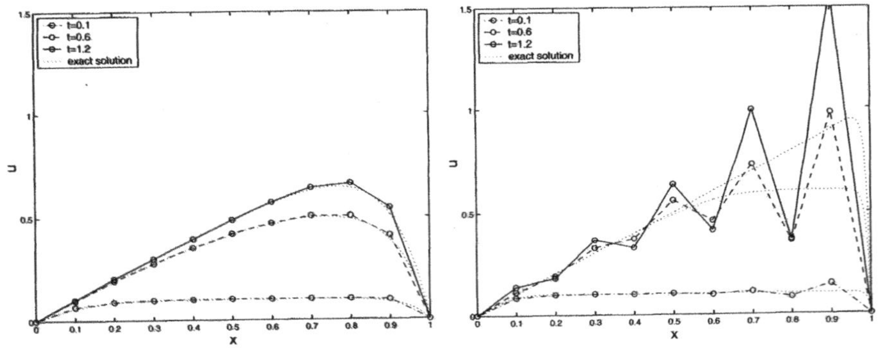

Fig. 5.5 Model problem for Crank–Nicolson at $Pe = 0.5$ (left) and $Pe = 5$ (right).

examples, if a smooth function is transported far from Dirichlet boundaries Galerkin will produce reasonable results. The previous accuracy analysis can be confirmed and, as seen in Chapter 3, Crank–Nicolson performs better (both in numerical damping and phase) than lower-order methods. Nevertheless, if the solution is confronted with Dirichlet boundary conditions developing a boundary layer, spatial instabilities (already seen in Chapter 2) will pollute the solution and spatial stabilization is required. A simple example using a uniform mesh of linear elements illustrates this point. It is the transient version of the example presented in Section 2.2.2, namely

$$\begin{cases} u_t + a\,u_x - \nu\,u_{xx} = 1 & \text{for } (x,t) \in \,]0,1[\times\mathbb{R}^+, \\ u(0,t) = 0 \quad \text{and} \quad u(1,t) = 0 & \text{for } t \in \mathbb{R}^+, \\ u(x,0) = 0 & \text{for } x \in \,]0,1[. \end{cases} \tag{5.17}$$

Figure 5.5 compares exact solution with results of the Galerkin/Crank–Nicolson formulation for different values of Pe and different instants.

5.4.3 Galerkin formulation of explicit Padé schemes

The Galerkin formulation of multistage explicit Padé methods, see (5.8), is

$$\begin{aligned}\left(w, u^{n+\beta_i}\right) = \left(w, u^n\right) + \beta_i\,\Delta t\Big[&\left(w, s^{n+\beta_{i-1}}\right) + \left(w, h^{n+\beta_{i-1}}\right)_{\Gamma_N} \\ &- c\left(\boldsymbol{a}; w, u^{n+\beta_{i-1}}\right) - a\left(w, u^{n+\beta_{i-1}}\right) - \left(w, \sigma\,u^{n+\beta_{i-1}}\right)\Big],\end{aligned}$$

for $i = 2, \ldots, \mathrm{n_{tg}}+1$. Recall that the first stage is trivial, $u^{n+\beta_1} = u^n$, thus $\beta_1 = 0$, and that $\beta_i = 1/(\mathrm{n_{tg}} + 2 - i)$, for $i = 2, \ldots, \mathrm{n_{tg}} + 1$.

The amplification factor associated with these schemes is evaluated by means of the Padé approximations presented in Table 5.1. The first row in this table defines the amplification factor:

$$G(\xi, C, d, c, r) = \mathrm{R}_{\mathrm{n_{tg}},0}(z) \quad \text{with } z := \frac{\mathcal{K}(\xi, d) - \mathcal{A}(\xi, C) - r\mathcal{M}(\xi)}{\mathcal{M}(\xi)}.$$

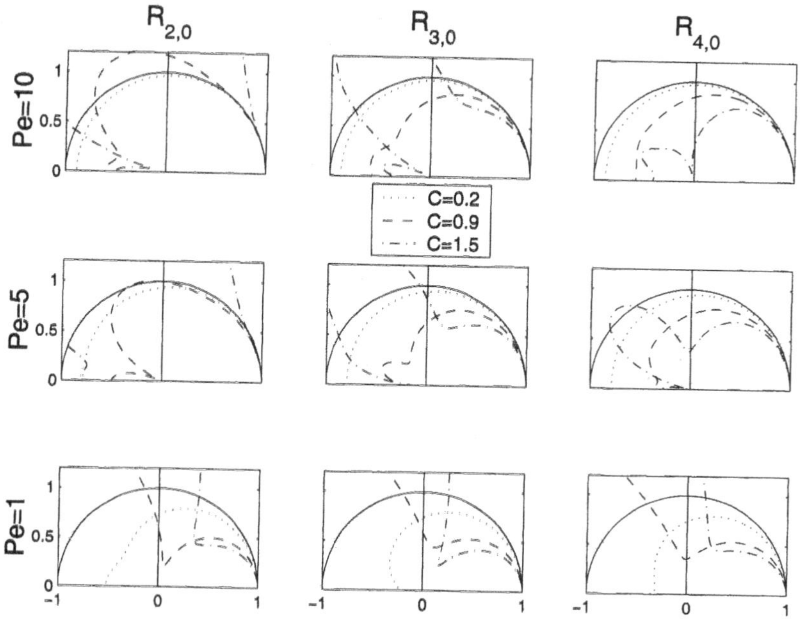

Fig. 5.6 Stability of multistage explicit Padé methods for different values of Péclet, Pe, and Courant, C, numbers.

These explicit methods are conditionally stable. Figure 5.1 shows the stability regions. Figure 5.6 presents the absolute value of the amplification factor in polar coordinates (ξ angle and $|G|$ radius) for different values of Pe and C. Note that diffusion strongly influences stability. For instance, the fourth-order method is stable for $C = 1.5$ at high Péclet numbers but becomes unstable as Pe decreases. This generates serious difficulties when explicit multistage methods are implemented in problems where convection is dominant in a region while diffusion dominates another, or simply when non-uniform meshes are needed due to geometrical requirements. From an accuracy point of view the same conclusions of Chapter 3 can be drawn here: higher-order methods are more accurate, in particular, with respect to phase errors.

5.4.4 Galerkin formulation of implicit multistage schemes

The time discretized equations (5.9) represent a strong form that must be solved at each time step. Note that any Runge–Kutta method, see (5.7), can also be expressed in the same form (with vectors and matrices of dimension $\mathrm{n}_{\mathrm{tg}} + 1$). The weighted residual form is

$$\left(\boldsymbol{w}, \frac{\Delta \boldsymbol{u}}{\Delta t}\right) - (\boldsymbol{w}, \mathbf{W}\, \Delta \boldsymbol{u}_t) = (\boldsymbol{w}, \mathbf{w}\, u_t^n),$$

where $\boldsymbol{w} \in [\mathcal{V}]^{\mathrm{n}}$, $\mathrm{n} = \mathrm{n}_{\mathrm{tg}} - 1$, for implicit multistage Padé schemes and $\mathrm{n} = \mathrm{n}_{\mathrm{tg}} + 1$, for standard implicit Runge–Kutta methods. Recall that $[\mathcal{V}]^{\mathrm{n}}$ imposes homogeneous conditions along the Dirichlet boundary. The solution $\Delta \boldsymbol{u} \in [\mathcal{S}]^{\mathrm{n}}$ and verifies the Dirichlet boundary conditions.

In the case of a linear operator $\mathcal{L}$, see (5.5), with constant coefficients and a source term s, see (5.11), the previous weighted residual equation can be rewritten as

$$\begin{aligned}\Big(\boldsymbol{w}, \frac{\Delta \boldsymbol{u}}{\Delta t}\Big) + c(\boldsymbol{a}; \boldsymbol{w}, \mathbf{W}\Delta\boldsymbol{u}) + a(\boldsymbol{w}, \mathbf{W}\Delta\boldsymbol{u}) + (\boldsymbol{w}, \sigma \mathbf{W}\Delta\boldsymbol{u}) \\ = (\boldsymbol{w}, \mathbf{w}\, s^n + \mathbf{W}\Delta \boldsymbol{s}) + (\boldsymbol{w}, \mathbf{w}\, h^n + \mathbf{W}\Delta \boldsymbol{h}) \\ - \Big(c(\boldsymbol{a}; \boldsymbol{w}, \mathbf{w}\, u^n) + a(\boldsymbol{w}, \mathbf{w}\, u^n) + (\boldsymbol{w}, \sigma \mathbf{w}\, u^n)\Big).\end{aligned}$$

From this weak form and Table 3.1 the equation which modifies each Fourier component is readily obtained

$$\begin{aligned}\Big(\mathcal{M}(\xi)\mathbf{I} - \theta\big(\mathcal{K}(\xi, d) - \mathcal{A}(\xi, C) - r\mathcal{M}(\xi)\big)\mathbf{W}\Big)\Delta\boldsymbol{u} \\ = \big(\mathcal{K}(\xi, d) - \mathcal{A}(\xi, C) - r\mathcal{M}(\xi)\big)u^n\mathbf{w}.\end{aligned}$$

Now the unknown is a vector with $\mathrm{n} = \mathrm{n_{tg}} - 1$ components for implicit multistage schemes. The amplification is the relation between the solution at t^{n+1} and t^n at a given interior point. To determine the numerical amplification factor, given the definition of $\Delta \boldsymbol{u}$, see (5.10), the following steps are necessary:

1. Solve the linear system of equations for $\boldsymbol{z} = (z_1, z_2, \dots, z_{\mathrm{n_{tg}}-1})^T$:

$$\begin{aligned}\Big(\mathcal{M}(\xi)\,\mathbf{I} - \big(\mathcal{K}(\xi, d) - \mathcal{A}(\xi, C) - r\mathcal{M}(\xi)\big)\mathbf{W}\Big)\boldsymbol{z} \\ = \big(\mathcal{K}(\xi, d) - \mathcal{A}(\xi, C) - r\mathcal{M}(\xi)\big)\mathbf{w}.\end{aligned}$$

2. And finally determine the amplification factor as

$$G(\xi, C, d, r) = 1 + z_1 + z_2 + \cdots + z_{\mathrm{n_{tg}}-1}.$$

Remark 5.7. This procedure to determine G is necessary for the stabilized methods shown next. In the case of a Galerkin formulation, similarly to the previous section, the Padé Table 5.1 allows us to directly compute the amplification factor. In this case implicit multistage schemes correspond to the diagonal of Table 5.1 and the amplification factor can be readily obtained as

$$G(\xi, C, d, c, r) = \mathsf{R}_{\mathrm{n_{tg}}-1, \mathrm{n_{tg}}-1}(z) \quad \text{with } z := \frac{\mathcal{K}(\xi, d) - \mathcal{A}(\xi, C) - r\mathcal{M}(\xi)}{\mathcal{M}(\xi)}.$$

These methods are unconditionally stable. They allow us to use high-order time-stepping schemes. Thus, accuracy can be improved drastically (obviously, for the solvable frequencies $0 \le \xi \le \pi/4$). Figure 5.7 shows relative errors in amplitude and phase for the fourth-order method, $\mathsf{R}_{2,2}$. But as previously observed for the θ family of methods, see Section 5.4.2, a Galerkin formulation presents spatial instabilities in the presence of boundary layers. The problem stated in (5.17) is solved again with a uniform mesh of 10 linear elements, see Figure 5.8. Since the $\mathsf{R}_{2,2}$ method has twice as many nodal unknowns, the Courant number is increased to $C = 3$.

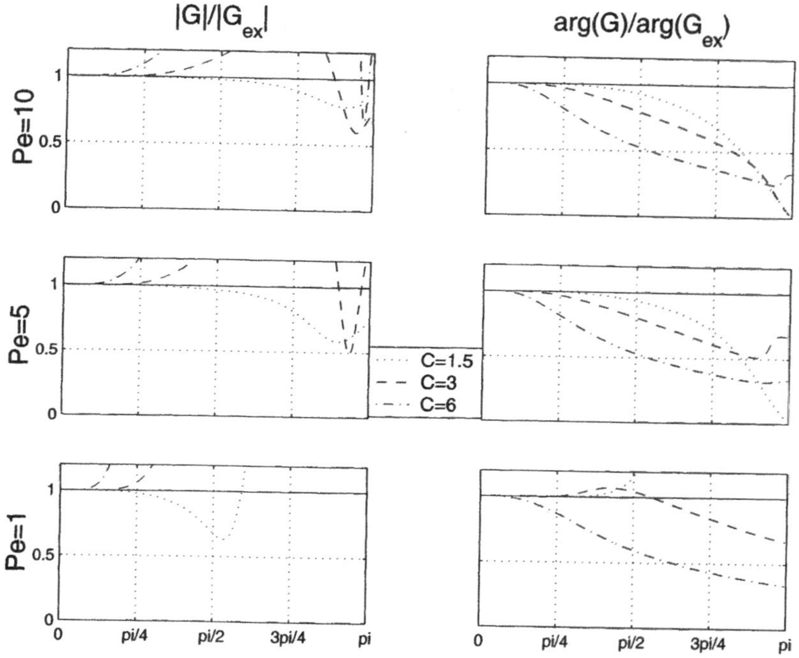

Fig. 5.7 Accuracy properties of implicit multistage $R_{2,2}$ at different values of Pe and C.

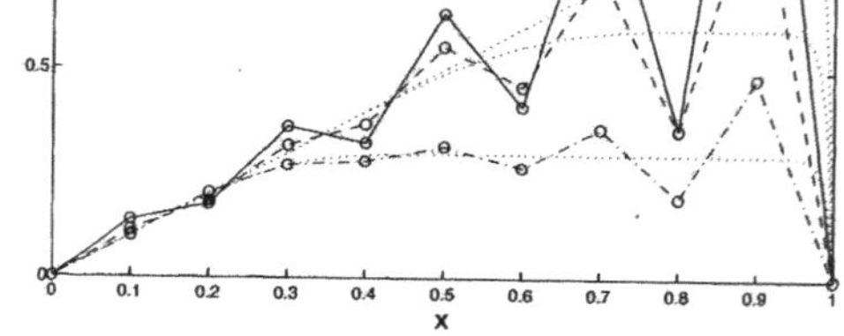

Fig. 5.8 Model problem for a Galerkin formulation of implicit multistage $R_{2,2}$.

5.4.5 Stabilization of the semi-discrete scheme

As amply discussed the Galerkin formulation lacks sufficient spatial stability when convective effects are important in the presence of boundary layers. In order to stabilize the convective term in a consistent manner, that is ensuring that the solution

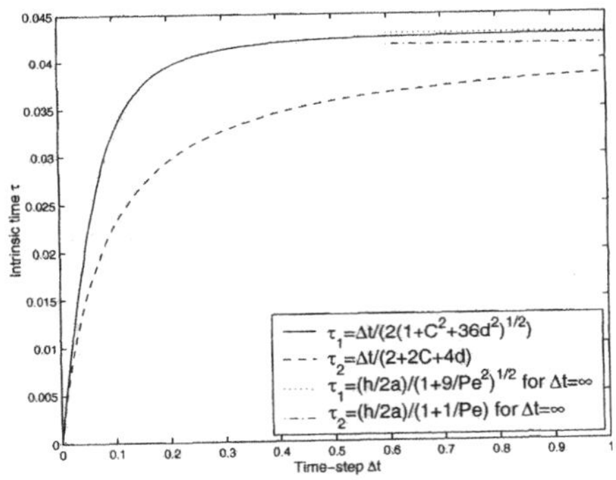

Fig. 5.9 Evolution of τ as Δt varies for the particular case: $a = 1, \nu = 10^{-2}$ and $h = 1/10$.

of the differential equation is also a solution of the weak form, Hughes and co-workers have proposed several techniques described in Chapter 2. These methods were subsequently extended to transient problems integrated with second-order time schemes or treated using space–time formulations.

A stabilized formulation of the convection–diffusion–reaction problem (5.1) can be stated as follows:

$$\big(w, u_t + \boldsymbol{a}\cdot\boldsymbol{\nabla} u + \sigma u\big) + a(w,u) + \sum_{e=1}^{n_{el}} \big(\mathcal{P}(w), \tau\mathcal{R}(u)\big)_{\Omega^e} = (w,s) + (w,h)_{\Gamma_N},$$

where the perturbation operator $\mathcal{P}$ characterizes the stabilization method, see Section 2.4. The stabilization term involves the residual $\mathcal{R}(u) = u_t + \boldsymbol{a}\cdot\boldsymbol{\nabla} u + \sigma u - \boldsymbol{\nabla}\cdot(\nu\boldsymbol{\nabla} u) - s$ of the governing equation, thus giving in principle a consistent formulation. Note that the residual includes the time derivative u_t of the unknown. This will result in a rather cumbersome implementation, except for the θ family of methods where u_t is replaced by $(u^{n+1} - u^n)/\Delta t$, or for space–time formulations.

Remark 5.8 (The transient stabilization parameter). When we are concerned by the steady solution it is usual to implement the parameters presented in Section 2.4.3. But if the transient solution is of interest we use the convection–diffusion–reaction extensions of the parameters described in Section 4.4.2.3. That is, for the θ family of methods, $\theta\in]0,1]$, following Shakib et al. (1991)

$$\tau = \left(\left(\frac{1}{\theta\Delta t}\right)^2 + \left(\frac{2a}{h}\right)^2 + 9\left(\frac{4\nu}{h^2}\right)^2 + \sigma^2\right)^{-1/2} = \frac{\Delta t}{2}\left(\left(\frac{1}{2\theta}\right)^2 + C^2 + 36d^2 + \left(\frac{r}{2}\right)^2\right)^{-1/2}$$

or from Soulaïmani and Fortin (1994), see also Codina (2000),

$$\tau = \left(\frac{1}{\theta\Delta t} + \frac{2a}{h} + \frac{4\nu}{h^2} + \sigma\right)^{-1} = \frac{\Delta t}{2}\left(\frac{1}{2\theta} + C + 2d + \frac{r}{2}\right)^{-1}.$$

Figure 5.9 shows, for $\theta = 1/2$, the variation of τ with Δt in a particular case. Note that in the limit $\Delta t = \infty$ both definitions are similar (the limits coincide in pure convection), but one definition reaches the limit much more rapidly.

5.4.6 Stabilization of multistage schemes

When accuracy requires us to go beyond second-order schemes and higher-order time integration schemes are employed, the standard stabilization of the semi-discrete equations is not trivial. In the context of transient problems, see Chapters 3 and 4, least-squares formulations can also be used to stabilize pure convection problems, see the appendix at the end of this chapter. However, their direct extension to convection–diffusion problems is also not trivial because of the presence of the second-order operator (the diffusion term). Huerta and Donea (2002) propose to combine spatial stabilization techniques with high-order implicit time-stepping schemes. Moreover, a least-squares (LS) formulation is also presented for these high-order time schemes combined with C^0 finite elements (in spite of the diffusion operator and without reducing the strong form to a system of first-order differential equations). Reference will be made to three stabilization techniques, namely, the *streamline-upwind Petrov–Galerkin* (SUPG), the *Galerkin/Least-squares* (GLS), and the *sub-grid scale* (SGS) methods, see Section 2.4.

In order to have a consistent stabilization of time discretized equations, such as those resulting from multistage time schemes, a residual must be defined. The residual, in the present case, is clearly chosen after time discretization. Thus, from the strong form (5.9) one gets

$$\mathcal{R}(\Delta \boldsymbol{u}) := \frac{\Delta \boldsymbol{u}}{\Delta t} - \mathbf{W}\,\Delta \boldsymbol{u}_t - \mathbf{w}\,\boldsymbol{u}_t^n.$$

In order to better visualize the spatial differential structure of the residual $\mathcal{R}(\Delta \boldsymbol{u})$, this equation is rewritten for the particular case of linear spatial operator $\mathcal{L}$ with constant coefficients, that is from (5.11),

$$\mathcal{R}(\Delta \boldsymbol{u}) = \frac{\Delta \boldsymbol{u}}{\Delta t} + \mathbf{W}\mathcal{L}(\Delta \boldsymbol{u}) - \mathbf{w}\big[s^n - \mathcal{L}(u^n)\big] - \mathbf{W}\Delta \boldsymbol{s}, \tag{5.18}$$

where, from the second term on the r.h.s., one can observe that the operator $\mathcal{L} = \boldsymbol{a}\cdot\boldsymbol{\nabla} - \nu\boldsymbol{\nabla}^2 + \sigma$ acts on each component of $\Delta \boldsymbol{u}$.

The consistently stabilized weak form of the time discretized problem is given by

$$\left(\boldsymbol{w}, \frac{\Delta \boldsymbol{u}}{\Delta t}\right) - \left(\boldsymbol{w}, \mathbf{W}\,\Delta \boldsymbol{u}_t\right) + \underbrace{\sum_e \big(\boldsymbol{\tau}\mathcal{P}(\boldsymbol{w}), \mathcal{R}(\Delta \boldsymbol{u})\big)_{\Omega^e}}_{\text{Stabilization term}} = \left(\boldsymbol{w}, \mathbf{w}\,\boldsymbol{u}_t^n\right). \tag{5.19}$$

Here again, the stabilization term is added to the Galerkin weak form. It contains the intrinsic time scale matrix $\boldsymbol{\tau}$ because (5.11), or (5.18), is a system of equations. The operator $\mathcal{P}$ characterizes the stabilization technique (i.e., SUPG, GLS, SGS or LS).

Remark 5.9 (On the stabilization matrix $\boldsymbol{\tau}$ for multistage schemes). When the convection–diffusion–reaction operator $\mathcal{L} = \boldsymbol{a}\cdot\boldsymbol{\nabla} - \nu\boldsymbol{\nabla}^2 + \sigma$ is replaced in the multistage form of the time-discretized equation, that is in (5.11), the system of equations that must be solved at each time step, namely

$$\frac{\Delta \boldsymbol{u}}{\Delta t} + \mathbf{W}(\boldsymbol{a} \cdot \boldsymbol{\nabla} \Delta \boldsymbol{u} - \nu \boldsymbol{\nabla}^2 \Delta \boldsymbol{u} + \sigma \Delta \boldsymbol{u}) = \mathbf{w}\big[s^n - (\boldsymbol{a} \cdot \boldsymbol{\nabla} u^n - \nu \boldsymbol{\nabla}^2 u^n + \sigma u^n)\big] + \mathbf{W} \Delta \boldsymbol{s},$$

can be reinterpreted as a system of *steady* convection–diffusion–reaction equations.

In this case, convection is controlled by $a_i \mathbf{W}$, $i = 1, \ldots, \mathrm{n_{sd}}$ (recall the Jacobian matrices in Section 4.4.2.3), the diffusion matrix is $\nu \mathbf{W}$ and the reaction matrix, $\mathbf{I}/\Delta t + \sigma \mathbf{W}$, includes the contribution form the transient term. Thus, in 1D, the stabilization matrix can be computed as

$$\begin{aligned} \boldsymbol{\tau} &= \left[\frac{\mathbf{W}^{-1}}{\Delta t} + \left(\frac{2a}{h} + \frac{4\nu}{h^2} + \sigma \right) \mathbf{I} \right]^{-T} \mathbf{W}^{-1} \\ &= \Delta t \big[\mathbf{W}^{-1} + (2C + 4d + r) \mathbf{I} \big]^{-T} \mathbf{W}^{-1}, \end{aligned}$$

when the analysis of Soulaïmani and Fortin (1994), see also Codina (2000), is particularized in this case. Note also that such a definition of the intrinsic matrix $\boldsymbol{\tau}$ is also valid for the scalar case and generalizes the previous one, see Remark 5.8.

In particular, for implicit Padé approximations, see the definitions of $\mathbf{W}$ in Section 5.3.3.5, $\mathbf{W}^{-1} = \frac{1}{2} \left(\begin{smallmatrix} 5 & 1 \\ -13 & 7 \end{smallmatrix} \right)$ for $\mathsf{R}_{2,2}$.

Finally, note that other definitions for $\boldsymbol{\tau}$ can also be used. For instance, if the steady solution is our main interest, as noted in Remark 5.8, it is usual to implement the parameters presented in Section 2.4.3, that is

$$\boldsymbol{\tau} = \frac{h}{2a} \left[\coth(\mathit{Pe}) - \frac{1}{\mathit{Pe}} \right] \mathbf{W}^{-1},$$

see the final example in the appendix at the end of this chapter.

Remark 5.10 (Consistent stabilization). The stabilization term involves the residual, which includes the second-order term $\nabla^2 u$. When linear finite elements are used this term vanishes or is largely under-represented, with the corresponding degradation in the consistency of the stabilized formulation. The lack of consistency leads to errors of order $\mathcal{O}(\tau)$, apart from the errors inherent in the time integration scheme.

In order to keep the convergence rates in time, several possibilities can be useful. The stabilization parameter τ can be defined to be asymptotically of order $\mathcal{O}(\Delta t^{2\mathrm{n}})$. That is, a specific intrinsic time τ should be designed for each one of the time integration schemes. Another possibility is to include flux jump terms across the element boundaries in the stabilized formulation to take into account the neglected terms, see Tezduyar and Osawa (2000) for details. In fact, Jansen, Collis, Whiting and Shakib (1999) show that when linear finite elements are used the lack of consistency due to the neglected terms leads also

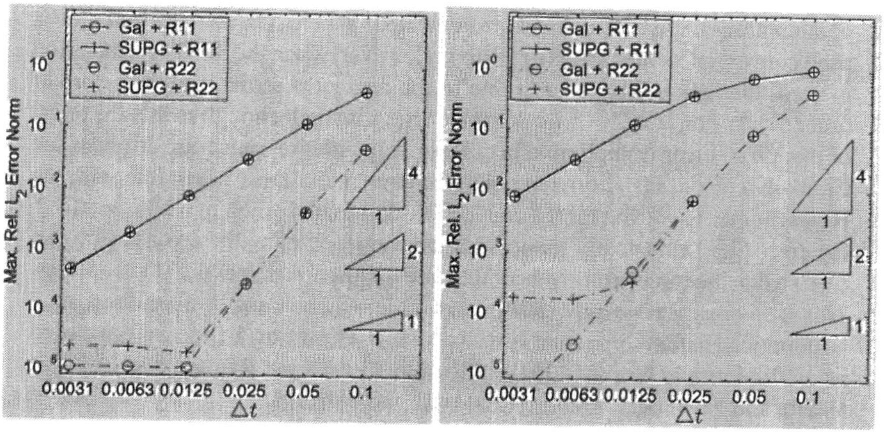

Fig. 5.10 Convergence results: $h = 0.001$, $a = 1$, $\nu = 10^{-2}$ (left) and $\nu = 10^{-4}$ (right).

to reduced convergence in space. For linear finite elements, they propose a global reconstruction of second derivatives. This method recovers the ability to approximate the residual in the stabilization term yielding a better consistency, through an iterative process. However, the increase in the computational cost is not negligible: a system of equations with global mass matrix must be solved at each iteration.

The use of high-order finite elements, such as quadratic elements, allows the inclusion of second derivatives of the approximation in the residual in the stabilization term, and thus consistent stabilized formulations can be defined. However, the computational cost and the implementation difficulties are highly increased due to the computation of second derivatives of the element mapping (see Jansen et al., 1999).

A numerical example is considered in order to experimentally assess the need for considering second derivatives in space. We solve the 1D convection–diffusion equation with constant coefficients,

$$u_t + au_x = \nu u_{xx}, \qquad (x,t) \in]0,2[\times]0,1[,$$

with homogeneous Dirichlet boundary conditions. The initial condition, at $t = 0$, is chosen such that the analytical solution is known

$$u(x,t) = \frac{\sigma_0}{\sigma} \exp\left\{-\frac{\left(x - (x_0 + at)\right)^2}{2\sigma^2}\right\}, \quad \sigma^2 = \sigma_0^2 + 2\nu t,$$

where $x_0 = 0.35$ and $\sigma_0 = 0.05$. Figure 5.10 shows the evolution of the error against the time step for linear finite elements with element size $h = 0.001$, and for two different values of the diffusion parameter ν. The error is evaluated in the $\mathcal{L}^2(\Omega)$ norm. Results are shown for Galerkin, `Gal`, and SUPG with Crank–Nicolson, $\mathsf{R}_{1,1}$, and a fourth-order Padé scheme, $\mathsf{R}_{2,2}$. In the stabilized computations second derivatives are neglected. When a Galerkin formulation is used convergence rates are as expected. However, when the SUPG formulation is used, the lack of consistency due to the neglected terms leads to errors of order $\mathcal{O}(\tau)$. For, $\nu = 10^{-2}$ the intrinsic time τ is small enough so that the effect of this $\mathcal{O}(\tau)$ error is negligible in comparison with the truncation errors of the Crank–Nicolson, $\mathsf{R}_{1,1}$, time-stepping scheme and almost negligible with the $\mathsf{R}_{2,2}$ scheme. However, for the convection-dominated problem with $\nu = 10^{-4}$, the error $\mathcal{O}(\tau)$ drastically reduces the convergence rate: the error is of order $\mathcal{O}(\tau)$ when the truncation errors of the time-stepping schemes are small enough. This is obviously more important in high-order schemes and its importance may be unnoticed in first- or second-order schemes. The intrinsic time τ is computed using the formula proposed by Shakib et al. (1991), see Remark 5.8. See also Huerta and Fernández-Méndez (2003) for more details.

5.4.6.1 *Streamline-upwind Petrov–Galerkin stabilization* The SUPG stabilization technique is defined by taking

$$\mathcal{P}(\boldsymbol{w}) := \mathbf{W}(\boldsymbol{a} \cdot \boldsymbol{\nabla})\boldsymbol{w}. \tag{5.20}$$

Note that matrix $\mathbf{W}$, which affects the convection term in (5.18), induces a *non-scalar* stabilization that is, each equation of the multistage time scheme is affected by different coefficients. The weak form for the SUPG method is obtained after substitution of the perturbation operator (5.20) in equation (5.19). The non-symmetric structure of the stabilization term induces some technical difficulties in the stability analysis of the SUPG method. This is avoided in the GLS stabilization technique, because it introduces a symmetric stabilization term in a consistent manner.

5.4.6.2 *Galerkin/Least-squares stabilization* The need for a previous time discretization is clear in this method. The GLS stabilization uses as perturbation operator $\mathcal{P}$ the spatial differential operator of the strong form, which in this case is affected by the time discretization. Note that the present approach differs from the standard space–time GLS formulation: no time derivatives appear in $\mathcal{P}$ and thus there is no need for multicorrector algorithms (see Shakib and Hughes, 1991; Codina, 1998). In the case of a linear convection–diffusion–reaction equation with constant coefficients, from (5.18) one deduces the operator $\mathcal{P}$, namely

$$\mathcal{P}(\boldsymbol{w}) := \frac{\boldsymbol{w}}{\Delta t} + \mathbf{W}\,\mathcal{L}(\boldsymbol{w}). \tag{5.21}$$

From a practical point of view there is no major difference between SUPG and GLS methods. Note that both methods are not identical, as in steady problems, for convection–diffusion (no reaction) and with linear elements (the second-order derivatives are zero in the element interiors). Moreover, the qualitative influence of each term in the definition of $\mathcal{P}$, equation (5.21), may be interpreted as follows:

$$\mathcal{P}(\boldsymbol{w}) = \frac{\boldsymbol{w}}{\Delta t} + \mathbf{W}\,\mathcal{L}(\boldsymbol{w}) = \underbrace{\frac{\boldsymbol{w}}{\Delta t}}_{\text{Galerkin}} + \mathbf{W}\big[\underbrace{(\boldsymbol{a}\cdot\boldsymbol{\nabla})\boldsymbol{w}}_{\text{SUPG}} - \underbrace{\boldsymbol{\nabla}\cdot(\nu\boldsymbol{\nabla}\boldsymbol{w})}_{0} + \underbrace{\sigma\boldsymbol{w}}_{\text{Galerkin}}\big].$$

The first term is a Galerkin weighting, the second term corresponds to the SUPG stabilization, the third term is zero for linear elements, and the fourth term is also a Galerkin weighting. Thus, for linear elements and a constant positive reaction, GLS is SUPG with the Galerkin term weighted $1 + \tau(\mathrm{C}\sigma + 1/\Delta t)$ times more (with C a constant related to $\mathbf{W}$). This implies that the instabilities introduced by Galerkin are slightly amplified in GLS compared with SUPG. This minor problem of the GLS stabilization technique is overcome in the simplest version of the sub-grid scale (SGS) method, see the next section.

In the case of the nonlinear convection–diffusion–reaction equation it is helpful to define the quasi-linear operator related to (5.5), see for instance the discussion by Hauke and Hughes (1998). This quasi-linear operator is then used in (5.21). Note, finally, that the stabilization term is symmetric but the complete weak form, that is (5.19) with (5.21), is in general non-symmetric.

5.4.6.3 Sub-grid scale stabilization The simplest version of the SGS stabilization assumes that the perturbation operator $\mathcal{P}(\boldsymbol{w})$ is minus the adjoint of the operator used in the GLS stabilization, that is

$$\mathcal{P}(\boldsymbol{w}) := -\frac{\boldsymbol{w}}{\Delta t} - \mathbf{W}\,\mathcal{L}^*(\boldsymbol{w}) = -\frac{\boldsymbol{w}}{\Delta t} + \mathbf{W}\big[(\boldsymbol{a}\cdot\boldsymbol{\nabla} + \sigma)\boldsymbol{w} + \boldsymbol{\nabla}\cdot(\nu\boldsymbol{\nabla}\boldsymbol{w})\big], \quad (5.22)$$

where $\mathcal{L}^*$ is the adjoint operator of $\mathcal{L}$. Following the same rationale as in the previous section to qualitatively determine the influence of the Galerkin term in SGS compared with SUPG, one gets that in SGS the Galerkin term is weighted $1 - \tau(\mathrm{C}\sigma + 1/\Delta t)$ times more than in SUPG. Thus it has less influence than in SUPG and GLS.

5.4.6.4 Least-squares stabilization The classical implementation of a standard LS formulation for the convection–diffusion–reaction problem (5.1) requires us to work in $\mathcal{H}^2$, unless a mixed LS formulation is used (see Park and Liggett, 1990; Carey, Shen and McLay, 1998). In fact, the least-squares minimization of the square of the residual of the governing equation invariably includes second spatial derivatives, thus requiring continuity of the unknown itself and of its first derivative.

In this section, an alternative procedure is proposed, which allows the use of standard $\mathcal{C}^0$ finite element interpolation and test functions. Furthermore, the proposed method does not increase the number of nodal unknowns (this is the case for a mixed formulation that yields a system of first-order equations or for higher-order interpolation methods). A standard LS formulation directly uses the spatial strong form to construct the integral equation. Here, since time discretization is already performed, equations (5.9) or (5.11) are used. Consequently, one gets the integral form

$$\left(\frac{\boldsymbol{w}}{\Delta t} + \mathbf{W}\,\mathcal{L}(\boldsymbol{w}), \mathcal{R}(\Delta\boldsymbol{u})\right) = 0, \quad (5.23)$$

where $\boldsymbol{w}$ and $\Delta\boldsymbol{u}$ are in subspaces of $[\mathcal{H}^2]^{\mathrm{n_{tg}}-1}$. However, an "equivalent" form following the same rationale as for standard stabilized methods, see equation (5.19), can be devised. It is equivalent in the sense that its unique solution is also the unique solution of (5.1) and also the solution of (5.23). The first argument in (5.23) is split by linearity and the term containing $\mathcal{L}(\boldsymbol{w})$ is only evaluated on the element interiors, namely

$$(\boldsymbol{w}, \mathcal{R}(\Delta\boldsymbol{u})) + \sum_e (\Delta t\, \mathbf{W}\, \mathcal{L}(\boldsymbol{w}), \mathcal{R}(\Delta\boldsymbol{u}))_{\Omega^e} = 0. \tag{5.24}$$

Now the interpolation and test functions can be taken in a subspace of $[\mathcal{H}^{1+}]^{n_{tg}-1}$,

$$\mathcal{H}^1 \subsetneq \mathcal{H}^{1+} := \{w \in \mathcal{H}^1(\Omega) \mid w|_{\Omega^e} \in \mathcal{H}^2(\Omega^e) \text{ for all element } \Omega^e\} \subsetneq \mathcal{H}^2.$$

Thus, standard $\mathcal{C}^0$ finite elements can be used. This approach can also be viewed as a standard stabilization technique: the first term accounts for the Galerkin part of the weak form, and the second one introduces the desired stabilization with the following definition of the operator $\mathcal{P}(\boldsymbol{w})$ and the intrinsic time τ:

$$\tau := \Delta t \quad \text{and} \quad \mathcal{P}(\boldsymbol{w}) := \mathbf{W}\, \mathcal{L}(\boldsymbol{w}). \tag{5.25}$$

In fact, the GLS formulation so-called "neglecting the inertia term" (i.e., $\boldsymbol{w}/dt$ is neglected) is used in common practice. It corresponds to the same perturbation operator (previous equation) and uses the standard τ instead of Δt.

5.4.6.5 Accuracy properties The Fourier analysis described in Section 5.4.4 for the Galerkin formulation can be repeated for each stabilized technique. For brevity we only present the GLS case; Huerta, Roig and Donea (2002) show further details. The equation associated with the GLS stabilization technique for a typical Fourier mode corresponding to a dimensionless wave number ξ is

$$\begin{aligned}
&\Big\{\mathcal{M}(\xi)\,\mathbf{I} - \big[\mathcal{K}(\xi,d) - \mathcal{A}(\xi,C) - r\mathcal{M}(\xi)\big]\mathbf{W} \\
&\quad - \frac{\tau}{\Delta t}\Big[-\mathcal{M}(\xi)\,\mathbf{I} + \mathcal{A}(\xi,C)(\mathbf{W}^T - \mathbf{W}) - r\mathcal{M}(\xi)(\mathbf{W}^T + \mathbf{W}) \\
&\qquad + \big(\mathcal{D}(\xi,C) - r^2\mathcal{M}(\xi)\big)\mathbf{W}^T\mathbf{W}\Big]\Big\}\Delta\boldsymbol{u} \\
&\qquad\quad = \Big\{\big[\mathcal{K}(\xi,d) - \mathcal{A}(\xi,C) - r\mathcal{M}(\xi)\big] \\
&\qquad\qquad - \frac{\tau}{\Delta t}\Big[\mathcal{A}(\xi,C) + r\mathcal{M}(\xi) - \big(\mathcal{D}(\xi,C) - r^2\mathcal{M}(\xi)\big)\mathbf{W}^T\Big]\Big\}u^n\ \mathbf{w}.
\end{aligned}$$

Note the nonlinear dependence on $\mathbf{W}$.

Recall that the intrinsic time scale can be defined in various ways, see Remark 5.8, although no major differences are observed for the usual definitions. Figures 5.11 and 5.12 show the relative errors for the numerical damping, $|G| \,/\, |G_{\text{ex}}|$, and the phase error, $\arg(G) \,/\, \arg(G_{\text{ex}})$, for different values of the Péclet number, $Pe = C/(2d)$. All the curves are plotted in polar coordinates, the angle is ξ and the radius is the relative error. From a precision point of view only the range $0 \leq \xi \leq \pi/4$ is of interest. Note that for Crank–Nicolson, Figure 5.11, the Courant numbers employed are 0.75, 1.5 and 3. For $\mathsf{R}_{2,2}$ larger values of the Courant number (1.5, 3 and 6) are used to

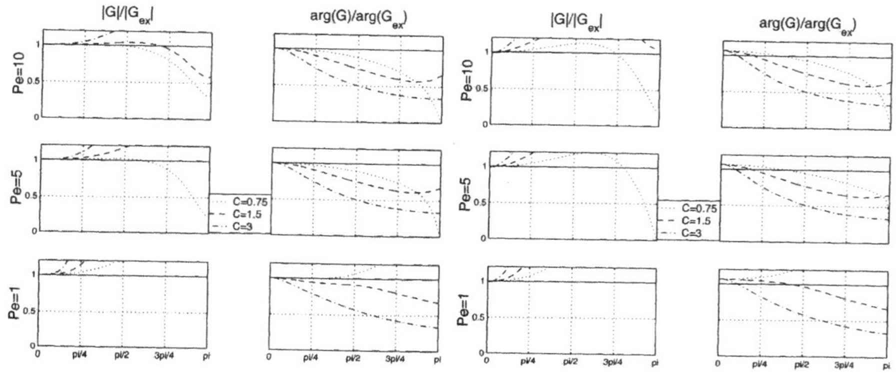

Fig. 5.11 Accuracy properties of Crank–Nicolson ($R_{1,1}$) with GLS stabilization: relative errors in amplitude and phase for $r = 0$ (left) and $r = 0.5$ (right) and different values of *Pe*.

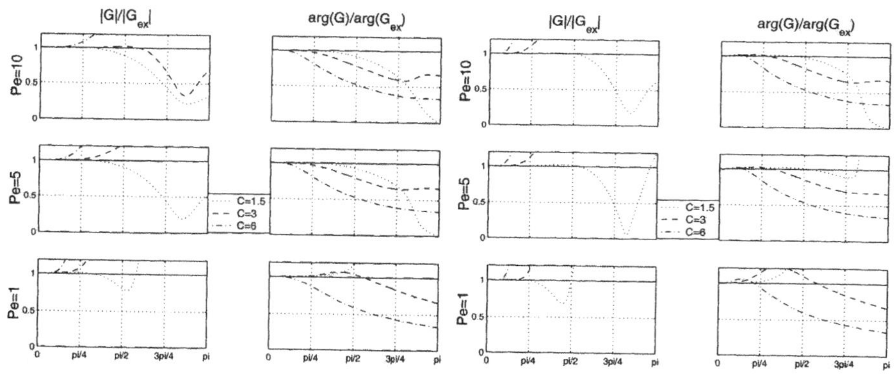

Fig. 5.12 Accuracy properties of $R_{2,2}$ with GLS stabilization: relative errors in amplitude and phase for $r = 0$ (left) and $r = 1$ (right) and different values of *Pe*.

highlight its range of precision which must compensate its extra cost (double number of unknowns per node).

Only the GLS stabilization results are shown, the other stabilization techniques present little differences. The LS formulation introduces, as expected, more numerical diffusion for $R_{2,2}$. Moreover, it is only noticeable in pure convection ($Pe = 0$) and this effect is beneficial in the presence of reaction. The phase accuracy is also affected by the LS stabilization but this only occurs for low Péclet numbers.

One can observe that, as expected, the stabilized methods introduce numerical damping for high frequencies (ξ close to π), which cannot be properly represented on the discrete mesh. Recall that Galerkin methods for pure convection do not introduce any numerical damping when combined with diagonal Padé approximations. Moreover, from an accuracy point of view (i.e., for dimensionless wave numbers such that $0 \leq \xi \leq \pi/4$), the response (for different Péclet, Courant and reaction values) is quantitatively the same as in the Galerkin formulation. Thus, the stabilization pro-

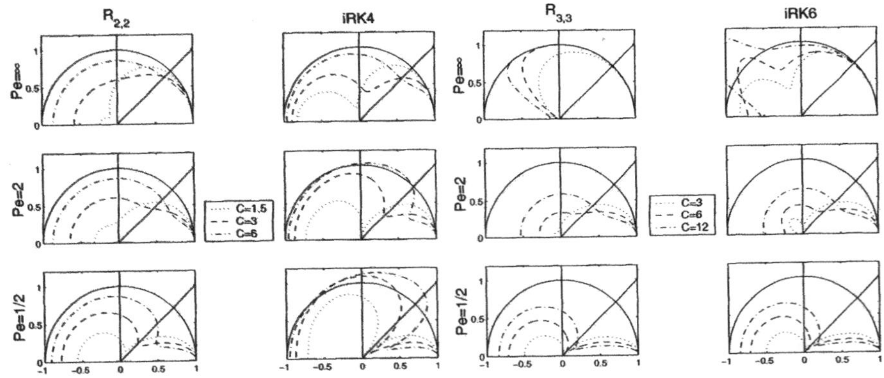

Fig. 5.13 Modulus of the amplification factor for stabilized Padé ($R_{n,n}$) and Lobatto IIIA implicit Runge–Kutta (iRKn) methods at different values of the Péclet number: fourth-order methods with LS (left) and sixth-order methods with GLS (right).

cedure introduces numerical diffusion where it is needed and does not compromise the accuracy. Moreover, from the point of view of the phase errors, the stabilized methods show a behavior identical to the Galerkin formulation. That is, phase errors are drastically reduced when the order of the temporal approximation is increased.

The reaction only influences the accuracy of the damping response; only for very large values of the reaction is the phase accuracy moderately degraded as r increases.

An important point to be made in concluding the accuracy analysis is that the use of non-overlapping increments in the multistage time scheme preserves unconditional stability when the method is combined with stabilized spatial formulations. By contrast, Figure 5.13 shows that the classical, fourth-order accurate, Lobatto IIIA implicit Runge–Kutta method combined with the LS stabilization loses stability for moderate to low Péclet numbers. It also shows that the sixth-order accurate Lobatto IIIA implicit Runge–Kutta method loses stability in pure convection when combined with the GLS stabilization. This is not the case for the corresponding multistage schemes derived from Padé approximations.

Now the stabilized formulation can be used in the model problem defined in (5.17) to preclude the usual global Galerkin instabilities. Figure 5.14 presents the results for Crank–Nicolson stabilized with SUPG, GLS, SGS and LS. In this case the Courant number is chosen equal to one because as seen in Figure 5.11 this value produces accurate results. Similarly, Figure 5.15 presents the results at the same instants for the stabilized implicit multistage Padé of fourth order, $R_{2,2}$. This scheme presents reasonable accuracy properties up to $C = 3$, see Figure 5.12. In this examples we use the definition of stabilization parameter for Crank–Nicolson proposed in Remark 5.8, and the definition of stabilization matrix for $R_{2,2}$ the given in Remark 5.9.

Note first that the solution is stabilized: oscillations, when they exist, are confined to the boundary layer and do not pollute the solution everywhere. Second, as already discussed in Section 2.4 (see Figure 2.14), the different stabilization (SUPG, GLS

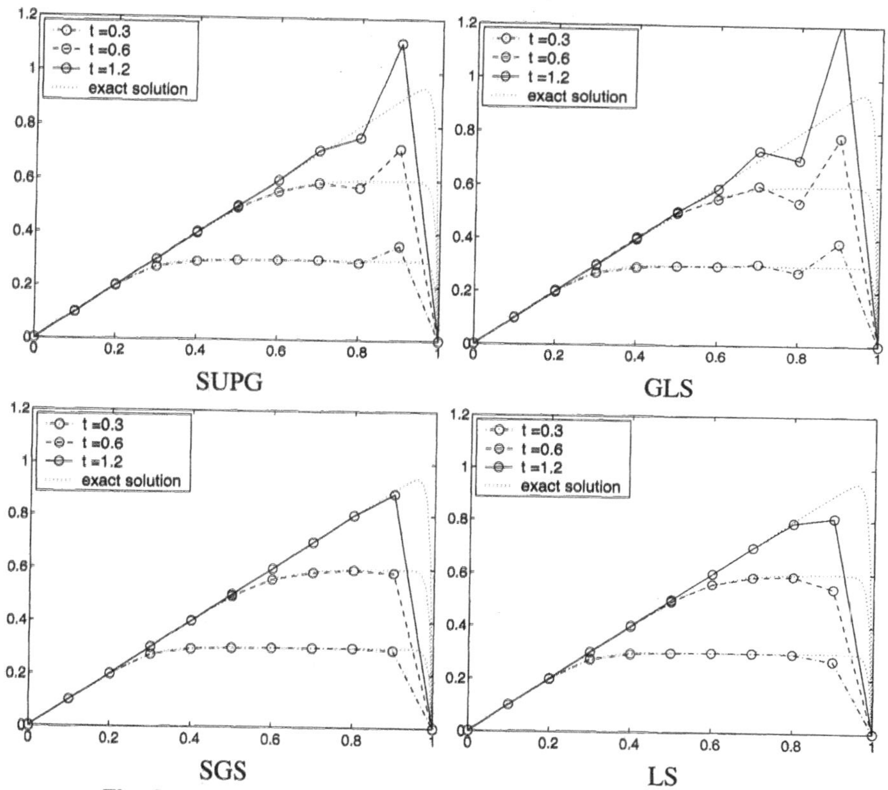

Fig. 5.14 Model problem with stabilized Crank–Nicolson and $C = 1$.

and SGS) techniques weight more or less the Galerkin term and thus the oscillation at the boundary layer is more or less amplified. Finally, these examples also show that LS stabilization techniques are more diffusive than SUPG, GLS or SGS. Moreover, by definition, see (5.25), the added diffusion is controlled by the time step ($\tau = \Delta t$). Therefore, the LS solution introduces more or less diffusion depending on the imposed Courant number.

5.5 STABILIZED SPACE–TIME FORMULATIONS

We conclude the discussion of spatial discretization procedures mentioning that stabilized methods for transient convection–diffusion–reaction can easily be extended to the space–time domain. Methods in this class were described for pure convection in Section 3.10 and we follow the notation introduced there. To give an example in the present context we consider the space–time Galerkin/Least-squares formulation, see Section 3.10.3 and in particular equation (3.65). We assume homogeneous Dirichlet boundary conditions. Due to the presence of the diffusion operator, least-squares

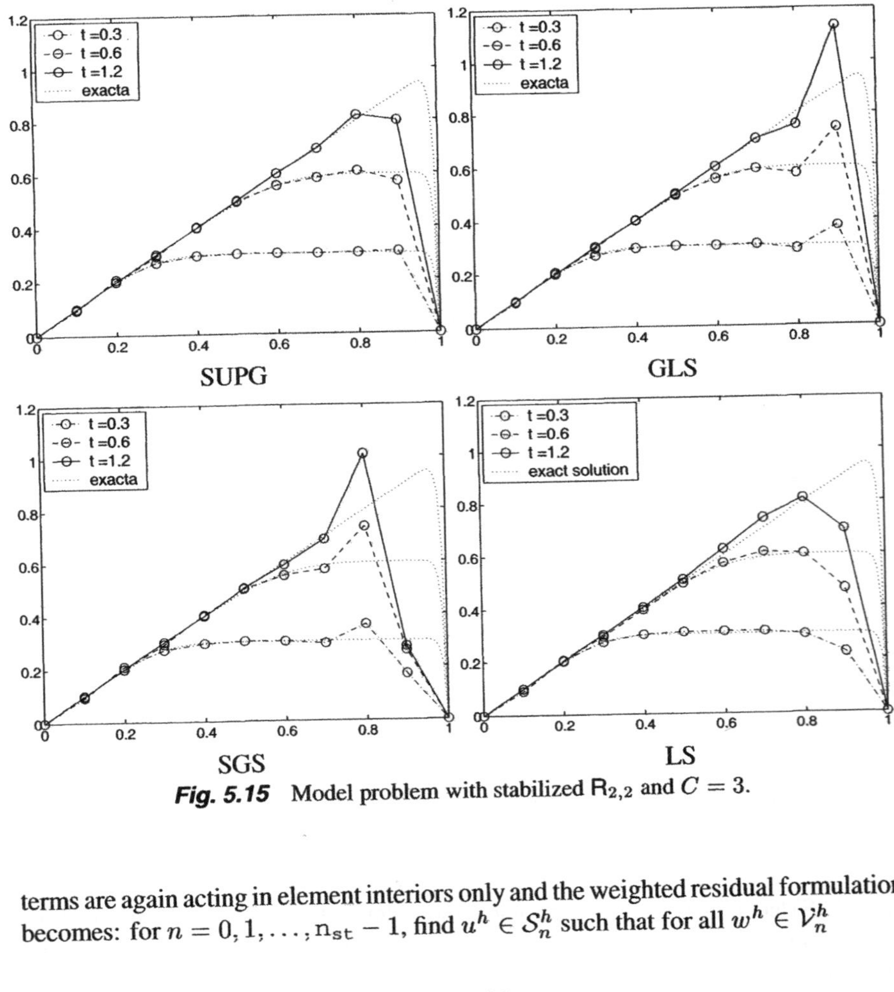

Fig. 5.15 Model problem with stabilized $\mathsf{R}_{2,2}$ and $C = 3$.

terms are again acting in element interiors only and the weighted residual formulation becomes: for $n = 0, 1, \ldots, \mathtt{n_{st}} - 1$, find $u^h \in \mathcal{S}_n^h$ such that for all $w^h \in \mathcal{V}_n^h$

$$\iint_{Q^n} w^h(u_t^h + \boldsymbol{a}\cdot\boldsymbol{\nabla}u^h + \sigma u^h)d\Omega\,dt + \iint_{Q^n} \boldsymbol{\nabla}w^h\cdot\nu\boldsymbol{\nabla}u^h\,d\Omega dt$$
$$+\sum_{e=1}^{\mathtt{n_{el}}}\iint_{Q_e^n}\left(w_t^h + \boldsymbol{a}\cdot\boldsymbol{\nabla}w^h + \sigma w^h - \boldsymbol{\nabla}\cdot(\nu\boldsymbol{\nabla}w^h)\right)$$
$$\tau\left(u_t^h + \boldsymbol{a}\cdot\boldsymbol{\nabla}u^h + \sigma u^h - \boldsymbol{\nabla}\cdot(\nu\boldsymbol{\nabla}u^h)\right)d\Omega\,dt$$
$$+\int_\Omega w^h(t_+^n)\left(u^h(t_+^n) - u^h(t_-^n)\right)d\Omega = 0, \quad (5.26)$$

with $u^h(t_-^0) = u_0$. The last integral is the jump condition. The third integral is the least-squares operator and parameter τ is the least-squares metric. Shakib and Hughes (1991) perform a Fourier stability and accuracy analysis of the space–time GLS method for constant-in-time and linear-in-time approximations.

5.6 SOLVED EXERCISES

5.6.1 Convection–diffusion of a Gaussian hill

The problem consists of solving the homogeneous linear convection–diffusion equation on the 1D domain $]0,1[$ with the initial condition

$$u(x,0) = \frac{5}{7}\exp\left\{-\left(\frac{x-x_0}{\ell}\right)^2\right\},$$

where $x_0 = 2/15$ and $\ell = 7\sqrt{2}/300$, and with boundary conditions: $u(0,t) = 0$, and $u(1,t) = 0$. The grid Péclet number is first taken as $Pe = 1$ and then increased to $Pe = 5$ and finally to $Pe = 100$. The convection velocity is $a = 1$. The problem is solved using a uniform mesh of linear elements of size $h = 1/150$ until time $t = 0.6$. The exact solution is given by

$$u(x,t) = \frac{5}{7\sigma(t)}\exp\left\{-\left(\frac{x-x_0-a\,t}{\ell\sigma(t)}\right)^2\right\}, \quad \text{where } \sigma(t) = \sqrt{1+4\nu\,t/(\ell^2)}.$$

Galerkin and Crank–Nicolson. We first want to highlight the fact that linear finite elements in the standard Galerkin formulation do not ideally combine with the second-order Crank–Nicolson time-stepping method in highly convective situations. The results reported in Figure 5.16 for a Courant number $C = 1$ show that the second-order time scheme performs well at low and moderate values of the Péclet number, but exhibits significant phase errors when the Péclet number is further increased. Moreover, the situation becomes worse when the time-step size corresponds to a Courant number larger than one, see the results for $C = 1.5$ and $Pe = 100$. As shown by the next tests of the Gaussian hill, the situation improves very much when passing to third- and fourth-order accurate time-stepping algorithms.

Galerkin and $\mathbf{R}_{2,2}$. Higher-order methods in time provide a gain in accuracy. Figure 5.17 shows, for a Courant number $C = 3$, that the fourth-order time scheme performs well for all values of the Péclet number. Obviously, results degrade when the time step is too large, see the results for $C = 4$ and $Pe = 100$.

Time-discontinuous Galerkin. The time-discontinuous Galerkin formulation introduced, for pure convection, in Section 3.10.1, see also the examples in Section 3.11.4, is used here. Linear finite element approximations are employed in both space and time, giving a third-order accurate and unconditionally stable method. Adapting the developments in Section 3.10.1 (pure convection) to the present convection–diffusion case, the following partitioned matrix system is obtained for the nodal unknowns $\mathbf{u}^{n+1}$ and $\mathbf{u}^{n^+}$:

$$\left(\mathbf{M} + \frac{2}{3}\Delta t\mathbf{C} + \frac{2}{3}\nu\Delta t\mathbf{K}\right)\mathbf{u}^{n+1} - \left(\mathbf{M} - \frac{1}{3}\Delta t\mathbf{C} - \frac{1}{3}\nu\Delta t\mathbf{K}\right)\mathbf{u}^{n^+} = 0$$

$$\left(\mathbf{M} + \frac{1}{3}\Delta t\mathbf{C} + \frac{1}{3}\nu\Delta t\mathbf{K}\right)\mathbf{u}^{n+1} + \left(\mathbf{M} + \frac{2}{3}\Delta t\mathbf{C} + \frac{2}{3}\nu\Delta t\mathbf{K}\right)\mathbf{u}^{n^+} = 2\mathbf{M}\,\mathbf{u}^{n^-}.$$

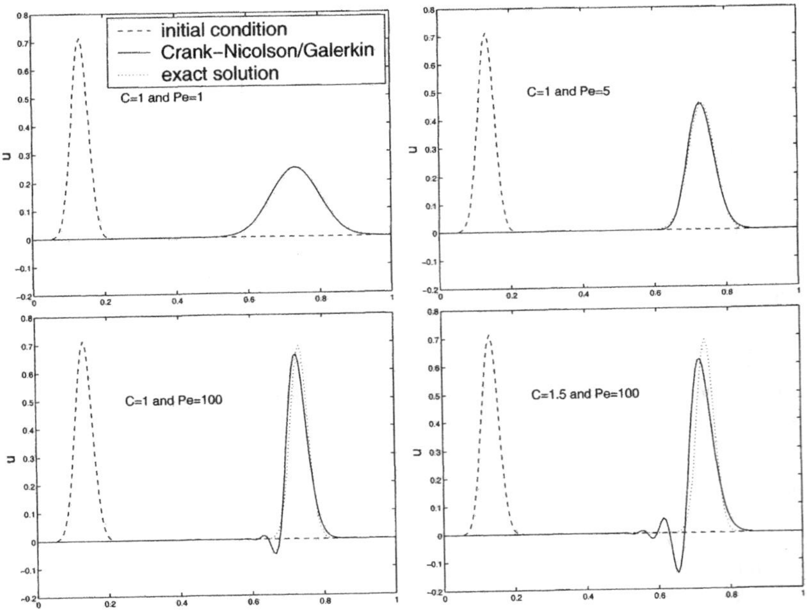

Fig. 5.16 Gaussian hill: standard Galerkin and Crank–Nicolson.

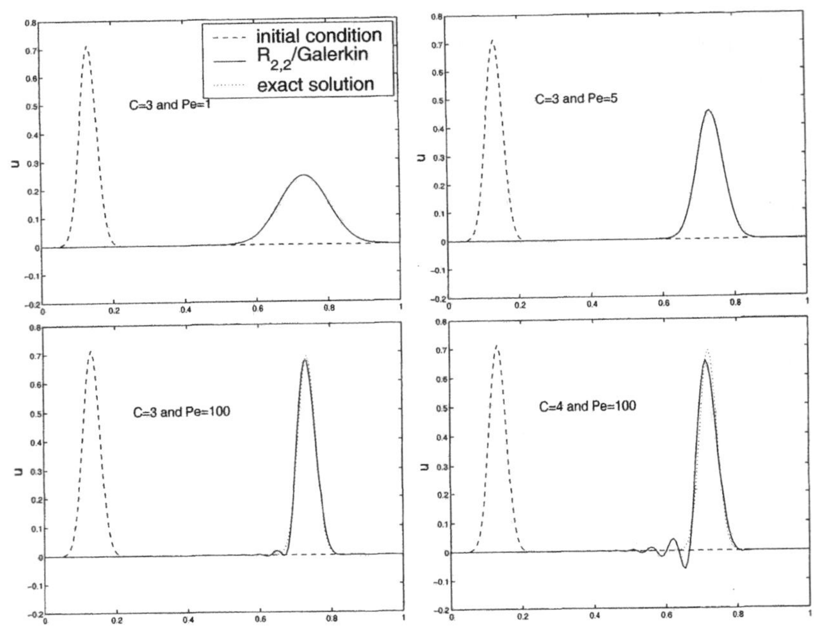

Fig. 5.17 Gaussian hill: standard Galerkin and $R_{2,2}$.

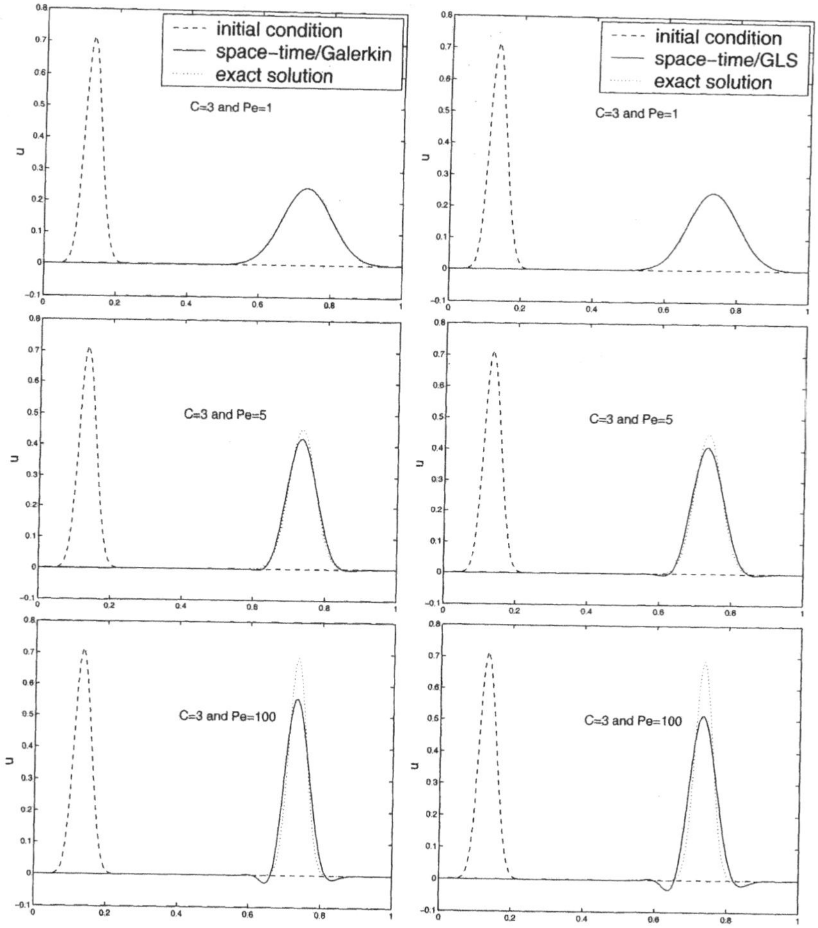

Fig. 5.18 Gaussian hill: space–time Galerkin (left) and Galerkin/Least-squares (right).

The matrices **M** (consistent mass), **C** (convection), and **K** (diffusion) are defined in (5.14). The conditions $u^{n+1}(0) = u^{n^+}(0) = 0$ and $u^{n+1}(1) = u^{n^+}(1) = 0$ are enforced to satisfy the boundary specifications.

Large time steps, corresponding to a Courant number $C = 3$, are used to test the capacity of this unconditionally stable scheme to deliver accurate results well beyond the stability limit of explicit methods for an equivalent computational cost (there are two unknowns per node). Figure 5.18 shows the results at $t = 0.6$ for the Galerkin formulation and three values of the Péclet number, $Pe = 1, 5$ and 100. Note that phase accuracy has improved with respect to the Crank–Nicolson results in Figure 5.16.

Time-discontinuous Galerkin/Least-squares. The formulation described in Section 5.5 is used. The partitioned matrix system for the nodal unknowns $\mathbf{u}^{n+1}$ and

$\mathbf{u}^{n^+}$ is now obtained in the form

$$\Big(\Big(1+\frac{2\tau}{\Delta t}\Big)\mathbf{M}+\frac{2}{3}\Delta t\mathbf{C}+\frac{2}{3}(\tau a^2+\nu)\Delta t\mathbf{K}\Big)\,\mathbf{u}^{n+1}$$
$$-\Big(\Big(1+\frac{2\tau}{\Delta t}\Big)\mathbf{M}-\Big(\frac{1}{3}\Delta t+2\tau\Big)\mathbf{C}-\frac{1}{3}(\tau a^2+\nu)\Delta t\mathbf{K}\Big)\,\mathbf{u}^{n^+}=0$$
$$\Big(\Big(1-\frac{2\tau}{\Delta t}\Big)\mathbf{M}+\Big(\frac{1}{3}\Delta t-2\tau\Big)\mathbf{C}+\frac{1}{3}(\tau a^2+\nu)\Delta t\mathbf{K}\Big)\,\mathbf{u}^{n+1}$$
$$+\Big(\Big(1+\frac{2\tau}{\Delta t}\Big)\mathbf{M}+\frac{2}{3}\Delta t\mathbf{C}+\frac{2}{3}(\tau a^2+\nu)\Delta t\mathbf{K}\Big)\,\mathbf{u}^{n^+}=2\mathbf{M}\,\mathbf{u}^{n^-}.$$

Remark 5.8 presents two expressions for the stabilization parameter τ. The first one is used here but other definitions can be employed. Numerical results at $t = 0.6$ are displayed in Figure 5.18. They confirm that the Galerkin/Least-squares space–time method can be operated with time increments much larger than the stability limit of explicit schemes. Nevertheless, the numerical results indicate that space–time methods exhibit a rather elevated numerical damping in smooth convection-dominated problems when a large value of the Courant number is employed. This is in agreement with the fact that τ increases with the time-step size.

5.6.2 Transient rotating pulse

In this 2D example both problems of the transient convection–diffusion equation are present: (1) accurate transport of the unknown is needed and (2) boundary layers appear in the solution due to the Dirichlet boundary conditions. Therefore, high-order time-stepping schemes and stabilized formulations are needed in order to obtain an accurate solution. The 2D convection–diffusion problem

$$\begin{cases} u_t + \boldsymbol{a}\cdot\boldsymbol{\nabla}u - \boldsymbol{\nabla}\cdot(\nu\boldsymbol{\nabla}u) = s & \text{in } \Omega =]0,1[\times]0,1[\\ u = 0 & \text{on } \partial\Omega \\ u = 0 & \text{at } t = 0 \end{cases}$$

is solved with small diffusion, $\nu = 10^{-5}$, on a uniform 40×40 bilinear mesh. The source term and the velocity field are defined as follows (see also Figure 5.19):

$$\boldsymbol{a} = (-y, x) \qquad s = \mathrm{e}^{-t^{10}} \begin{cases} \cos(\pi/2\sqrt{x^2+y^2}) & \text{if } \sqrt{x^2+y^2} \le 1, \\ 0 & \text{otherwise.} \end{cases}$$

Figures 5.20 to 5.24 show the numerical results obtained at $t = \pi$ and $t = 5\pi$ for the Galerkin, SUPG and LS methods. Crank–Nicolson and $\mathsf{R}_{2,2}$ time integration schemes have been used with Courant 1 and 3, respectively.

Boundary layers are present in the solution due to the convective character of the equation and the homogeneous Dirichlet boundary conditions. Thus, the typical instabilities of the Galerkin formulation soon appear. The numerical solution is clearly improved for the stabilized formulations: oscillations are alleviated and almost

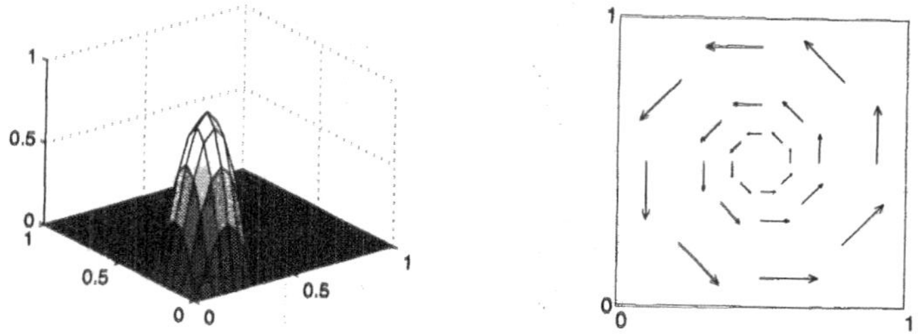

Source term s Convection velocity $\mathbf{a}$

Fig. 5.19 Source term and convection velocity

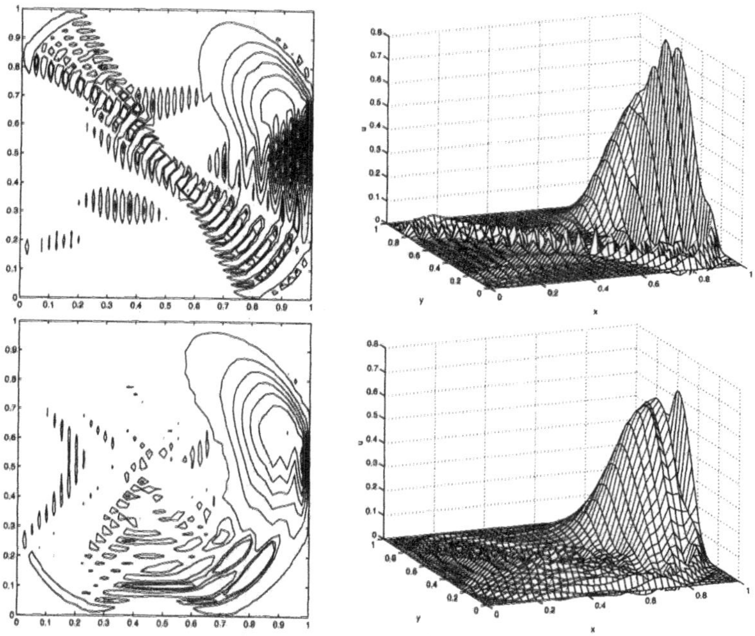

Fig. 5.20 Galerkin with Crank–Nicolson results at $t = \pi$ (top) and $t = 5\pi$ (bottom).

suppressed in the whole domain. However, important phase and amplitude errors can be observed in the numerical solution obtained for the Crank–Nicolson time-stepping scheme. Note that Crank–Nicolson results present negative values, with no physical sense, lagging behind the pulse.

It is also noticeable that both stabilization techniques produce similar results. SUPG presents more oscillations at the boundary layer. But results can easily be improved with the corresponding element-by-element definition of τ. Here the first formula presented in Remark 5.8 is employed. The least-squares results diffuse the spurious oscillations more rapidly but can easily become overly diffusive if Δt or the element size are increased.

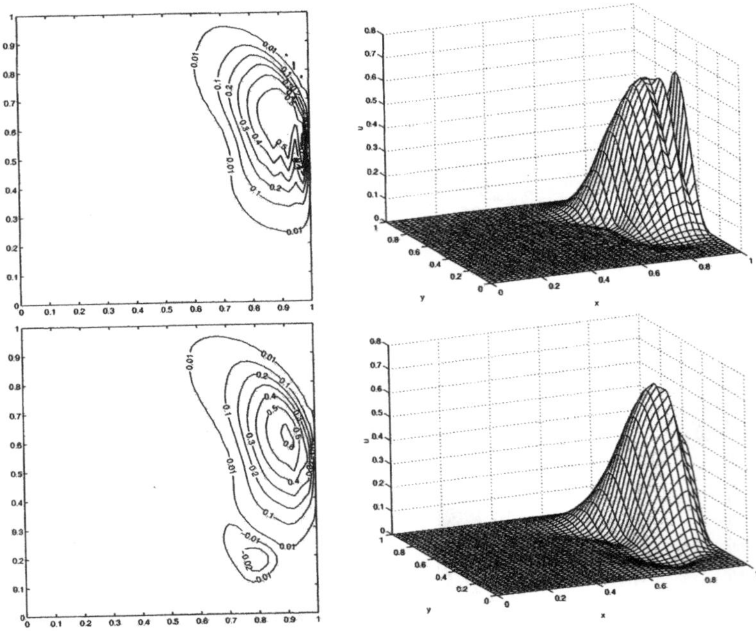

Fig. 5.21 SUPG with Crank–Nicolson results at $t = \pi$ (top) and $t = 5\pi$ (bottom).

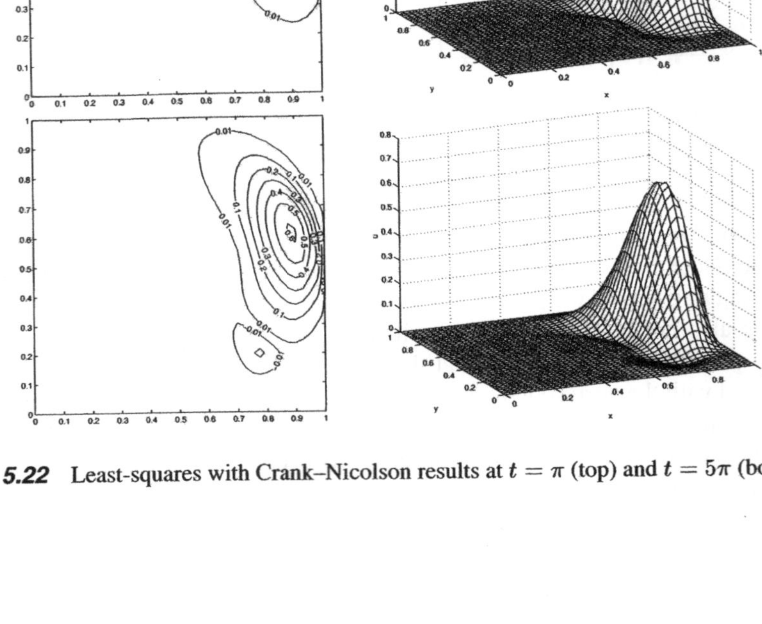

Fig. 5.22 Least-squares with Crank–Nicolson results at $t = \pi$ (top) and $t = 5\pi$ (bottom).

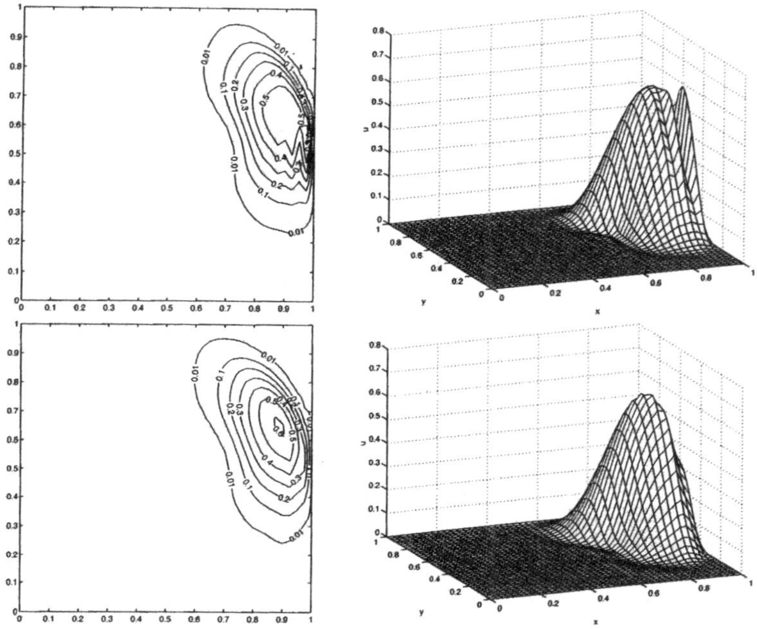

Fig. 5.23 SUPG with $\mathsf{R}_{2,2}$ results at $t = \pi$ (top) and $t = 5\pi$ (bottom).

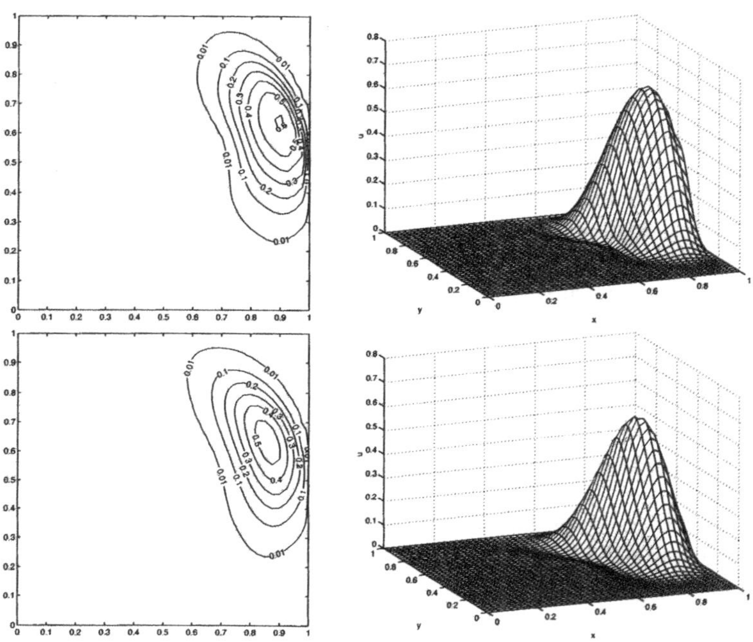

Fig. 5.24 Least-squares with $\mathsf{R}_{2,2}$ results at $t = \pi$ (top) and $t = 5\pi$ (bottom).

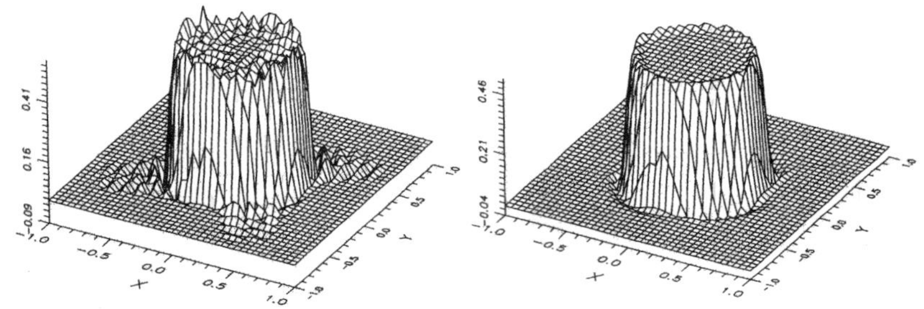

Fig. 5.25 Steady-state solution of the rotating pulse problem using $R_{2,2}$ with $C = 2$ with Galerkin (left) and least-squares (right) formulations.

5.6.3 Steady rotating pulse problem

This example illustrates the ability of the stabilized Padé schemes discussed in Section 5.4.6 to accurately solve problems with internal boundary layers. The linear convection–diffusion–reaction equation (5.1a) is solved with $R_{2,2}$ in the square domain $\Omega =]-1,1[\times]-1,1[$ with homogeneous Dirichlet boundary conditions and the following definitions: $\sigma = 2$,

$$\boldsymbol{a} = \phi(\varrho)\begin{pmatrix}-y\\ x\end{pmatrix}, \phi(r) = \begin{cases}1-\varrho^2 & \text{if } \varrho \le 1,\\ 0 & \text{otherwise,}\end{cases} \quad \text{and } s = \begin{cases}1 & \text{if } \varrho \le 1/2,\\ 0 & \text{otherwise,}\end{cases}$$

where $\varrho = \sqrt{x^2+y^2}$. The asymptotic steady-state solution presents a clear pattern with $u \approx 1/2$ if $\varrho \le 1/2$ and $u \approx 0$ otherwise; the boundary layer is $\mathcal{O}(\sqrt{\nu})$ along the circle $\varrho = 1/2$. A uniform mesh of 40×40 bilinear elements is used. Figure 5.25 shows the results obtained with the Courant number $C = 2$ for a Péclet number of 50000. While the Galerkin formulation fails to deliver stable results, one notes that the least-squares stabilization succeeds in eliminating the spurious oscillations characteristic of the Galerkin approach. Nevertheless, some oscillations remain near sharp solution gradients. Nonlinear viscosity of the type discussed in Chapter 4 (shock-capturing schemes) should be locally added to suppress such residual oscillations. Results for other stabilized methods or for higher-order time-stepping schemes ($R_{3,3}$) present negligible differences.

5.6.4 Nonlinear propagation of a step

This example studies the influence of the numerical time accuracy in the propagation of a vertical front over a nonuniform mesh. Crank–Nicolson and $R_{2,2}$ are compared and Burgers' equation,

$$u_t + u\,u_x = \nu\,u_{xx},$$

is solved in $]0,1[$ with a Dirichlet condition $u(0,t) = 1$ and a homogeneous Neumann condition of zero total flux at $x = 1$. Note that now we are confronted with a nonlinear

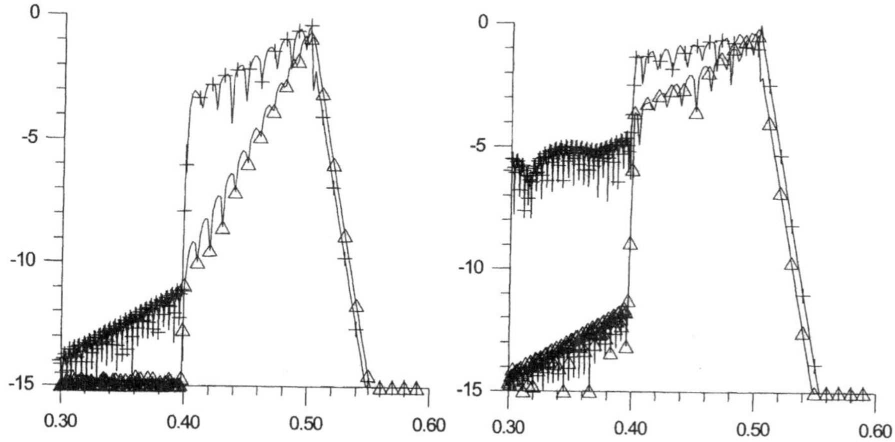

Fig. 5.26 Logarithm of the absolute error at time $t = 1$ for the LS (left) and GLS (right) stabilization. Crank–Nicolson is used with $C = 0.75$ (+) and $\mathsf{R}_{2,2}$ with $C = 3.0$ ($\triangle$).

partial differential equation. The analytical solution given by Sachdev (1987) is

$$u(x,t) = \left(1 + \exp\left(\frac{x - 0.5t}{2\nu}\right)\right)^{-1}.$$

The initial condition is imposed at $t = 0.02$ and a (maximum) Péclet number of 1000 is considered. Thus, the variation of u is extremely sharp and stabilized formulations are required. Here, we also compare the GLS and LS the formulations discussed in Sections 5.4.6.2 and 5.4.6.4, respectively. A piecewise uniform mesh of 1000 linear elements is used to minimize the error due to space discretization and highlight temporal errors. The mesh size is $h = 0.002$ on $[0, 0.2] \cup [0.4, 1]$ and $h/10$ on $]0.2, 0.4[$. As shown in Figure 5.26, the response obtained at time $t = 1$ with the stabilized/Crank–Nicolson is much worse than the one given by the stabilized/$\mathsf{R}_{2,2}$. Since the spatial mesh size is very small, this is clearly due to the inferior phase accuracy of the second-order Crank–Nicolson method. Notice that in both time-stepping schemes the added dissipation introduced by the LS approach increases with the size of the time step.

5.6.5 Burgers' equation in 1D

Burgers' equation is now solved in $]0, 1[$ with homogeneous Dirichlet boundary conditions. The problem is defined by

$$\begin{cases} u_t + uu_x = \nu u_{xx}, \\ u(x,0) = \sin(\pi x), \\ u(0,t) = u(1,t) = 0 \end{cases}$$

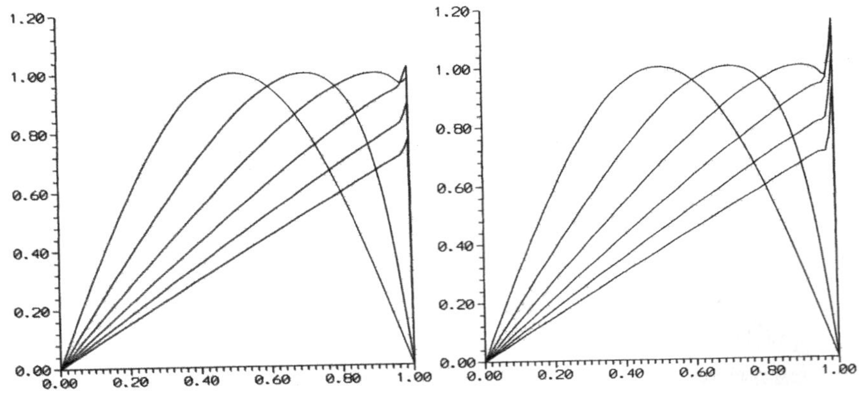

Fig. 5.27 Burgers' equation in 1D: least-squares/$R_{2,2}$ with $C = 2$ for $Pe = 5$ (left) and $Pe = 5000$ (right). Results are shown for $t = 0, 0.2, 0.4, 0.6, 0.8, 1$.

and presents a boundary layer at $x = 1$. The numerical results in Figure 5.27 were obtained using a uniform mesh of 100 linear elements.

A central spatial discretization, such as the Galerkin formulation, is unable to reproduce the solution to this problem for high values of the Péclet number. We note from Figure 5.27 that the LS/$R_{2,2}$ scheme produces an excellent response, except in the boundary layer where it presents a localized oscillation. The amplitude of this oscillation is controlled by the Courant number. A Newton–Raphson procedure has been used to iteratively solve the algebraic system governing the nodal values of the solution and good accuracy was obtained in few iterations.

5.6.6 Two-dimensional Burgers' equation

A 2D Burgers' problem is now considered over the square domain $\Omega =]0,1[\times]0,1[$. The problem is defined by the coupled equations

$$\begin{cases} u_t + u\,u_x + v\,u_y = \nu\nabla^2 u, \\ v_t + u\,v_x + v\,v_y = \nu\nabla^2 v, \end{cases}$$

and by the following initial and boundary conditions:

$$\begin{aligned} u(x,y,0) &= \sin(\pi x)\,\cos(\pi y), & v(x,y,0) &= \cos(\pi x)\,\sin(\pi y), \\ u(0,y,t) &= u(1,y,t) = 0, & v(x,0,t) &= v(x,1,t) = 0, \\ \frac{\partial u}{\partial n}(x,0,t) &= \frac{\partial u}{\partial n}(x,1,t) = 0, & \frac{\partial v}{\partial n}(0,y,t) &= \frac{\partial v}{\partial n}(1,y,t) = 0. \end{aligned}$$

A mesh of 30×30 bilinear elements is used for the spatial discretization. The problem exhibits various symmetries:

$$u(x,y,t) = v(y,x,t) \quad \text{and} \quad u(x,y,t) = -u(1-x,1-y,t).$$

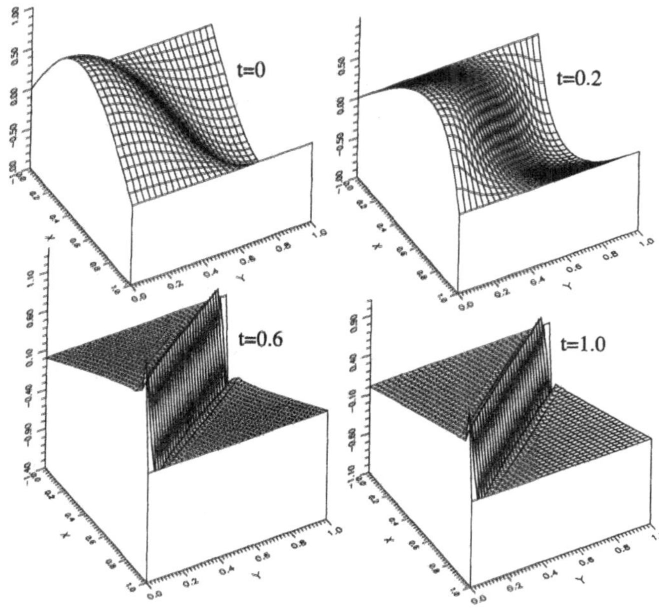

Fig. 5.28 Burgers' equation in 2D: least-squares and $\mathsf{R}_{2,2}$ for $Pe = 5000$ and $C = 2$.

Along the transverse section of the domain they imply that

$$u(x, x, t) = v(x, x, t) \quad \text{and} \quad u(x, x, t) = -u(1 - x, 1 - x, t).$$

When convection dominates the nonlinear transport, the solution includes the formation of a discontinuity along the diagonal of the domain passing through the points $(0, 1)$ and $(1, 0)$.

As in the 1D Burgers' problem, the Galerkin spatial discretization fails to give acceptable results for high values of the Péclet number. Figure 5.28 shows that the LS finite element scheme combined with Padé approximation $\mathsf{R}_{2,2}$ succeeds in capturing the solution with good accuracy. Since the scheme is not monotone, it exhibits residual oscillations near sharp gradients. It does, however, not degrade the solution outside the zones of strong solution gradient.

Appendix Least-squares in transient/relaxation problems

The LS formulation is intrinsically ill-posed for steady convection–diffusion problems. This is not the case of other well-known stabilization techniques such as the SUPG or GLS formulations. For this reason, LS methods were not introduced in Chapter 2. However, LS techniques are commonly used in transient pure convection, see Chapters 3 and 4 and have been extended to transient convection–diffusion in this chapter. Are the intrinsic problems of least-squares in steady convection–diffusion hidden in the transient case? Do they corrupt the transient solution or its steady-state?

Here a simple example/exercise is used to recall the difficulties of LS in a steady-state (elliptic) formulation. We also show that these difficulties do not exist in transient situations. Moreover, the steady solution obtained as the result of a relaxation technique with LS is well-behaved and lacks the inconsistencies of the pure steady solution. Thus, LS can be an alternative to SUPG or GLS in transient (or relaxation) problems because it is also well-posed and can be implemented with standard C^0 finite elements without introducing additional variables (Huerta and Donea, 2002).

The steady-state convection–diffusion equation. The model problem, which is studied in order to show the inherent difficulties of the LS formulation in steady-state problems, is the homogenous 1D convection–diffusion equation,

$$a\,u_x - \nu\,u_{xx} = 0. \tag{A.1}$$

A particular example suffices to illustrate the inconsistency of the LS formulation in steady-state problems, in particular,

$$\begin{cases} a\,u_x - \nu\,u_{xx} = 0 & \text{in } \Omega :=]0,1[, \\ u(0) = 0 \text{ and } u(1) = 1. \end{cases} \tag{A.2}$$

The exact solution of (A.2) is:

$$u_{\texttt{exact}}(x) = \frac{1 - \exp(ax/\nu)}{1 - \exp(a/\nu)}, \tag{A.3}$$

where, as expected, a/ν controls the relative importance of the convection and diffusion term, see Figure A.1.

The variational form for the LS formulation of (A.2) is: find u in the appropriate trial space such that

$$\big(a\,v_x - \nu\,v_{xx}, a\,u_x - \nu\,u_{xx}\big) = 0 \qquad \forall v \in \mathcal{H}_0^2(\Omega). \tag{A.4}$$

Remark A.1 (The trial and test spaces). The variational form involves integrals of second derivatives of both u and v. Thus, it is necessary that the spaces of trial and test functions are subspaces of $\mathcal{H}^2$. On one hand, the space of the test (weighting) functions is, as usual, chosen such that homogenous conditions are verified on the Dirichlet portion of the boundary, Γ_D. Thus, in general, the test functions belong to

$$\mathcal{H}^2_{\Gamma_D}(\Omega) := \{v \in \mathcal{H}^2(\Omega) \mid v = 0 \text{ on } \Gamma_D\}.$$

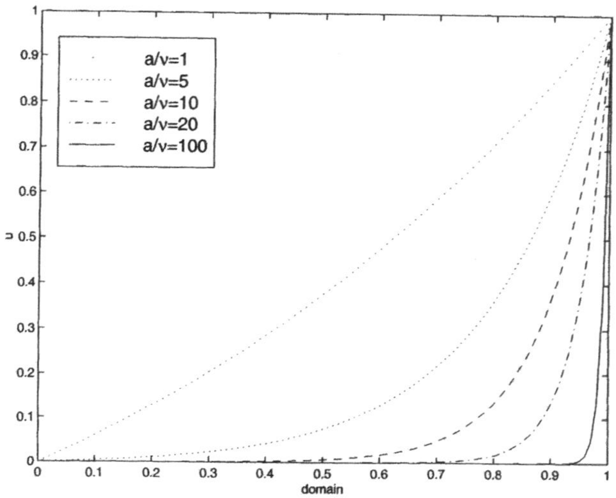

Fig. A.1 Exact solution of problem (A.2) for increasing values of a/ν.

In the particular case studied here, see (A.2), only essential boundary conditions are imposed, consequently,

$$\mathcal{H}^2_{\Gamma_D} = \mathcal{H}^2_0.$$

On the other hand, the trial solutions, u, must satisfy the Dirichlet (forced) boundary conditions; thus, the space of trial solutions is, in general,

$$\{u \in \mathcal{H}^2(\Omega) \mid u = u_D \text{ on } \Gamma_D\} \equiv \mathcal{H}^2_0(\Omega) + \{\overline{u}_D\},$$

where u_D are the Dirichlet boundary data on Γ_D, and $\overline{u}_D$ is any function of $\mathcal{H}^2(\Omega)$ satisfying the Dirichlet boundary conditions.

Remark A.2 (The actual steady problem). The extra regularity in the LS weak form also implies extra boundary conditions for the strong form of the problem. These extra boundary conditions are naturally enforced in (A.4). In fact, if equation (A.4) is integrated by parts recalling that $v \in \mathcal{H}^2_0$, the Euler–Lagrange equation and the natural boundary conditions associated with (A.4) become apparent, namely

$$\begin{cases} a^2\, u_{xx} - \nu^2\, u_{xxxx} = 0 & \text{in } \Omega :=]0,1[, \\ u(0) = 0 \text{ and } u(1) = 1, & \\ \nu(a\, u_x - \nu u_{xx}) = 0 & \text{at } x = 0 \text{ and } x = 1. \end{cases} \tag{A.5}$$

Note that this problem is consistent with the original one, see (A.2).

Remark A.3. It is well known that the LS formulation is formally equivalent to a higher-order problem. This has originated a debate on boundary conditions in LS problems, see for instance the books by Zienkiewicz and Morgan (1983, Sec. 6.8.) and Jiang (1998, p 25). Here it is clear that, in finite elements, the extra boundary conditions are natural conditions which do not appear explicitly in the weak form. But they are imposed implicitly! If other conditions, different from these natural ones, are imposed on the boundary spurious solutions may be obtained, as noted by Zienkiewicz and Morgan (1983). Thus, as observed by Jiang (1998), if other numerical techniques are used, such as finite differences, the extra (natural) boundary conditions must be enforced explicitly.

The objective now is to study the stabilization effects of the LS formulation for convection-dominated situations. Thus, the limit when ν approaches 0 is analyzed, in this case, the variational form (A.4) becomes: find $u \in \mathcal{H}_0^1(\Omega) + \{\overline{u}_D\}$, such that

$$\big(v_x, u_x\big) = 0, \qquad \forall v \in \mathcal{H}_0^1(\Omega). \tag{A.6}$$

However, the equivalent strong form (i.e., the Euler–Lagrange equation associated with the variational form and the boundary conditions induced by the functional spaces) associated with (A.6) is

$$\begin{cases} u_{xx} = 0 & \text{in } \Omega =]0,1[\\ u(0) = 0 \text{ and } u(1) = 1. \end{cases} \tag{A.7}$$

The unique solution of this problem is $u(x) = x$, which is obviously non consistent with the limit case of (A.3) for ν approaching zero. In fact, this solution coincides with $\lim_{\nu\to\infty} u_{\text{exact}}(x) = x$. Thus, the steady LS solution will converge in the limit case $\nu = 0$ to the exact solution for the limit case $\nu \to \infty$. This is intrinsic to the LS formulation in steady problems. This is why LS are not used as a stabilization technique for the steady convection–diffusion equation.

It is important to note that the problem defined by (A.7) corresponds to the limit problem of (A.5). Thus, the difficulties of the LS formulation are also present for small values of ν as it will be seen in the following numerical examples.

These conclusions can be confirmed numerically. Figure A.2 shows numerical solutions of (A.4) for a uniform mesh of 10 $\mathcal{C}^1$ finite elements (Hermite elements of degree 3) as a/ν grows. The Galerkin approximations with the same mesh are also plotted for comparison. As expected, when convection becomes dominant, the Galerkin formulation becomes unstable and the LS formulation approaches the "non-physical" solution $u = x$.

Figure A.3 presents the influence of the element size in this LS approximation. As expected, as the number of elements increases the Péclet number decreases and the approximation improves. Note however the poor results obtained with 100 elements (200 degrees of freedom).

If standard stabilization techniques for steady problems, such as SUPG and GLS, are used, this intrinsic problem in the limit $\nu \to 0$ does not appear, see Chapter 2. The model problem presented in (A.2), can be rewritten in weak form using the SUPG

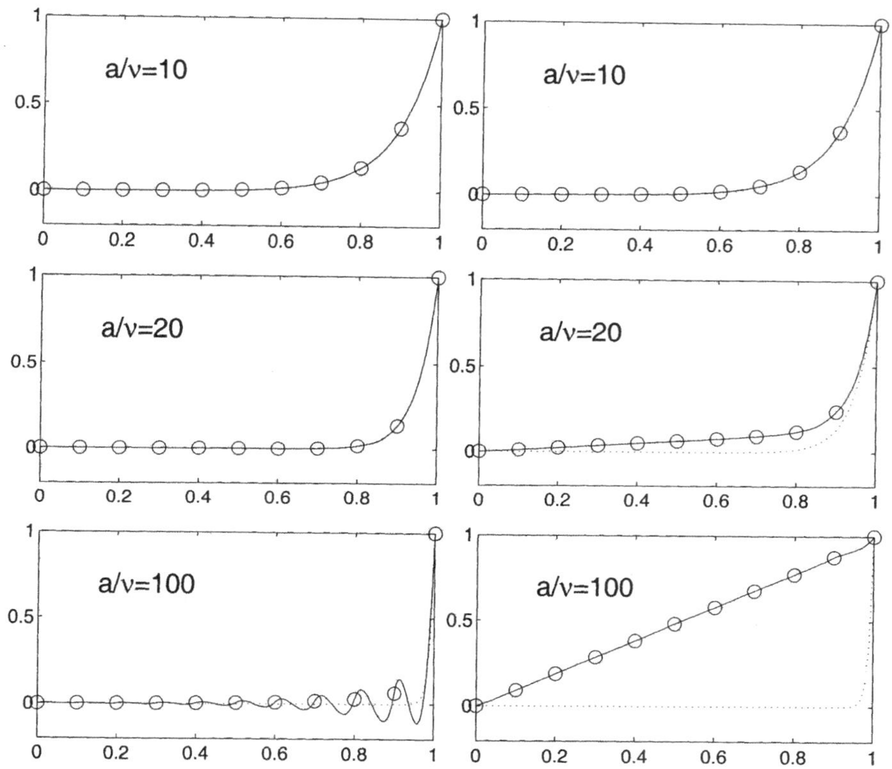

Fig. A.2 Galerkin (left) and Least-Squares (right) approximations (solid lines) to the exact (dotted) solution of the model problem for increasing values of a/ν.

formulation as, find u such that

$$\underbrace{(v, a\,u_x) + (v_x, \nu\,u_x)}_{\text{Galerkin}} + \underbrace{\sum_e \tau\big(a\,v_x, a\,u_x - \nu\,u_{xx}\big)_{\Omega^e}}_{\text{SUPG Stabilization}} = 0 \quad \forall v \in \mathcal{H}_0^1(\Omega), \tag{A.8}$$

where the Galerkin term and the stabilization term are easily recognized.

Remark A.4. Note that u must have the first derivative integrable in the whole domain and the second derivative integrable inside the elements. Thus, u belongs to

$$\{u \in \mathcal{H}^1(\Omega) \mid u = u_D \text{ on } \Gamma_D,\ u|_{\Omega^e} \in \mathcal{H}^2(\Omega^e) \text{ for every element } \Omega^e\}$$
$$\equiv \mathcal{H}_0^{1+}(\Omega) + \{\overline{u}_D\},$$

where

$$\mathcal{H}_0^{1+}(\Omega) = \{v \in \mathcal{H}_0^1(\Omega) \mid v|_{\Omega^e} \in \mathcal{H}^2(\Omega^e) \text{ for every element } \Omega^e\} \tag{A.9}$$

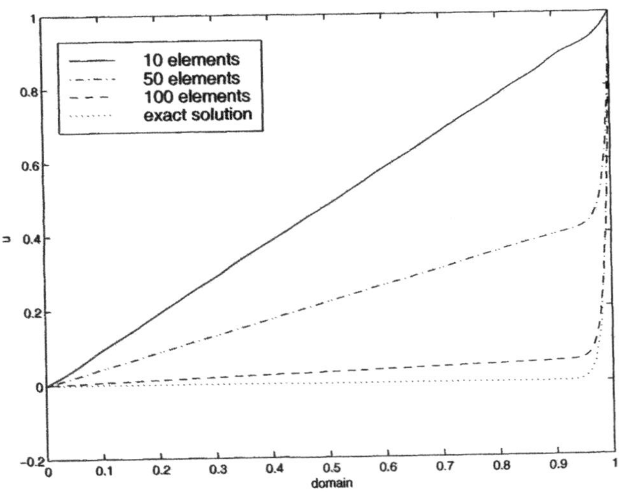

Fig. A.3 Influence of the element size on the LS approximations for $a/\nu = 100$.

and $\overline{u}_D$ is any function of $\mathcal{H}^{1+}(\Omega)$ satisfying the Dirichlet boundary conditions.

As previously discussed for the LS formulation, the limit $\nu \to 0$ is analyzed. In this case, the previous weak form becomes

$$\big(v, a\,u_x\big) + \tau\big(a\,v_x, a\,u_x\big) = 0 \quad \forall v \in \mathcal{H}_0^1(\Omega), \tag{A.10}$$

which can be interpreted as the Galerkin discretization of

$$\begin{cases} a\,u_x - a^2\tau\,u_{xx} = 0 & \text{in } \Omega \\ u(0) = 0 \text{ and } u(1) = 1. \end{cases} \tag{A.11}$$

As expected, the SUPG formulation stabilizes the solution. It introduces an artificial diffusion (viscosity) $\bar{\nu} = a^2\tau$. Figure A.4 shows the SUPG approximation for 10 $\mathcal{C}^1$ finite elements, $a/\nu = 100$, and the optimal intrinsic time for linear elements, $\tau = (\coth(Pe) - 1/Pe)h/(2a)$, which, as expected, is not optimal in this case because Hermite elements of degree 3 are used.

Remark A.5 (The GLS formulation). Similar conclusions are drawn if a GLS formulation is employed. Equation (A.8) becomes

$$\underbrace{\big(v, a\,u_x\big) + \big(v_x, \nu\,u_x\big)}_{\text{Galerkin}} + \underbrace{\sum_e \tau\big(a\,v_x - \nu\,v_{xx}, a\,u_x - \nu\,u_{xx}\big)_{\Omega^e}}_{\text{GLS Stabilization}} = 0,$$

for all $v \in \mathcal{H}_0^{1+}(\Omega)$, which induces exactly the same weak, (A.10), and strong, (A.11), forms as SUPG in the limiting case of zero viscosity.

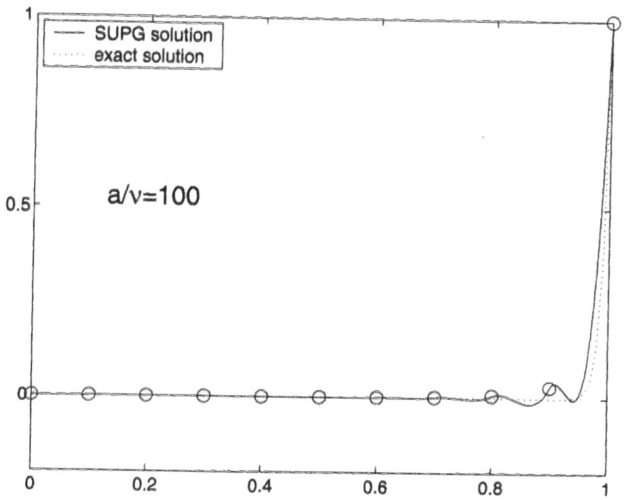

Fig. A.4 SUPG approximation for $a/\nu = 100$.

The transient convection–diffusion equation. It is well known that for the transient convection–diffusion equation the LS formulation works properly. Recall that this is also the case for both SUPG and GLS stabilization techniques. In order to understand this behavior, the variational form obtained for each method will be presented. A homogenous 1D transient convection–diffusion equation,

$$u_t + a\,u_x - \nu\,u_{xx} = 0, \tag{A.12}$$

is used to study the different stabilization techniques. The strong form of the model problem in the transient case is simply

$$\begin{cases} u_t + a\,u_x - \nu\,u_{xx} = 0 & \text{in } \Omega \times \mathbb{R}^+ :=]0,1[\times]0,\infty[, \\ u(0,t) = u_0 \text{ and } u(1,t) = u_1 & \text{for } t \in \mathbb{R}^+ \text{ (boundary conditions)}, \\ u(x,0) = f(x) & \text{for } x \in \Omega \text{ (initial condition)}, \end{cases} \tag{A.13}$$

where the initial condition $f(x)$ must be a smooth function.

In order to implement a stabilized formulations (LS, SUPG or GLS) for transient problems we perform first the time discretization of (A.13). For simplicity, a single-step time-integration scheme is used, but higher-order in time schemes can also be implemented. After time discretization the spatial equation that must be solved at each time step is

$$\frac{\Delta u}{\Delta t} + \theta(a\,\Delta u_x - \nu\,\Delta u_{xx}) = -(a\,u_x^n - \nu\,u_{xx}^n) \tag{A.14}$$

with $\Delta u = u^{n+1} - u^n$. Thus, the spatial operator to be used for the LS approach is obtained from the l.h.s. of the previous equation, namely

$$\mathcal{L} = 1/\Delta t + \theta(a\,\partial_x - \nu\,\partial_{xx}). \tag{A.15}$$

Once the spatial operator is known the variational form induced by the LS formulation can be defined, that is, $\forall v \in \mathcal{H}_0^2(\Omega)$,

$$\begin{aligned} 0 &= \big(\mathcal{L}(v), u_t + a\,u_x - \nu\,u_{xx}\big) \\ &= \Big(\frac{v}{\Delta t} + \theta(a\,v_x - \nu\,v_{xx}), u_t + a\,u_x - \nu\,u_{xx}\Big). \end{aligned} \tag{A.16}$$

Remark A.6. Note that, as usual, the test space is independent of time and the trial space is in this case defined as

$$\begin{aligned} &\{u \mid u(\cdot,t) \in \mathcal{H}^2(\Omega), t \in \mathbb{R}^+ \text{ and } u(x,t) = u_D \text{ for } x \text{ on } \Gamma_D\} \\ &\qquad\qquad \equiv \{u \mid u(\cdot,t) \in \mathcal{H}_0^2(\Omega) + \{\overline{u}_D\}, t \in \mathbb{R}^+\} \end{aligned}$$

As expected, the LS formulation still requires higher regularity for the solution because of the presence of the second-order spatial derivatives.

Remark A.7 (The actual transient problem). As previously observed in Remark A.2, the extra regularity in the LS weak form also implies extra boundary conditions for the strong form of the problem. This extra boundary conditions are naturally enforced in (A.16). The Euler–Lagrange equation and the boundary conditions associated with (A.16) are

$$\begin{cases} u_t + a\,u_x - \nu\,u_{xx} - a\theta\Delta t(u_t + a\,u_x)_x & \\ \qquad\qquad - \nu\theta\Delta t(u_t - \nu\,u_{xx})_{xx} = 0 & \text{in } \Omega \times \mathbb{R}^+, \\ u(0,t) = u_0 \text{ and } u(1,t) = u_1 & \text{for } t \in \mathbb{R}^+, \\ \nu\theta\Delta t(u_t + a\,u_x - \nu\,u_{xx})\big|_{x=0} = 0 & \text{for } t \in \mathbb{R}^+, \\ \nu\theta\Delta t(u_t + a\,u_x - \nu\,u_{xx})\big|_{x=1} = 0 & \text{for } t \in \mathbb{R}^+, \\ u(x,0) = f(x) & \text{on } x \in \Omega. \end{cases} \tag{A.17}$$

Note that this problem is also consistent with the original one, namely (A.13), hint: use the transient differential equation (A.12) to verify the Euler–Lagrange equation.

In the limit case when $\nu \to 0$ the weak form (A.16) becomes

$$\big(v, u_t + a\,u_x\big)_\Omega + a\,\varepsilon\big(v_x, u_t + a\,u_x\big)_\Omega = 0 \qquad \forall v \in \mathcal{H}_0^1(\Omega), \tag{A.18}$$

where, for convenience, a parameter, $\varepsilon := \theta\Delta t$, is defined. Note that, as previously observed, the regularity conditions are relaxed in this limit case.

The Euler–Lagrange equation associated with the previous weak form is

$$\big[1 - a\,\varepsilon\,\partial_x\big](u_t + a\,u_x) = 0, \tag{A.19}$$

and the strong form induced by (A.18) is

$$\begin{cases} \big[1 - a\,\varepsilon\,\partial_x\big](u_t + a\,u_x) = 0 & \text{in } \Omega \times \mathbb{R}^+ :=]0,1[\times]0,\infty[\\ u(0,t) = u_0 \text{ and } u(1,t) = u_1 & \text{for } t \in \mathbb{R}^+, \\ u(x,0) = f(x) & \text{on } x \in \Omega. \end{cases} \tag{A.20}$$

That is, (A.18) can be interpreted as the Galerkin discretization of this strong form. It is important to note, for consistency, that the strong form defined by (A.20) corresponds to the limit, when ν approaches 0, of problem (A.17).

Remark A.8. Note that in the limit case when ν approaches 0, (A.12) becomes $u_t + a\,u_x = 0$, whose solutions also satisfy (A.19).

Remark A.9 (Stabilizing effect of the LS formulation). At steady-state, that is when $u_t \to 0$, the problem defined by (A.20) reduces to

$$\begin{cases} a\,u_x - a^2\varepsilon\,u_{xx} = 0, & \text{in } \Omega \\ u(0) = u_0 \text{ and } u(1) = u_1, \end{cases}$$

which clearly shows both the stabilizing effect of the LS formulation and the fact that the steady solution of the transient LS formulation is well-posed. Note the equivalence with (A.11) and that the artificial diffusion is controlled, as expected, by the time increment, $\bar{\nu} = a^2\,\varepsilon = a^2\theta\Delta t$.

However, this problem is the transient counterpart to (A.7), whose solution was clearly "non-physical". Thus, the obvious goal now is to determine if (A.20) presents physical solutions, or not. First, however, we wonder if (A.20) is well-posed. Note that this is a hyperbolic equation with prescribed Dirichlet conditions on the whole boundary. Does (A.20) present feasible solutions? These questions will be answered. Before, however, the SUPG and GLS stabilization techniques are studied in order to obtain the transient counterpart to (A.11).

In SUPG and GLS formulations the extra term added to the Galerkin weak form is a function of the residual to ensure consistency of the formulation. The residual of (A.14) is simply

$$\mathcal{R}(\Delta u) := \frac{\Delta u}{\Delta t} + \theta(a\,\Delta u_x - \nu\,\Delta u_{xx}) + a\,u_x^n - \nu\,u_{xx}^n$$

and the stabilized weak form is

$$\big(v, \mathcal{R}(\Delta u)\big) + \sum_e \tau\big(\mathcal{P}(v), \mathcal{R}(\Delta u)\big)_{\Omega^e} = 0 \quad \forall v \in \mathcal{H}_0^{1+}(\Omega). \tag{A.21}$$

In 1D and for SUPG, see Chapter 2, this operator is defined as $\mathcal{P}(v) := \theta a\,v_x$. Thus, the variational form associated with SUPG is, $\forall v \in \mathcal{H}_0^1(\Omega)$,

$$\big(v, u_t + a\,u_x - \nu\,u_{xx}\big) + \sum_e \tau\big(\theta a\,v_x, u_t + a\,u_x - \nu\,u_{xx}\big)_{\Omega^e} = 0. \tag{A.22}$$

Moreover, in the limit case when ν approaches 0 this form becomes

$$\big(v, u_t + a\,u_x\big) + \tau\big(\theta a\,v_x, u_t + a\,u_x\big) = 0 \quad \forall v \in \mathcal{H}_0^1(\Omega), \tag{A.23}$$

where the sum over element interiors is no longer necessary because only first derivatives are present (second derivatives disappear in the limit $\nu \to 0$).

This equation can be interpreted as the standard Galerkin discretization of

$$\big[1 - a\theta\tau\,\partial_x\big](u_t + a\,u_x) = 0,$$

which corresponds to (A.19) for $\varepsilon := \theta\tau$. Thus, the strong form corresponding to the stabilized formulation is identical to (A.20) (where ε now depends on the intrinsic time τ) and the analytical solution developed next will also be valid for the SUPG stabilization.

Remark A.10 (GLS stabilisation). In the GLS formulation, the perturbation operator is defined as $\mathcal{P} := \mathcal{L}$, see equation (A.15). Then, the variational form obtained for the limit case $\nu \to 0$ can be interpreted as the Galerkin discretization of

$$\Big[\big(1 + \frac{\tau}{\Delta t}\big) - a\theta\tau\,\partial_x\Big](u_t + a\,u_x) = 0,$$

which also corresponds to (A.19) where, in this case, $\varepsilon := \theta\tau\Delta t/(\tau + \Delta t)$.

The analytical solution of the strong form associated with the stabilized formulations is obtained from the general solution of (A.19), which is

$$u(x,t) = F(t)\exp(x/a\,\varepsilon) + G(at - x), \tag{A.24}$$

where functions F and G are be determined imposing the boundary and initial conditions. In this particular case, F and G are piecewise defined functions, for $(n-1)/a \le t \le n/a$

$$F(t) = u_1 \frac{1 - \exp(\frac{-n}{a\,\varepsilon})}{\exp(\frac{1}{a\,\varepsilon}) - 1} - u_0 \frac{1 - \exp\big(\frac{-n+1}{a\,\varepsilon}\big)}{\exp(\frac{1}{a\,\varepsilon}) - 1} - f(n - at)\exp(\frac{-n}{a\,\varepsilon}),$$

and for $(n-1) \le z \le n$

$$G(z) = u_0\left(1 + \frac{1 - \exp\big(\frac{-n+1}{a\,\varepsilon}\big)}{\exp(\frac{1}{a\,\varepsilon}) - 1}\right) - u_1 \frac{1 - \exp(\frac{-n}{a\,\varepsilon})}{\exp(\frac{1}{a\,\varepsilon}) - 1} + f(n - z)\exp\Big(\frac{-n}{a\,\varepsilon}\Big).$$

Note that the analytical steady-state solution (i.e., when $t \to \infty$) is

$$u(x) = (u_1 - u_0)\frac{\exp(x/a\,\varepsilon) - 1}{\exp(1/a\,\varepsilon) - 1} + u_0.$$

This equation can be rewritten in terms of an artificial viscosity $\bar{\nu} = a^2\,\varepsilon$, namely

$$u(x) = (u_1 - u_0)\frac{\exp(ax/\bar{\nu}) - 1}{\exp(a/\bar{\nu}) - 1} + u_0, \tag{A.25}$$

which clearly shows the boundary layer structure of the solution, see (A.3), and, as expected, only depends on the boundary conditions (not on the initial condition).

Therefore, in the transient case the strong form of the limit problem (i.e., when $\nu \to 0$) is well-posed, has an analytical solution and the steady-state solution presents the desired structure.

The model problem (A.13) is now numerically solved for the particular case

$$\begin{cases} u_t + a\,u_x - \nu\,u_{xx} = 0 & \text{in } \Omega \times \mathbb{R}^+ :=]0,1[\times]0,\infty[, \\ u(0,t) = 0 \text{ and } u(1,t) = 1 & \text{for } t \in \mathbb{R}^+, \\ u(x,0) = x & \text{for } x \in \Omega. \end{cases}$$

In this particular case, the analytical steady solution of the stabilized problem, equation (A.25), is

$$\lim_{t\to\infty} u(x,t) = \frac{\exp(ax/\bar{\nu}) - 1}{\exp(a/\bar{\nu}) - 1}.$$

Note that, the boundary layer is governed by numerical parameters, $\bar{\nu} = a^2\theta\Delta t$ for LS, $\bar{\nu} = a^2\theta\tau$ for SUPG and $\bar{\nu} = a^2\theta\tau\Delta t/(\tau + \Delta t)$ for GLS.

Before presenting the numerical results, the formulation proposed in Section 5.4.6.4 is recalled, see also Huerta and Donea (2002). It allows us to use standard $\mathcal{C}^0$ finite elements. The weak problem is as follows: find $u(\cdot,t) \in \mathcal{H}_0^{1+}(\Omega) + \{\overline{u}_D\}$, for all $v \in \mathcal{H}_0^{1+}(\Omega)$, such that

$$\big(v, u_t + a\,u_x\big) + \big(v_x, \nu\,u_{xx}\big) + \sum_e \theta\Delta t\big(a\,v_x - \nu\,v_{xx}, u_t + a\,u_x - \nu\,u_{xx}\big) = 0.$$

Note that this formulation, as well as the standard LS in $\mathcal{H}^2$, is symmetric, this is not the case of SUPG and GLS.

The model problem is solved for $a/\nu = 100$. Figure A.5 shows numerical results at different instants for LS, SUPG and GLS with a uniform mesh of 10 $\mathcal{C}^0$ elements (left) and a Crank–Nicolson scheme with $\Delta t = 1$ (i.e., $C = 1$). The exact solution is also displayed for comparison. For SUPG and GLS, the intrinsic time is chosen equal to the optimal stabilization parameter for linear elements in the steady convection–diffusion equation, namely

$$\tau = \frac{h}{2a}\left[\coth(Pe) - \frac{1}{Pe}\right]\frac{1}{\theta},$$

where, in this case, $\theta = 1/2$, see also Remark 5.9. Thus the exact nodal solution must be recovered for SUPG at steady-state. The same figure also shows results for 10 $\mathcal{C}^1$ elements (right). As expected both $\mathcal{C}^0$ and $\mathcal{C}^1$ elements produce reasonable results in all cases, in particular, with the LS stabilization.

In summary, standard LS stabilization (in $\mathcal{H}^2$) can not be used directly in steady convection–diffusion problems because it produces non-physical solutions in the pure convection limit. But the transient convection–diffusion equation can be stabilized with LS.

It should also be observed that the LS formulation proposed can be implemented with standard $\mathcal{C}^0$ finite elements and does not necessitate to introduce additional nodal

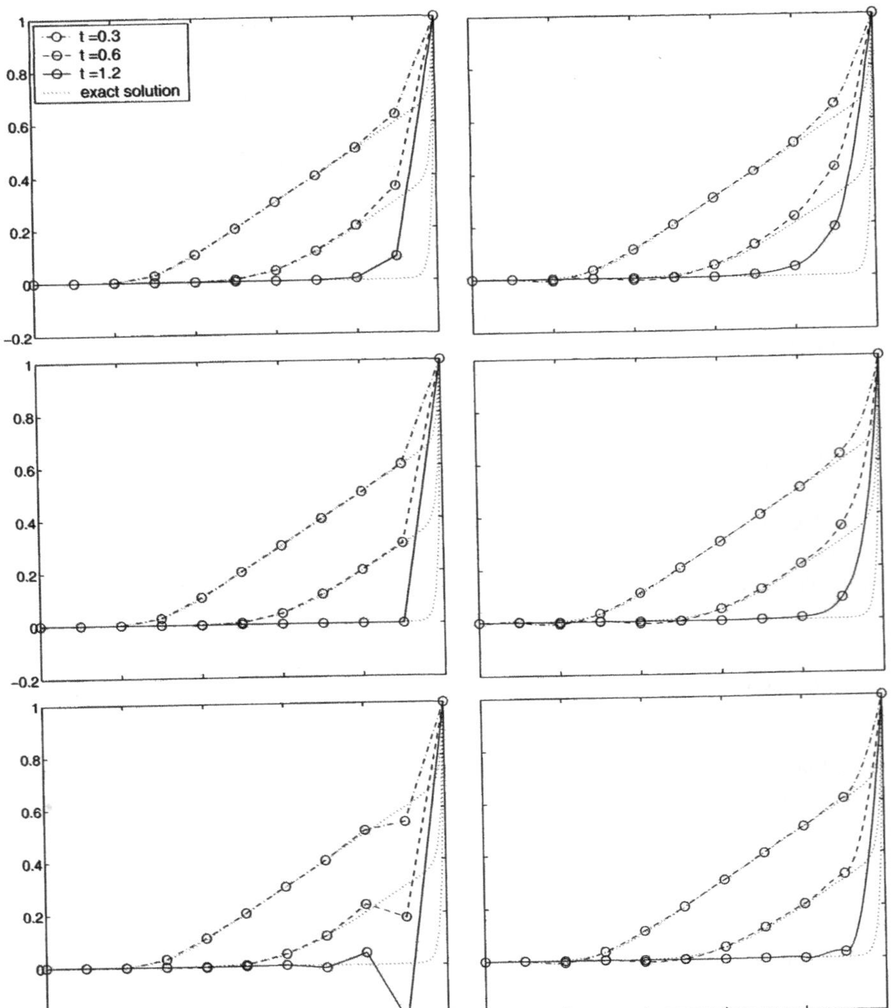

Fig. A.5 Least-Squares (top) SUPG (center) and GLS (bottom) approximations using $\mathcal{C}^0$ (left) and $\mathcal{C}^1$ (right) elements.

variables. In fact, if the steady-state solution is desired, LS produces correct results when a relaxation technique is employed.

This intrinsic difficulty of LS in pure steady-state problems is not present in other stabilization techniques, such as SUPG or GLS, which stabilize properly both steady and transient equations.

6

Viscous incompressible flows

This chapter is concerned with the finite element treatment of viscous incompressible flows governed by the Navier–Stokes equations. Solving these equations numerically meets two difficult problems. The first one is related to the discretization of the nonlinear convective terms, it requires the use of stabilized finite element formulations to properly treat high Reynolds number flows. This issue has been already discussed in previous chapters. The second difficulty is related to the numerical treatment of the saddle-point problem which arises from the variational formulation of the incompressible flow equations with the pressure acting as a Lagrangian multiplier of the incompressibility constraint. Different methods have been devised for the solution of the above numerical difficulties and some of them are discussed in this chapter.

6.1 INTRODUCTION

In this last chapter we consider the finite element modeling of steady and transient viscous, incompressible flows governed by the Navier–Stokes equations. Section 6.2 presents a review of the basic continuum mechanics concepts that are needed for the finite element formulation of incompressible flow problems. The Navier–Stokes equations governing viscous incompressible flows in the laminar regime are then introduced, together with the associated initial and boundary conditions.

The following sections are devoted to the application of finite element techniques to model incompressible flows in the Eulerian description. As shown in Section 6.3, there are two major sources of numerical difficulty in the use of the standard Galerkin finite element method. The first is related to the incompressibility of the fluid and manifests itself when an inappropriate combination of element interpolation functions for the velocity and pressure is employed. As a consequence, instabilities in the pressure field may appear, and this is independent of the Reynolds number. That is, instabilities may occur even at very "slow" flows (low Reynolds number). A proper combination of interpolation spaces (for velocity and pressure) is needed unless a specific formulation circumventing these instabilities is employed.

A second source of numerical difficulty is due to the presence of nonlinear convective terms in the Navier–Stokes equations. As seen in the preceding chapters, the Galerkin formulation typically lacks stability when convective effects dominate and alternative spatial discretization procedures must be advocated to restore stability without compromising the accuracy. In the present chapter attention will therefore be focused on situations in which convective effects are important, as well as on the numerical difficulties arising from the fluid incompressibility.

In Section 6.4 we define the function spaces that are needed for the approximation of the velocity and pressure fields. Section 6.5 is concerned with the finite element formulation of stationary Stokes problems, thus leaving aside the problems arising from the convective terms in the full Navier–Stokes equations. Emphasis is placed on mixed finite element formulations in which both velocity and pressure are retained as unknowns. Conditions to be satisfied that produce stable approximations with mixed methods in the standard Galerkin formulation are discussed in some detail. Stabilized formulations are then introduced which provide a remedy for the deficiencies of the Galerkin approach. We also discuss the penalty method which uncouples the determination of the velocity and pressure fields through a relaxation of the incompressibility constraint.

Section 6.6 deals with steady Navier–Stokes problems. We underline the additional difficulty introduced by the nonlinear convective terms. Fortunately, stabilization techniques provide a remedy for the deficiencies of the Galerkin approach arising from both the convective terms and the incompressibility constraint.

Transient problems are introduced in Section 6.7. Here, emphasis is placed on the advantages of a fractional-step projection method for the time integration of the Navier–Stokes equations. Simple test problems easily solvable by the interested reader are presented in Section 6.8. In particular, as an illustration, an application of the fractional-step method for natural convection problems is discussed in Section 6.8.4.

Several excellent texts devoted to the finite element modeling of incompressible flow problems are available as complements to the present introductory chapter. Particularly noteworthy are Glowinski (1984), Carey and Oden (1986), Girault and Raviart (1986), Glowinski and Le Tallec (1989), Gresho and Sani (2000), Gunzburger (1989), Pironneau (1989), Quartapelle (1993), Quarteroni and Valli (1994) and Temam (2001). The finite element books by Baker (1983), Jiang (1998), Hughes (2000), Zienkiewicz and Taylor (2000b) and Reddy and Gartling (2001) also contain interesting chapters on incompressible problems.

6.2 BASIC CONCEPTS

6.2.1 Strain rate and spin tensors

The analysis of the relative motion of neighboring particles within a fluid is similar to the theory of deformation of an elastic solid: the rate of strain and the rate of rotation in the fluid take the place of the strain and rotation of the solid. An important variable in the characterization of fluid motion is the *velocity gradient.* The velocity gradient is a second-order tensor defined in a Cartesian coordinate system by

$$\nabla \boldsymbol{v} = \begin{pmatrix} \frac{\partial v_1}{\partial x_1} & \frac{\partial v_1}{\partial x_2} & \frac{\partial v_1}{\partial x_3} \\ \frac{\partial v_2}{\partial x_1} & \frac{\partial v_2}{\partial x_2} & \frac{\partial v_2}{\partial x_3} \\ \frac{\partial v_3}{\partial x_1} & \frac{\partial v_3}{\partial x_2} & \frac{\partial v_3}{\partial x_3} \end{pmatrix}.$$

It may be decomposed into its symmetric and skew-symmetric parts according to

$$\frac{\partial v_i}{\partial x_j} = \frac{1}{2}\Big(\frac{\partial v_i}{\partial x_j} + \frac{\partial v_j}{\partial x_i}\Big) + \frac{1}{2}\Big(\frac{\partial v_i}{\partial x_j} - \frac{\partial v_j}{\partial x_i}\Big),$$

for $i, j = 1, \dots, \mathrm{n_{sd}}$, or

$$\nabla \boldsymbol{v} = \nabla^{\mathrm{S}} \boldsymbol{v} + \nabla^{\mathrm{W}} \boldsymbol{v}, \quad \text{where} \quad \begin{cases} \nabla^{\mathrm{S}} := \frac{1}{2}(\nabla + \nabla^T), & \text{and} \\ \nabla^{\mathrm{W}} := \frac{1}{2}(\nabla - \nabla^T) \end{cases} \tag{6.1}$$

The symmetric tensor $\nabla^{\mathrm{S}} \boldsymbol{v}$ is called the *rate of deformation (or strain rate) tensor.* We shall frequently use the alternative notation

$$v_{(i,j)} := \frac{1}{2}\Big(\frac{\partial v_i}{\partial x_j} + \frac{\partial v_j}{\partial x_i}\Big) \quad \text{for } i, j = 1, \dots, \mathrm{n_{sd}}$$

to denote the components $[\nabla^{\mathrm{S}} \boldsymbol{v}]_{ij}$ of the strain rate tensor. The skew-symmetric tensor $\nabla^{\mathrm{W}} \boldsymbol{v}$ is called *vorticity tensor* (or *spin tensor*). Its components are defined by

$$[\nabla^{\mathrm{W}} \boldsymbol{v}]_{ij} = \frac{1}{2}\Big(\frac{\partial v_i}{\partial x_j} - \frac{\partial v_j}{\partial x_i}\Big).$$

If the rate of deformation tensor $\nabla^{\mathrm{S}} \boldsymbol{v}$ at a given point is identically zero, the motion in the neighborhood of that point is a rigid body rotation.

Associated with the vorticity tensor is the vorticity vector $\boldsymbol{\omega}$ defined by

$$\boldsymbol{\omega} = \nabla \times \boldsymbol{v},$$

or in component form

$$\omega_1 = \frac{\partial v_3}{\partial x_2} - \frac{\partial v_2}{\partial x_3}, \qquad \omega_2 = \frac{\partial v_1}{\partial x_3} - \frac{\partial v_3}{\partial x_1}, \qquad \omega_3 = \frac{\partial v_2}{\partial x_1} - \frac{\partial v_1}{\partial x_2}.$$

Note that, a velocity field is said to be irrotational if vorticity vanishes everywhere within the flow field.

Remark 6.1. In 3D, the relation between the vorticity vector $\boldsymbol{\omega}$ and the vorticity tensor $\boldsymbol{\nabla}^{\mathrm{W}}\boldsymbol{v}$ is explicitly expressed by

$$\boldsymbol{\omega} = -\mathbf{P} : \boldsymbol{\nabla}^{\mathrm{W}}\boldsymbol{v}, \quad \text{i.e.,}\ \omega_i = -\sum_{j,k=1}^{3} P_{ijk}[\boldsymbol{\nabla}^{\mathrm{W}}\boldsymbol{v}]_{jk};$$

where $\mathbf{P}$ is the unit antisymmetric multilinear form of order 3 (see van der Waerden, 1970, p. 76). $\mathbf{P}$ is defined such that $P_{ijk} = 0$ if one of the indices is repeated and for i, j and k different

$$P_{ijk} = \begin{cases} 1 & \text{if } \pi \text{ is even} \\ -1 & \text{if } \pi \text{ is odd} \end{cases}$$

where π is the permutation taking 123 into ijk.

6.2.2 The stress tensor in a Newtonian fluid

In a general fluid at rest, only normal stresses are present and the stress tensor has the isotropic form

$$\sigma_{ij} = -p\,\delta_{ij},$$

where p is the static fluid pressure and δ_{ij} the Kronecker delta. The situation is different for a fluid in motion. Then, in general, tangential stresses are non-zero, and the normal component of the stress acting across a surface element depends on the direction of the normal to the element. The quantity $-\frac{1}{3}\sigma_{ii}$ (sum on i), which is invariant under rotation of the reference axes and reduces to the static fluid pressure when the fluid is at rest, is used to define the *pressure at a point in a moving fluid*:

$$p = -\frac{1}{3}\,\sigma_{ii}.$$

Note that this is a purely mechanical definition of pressure, which is thus not connected to the usual definition of pressure in thermodynamics.

It is convenient to decompose the Cauchy stress tensor σ_{ij} into the sum of an isotropic part $-p\,\delta_{ij}$ and a remaining non-isotropic part s_{ij}, the *deviatoric stress tensor*:

$$\sigma_{ij} = -p\,\delta_{ij} + s_{ij}.$$

For a *Newtonian fluid*, it is assumed that the stress tensor and the strain rate tensor are linearly related. The stress–strain rate relationship is given by

$$\sigma_{ij} = -p\,\delta_{ij} + s_{ij} = -p\,\delta_{ij} + \mu\Big(\frac{\partial v_i}{\partial x_j} + \frac{\partial v_j}{\partial x_i}\Big) + \lambda\,\frac{\partial v_k}{\partial x_k}\delta_{ij};$$

where μ is the fluid dynamic viscosity and λ the so-called second coefficient of viscosity. For an incompressible fluid one has $\boldsymbol{\nabla} \cdot \boldsymbol{v} = 0$ and consequently the above relation reduces to *Stokes' law*

$$\sigma_{ij} = -p\,\delta_{ij} + \mu\Big(\frac{\partial v_i}{\partial x_j} + \frac{\partial v_j}{\partial x_i}\Big) = -p\,\delta_{ij} + 2\mu\, v_{(i,j)}.$$

In compact form Stokes' law reads

$$\boldsymbol{\sigma} = -p\,\mathbf{I} + 2\mu \boldsymbol{\nabla}^{\mathrm{S}} \boldsymbol{v}, \tag{6.2}$$

or explicitly in 3D:

$$\begin{pmatrix} \sigma_{11} & \sigma_{12} & \sigma_{13} \\ \sigma_{21} & \sigma_{22} & \sigma_{23} \\ \sigma_{31} & \sigma_{32} & \sigma_{33} \end{pmatrix} = -p \begin{pmatrix} 1 & 0 & 0 \\ 0 & 1 & 0 \\ 0 & 0 & 1 \end{pmatrix}$$
$$+\, 2\mu \begin{pmatrix} \dfrac{\partial v_1}{\partial x_1} & \dfrac{1}{2}\Big(\dfrac{\partial v_1}{\partial x_2} + \dfrac{\partial v_2}{\partial x_1}\Big) & \dfrac{1}{2}\Big(\dfrac{\partial v_1}{\partial x_3} + \dfrac{\partial v_3}{\partial x_1}\Big) \\ \dfrac{1}{2}\Big(\dfrac{\partial v_2}{\partial x_1} + \dfrac{\partial v_1}{\partial x_2}\Big) & \dfrac{\partial v_2}{\partial x_2} & \dfrac{1}{2}\Big(\dfrac{\partial v_2}{\partial x_3} + \dfrac{\partial v_3}{\partial x_2}\Big) \\ \dfrac{1}{2}\Big(\dfrac{\partial v_3}{\partial x_1} + \dfrac{\partial v_1}{\partial x_3}\Big) & \dfrac{1}{2}\Big(\dfrac{\partial v_3}{\partial x_2} + \dfrac{\partial v_2}{\partial x_3}\Big) & \dfrac{\partial v_3}{\partial x_3} \end{pmatrix}.$$

Remark 6.2. It is easy to prove the identity

$$[\boldsymbol{\nabla} \boldsymbol{v}]_{ij}\, \sigma_{ij} = [\boldsymbol{\nabla}^{\mathrm{S}} \boldsymbol{v}]_{ij}\, \sigma_{ij} = v_{(i,j)}\, \sigma_{ij},$$

which will be repeatedly used in the derivation of the variational form of Stokes and Navier–Stokes problems. The proof (see for instance Hughes, 2000, p. 79) uses the additive decomposition (6.1) and the fact that

$$[\boldsymbol{\nabla}^{\mathrm{W}} \boldsymbol{v}]_{ij}\, \sigma_{ij} = -[\boldsymbol{\nabla}^{\mathrm{W}} \boldsymbol{v}]_{ji}\, \sigma_{ij} = -[\boldsymbol{\nabla}^{\mathrm{W}} \boldsymbol{v}]_{ji}\, \sigma_{ji} = -[\boldsymbol{\nabla}^{\mathrm{W}} \boldsymbol{v}]_{ij}\, \sigma_{ij},$$

in view of the skew-symmetry of $\boldsymbol{\nabla}^{\mathrm{W}} \boldsymbol{v}$ and the symmetry of $\boldsymbol{\sigma}$.

6.2.3 The Navier–Stokes equations

A number of important phenomena in fluid mechanics are described by the Navier–Stokes equations. They are a statement of the dynamical effect of the externally applied forces and the internal forces of a fluid that we shall assume Newtonian. The internal forces are due to the pressure and the viscosity of the fluid. We consider a flow region $\Omega \in \mathbb{R}^{\mathrm{n_{sd}}}$, where $\mathrm{n_{sd}} = 2$ or 3. The domain Ω occupied by the fluid will be assumed bounded (finite size). The boundary $\Gamma = \partial\Omega$ of the fluid domain is assumed to be Lipschitz continuous, meaning that it is a closed and sufficiently regular surface. Then, the time-dependent flow of a viscous incompressible fluid is governed by the following form of the momentum equation (1.15) and the mass-conservation equation (1.11), called the *Navier–Stokes equations*:

$$\rho\,(\boldsymbol{v}_t + (\boldsymbol{v}\cdot\boldsymbol{\nabla})\boldsymbol{v}) = \boldsymbol{\nabla}\cdot\boldsymbol{\sigma} + \rho\boldsymbol{b} \qquad \text{in } \Omega\times]0,T[, \tag{6.3a}$$

$$\boldsymbol{\nabla}\cdot\boldsymbol{v} = 0 \qquad \text{in } \Omega\times]0,T[. \tag{6.3b}$$

Here, ρ is the fluid density and $\boldsymbol{b}$ the volume force per unit mass of fluid. Using Stokes' law (6.2), the equation of motion (6.3a) can be expressed in the more convenient form

$$\boldsymbol{v}_t + (\boldsymbol{v}\cdot\boldsymbol{\nabla})\boldsymbol{v} - 2\nu\boldsymbol{\nabla}\cdot\boldsymbol{\nabla}^{\mathrm{S}}\boldsymbol{v} + \boldsymbol{\nabla}p = \boldsymbol{b},$$

where $\nu = \mu/\rho$ is the fluid kinematic viscosity and p the kinematic pressure, that is pressure divided by density. Usually the previous equation is rewritten as

$$\boldsymbol{v}_t + (\boldsymbol{v}\cdot\boldsymbol{\nabla})\boldsymbol{v} - \nu\boldsymbol{\nabla}^2\boldsymbol{v} - \nu\boldsymbol{\nabla}(\boldsymbol{\nabla}\cdot\boldsymbol{v}) + \boldsymbol{\nabla}p = \boldsymbol{b} \tag{6.4}$$

and referred as the velocity–pressure stress-divergence form.

The incompressibility condition (6.3b) is a consequence of the fact that in an incompressible continuum the rate of change of the mass density following the motion is zero, see equation (1.12). Under the incompressibility condition (6.3b), the momentum equation (6.3a), or (6.4), can be transformed to

$$\boldsymbol{v}_t + (\boldsymbol{v}\cdot\boldsymbol{\nabla})\boldsymbol{v} - \nu\boldsymbol{\nabla}^2\boldsymbol{v} + \boldsymbol{\nabla}p = \boldsymbol{b} \qquad \text{in } \Omega\times]0,T[, \tag{6.5}$$

where $\boldsymbol{\nabla}^2\boldsymbol{v}$ denotes the Laplacian operator applied to the velocity vector. Its components are defined by

$$[\boldsymbol{\nabla}^2\boldsymbol{v}]_i = \sum_{j=1}^{\mathrm{n_{sd}}} \frac{\partial^2 v_i}{\partial x_j^2}.$$

The momentum equation is usually expressed in form (6.5), because, apart from the convective term $(\boldsymbol{v}\cdot\boldsymbol{\nabla})\boldsymbol{v}$, these equations uncouple the velocity components.

The Navier–Stokes problem must be completed with suitable initial and boundary conditions to form a well-posed initial boundary value problem. Typical boundary conditions consist of prescribing the value $\boldsymbol{v}_D$ of the velocity on a portion Γ_D of the boundary:

$$\boldsymbol{v}(\boldsymbol{x},t) = \boldsymbol{v}_D(\boldsymbol{x},t), \qquad \boldsymbol{x}\in\Gamma_D, \quad t\in]0,T[, \tag{6.6}$$

and boundary traction $\boldsymbol{t}$ on the complementary portion Γ_N:

$$\boldsymbol{n}\cdot\boldsymbol{\sigma}(\boldsymbol{x},t) = \boldsymbol{t}(\boldsymbol{x},t), \qquad \boldsymbol{x}\in\Gamma_N, \quad t\in]0,T[, \tag{6.7}$$

where vector $\boldsymbol{n}$ denotes the unit outer normal to the boundary.

Remark 6.3 (Imposing boundary tractions). Note that for fluids obeying Stokes' law, equation (6.7) is equivalent to

$$\boldsymbol{n}\cdot\boldsymbol{\sigma} = -p\boldsymbol{n} + 2\nu\,\boldsymbol{n}\cdot\boldsymbol{\nabla}^{\mathrm{S}}\boldsymbol{v} = \boldsymbol{t}. \tag{6.8}$$

In 2D and with reference to a local system of Cartesian axes (n, τ), it becomes

$$-p + 2\nu \frac{\partial v_n}{\partial n} = t_n,$$
$$\nu\Big(\frac{\partial v_\tau}{\partial n} + \frac{\partial v_n}{\partial \tau}\Big) = t_\tau.$$

The first condition involves both pressure p and the velocity gradient. It represents a condition of applied normal stress. The second one models applied tangential stresses. Such boundary conditions are frequently employed, in particular, to simulate a no-stress condition at outflow boundaries.

In the case of a time-dependent problem, the value of the velocity field at the initial time $t = 0$ must be given in Ω:

$$\boldsymbol{v}(\boldsymbol{x}, 0) = \boldsymbol{v}_0(\boldsymbol{x}), \qquad \boldsymbol{x} \in \Omega.$$

Moreover, the initial velocity field $\boldsymbol{v}_0$ must be divergence free, that is

$$\boldsymbol{\nabla} \cdot \boldsymbol{v}_0 = 0 \qquad \text{in } \Omega.$$

Remark 6.4. No initial condition needs to be specified for the fluid pressure. This is a consequence of the fact that no time derivative of pressure appears in the governing equations. When Dirichlet conditions are imposed everywhere on the boundary, $\Gamma_N = \emptyset$, pressure is only present by its gradient in the Navier–Stokes equations, and thus it is determined only up to an arbitrary constant. In this case, it is usual to impose the pressure average or the value of the pressure at one point to uniquely define the pressure field.

Remark 6.5. In the case of highly viscous flow, the convective terms in the Navier–Stokes equations can often be neglected if compared with the dominant viscous terms. The resulting equations are identical to the equations of isotropic incompressible elasticity and are called *equations of Stokes flow*:

$$\begin{cases} \boldsymbol{v}_t - \nu \boldsymbol{\nabla}^2 \boldsymbol{v} + \boldsymbol{\nabla} p = \boldsymbol{b} \\ \boldsymbol{\nabla} \cdot \boldsymbol{v} = 0 \end{cases} \qquad \text{in } \Omega \times]0, T[.$$

Remark 6.6. A dimensionless form of the Navier–Stokes equations is obtained by replacing the kinematic viscosity ν by the inverse of the flow Reynolds number defined as $Re = VL/\nu$, where V and L stand, respectively, for a characteristic velocity and a characteristic length of the flow. Reynolds number Re characterizes the ratio between the inertia forces and the viscous forces. In the absence of body forces, the Navier–Stokes equations in primitive variables read

$$\begin{cases} \boldsymbol{v}_t + (\boldsymbol{v} \cdot \boldsymbol{\nabla})\boldsymbol{v} - \dfrac{1}{Re} \boldsymbol{\nabla}^2 \boldsymbol{v} + \boldsymbol{\nabla} p = 0 \\ \boldsymbol{\nabla} \cdot \boldsymbol{v} = 0 \end{cases} \qquad \text{in } \Omega \times]0, T[.$$

Remark 6.7. In the case of 2D motion, $\boldsymbol{v} = (v_1, v_2, 0)$ and v_1 and v_2 are independent of x_3. From the incompressibility condition, $\boldsymbol{\nabla} \cdot \boldsymbol{v} = \partial v_1/\partial x_1 + \partial v_2/\partial x_2 = 0$, we may pose

$$v_1 = \frac{\partial \psi}{\partial x_2}, \qquad v_2 = -\frac{\partial \psi}{\partial x_1},$$

where the unknown scalar function $\psi(\boldsymbol{x}, t)$, called *stream function*, is defined by

$$\psi(\boldsymbol{x}, t) - \psi_O = \int_O^P \boldsymbol{\nabla}\psi \cdot d\boldsymbol{x} = \int_O^P [\boldsymbol{v} \times d\boldsymbol{x}]_3,$$

where ψ_O is a constant and the line integral is taken along an arbitrary curve joining some reference point O to the point P with coordinates $\boldsymbol{x}$; note that $d\boldsymbol{x}$ represents a tangent vector to the arbitrary curve. It is common practice in computational fluid dynamics to provide a picture of a flow field by drawing a family of streamlines (curves at constant ψ and $\boldsymbol{v}$ tangent). In the finite element context, contours of streamlines, with specified intervals in ψ between pairs of neighboring streamlines, are obtained by numerical integration along the element sides or through the solution of a Poisson equation.

6.3 MAIN ISSUES IN INCOMPRESSIBLE FLOW PROBLEMS

Before introducing finite element techniques for the numerical solution of the Navier–Stokes equations, we wish to underline the main difficulties involved in the numerical simulation of incompressible flow problems.

A first difficulty is due to the presence of nonlinear and non-symmetric convective terms in the momentum equation (6.3a). Such difficulty increases with the value of the flow Reynolds number. High Reynolds number flows are convection dominated and, as repeatedly mentioned in the previous chapters, the standard Galerkin formulation is unstable. Stabilization techniques, such as SUPG, GLS, SGS or LS must be used to provide meaningful finite element solutions at high Reynolds numbers. These issues have been discussed in detail in previous chapters.

Another source of numerical difficulty is the incompressibility condition. The continuity equation for an incompressible fluid takes the peculiar form expressed in equation (6.3b). It consists of a constraint on the velocity field which must be divergence free. Then, the pressure has to be considered as a variable not related to any constitutive equation. Its presence in the momentum equation has the purpose of introducing an additional degree of freedom needed to satisfy the incompressibility constraint. The role of the pressure variable is thus to adjust itself instantaneously in order to satisfy the condition of divergence-free velocity. That is, the pressure is acting as a Lagrangian multiplier of the incompressibility constraint and thus there is a coupling between the velocity and the pressure unknowns.

Various formulations have been proposed in the literature to deal with incompressible flow problems. Here emphasis is placed on primitive variable formulations (retaining velocity and pressure as unknowns). Penalty methods are also discussed because they allow us to uncouple the determination of the velocity and pressure fields.

The primitive variable formulation, with both velocity and pressure unknowns, leads to so-called *mixed finite element methods*. Such methods present numerical difficulties caused by the *saddle-point* nature of the resulting variational problem; recall that pressure acts as a Lagrangian multiplier of the incompressibility constraint. Then, the algebraic system for the nodal values of velocity and pressure in a Galerkin formulation is governed by a partitioned matrix with a null submatrix on the diagonal. Solvability of the algebraic system depends on a proper choice of finite element spaces for velocity and pressure. They must satisfy a compatibility condition, the so-called LBB condition. There are, however, alternative finite element formulations that allow us to circumvent the LBB condition and enable the use of velocity–pressure pairs that are unstable in the standard Galerkin formulation.

The *penalty formulation* allows the elimination of the pressure variable from the Navier–Stokes problem through a relaxation of the incompressibility condition. The constraint $\boldsymbol{\nabla} \cdot \boldsymbol{v} = 0$ is replaced by $\boldsymbol{\nabla} \cdot \boldsymbol{v}^{(\lambda)} = -p^{(\lambda)}/\lambda$ where λ is a large parameter. This substitution eliminates the pressure gradient term from the momentum equation. Since it involves only velocities, the penalty method is computationally very attractive. A disadvantage is the presence of the penalty parameter λ, which may cause a loss of accuracy for excessively large values of λ, and prevent convergence to the actual solution for insufficiently large parameters.

As a starting point for the development of finite element models for incompressible flows, we introduce in the next section finite element spaces for velocity and pressure.

6.4 TRIAL SOLUTIONS AND WEIGHTING FUNCTIONS

The weak forms of Stokes and Navier–Stokes problems requires the introduction of classes of functions for the velocity field and the pressure field. With respect to velocity, $\boldsymbol{v}$, the space of trial solutions is denoted by $\boldsymbol{\mathcal{S}}$. As discussed in Section 1.5.2, candidate approximating functions must satisfy a priori Dirichlet boundary conditions, see (6.6), on Γ_D. The trial solution space $\boldsymbol{\mathcal{S}}$ containing the approximating functions for the velocity is thus characterized as follows:

$$\boldsymbol{\mathcal{S}} := \left\{ \boldsymbol{v} \in \boldsymbol{\mathcal{H}}^1(\Omega) \mid \boldsymbol{v} = \boldsymbol{v}_D \text{ on } \Gamma_D \right\} \quad \text{(trial solutions)}, \tag{6.9a}$$

where bold spaces contain vector functions such that each component is in the corresponding space of scalar functions. The weighting functions of the velocity, $\boldsymbol{w}$, belong to $\boldsymbol{\mathcal{V}}$. Functions in this class have the same characteristics as those in class $\boldsymbol{\mathcal{S}}$, except that the weighting functions are required to vanish on Γ_D where the velocity is prescribed. The class $\boldsymbol{\mathcal{V}}$ is thus symbolically defined by

$$\boldsymbol{\mathcal{V}} := \boldsymbol{\mathcal{H}}^1_{\Gamma_D}(\Omega) = \left\{ \boldsymbol{w} \in \boldsymbol{\mathcal{H}}^1(\Omega) \mid \boldsymbol{w} = \boldsymbol{0} \text{ on } \Gamma_D \right\} \quad \text{(weighting functions)}. \tag{6.9b}$$

Finally, we introduce a space of functions, denoted $\mathcal{Q}$, for the pressure. As we shall see, spatial derivatives of pressure do not appear in the weak form of the (Navier–

Stokes) problem; thus functions in $\mathcal{Q}$ are simply required to be square-integrable. Moreover, since there are no explicit boundary conditions on pressure, the space $\mathcal{Q}$,

$$\mathcal{Q} := \mathcal{L}_2(\Omega) \quad \text{(pressure space)}, \tag{6.9c}$$

suffices as the trial solution space and as the weighting function space. Note however that in the case of purely Dirichlet velocity boundary conditions, the pressure is defined up to a constant, see Remark 6.4. In such a case, its value must be prescribed at a given point of the domain Ω and the pressure space $\mathcal{Q}$ is thus replaced by $\mathcal{L}_2(\Omega)/\mathbb{R}$.

Remark 6.8 (Construction of a solenoidal velocity field). Additional subspaces accounting for the incompressible constraint are required for the analysis of Stokes and Navier–Stokes problems. They are subspaces of

$$\boldsymbol{\mathcal{H}}(\text{div};\Omega) := \{\boldsymbol{v} \in \boldsymbol{\mathcal{L}}_2(\Omega) \mid \boldsymbol{\nabla}\cdot\boldsymbol{v} \in \mathcal{L}_2(\Omega)\}, \tag{6.10}$$

which is a Hilbert space endowed with the following norm:

$$\|\boldsymbol{v}\|^2_{\text{div};\Omega} = \int_\Omega \boldsymbol{v}\cdot\boldsymbol{v}\, d\Omega + \int_\Omega (\boldsymbol{\nabla}\cdot\boldsymbol{v})^2\, d\Omega = (\boldsymbol{v},\boldsymbol{v}) + (\boldsymbol{\nabla}\cdot\boldsymbol{v},\boldsymbol{\nabla}\cdot\boldsymbol{v}).$$

The inner product for vector-valued functions was introduced in Section 1.5.1.2. Consider, for instance, the space $\mathcal{D}(\Omega)$, or $C_0^\infty(\Omega)$, of infinitely differentiable functions having compact support in Ω. Then, the set of solenoidal vector fields of $\boldsymbol{\mathcal{D}}(\Omega)$ is defined as

$$\boldsymbol{\mathcal{J}}(\Omega) := \{\boldsymbol{v} \in \boldsymbol{\mathcal{D}}(\Omega) \mid \boldsymbol{\nabla}\cdot\boldsymbol{v} = 0\}.$$

Likewise, the set of divergence-free velocity fields in $\boldsymbol{\mathcal{H}}_0^1(\Omega)$ is

$$\boldsymbol{\mathcal{J}}_0^1(\Omega) := \{\boldsymbol{v} \in \boldsymbol{\mathcal{H}}_0^1(\Omega) \mid \boldsymbol{\nabla}\cdot\boldsymbol{v} = 0\}.$$

For instance, to this subspace belongs the solution of the weak form of the Stokes problem with homogenous boundary conditions. In particular, if the domain Ω is bounded and its boundary Γ is Lipschitz continuous, we can define

$$\boldsymbol{\mathcal{J}}_0(\Omega) := \{\boldsymbol{v} \in \boldsymbol{\mathcal{L}}_2(\Omega) \mid \boldsymbol{\nabla}\cdot\boldsymbol{v} = 0 \text{ in } \Omega,\ \boldsymbol{n}\cdot\boldsymbol{v} = 0 \text{ on } \Gamma\}.$$

Note that $\boldsymbol{\mathcal{H}}(\text{div};\Omega)$ is the closure of $\boldsymbol{\mathcal{D}}(\overline{\Omega})$ with respect to the norm $\|\cdot\|_{\text{div};\Omega}$ and $\boldsymbol{\mathcal{J}}_0(\Omega)$ is the closure of $\boldsymbol{\mathcal{J}}(\Omega)$ with respect to the $\boldsymbol{\mathcal{L}}_2(\Omega)$ norm, see Section 1.5.1.1. Since $\boldsymbol{\mathcal{J}}_0(\Omega)$ is a closed subspace of $\boldsymbol{\mathcal{L}}_2(\Omega)$, we can define the decomposition

$$\boldsymbol{\mathcal{L}}_2(\Omega) = \boldsymbol{\mathcal{J}}_0(\Omega) \oplus \boldsymbol{\mathcal{J}}_0^\perp(\Omega),$$

where the characterization of $\boldsymbol{\mathcal{J}}_0^\perp(\Omega)$ derives from a theorem due to Ladyzhenskaya (1969) which states that

$$\boldsymbol{\mathcal{J}}_0^\perp(\Omega) := \{\boldsymbol{w} \in \boldsymbol{\mathcal{L}}_2(\Omega) \mid \boldsymbol{w} = \boldsymbol{\nabla}p,\ p \in \mathcal{H}^1(\Omega)\}.$$

This result is a consequence of the Helmholtz decomposition principle (see for instance Temam, 2001, Chap. I, Sec. 1) which states that any vector field $\boldsymbol{v}$ defined in Ω admits a unique orthogonal decomposition into the sum of a solenoidal field and the gradient of a scalar function. This characterization of $\mathcal{J}_0^{\perp}(\Omega)$ implies, in particular, that for all $\boldsymbol{w} \in \mathcal{L}_2(\Omega)$ orthogonal to any $\boldsymbol{u} \in \mathcal{J}_0(\Omega)$, that is $(\boldsymbol{w}, \boldsymbol{u}) = 0$, there exist p such that $\boldsymbol{w} = \boldsymbol{\nabla} p$. Note that the reciprocal holds.

As shown in Section 6.7, the projection of $\mathcal{L}_2(\Omega)$ onto $\mathcal{J}_0(\Omega)$, denoted by $\mathbb{P}$, is of prime importance for the construction of a solenoidal velocity field in the so-called fractional-step projection methods for solving incompressible flow problems.

6.5 STATIONARY STOKES PROBLEM

Before studying the Navier–Stokes equations, we shall study the steady Stokes problem. That is, we neglect the time-dependent and convective terms of the full Navier–Stokes equations. Two distinct formulations of Stokes problem are considered:

First, the momentum equation is written in terms of the Cauchy stress (also known as stress-divergence form). The Stokes' constitutive law is only invoked after setting the weak form. The advantage of this approach is that it can readily treat problems with fluid constitutive equations more general than the linear Stokes' law.

Second, the problem is directly formulated in terms of velocity and pressure. Use is made of Stokes' law and of the incompressibility condition to express the viscous term as the Laplacian of velocity. This is the standard form of the Stokes equations.

6.5.1 Formulation in terms of Cauchy stress

6.5.1.1 *Strong form* In differential form, a steady Stokes problem is stated as follows in terms of Cauchy stress: given the body force $\boldsymbol{b}$, prescribed velocities $\boldsymbol{v}_D$ on portion Γ_D of the boundary and imposed boundary tractions $\boldsymbol{t}$ on the remaining portion Γ_N, determine the velocity field $\boldsymbol{v}$ and the pressure field p such that

$$
\begin{aligned}
-\boldsymbol{\nabla}\cdot\boldsymbol{\sigma} &= \boldsymbol{b} && \text{in } \Omega && \text{(equilibrium)}, && (6.11\text{a})\\
\boldsymbol{\nabla}\cdot\boldsymbol{v} &= 0 && \text{in } \Omega && \text{(incompressibility)}, && (6.11\text{b})\\
\boldsymbol{v} &= \boldsymbol{v}_D && \text{on } \Gamma_D && \text{(Dirichlet b.c.)}, && (6.11\text{c})\\
\boldsymbol{n}\cdot\boldsymbol{\sigma} &= \boldsymbol{t} && \text{on } \Gamma_N && \text{(Neumann b.c.)}. && (6.11\text{d})
\end{aligned}
$$

Note, that a constitutive equation is needed to close the problem. That is, the Cauchy stress, $\boldsymbol{\sigma}$, must be related to velocity, $\boldsymbol{v}$, and pressure, p, that is $\boldsymbol{\sigma} = \boldsymbol{\sigma}(p, \boldsymbol{v})$, for instance by the linear Stokes' law (6.2).

6.5.1.2 *Weak form* The Stokes equations (6.11) define a *constrained equilibrium problem.* Techniques from optimization theory are available to treat such problems. In the finite element context standard methods are Lagrange multipliers, penalty, augmented Lagrangian and perturbed Lagrangian methods. A detailed account of such methods, including implementation details, can be found elsewhere (for instance, Glowinski, 1984; Carey and Oden, 1986; Glowinski and Le Tallec, 1989; Brezzi and Fortin, 1991; Belytschko et al., 2000; Quarteroni and Valli, 1994; Hughes, 2000; Gresho and Sani, 2000).

Here, the Lagrange multiplier method is used to solve the Stokes problem. In this method, the minimization of an objective function (the total potential energy) subject to constraints (the incompressibility) corresponds to the stationary points of the sum of the objective function and the constraints weighted by the Lagrange multipliers. This saddle point problem is further discussed in Remark 6.9.

The weak form can also be obtained multiplying the equation of motion (6.11a) by the velocity test function $\boldsymbol{w}$ and integrating by parts the stress term, thereby generating the natural boundary condition (6.11d) on Γ_N. Similarly, the incompressibility condition (6.11b) is multiplied by the pressure test function q and the result integrated over the computational domain Ω. Thus the weak form of Stokes problem becomes: given $\boldsymbol{b}$, $\boldsymbol{v}_D$ and the boundary traction $\boldsymbol{t}$, find the velocity field $\boldsymbol{v} \in \boldsymbol{\mathcal{S}}$ and the pressure field $p \in \mathcal{Q}$, such that for all velocity test functions $\boldsymbol{w} \in \boldsymbol{\mathcal{V}}$ and all pressure test functions $q \in \mathcal{Q}$,

$$\begin{cases} \displaystyle\int_\Omega \boldsymbol{\nabla w} : \boldsymbol{\sigma}\, d\Omega = \int_\Omega \boldsymbol{w} \cdot \boldsymbol{b}\, d\Omega + \int_{\Gamma_N} \boldsymbol{w} \cdot \boldsymbol{t}\, d\Gamma \\ \displaystyle\int_\Omega q\, \boldsymbol{\nabla} \cdot \boldsymbol{v}\, d\Omega = 0. \end{cases} \tag{6.12}$$

To further clarify which equations are satisfied by the the weak form (6.12), we integrate by parts the term involving $\boldsymbol{\sigma}$ and use the divergence theorem. This gives, adding both equations,

$$0 = \int_\Omega \boldsymbol{w} \cdot \underbrace{(\boldsymbol{\nabla} \cdot \boldsymbol{\sigma} + \boldsymbol{b})}_{\text{equilibrium}}\, d\Omega + \int_\Omega q \underbrace{\boldsymbol{\nabla} \cdot \boldsymbol{v}}_{\text{incomp.}}\, d\Omega - \int_{\Gamma_N} \boldsymbol{w} \cdot \underbrace{(\boldsymbol{n} \cdot \boldsymbol{\sigma} - \boldsymbol{t})}_{\text{Neumann b.c.}}\, d\Gamma.$$

The weighting functions $\boldsymbol{w}$ and q are arbitrary, thus the solution of the variational problem (6.12) verifies the strong form (6.11) of the steady Stokes problem.

If the fluid constitutive equation can be expressed as

$$\sigma_{ij} = -p\,\delta_{ij} + s_{ij}(\boldsymbol{v}),$$

where the deviatoric part of $\boldsymbol{\sigma}$ is denoted by s_{ij}. The variational problem (6.12) can be rewritten in compact form in terms of velocity and pressure, find $(\boldsymbol{v}, p) \in \boldsymbol{\mathcal{S}} \times \mathcal{Q}$, such that

$$\begin{cases} a(\boldsymbol{w}, \boldsymbol{v}) + b(\boldsymbol{w}, p) = (\boldsymbol{w}, \boldsymbol{b}) + (\boldsymbol{w}, \boldsymbol{t})_{\Gamma_N} & \forall \boldsymbol{w} \in \boldsymbol{\mathcal{V}} \\ b(\boldsymbol{v}, q) = 0 & \forall q \in \mathcal{Q} \end{cases} \tag{6.13}$$

or equivalently, find $(\boldsymbol{v}, p) \in \boldsymbol{\mathcal{S}} \times \mathcal{Q}$, such that

$$a(\boldsymbol{w}, \boldsymbol{v}) + b(\boldsymbol{w}, p) + b(\boldsymbol{v}, q) = (\boldsymbol{w}, \boldsymbol{b}) + (\boldsymbol{w}, \boldsymbol{t})_{\Gamma_N} \quad \forall (\boldsymbol{w}, q) \in \boldsymbol{\mathcal{V}} \times \mathcal{Q}$$

with the following definitions of the forms,

$$a(\boldsymbol{w},\boldsymbol{v}) = \int_\Omega w_{(i,j)}\, s_{ij}\, d\Omega \quad \text{and} \quad b(\boldsymbol{v},q) = -\int_\Omega q \boldsymbol{\nabla}\cdot\boldsymbol{v}\, d\Omega,$$

see also (1.27). Note that the symmetry of the stress tensor has been used

$$\boldsymbol{\nabla w} : \boldsymbol{\sigma} = \sum_{i,j=1}^{n_{sd}} \frac{\partial w_i}{\partial x_j}\sigma_{ij} = \sum_{i,j=1}^{n_{sd}} w_{(i,j)}\,\sigma_{ij}.$$

Moreover, if the Cauchy stress $\boldsymbol{\sigma}$ is assumed given by the linear Stokes' law (6.2),

$$s_{ij} = C_{ijkl}\, v_{(k,l)}, \quad \text{where } C_{ijkl} = \nu\,(\delta_{ik}\,\delta_{jl} + \delta_{il}\,\delta_{jk}),$$

and this allows us to write the bilinear form $a(\boldsymbol{w},\boldsymbol{v})$ as

$$a(\boldsymbol{w},\boldsymbol{v}) = \int_\Omega w_{(i,j)}\, C_{ijkl}\, v_{(k,l)}\, d\Omega. \tag{6.14}$$

Remark 6.9 (Saddle-point nature of the solution). For homogeneous Dirichlet boundary conditions, $\boldsymbol{v} = 0$ on $\Gamma_D = \partial\Omega$ and $\Gamma_N = \emptyset$, equation (6.13) reads

$$a(\boldsymbol{w},\boldsymbol{v}) + b(\boldsymbol{w},p) + b(\boldsymbol{v},q) = (\boldsymbol{w},\boldsymbol{b}), \qquad \forall(\boldsymbol{w},q) \in \boldsymbol{\mathcal{V}} \times \mathcal{Q}.$$

It can be shown (see for instance Temam, 2001) that the solution $(\boldsymbol{v},p)$ of the previous Stokes problem is a *saddle point* of the Lagrangian functional

$$I(\boldsymbol{w},q) := \frac{1}{2}a(\boldsymbol{w},\boldsymbol{w}) + b(\boldsymbol{w},q) - (\boldsymbol{w},\boldsymbol{b}) \qquad \forall(\boldsymbol{w},q) \in \boldsymbol{\mathcal{V}} \times \mathcal{Q};$$

that is,

$$I(\boldsymbol{v},q) \le I(\boldsymbol{v},p) \le I(\boldsymbol{w},p) \qquad \forall(\boldsymbol{w},q) \in \boldsymbol{\mathcal{V}} \times \mathcal{Q},$$

or equivalently,

$$I(\boldsymbol{v},p) = \min_{\boldsymbol{w}\in\boldsymbol{\mathcal{V}}} \max_{q\in\mathcal{Q}} I(\boldsymbol{w},q)$$

Remark 6.10 (Stress–strain form). It is sometimes interesting to express the bilinear form $a(\boldsymbol{w},\boldsymbol{v})$ in equation (6.14) in terms of the *strain rate vector*

$$\dot{\boldsymbol{\epsilon}}(\boldsymbol{v})^T = \left(\frac{\partial v_1}{\partial x_1}, \frac{\partial v_2}{\partial x_2}, \frac{\partial v_3}{\partial x_3}, \frac{\partial v_1}{\partial x_2} + \frac{\partial v_2}{\partial x_1}, \frac{\partial v_2}{\partial x_3} + \frac{\partial v_3}{\partial x_2}, \frac{\partial v_3}{\partial x_1} + \frac{\partial v_1}{\partial x_3}\right),$$

containing the six strain rate components relevant in 3D analysis. To express $a(\boldsymbol{w},\boldsymbol{v})$ in terms of the strain rate vector $\dot{\boldsymbol{\epsilon}}(\boldsymbol{v})$, we define for 3D the constitutive matrix

$$\mathbf{C}_\nu = \nu \begin{pmatrix} 2 & 0 & 0 & 0 & 0 & 0 \\ 0 & 2 & 0 & 0 & 0 & 0 \\ 0 & 0 & 2 & 0 & 0 & 0 \\ 0 & 0 & 0 & 1 & 0 & 0 \\ 0 & 0 & 0 & 0 & 1 & 0 \\ 0 & 0 & 0 & 0 & 0 & 1 \end{pmatrix}. \tag{6.15}$$

It can then be shown that $w_{(i,j)}\, C_{ijkl}\, v_{(k,l)} = \dot{\boldsymbol{\epsilon}}(\boldsymbol{w})^T\, \mathbf{C}_\nu\, \dot{\boldsymbol{\epsilon}}(\boldsymbol{v})$, so that the bilinear form (6.14) can be rewritten as

$$a(\boldsymbol{w},\boldsymbol{v}) = \int_\Omega \dot{\boldsymbol{\epsilon}}(\boldsymbol{w})^T\, \mathbf{C}_\nu\, \dot{\boldsymbol{\epsilon}}(\boldsymbol{v})\, d\Omega.$$

Note that $\mathbf{C}_\nu\, \dot{\boldsymbol{\epsilon}}(\boldsymbol{v})$ is the deviatoric stress vector $\boldsymbol{s}(\boldsymbol{v})$ with components

$$\boldsymbol{s}(\boldsymbol{v})^T = (s_{11}, s_{22}, s_{33}, s_{12}, s_{23}, s_{31})\,.$$

It follows that

$$a(\boldsymbol{w},\boldsymbol{v}) = \int_\Omega \dot{\boldsymbol{\epsilon}}(\boldsymbol{w})^T\, \boldsymbol{s}(\boldsymbol{v})\, d\Omega$$

in complete analogy with linear elasticity (see the excellent presentation by Belytschko et al., 2000, App. 1).

6.5.2 Formulation in terms of velocity and pressure

6.5.2.1 Strong form The formulation of the Stokes problem in terms of Cauchy stress has the advantage that it is applicable to arbitrary fluid constitutive relations. When the linear Stokes' law can be invoked, it is preferable to start from a strong form of written in terms of velocity and pressure because in this form the velocity components are uncoupled. The steady Stokes problem is usually restated as follows:

$$-\nu\boldsymbol{\nabla}^2\boldsymbol{v} + \boldsymbol{\nabla}p = \boldsymbol{b} \quad \text{in } \Omega \quad \text{(equilibrium),} \tag{6.16a}$$

$$\boldsymbol{\nabla}\cdot\boldsymbol{v} = 0 \quad \text{in } \Omega \quad \text{(incompressibility),} \tag{6.16b}$$

$$\boldsymbol{v} = \boldsymbol{v}_D \quad \text{on } \Gamma_D \quad \text{(Dirichlet b.c.),} \tag{6.16c}$$

$$-p\boldsymbol{n} + \nu(\boldsymbol{n}\cdot\boldsymbol{\nabla})\boldsymbol{v} = \boldsymbol{t} \quad \text{on } \Gamma_N \quad \text{(Neumann b.c.).} \tag{6.16d}$$

Having used the kinematic viscosity $\nu = \mu/\rho$ and p denoting the dynamic pressure, the ("dynamic" or "scaled") Cauchy stress tensor (normalized by density), if needed, is also given by Stokes' law (6.2), $\boldsymbol{\sigma} = -p\mathbf{I} + 2\nu\,\boldsymbol{\nabla}^{\mathrm{S}}\boldsymbol{v}$.

The equilibrium equation (6.16a) is obtained from (6.11a) using the Stokes' law (6.2) and the incompressibility condition (6.11b), see also the velocity–pressure stress-divergence form (6.4).

In this case, however, $\boldsymbol{t}$ does not correspond to boundary tractions. It is usually called a "pseudo-traction".

Remark 6.11 (Laplace or stress-divergence forms). As noted previously, in the Neumann boundary condition defined by (6.16d) $\boldsymbol{t}$ is not a boundary traction but a "pseudo-traction". Boundary tractions are clearly enforced if the formulation is in terms of Cauchy stress, see for instance (6.11d) or Remark 6.3. In fact, (6.8) is different from (6.16d), that is

$$\boldsymbol{n}\cdot\boldsymbol{\sigma} = -p\boldsymbol{n} + 2\nu\,\boldsymbol{n}\cdot\boldsymbol{\nabla}^{\mathrm{S}}\boldsymbol{v} \neq -p\boldsymbol{n} + \nu(\boldsymbol{n}\cdot\boldsymbol{\nabla})\boldsymbol{v}.$$

This inequality becomes an equality if $(\partial v_j/\partial x_i)n_j = 0$ for $i = 1,\dots,\mathrm{n_{sd}}$, which, in the case of incompressible fluids, is equivalent to $(\boldsymbol{\tau}\cdot\boldsymbol{\nabla})\boldsymbol{v} = \boldsymbol{0}$ for all vectors $\boldsymbol{\tau}$ orthogonal to $\boldsymbol{n}$, that is $(\partial v_i/\partial x_j)\tau_j = 0$ for $i = 1,\dots,\mathrm{n_{sd}}$.

Nevertheless, "pseudo-tractions" are the natural boundary conditions for a weak (velocity–pressure) Laplace formulation and thus, it is often a convenient form to impose open/artificial boundary conditions (see Section 6.8.3).

6.5.2.2 *Weak form*

Applying the divergence theorem to the pressure gradient term and to the second-derivative (viscous) term, the weak formulation of Stokes problem (6.16) is also given by (6.13), namely find $(\boldsymbol{v},p) \in \boldsymbol{\mathcal{S}} \times \mathcal{Q}$, such that

$$a(\boldsymbol{w},\boldsymbol{v}) + b(\boldsymbol{w},p) + b(\boldsymbol{v},q) = (\boldsymbol{w},\boldsymbol{b}) + (\boldsymbol{w},\boldsymbol{t})_{\Gamma_N} \quad \forall(\boldsymbol{w},q) \in \boldsymbol{\mathcal{V}} \times \mathcal{Q},$$

with the following new definition of the viscous bilinear term:

$$a(\boldsymbol{w},\boldsymbol{v}) = \int_\Omega \boldsymbol{\nabla w} : \nu \boldsymbol{\nabla v}\, d\Omega. \tag{6.17}$$

Remark 6.12. Depending on the definition of the bilinear form $a(\cdot,\cdot)$, either (6.14) or (6.17), the interpretation of $\boldsymbol{t}$ may be different. If the stress-divergence form is employed, see (6.4) or (6.11a), the viscous bilinear form for an incompressible Newtonian fluid is (6.14) and $\boldsymbol{t}$ are boundary tractions. If the Laplace formulation for the viscous term is preferred, (6.17) defines the viscous bilinear form and $\boldsymbol{t}$ are "pseudo-tractions", see Remark 6.11. Note that these remarks are also pertinent to the Navier-Stokes equations.

6.5.3 Galerkin formulation

The Galerkin formulation of the Stokes problem leads to a *mixed finite element method.* We need to introduce local approximations for both the velocity components v_i^h and pressure p^h, as well as for their associated weighting functions w_i^h and q^h. We denote by $\boldsymbol{\mathcal{S}}^h$ and $\boldsymbol{\mathcal{V}}^h$ the finite dimensional subspaces of $\boldsymbol{\mathcal{S}}$ and $\boldsymbol{\mathcal{V}}$, and $\mathcal{Q}^h$ the finite dimensional subspace of $\mathcal{Q}$.

The velocity approximation $\boldsymbol{v}^h \in \boldsymbol{\mathcal{S}}^h$ admits the representation $\boldsymbol{v}^h = \boldsymbol{u}^h + \boldsymbol{v}_D^h$, where the field $\boldsymbol{v}_D^h$ satisfies (approximates) the Dirichlet boundary condition, see (6.11c), on Γ_D. Thus, the auxiliary velocity $\boldsymbol{u}^h$ belongs to the same space as the test function $\boldsymbol{w}^h$, namely $\boldsymbol{\mathcal{V}}^h$. The Galerkin formulation of the Stokes problem, see (6.13), may then be stated as follows: given $\boldsymbol{b}$, $\boldsymbol{v}_D$ and $\boldsymbol{t}$, find the auxiliary velocity $\boldsymbol{u}^h \in \boldsymbol{\mathcal{V}}^h$ and the pressure field $p^h \in \mathcal{Q}^h$ for all $(\boldsymbol{w}^h,q^h) \in \boldsymbol{\mathcal{V}}^h \times \mathcal{Q}^h$, such that

$$\left\{\begin{aligned} &a(\boldsymbol{w}^h,\boldsymbol{u}^h) + b(\boldsymbol{w}^h,p^h) = (\boldsymbol{w}^h,\boldsymbol{b}^h) + (\boldsymbol{w}^h,\boldsymbol{t}^h)_{\Gamma_N} - a(\boldsymbol{w}^h,\boldsymbol{v}_D^h)\\ &b(\boldsymbol{u}^h,q^h) = -b(\boldsymbol{v}_D^h,q^h). \end{aligned}\right. \tag{6.18}$$

As will become clear in Section 6.5.6, the success of a mixed finite element formulation crucially depends on a proper choice of the local interpolations of the velocity and the pressure. Since the pressure gradient does not enter the weak form, pressure is not required to be continuous at the interface between elements, a condition which is mandatory for the interpolation of velocity.

The next step in the Galerkin formulation consists in approximating the velocity components $v_i^h = u_i^h + v_{Di}^h$ in terms of shape functions and associated nodal values. Following the terminology introduced in Section 1.5.5, we denote by $\eta = \{1, 2, \ldots, \mathrm{n_{np}}\}$ the set of global velocity node numbers in the finite element mesh. Furthermore, we denote by $\eta_{Di} \subset \eta$ the subset of velocity nodes belonging to the Dirichlet portion of the boundary where component i of the velocity is prescribed.

The velocity components are then approximated as follows:

$$\left.\begin{aligned} u_i^h(\boldsymbol{x}) &= \sum_{A \in \eta \setminus \eta_{Di}} N_A(\boldsymbol{x})\, \mathrm{u}_{iA} \\ v_{Di}^h(\boldsymbol{x}) &= \sum_{A \in \eta_{Di}} N_A(\boldsymbol{x})\, v_{Di}(\boldsymbol{x}_A) \end{aligned}\right\} \quad \text{(no sum on } i\text{)}, \tag{6.19a}$$

where N_A is the shape function associated with global node number A, and u_{iA} the value of u_i^h at node number A. Recall that each velocity node has $\mathrm{n_{sd}}$ degrees of freedom. Thus, in the Galerkin formulation the test functions are defined such that

$$w_i^h \in \mathcal{V}_i^h := \operatorname*{span}_{A \in \eta \setminus \eta_{Di}} \{N_A\}.$$

The vector version of (6.19a) is defined with the aid of the canonical basis of $\mathbb{R}^{\mathrm{n_{sd}}}$, namely $\{\boldsymbol{e}_1, \ldots, \boldsymbol{e}_{\mathrm{n_{sd}}}\}$. For $\mathrm{n_{sd}} = 3$ the expressions of $\boldsymbol{e}_1$, $\boldsymbol{e}_2$ and $\boldsymbol{e}_3$ are the following:

$$\boldsymbol{e}_1 = (1, 0, 0)^T, \quad \boldsymbol{e}_2 = (0, 1, 0)^T, \quad \boldsymbol{e}_3 = (0, 0, 1)^T.$$

The vector version of the interpolation, (6.19a), and test functions are

$$\boldsymbol{u}^h(\boldsymbol{x}) = \sum_{i=1}^{\mathrm{n_{sd}}} u_i^h(\boldsymbol{x})\, \boldsymbol{e}_i = \sum_{i=1}^{\mathrm{n_{sd}}} \sum_{A \in \eta \setminus \eta_{Di}} N_A(\boldsymbol{x})\, \mathrm{u}_{iA}\, \boldsymbol{e}_i;$$

$$\boldsymbol{w}^h(\boldsymbol{x}) = \sum_{i=1}^{\mathrm{n_{sd}}} w_i^h(\boldsymbol{x})\, \boldsymbol{e}_i.$$

The pressure field is interpolated using a possibly different set of pressure nodes denoted by $\hat{\eta}$ and the shape functions $\hat{N}_{\hat{A}}$, as

$$p^h(\boldsymbol{x}) = \sum_{\hat{A} \in \hat{\eta}} \hat{N}_{\hat{A}}(\boldsymbol{x})\, \mathrm{p}_{\hat{A}},$$

where $\hat{A}$ is the global pressure node number and $\mathrm{p}_{\hat{A}}$ is the pressure value at $\hat{A}$. As previously done for the velocity, we shall use small letters, namely $\hat{a}$ and $\hat{b}$, to denote element pressure node numbers; they range from 1 to $\hat{\mathrm{n}}_{\mathrm{en}}$ pressure nodes. Similarly, the weighting function q^h for the pressure is expressed as

$$q^h \in \mathcal{Q}^h := \operatorname*{span}_{\hat{A} \in \hat{\eta}} \{\hat{N}_{\hat{A}}\}. \tag{6.19b}$$

Table 6.1 Correspondence between matrices and vectors in (6.21) and the variational forms.

Matrix/Vector	Corresponding term of equation (6.13) or (6.18)	
$\mathbf{K}$	$a(\boldsymbol{w},\boldsymbol{u})$	$\int_\Omega \boldsymbol{\epsilon}(\boldsymbol{w})^T \mathbf{C}_\nu \, \boldsymbol{\epsilon}(\boldsymbol{v})\, d\Omega$ or $\int_\Omega \boldsymbol{\nabla w} : \nu \boldsymbol{\nabla v}\, d\Omega$
$\mathbf{G}$	$b(\boldsymbol{w},p)$	$-\int_\Omega p \boldsymbol{\nabla}\cdot \boldsymbol{w}\, d\Omega$
$\mathbf{G}^T$	$b(\boldsymbol{u},q)$	$-\int_\Omega q \boldsymbol{\nabla}\cdot \boldsymbol{u}\, d\Omega$
$\mathbf{f}$		$(\boldsymbol{w},\boldsymbol{b}) + (\boldsymbol{w},\boldsymbol{t})_{\Gamma_N} - a(\boldsymbol{w},\boldsymbol{v}_D)$
$\mathbf{h}$		$-b(\boldsymbol{v}_D,q)$

6.5.4 Matrix problem

To derive the matrix problem governing Stokes flow in the Galerkin formulation, we introduce the above expressions for the trial and weighting functions into the Galerkin form (6.18). This results in the following set of nodal equations for the unknown components of the velocity field: for each $A \in \eta \setminus \eta_{Di}$ and $1 \leq i \leq \mathrm{n_{sd}}$

$$\sum_{j=1}^{\mathrm{n_{sd}}}\Big\{\sum_{B\in\eta\setminus\eta_{Dj}} a(N_A\,\boldsymbol{e}_i, N_B\,\boldsymbol{e}_j)\mathrm{u}_{jB}\Big\} + \sum_{\hat{A}\in\hat{\eta}} b(N_A\,\boldsymbol{e}_i, \hat{N}_{\hat{A}})\,\mathrm{p}_{\hat{A}}$$
$$= (N_A\,\boldsymbol{e}_i, \boldsymbol{b}^h) + (N_A\,\boldsymbol{e}_i, \boldsymbol{t}^h)_{\Gamma_N} - \sum_{j=1}^{\mathrm{n_{sd}}}\Big\{\sum_{B\in\eta_{Dj}} a(N_A\,\boldsymbol{e}_i, N_B\,\boldsymbol{e}_j)\, v_{Dj}\Big\}. \quad (6.20)$$

Note that the viscous bilinear form can be defined either by (6.14) or by (6.17). Similarly, the following set of discrete equations corresponding to the incompressibility constraint is obtained: for every $\hat{A} \in \hat{\eta}$

$$\sum_{i=1}^{\mathrm{n_{sd}}}\Big\{\sum_{B\in\eta\setminus\eta_{Di}} b(N_B\,\boldsymbol{e}_i, \hat{N}_{\hat{A}})\mathrm{u}_{iB}\Big\} = -\sum_{i=1}^{\mathrm{n_{sd}}}\Big\{\sum_{B\in\eta_{Di}} b(N_B\,\boldsymbol{e}_i, \hat{N}_{\hat{A}}) v_{Di}\Big\}.$$

From these equations one finds that the matrix system which governs the discrete Stokes problem assumes the following partitioned form:

$$\begin{pmatrix} \mathbf{K} & \mathbf{G} \\ \mathbf{G}^T & \mathbf{0} \end{pmatrix} \begin{pmatrix} \mathbf{u} \\ \mathbf{p} \end{pmatrix} = \begin{pmatrix} \mathbf{f} \\ \mathbf{h} \end{pmatrix}. \quad (6.21)$$

The matrices and vectors in this algebraic system of equations and the corresponding terms in the variational equation (6.18) are identified in Table 6.1.

Matrix $\mathbf{K}$ is the *viscosity matrix* and results from the discretization of $a(\cdot,\cdot)$, see (6.14) or (6.17). It is obtained, as usual, from the assembly of element contributions,

$$\mathbf{K} = \mathop{\mathsf{A}}^{e} \mathbf{K}^e.$$

Each velocity node has as many degrees of freedom as spatial dimensions. Thus, $\mathbf{K}^e$ is a square matrix whose dimension is given by $\mathrm{n_{ee}} = \mathrm{n_{en}} \cdot \mathrm{n_{sd}}$, where $\mathrm{n_{en}}$ is the number of element nodes and $\mathrm{n_{sd}}$ is the number of space dimensions. Each element of this matrix, K^e_{rs}, is characterized by indices r and s (element equation numbers) related to the element node numbers, a and b, and the components i and j of the velocity:

$$\begin{matrix} r = \mathrm{n_{sd}}(a-1)+i \\ s = \mathrm{n_{sd}}(b-1)+j \end{matrix} \quad \text{with} \quad \begin{matrix} 1 \le r,s \le \mathrm{n_{ee}} \\ 1 \le a,b \le \mathrm{n_{en}} \\ 1 \le i,j \le \mathrm{n_{sd}}. \end{matrix} \tag{6.22}$$

Each element of the viscosity matrix is defined as $K^e_{rs} = \boldsymbol{e}_i^T \mathbf{K}^e_{ab}\, \boldsymbol{e}_j$, where the particular expression of $\mathbf{K}^e_{ab}$ depends on the formulation chosen (recall the relation between r and s and a, b, i and j given in the previous equation). In 3D, a formulation based on the Cauchy stress gives

$$\mathbf{K}^e_{ab} = \int_{\Omega^e} \mathbf{B}_a^T \mathbf{C}_\nu \mathbf{B}_b \, d\Omega, \quad \text{with}$$
$$\mathbf{B}_a^T = \begin{pmatrix} \frac{\partial N_a}{\partial x_1} & 0 & 0 & \frac{\partial N_a}{\partial x_2} & 0 & \frac{\partial N_a}{\partial x_3} \\ 0 & \frac{\partial N_a}{\partial x_2} & 0 & \frac{\partial N_a}{\partial x_1} & \frac{\partial N_a}{\partial x_3} & 0 \\ 0 & 0 & \frac{\partial N_a}{\partial x_3} & 0 & \frac{\partial N_a}{\partial x_2} & \frac{\partial N_a}{\partial x_1} \end{pmatrix}, \tag{6.23}$$

where the strain rate–velocity matrix $\mathbf{B}$ is identical to the strain–displacement matrix in elasticity, and the matrix $\mathbf{C}_\nu$ is defined in (6.15). In fact, the viscosity matrix is equivalent to the stiffness matrix in elasticity; it is symmetric and positive definite. A velocity–pressure formulation induces a matrix with the same properties and the following expression:

$$\mathbf{K}^e_{ab} = \int_{\Omega^e} \mathbf{B}_a^T \nu \, \mathbf{B}_b \, d\Omega, \quad \text{with}$$
$$\mathbf{B}_a^T = \begin{pmatrix} \frac{\partial N_a}{\partial x_1} & \frac{\partial N_a}{\partial x_2} & \frac{\partial N_a}{\partial x_3} & 0 & 0 & 0 & 0 & 0 & 0 \\ 0 & 0 & 0 & \frac{\partial N_a}{\partial x_1} & \frac{\partial N_a}{\partial x_2} & \frac{\partial N_a}{\partial x_3} & 0 & 0 & 0 \\ 0 & 0 & 0 & 0 & 0 & 0 & \frac{\partial N_a}{\partial x_1} & \frac{\partial N_a}{\partial x_2} & \frac{\partial N_a}{\partial x_3} \end{pmatrix}. \tag{6.24}$$

Matrix $\mathbf{G}$ is the discrete *gradient operator*, and $\mathbf{G}^T$ is the discrete *divergence operator*. Matrix $\mathbf{G}$ arises from the discretization of the term $b(\boldsymbol{w}^h, p^h)$ in the Galerkin variational form (6.18):

$$\mathbf{G} = \mathop{\mathbf{A}}_{e} \mathbf{G}^e, \qquad G^e_{r\hat{a}} = b(N_a\, \boldsymbol{e}_i, \hat{N}_{\hat{a}})_{\Omega^e} = -(\hat{N}_{\hat{a}}, \boldsymbol{\nabla}\cdot(N_a\, \boldsymbol{e}_i))_{\Omega^e},$$

where $\hat{a}$ is an element pressure node number and r an element equation number. Note that $\boldsymbol{\nabla}\cdot(N_a \boldsymbol{e}_i) = \partial N_a/\partial x_i$.

Vectors $\mathbf{f}$ and $\mathbf{h}$ incorporate the effect of the velocity $\boldsymbol{v}_D$ prescribed on the Dirichlet portion Γ_D of the boundary. Vector $\mathbf{f}$ also includes the contributions emanating from the applied body force $\boldsymbol{b}$ and the prescribed traction $\boldsymbol{t}$ on the Neumann portion of the boundary. They take the form

$$\begin{aligned} \mathbf{f} &= \mathop{\mathsf{A}}^{e} \mathbf{f}^{e}, \quad f_r^e = \left(N_a, b_i^h\right)_{\Omega^e} + \left(N_a, t_i^h\right)_{\Gamma_N^e} - \sum_{q=1}^{\mathrm{n_{ee}}} K_{rs}^e v_{Ds}^e, \\ \mathbf{h} &= \mathop{\mathsf{A}}^{e} \mathbf{h}^{e}, \quad h_{\hat{a}}^e = \sum_{p=1}^{\mathrm{n_{ee}}} \left(\hat{N}_{\hat{a}}, \boldsymbol{\nabla} \cdot (N_a \boldsymbol{e}_i)\right)_{\Omega^e} v_{Dr}^e. \end{aligned} \tag{6.25}$$

Here, $v_{Ds}^e = v_{Djb}^e$ if v_{Dj} is prescribed at node b, and equals zero otherwise.

6.5.5 Solvability condition and solution procedure

The partitioned system (6.21) could in principle be solved in several ways. These include global methods in which the original system is solved iteratively, for instance using the Uzawa method (Brezzi and Fortin, 1991), as well as the so-called pressure-matrix method in which an independent linear system is generated for the pressure.

However, first of all, given the peculiar form of the Stokes system (6.21) with a null submatrix on the diagonal, the following critical question can be raised: under which condition can the algebraic system governing the velocity and pressure be safely solved? It can be shown that, provided the kernel (null space) of matrix $\mathbf{G}$ is zero, the global matrix (6.21) is non-singular, that is $\mathbf{u}$ and $\mathbf{p}$ are uniquely defined. Recall that the kernel of $\mathrm{n_{eq}} \times \hat{\mathrm{n}}_{\mathrm{eq}}$ matrix $\mathbf{G}$ is the set of all vectors $\mathbf{q}$ ($\hat{\mathrm{n}}_{\mathrm{eq}}$ components) such that $\mathbf{Gq} = \mathbf{0}$, namely

$$\ker \mathbf{G} := \{\mathbf{q} \mid \mathbf{q} \in \mathbb{R}^{\hat{\mathrm{n}}_{\mathrm{eq}}} \text{ and } \mathbf{Gq} = \mathbf{0}\},$$

where $\mathrm{n_{eq}}$ is the number of velocity unknowns, $\mathrm{n_{eq}} = \mathrm{n_{np}} \cdot \mathrm{n_{sd}} - \sum_{i=1}^{\mathrm{n_{sd}}} \dim(\eta_{Di})$, and $\hat{\mathrm{n}}_{\mathrm{eq}}$ is the number of pressure unknowns, $\hat{\mathrm{n}}_{\mathrm{eq}} = \dim(\hat{\eta}) = \dim(\mathcal{Q}^h)$.

In turn, to have $\ker \mathbf{G} = \{\mathbf{0}\}$, the velocity and pressure interpolations must satisfy a compatibility condition, called the LBB condition, to be discussed in the next section. Inappropriate combinations of velocity and pressure interpolations, namely pairs of spaces not satisfying the LBB compatibility condition, may render the discrete divergence matrix, $\mathbf{G}^T$, rank deficient.

Provided the kernel of matrix $\mathbf{G}$ is zero, the following pressure-matrix method can be employed to solve the partitioned Stokes system. From the first equation in (6.21), one obtains

$$\mathbf{u} = \mathbf{K}^{-1}\left(\mathbf{f} - \mathbf{Gp}\right). \tag{6.26}$$

The introduction of this result in the second equation of (6.21) yields an algebraic system for the pressure based on the so-called Schur complement matrix:

$$\left(\mathbf{G}^T \mathbf{K}^{-1} \mathbf{G}\right) \mathbf{p} = \mathbf{G}^T \mathbf{K}^{-1} \mathbf{f} - \mathbf{h}. \tag{6.27}$$

Note that the pressure matrix $\left(\mathbf{G}^T \mathbf{K}^{-1} \mathbf{G}\right)$ is symmetric, but full due to the presence of $\mathbf{K}^{-1}$. It is positive definite if $\ker \mathbf{G} = \{\mathbf{0}\}$. Once the pressure nodal values, $\mathbf{p}$, are determined from (6.27), the velocity field is obtained using equation (6.26). Usually this approach is combined with an iterative solver for the resulting linear systems and therefore ends up with two nested iterative loops which make the solution strategy quite expensive. To reduce the solution cost, alternative solution strategies have been proposed. They are described in detail by Quarteroni and Valli (1994) and Gresho and Sani (2000). Some of them are based on splitting the original problem into successive subproblems cheaper to solve. An example of an alternative solution strategy is presented in Section 6.7 in connection with the modeling of time-dependent Navier–Stokes problems.

6.5.6 The LBB compatibility condition

We have seen that the symmetric pressure matrix $(\mathbf{G}^T\mathbf{K}^{-1}\mathbf{G})$ is positive definite only if $\ker \mathbf{G} = \{\mathbf{0}\}$. If this is the case, the partitioned matrix (6.21) is non-singular and delivers uniquely defined velocity and pressure fields. If this is not the case, a stable and convergent velocity field might be obtained, but the pressure field is likely to present spurious and oscillatory results.

Ladyzhenskaya (1969), Babuška (1970/71) and Brezzi (1974) have determined the compatibility condition, known as the LBB (or inf-sup) condition that continuous and discrete spaces must satisfy to guarantee the stability of a mixed method. A complete discussion of the theory of mixed methods is beyond the scope of the present text, see Girault and Raviart (1986) or Brezzi and Fortin (1991) for more details. We shall limit ourselves to give a flavor of the numerical difficulties associated with these methods.

The LBB condition states that velocity and pressure spaces cannot be chosen arbitrarily, a link between them is necessary. To illustrate this let us consider once more the partitioned matrix system (6.21) governing steady Stokes flow, namely

$$\begin{pmatrix} \mathbf{K} & \mathbf{G} \\ \mathbf{G}^T & \mathbf{0} \end{pmatrix} \begin{pmatrix} \mathbf{u} \\ \mathbf{p} \end{pmatrix} = \begin{pmatrix} \mathbf{f} \\ \mathbf{h} \end{pmatrix},$$

where for the present purposes $\mathbf{K}$ is a square matrix ($\mathrm{n_{eq}} \times \mathrm{n_{eq}}$), $\mathbf{G}$ a rectangular matrix ($\mathrm{n_{eq}} \times \hat{\mathrm{n}}_{\mathrm{eq}}$), and the vectors $\mathbf{u}$, $\mathbf{p}$, $\mathbf{f}$ and $\mathbf{h}$ have the corresponding dimensions. Note that to ensure a unique solution, the matrix in the previous system, equation (6.21), should have a *full rank*. Recall that the rank of a matrix is the order of the largest square array within that matrix, formed by deleting certain rows and columns, whose determinant does not vanish. In the present case, since $\mathbf{K}$ is regular ($\operatorname{rank}\mathbf{K} = \mathrm{n_{eq}}$), the row vectors of $(\mathbf{G}^T, \mathbf{0})$ must be linearly independent, that is $\operatorname{rank}\mathbf{G}^T = \hat{\mathrm{n}}_{\mathrm{eq}}$. Matrix $\mathbf{G}$ has $\mathrm{n_{eq}}$ rows. Thus, a necessary condition for $\operatorname{rank}\mathbf{G} = \hat{\mathrm{n}}_{\mathrm{eq}}$ is that $\hat{\mathrm{n}}_{\mathrm{eq}} \leq \mathrm{n_{eq}}$. This means that in order for $\mathbf{u}$ and $\mathbf{p}$ to be uniquely determined from (6.21), a necessary, but not sufficient, condition is that

$$\dim \mathcal{Q}^h \leq \dim \boldsymbol{\mathcal{V}}^h.$$

Thus we have found a necessary relation between the dimension of the discrete velocity and pressure spaces.

The sufficient condition linking these spaces is given by the Ladyzhenskaya–Babuška–Brezzi (LBB) compatibility condition, which states that: *The existence of a stable finite element approximate solution* $(\mathbf{u}^h, \mathbf{p}^h)$ *to the steady Stokes problem depends on choosing a pair of spaces* $\mathcal{V}^h$ *and* $\mathcal{Q}^h$, *such that the following* inf-sup condition *holds*:

$$\inf_{q^h \in \mathcal{Q}^h} \sup_{\boldsymbol{w}^h \in \mathcal{V}^h} \frac{\left(q^h, \boldsymbol{\nabla} \cdot \boldsymbol{w}^h\right)}{\|q\|_0 \|\, \boldsymbol{w}^h\|_1} \geq \alpha > 0,$$

where α is independent of the mesh size h. If the LBB compatibility condition is satisfied, then there exists a unique $\boldsymbol{u}^h \in \mathcal{V}^h$ and a $p^h \in \mathcal{Q}^h$ (determined up to an arbitrary constant in the case of purely Dirichlet boundary conditions).

If the velocity–pressure pair satisfies the LBB condition, the discrete gradient operator $\mathbf{G}$ in the partitioned matrix (6.21) is such that $\ker \mathbf{G} = \{\mathbf{0}\}$. Hence, the pressure matrix $(\mathbf{G}^T\mathbf{K}^{-1}\mathbf{G})$ in (6.27) is positive definite and the partitioned matrix (6.21) is non-singular. Existence and uniqueness of the solution are thus guaranteed.

6.5.7 Some popular velocity–pressure couples

In the finite element context, it is by no means easy to prove whether or not a given velocity–pressure pair satisfies the LBB compatibility condition. Although several numerical techniques can help, from the simplest constraint ratio (see for instance Hughes, 2000, Sec. 4.3.7) to the more recent numerical inf-sup testing by Brezzi and Bathe (1990) or Bathe et al. (2000) (see also Brezzi and Fortin, 1991, Sec. II.3.2). Possible combinations of the velocity and pressure approximations leading to stable results are given in Figure 6.1, together with seemingly natural combinations which fail to satisfy the LBB condition.

Among the stable elements listed in Figure 6.1, the triangular and quadrilateral Taylor–Hood elements (Taylor and Hood, 1973) appear to be the most natural ones, in that they only make use of the base element nodes. The velocity and pressure interpolations are both piecewise continuous, the former being quadratic and the latter linear. In addition to being stable, the Taylor–Hood elements exhibit optimal quadratic convergence. The mini element proposed by Arnold, Brezzi and Fortin (1984) and the elements developed by Crouzeix and Raviart (1973) are perhaps less appealing from the viewpoint of computer implementation because they include bubble functions. In the mini element, the piecewise linear velocity field is enriched with a bubble function, which is actually a cubic function in 2D (product of all barycentric coordinates) that vanishes on the boundary. The pressure is continuous and piecewise linear. The mini element exhibits a linear convergence and gives poor pressure approximations in 3D.

The Crouzeix–Raviart triangular element in Figure 6.1 is actually part of a family of stable elements developed by the same authors. It is based on a piecewise linear, but discontinuous, pressure representation. The velocity field is described by a continuous quadratic polynomial enriched with a bubble (as in the mini element product of the barycentric coordinates, i.e. cubic in 2D, quartic in 3D). In spite of its extra cost (certain versions allow more economical implementations), it is considered an accurate and effective element.

A few other simple elements, not listed in Figure 6.1, do satisfy the LBB condition. These include the quadrilateral element with continuous piecewise biquadratic velocity interpolation and discontinuous linear pressure (as in the Crouzeix–Raviart

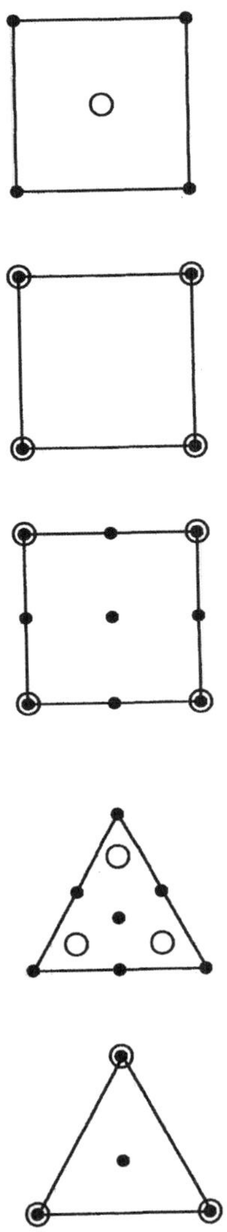

Q1P0 element:
Continuous bilinear velocity,
Discontinuous constant pressure,
Does not satisfy LBB condition.
(same for linear/constant triangle)

Q1Q1 element:
Continuous bilinear velocity,
Continuous bilinear pressure,
Does not satisfy LBB condition.
(same for linear/linear triangle)

Q2Q1 element:
(Taylor–Hood element)
Continuous biquadratic velocity,
Continuous bilinear pressure,
Satisfies LBB condition,
Quadratic convergence.
(same for quadratic/linear triangle)

Crouzeix–Raviart element:
Velocity: continuous quadratic
+ cubic bubble function,
Pressure: discontinuous linear,
Satisfies LBB condition,
Quadratic convergence.

Mini element:
Velocity: continuous linear
+ cubic bubble function,
Pressure: continuous linear,
Satisfies LBB condition,
Linear convergence.

Nodes: • Velocity

○ Pressure

Fig. 6.1 Examples of 2D stable and unstable velocity and pressure interpolations.

triangular element). This element enjoys quadratic convergence properties. Also stable is the triangular element with continuous piecewise quadratic velocity and elementwise constant pressure. This element has a linear convergence rate.

6.5.8 Stabilization of the Stokes problem

In recent years, the efforts of researchers in the area of mixed methods have been directed towards circumventing the LBB condition, thus opening the way to the use of velocity–pressure pairs which are not stable in the standard Galerkin formulation.

The basic idea behind stabilization procedures is to enforce the positive definiteness of matrix $\left(\begin{smallmatrix}\mathbf{K} & \mathbf{G}\\ \mathbf{G}^T & \mathbf{0}\end{smallmatrix}\right)$, see equation (6.21). This can, for instance, be accomplished through a modification of the weak form of the incompressibility condition in order to render non-zero the diagonal term resulting from the incompressibility condition.

It is now standard to stabilize incompressible flow problems using techniques inspired from stabilized formulations, such as SUPG and Galerkin/Least-squares (GLS), introduced in Chapter 2.

Let us illustrate the GLS stabilization proposed for the Stokes problem by Hughes and Franca (1987). The method consists of modifying the variational form of the Stokes problem (6.16), that is equations (6.13), by the addition of the terms emanating from the minimization of the least-squares form

$$L_S(\boldsymbol{v},p) := \big(-\nu\boldsymbol{\nabla}^2\boldsymbol{v} + \boldsymbol{\nabla}p - \boldsymbol{b}, -\nu\boldsymbol{\nabla}^2\boldsymbol{v} + \boldsymbol{\nabla}p - \boldsymbol{b}\big),$$

that is, the square of the residual of the momentum equation, see (6.16a). The stationarity condition of $L_S(\boldsymbol{v},p)$ implies that

$$\frac{dL_S(\boldsymbol{v}+\epsilon\boldsymbol{w}, p+\epsilon q)}{d\epsilon}\bigg|_{\epsilon=0} = 0$$

for all values of $\boldsymbol{w}$ and q, variations of the velocity and of the pressure, respectively. The derivation foreseen in the previous equation implies that

$$\big(-\nu\boldsymbol{\nabla}^2\boldsymbol{w} + \boldsymbol{\nabla}q, -\nu\boldsymbol{\nabla}^2\boldsymbol{v} + \boldsymbol{\nabla}p - \boldsymbol{b}\big) = 0 \qquad \forall(\boldsymbol{w},q) \in \boldsymbol{\mathcal{V}} \times \mathcal{Q},$$

or equivalently,

$$\begin{cases} \big(-\nu\boldsymbol{\nabla}^2\boldsymbol{w}, -\nu\boldsymbol{\nabla}^2\boldsymbol{v} + \boldsymbol{\nabla}p - \boldsymbol{b}\big) = 0 & \forall\boldsymbol{w} \in \boldsymbol{\mathcal{V}}\\ \big(\boldsymbol{\nabla}q, -\nu\boldsymbol{\nabla}^2\boldsymbol{v} + \boldsymbol{\nabla}p - \boldsymbol{b}\big) = 0 & \forall q \in \mathcal{Q}. \end{cases}$$

The stabilization of the Stokes problem is obtained by adding to the Galerkin weak form (6.13) the previous equations emanating from the least-squares form. This entails a modification of both the momentum and the continuity equations. To avoid additional continuity requirements due to the presence of second spatial derivatives, the terms added to the Galerkin weak form act on the element interiors only. These terms depend on the residual of the momentum equation and therefore ensure the consistency of the stabilized formulation. The stabilized discrete problem is then formulated as: find $\boldsymbol{v}^h \in \boldsymbol{\mathcal{S}}^h$ and $p^h \in \mathcal{Q}^h$, for all $(\boldsymbol{w}^h, q^h) \in \boldsymbol{\mathcal{V}}^h \times \mathcal{Q}^h$, such that

$$\begin{cases} a(\boldsymbol{w}^h, \boldsymbol{v}^h) + b(\boldsymbol{w}^h, p^h) + \sum_{e=1}^{n_{el}} \tau_e \big(-\nu \boldsymbol{\nabla}^2 \boldsymbol{w}^h, -\nu \boldsymbol{\nabla}^2 \boldsymbol{v}^h + \boldsymbol{\nabla} p^h - \boldsymbol{b}^h\big)_{\Omega^e} \\ \qquad\qquad = (\boldsymbol{w}^h, \boldsymbol{b}^h) + (\boldsymbol{w}^h, \boldsymbol{t}^h)_{\Gamma_N}, \\ b(\boldsymbol{v}^h, q^h) - \sum_{e=1}^{n_{el}} \tau_e \big(\boldsymbol{\nabla} q^h, -\nu \boldsymbol{\nabla}^2 \boldsymbol{v}^h + \boldsymbol{\nabla} p^h - \boldsymbol{b}^h\big)_{\Omega^e} = 0, \end{cases}$$

where τ_e is the stabilization parameter. Note that the presence in the second equation of the term $(\boldsymbol{\nabla} q^h, \boldsymbol{\nabla} p^h)$ introduces a non-zero diagonal term in the partitioned matrix resulting from the spatial discretization of the above GLS weak form. This produces the desired stabilization of the pressure field.

For linear elements the GLS stabilization does not affect the weak form of the momentum equation because the terms involving the second derivatives of the weighting function $\boldsymbol{w}$ vanish. The GLS weak formulation then reduces to the following variational problem: find $\boldsymbol{v}^h \in \boldsymbol{\mathcal{S}}^h$ and $p^h \in \mathcal{Q}^h$, such that, for all $(\boldsymbol{w}^h, q^h) \in \boldsymbol{\mathcal{V}}^h \times \mathcal{Q}^h$,

$$\begin{cases} a(\boldsymbol{w}^h, \boldsymbol{v}^h) + b(\boldsymbol{w}^h, p^h) = (\boldsymbol{w}^h, \boldsymbol{b}^h) + (\boldsymbol{w}^h, \boldsymbol{t}^h)_{\Gamma_N}, \\ b(\boldsymbol{v}^h, q^h) - \sum_{e=1}^{n_{el}} \tau_e \big(\boldsymbol{\nabla} q^h, \boldsymbol{\nabla} p^h\big)_{\Omega^e} = -\sum_{e=1}^{n_{el}} \tau_e \big(\boldsymbol{\nabla} q^h, \boldsymbol{b}^h\big)_{\Omega^e}. \end{cases} \tag{6.28}$$

Note that the second term in the second equation indicates that a Poisson equation has been generated for the pressure field. An interesting consequence of the GLS stabilization of the Stokes problem is that elements with equal order interpolations, which are unstable in the Galerkin formulation, now become stable. This is, in particular, the case for the quadrilateral element in Figure 6.1 with continuous piecewise bilinear interpolations, as well as for the linear/linear three-node triangle. The stabilization parameter is chosen as

$$\tau_e = \alpha_0 \frac{h_e^2}{4\nu},$$

where h_e is a measure of the element size. Since the Stokes equations do not contain convective terms, the only component of τ_e is the viscous contribution. The choice $\alpha_0 = 1/3$ appears to be optimal for linear elements.

6.5.9 Penalty method

The penalty method may be interpreted as enabling a relaxation of the incompressibility constraint in the sense that the incompressible problem is approximated by means of a slightly compressible formulation.

In the mixed finite element approach pressure is an unknown. However, the penalty method allows its elimination and thus reduces the size of the matrix problem. There is a vast literature on penalty methods for incompressible flow problems. In the finite element framework, initial studies of the method were performed, among others, by Temam (2001), Girault and Raviart (1986), Malkus and Hughes (1978), Zienkiewicz and Godbole (1975), Bercovier (1978) and Bercovier and Engelman (1979). Articles by Hughes, Liu and Brooks (1979) and Heinrich, Marshall and Zienkiewicz (1978) also provide a detailed presentation of the penalty formulation for the Stokes problem, including computer implementation details.

The starting point for the penalty method consists in replacing the incompressibility constraint $\boldsymbol{\nabla}\cdot\boldsymbol{v} = 0$ by

$$\boldsymbol{\nabla}\cdot\boldsymbol{v}^{(\lambda)} = -p^{(\lambda)}/\lambda, \tag{6.29}$$

where λ is a mesh-size- and problem-independent parameter taken of the order of $10^7{\sim}10^8$ in double-precision calculations. The fluid constitutive equation (6.2) is then replaced by the following relation:

$$\sigma_{ij}^{(\lambda)} = -p^{(\lambda)}\delta_{ij} + 2\nu\, v_{(i,j)}^{(\lambda)},$$

where pressure is now defined by (6.29), namely $p^{(\lambda)} = -\lambda\boldsymbol{\nabla}\cdot\boldsymbol{v}^{(\lambda)}$.

The strict incompressibility constraint (6.11b) is thus abandoned. More precisely, it is introduced in the momentum equation as a penalized term emanating from relation (6.29). Consequently, the boundary value problem for Stokes flow is formulated as:

$$\begin{cases} -\boldsymbol{\nabla}\cdot\boldsymbol{\sigma}^{(\lambda)} = \boldsymbol{b} & \text{in } \Omega, \\ \boldsymbol{v}^{(\lambda)} = \boldsymbol{v}_D & \text{on } \Gamma_D, \\ \boldsymbol{n}\cdot\boldsymbol{\sigma}^{(\lambda)} = \boldsymbol{t} & \text{on } \Gamma_N. \end{cases}$$

The weak form for the penalty method is easily obtained in the primitive variable, $\boldsymbol{v}$,

$$a(\boldsymbol{w},\boldsymbol{v}) + \lambda\,(\boldsymbol{\nabla}\cdot\boldsymbol{w},\boldsymbol{\nabla}\cdot\boldsymbol{v}) = (\boldsymbol{w},\boldsymbol{b}) + (\boldsymbol{w},\boldsymbol{t})_{\Gamma_N} \qquad \forall \boldsymbol{w}\in\boldsymbol{\mathcal{V}}.$$

Note the presence of a new viscous-type term associated with the penalty parameter. The incompressibility equation is not needed because the pressure in the momentum equation is replaced using expression (6.29). As in Section 6.5.3, we use the representation $\boldsymbol{v}^h = \boldsymbol{u}^h + \boldsymbol{v}_D^h$ for the finite dimensional approximation of the velocity. The discrete formulation is then given as: find $\boldsymbol{u}^h \in \boldsymbol{\mathcal{V}}^h$, for all $\boldsymbol{w}^h \in \boldsymbol{\mathcal{V}}^h$, such that

$$\begin{aligned} a(\boldsymbol{w}^h,\boldsymbol{u}^h) + \lambda\,(\boldsymbol{\nabla}\cdot\boldsymbol{w}^h,\boldsymbol{\nabla}\cdot\boldsymbol{u}^h) = (\boldsymbol{w}^h,\boldsymbol{b}^h) &+ (\boldsymbol{w}^h,\boldsymbol{t}^h)_{\Gamma_N} \\ &- a(\boldsymbol{w}^h,\boldsymbol{v}_D^h) - \lambda\,(\boldsymbol{\nabla}\cdot\boldsymbol{w}^h,\boldsymbol{\nabla}\cdot\boldsymbol{v}_D^h). \end{aligned}$$

After spatial discretization, this equation produces the following set of nodal equations for the components of the auxiliary velocity: for each $A\in\eta\setminus\eta_{Di}$ and $1\leq i\leq \mathrm{n_{sd}}$

$$\begin{aligned} &\sum_{j=1}^{\mathrm{n_{sd}}}\Big\{\sum_{B\in\eta\setminus\eta_{Dj}} \big[a(N_A\,\boldsymbol{e}_i, N_B\,\boldsymbol{e}_j) + \lambda(\boldsymbol{\nabla}\cdot(N_A\boldsymbol{e}_i),\boldsymbol{\nabla}\cdot(N_B\boldsymbol{e}_j))\big]\,\mathrm{u}_{jB}\Big\} \\ &\quad = (N_A\,\boldsymbol{e}_i,\boldsymbol{b}^h) + (N_A\,\boldsymbol{e}_i,\boldsymbol{t}^h)_{\Gamma_N} \\ &\qquad - \sum_{j=1}^{\mathrm{n_{sd}}}\Big\{\sum_{B\in\eta_{Dj}} \big[a(N_A\,\boldsymbol{e}_i, N_B\,\boldsymbol{e}_j) + \lambda\,(\boldsymbol{\nabla}\cdot(N_A\boldsymbol{e}_i),\boldsymbol{\nabla}\cdot(N_B\boldsymbol{e}_j))\big]\,v_{Dj}\Big\}. \end{aligned}$$

From this set of algebraic equations, the following matrix system is obtained:

$$\left(\mathbf{K} + \mathbf{K}^{\lambda}\right) \mathbf{u}^{(\lambda)} = \mathbf{f}, \tag{6.30}$$

which governs the penalty formulation of the Stokes problem. $\mathbf{K}$ is the viscosity matrix arising from the term $a(\boldsymbol{w}^h, \boldsymbol{u}^h)$. It is identical to the viscosity matrix obtained in Section 6.5.4 for the mixed method. $\mathbf{K}^{\lambda}$ is the so-called *penalty matrix*, which has the same structure as $\mathbf{K}$ and is defined as

$$\mathbf{K}^{\lambda} = \mathop{\mathsf{A}}^{e} [\mathbf{K}^{\lambda}]^e, \qquad [K^{\lambda}]^e_{rs} = \lambda \left(\boldsymbol{\nabla} \cdot (N_a \boldsymbol{e}_i), \boldsymbol{\nabla} \cdot (N_b \boldsymbol{e}_j)\right)_{\Omega^e},$$

recall the definition of the indices given by (6.22). Thus, it is identical to expression (6.24) but with λ replacing ν. Finally, the nodal vector $\mathbf{f}$ is defined in (6.25) (where K^e_{rs} now accounts for the sum of the viscosity and penalty matrices).

Observe that the global matrices $\mathbf{K}$ and $\mathbf{K}^{\lambda}$ are proportional to ν and λ, respectively. In order to impose incompressibility, parameter λ must be selected very large. However, when λ is very large, the penalty matrix plays a dominant role in system (6.30). Then, if matrix $\mathbf{K}^{\lambda}$ is regular, only the trivial solution $\mathbf{u}^{(\lambda)} = \mathbf{0}$ of the homogeneous form of (6.30) is possible. It follows that matrix $\mathbf{K}^{\lambda}$ must be singular to obtain a meaningful solution, that is

$$\mathbf{K}^{\lambda}\, \mathbf{u}^{(\lambda)} = \mathbf{0} \qquad \text{and} \qquad \mathbf{u}^{(\lambda)} \neq \mathbf{0}.$$

Standard numerical integration rules used to evaluate finite element matrices are such that, with conforming elements, they practically always result in a regular $\mathbf{K}^{\lambda}$ matrix. To render the penalty matrix singular, one needs to lower the order of the quadrature used to evaluate the element integrals involving the λ term. Note that the rank of system (6.30) is conserved using the required (full) quadrature for the viscous diffusion term $\mathbf{K}^{\nu}$. Figure 6.2, reproduced from the book by Hughes (2000), gives the Gaussian integration rules to be used for the viscous and penalty terms with some popular elements in 2D and 3D.

The convergence of the solution of the Stokes problem obtained by the penalty method has been proved by Temam (2001). The main advantage of the penalty formulation is the uncoupling of the velocity and pressure solutions and the elimination of the incompressibility condition from the variational formulation. The reduction in the number of variables renders the penalty method quite attractive. However, an improper choice of the value of parameter λ might cause numerical problems. If λ is too small compressibility and pressure errors will occur. An excessively large value may result in numerical ill conditioning. For Stokes flow, Hughes et al. (1979) suggest selecting λ according to the relation $\lambda = c\mu$, where μ is the fluid dynamic viscosity and c is a constant of the order of 10^7 for double-precision calculations.

Remark 6.13 (The need for under-integration of the penalty term). As an illustration, consider the case of a local approximation of the velocity $\boldsymbol{u}^h$ based upon a nine-node biquadratic interpolation in the plane (x_1, x_2). The components (u_1^h, u_2^h) of the velocity are polynomials of the form

$$u_j^h = a_{1,j} + a_{2,j}\,x_1 + a_{3,j}\,x_2 + a_{4,j}\,x_1x_2 + a_{5,j}\,x_1^2 + a_{6,j}\,x_2^2 + a_{7,j}\,x_1^2x_2 + a_{8,j}\,x_1x_2^2 + a_{9,j}\,x_1^2x_2^2,$$

where $a_{i,j}, i = 1, \ldots, 9$, are the nine coefficients of the biquadratic polynomial for each component $(j = 1, 2)$ of the local velocity within the element. When λ is very large the flow is incompressible and the divergence-free condition $\partial u_1^h/\partial x_1 + \partial u_2^h/\partial x_2 = 0$ becomes

$$a_{2,1} + a_{4,1}\,x_2 + 2a_{5,1}\,x_1 + 2a_{7,1}\,x_1x_2 + a_{8,1}\,x_2^2 + 2a_{9,1}\,x_1x_2^2 + a_{3,2} + a_{4,2}\,x_1 + 2a_{6,2}\,x_2 + a_{7,2}\,x_1^2 + 2a_{8,2}\,x_1x_2 + 2a_{9,2}\,x_1^2x_2 = 0.$$

If the penalty term is integrated exactly the above relationship must be satisfied at every point (x_1, x_2) of the element and a total of *eight constraints* are thus imposed for each element:

$$\begin{array}{llll} a_{2,1} + a_{3,2} = 0, & a_{4,1} + 2a_{6,2} = 0, & a_{8,1} = 0, & a_{7,2} = 0, \\ 2a_{5,1} + a_{4,2} = 0, & 2a_{7,1} + 2a_{8,2} = 0, & a_{9,1} = 0, & a_{9,2} = 0. \end{array}$$

This coincides with the number of velocity degrees of freedom per element in an infinite plane mesh,

$$\lim_{n\to\infty} \frac{2\,(2n+1)^2}{n^2} = 8.$$

Thus, in the case of full integration of the penalty term, the constraints are seen to consume a great number of degrees of freedom and too few are left to properly simulate the flow problem at hand. To cure this situation a 2×2 Gaussian integration rule must be used to evaluate the penalty term, while the normal 3×3 rule for the biquadratic element should be used for the viscous diffusion term. In this way, only four constraints are imposed per element. Incompressibility is therefore satisfied in the mean, but now four degrees of freedom per element are available to satisfy the equations of motion.

Success of the penalty method crucially depends on an appropriate reduction of the number of constraints introduced by the penalty term. The use of a reduced integration rule for the penalty term leads to a proper balance between the number of degrees available to satisfy the momentum equations and those consumed to approximate the incompressibility of the flow.

Remark 6.14 (Equivalence of penalty and mixed methods). Malkus and Hughes (1978) (see also Hughes, 2000) have established this important result. They demonstrate the *equivalence*, for incompressible problems, between the reduced integration of the penalty term and a mixed finite element method if the pressure nodes coincide with the integration points of the reduced rule. The equivalence between mixed elements and penalty-type elements with selective integration is illustrated in Figure 6.2.

	Mixed elements		Equivalent elements with selective integration	
	Velocity interpolation	Pressure interpolation	Standard integration (Term $\mathbf{K}$)	Reduced integration (Term $\mathbf{K}^{\lambda}$)
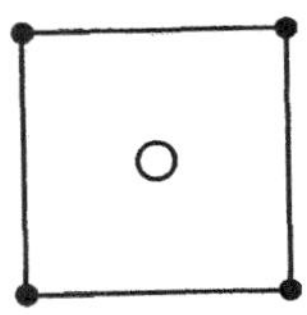	Bilinear	Constant	2×2	1 point
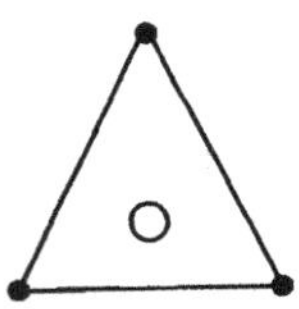	Linear	Constant	1 point	1 point
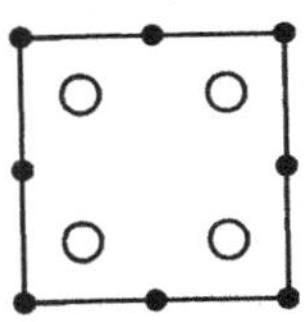	Serendipity	Bilinear	3×3	2×2
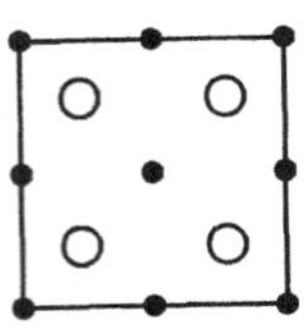	Biquadratic	Bilinear	3×3	2×2
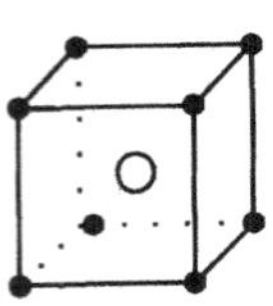	Trilinear	Constant	2×2×2	1 point

Fig. 6.2 Equivalence of mixed and reduced/selective integration elements.

6.6 STEADY NAVIER–STOKES PROBLEM

We shall now consider the Navier–Stokes equations for steady flow. From the definitions of Section 6.2.3 and, in particular, equation (6.5) the strong form of the boundary value problem is stated as follows: find the velocity field $\boldsymbol{v}$ and the pressure field p, such that

$$
\begin{aligned}
-\nu\boldsymbol{\nabla}^2\boldsymbol{v} + (\boldsymbol{v}\cdot\boldsymbol{\nabla})\boldsymbol{v} + \boldsymbol{\nabla}p &= \boldsymbol{b} && \text{in } \Omega, && (6.31\text{a})\\
\boldsymbol{\nabla}\cdot\boldsymbol{v} &= 0 && \text{in } \Omega, && (6.31\text{b})\\
\boldsymbol{v} &= \boldsymbol{v}_D && \text{on } \Gamma_D, && (6.31\text{c})\\
-p\boldsymbol{n} + \nu(\boldsymbol{n}\cdot\boldsymbol{\nabla})\boldsymbol{v} &= \boldsymbol{t} && \text{on } \Gamma_N. && (6.31\text{d})
\end{aligned}
$$

Note that the Cauchy stress is again assumed to be given by Stokes' law.

6.6.1 Weak form and Galerkin formulation

The Navier–Stokes problem differs from the Stokes problem because of the presence of the nonlinear convective term $(\boldsymbol{v}\cdot\boldsymbol{\nabla})\boldsymbol{v}$. Considering the same spaces $\boldsymbol{\mathcal{S}}$, $\boldsymbol{\mathcal{V}}$ and $\mathcal{Q}$ defined by (6.9), the weak formulation of problem (6.31) becomes: find $\boldsymbol{v}\in\boldsymbol{\mathcal{S}}$ and $p\in\mathcal{Q}$, such that

$$
\begin{cases}
a(\boldsymbol{w},\boldsymbol{v}) + c(\boldsymbol{v};\boldsymbol{w},\boldsymbol{v}) + b(\boldsymbol{w},p) = (\boldsymbol{w},\boldsymbol{b}) + (\boldsymbol{w},\boldsymbol{t})_{\Gamma_N} & \forall\boldsymbol{w}\in\boldsymbol{\mathcal{V}},\\
b(\boldsymbol{v},q) = 0 & \forall q\in\mathcal{Q},
\end{cases}
$$

where, apart from the bilinear form used in the weak form (6.16) of the Stokes problem, we have introduced the trilinear form, see Section 1.5.3,

$$
c(\boldsymbol{a};\boldsymbol{w},\boldsymbol{v}) = (\boldsymbol{w},(\boldsymbol{a}\cdot\boldsymbol{\nabla})\boldsymbol{v}) = \int_\Omega \boldsymbol{w}\cdot(\boldsymbol{a}\cdot\boldsymbol{\nabla})\boldsymbol{v}\,d\Omega,
$$

which, in this case, is associated with the nonlinear convective term in the momentum equation. Introducing the finite dimensional subspaces $\boldsymbol{\mathcal{S}}^h$, $\boldsymbol{\mathcal{V}}^h$ and $\mathcal{Q}^h$, and proceeding as for Stokes flow, we obtain the Galerkin counterpart of the previous weak formulation. Define $\boldsymbol{v}_D^h\in\boldsymbol{\mathcal{S}}^h$ such that $\boldsymbol{v}^h = \boldsymbol{u}^h + \boldsymbol{v}_D^h$, then find the auxiliary velocity field $\boldsymbol{u}^h\in\boldsymbol{\mathcal{V}}^h$ and the pressure $p^h\in\mathcal{Q}^h$, such that, for all $(\boldsymbol{w}^h,q^h)\in\boldsymbol{\mathcal{V}}^h\times\mathcal{Q}^h$,

$$
\begin{cases}
a(\boldsymbol{w}^h,\boldsymbol{u}^h) + c(\boldsymbol{v}^h;\boldsymbol{w}^h,\boldsymbol{u}^h) + b(\boldsymbol{w}^h,p^h) = (\boldsymbol{w}^h,\boldsymbol{b}^h) + (\boldsymbol{w}^h,\boldsymbol{t}^h)_{\Gamma_N}\\
\qquad\qquad\qquad - a(\boldsymbol{w}^h,\boldsymbol{v}_D^h) - c(\boldsymbol{v}^h;\boldsymbol{w}^h,\boldsymbol{v}_D^h)\\
b(\boldsymbol{u}^h,q^h) = -b(\boldsymbol{v}_D^h,q^h).
\end{cases}
$$

Note that now, the r.h.s. term depends on the unknown because $\boldsymbol{v}^h = \boldsymbol{u}^h + \boldsymbol{v}_D^h$.

6.6.2 Matrix problem

We then express the finite dimensional approximations in terms of the velocity and pressure shape functions defined in (6.19). Proceeding as in Section 6.5.4, the matrix system governing the discrete Navier–Stokes problem is obtained in the following partitioned form:

$$\begin{pmatrix} \mathbf{K}+\mathbf{C}(\boldsymbol{v}) & \mathbf{G} \\ \mathbf{G}^T & \mathbf{0} \end{pmatrix} \begin{pmatrix} \mathbf{u} \\ \mathbf{p} \end{pmatrix} = \begin{pmatrix} \mathbf{f}(\boldsymbol{v}) \\ \mathbf{h} \end{pmatrix}, \tag{6.32}$$

where, as in the Stokes problem, matrices $\mathbf{K}$ and $\mathbf{G}$ are, respectively, the viscosity matrix and the discrete gradient operator. $\mathbf{C} = \mathbf{C}(\boldsymbol{v})$ is the convection matrix defined by

$$\mathbf{C} = \mathop{\mathbf{A}}^{e} \mathbf{C}^e, \qquad \text{with} \quad C^e_{rs} = \boldsymbol{e}_i^T \int_{\Omega^e} N_a\, \boldsymbol{v}^h \cdot \boldsymbol{\nabla} N_b \, d\Omega \ \boldsymbol{e}_j. \tag{6.33}$$

Vector $\mathbf{f}$ is the same as for the Stokes problem, see (6.25), plus an added term: the product of the previous matrix by the Dirichlet velocities. Recall the definition of the indices given by (6.22).

Here again, the saddle-point nature of the problem, with pressure acting as a Lagrangian multiplier of the incompressibility constraint, yields a coupling between velocity and pressure. System (6.32) is nonlinear and non-symmetric (due to the convective terms), an appropriate iterative solution technique must be used. An in-depth discussion of solution algorithms for the nonlinear Navier–Stokes equations may be found in the textbooks mentioned in the introduction to this chapter.

There are two potential sources of numerical instability in the Galerkin finite element solution of steady Navier–Stokes problems. The first is due to the treatment of the convective term and manifests itself in high Reynolds number flows when unresolved internal or boundary layers are present in the solution. The second source of potential instability, already encountered in Stokes flow, is an inappropriate combination of interpolation functions for velocity and pressure. Fortunately, stabilization procedures for the Navier–Stokes equations capable of curing both types of numerical instability are available. They are presented in Section 6.7.2.

6.7 UNSTEADY NAVIER–STOKES EQUATIONS

Let us now consider the case of unsteady viscous incompressible flows. Here, emphasis will be placed on stabilized finite element formulations and on the advantages of a fractional-step projection method for time integration of the Navier–Stokes equations. Again, using the definitions of Section 6.2.3, the governing equations and associated initial/boundary conditions are:

$$\boldsymbol{v}_t - \nu\boldsymbol{\nabla}^2\boldsymbol{v} + (\boldsymbol{v}\cdot\boldsymbol{\nabla})\boldsymbol{v} + \boldsymbol{\nabla}p = \boldsymbol{b} \qquad \text{in } \Omega\times]0,T[, \tag{6.34a}$$

$$\boldsymbol{\nabla}\cdot\boldsymbol{v} = 0 \qquad \text{in } \Omega\times]0,T[, \tag{6.34b}$$

$$\boldsymbol{v} = \boldsymbol{v}_D \qquad \text{on } \Gamma_D\times]0,T[, \tag{6.34c}$$

$$-p\boldsymbol{n} + \nu(\boldsymbol{n}\cdot\boldsymbol{\nabla})\boldsymbol{v} = \boldsymbol{t} \qquad \text{on } \Gamma_N\times]0,T[, \tag{6.34d}$$

$$\boldsymbol{v}(\boldsymbol{x},0) = \boldsymbol{v}_0(\boldsymbol{x}) \qquad \text{in } \Omega. \tag{6.34e}$$

Note that the initial velocity field is assumed solenoidal: $\boldsymbol{\nabla}\cdot\boldsymbol{v}_0 = 0$.

6.7.1 Weak formulation and spatial discretization

The weak formulation is obtained, as usual, by projection of the equations (6.34) onto a space of weighting functions $\boldsymbol{w} \in \boldsymbol{\mathcal{V}}$ for the momentum equation and $q \in \mathcal{Q}$ for the incompressibility condition. The end result is the following variational problem: given $\boldsymbol{b}, \boldsymbol{v}_D, \boldsymbol{t}$ and $\boldsymbol{v}_0$, find $\boldsymbol{v}(\boldsymbol{x}, t) \in \boldsymbol{\mathcal{S}} \times]0, T[$ and $p(\boldsymbol{x}, t) \in \mathcal{Q} \times]0, T[$, such that, for all $(\boldsymbol{w}, q) \in \boldsymbol{\mathcal{V}} \times \mathcal{Q}$,

$$\begin{cases} (\boldsymbol{w}, \boldsymbol{v}_t) + a(\boldsymbol{w}, \boldsymbol{v}) + c(\boldsymbol{v}; \boldsymbol{w}, \boldsymbol{v}) + b(\boldsymbol{w}, p) = (\boldsymbol{w}, \boldsymbol{b}) + (\boldsymbol{w}, \boldsymbol{t})_{\Gamma_N}, \\ b(\boldsymbol{v}, q) = 0 \end{cases}$$

with $\boldsymbol{v}(0) = \boldsymbol{v}_0$.

The Galerkin spatial discretization of this time-dependent problem proceeds as previously. For each $t \in]0, T[$, we define $\boldsymbol{v}_D^h(t) \in \boldsymbol{\mathcal{S}}^h$ such that $\boldsymbol{v}^h(t) = \boldsymbol{u}^h(t) + \boldsymbol{v}_D^h(t)$. Then, we seek the auxiliary velocity field $\boldsymbol{u}^h(\cdot, t) \in \boldsymbol{\mathcal{V}}^h$ and pressure $p^h(\cdot, t) \in \mathcal{Q}^h$, such that, for all $(\boldsymbol{w}^h, q^h) \in \boldsymbol{\mathcal{V}}^h \times \mathcal{Q}^h$,

$$\begin{cases} (\boldsymbol{w}^h, \boldsymbol{u}_t^h) + a(\boldsymbol{w}^h, \boldsymbol{u}^h) + c(\boldsymbol{v}^h; \boldsymbol{w}^h, \boldsymbol{u}^h) + b(\boldsymbol{w}^h, p^h) \\ \qquad\qquad = (\boldsymbol{w}^h, \boldsymbol{b}^h) + (\boldsymbol{w}^h, \boldsymbol{t}^h)_{\Gamma_N} - a(\boldsymbol{w}^h, \boldsymbol{v}_D^h) - c(\boldsymbol{v}^h; \boldsymbol{w}^h, \boldsymbol{v}_D^h) \\ b(\boldsymbol{u}^h, q^h) = -b(\boldsymbol{v}_D^h, q^h) \end{cases}$$

with $\boldsymbol{v}^h(0) = \boldsymbol{v}_0^h$.

Finally, the finite element discretization of this weak form yields the system of semi-discrete equations for $t \in]0, T[$

$$\begin{cases} \mathbf{M}\,\dot{\mathbf{u}}(t) + \left[\mathbf{K} + \mathbf{C}(\boldsymbol{v}(t))\right]\mathbf{u}(t) + \mathbf{G}\,\mathbf{p}(t) = \mathbf{f}(t, \boldsymbol{v}(t)) \\ \mathbf{G}^T\mathbf{u}(t) = \mathbf{h}(t) \\ \mathbf{u}(0) = \mathbf{v}_0 - \mathbf{v}_D(0) \end{cases} \tag{6.35}$$

where $\mathbf{M}$ is the standard finite element mass matrix.

To trace the transient response, this system of semi-discrete equations can be advanced in time by suitable finite difference schemes such as the θ family methods introduced in Section 3.4.2.3. Note that a fully implicit method requires the solution of a nonlinear algebraic system at each time step. Semi-implicit methods in which the convection matrix $\mathbf{C}(\boldsymbol{v}(t))$ and $\mathbf{f}(t, \boldsymbol{v}(t))$ are treated explicitly are thus generally preferred. In Section 6.7.3 we shall consider the time discretization of the unsteady Navier–Stokes equations by means of a fractional-step procedure.

Remark 6.15. An in-depth analysis of the semi-discrete system induced by the unsteady Navier–Stokes equations can be found in the series of papers by Heywood and Rannacher (1982; 1986; 1988; 1990). In order to analyze stability and obtain an energy estimate, these authors use the nowadays standard skew-symmetric form of the convection term, see for instance Temam (2001). That is, the trilinear form $c(\boldsymbol{v}; \boldsymbol{w}, \boldsymbol{u})$ is replaced by

$$\widehat{c}(\boldsymbol{v}; \boldsymbol{w}, \boldsymbol{u}) = \frac{1}{2}\Big(c(\boldsymbol{v}; \boldsymbol{w}, \boldsymbol{u}) - c(\boldsymbol{v}; \boldsymbol{u}, \boldsymbol{w})\Big),$$

which is skew-symmetric, i.e. $\widehat{c}(\boldsymbol{v};\boldsymbol{w},\boldsymbol{w}) = 0$. This modified trilinear term is consistent with the original unsteady incompressible Navier–Stokes problem. It is easy to verify that the replacement of the original trilinear term, $c(\cdot;\cdot,\cdot)$, by the new skew-symmetric form is equivalent to the replacement of the original convective term, $(\boldsymbol{v}\cdot\boldsymbol{\nabla})\boldsymbol{v}$, by $(\boldsymbol{v}\cdot\boldsymbol{\nabla})\boldsymbol{v}+\frac{1}{2}(\boldsymbol{\nabla}\cdot\boldsymbol{v})\boldsymbol{v}$. Note that for a divergence-free velocity field, such as the incompressible solution of (6.34), this modification is legitimate.

Finally, it is important to remark that the skew-symmetric form of the convection term is also exploited in the computational schemes in order to ensure unconditional time stability, see Remark 6.18.

6.7.2 Stabilized finite element formulation

Recall that the Galerkin finite element method leads to central approximations of the convective terms and is thus not optimal when convection dominate diffusion (the viscosity effects), that is for high Reynolds number flows. In such cases, use should be made of a stabilized finite element formulation to obtain reliable numerical solutions. In addition, as was the case for Stokes flow, the stability of the Galerkin method applied to the incompressible Navier–Stokes equations depends on satisfying of the LBB condition. It is nevertheless possible to circumvent this condition, as already seen for the Stokes problem, by making use of a stabilization technique. There are therefore two major reasons for stabilizing the incompressible Navier–Stokes equations.

The extension of the stabilized formulation of the Stokes problem discussed in Section 6.5.8 to the incompressible Navier–Stokes equations (6.34) has been studied in a series of papers by Johnson and Saranen (1986), Hughes, Franca and Hulbert (1989), Hansbo and Szepessy (1990), Franca et al. (1992), Franca and Frey (1992), Franca and Hughes (1993), Tezduyar, Mittal, Ray and Shih (1992) and Tezduyar and Osawa (2000), among others. See also the review paper by Tezduyar (1992) on finite element stabilization methods for incompressible flow computations.

The stabilized finite element formulation of the Navier–Stokes problem (6.34) proposed by Tezduyar and Osawa (2000) is given as follows: find $\boldsymbol{v}^h \in \boldsymbol{\mathcal{S}}^h\times]0,T[$ and $p^h \in \mathcal{Q}^h\times]0,T[$, for all $(\boldsymbol{w}^h, q^h) \in \boldsymbol{\mathcal{V}}^h \times \mathcal{Q}^h$, such that

$$\left\{\begin{aligned}
&(\boldsymbol{w}^h,\boldsymbol{v}_t^h)+a(\boldsymbol{w}^h,\boldsymbol{v}^h)+c(\boldsymbol{v}^h;\boldsymbol{w}^h,\boldsymbol{v}^h)+b(\boldsymbol{w}^h,p^h)-(\boldsymbol{w}^h,\boldsymbol{b}^h)-(\boldsymbol{w}^h,\boldsymbol{t}^h)_{\Gamma_N}\\
&\qquad+\sum_{e=1}^{n_{\mathrm{el}}}\tau_{\mathrm{SUPG}}\big((\boldsymbol{v}^h\cdot\boldsymbol{\nabla})\boldsymbol{w}^h,\mathcal{R}(\boldsymbol{v}^h)\big)_{\Omega^e}+\tau_{\mathrm{LSIC}}\big(\boldsymbol{\nabla}\cdot\boldsymbol{w}^h,\boldsymbol{\nabla}\cdot\boldsymbol{v}^h\big)_{\Omega^e}=0,\\
&b(\boldsymbol{v}^h,q^h)+\sum_{e=1}^{n_{\mathrm{el}}}\tau_{\mathrm{PSPG}}\big(\boldsymbol{\nabla}q^h,\mathcal{R}(\boldsymbol{v}^h)\big)_{\Omega^e}=0,
\end{aligned}\right.$$

where

$$\mathcal{R}(\boldsymbol{v}^h)=\boldsymbol{v}_t^h+(\boldsymbol{v}^h\cdot\boldsymbol{\nabla})\boldsymbol{v}^h-\nu\boldsymbol{\nabla}^2\boldsymbol{v}^h+\boldsymbol{\nabla}p^h-\boldsymbol{b}^h$$

is the residual of the momentum equation.

In the variational form of the momentum equation, the first line represents the standard Galerkin formulation. Similarly, the first term in the weak form of the incompressibility constraint is the usual Galerkin term.

The terms involving the element-level integrals are the added stabilization terms. These terms depend on three stabilization parameters. The terms involving parameter τ_{SUPG} stabilize the Galerkin formulation in the presence of a dominating convective term in the momentum equation, see Sections 2.4, 5.4.5 or 5.4.6. Note that other stabilization techniques such as GLS, SGS or LS can also be employed.

Parameter τ_{LSIC} is in fact an artificial diffusion. Its dimension is length square over time (same as the kinematic viscosity ν), and a typical choice is $\|\boldsymbol{v}^h\| h_e/2$. Tezduyar and Osawa (2000) propose to use this term as a least-squares stabilization on the incompressibility constraint. It provides additional stability for flows at large Reynolds numbers.

The terms associated with parameter τ_{PSPG} (pressure-stabilizing/Petrov–Galerkin) allow the use of mixed elements with equal-order interpolations for the velocity and pressure, see Section 6.5.8, in particular equations (6.28). Note that all stabilization terms are weighted residuals, therefore ensuring the consistency of the formulation.

Tezduyar and Osawa (2000) suggest evaluating the stabilization parameters using element-level matrices and vectors which automatically account for the local length scales, advection field and flow Reynolds number Re. The reader interested in the actual construction of element-based stability parameters should consult the above-mentioned article. Note that alternative definitions of τ_{SUPG} have been proposed by Codina (2000) and Shakib et al. (1991), see Sections 2.4.3, 5.4.5 and 5.4.6.

6.7.3 Time discretization by fractional-step methods

A popular method for the time discretization of the unsteady Navier–Stokes equations is the fractional-step method in which, as already mentioned in Section 5.3.2, the time advancement is decomposed into a sequence of two or more steps. Fractional-step methods for the incompressible Navier–Stokes equations were originated independently by Chorin (1968; 1969) and Temam (1969; 2001).

As we shall see, the fractional-step approach to time integration allows us to alleviate the numerical difficulties related to the saddle-point problem which arises from the variational formulation of the Navier–Stokes equations. The basic idea is to split the numerical treatment of the various operators in the equations, thus decomposing the initially difficult problem into relatively easier substeps. There are several ways to perform such splitting and therefore a variety of fractional-step methods for the unsteady Navier–Stokes equations do exist. The books by Quartapelle (1993), Quarteroni and Valli (1994), Gresho and Sani (2000), and references therein, should be consulted for a detailed exposition of fractional-step methods.

Fractional-step methods described here perform time discretization before the spacial discretization. When this approach is adopted, a controversy arises on which boundary conditions must be imposed at each step, because the intermediate semi-discrete problems must be well-posed. Another important feature in fractional-step methods is the overall order of accuracy with respect to time discretization. Most methods are first-order accurate, but some second-order accurate methods have also been developed. Both types of methods will be considered in this section.

While the first step in the time discretization which treats the convective and viscous terms in the Navier–Stokes equations might be treated by an explicit algorithm, the final step describing the pressure/incompressibility phase must necessarily be treated by an implicit time integration scheme. Here, an approach is adopted which gives the choice of either an explicit, a semi-implicit, or a fully implicit time-stepping scheme for the first step. Note, however, that adopting a fully implicit scheme for the first step considerably increases the computational burden due to the nonlinearity of the convective terms in the Navier–Stokes equations. Semi-implicit methods are therefore generally preferred in large-scale computations.

6.7.3.1 *Chorin–Temam projection method* The principle of the projection method is to compute the velocity and pressure fields separately through the computation of an intermediate velocity, which is then projected onto the subspace of the solenoidal vector functions. A detailed account of the method can be found in the textbook by Temam (2001). Basic to the derivation of projection methods is a theorem of orthogonal decomposition due to Ladyzhenskaya (1969), which is based on the Helmholtz decomposition principle. The theorem states that any vector field $\boldsymbol{w}$ in Ω admits the unique orthogonal decomposition

$$\boldsymbol{w} = \boldsymbol{v} + \boldsymbol{\nabla}\phi \tag{6.36}$$

into a solenoidal field, $\boldsymbol{v}$, with zero normal component on the domain boundary (i.e., $\boldsymbol{\nabla} \cdot \boldsymbol{v} = 0$ and $\boldsymbol{n} \cdot \boldsymbol{v} = 0$ on Γ) and the gradient of some scalar function ϕ, see also the Remark 6.8. In the present context, an intermediate velocity field, $\boldsymbol{v}_{\text{int}}^{n+1}$, is decomposed into the sum of a solenoidal velocity field, $\boldsymbol{v}^{n+1}$, and the gradient of a scalar function proportional to the unknown pressure, namely, $\boldsymbol{\nabla} p^{n+1}$.

For simplicity, we shall consider a purely Dirichlet problem: that is, we prescribe the condition $\boldsymbol{v} = \boldsymbol{v}_D$ on the boundary Γ of the computational domain Ω. The Chorin–Temam projection method includes two basic steps as follows.

The *first step* includes the viscous and convective terms in the Navier–Stokes equations (6.34) and, given the previous time-step velocity field $\boldsymbol{v}^n$, consists of finding an intermediate velocity field, $\boldsymbol{v}_{\text{int}}^{n+1}$, such that

$$\begin{cases} \dfrac{\boldsymbol{v}_{\text{int}}^{n+1} - \boldsymbol{v}^n}{\Delta t} + (\boldsymbol{v}^* \cdot \boldsymbol{\nabla})\,\boldsymbol{v}^{**} - \nu \boldsymbol{\nabla}^2 \boldsymbol{v}^{**} = \boldsymbol{b}^{n+1} & \text{in } \Omega, \\ \boldsymbol{v}_{\text{int}}^{n+1} = \boldsymbol{v}_D^{n+1} & \text{on } \Gamma, \end{cases} \tag{6.37}$$

where the velocities $\boldsymbol{v}^*$ and $\boldsymbol{v}^{**}$ must be chosen suitably for the treatment of the nonlinear convective term, possible options are

$$\begin{aligned} &\boldsymbol{v}^* = \boldsymbol{v}^{**} = \boldsymbol{v}^n && \text{for the explicit Euler method,} \\ &\boldsymbol{v}^* = \boldsymbol{v}^n \quad \text{and} \quad \boldsymbol{v}^{**} = \boldsymbol{v}_{\text{int}}^{n+1} && \text{for a semi-implicit method,} \\ &\boldsymbol{v}^* = \boldsymbol{v}^{**} = \boldsymbol{v}_{\text{int}}^{n+1} && \text{for the implicit Euler method.} \end{aligned}$$

Note that the complete Dirichlet boundary conditions are imposed in this first step. This is due to the fact that this step includes the viscous term.

To construct a finite element version of the fractional-step method, a weak form of the first-step equations (6.37) is necessary. The problem requires to find the intermediate velocity $\boldsymbol{v}_{\text{int}}^{n+1} \in \boldsymbol{\mathcal{S}}_{\text{int}}$, such that for all $\boldsymbol{w} \in \boldsymbol{\mathcal{V}}_{\text{int}}$

$$\left(\boldsymbol{w}, \frac{\boldsymbol{v}_{\text{int}}^{n+1} - \boldsymbol{v}^n}{\Delta t}\right) + c(\boldsymbol{v}^*; \boldsymbol{w}, \boldsymbol{v}^{**}) + a(\boldsymbol{w}, \boldsymbol{v}^{**}) = (\boldsymbol{w}, \boldsymbol{b}^{n+1}), \tag{6.38}$$

where the trilinear and bilinear forms have already been defined, see for instance Section 6.6.1 and equation 6.17. Note that the functional spaces $\boldsymbol{\mathcal{S}}_{\text{int}}$ and $\boldsymbol{\mathcal{V}}_{\text{int}}$ are such that the complete Dirichlet boundary conditions are verified, namely $\boldsymbol{v}_{\text{int}}^{n+1} = \boldsymbol{v}_D^{n+1}$ on Γ.

For the semi-implicit and fully implicit cases, the algebraic system resulting from the finite element discretization is

$$\mathbf{M}_1 \left(\frac{\mathbf{v}_{\text{int}}^{n+1} - \mathbf{v}^n}{\Delta t}\right) + \big(\mathbf{C}(\boldsymbol{v}^*) + \mathbf{K}\big)\, \mathbf{v}_{\text{int}}^{n+1} = \mathbf{f}^{n+1}, \tag{6.39}$$

where $\mathbf{M}_1$ is the consistent mass matrix, $\mathbf{C}$ is the convection matrix defined in (6.33), $\mathbf{K}$ is the viscosity matrix identical to the one already defined in Section 6.5.4, and vector $\mathbf{f}^{n+1}$ accounts for the applied body force $\boldsymbol{b}$ and the Dirichlet boundary conditions.

Note the computational complexity of the fully implicit option, $\boldsymbol{v}^* = \boldsymbol{v}_{\text{int}}^{n+1}$, for time integration, which requires repeated computations of the inverse of the nonlinear and non-symmetric matrix $\mathbf{M}_1 + \Delta t\, \big(\mathbf{C}(\boldsymbol{v}_{\text{int}}^{n+1}) + \mathbf{K}\big)$. In this case, predictor–corrector methods are usually employed to solve (6.39). In the semi-implicit case, $\boldsymbol{v}^* = \boldsymbol{v}^n$, a modification of the convective term is required to maintain unconditional stability as explained in Remark 6.18.

The *second step* of the Chorin–Temam method determines the end-of-step velocity $\boldsymbol{v}^{n+1}$ and pressure p^{n+1} solving

$$\begin{cases} \dfrac{\boldsymbol{v}^{n+1} - \boldsymbol{v}_{\text{int}}^{n+1}}{\Delta t} + \boldsymbol{\nabla} p^{n+1} = 0 & \text{in } \Omega, \\ \boldsymbol{\nabla} \cdot \boldsymbol{v}^{n+1} = 0 & \text{in } \Omega, \\ \boldsymbol{n} \cdot \boldsymbol{v}^{n+1} = \boldsymbol{n} \cdot \boldsymbol{v}_D^{n+1} & \text{on } \Gamma. \end{cases} \tag{6.40}$$

Note that this second step includes the remaining term (pressure) and equation (incompressibility) of the Navier–Stokes equations. Now, the boundary condition only prescribes the normal component of the velocity (not the tangential components). This is a crucial aspect of this fractional-step method: the tangential components of the velocity cannot be controlled on the boundary in accordance with the Helmholtz decomposition principle, which only allows us to prescribe a condition on the normal component of the velocity. The first equation in (6.40) is

$$\boldsymbol{v}_{\text{int}}^{n+1} = \boldsymbol{v}^{n+1} + \Delta t\, \boldsymbol{\nabla} p^{n+1}, \quad \text{or} \quad \boldsymbol{v}^{n+1} = \boldsymbol{v}_{\text{int}}^{n+1} - \Delta t\, \boldsymbol{\nabla} p^{n+1}. \tag{6.41}$$

In the particular case of homogeneous boundary values for the normal component of the velocity (i.e., $\boldsymbol{n}\cdot\boldsymbol{v}_D = 0$), this is precisely the orthogonal decomposition applied to the intermediate velocity $\boldsymbol{v}_{\text{int}}^{n+1} = \boldsymbol{w}$ in (6.36). Thus, $\boldsymbol{\nabla}\cdot\boldsymbol{v}^{n+1} = 0$ and $\boldsymbol{n}\cdot\boldsymbol{v}^{n+1} = 0$ on Γ. This shows that the fractional-step method involves an orthogonal projection operator $\mathbb{P}$, see Remark 6.8, such that

$$\boldsymbol{v}^{n+1} = \mathbb{P}\,\boldsymbol{v}_{\text{int}}^{n+1} \quad \text{and} \quad \Delta t\,\boldsymbol{\nabla} p^{n+1} = (\mathbf{I}-\mathbb{P})\,\boldsymbol{v}_{\text{int}}^{n+1}.$$

The weak form of the second-step equations (6.40) is given as follows: find the end-of-step velocity $\boldsymbol{v}^{n+1}\in\boldsymbol{\mathcal{S}}$ and the pressure $p^{n+1}\in\mathcal{Q}$, such that, for all $(\boldsymbol{w},q)\in\boldsymbol{\mathcal{V}}\times\mathcal{Q}$,

$$\begin{cases} \left(\boldsymbol{w}, \dfrac{\boldsymbol{v}^{n+1}-\boldsymbol{v}_{\text{int}}^{n+1}}{\Delta t}\right) + b(\boldsymbol{w},p^{n+1}) = 0, \\ b(\boldsymbol{v}^{n+1},q) = 0. \end{cases} \tag{6.42}$$

Now, the functional spaces $\boldsymbol{\mathcal{S}}$ and $\boldsymbol{\mathcal{V}}$ are such that the solution verifies the prescribed boundary conditions, $\boldsymbol{n}\cdot\boldsymbol{v}^{n+1} = \boldsymbol{n}\cdot\boldsymbol{v}_D^{n+1}$ on Γ.

The discrete equations emanating from the discretization of (6.42) induce the following system of algebraic equations:

$$\begin{cases} \mathbf{M}_2\left(\dfrac{\mathbf{v}^{n+1}-\mathbf{v}_{\text{int}}^{n+1}}{\Delta t}\right) + \mathbf{G}\,\mathbf{p}^{n+1} = 0, \\ \mathbf{G}^T\mathbf{v}^{n+1} = 0, \end{cases}$$

or equivalently,

$$\begin{pmatrix} \mathbf{M}_2/\Delta t & \mathbf{G} \\ \mathbf{G}^T & \mathbf{0} \end{pmatrix}\begin{pmatrix} \mathbf{v}^{n+1} \\ \mathbf{p}^{n+1} \end{pmatrix} = \begin{pmatrix} \mathbf{M}_2\mathbf{v}_{\text{int}}^{n+1}/\Delta t \\ \mathbf{0} \end{pmatrix}.$$

Note that this system has the same structure as the one obtained for the Stokes problem, see (6.21), where the viscous matrix is now replaced by a mass matrix. As already discussed in Section 6.5.5 this system can be solved in two steps, first compute the pressure field from

$$(\mathbf{G}^T\mathbf{M}_2^{-1}\mathbf{G})\,\mathbf{p}^{n+1} = \frac{1}{\Delta t}\,\mathbf{G}^T\,\mathbf{v}_{\text{int}}^{n+1}$$

and then compute the end-of-step velocity from

$$\mathbf{M}_2\mathbf{v}^{n+1} = \mathbf{M}_2\mathbf{v}_{\text{int}}^{n+1} - \Delta t\,\mathbf{G}\mathbf{p}^{n+1},$$

see also (6.41). Recall that when Dirichlet conditions are precibed on the whole boundary, a reference value of pressure must be specified at an arbitrary pressure node because $\mathcal{Q} = \mathcal{L}_2(\Omega)/\mathbb{R}$.

The solvability issues discussed in Section 6.5.5 are automatically verified here. That is, this algorithm provides a unique solution which converges to the exact one for any velocity–pressure pair provided that the boundary is smooth (the intermediate velocity also converges to the exact one). This property and the fact that velocity and pressure are uncoupled (recall that the mass matrix is adequately approximated by the lumped one) have contributed to the popularity of this method.

Nevertheless, this scheme is first order in time; although modified schemes, for instance with iterative procedures see Bell, Colella and Glaz (1989), have been developed for higher-order methods. Another drawback is that the end-of-step velocity $\boldsymbol{v}^{n+1}$ does not satisfy exactly the Dirichlet boundary conditions, this may cause an non-physical boundary layer of $\mathcal{O}\left(\sqrt{\nu\Delta t}\right)$, see Remark 6.20 and the papers by Gresho (1990) and Temam (1991).

Finally, it is important to note that the condition on the smoothness of boundary Γ is, in some cases, crucial. For instance, the cavity flow problem, which is studied in Section 6.8.2, does not present a smooth boundary. One can verify that the bilinear equal-interpolation for velocity and pressure, that is the Q1Q1 element (see Figure 6.1), induces a non regular $\mathbf{G}^T\mathbf{M}_2^{-1}\mathbf{G}$ matrix. Note that this is not the case for the same problem and another element not passing the LBB condition: the Q1P0 element.

Remark 6.16 (Incremental projection scheme). The original Chorin–Temam method can be modified to incorporate the pressure increment $p^{n+1}-p^n$, instead of the total pressure, in the pressure/incompressibility phase. This leads to a projection method with improved convergence properties. The first step (6.37) of the incremental projection scheme is modified to include the pressure gradient term $\boldsymbol{\nabla}p^n$ and becomes

$$\begin{cases} \dfrac{\boldsymbol{v}_{\text{int}}^{n+1} - \boldsymbol{v}^n}{\Delta t} + (\boldsymbol{v}^* \cdot \boldsymbol{\nabla})\,\boldsymbol{v}^{**} - \nu \boldsymbol{\nabla}^2 \boldsymbol{v}^{**} = \boldsymbol{b}^{n+1} - \boldsymbol{\nabla}p^n & \text{in } \Omega, \\ \boldsymbol{v}_{\text{int}}^{n+1} = \boldsymbol{v}_D^{n+1} & \text{on } \Gamma, \end{cases}$$

while the projection step is replaced by

$$\begin{cases} \dfrac{\boldsymbol{v}^{n+1} - \boldsymbol{v}_{\text{int}}^{n+1}}{\Delta t} + \boldsymbol{\nabla}(p^{n+1} - p^n) = 0 & \text{in } \Omega, \\ \boldsymbol{\nabla}\cdot\boldsymbol{v}^{n+1} = 0 & \text{in } \Omega, \\ \boldsymbol{n}\cdot\boldsymbol{v}^{n+1} = \boldsymbol{n}\cdot\boldsymbol{v}_D^{n+1} & \text{on } \Gamma. \end{cases}$$

Remark 6.17 (Treatment of Neumann-type boundary conditions). When Neumann boundary conditions, see (6.34d), are present, the condition

$$-p\boldsymbol{n} + \nu(\boldsymbol{n}\cdot\boldsymbol{\nabla})\boldsymbol{v} = \boldsymbol{t}$$

is decomposed in order to include the viscous part of the prescribed boundary traction in the first step and the pressure contribution in the second step. The process is illustrated in a paper by Laval and Quartapelle (1990). This splitting is also feasible when a velocity–pressure stress-divergence formulation is employed, recall Remarks 6.3, 6.11 and 6.12.

Remark 6.18 (Treatment of convective terms). As previously discussed, in the first step, convective terms can be treated in various ways, from fully explicit to fully implicit. The drawback of implicit integration is the need to solve nonlinear and non-symmetric systems at each time step. Despite their conditional stability, explicit schemes are advantageous because the linear algebraic systems that must be solved are symmetric and positive definite. Thus, effective iterative solvers can be employed.

The interesting features of a semi-implicit approach are the linearization of the convective term, the possible unconditional stability of the scheme, and, as shown by Guermond and Quartapelle (1998b), the possibility of eliminating the velocity update in the incompressibility step. To guarantee unconditional stability in the case of a semi-implicit time integration, the skew-symmetric form of the convective term must be used, see Remark 6.15.

Remark 6.19 (Elimination of the end-of-step velocity). Since the intermediate velocity also converges to the exact one, using the skew-symmetric form of the convective term, Guermond and Quartapelle (1997; 1998b) rewrite the first step, see (6.37), in the semi-implicit form

$$\begin{cases} \dfrac{\boldsymbol{v}_{\text{int}}^{n+1} - \boldsymbol{v}^n}{\Delta t} + (\boldsymbol{v}_{\text{int}}^n \cdot \boldsymbol{\nabla})\boldsymbol{v}_{\text{int}}^{n+1} + \dfrac{1}{2}(\boldsymbol{\nabla} \cdot \boldsymbol{v}_{\text{int}}^n)\boldsymbol{v}_{\text{int}}^{n+1} - \nu \boldsymbol{\nabla}^2 \boldsymbol{v}_{\text{int}}^{n+1} = \boldsymbol{b}^{n+1} & \text{in } \Omega, \\ \boldsymbol{v}_{\text{int}}^{n+1} = \boldsymbol{v}_D^{n+1} & \text{on } \Gamma, \end{cases}$$

where the convective velocity is $\boldsymbol{v}_{\text{int}}^n$ instead of $\boldsymbol{v}^n$. Using now the momentum equation of the second step, equation (6.41), at t^n (i.e., $\boldsymbol{v}^n = \boldsymbol{v}_{\text{int}}^n - \Delta t\, \boldsymbol{\nabla} p^n$) the first equation can be expressed in terms of the intermediate velocity only:

$$\frac{\boldsymbol{v}_{\text{int}}^{n+1} - \boldsymbol{v}_{\text{int}}^n}{\Delta t} + (\boldsymbol{v}_{\text{int}}^n \cdot \boldsymbol{\nabla})\,\boldsymbol{v}_{\text{int}}^{n+1} + \frac{1}{2}(\boldsymbol{\nabla} \cdot \boldsymbol{v}_{\text{int}}^n)\boldsymbol{v}_{\text{int}}^{n+1} - \nu \boldsymbol{\nabla}^2 \boldsymbol{v}_{\text{int}}^{n+1} = \boldsymbol{b}^{n+1} - \boldsymbol{\nabla} p^n.$$

Since this algorithm has eliminated the end-of-step velocity completely, if needed it is computed from (6.41). Finally, observe that this fractional-step scheme requires the evaluation of pressure at each step. This can be done with a pressure Poisson equation, see the next remark, with no evaluation of the end-of-step velocity.

Remark 6.20 (Pressure Poisson equation). Observe that the equations of the second step can be reformulated in terms of a Poisson equation for the pressure. In fact, applying the divergence operator to the first equation in (6.40), since $\boldsymbol{\nabla} \cdot \boldsymbol{v}^{n+1} = 0$, we obtain the following Neumann problem for the pressure:

$$\begin{cases} \boldsymbol{\nabla}^2 p^{n+1} = \dfrac{1}{\Delta t} \boldsymbol{\nabla} \cdot \boldsymbol{v}_{\text{int}}^{n+1} & \text{in } \Omega, \\ \boldsymbol{n} \cdot \boldsymbol{\nabla} p^{n+1} = 0 & \text{on } \Gamma. \end{cases}$$

Once the end-of-step pressure p^{n+1} is determined, the end-of-step velocity $\boldsymbol{v}^{n+1}$, if needed, can be computed by the explicit relation (6.41), that is the first equation in (6.40).

The Neumann boundary conditions for the end-of-step pressure are obtained using (6.41) in the boundary condition of the second step, namely $\boldsymbol{n}\cdot\boldsymbol{v}^{n+1} = \boldsymbol{n}\cdot\boldsymbol{v}_D^{n+1}$, and using the fact that $\boldsymbol{v}_{\text{int}}^{n+1} = \boldsymbol{v}_D^{n+1}$ from the first step. There has been some controversy on these non-physical boundary conditions for pressure. Note that the exact pressure for sufficiently smooth conditions does not verify homogeneous boundary conditions (see Gresho and Sani, 1987). Together with the weak representation of the tangential velocity on the boundary, the non-physical boundary conditions for the pressure represent a potential drawback of the Chorin–Temam method.

Remark 6.21 (Viscosity splitting fractional-step method). To alleviate the difficulties regarding the imposition of Dirichlet boundary conditions in the second step of the Chorin–Temam projection method, Blasco, Codina and Huerta (1997; 1998) introduced a viscosity splitting fractional-step method in which the second step avoids using the projection idea. Instead, they introduce a diffusion term in the momentum equation of the second step, which consequently loses its inviscid character responsible for preventing control of the prescribed tangential component of the velocity at the boundary. The second step then consists of determining the end-of-step velocity $\boldsymbol{v}^{n+1}$ and pressure p^{n+1} as being the solution of

$$\begin{cases} \dfrac{\boldsymbol{v}^{n+1} - \boldsymbol{v}_{\text{int}}^{n+1}}{\Delta t} - \nu\,\boldsymbol{\nabla}^2\left(\boldsymbol{v}^{n+1} - \boldsymbol{v}_{\text{int}}^{n+1}\right) + \boldsymbol{\nabla} p^{n+1} = 0 & \text{in } \Omega, \\ \boldsymbol{\nabla}\cdot\boldsymbol{v}^{n+1} = 0 & \text{in } \Omega, \\ \boldsymbol{v}^{n+1} = \boldsymbol{v}_D^{n+1} & \text{on } \Gamma. \end{cases}$$

When combined with the equations (6.37) of the first step, this formulation of the second step allows the imposition of the original Dirichlet boundary conditions in both phases of the fractional-step method. This method, however, requires finite elements passing the LBB condition.

Remark 6.22 (The LBB condition). It was initially thought that Poisson-based projection techniques could be used with velocity–pressure pairs not satisfying the LBB condition and that this family of methods could thus be considered as pertaining to the class of stabilized finite element methods. This is unfortunately not always true. Guermond and Quartapelle (1998a; 1998b) performed an in-depth study of the stability and convergence properties of fractional-step projection methods in which the pressure/incompressibility step is recast in terms of a Poisson equation for the pressure. Their conclusion is that in the case of the incremental projection scheme, the velocity–pressure pairs must satisfy the LBB compatibility condition to obtain non-oscillatory numerical results. By contrast, when the total pressure is employed in the second step, they found that equal-order interpolations could be safely used, provided the time step is not too small with respect to the spatial mesh size, in the sense that $\Delta t \geq c\,h^k$, k being the degree of the velocity interpolation and h a measure of the mesh size. Unfortunately, the non-incremental projection method has inferior convergence properties with respect to the incremental scheme. Minev (2001) presents a discussion on which fractional-step methods require an LBB compliant approximation and which schemes do not.

6.7.3.2 *Algebraic splitting* The Chorin–Temam projection method, which is based on the Helmholtz decomposition principle, can be viewed as a *physical splitting* of the original incompressible Navier–Stokes equations. As an alternative method for solving the Navier–Stokes problem in a sequence of simpler steps, Perot (1993) and Abdallah (1995) in a finite volume context and more recently Quarteroni, Saleri and Veneziani (2000) in the finite element context introduced an *algebraic splitting*. It is based on an incomplete (or approximate) block LU factorization of the original partitioned matrix system arising from the space–time discretization of the unsteady Navier–Stokes equations (6.35), namely

$$\begin{pmatrix} \mathbf{B} & \mathbf{G} \\ \mathbf{G}^T & \mathbf{0} \end{pmatrix} \begin{pmatrix} \mathbf{u}^{n+1} \\ \mathbf{p}^{n+1} \end{pmatrix} = \begin{pmatrix} \mathbf{f}^* \\ \mathbf{h} \end{pmatrix}, \tag{6.43}$$

where $\mathbf{f}^*$ accounts for the r.h.s. term in the first equation in (6.35) and all the known terms (evaluated at t^n) resulting from the time discretization of the momentum equation. To simplify the notation, we have posed

$$\mathbf{B} = \frac{1}{\Delta t}\mathbf{M} + \big(\mathbf{K} + \mathbf{C}(\boldsymbol{v}^{n+1})\big). \tag{6.44}$$

As already mentioned in Section 6.5.5 for the Stokes problem, the partitioned system (6.43) could in principle be solved in two steps (6.26) and (6.27), namely

$$\begin{cases} (\mathbf{G}^T\mathbf{B}^{-1}\mathbf{G})\,\mathbf{p}^{n+1} = \mathbf{G}^T\mathbf{B}^{-1}\mathbf{f}^* - \mathbf{h}, \\ \mathbf{B}\mathbf{u}^{n+1} = \mathbf{f}^* - \mathbf{G}\,\mathbf{p}^{n+1}. \end{cases}$$

The second equation is obtained making the first equation in (6.43) explicit with respect to $\mathbf{u}^{n+1}$ and the pressure equation, which is solved first, is obtained after this expression for $\mathbf{u}^{n+1}$ is introduced in the second equation of (6.43).

Unfortunately, this strategy is computationally unaffordable in engineering applications. The reduction in computational complexity is usually limited to replace $\mathbf{B}^{-1}$ by a simpler matrix. From (6.44) we have

$$\mathbf{B}^{-1} = \Delta t\Big(\mathbf{I} + \Delta t\mathbf{M}^{-1}\big(\mathbf{K} + \mathbf{C}(\boldsymbol{v}^{n+1})\big)\Big)^{-1}\mathbf{M}^{-1}.$$

The standard approach is to use the first-order approximation of matrix $\mathbf{B}^{-1}$ with respect to Δt given by

$$\mathbf{B}^{-1} \approx \mathbf{H} = \Delta t\mathbf{M}^{-1}.$$

Since, by contrast with matrix $\mathbf{B}^{-1}$, matrix $\mathbf{M}^{-1}$ remains invariant during the transient calculation, a significant reduction of the computational effort is obtained by this approximation. Higher-order expressions can also be devised, the second-order approximation is

$$\mathbf{H} = \Delta t\big(\mathbf{I} - \Delta t\mathbf{M}^{-1}(\mathbf{K} + \mathbf{C})\big)\mathbf{M}^{-1}$$

and the third-order one becomes

$$\mathbf{H} = \Delta t\Big(\mathbf{I} - \Delta t\mathbf{M}^{-1}(\mathbf{K}+\mathbf{C}) + \Delta t^2\big(\mathbf{M}^{-1}(\mathbf{K}+\mathbf{C})\big)^2\Big)\mathbf{M}^{-1},$$

where $\mathbf{C}$ is non-symmetric and varies with time. Note that only the inverse of the mass matrix is required (recall that the lumped mass matrix is a good approximation of the consistent one).

Perot (1993) and Quarteroni et al. (2000) generalize this approach using the exact LU factorization of the matrix in (6.43),

$$\begin{pmatrix} \mathbf{B} & \mathbf{G} \\ \mathbf{G}^T & \mathbf{0} \end{pmatrix} = \begin{pmatrix} \mathbf{B} & \mathbf{0} \\ \mathbf{G}^T & -\mathbf{G}^T\mathbf{B}^{-1}\mathbf{G} \end{pmatrix} \begin{pmatrix} \mathbf{I} & \mathbf{B}^{-1}\mathbf{G} \\ \mathbf{0} & \mathbf{I} \end{pmatrix}.$$

This process is indeed equivalent to a direct solution of system (6.43). Its interest lies in the fact that suitable approximations of matrix $\mathbf{B}^{-1}$ allow the construction of more economic solution schemes. Now $\mathbf{B}^{-1}$ can be replaced by two simpler matrices, one in the L-block, $\mathbf{H}_1$, and one in the U-block, $\mathbf{H}_2$. Thus, the original matrix is approximated by

$$\begin{pmatrix} \mathbf{B} & \mathbf{0} \\ \mathbf{G}^T & -\mathbf{G}^T\mathbf{H}_1\mathbf{G} \end{pmatrix} \begin{pmatrix} \mathbf{I} & \mathbf{H}_2\mathbf{G} \\ \mathbf{0} & \mathbf{I} \end{pmatrix} = \begin{pmatrix} \mathbf{B} & \mathbf{B}\mathbf{H}_2\mathbf{G} \\ \mathbf{G}^T & \mathbf{G}^T(\mathbf{H}_2 - \mathbf{H}_1)\mathbf{G} \end{pmatrix}. \tag{6.45}$$

From this inexact factorization three strategies become apparent.

First, if $\mathbf{H} = \mathbf{H}_1 = \mathbf{H}_2$ the previously discussed methodology is recovered. Note that this strategy is the only one that preserves mass because the lower-left block of the inexact factorization is a null matrix, see (6.45). The momentum equation, however, is modified by the fact that $\mathbf{H}_2 \neq \mathbf{B}^{-1}$. Projection fractional-step methods, such as the ones discussed in the previous section, belong to this strategy.

Second, if $\mathbf{H}_2 = \mathbf{B}^{-1}$ and $\mathbf{H}_1 \neq \mathbf{H}_2$ momentum is preserved because the momentum equation is not modified but the mass conservation equation is modified. Quasi-compressible (i.e., pressure-stabilized, penalty, etc.) methods belong to this category, see for instance (6.28) and (6.29).

Third, if $\mathbf{H}_2 \neq \mathbf{B}^{-1}$ and $\mathbf{H}_1 \neq \mathbf{H}_2$ a more general strategy is devised where both the momentum and continuity equations are perturbed, see for instance Section 6.7.2.

The inexact factorization (6.45) in equation (6.43) allows the computation of $\mathbf{u}^{n+1}$ and $\mathbf{p}^{n+1}$ by means of the following sequence of steps:

$$\text{L-step}\ \begin{cases} \mathbf{B}\,\mathbf{u}_{\text{int}}^{n+1} = \mathbf{f}^*, \\ \mathbf{G}^T\,\mathbf{u}_{\text{int}}^{n+1} - \mathbf{G}^T\mathbf{H}_1\mathbf{G}\,\mathbf{p}_{\text{int}}^{n+1} = \mathbf{h}, \end{cases} \qquad \text{U-step}\ \begin{cases} \mathbf{u}^{n+1} + \mathbf{H}_2\mathbf{G}\,\mathbf{p}^{n+1} = \mathbf{u}_{\text{int}}^{n+1}, \\ \mathbf{p}^{n+1} = \mathbf{p}_{\text{int}}^{n+1}, \end{cases}$$

or, equivalently

$$\begin{cases} \mathbf{B}\,\mathbf{u}_{\text{int}}^{n+1} = \mathbf{f}^* & \text{(intermediate velocity computation)}, \\ \mathbf{G}^T\mathbf{H}_1\mathbf{G}\,\mathbf{p}^{n+1} = \mathbf{G}^T\,\mathbf{u}_{\text{int}}^{n+1} - \mathbf{h} & \text{(pressure computation)}, \\ \mathbf{u}^{n+1} = \mathbf{u}_{\text{int}}^{n+1} - \mathbf{H}_2\mathbf{G}\,\mathbf{p}^{n+1} & \text{(end-of-step velocity computation)}. \end{cases}$$

A number of other schemes based on the inexact factorization (6.45) are investigated by Quarteroni et al. (2000).

6.8 APPLICATIONS AND SOLVED EXERCICES

6.8.1 Stokes flow with analytical solution

In order to illustrate the behavior of selected mixed finite elements in the solution of stationary Stokes flow, see Section 6.5.2, we consider a two-dimensional problem in the square domain $\Omega =]0,1[\times]0,1[$, which possesses a closed-form analytical solution. The problem consists of determining the velocity field $\boldsymbol{v} = (v_1, v_2)$ and the pressure p such that

$$\begin{cases} -\nu\boldsymbol{\nabla}^2\boldsymbol{v} + \boldsymbol{\nabla}p = \boldsymbol{b} & \text{in } \Omega, \\ \boldsymbol{\nabla}\cdot\boldsymbol{v} = 0 & \text{in } \Omega, \\ \boldsymbol{v} = 0 & \text{on } \Gamma, \end{cases}$$

where the fluid viscosity is taken as $\nu = 1$. The components of the body force $\boldsymbol{b}$ are prescribed as

$$\begin{aligned} b_1 &= (12 - 24\,y)\,x^4 + (-24 + 48\,y)\,x^3 + (-48\,y + 72\,y^2 - 48\,y^3 + 12)\,x^2 \\ &\quad + (-2 + 24\,y - 72\,y^2 + 48\,y^3)\,x + 1 - 4\,y + 12\,y^2 - 8\,y^3, \\ b_2 &= (8 - 48\,y + 48\,y^2)\,x^3 + (-12 + 72\,y - 72\,y^2)\,x^2 \\ &\quad + (4 - 24\,y + 48\,y^2 - 48\,y^3 + 24\,y^4)\,x - 12\,y^2 + 24\,y^3 - 12\,y^4, \end{aligned}$$

where, for simplicity, we have used in this 2D problem the notation $(x, y) := (x_1, x_2)$. With this prescribed body force, the exact solution is

$$\begin{aligned} v_1(x,y) &= x^2\,(1-x)^2\,(2\,y - 6\,y^2 + 4\,y^3), \\ v_2(x,y) &= -y^2\,(1-y)^2\,(2\,x - 6\,x^2 + 4\,x^3), \\ p(x,y) &= x\,(1-x). \end{aligned}$$

The square domain is discretized with different uniform meshes: 20×20 Q1P0 elements, 20×20 Q1Q1 elements, 40×40 `Mini` elements, and finally 10×10 Q2Q1 elements (the characteristic distance between velocity nodes is kept constant). Figures 6.3 and 6.4 show the pressure field and velocity vectors obtained with the various elements and a Galerkin formulation. Clearly, elements Q1P0 and Q1Q1, which do not satisfy the LBB compatibility condition, exhibit a spurious pressure response. However, one should note that they deliver an acceptable velocity response, see the excellent discussion by Brezzi and Fortin (1991, Sec. II.3.3) entitled "Is the inf-sup condition so important?". The results obtained with the `Mini` element and the Q2Q1 element, which satisfy the LBB condition, deliver good answers for both velocity and pressure.

These conclusions are clarified in the convergence plots, see Figures 6.5 and 6.6, where uniform meshes have been used with characteristic lengths: $h = 1/10$, $1/20$, $1/30$, and $1/40$. The velocity field converges to the exact solution for all the elements as shown in Figure 6.5 and the convergence rates coincide with the theoretical ones. This is not the case for the pressure field. Figure 6.6 shows two plots, the left one

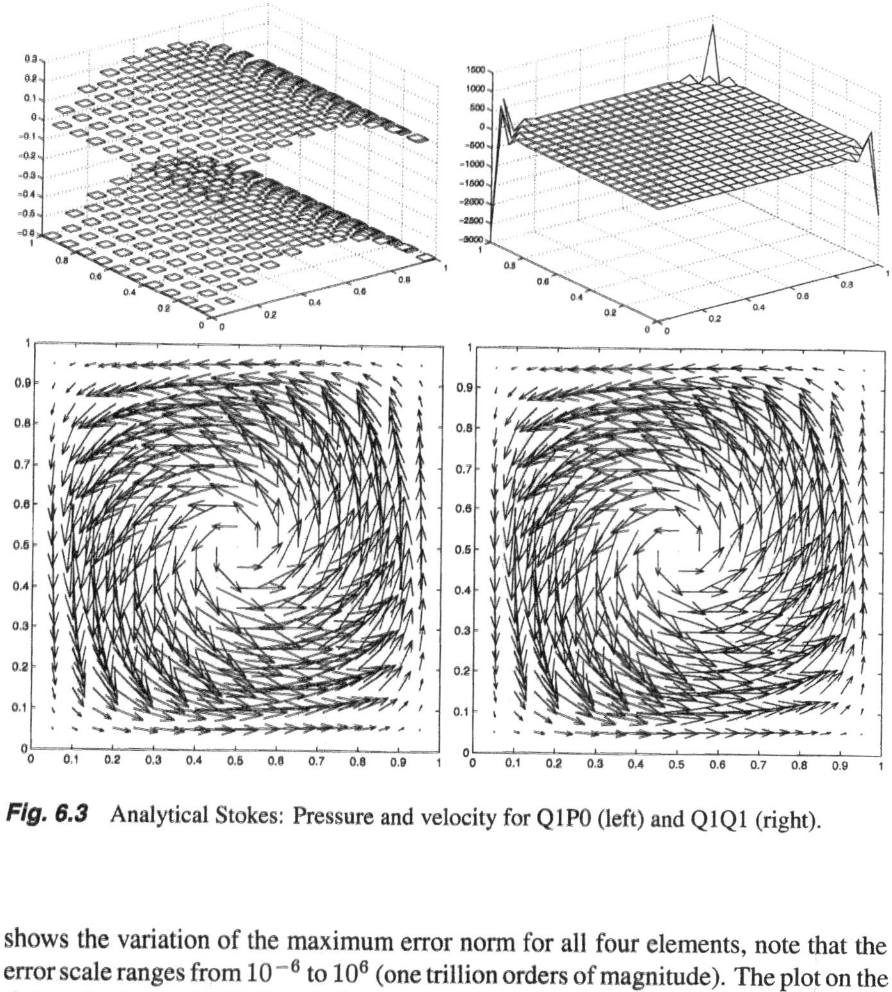

Fig. 6.3 Analytical Stokes: Pressure and velocity for Q1P0 (left) and Q1Q1 (right).

shows the variation of the maximum error norm for all four elements, note that the error scale ranges from 10^{-6} to 10^{6} (one trillion orders of magnitude). The plot on the right only shows the LBB compliant elements and clearly indicates that the pressure field converges to the exact one.

6.8.2 Cavity flow problem

This example has become a standard benchmark test for incompressible flows. We will show here results for the Stokes and Navier–Stokes problems. Figure 6.7 shows a schematic representation of the problem statement. It models a plane flow of an isothermal fluid in a square lid-driven cavity. The upper side of the cavity moves in its own plane at unit speed, while the other sides are fixed.

The boundary conditions are indicated in Figure 6.7. There is a discontinuity in the boundary conditions at the two upper corners of the cavity. Two cases can be envisioned: the two upper corners are either considered as belonging to the top mobile side (leaky cavity), or they are assumed to belong to the fixed vertical walls

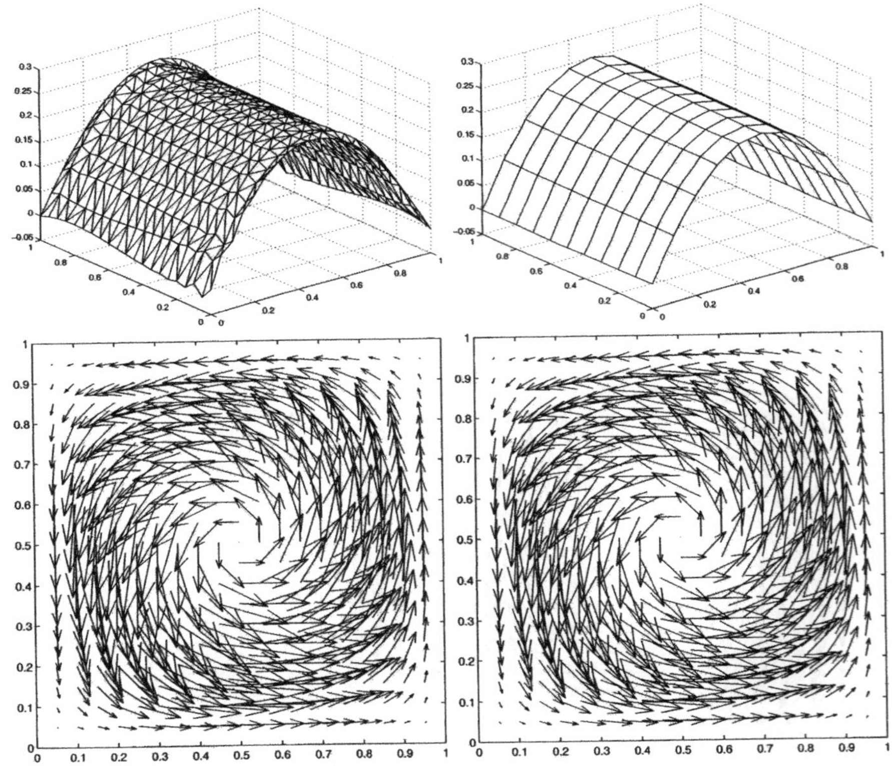

Fig. 6.4 Analytical Stokes: Pressure and velocity for `Mini` (left) and Q2Q1 (right).

(non-leaky). The former case is adopted here. It introduces a singulary in the pressure field precisely at those two upper corners.

Finally, it should be noticed that Dirichlet boundary conditions are imposed on every boundary in this example. As commented earlier this implies that pressure is known up to a constant ($p \in \mathcal{L}_2(\Omega)/\mathbb{R}$). Thus at an arbitrary point, the lower left corner of the cavity, the reference value $p = 0$ is prescribed.

First, we solve the lid-driven cavity for the Stokes problem and the standard Galerkin formulation. The main features in this case are the symmetry with respect to the vertical centerline and the pressure singularity at the two upper corners. In fact, no shear layers are present in the Stokes problem, but results (the pressure jump between both corners) improve if a nonuniform mesh is employed. The cavity is discretized with a nonuniform mesh of 30×30 Q1P0 elements, 30×30 Q1Q1 elements, 60×60 `Mini` elements, and finally 15×15 Q2Q1 elements. Thus, all these meshes have the same characteristic element size: $h = 1/30$.

Figure 6.8 shows the symmetric streamlines for the Q2Q1 element. This distribution of streamlines is very similar to the distributions obtained for the other elements. Figure 6.9 shows the pressure field for the non LBB compliant elements.

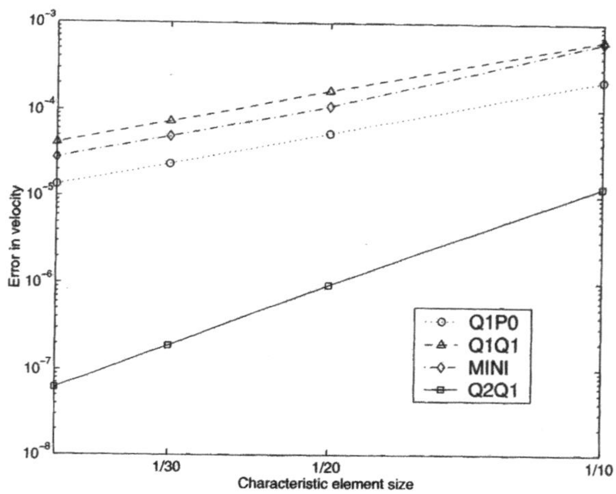

Fig. 6.5 Analytical Stokes: velocity error versus element size.

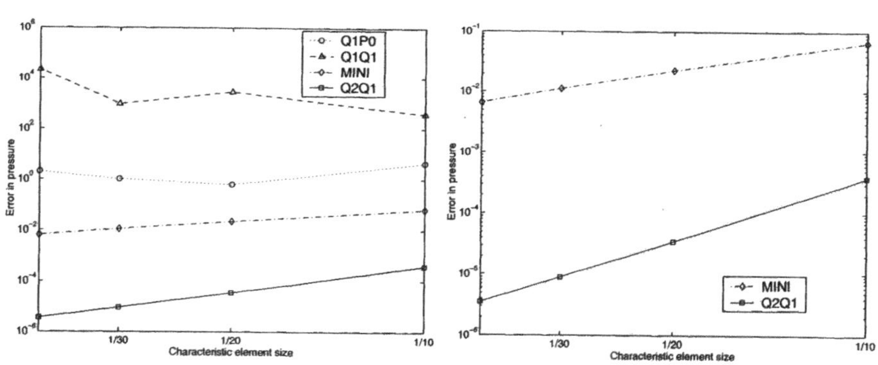

Fig. 6.6 Analytical Stokes: pressure error versus element size.

As expected they present inaccurate pressure results. Both the Q1P0 and the Q1Q1 elements present oscillations which are more pronounced in the corners. The element-to-element oscillations are more obvious on uniform meshes. Both the `Mini` and the Q2Q1, which are LBB compliant, show, as expected, reasonable results for pressure, see the corresponding pressure fields in Figure 6.10.

The stationary Navier–Stokes solution, which is the objective now, is entirely characterized by the Reynolds number,

$$Re = \frac{V_{\text{ref}} \, L_{\text{ref}}}{\nu},$$

ν being the kinematic viscosity of the fluid. The reference velocity used in the Reynolds number is the velocity of the mobile side: $V_{\text{ref}} = 1$. The reference length is the side of the cavity: $L_{\text{ref}} = 1$. The influence of the Reynolds number can be clearly

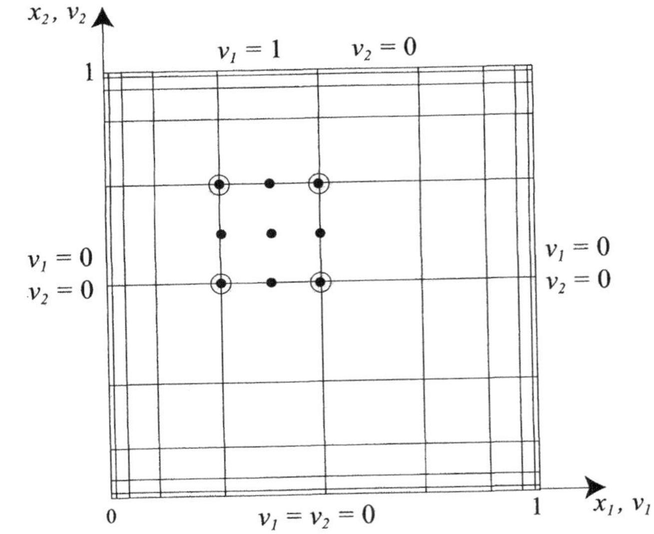

Fig. 6.7 Lid-driven cavity: problem statement with boundary conditions and schematic discretization with Q2Q1 elements. Dots indicate velocity nodes and circles denote pressure nodes.

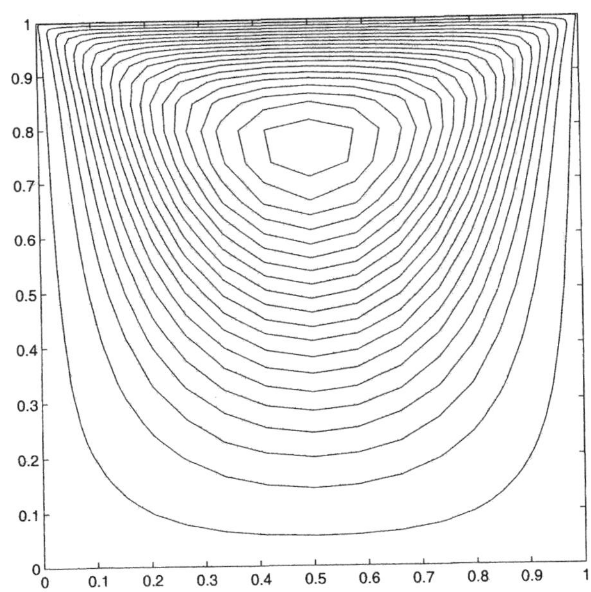

Fig. 6.8 Stokes cavity flow: streamlines for the Q2Q1 element.

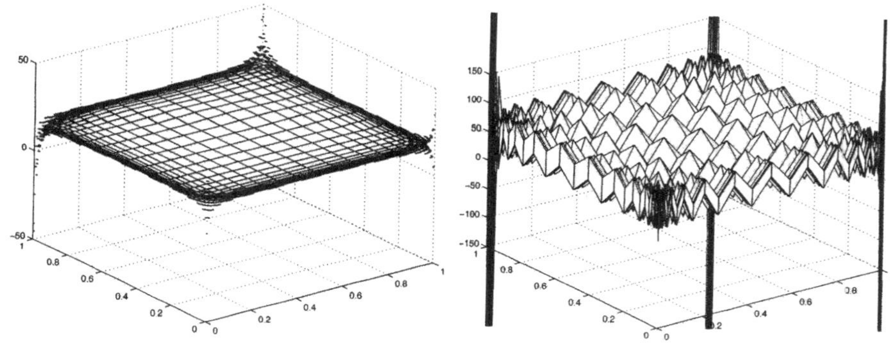

Fig. 6.9 Stokes cavity flow: pressure field for Q1P0 (left) and Q1Q1 (right) elements.

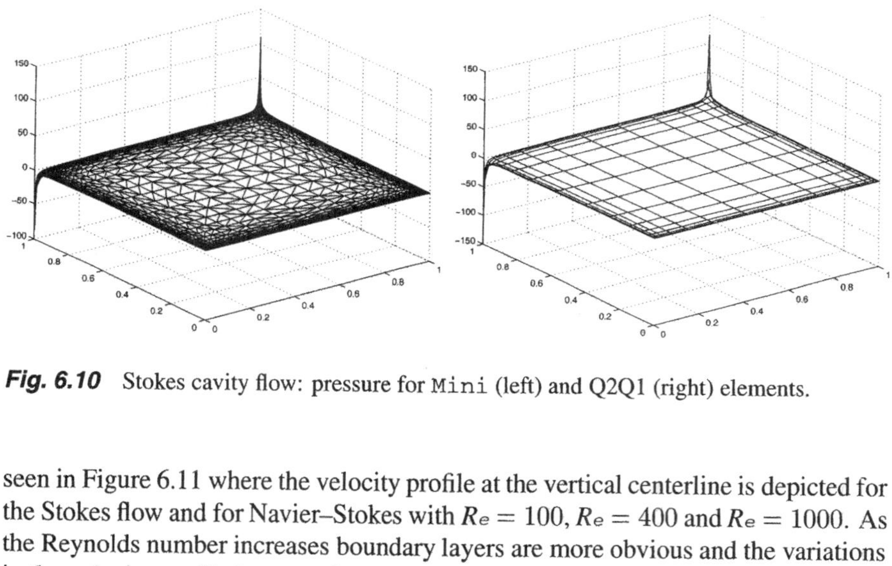

Fig. 6.10 Stokes cavity flow: pressure for `Mini` (left) and Q2Q1 (right) elements.

seen in Figure 6.11 where the velocity profile at the vertical centerline is depicted for the Stokes flow and for Navier–Stokes with $Re = 100$, $Re = 400$ and $Re = 1000$. As the Reynolds number increases boundary layers are more obvious and the variations in the velocity profile become sharper.

We show velocity and pressure results for Reynolds numbers of 100 and 1000. Note that computations are performed using the standard Galerkin finite element method. Stabilized formulations, see Section 6.7.2, must be employed for larger values of the Reynolds number or coarser meshes. An iterative technique (Picard or Newton-Raphson methods) must be employed to iteratively solve the resulting system of nonlinear algebraic equations, equations (6.32).

Results for the cavity flow are displayed in graphical form in Figures 6.12 and 6.13 which allow us to visualize the streamlines and the pressure response. The value and position of the main vortex are indicated in Table 6.2; a comparison with some reference solutions from the literature is also indicated. A satisfactory agreement is observed for all values of the Reynolds number.

As can be seen in these figures and in the table, the position of the main vortex moves towards the center of the cavity when the Reynolds number increases. The

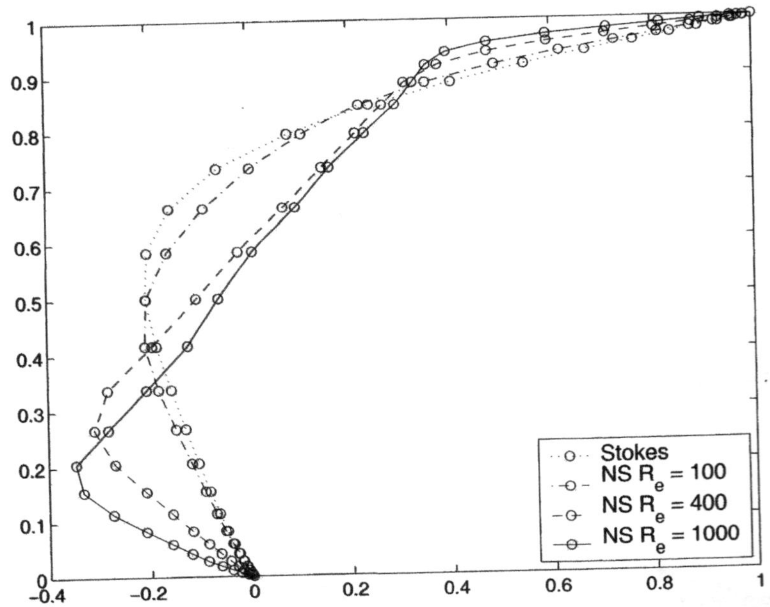

Fig. 6.11 Stokes cavity flow: velocity profiles at the vertical centerline for Stokes and Navier–Stokes with different values of Re.

Table 6.2 Position and strength of the cavity main vortex as a function of Reynolds number.

Square cavity		x_1	x_2	Stream Function
$Re = 100$	Present simulation	0.62	0.74	0.103
	Burggraf (1966)	0.62	0.74	0.101
	Tuann and Olson (1978)	0.61	0.722	0.104
$Re = 400$	Present simulation	0.568	0.606	0.110
	Burggraf (1966)	0.560	0.620	0.101
	Tuann and Olson (1978)	0.506	0.583	0.1213
	Ozawa (1975)	0.559	0.614	0.1083
$Re = 1000$	Present simulation	0.540	0.573	0.110
	Ozawa (1975)	0.533	0.569	0.118
	Goda (1979)	0.538	0.575	—

development of a secondary vortex in the right bottom corner of the cavity becomes progressively apparent and a third vortex appears at the lower left corner.

Elevated velocity gradients develop near the cavity walls for large values of the flow Reynolds number. This generates non-physical oscillations in the Galerkin solution for the velocity. A stabilized formulation would then be required, see Section 6.7.2. Solutions of the lid-driven cavity problem obtained with stabilized finite element

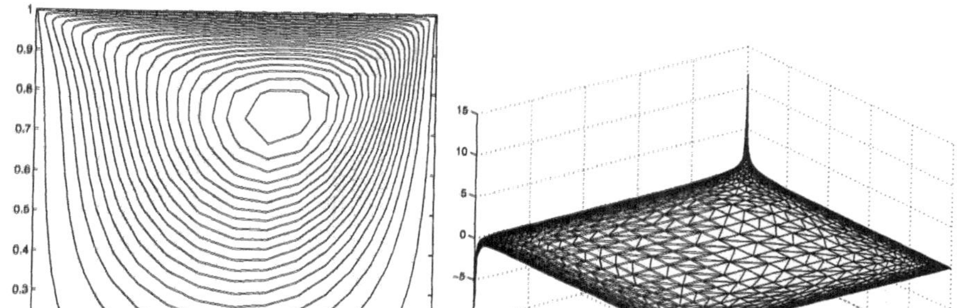

Fig. 6.12 Cavity: `Mini` element, streamlines and pressure for $Re = 100$ (top) and 1000 (bottom).

formulations are reported by Tezduyar et al. (1992), Franca and Frey (1992), Hannani, Stanislas and Dupont (1995), among others.

6.8.3 Plane jet simulation

As an example of a truly transient situation, a plane jet problem is now considered in which the flow domain is the right half-space, namely $\{\boldsymbol{x} \mid x_1 > 0, -\infty < x_2 < \infty\}$. The computational domain is limited to the square defined by $]0,1[\times]0,1[$, which is discretized by a 16×16 uniform mesh of Q2Q1 elements.

The jet aperture is centered at $\boldsymbol{x} = (0.0, 0.5)$ and is $1/16$ wide. The jet profile is parabolic with a maximum velocity of one. The fluid viscosity is taken to be $\nu = 5 \times 10^{-4}$ and its density $\rho = 1$ (non-dimensional variables are used). The boundary conditions simulating the actual flow domain are as follows:

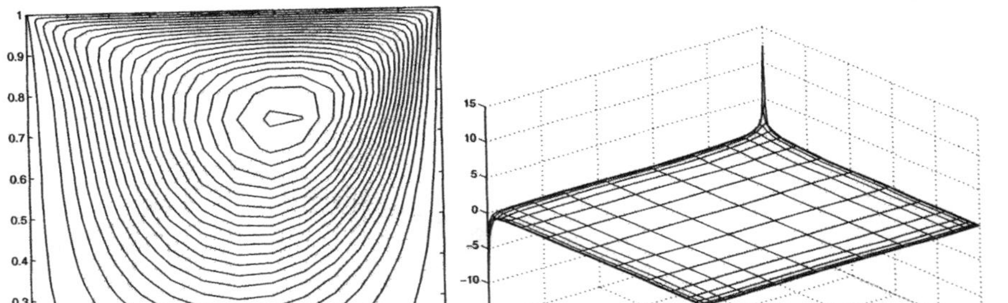

Fig. 6.13 Cavity: Q2Q1, streamlines and pressure for $Re = 100$ (top) and 1000 (bottom).

- On the inflow side, $x_1 = 0$, Dirichlet conditions for the velocity are imposed. The vertical component of the velocity is prescribed to be zero, in addition to the obvious conditions on the horizontal component.
- On the other sides, we want to simulate the situation at the outlet of the computational domain, that is *open/artificial boundary conditions*. The issue of open boundary conditions for the Navier–Stokes equations is still open and has been avoided in this text. It is however accepted, in general, that a good alternative is to impose on the primary variables the condition $\boldsymbol{n} \cdot \boldsymbol{\nabla}(\cdot) = 0$. In the context of the fractional-step method, such a boundary condition is easy to implement. In the first step the homogenous natural boundary condition is simply $\nu(\boldsymbol{n} \cdot \boldsymbol{\nabla})\boldsymbol{v} = \boldsymbol{0}$, which imposes the desired open boundary condition on the velocity. Note that in 2D and with reference to a local system of Cartesian axes $(\boldsymbol{n}, \boldsymbol{\tau})$, this is simply

$$\frac{\partial v_n}{\partial n} = 0 \qquad \text{and} \qquad \frac{\partial v_\tau}{\partial n} = 0.$$

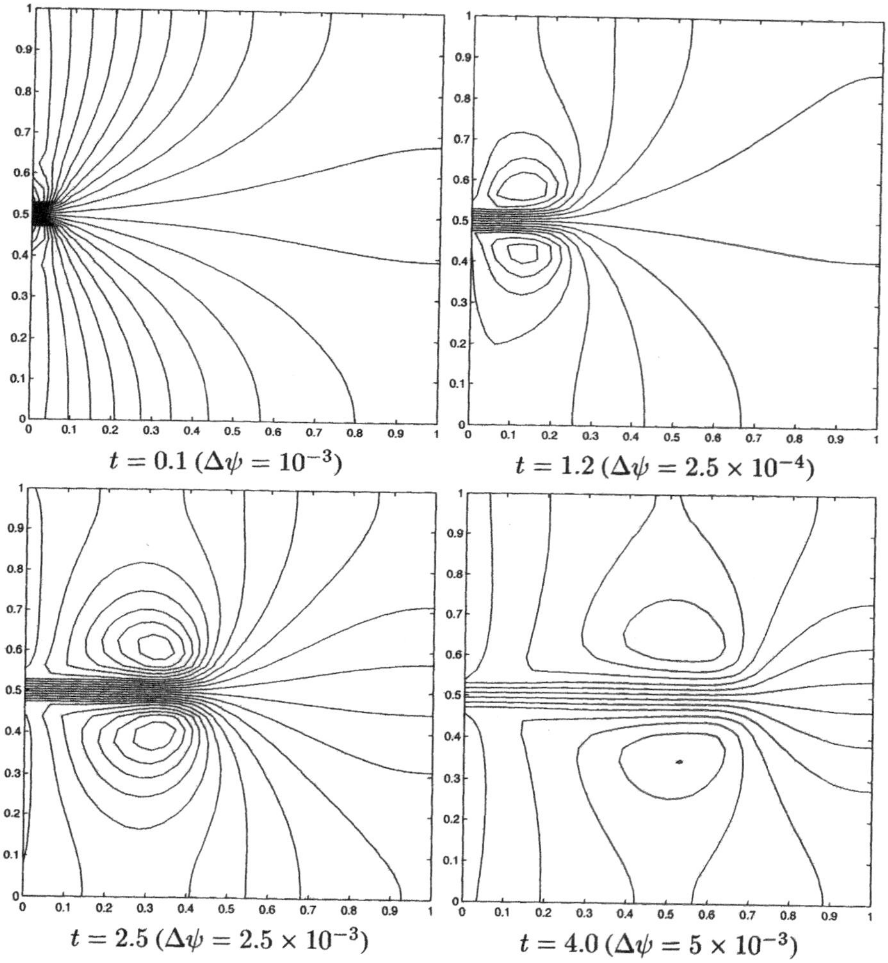

$t = 0.1$ ($\Delta\psi = 10^{-3}$)

$t = 1.2$ ($\Delta\psi = 2.5 \times 10^{-4}$)

$t = 2.5$ ($\Delta\psi = 2.5 \times 10^{-3}$)

$t = 4.0$ ($\Delta\psi = 5 \times 10^{-3}$)

Fig. 6.14 Streamlines for plane jet problem

If the pressure Poisson equation is employed in the second step, see Remark 6.20, again the homogenous natural boundary condition, $\boldsymbol{n} \cdot \boldsymbol{\nabla} p = 0$, imposes the desired open boundary condition on the pressure.

The Chorin–Temam projection method described in Section 6.7.3.1 is used for marching in time. The idea is to present a solution strategy that could easily be programmed by the interested reader. With this in mind, the computational schemes have been selected so as to be characterized by a very simple algorithmic structure. The streamlines at different instants are shown in Figure 6.14.

The fluid is at rest at $t = 0$. The explicit Euler scheme is used for time integration of the first-step equations (although a backward Euler method can also be used to avoid instabilities and would induce similar results). To make explicit schemes

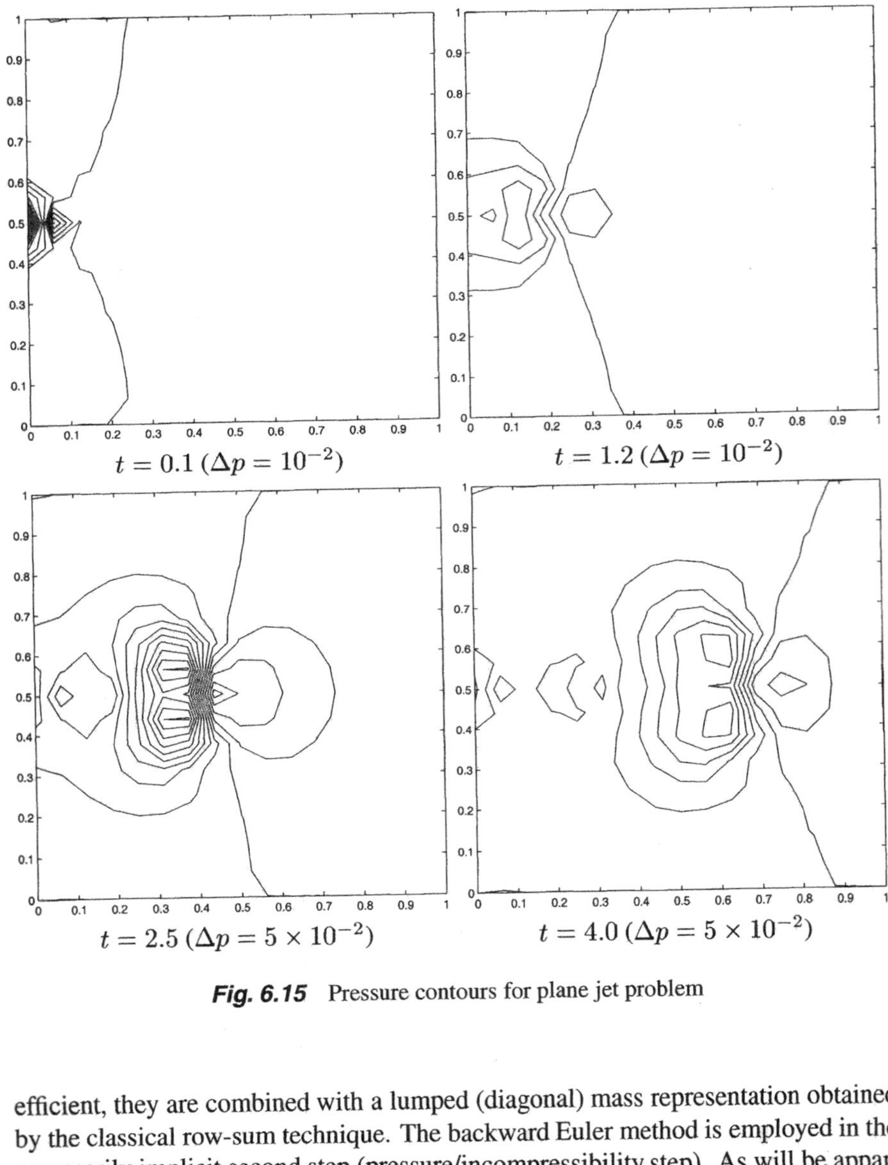

Fig. 6.15 Pressure contours for plane jet problem

efficient, they are combined with a lumped (diagonal) mass representation obtained by the classical row-sum technique. The backward Euler method is employed in the necessarily implicit second step (pressure/incompressibility step). As will be apparent from the numerical results, in addition to its simplicity, the proposed strategy is capable of producing accurate results. The time step is taken to be $\Delta t = 0.01$ as done by Laval and Quartapelle (1990).

The flow pattern, particularly the vortex creation close to the jet aperture, is represented at different instants by the streamlines in Figure 6.14. The corresponding pressure contours are shown in Figure 6.15. The results are in good agreement with those obtained by Laval and Quartapelle (1990) using a fractional-step Taylor–Galerkin method. Their method consists of three phases as follows. A pure convection phase is solved first using a second-order explicit Taylor–Galerkin method. This is followed by a viscous diffusion phase solved using the forward Euler method. Finally, the pressure/incompressibility phase is solved using the backward Euler scheme. The present results can also be compared with those obtained by Blasco, Codina and Huerta (1997) using a predictor-multicorrector algorithm.

6.8.4 Natural convection in a square cavity

To illustrate the use of the Chorin–Temam fractional-step method in another incompressible flow problem, we shall describe the splitting-up approximate solution of unsteady natural convection problems. Such problems involve a coupling between the Navier–Stokes equations describing the fluid motion and the thermal energy equation governing the space–time evolution of the temperature. The forces which induce natural convection are in fact spatially variable gravity forces generated by (buoyancy) density variations in the fluid due to the non-uniformity of the temperature.

6.8.4.1 The Navier–Stokes and temperature equations In natural convection problems the mathematical description of fluid motion is given by the following form of the Navier–Stokes equations:

$$\varrho\big(\boldsymbol{v}_t + (\boldsymbol{v}\cdot\boldsymbol{\nabla})\boldsymbol{v}\big) = \boldsymbol{\nabla}\cdot\boldsymbol{\sigma} + \varrho\boldsymbol{b} \qquad \text{in } \Omega\times]0,T[, \tag{6.46}$$

where

$$\boldsymbol{b} = \big(1 - \beta(\mathsf{T} - \mathsf{T}_0)\big)\,\boldsymbol{g} \tag{6.47}$$

is the gravity force per unit mass derived on the basis of Boussinesq approximation, vector $\boldsymbol{g}$ denotes the gravity field, β is the coefficient of thermal expansion of the fluid, T is the temperature field, T_0 is the reference temperature, and ϱ is the reference density of the fluid (i.e., the fluid density corresponding to the reference temperature, when $\mathsf{T} = \mathsf{T}_0$).

The fluid is assumed incompressible according to Boussinesq approximation. That is, its density is assumed constant, except in the gravity force term where it depends on temperature according to the indicated linear law, see (6.47).

Under the usual assumption of a Newtonian fluid, the components σ_{ij} of Cauchy stress are linearly related to the strain rate by Stokes' law. The incompressibility condition is expressed in the standard form $\boldsymbol{\nabla}\cdot\boldsymbol{v} = 0$, and the Navier–Stokes equations (6.46) are completed with the usual initial and boundary conditions.

The Boussinesq body forces, see (6.47), introduce a coupling between the Navier–Stokes equations and the thermal energy equation

$$\varrho c\big(\mathsf{T}_t + \boldsymbol{v}\cdot\boldsymbol{\nabla}\mathsf{T}\big) = \boldsymbol{\nabla}\cdot(k\boldsymbol{\nabla}\mathsf{T}) + Q \qquad \text{in } \Omega\times]0,T[. \tag{6.48}$$

Here, c denotes the specific heat of the fluid at constant volume, k is the thermal conductivity (assumed isotropic in this case) and Q represents volumetric heat generation.

Equation (6.48) combined with boundary conditions such as

$$
\begin{aligned}
\mathsf{T}(\boldsymbol{x}, t) &= \mathsf{T}_D(\boldsymbol{x}, t) && \text{on } \Gamma_D^{\mathsf{T}} \times]0, T[, \\
k\frac{\partial \mathsf{T}(\boldsymbol{x}, t)}{\partial n} &= q_N(\boldsymbol{x}, t) && \text{on } \Gamma_N^{\mathsf{T}} \times]0, T[,
\end{aligned}
$$

where T_D represents a prescribed temperature, while q_N denotes a prescribed normal heat flux. The initial condition $\mathsf{T}(\boldsymbol{x}, 0) = \mathsf{T}_0(\boldsymbol{x})$ for all $\boldsymbol{x} \in \Omega$, completes the initial boundary value problem for the temperature.

6.8.4.2 A fractional-step method for Navier–Stokes and temperature

The implicit character of pressure in an incompressible fluid precludes the use of purely explicit time integration algorithms in the Navier–Stokes equations. As seen in Section 6.7.3, it is nevertheless possible to take advantage of the algorithmic simplicity of explicit time-stepping schemes by using a fractional-step approach. That is, the convective and Boussinesq terms could be evaluated explicitly. Here a simple example is shown and to simplify the exposition, we assume that Dirichlet conditions for the velocity are prescribed on the whole boundary of the fluid domain.

The first step considers the convective, viscous and Boussinesq terms of equation (6.46), see also (6.37). Thus, the first-step equation reads

$$\boldsymbol{v}_t + (\boldsymbol{v} \cdot \boldsymbol{\nabla})\boldsymbol{v} - \boldsymbol{\nabla} \cdot (2\nu \boldsymbol{\nabla}^{\mathsf{S}} \boldsymbol{v}) = \boldsymbol{b}.$$

Note that this equation is normalized by the reference density, ϱ, of the fluid. The simplest option for this first step is an explicit time integration scheme. For instance, we can use the explicit Euler method or the second-order Adams–Bashforth method described in Section 5.3.1. The result is an intermediate velocity field $\boldsymbol{v}_{\text{int}}^{n+1}$, satisfying the Dirichlet boundary conditions, that is $\boldsymbol{v}_{\text{int}}^{n+1} = \boldsymbol{v}_D^{n+1}$.

The second step of the time integration of the Navier–Stokes equations is the pressure/incompressibiblity step. The end-of-step velocity field $\boldsymbol{v}^{n+1}$ is determined adding the intermediate velocity $\boldsymbol{v}_{\text{int}}^{n+1}$ and the effect of the pressure gradient. Recall that pressure is such that the end-of-step velocity is solenoidal. Time integration is necessarily implicit and the simplest algorithm is given by the backward Euler method, see equations (6.40).

A two-step procedure can also be used to update temperature from T^n to T^{n+1}. The convective term is integrated in the first step and the diffusion and source terms in the second one, see equation (6.48). Thus, the first step reduces to the hyperbolic equation

$$\mathsf{T}_t + \boldsymbol{v} \cdot \boldsymbol{\nabla} \mathsf{T} = 0.$$

The third-order explicit Taylor–Galerkin method, see Section 3.6.2, can be used for the time integration of this equation. An intermediate temperature field, T_{int}, is then obtained. Since this is a pure convection equation Dirichlet boundary condition $\mathsf{T}_{\text{int}} = \mathsf{T}_D^{n+1}$ are only prescribed on the inflow portion of Γ_D^{T} (i.e., where $\boldsymbol{n} \cdot \boldsymbol{v} < 0$).

The second step, which accounts for the diffusion and source terms, reads

$$\varrho c\, \mathsf{T}_t = \boldsymbol{\nabla} \cdot (k \boldsymbol{\nabla} \mathsf{T}) + Q.$$

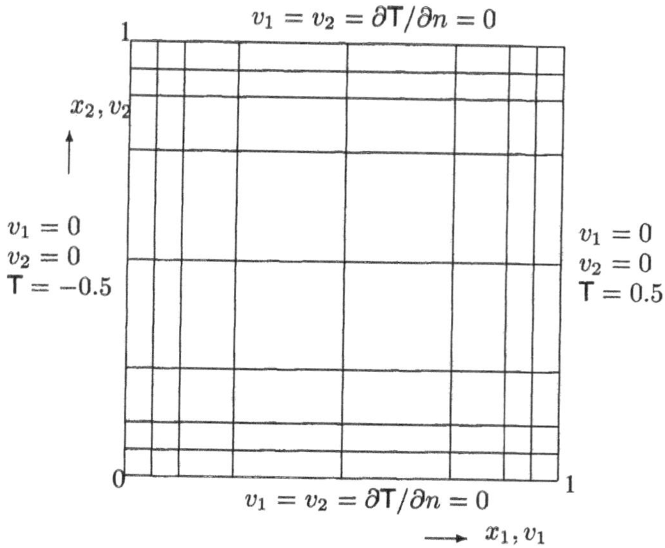

Fig. 6.16 Natural convection: problem statement and discretization.

The complete (Dirichlet and Neumann) boundary conditions are imposed. In this parabolic phase an implicit time-stepping algorithm is to be preferred since the stable time step for explicit methods decreases with the square of the mesh size. For instance, the second-order Crank–Nicolson method can be a reasonable choice.

6.8.4.3 The numerical example The classical test problem of a differentially heated square cavity is chosen to illustrate the use of the fractional-step method for solving natural convection flows in enclosures.

The problem statement is depicted in Figure 6.16. At the initial time, a fluid at uniform temperature $T_0 = 0$ is at rest in a square cavity with unit sides. At the start of the calculation, the temperature of the left wall of the cavity is suddenly lowered by 0.5, while the temperature of the right wall is increased by the same quantity. The horizontal walls are assumed to be thermally insulated (zero heat flux). No-slip and no penetration velocity boundary conditions are prescribed on all walls.

The motion induced by natural convection is governed by two dimensionless numbers, the Rayleigh and the Prandtl numbers defined by

$$Ra = \frac{g\beta L_{\text{ref}}^3\,\Delta T}{\nu\,\kappa} \qquad \text{and} \qquad Pr = \frac{\nu}{\kappa},$$

where g is the gravitational acceleration, β the coefficient of thermal expansion, ΔT the temperature difference between the hot and cold walls, ν denotes the kinematic viscosity and κ is the thermal diffusivity of the fluid. The reference length is $L_{\text{ref}} = 1$, the length of the cavity side.

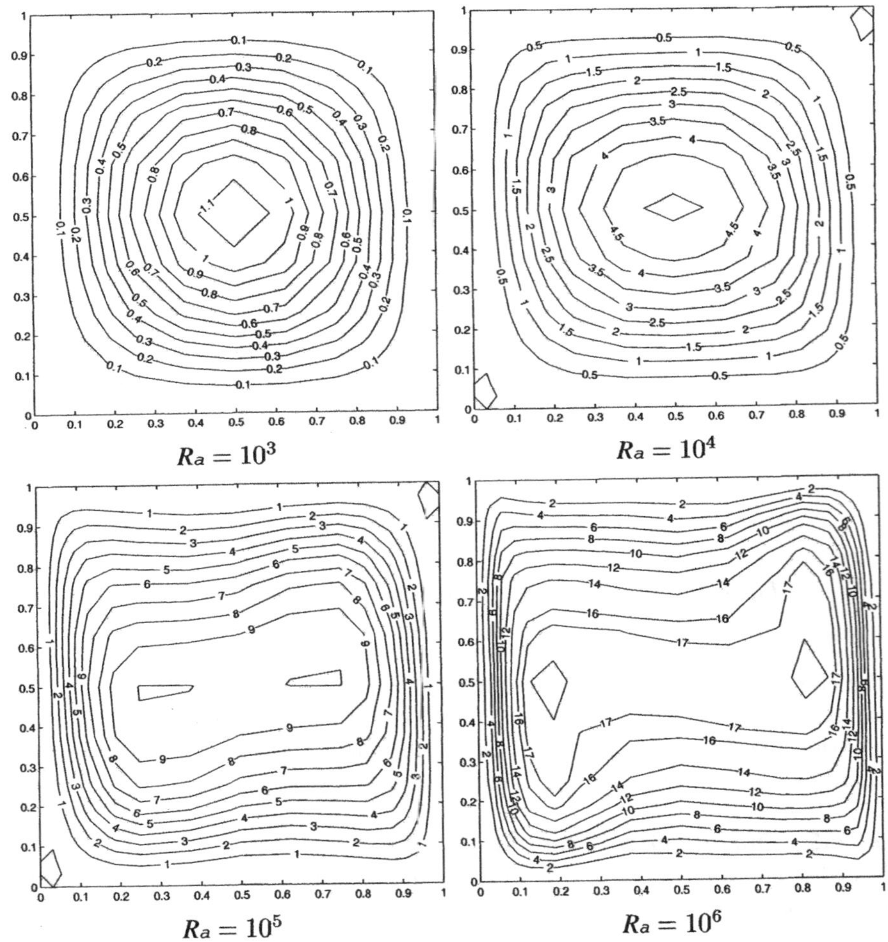

Fig. 6.17 Natural convection: streamlines in the thermal cavity for different Rayleigh numbers and Pr = 1.

The problem is symmetric with respect to the center of the cavity. This property can be exploited to reduce the cost of the simulation or to test the quality of the numerical results. We select the second option. The finite element computation is performed using the non-uniform mesh of 8×8 Q2Q1 elements for the mixed interpolation of velocity and pressure, see Figure 6.16. A biquadratic interpolation is used for the temperature. The time integration based upon the fractional-step approach described in Section 6.8.4.2 is used until a steady state is reached.

The results are for a Prandtl number of one and Rayleigh numbers from 10^3 to 10^6. The development and the structure of the flow vary considerably with the Rayleigh number. The main characteristics of the flow are illustrated in Figures 6.17 and 6.18:

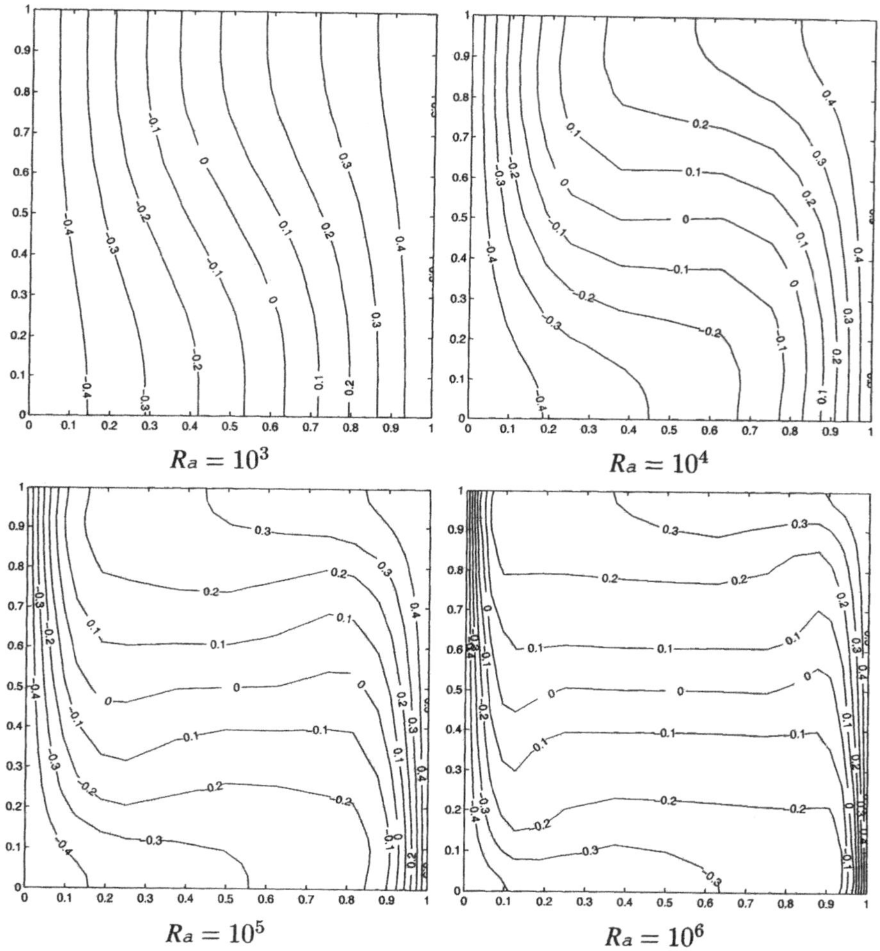

Fig. 6.18 Natural convection: isotherms in the thermal cavity for different Rayleigh numbers and Pr $= 1$.

1. A secondary flow develops at a Rayleigh number situated between 10^4 and 10^5.
2. A boundary layer appears progressively in the vicinity of the vertical walls of the cavity.
3. The secondary flow evolves at the center of the cavity in the form of "eyes of a cat".

The computed streamlines are plotted in Figure 6.17. For $Ra < 10^5$, the flow is mono-cellular. The fluid rises along the hot side of the cavity (right wall) and comes down along the cold wall. For $Ra = 10^5$, the development of a secondary flow starts in the central portion of the cavity. The circulation direction of the secondary flow is the same as that of the base flow. The value of the stream function at the center

Table 6.3 Stream function values at the center of the thermal cavity

Rayleigh number		10^3	10^4	10^5	10^6
Present simulation	(8 × 8)	1.19	5.15	9.74	17.99
Heinrich et al. (1978)	(4 × 4)	1.18	5.13	9.44	—
Marshall et al. (1978)	(8 × 8)	—	5.12	9.54	17.32

of the cavity gives a measure of the importance of the recirculation. The results are displayed in Table 6.3 in comparison with other simulations from the literature.

The isotherms in the cavity are displayed in Figure 6.18. At low values of the Rayleigh number they are practically straight lines, in accordance with a diffusion-dominated heat transfer process within the cavity. When the Rayleigh number increases, the convective effects become more and more important. The isotherms deform progressively and take an S shape. The distortion of the isotherms generates high temperature gradients near the vertical walls of the cavity.

References

Abdallah, S. (1995), 'Comments on the fractional step method: "An analysis of the fractional step method" [J. Comput. Phys. **108** (1993), no. 1, 51–58] by J. B. Perot', *J. Comput. Phys.* **117**(1), 179–180.

Adams, R. A. (1975), *Sobolev spaces*, Vol. 65 of *Pure and Applied Mathematics*, Academic Press, New York.

Ainsworth, M. (2001), 'Essential boundary conditions and multi-point constraints in finite element analysis', *Comput. Methods Appl. Mech. Eng.* **190**(48), 6323–6339.

Ainsworth, M. and Oden, J. T. (2000), *A posteriori error estimation in finite element analysis*, John Wiley & Sons, Chichester.

Allievi, A. and Bermejo, R. (2000), 'Finite element modified method of characteristics for the Navier-Stokes equations', *Int. J. Numer. Methods Fluids* **32**(4), 439–464.

Ames, W. F. (1992), *Numerical methods for partial differential equations*, third edn, Academic Press, New York.

Arnold, D. N., Brezzi, F. and Fortin, M. (1984), 'A stable finite element for the Stokes equations', *Calcolo* **21**(4), 337–344.

Babuška, I. (1970/71), 'Error-bounds for finite element method', *Numer. Math.* **16**, 322–333.

Baiocchi, C., Brezzi, F. and Franca, L. (1993), 'Virtual bubbles and the Galerkin-least-squares method', *Comput. Methods Appl. Mech. Eng.* **105**(1), 125–141.

Baker, A. J. (1983), *Finite element computational fluid mechanics*, Hemisphere, Washington, D.C.

Batchelor, G. K. (1999), *An introduction to fluid dynamics*, Cambridge University Press, Cambridge.

Bathe, K.-J., Hendriana, D., Brezzi, F. and Sangalli, G. (2000), 'Inf-sup testing of upwind methods', *Int. J. Numer. Methods Eng.* **48**(5), 745–760.

Baumann, C. E. and Oden, J. T. (1999), 'A discontinuous hp finite element method for convection-diffusion problems', *Comput. Methods Appl. Mech. Eng.* **175**(3–4), 311–341.

Baumann, C. E. and Oden, J. T. (2000), 'An adaptive-order discontinuous Galerkin method for the solution of the Euler equations of gas dynamics', *Int. J. Numer. Methods Eng.* **47**(1–3), 61–73.

Bell, J. B., Colella, P. and Glaz, H. M. (1989), 'A second-order projection method for the incompressible Navier-Stokes equations', *J. Comput. Phys.* **85**(2), 257–283.

Belytschko, T. (1983), 'An overview of semidiscretization and time integration procedures', *in* T. Belytschko and T. J. R. Hughes, eds, *Computational methods for transient analysis*, Vol. 1 of *Mechanics and Mathematical Methods. A Series of Handbooks. Subseries: Computational Methods in Mechanics*, Elsevier Scientific, Amsterdam, Chap. 1, pp. 1–65.

Belytschko, T. and Eldib, I. (1979), 'Analysis of a finite element upwind scheme', *in* T. J. R. Hughes, ed., *Finite element methods for convection dominated flows*, AMD – Vol. 34, Presented at the Winter Annual Meeting of the ASME, Amer. Soc. Mech. Engrs. (ASME), New York, pp. 195–200.

Belytschko, T. and Kennedy, J. M. (1978), 'Computer methods for subassembly simulation', *Nucl. Eng. Des.* **49**, 17–38.

Belytschko, T., Kennedy, J. M. and Schoeberle, D. F. (1978), 'Quasi-Eulerian finite element formulation for fluid-structure interaction', Proceedings of the Joint ASME/CSME Pressure Vessels and Piping Conference, Amer. Soc. Mech. Engrs. (ASME), New York, p. 13. ASME paper 78-PVP-60.

Belytschko, T., Krongauz, Y., Organ, D., Fleming, M. and Krysl, P. (1996), 'Meshless methods: an overview and recent developments', *Comput. Methods Appl. Mech. Eng.* **139**(1–4), 3–48.

Belytschko, T., Liu, W. K. and Moran, B. (2000), *Nonlinear finite elements for continua and structures*, John Wiley & Sons, Chichester.

Belytschko, T., Lu, Y. Y. and Gu, L. (1994), 'Element-free Galerkin methods', *Int. J. Numer. Methods Eng.* **37**(2), 229–256.

Benqué, J. P., Ibler, B., Keramsi, A. and Labadie, G. (1980), 'A finite element method for the Navier–Stokes equations', Proceedings from the Third International Conference on Finite Elements in Flow Problems, Banff, Alberta, Canada, June 10–13, 1980, Vol. I, pp. 110–120.

Benqué, J. P., Labadie, G. and Ronat, J. (1982), 'A new finite element method for the Navier–Stokes equations coupled with a temperature equation', in T. Kawai, ed., Finite element flow analysis, Proceedings of the Fourth International Symposium, (Tokyo 1982), North-Holland, Amsterdam, pp. 295–302.

Bercovier, M. (1978), 'Perturbation of mixed variational problems. Application to mixed finite element methods', *RAIRO Anal. Numér.* **12**(3), 211–236.

Bercovier, M. and Engelman, M. (1979), 'A finite element for the numerical solution of viscous incompressible flows', *J. Comput. Phys.* **30**(2), 181–201.

Bermejo, R. (1995), 'A Galerkin-characteristic algorithm for transport-diffusion equations', *SIAM J. Numer. Anal.* **32**(2), 425–454.

Blasco, J., Codina, R. and Huerta, A. (1997), 'Analysis of fractional step finite element methods for the incompressible Navier-Stokes equations', Technical Report 38, International Center for Numerical Methods in Engineering (CIMNE), Barcelona.

Blasco, J., Codina, R. and Huerta, A. (1998), 'A fractional-step method for the incompressible Navier-Stokes equations related to a predictor-multicorrector algorithm', *Int. J. Numer. Methods Fluids* **28**(10), 1391–1419.

Bonet, J. and Wood, R. D. (1997), *Nonlinear continuum mechanics for finite element analysis*, Cambridge University Press, Cambridge.

Boris, J. P. and Book, D. L. (1997), 'Flux-corrected transport. I. SHASTA, a fluid transport algorithm that works [J. Comput. Phys. **11** (1973), no. 1, 38–69]', *J. Comput. Phys.* **135**(2), 170–186. With an introduction by Steven T. Zalesak, Commemoration of the 30th anniversary of *J. Comput. Phys.*

Brezzi, F. (1974), 'On the existence, uniqueness and approximation of saddle-point problems arising from Lagrangian multipliers', *RAIRO Anal. Numér.* **8**(R-2), 129–151.

Brezzi, F. and Bathe, K.-J. (1990), 'A discourse on the stability conditions for mixed finite element formulations', *Comput. Methods Appl. Mech. Eng.* **82**(1-3), 27–57.

Brezzi, F. and Fortin, M. (1991), *Mixed and hybrid finite element methods*, Vol. 15 of *Springer Series in Computational Mathematics*, Springer-Verlag, New York.

Brezzi, F., Franca, L. P. and Russo, A. (1998), 'Further considerations on residual-free bubbles for advective-diffusive equations', *Comput. Methods Appl. Mech. Eng.* **166**(1–2), 25–33.

Brooks, A. N. and Hughes, T. J. R. (1982), 'Streamline upwind/Petrov-Galerkin formulations for convection dominated flows with particular emphasis on the incompressible Navier-Stokes equations', *Comput. Methods Appl. Mech. Eng.* **32**(1–3), 199–259.

Bung, H., Casadei, F., Halleux, J.-P. and Lepareux, M. (1989), 'PLEXIS-3C: a computer code for fast dynamic problems in structures and fluids', Proceedins of the 10th SMIRT Conference, Anaheim, USA.

Burggraf, O. R. (1966), 'Analytical and numerical studies of the structure of steady separated flows', *J. Fluid Mechanics* **24**(1), 113–152.

Carette, J.-C. (1997), 'Adaptive unstructured mesh algorithms and SUPG finite element method for compressible high Reynolds number flows', PhD thesis, Université Libre de Bruxelles, Belgium.

Carey, G. F. and Jiang, B. N. (1988), 'Least-squares finite elements for first-order hyperbolic systems', *Int. J. Numer. Methods Eng.* **26**(1), 81–93.

Carey, G. F. and Oden, J. T. (1983), *Finite elements. A second course*, Vol. II of *The Texas Finite Element Series*, Prentice Hall, Englewood Cliffs, NJ.

Carey, G. F. and Oden, J. T. (1986), *Finite elements. Fluid mechanics*, Vol. VI of *The Texas Finite Element Series*, Prentice Hall, Englewood Cliffs, NJ.

Carey, G. F., Shen, Y. and McLay, R. T. (1998), 'Parallel conjugate gradient performance for least-squares finite elements and transport problems', *Int. J. Numer. Methods Fluids* **28**(10), 1421–1440.

Casadei, F. and Halleux, J.-P. (1995), 'An algorithm for permanent fluid-structure interaction in explicit transient dynamics', *Comput. Methods Appl. Mech. Eng.* **128**(3–4), 231–289.

Casadei, F., Halleux, J.-P., Sala, A. and Chillè, F. (2001), 'Transient fluid-structure interaction algorithms for large industrial applications', *Comput. Methods Appl. Mech. Eng.* **190**(24–25), 3081–3110.

Casadei, F. and Sala, A. (1999), 'Finite element and finite volume simulation of industrial fast transient fluid-structure interactions', European Conference on Computational Mechanics, ECCM'99, Munich.

Chorin, A. J. (1968), 'Numerical solution of the Navier-Stokes equations', *Math. Comput.* **22**, 745–762.

Chorin, A. J. (1969), 'On the convergence of discrete approximations to the Navier-Stokes equations', *Math. Comput.* **23**, 341–353.

Christie, I. (1985), 'Upwind compact finite difference schemes', *J. Comput. Phys.* **59**(2), 353–368.

Christie, I., Griffiths, D. F., Mitchell, A. R. and Zienkiewicz, O. C. (1976), 'Finite element methods for second order differential equations with significant first derivatives', *Int. J. Numer. Methods Eng.* **10**(6), 1389–1396.

Christon, M. A. (1991), 'The influence of the mass matrix on the dispersive nature of the semi-discrete, second-order wave equation', *Comput. Methods Appl. Mech. Eng.* **173**(1–2), 147–166.

Ciarlet, P. G. (1978), *The finite element method for elliptic problems*, Vol. 4 of *Studies in Mathematics and its Applications*, North-Holland, Amsterdam.

Cockburn, B. (1998), 'An introduction to the discontinuous Galerkin method for convection-dominated problems', *in* A. Quarteroni, ed., *Advanced numerical approximation of nonlinear hyperbolic equations*, Vol. 1697 of *Lecture Notes in Mathematics*, Springer-Verlag, Berlin, pp. 151–268. Papers from the C.I.M.E. Summer School held in Cetraro, June 23–28, 1997.

Cockburn, B. and Shu, C.-W. (2001), 'Runge-Kutta discontinuous Galerkin methods for convection-dominated problems', *J. Sci. Comput.* **16**(3), 173–261.

Codina, R. (1993a), 'A discontinuity-capturing crosswind-dissipation for the finite element solution of the convection-diffusion equation', *Comput. Methods Appl. Mech. Eng.* **110**(3–4), 325–342.

Codina, R. (1993b), 'A finite element formulation for the numerical solution of the convection-diffusion equation', Technical Report 14, International Center for Numerical Methods in Engineering (CIMNE), Barcelona.

Codina, R. (1998), 'Comparison of some finite element methods for solving the diffusion-convection-reaction equation', *Comput. Methods Appl. Mech. Eng.* **156**(1–4), 185–210.

Codina, R. (2000), 'On stabilized finite element methods for linear systems of convection-diffusion-reaction equations', *Comput. Methods Appl. Mech. Eng.* **188**(1–3), 61–82.

Codina, R., Vázquez, M. and Zienkiewicz, O. C. (1998), 'A general algorithm for compressible and incompressible flows. III. The semi-implicit form', *Int. J. Numer. Methods Fluids* **27**(1–4, Special Issue), 13–32.

Courant, R., Friedrichs, K. and Lewy, H. (1967), 'On the partial difference equations of mathematical physics', *IBM J. Res. Develop.* **11**, 215–234. English translation of an article originally published in German in *Math. Ann.* **100** (1928), 32–74.

Crouzeix, M. and Raviart, P.-A. (1973), 'Conforming and nonconforming finite element methods for solving the stationary Stokes equations. I', *RAIRO Anal. Numér.* **7**(R-3), 33–75.

Dahmen, W., Kurdila, A. J. and Oswald, P. (eds) (1997), *Multiscale wavelet methods for partial differential equations*, Academic Press, San Diego, CA.

Davis, G. D. V. and Mallinson, G. (1976), 'An evaluation of upwind and central difference approximations by a study of recirculating flow', *Comput. Fluids* **4**, 29–43.

Deconinck, H., Hirsch, C. and Peuteman, J. (1986), 'Characteristic decomposition methods for the multidimensional Euler equations', *Tenth international conference on numerical methods in fluid dynamics (Beijing, 1986)*, Vol. 264 of *Lecture Notes in Phys.*, Springer-Verlag, Berlin, pp. 216–221.

Donea, J. (1983), 'Arbitrary Lagrangian-Eulerian finite element methods', *in* T. Belytschko and T. J. R. Hughes, eds, *Computational methods for transient analysis*, Vol. 1 of *Mechanics and Mathematical Methods. A Series of Handbooks. Subseries: Computational Methods in Mechanics*, Elsevier Scientific, Amsterdam, Chap. 10, pp. 473–516.

Donea, J. (1984), 'A Taylor-Galerkin method for convective transport problems', *Int. J. Numer. Methods Eng.* **20**(24), 101–120.

Donea, J., Belytschko, T. and Smolinski, P. (1985), 'A generalized Galerkin method for steady convection–diffusion problems with application to quadratic shape functions', *Comput. Methods Appl. Mech. Eng.* **48**(1), 25–43.

Donea, J., Fasoli-Stella, P. and Giuliani, S. (1977), 'Lagrangian and Eulerian finite element techniques for transient fluid-structure interaction problems', Transactions of the 4th SMIRT Conference, Vol. B. paper B1/2, San Francisco, 15-19 August.

Donea, J. and Giuliani, S. (1981), 'A simple method to generate high-order accurate convection operators for explicit schemes based on linear finite elements', *Int. J. Numer. Methods Fluids* **1**(1), 63–79.

Donea, J., Giuliani, S. and Halleux, J.-P. (1982), 'An Arbitrary Lagrangian-Eulerian finite element method for transient dynamic fluid-structure interactions', *Comput. Methods Appl. Mech. Eng.* **33**, 689–723.

Donea, J., Giuliani, S., Laval, H. and Quartapelle, L. (1982), 'Finite element solution of the unsteady Navier-Stokes equations by a fractional step method', *Comput. Methods Appl. Mech. Eng.* **30**, 53–73.

Donea, J. and Quartapelle, L. (1992), 'An introduction to finite element methods for transient advection problems', *Comput. Methods Appl. Mech. Eng.* **95**(2), 169–203.

Donea, J., Quartapelle, L. and Selmin, V. (1987), 'An analysis of time discretization in the finite element solution of hyperbolic problems', *J. Comput. Phys.* **70**(2), 463–499.

Donea, J., Roig, B. and Huerta, A. (1998), 'High-order accurate time-stepping schemes for convection-diffusion problems', Technical Report 42, International Center for Numerical Methods in Engineering (CIMNE), Barcelona.

Donea, J., Roig, B. and Huerta, A. (2000), 'High-order accurate time-stepping schemes for convection-diffusion problems', *Comput. Methods Appl. Mech. Eng.* **182**(3–4), 249–275.

Donea, J., Selmin, V. and Quartapelle, L. (1988), 'Recent developments of the Taylor-Galerkin method for the numerical solution of hyperbolic problems', *Numerical methods for fluid dynamics, III (Proceedings of the 3rd Conference, Oxford, 1988)*, Vol. 17 of *Inst. Math. Appl. Conf. Ser. New Ser.*, Oxford Univ. Press, New York, pp. 171–185.

Douglas, Jr, J. and Russell, T. F. (1982), 'Numerical methods for convection-dominated diffusion problems based on combining the method of characteristics with finite element or finite difference procedures', *SIAM J. Numer. Anal.* **19**(5), 871–885.

Duarte, C. A. and Oden, J. T. (1996), 'H-p clouds—an h-p meshless method', *Numer. Meth. Part. Differ. Equ.* **12**(6), 673–705.

Felippa, C. A. (2001), 'A historical outline of matrix structural analysis: a play in three acts', *Comput. Struct.* **79**(14), 1313–1324.

Franca, L. P. and Frey, S. L. (1992), 'Stabilized finite element methods: II. The incompressible Navier-Stokes equations', *Comput. Methods Appl. Mech. Eng.* **99**(2–3), 209–233.

Franca, L. P., Frey, S. L. and Hughes, T. J. R. (1992), 'Stabilized finite element methods. I. Application to the advective-diffusive model', *Comput. Methods Appl. Mech. Eng.* **95**(2), 253–276.

Franca, L. P. and Hughes, T. J. R. (1993), 'Convergence analyses of Galerkin least-squares methods for symmetric advective-diffusive forms of the Stokes and incompressible Navier-Stokes equations', *Comput. Methods Appl. Mech. Eng.* **105**(2), 285–298.

Franca, L. P., Nesliturk, A. and Stynes, M. (1998), 'On the stability of residual-free bubbles for convection-diffusion problems and their approximation by a two-level finite element method', *Comput. Methods Appl. Mech. Eng.* **166**(1–2), 35–49.

Galeão, A. C. and do Carmo, E. G. D. (1988), 'A consistent approximate upwind Petrov-Galerkin method for convection dominated problems', *Comput. Methods Appl. Mech. Eng.* **68**(1), 83–95.

Gear, C. W. (1971), *Numerical initial value problems in ordinary differential equations*, Prentice Hall, Englewood Cliffs, NJ.

Giles, M. B. (1997), 'Stability analysis of a Galerkin/Runge-Kutta Navier-Stokes discretisation on unstructured tetrahedral grids', *J. Comput. Phys.* **132**(2), 201–214.

Girault, V. and Raviart, P.-A. (1986), *Finite element methods for Navier-Stokes equations. Theory and algorithms*, Springer-Verlag, Berlin.

Giuliani, S. (1982), 'An algorithm for continuous rezoning of the hydrodynamic grid in Arbitrary Lagrangian-Eulerian computer codes', *Nucl. Eng. Des.* **72**, 205–212.

Glowinski, R. (1984), *Numerical methods for nonlinear variational problems*, Springer-Verlag, New York.

Glowinski, R. and Le Tallec, P. (1989), *Augmented Lagrangian and operator-splitting methods in nonlinear mechanics*, Society for Industrial and Applied Mathematics (SIAM), Philadelphia, PA.

Godlewski, E. and Raviart, P.-A. (1996), *Numerical Approximation of Hyperbolic Systems of Conservation Laws*, Vol. 118 of *Springer Series in Applied Mathematical Sciences*, Springer-Verlag, Berlin.

Godunov, S. K. (1959), 'A difference method for numerical calculation of discontinuous solutions of the equations of hydrodynamics', *Mat. Sb. (N.S.)* **47 (89)**, 271–306.

Gresho, P. M. (1990), 'On the theory of semi-implicit projection methods for viscous incompressible flow and its implementation via a finite element method that also introduces a nearly consistent mass matrix. I. Theory', *Int. J. Numer. Methods Fluids* **11**(5), 587–620.

Gresho, P. M. and Lee, R. L. (1979), 'Don't suppress the wiggles—they're telling you something!', *in* T. J. R. Hughes, ed., *Finite element methods for convection dominated flows*, AMD – Vol. 34, Presented at the Winter Annual Meeting of the ASME, Amer. Soc. Mech. Engrs. (ASME), New York, pp. 37–61.

Gresho, P. M. and Sani, R. L. (1987), 'On pressure boundary conditions for the incompressible Navier-Stokes equations', *Int. J. Numer. Methods Fluids* **7**(10), 1111–1145.

Gresho, P. M. and Sani, R. L. (2000), *Incompressible flow and the finite element method. Vol. 1: Advection diffusion. Vol. 2: Isothermal laminar flow*, John Wiley & Sons, Chichester.

Griffiths, D. F. and Mitchell, A. R. (1979), 'On generating upwind finite element methods', *in* T. J. R. Hughes, ed., *Finite element methods for convection dominated flows*, AMD – Vol. 34, Presented at the Winter Annual Meeting of the ASME, Amer. Soc. Mech. Engrs. (ASME), New York, pp. 91–104.

Guermond, J.-L. (1999a), 'Stabilisation par viscosité de sous-maille pour l'approximation de Galerkin des opérateurs linéaires monotones', *Comptes Rendus Acad. Sci. Ser. I-Math.* **328**(7), 617–622.

Guermond, J.-L. (1999b), 'Stabilization of Galerkin approximations of transport equations by subgrid modeling', *ESAIM-Math. Model. Numer. Anal.* **33**(6), 1293–1316.

Guermond, J.-L. and Quartapelle, L. (1997), 'Calculation of incompressible viscous flow by an unconditionally stable projection FEM', *J. Comput. Phys.* **132**(1), 12–33.

Guermond, J.-L. and Quartapelle, L. (1998a), 'On stability and convergence of projection methods based on pressure Poisson equation', *Int. J. Numer. Methods Fluids* **26**(9), 1039–1053.

Guermond, J.-L. and Quartapelle, L. (1998b), 'On the approximation of the unsteady Navier-Stokes equations by finite element projection methods', *Numer. Math.* **80**(2), 207–238.

Gunzburger, M. D. (1989), *Finite element methods for viscous incompressible flows. A guide to theory, practice, and algorithms*, Academic Press, Boston, MA.

Hairer, E., Nørsett, S. P. and Wanner, G. (1993), *Solving ordinary differential equations. Vol. I. Nonstiff problems*, second edn, Springer-Verlag, Berlin.

Hairer, E. and Wanner, G. (1996), *Solving ordinary differential equations. Vol. II. Stiff and differential–algebraic problems*, second edn, Springer-Verlag, Berlin.

Hannani, S. K., Stanislas, M. and Dupont, P. (1995), 'Incompressible Navier-Stokes computations with SUPG and GLS formulations', *Comput. Methods Appl. Mech. Eng.* **124**(1–2), 153–170.

Hansbo, P. (1993), 'Explicit streamline diffusion finite element methods for the compressible Euler equations in conservation variables', *J. Comput. Phys.* **109**(2), 274–288.

Hansbo, P. and Johnson, C. (1991), 'Adaptive streamline diffusion methods for compressible flows using conservation variables', *Comput. Methods Appl. Mech. Eng.* **87**(2–3), 267–280.

Hansbo, P. and Szepessy, A. (1990), 'A velocity-pressure streamline diffusion finite element method for the incompressible Navier-Stokes equations', *Comput. Methods Appl. Mech. Eng.* **84**(2), 175–192.

Harari, I., Frey, S. and Franca, L. P. (2002), 'A note on a recent study of stabilized finite element computations for heat conduction', *Comput. Mech.* **28**(1), 63–65.

Harari, I. and Hughes, T. J. R. (1994), 'Stabilized finite element methods for steady advection-diffusion with production', *Comput. Methods Appl. Mech. Eng.* **115**(1–2), 165–191.

Harten, A. (1983), 'High resolution schemes for hyperbolic conservation laws', *J. Comput. Phys.* **49**(3), 357–393.

Harten, A. (1997), 'High resolution schemes for hyperbolic conservation laws', *J. Comput. Phys.* **135**(2), 259–278. With an introduction by Peter Lax, Commemoration of the 30th anniversary of *J. Comput. Phys.*

Harten, A., Hyman, J. M. and Lax, P. D. (1976), 'On finite-difference approximations and entropy conditions for shocks', *Commun. Pure Appl. Math.* **29**(3), 297–322. With an appendix by B. Keyfitz.

Harten, A. and Tal-Ezer, H. (1981), 'On a fourth order accurate implicit finite difference scheme for hyperbolic conservation laws. I. Nonstiff strongly dynamic problems', *Math. Comput.* **36**(154), 353–373.

Hauke, G. and Hughes, T. J. R. (1998), 'A comparative study of different sets of variables for solving compressible and incompressible flows', *Comput. Methods Appl. Mech. Eng.* **153**(1–2), 1–44.

Heinrich, J. C., Huyakorn, P. S., Zienkiewicz, O. C. and Mitchell, A. R. (1977), 'An "upwind" finite element scheme for two-dimensional convective transport equation', *Int. J. Numer. Methods Eng.* **11**(1), 131–143.

Heinrich, J. C., Marshall, R. S. and Zienkiewicz, O. C. (1978), 'Penalty function solution of coupled convective and conductive heat transfer', *in* C. Taylor, K. Morgan and C. Brebbia, eds, *Numerical Methods in Laminar and Turbulent Flows*, Pentech Press, pp. 435–447.

Heinrich, J. C. and Zienkiewicz, O. C. (1979), 'The finite element method and "upwinding" techniques in the numerical solution of convection dominated flow problems', *in* T. J. R. Hughes, ed., *Finite element methods for convection dominated flows*, AMD – Vol. 34, Presented at the Winter Annual Meeting of the ASME, Amer. Soc. Mech. Engrs. (ASME), New York, pp. 105–136.

Heywood, J. G. and Rannacher, R. (1982), 'Finite element approximation of the nonstationary Navier-Stokes problem. I. Regularity of solutions and second-order error estimates for spatial discretization', *SIAM J. Numer. Anal.* **19**(2), 275–311.

Heywood, J. G. and Rannacher, R. (1986), 'Finite element approximation of the nonstationary Navier-Stokes problem. II. Stability of solutions and error estimates uniform in time', *SIAM J. Numer. Anal.* **23**(4), 750–777.

Heywood, J. G. and Rannacher, R. (1988), 'Finite element approximation of the nonstationary Navier-Stokes problem. III. Smoothing property and higher order error estimates for spatial discretization', *SIAM J. Numer. Anal.* **25**(3), 489–512.

Heywood, J. G. and Rannacher, R. (1990), 'Finite-element approximation of the nonstationary Navier-Stokes problem. IV. Error analysis for second-order time discretization', *SIAM J. Numer. Anal.* **27**(2), 353–384.

Hirsch, C. (1990), *Numerical computation of internal and external flows. Vol. 2: Computational methods for inviscid and viscous flows*, Wiley Series in Numerical Methods in Engineering, John Wiley & Sons, Chichester.

Hirt, C. W., Amsden, A. A. and Cook, J. L. (1974), 'An Arbitrary Lagrangian-Eulerian computing method for all flow speeds', *J. Comput. Phys.* **14**(3), 227–253.

Hirt, C. W., Amsden, A. A. and Cook, J. L. (1997), 'An Arbitrary Lagrangian-Eulerian computing method for all flow speeds', *J. Comput. Phys.* **135**(2), 203–216. Reprinted from ibid. **14**, 227–253 (1974).

Huerta, A. and Donea, J. (2002), 'Time-accurate solution of stabilized convection-diffusion-reaction equations: I — Time and space discretization', *Commun. Numer. Methods Eng.* **18**(8), 565–573.

Huerta, A. and Fernández-Méndez, S. (2000), 'Enrichment and coupling of the finite element and meshless methods', *Int. J. Numer. Methods Eng.* **48**(11), 1615–1636.

Huerta, A. and Fernández-Méndez, S. (2003), 'Time accurate consistently stabilized mesh-free methods for convection dominated problems', *Int. J. Numer. Methods Eng.* **59**(9), 1225–1242.

Huerta, A. and Liu, W. K. (1988), 'Viscous flow with large free surface motion', *Comput. Methods Appl. Mech. Eng.* **69**(3), 277–324.

Huerta, A., Rodríguez-Ferran, A., Díez, P. and Sarrate, J. (1999), 'Adaptive finite element strategies based on error assessment', *Int. J. Numer. Methods Eng.* **46**(10), 1803–1818.

Huerta, A., Roig, B. and Donea, J. (2002), 'Time-accurate solution of stabilized convection-diffusion-reaction equations: II — Accuracy analysis and examples', *Commun. Numer. Methods Eng.* **18**(8), 575–584.

Hughes, T. J. R. (1978), 'A simple scheme for developing "upwind" finite elements', *Int. J. Numer. Methods Eng.* **12**(9), 1359–1365.

Hughes, T. J. R. (1995), 'Multiscale phenomena: Green's functions, the Dirichlet-to-Neumann formulation, subgrid scale models, bubbles and the origins of stabilized methods', *Comput. Methods Appl. Mech. Eng.* **127**(1–4), 387–401.

Hughes, T. J. R. (2000), *The finite element method: linear static and dynamic finite element analysis*, Dover Publications, New York. Corrected reprint of the 1987 original [Prentice Hall, Englewood Cliffs, NJ].

Hughes, T. J. R. and Brooks, A. (1979), 'A multidimensional upwind scheme with no crosswind diffusion', *in* T. J. R. Hughes, ed., *Finite element methods for convection dominated flows*, AMD – Vol. 34, Presented at the Winter Annual Meeting of the ASME, Amer. Soc. Mech. Engrs. (ASME), New York, pp. 19–35.

Hughes, T. J. R. and Brooks, A. N. (1982), 'A theoretical framework for Petrov-Galerkin methods with discontinuous weighting functions: application to the streamline-upwind procedure', *in* R. H. Gallagher, D. H. Norrie, J. T. Oden and O. C. Zienkiewicz, eds, *Finite Elements in Fluids*, Vol. 4, Selected papers from the Third International Conference on Finite Elements in Flow Problems, Banff, Alberta, Canada, June 10–13, 1980, John Wiley & Sons, New York, pp. 47–65.

Hughes, T. J. R., Feijóo, G. R., Mazzei, L. and Quincy, J.-B. (1998), 'The variational multiscale method - a paradigm for computational mechanics', *Comput. Methods Appl. Mech. Eng.* **166**(1–2), 3–24.

Hughes, T. J. R. and Franca, L. P. (1987), 'A new finite element formulation for computational fluid dynamics. VII. The Stokes problem with various well-posed boundary conditions: symmetric formulations that converge for all velocity/pressure spaces', *Comput. Methods Appl. Mech. Eng.* **65**(1), 85–96.

Hughes, T. J. R., Franca, L. P. and Hulbert, G. M. (1989), 'A new finite element formulation for computational fluid dynamics. VIII. The Galerkin/least-squares method for advective-diffusive equations', *Comput. Methods Appl. Mech. Eng.* **73**(2), 173–189.

Hughes, T. J. R., Franca, L. P. and Mallet, M. (1987), 'A new finite element formulation for computational fluid dynamics. VI. Convergence analysis of the generalized SUPG formulation for linear time-dependent multidimensional advective-diffusive systems', *Comput. Methods Appl. Mech. Eng.* **63**(1), 97–112.

Hughes, T. J. R., Liu, W. K. and Brooks, A. (1979), 'Finite element analysis of incompressible viscous flows by the penalty function formulation', *J. Comput. Phys.* **30**(1), 1–60.

Hughes, T. J. R., Liu, W. K. and Zimmermann, T. K. (1978), 'Lagrangian-Eulerian finite element formulation for incompressible viscous flows', U.S.-Japan Seminar on Interdisciplinary Finite Element Analysis, Cornell University, Ithaca, NY.

Hughes, T. J. R. and Mallet, M. (1986a), 'A new finite element formulation for computational fluid dynamics. III. The generalized streamline operator for multidimensional advective-diffusive systems', *Comput. Methods Appl. Mech. Eng.* **58**(3), 305–328.

Hughes, T. J. R. and Mallet, M. (1986b), 'A new finite element formulation for computational fluid dynamics. IV. A discontinuity-capturing operator for multidimensional advective-diffusive systems', *Comput. Methods Appl. Mech. Eng.* **58**(3), 329–336.

Hughes, T. J. R., Mazzei, L. and Jansen, K. E. (2000), 'Large eddy simulation and variational multiscale method', *Computing and Visualization in Science* **3**(1/2), 47–59.

Hughes, T. J. R., Mazzei, L., Oberai, A. O. and Wray, A. A. (2001), 'The multiscale formulation of large eddy simulation: decay of homogeneous isotropic turbulence', *Phys. Fluids* **13**(2), 505–512.

Hughes, T. J. R. and Tezduyar, T. E. (1984), 'Finite element methods for first-order hyperbolic systems with particular emphasis on the compressible Euler equations', *Comput. Methods Appl. Mech. Eng.* **45**(1–3), 217–284.

Hughes, T. J. R., Tezduyar, T. E. and Brooks, A. N. (1982), 'A Petrov Galerkin finite element formulation for systems of conservation law with special reference to the compressible Euler equations', *in* K. W. Morton and M. J. Baines, eds, *Numerical methods for fluid dynamics, (Reading 1982)*, Academic Press, London, pp. 97–125.

Iannelli, G. S. and Baker, A. J. (1991), 'A globally well-posed finite element algorithm for aerodynamic applications', *Int. J. Numer. Methods Fluids* **12**(5), 407–441.

Idelsohn, S., Nigro, N., Storti, M. and Buscaglia, G. (1996), 'A Petrov-Galerkin formulation for advection-reaction-diffusion problems', *Comput. Methods Appl. Mech. Eng.* **136**(1–2), 27–46.

Ilinca, F., Hétu, J.-F. and Peletier, D. (2000), 'On stabilized finite element formulations for incompressible advective-diffusive transport and fluid flow problems', *Comput. Methods Appl. Mech. Eng.* **188**(1–3), 235–255.

Isaacson, E. and Keller, H. B. (1994), *Analysis of numerical methods*, Dover Publications, New York. Corrected reprint of the 1966 original [John Wiley & Sons, New York].

Jameson, A. (1985), 'Numerical solution of the Euler equations for compressible inviscid fluids', *in* F. Angrand, A. Dervieux, J. A. Desideri and R. Glowinski, eds, *Numerical methods for the Euler equations of fluid dynamics*, SIAM, Philadelphia, pp. 199–245. Proceedings of the INRIA Workshop, Rocquencourt, 1983.

Jameson, A., Schmidt, W. and Turkel, E. (1981), 'Numerical solutions of the Euler equations by finite volume methods using Runge-Kutta time stepping schemes', AIAA Computational Fluid Dynamics Conference, San Diego. Paper 81-1259.

Jamet, P. (1978), 'Galerkin-type approximations which are discontinuous in time for parabolic equations in a variable domain', *SIAM J. Numer. Anal.* **15**(5), 912–928.

Jansen, K. E., Collis, S. S., Whiting, C. and Shakib, F. (1999), 'A better consistency for low-order stabilized finite element methods', *Comput. Methods Appl. Mech. Eng.* **174**(1–2), 153–170.

Jiang, B.-n. (1998), *The least-squares finite element method. Theory and applications in computational fluid dynamics and electromagnetics*, Springer-Verlag, Berlin.

Jiang, C. B. and Kawahara, M. (1993), 'The analysis of unsteady incompressible flows by a three-step finite element method', *Int. J. Numer. Methods Fluids* **16**(9), 793–811.

John, F. (1991), *Partial differential equations*, fourth edn, Springer-Verlag, New York.

Johnson, C. (1987), *Numerical solution of partial differential equations by the finite element method*, Cambridge University Press, Cambridge.

Johnson, C., Nävert, U. and Pitkäranta, J. (1984), 'Finite element methods for linear hyperbolic equations', *Comput. Methods Appl. Mech. Eng.* **45**(1–3), 285–312.

Johnson, C. and Saranen, J. (1986), 'Streamline diffusion methods for the incompressible Euler and Navier-Stokes equations', *Math. Comput.* **47**(175), 1–18.

Johnson, C., Szepessy, A. and Hansbo, P. (1990), 'On the convergence of shock-capturing streamline diffusion finite element methods for hyperbolic conservation laws', *Math. Comput.* **54**(189), 107–129.

Kehr-Candille, V. and Ohayon, R. (1992), 'Elastoacoustic damped vibrations. Finite element and modal reduction methods', *in* P. Ladevèze and O. Zienkiewicz, eds, *New advances in computational structural mechanics (Giens, 1991)*, Vol. 32 of *Studies in Applied Mechanics*, Elsevier, Amsterdam, pp. 321–334.

Kelly, D. W., Nakazawa, S., Zienkiewicz, O. C. and Heinrich, J. C. (1980), 'A note of upwinding and anisotropic balancing dissipation in finite element approximation to convective diffusion problems', *Int. J. Numer. Methods Eng.* **15**(9), 1705–1711.

Ladevèze, P. and Pelle, J.-P. (2001), *La maîtrise du calcul en mécanique linéaire et non linéaire*, Hermes Science, Paris.

Ladyzhenskaya, O. A. (1969), *The mathematical theory of viscous incompressible flow*, Gordon and Breach Science, New York.

Lamb, H. (1993), *Hydrodynamics*, sixth edn, Cambridge University Press, Cambridge. Reprint of the 1932 edition.

Lambert, J. D. (1991), *Numerical methods for ordinary differential equations. The initial value problem*, John Wiley & Sons, Chichester.

Landau, L. D. and Lifshitz, E. M. (1959), *Fluid mechanics*, Pergamon Press, London.

Lapidus, A. (1967), 'A detached shock calculation by second order finite differences.', *J. Comput. Phys.* **2**, 154–177.

Lasaint, P. and Raviart, P.-A. (1974), 'On a finite element method for solving the neutron transport equation', *Mathematical aspects of finite elements in partial differential equations*, Math. Res. Center, University of Wisconsin-Madison, Academic Press, New York, pp. 89–123. Publication No. 33. Proceedings of the Symposium, Madison, 1974.

Laval, H. (1988), 'Taylor–Galerkin solution of the time-dependent Navier–Stokes equations', *in* H. Niki and M. Kawahara, eds, *Computational Methods in Flow Analysis*, Okayama University of Science, pp. 414–421.

Laval, H. and Quartapelle, L. (1990), 'A fractional-step Taylor–Galerkin method for unsteady incompressible flows', *Int. J. Numer. Methods Fluids* **11**(5), 501–513.

Leonard, B. P. (1979), 'A survey of finite differences of opinion on numerical muddling of the incomprehensible defective confusion equation', *in* T. J. R. Hughes, ed., *Finite element methods for convection dominated flows*, AMD – Vol. 34, Presented at the Winter Annual Meeting of the ASME, Amer. Soc. Mech. Engrs. (ASME), New York, pp. 1–17.

LeVeque, R. J. (1992), *Numerical methods for conservation laws*, second edn, Birkhäuser-Verlag, Basel.

Li, C. W. (1990), 'Least-squares characteristics and finite elements for advection-dispersion simulation', *Int. J. Numer. Methods Eng.* **29**(6), 1343–1358.

Liu, F. and Jameson, A. (1993), 'Multigrid Euler calculations for three-dimensional cascades', *AIAA J.* **31**(10), 1785–1791.

Liu, W. K. and Chang, H. (1986), 'On a numerical method for liquid filled systems', *Comput. Struct.* **23**(5), 671–677.

Liu, W. K. and Chang, H. G. (1985), 'A method of computation for fluid structure interaction', *Comput. Struct.* **20**(1–3), 311–320.

Liu, W. K. and Gvildys, J. (1986), 'Fluid-structure interaction of tanks with an eccentric core barrel', *Comput. Methods Appl. Mech. Eng.* **58**(1), 51–77.

Liu, W. K., Jun, S., Li, S., Adee, J. and Belytschko, T. (1995), 'Reproducing kernel particle methods for structural dynamics', *Int. J. Numer. Methods Eng.* **38**(10), 1655–1679.

Löhner, R., Morgan, K. and Peraire, J. (1985), 'A simple extension to multidimensional problems of the artificial viscosity due to Lapidus', *Commun. Numer. Methods Eng.* **1**(4), 141–147.

Malkus, D. S. and Hughes, T. J. R. (1978), 'Mixed finite element methods - reduced and selective integration techniques: a unification of concepts', *Comput. Methods Appl. Mech. Eng.* **15**(1), 63–81.

Marchuk, G. I. (1982), *Methods of numerical mathematics*, second edn, Springer-Verlag, New York. Translated from the Russian by Arthur A. Brown.

Marchuk, G. I. (1990), 'Splitting and alternating direction methods', *in* P. G. Ciarlet and J.-L. Lions, eds, *Handbook of numerical analysis, Vol. I*, North-Holland, Amsterdam, pp. 197–462.

Marshall, R. S., Heinrich, J. C. and Zienkiewicz, O. C. (1978), 'Natural convection in a square enclosure by a finite element penalty function method using primitive fluid variables', *J. Num. Methods in Heat Transfer* **1**, 315–330.

Masud, A. and Hughes, T. J. R. (1997), 'A space-time Galerkin/least-squares finite element formulation of the Navier-Stokes equations for moving domain problems', *Comput. Methods Appl. Mech. Eng.* **146**(1–2), 91–126.

Melenk, J. M. and Babuška, I. (1996), 'The partition of unity finite element method: basic theory and applications', *Comput. Methods Appl. Mech. Eng.* **139**(1–4), 289–314.

Minev, P. D. (2001), 'A stabilized incremental projection scheme for the incompressible Navier-Stokes equations', *Int. J. Numer. Methods Fluids* **36**(4), 441–464.

Mitchell, A. R. and Griffiths, D. F. (1980), *The finite difference method in partial differential equations*, John Wiley & Sons, Chichester.

Moran, H. J. P. and Ohayon, R. (1979), 'Substructure variational analysis of the vibrations of coupled fluid-structure systems. Finite element results', *Int. J. Numer. Methods Eng.* **14**(5), 741–755.

Morton, K. W. (1982), 'Schock capturing, fitting and recovery', *in* E. Krause, ed., *Eigth International Conference on Numerical Methods in Fluid Dynamics*, Vol. 170 of *Lecture Notes in Phys.*, Springer-Verlag, Berlin, pp. 77–93.

Morton, K. W. (1983), 'Characteristing Galerkin methods for hyperbolic problems', in M. Pandolfi and R. Piva, eds, Fifth GAMM Conference on Numerical Methods in Fluid Dynamics, Vieweg, Braunschweig, pp. 243–250.

Morton, K. W. (1985), 'Generalised Galerkin methods for hyperbolic problems', *Comput. Methods Appl. Mech. Eng.* **52**(1–3), 847–871.

Morton, K. W. (1996), *Numerical Solution of Convection-Diffusion Problems*, Vol. 12 of *Applied mathematics and mathematical computation*, R.J. Knops and K.W. Morton, eds, Chapman & Hall, London.

Morton, K. W. and Parrott, A. K. (1980), 'Generalised Galerkin methods for first-order hyperbolic equations', *J. Comput. Phys.* **36**(2), 249–270.

Nayroles, B., Touzot, G. and Villon, P. (1992), 'Generalizing the finite element method: diffuse approximation and diffuse elements', *Comput. Mech.* **10**(5), 307–318.

Nguyen, H. and Reynen, J. (1984), 'A space–time least-squares finite element scheme for advection–diffusion equations', *Comput. Methods Appl. Mech. Eng.* **42**(3), 331–342.

Noh, W. F. (1964), 'Cel: A time-dependent, two-space-dimensional, coupled eulerian-lagrange code', *in* B. Alder, S. Fernbach and M. Rotenberg, eds, *Methods in computational physics. Advances in research and applications. Fundamental methods in hydrodynamics*, Vol. 3, Academic Press, New York, pp. 117–179.

Oden, J. T. and Reddy, J. N. (1976), *An introduction to the mathematical theory of finite elements*, Pure and Applied Mathematics, John Wiley & Sons, New York.

Oleĭnik, O. (1957), 'Discontinuous solutions of non-linear differential equations', *Usp. Mat. Nauk (N.S.)* **12**, 3–73 (in Russian), translated in *Am. Math. Soc. Transl. (Ser. 2)* **26** (1963), 95–172.

Oñate, E. (1998), 'Derivation of stabilized equations for numerical solution of advective-diffusive transport and fluid flow problems', *Comput. Methods Appl. Mech. Eng.* **151**(1–2), 233–265.

Oñate, E. and Manzán, M. (1999), 'A general procedure for deriving stabilized space-time finite element methods for advective-diffusive problems', *Int. J. Numer. Methods Fluids* **31**(1), 203–221.

Ozawa, S. (1975), 'Numerical studies of steady flow in a two-dimensional square cavity at high reynolds numbers', *J. Phys. Soc. Jpn.* **38**(3), 889–895.

Paillère, H. (1995), 'Multidimensional upwind residual distribution schemes for the Euler and Navier-Stokes equations on unstructured grids', PhD thesis, Universit é Libre de Bruxelles and von Karman Institute, Belgium.

Park, N.-S. and Liggett, J. A. (1990), 'Taylor-least-squares finite element for two-dimensional advection-dominated unsteady advection-diffusion problems', *Int. J. Numer. Methods Fluids* **11**(1), 21–38.

Park, N. S. and Liggett, J. A. (1991), 'Application of Taylor-least squares finite element to three dimensional advection-diffusion equation', *Int. J. Numer. Methods Fluids* **13**(6), 759–773.

Peraire, J. (1986), 'A finite element method for convection dominated flows', PhD thesis, University College of Swansea, Wales.

Peraire, J., Zienkiewicz, O. C. and Morgan, K. (1986), 'Shallow water problems: a general explicit formulation', *Int. J. Numer. Methods Eng.* **22**(3), 547–574.

Pereira, J. M. C., Kobayashi, M. H. and Pereira, J. C. F. (2001), 'A fourth-order-accurate finite volume compact method for the incompressible Navier-Stokes solutions', *J. Comput. Phys.* **167**(1), 217–243.

Perot, J. B. (1993), 'An analysis of the fractional step method', *J. Comput. Phys.* **108**(1), 51–58.

Perrochet, P. and Azérad, P. (1995), 'Space-time integrated least-squares: solving a pure advection equation with a pure diffusion operator', *J. Comput. Phys.* **117**(2), 183–193.

Pironneau, O. (1981/82), 'On the transport-diffusion algorithm and its applications to the Navier-Stokes equations', *Numer. Math.* **38**(3), 309–332.

Pironneau, O. (1989), *Finite element methods for fluids*, John Wiley & Sons, Chichester.

Pudykiewicz, J. and Staniforth, A. (1984), 'Some properties and comparative performance of semi-Lagrangian method of Robert in the solution of the advection–diffusion equation', *Atmos.-Ocean* **22**(3), 283–308.

Purnell, D. K. (1976), 'Solution of the advective equation by upstream interpolation with a cubic spline', *Mon. Weather Rev.* **104**, 42–48.

Quartapelle, L. (1993), *Numerical solution of the incompressible Navier-Stokes equations*, Vol. 113 of *International Series of Numerical Mathematics*, Birkhäuser-Verlag, Basel.

Quartapelle, L. and Rebay, S. (1990), 'Numerical solution of two-point boundary value problems', *J. Comput. Phys.* **86**(2), 314–354.

Quarteroni, A., Saleri, F. and Veneziani, A. (2000), 'Factorization methods for the numerical approximation of Navier-Stokes equations', *Comput. Methods Appl. Mech. Eng.* **188**(1–3), 505–526.

Quarteroni, A. and Valli, A. (1994), *Numerical Approximation of Partial Differential Equations*, Vol. 23 of *Springer Series in Computational Mathematics*, Springer-Verlag, Berlin.

Reddy, J. N. and Gartling, D. K. (2001), *The finite element method in heat transfer and fluid dynamics*, second edn, CRC Press, Boca Raton, FL.

Richtmyer, R. D. and Morton, K. W. (1967), *Difference Methods for Initial-Value Problems*, Vol. 4 of *Interscience tracts in pure and applied mathematics*, second edn, L. Bers, R. Courant and J.J. Stoker, eds, John Wiley & Sons, New York.

Robert, A. (1981), 'A stable numerical integration scheme for the primitive meteorological equations', *Atmos.-Ocean* **19**(1), 35–46.

Robert, A. (1982), 'A semi-Lagrangian and semi-implicit numerical integration scheme for the primitive meteorological equations', *J. Meteorol. Soc. Jpn* **60**, 319–325.

Rodríguez-Ferran, A. and Huerta, A. (1999), 'Adapting Broyden method to handle linear constraints imposed via Lagrange multipliers', *Int. J. Numer. Methods Eng.* **46**(12), 2011–2026.

Roe, P. L. (1981), 'Approximate Riemann solvers, parameter vectors, and difference schemes', *J. Comput. Phys.* **43**(2), 357–372.

Roe, P. L. (1984), 'Generalized formulation of TVD Lax-Wendroff schemes', Technical Report n. 84-53, ICASE, Hampton.

Roe, P. L. (1985), 'Some contributions to the modelling of discontinuous flows', in B. Engquist, S. Osher and R. Somerville, eds, Large-scale computations in fluid mechanics, Part 2 (La Jolla, Calif., 1983), American Mathematical Society, Providence, RI, pp. 163–193.

Roe, P. L. (1986), 'Discrete models for the numerical analysis of time-dependent multidimensional gas dynamics', *J. Comput. Phys.* **63**(2), 458–476.

Sachdev, P. L. (1987), *Nonlinear diffusive waves*, Cambridge University Press, Cambridge.

Sampaio, P. A. B. D. and Moreira, M. L. (2000), 'A new finite element formulation for both compressible and nearly incompressible fluid dynamics', *Int. J. Numer. Methods Fluids* **32**(1), 51–78.

Selmin, V. (1987), 'Third-order finite element schemes for the solution of hyperbolic problems', Technical Report 707, INRIA, France.

Selmin, V., Donéa, J. and Quartapelle, L. (1985), 'Finite element methods for non-linear advection', *Comput. Methods Appl. Mech. Eng.* **52**(1–3), 817–845.

Shakib, F. (1989), 'Finite element analysis of the compressible Euler and Navier-Stokes equations', PhD thesis, Stanford University, USA.

Shakib, F. and Hughes, T. J. R. (1991), 'A new finite element formulation for computational fluid dynamics. IX. Fourier analysis of space-time Galerkin/least-squares algorithms', *Comput. Methods Appl. Mech. Eng.* **87**(1), 35–58.

Shakib, F., Hughes, T. J. R. and Johan, Z. (1991), 'A new finite element formulation for computational fluid dynamics. X. The compressible Euler and Navier-Stokes equations', *Comput. Methods Appl. Mech. Eng.* **89**(1–3), 141–219.

Sod, G. A. (1978), 'A survey of several finite difference methods for systems of nonlinear hyperbolic conservation laws', *J. Comput. Phys.* **27**(1), 1–31.

Soulaïmani, A. and Fortin, M. (1994), 'Finite element solution of compressible viscous flows using conservative variables', *Comput. Methods Appl. Mech. Eng.* **118**(3–4), 319–350.

Staniforth, A. and Côté, J. (1991), 'Semi-Lagrangian integration schemes for atmospheric models - A review', *Mon. Weather Rev.* **119**(9), 2206–2223.

Staniforth, A. and Pudykiewicz, J. (1985), 'Reply to comments on and addenda to "Some properties and comparative performance of semi-Lagrangian method of Robert in the solution of the advection–diffusion equation"', *Atmos.-Ocean* **23**, 195–200.

Steger, J. L. and Warming, R. F. (1981), 'Flux vector splitting of the inviscid gasdynamic equations with application to finite-difference methods', *J. Comput. Phys.* **40**(2), 263–293.

Stein, L. R., Gentry, R. A. and Hirt, C. (1977), 'Computational simulation of transient blast loading on three-dimensional structures', *Comput. Methods Appl. Mech. Eng.* **11**, 57–74.

Strang, G. and Fix, G. J. (1973), *An analysis of the finite element method*, Prentice Hall Series in Automatic Computation, Prentice Hall, Englewood Cliffs, NJ.

Strouboulis, T., Copps, K. and Babuška, I. (2000), 'The generalized finite element method: an example of its implementation and illustration of its performance', *Int. J. Numer. Methods Eng.* **47**(8), 1401–1417.

Sweby, P. K. (1984), 'High resolution schemes using flux limiters for hyperbolic conservation laws', *SIAM J. Numer. Anal.* **21**(5), 995–1011.

Tanguay, M., Simard, A. and Staniforth, A. (1989), 'A three-dimensional semi-Lagrangian integration scheme for the Canadian regional finite-element forecast model', *Mon. Weather Rev.* **117**, 1861–1871.

Taylor, C. and Hood, P. (1973), 'A numerical solution of the Navier-Stokes equations using the finite element technique', *Comput. Fluids* **1**(1), 73–100.

Temam, R. (1969), 'Sur l'approximation de la solution des équations de Navier-Stokes par la méthode des pas fractionnaires. II', *Arch. Ration. Mech. Anal.* **33**, 377–385.

Temam, R. (1991), 'Remark on the pressure boundary condition for the projection method', *Theor. Comput. Fluid Dyn.* **3**(3), 181–184.

Temam, R. (2001), *Navier-Stokes equations. Theory and numerical analysis*, AMS Chelsea Publishing, Providence, RI. Corrected reprint of the 1984 edition [North-Holland, Amsterdam, 1984].

Tezduyar, T. E. (1992), 'Stabilized finite element formulations for incompressible flow computations', *Adv. Appl. Mech.* **28**, 1–44.

Tezduyar, T. E. and Ganjoo, D. K. (1986), 'Petrov-Galerkin formulations with weighting functions dependent upon spatial and temporal discretization: applications to transient convection-diffusion problems', *Comput. Methods Appl. Mech. Eng.* **59**(1), 49–71.

Tezduyar, T. E., Mittal, S., Ray, S. E. and Shih, R. (1992), 'Incompressible flow computations with stabilized bilinear and linear equal-order-interpolation velocity-pressure elements', *Comput. Methods Appl. Mech. Eng.* **95**(2), 221–242.

Tezduyar, T. E., Mittal, S. and Shih, R. (1991), 'Time-accurate incompressible flow computations with quadrilateral velocity-pressure elements', *Comput. Methods Appl. Mech. Eng.* **87**(2–3), 363–384.

Tezduyar, T. E. and Osawa, Y. (2000), 'Finite element stabilization parameters computed from element matrices and vectors', *Comput. Methods Appl. Mech. Eng.* **190**(3–4), 411–430.

Tezduyar, T. E. and Park, Y. (1986), 'Discontinuity-capturing finite element formulations for nonlinear convection-diffusion-reaction equations', *Comput. Methods Appl. Mech. Eng.* **59**(3), 307–325.

Trulio, J. G. (1966), 'Theory and structure of the AFTON codes', Technical Report AFWL-TR-66-19, Air Force Weapons Lab.

Tuann, S. Y. and Olson, M. D. (1978), 'Review of computing methods for recirculating flows', *J. Comput. Phys.* **29**(1), 1–19.

van der Waerden, B. L. (1970), *Algebra. Vol. 1*, Frederick Ungar, New York.

van Leer, B. (1974), 'Towards the ultimate conservative difference scheme II. Monotonicity and conservation combined in a second order scheme', *J. Comput. Phys.* **14**, 361–370.

van Leer, B. (1982), 'Flux-vector splitting for the Euler equations', *in* E. Krause, ed., *Eigth International Conference on Numerical Methods in Fluid Dynamics*, Vol. 170 of *Lecture Notes in Phys.*, Springer-Verlag, Berlin, pp. 507–512.

von Neumann, J. and Richtmyer, R. D. (1950), 'A method for the numerical calculation of hydrodynamic shocks', *J. Appl. Phys.* **21**, 232–237.

Wait, R. and Mitchell, A. R. (1985), *Finite element analysis and applications*, John Wiley & Sons, New York.

Warming, R. F. and Hyett, B. J. (1974), 'The modified equation approach to the stability and accuracy analysis of finite-difference method', *J. Comput. Phys.* **14**, 159–179.

Wilkins, M. L. (1969), 'Calculation of elastic-plastic flow', Technical Report UCRL-7322 Rev 1, University of California.

Wong, J. S., Darmofal, D. L. and Peraire, J. (2001), 'The solution of the compressible Euler equations at low Mach numbers using a stabilized finite element algorithm', *Comput. Methods Appl. Mech. Eng.* **190**(43–44), 5719–5737.

Woodward, P. and Colella, P. (1984), 'The numerical simulation of two-dimensional fluid flow with strong shocks', *J. Comput. Phys.* **54**(1), 115–173.

Yanenko, N. N. (1971), *The method of fractional steps. The solution of problems of mathematical physics in several variables*, Springer-Verlag, Berlin. Translated from the Russian by T. Cheron. English translation edited by M. Holt.

Yee, H. C. (1987), 'Construction of explicit and implicit symmetric TVD schemes and their applications', *J. Comput. Phys.* **68**(1), 151–179.

Yosida, K. (1995), *Functional analysis*, Springer-Verlag, Berlin. Reprint of the sixth (1980) edition.

Zienkiewicz, O. C. and Codina, R. (1995), 'A general algorithm for compressible and incompressible flow. I. The split, characteristic-based scheme', *Int. J. Numer. Methods Fluids* **20**(8–9), 869–885.

Zienkiewicz, O. C. and Godbole, P. N. (1975), 'Viscous, incompressible flow with special reference to non-Newtonian (plastic) fluids', *in* R. H. Gallagher, J. T. Oden, C. Taylor and O. C. Zienkiewicz, eds, *Finite elements in fluids, Vol. 1: Viscous Flow and Hydrodynamics*, International Symposium on the Finite Element Method in Flow Problems held at Swansea, Wales, John Wiley & Sons, Chichester, pp. 25–55.

Zienkiewicz, O. C. and Morgan, K. (1983), *Finite elements and approximation*, John Wiley & Sons, New York.

Zienkiewicz, O. C., Morgan, K., Satya Sai, B. V. K., Codina, R. and Vázquez, M. (1995), 'A general algorithm for compressible and incompressible flow. II. Tests on the explicit form', *Int. J. Numer. Methods Fluids* **20**(8–9), 887–913.

Zienkiewicz, O. C. and Taylor, R. L. (2000a), *The finite element method. Vol. 1 The basis*, fifth edn, Butterworth Heinemann, Oxford.

Zienkiewicz, O. C. and Taylor, R. L. (2000b), *The finite element method. Vol. 3 Fluid dynamics*, fifth edn, Butterworth Heinemann, Oxford.

Index

T

U

Printed and bound by CPI Group (UK) Ltd, Croydon, CR0 4YY
16/09/2026
14963471-0007